13th International Conference on Nanomaterials - Research and Application (NANOCON 2021)

Brno, Czech Republic
20 - 22 October 2021

ISBN: 978-1-7138-4379-5

Printed from e-media with permission by:

Curran Associates, Inc.
57 Morehouse Lane
Red Hook, NY 12571

Some format issues inherent in the e-media version may also appear in this print version.

Printed with permission by Curran Associates, Inc. (2022)

For permission requests, please contact Tanger Ltd
at the address below.

Tanger Ltd
Keltickova 62
710 00 Slezka Ostrava
Czech Republic

www.tanger.cz

Additional copies of this publication are available from:

Curran Associates, Inc.
57 Morehouse Lane
Red Hook, NY 12571 USA
Phone: 845-758-0400
Fax: 845-758-2633
Email: curran@proceedings.com
Web: www.proceedings.com

TANGER Ltd.
The Czech Society for New Materials and Technologies
Regional Centre of Advanced Technologies and Materials

2021
NANOCON®

13th International Conference on Nanomaterials - Research & Application

CONFERENCE PROCEEDINGS

October 20 - 22, 2021
Brno, Czech Republic, EU

ISSN 2694-930X

NANOCON 2021 **Conference Proceedings**

Different Authors

October 20-22, 2021 Orea Congress Hotel Brno, Czech Republic, EU

Issued by: TANGER Ltd., Keltickova 62, 710 00 Ostrava, Czech Republic, EU

Edition: 1st Edition, 2021

Number of pages: 470

SESSION A - NANOMATERIALS FOR ELECTRONIC, MAGNETIC AND OPTIC APPLICATIONS. CARBON NANOSTRUCTURES, QUANTUM DOTS

SESSION E - ADVANCED METHODS OF PREPARATION AND CHARACTERIZATION OF NANOMATERIALS

SESSION A
NANOMATERIALS FOR ELECTRONIC, MAGNETIC AND OPTIC APPLICATIONS. CARBON NANOSTRUCTURES, QUANTUM DOTS

Chairmen

Prof. Dr. Eduard HULICIUS	Institute of Physics of the CAS, Prague, Czech Republic, EU
Assoc. Prof. Dr. Martin KALBAC	J. Heyrovsky Institute of Physical Chemistry of the CAS, Prague, Czech Republic, EU
Dr. Oldrich SCHNEEWEISS	Institute of the Physics of Materials of the CAS, Brno, Czech Republic, EU
Prof. Dr. Bohuslav REZEK	Czech Technical University in Prague, Czech Republic, EU
Prof. Dr. Zdenek SOFER	University of Chemistry and Technology, Prague, Czech Republic, EU

ANOMALOUS PRESSURE-INDUCED CHANGES IN MAGNETISM OF FERRIMAGNETIC Ni-Mn-Sn STRUCTURES

[1]Martin FRIÁK, [1,2]Martina MAZALOVÁ, [1]Ivana MIHÁLIKOVÁ, [2,1]Mojmír ŠOB

[1]*Institute of Physics of Materials, v.v.i., Czech Academy of Sciences, Brno, Czech Republic, EU,*
friak@ipm.cz, 394206@mail.muni.cz, mihalikova@ipm.cz

[2]*Department of Chemistry, Faculty of Science, Masaryk University, Brno, Czech Republic, EU*
mojmir@ipm.cz

https://doi.org/10.37904/nanocon.2021.4306

Abstract

We have performed a quantum-mechanical study of two ferrimagnetic Ni_2MnSn structures, a stoichiometric one with Mn-Ni swaps and an off-stoichiometric ($Ni_7Mn_8Sn_1$) alloy. We used 16-atom supercells with the atomic positions related to those in the full Heusler structure, i.e. austenite phase (unless distorted by internal distribution of point defects as in the case of Ni_2MnSn with swaps). We have determined thermodynamic, magnetic and structural properties of both pressure-free states as well as those corresponding to hydrostatic pressures of a few GPa. The atomic and magnetic configurations of the studied states are found to exhibit anomalous pressure-induced changes in the total magnetic moment. In particular, the total magnetic moments per 16-atom supercells increase with increasing hydrostatic pressure. Despite this peculiar trend in the total magnetic moment, the magnitudes of local magnetic moments of Mn atoms, that are decisive for the value of the total magnetic moment, decrease with increasing hydrostatic pressure (as is common in majority of magnetic systems). The identified phenomena may be related to an interplay of a few contributing mechanisms. First, the magnetic moments of Mn atoms, that are either parallel or antiparallel to the orientation of the total magnetic moment of the supercells, nearly compensate each other due to the ferrimagnetic nature of the studied magnetic states. Second, the swapped and off-stoichiometric atoms lead to different local atomic environments of Mn atoms and, consequently, to different local magnetic moments of these atoms as well as their different response to hydrostatic pressures. Importantly, the local magnetic moments of Mn atoms, that are antiparallel to the orientation of the total magnetic moment, are more sensitive to the applied pressures. Regarding thermodynamics, the studied states are excited ones.

Keywords: Ni-Mn-Sn, magnetism, pressure, ferrimagnetic, austenite, quantum-mechanical, defects

1. INTRODUCTION

One of the most intensively studied materials is ternary X_2YZ materials with Heusler-type crystal lattice [1]. The Heusler alloys include numerous combinations of different chemical elements and some of them possess an extraordinary magneto-structural transitions that are often accompanied by significant changes of their properties [2]. In particular, the off-stoichiometric Ni_2MnX (X = Ga, In, Sn) compounds exhibit unusual magnetic behavior stemming from competing antiferromagnetic and ferromagnetic interactions between magnetic sublattices with Mn atoms [2-5]. The stoichiometric austenite Ni_2Mn-based alloys possess a long range ferromagnetic arrangement of moments below the Curie temperature, see, e.g. Ref. [7]. In contrast, the Ni-Mn-X compounds undergo several magnetic and magneto-structural transition [8-10] and some of these interesting phenomena are not fully understood yet. Our study was motivated by the current interest in pressure-induced changes in Ni-Mn-Sn structures [7] when an unexpected complexity is often found depending on the type and distribution of internal defects as well as their magnetic state, see, e.g. [11].

Before we proceed to describing two ferrimagnetic Ni-Mn-Sn structures, that we have studied, it is worth making a few comments on the structure and properties of the stoichiometric austenitic Ni_2MnSn without any defects. Its atomic distribution is that of the full Heusler structure, see **Figure 1**. The lowest energy configuration exhibits local magnetic moments of atoms as listed in **Figure 1(a)**; further details can be found in our recent paper [11]. The magnetic state is, in principle, ferrimagnetic due to the orientation of local magnetic moment of Sn atoms that is antiparallel to those of Mn and Ni atoms. But the local magnetic moments of Sn atoms are equal only to 0.05 μ_B, i.e. nearly an order of magnitude smaller than the magnitudes of local magnetic moments of Ni atoms and nearly two orders of magnitude smaller than those of Mn atoms. In fact, the magnitudes and orientations of Mn atoms are often decisive for the total magnetic moment of Ni-Mn-Sn systems. Therefore, the state shown in **Figure 1(a)** is often called ferromagnetic reflecting the ferromagnetic orientation of local magnetic moments of Mn atoms, as well as of Ni atoms, with respect to the orientation of the total magnetic moment. Our previous study [11] of various magnetic and structural variants of stoichiometric Ni_2MnSn with different types of swapped atoms revealed that the response to the hydrostatic pressures is highly sensitive to both atomic configuration and magnetic state.

2. COMPUTATIONAL METHODOLOGY

As a continuation of our previous research [11], we have computed pressure-induced changes in the case of two ferrimagnetic Ni-Mn-Sn states. One was the stoichiometric Ni_2MnSn case with two Ni-Mn swaps and the second quite off-stoichiometric Mn-rich and Sn-poor alloy with the stoichiometry $Ni_7Mn_8Sn_1$. Both were modeled using specific variants of the 16-atom supercell shown in **Figure 1(b)**. The supercell of the former state has a tetragonal symmetry while the latter one possesses a cubic one (details will be discussed below).

The Vienna Ab initio Simulation package (VASP) [12,13] implementing the density functional theory (DFT) [14,15] was used to perform our *ab initio* calculations. We have employed projector augmented wave (PAW) [16,17] pseudopotentials (Ni_pv, Mn_pv and Sn_d versions from the potpaw_PBE.52 VASP database) and the generalized gradient approximation (GGA) as parametrized by Perdew, Burke and Ernzerhof (PBE'96), see Ref. [18]. In the case of 16-atom supercells, the plane-wave energy cut-off was set to 700 eV and the 8x8x8 k-point mesh with its origin shifted to (1/2, 1/2, 1/2) was used. We applied the Methfessel-Paxton order 1 smearing with the smearing parameter set to 0.23. Supercells modeling the impact of hydrostatic pressures were partly relaxed, i.e., their total energy was minimized with respect to the atomic positions and the cell shape for a given volume corresponding to a certain applied hydrostatic pressure.

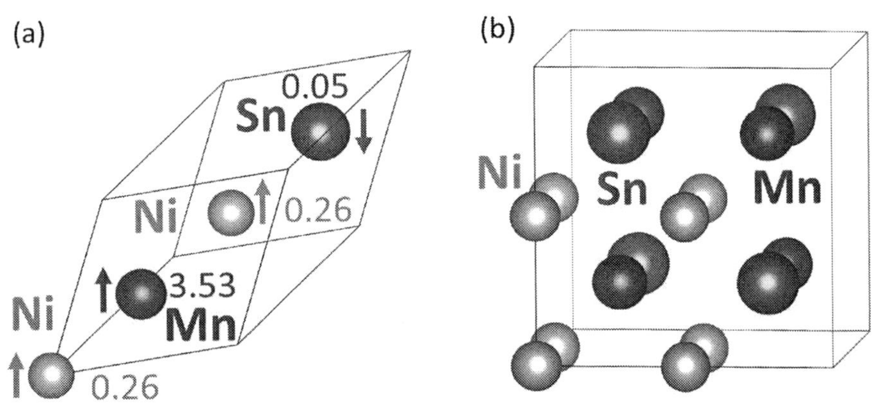

Figure 1 Schematic visualizations of (a) 4-atom rhombohedral primitive unit cell and (b) 16-atom cube-shaped conventional supercell of Heusler-structure (i.e. austenitic) Ni_2MnSn without any defects. The latter supercell was used for our modeling of states with a pair of Mn-Ni swaps and an off-stoichiometric $Ni_7Mn_8Sn_1$ alloy. Sub-figure (a) contains also computed values of local magnetic moments of atoms in Bohr magnetons and their orientation indicated by arrows, for details see Ref. [11].

3. RESULTS

Figure 2(a) visualizes the atomic configuration of the stoichiometric allotrope of Ni_2MnSn with two Ni-Mn swaps per 16-atom supercell. We focused on a ferrimagnetic configuration when the swapped Mn atoms on the Ni sublattice have their local magnetic moments antiparallel to local magnetic moments of the two remaining Mn atoms on the Mn sublattice as presented in **Figure 2(b)**. When minimizing the energy for this configuration, i.e. the state without any hydrostatic pressure, the swapped Mn atoms on the Ni sublattice turned out to have the magnitude of the local magnetic moments (2.78 μ_B) lower than the Mn atoms on the Mn sublattice (3.49 μ_B) by 0.71 Bohr magneton, see **Figure 2(b)**. Contrary to this trend, the Ni atoms on the Mn sublattice have the magnitude of their local magnetic moments (0.25 μ_B) higher than those of Ni atoms on the Ni sublattice (0.9-0.14 μ_B). The orientation of Ni atoms on the Mn sublattice is antiparallel to that of local magnetic moments of Ni atoms on the Ni sublattice (and this orientation of the total magnetic moment is also that of the whole 16-atom supercell). The magnitude of the total magnetic moment is only 1.77 μ_B per 16-atom supercell due to a nearly complete compensation of parallel and antiparallel local magnetic moments of atoms. The pressure-dependence of the total magnetic moment of the whole supercell, see **Figure 2(c)**, has an anomalous trend when it increases for increasing pressure, i.e. decreasing volume.

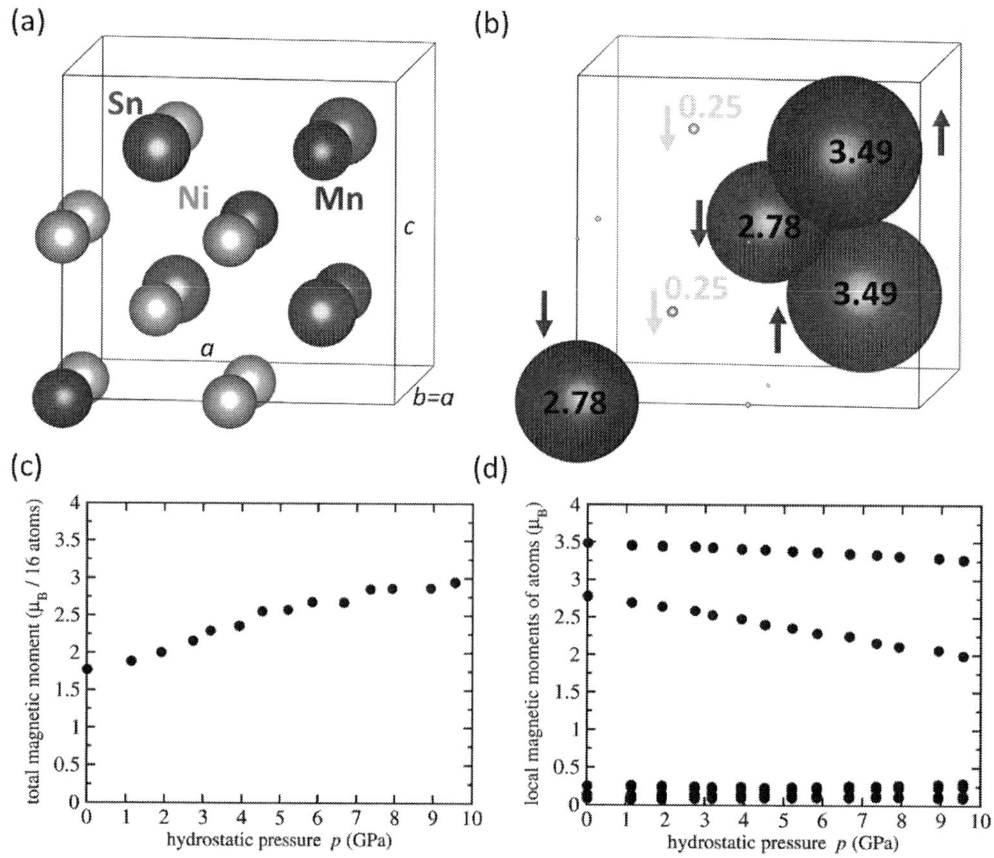

Figure 2 A visualization of 16-atom supercell of austenitic stoichiometric Ni_2MnSn with two Mn-Ni swaps (a). Part (b) includes the values of local magnetic moments of Mn and Ni atoms that are numerically listed in Bohr magnetons, indicated by the diameter of spheres representing the atoms and their orientation is shown by the arrows. The calculated pressure dependence of the total magnetic moment of the 16-atom supercell is shown sub-figure (c) and those of the local magnetic moments of constituent atoms are exhibited in sub-figure (d). The local magnetic moments lower than 0.5 μ_B are those of Ni atoms while values above 1.9 μ_B represent those of Mn atoms. Magnetic moments of Sn atoms are lower than 0.1 μ_B and are not shown.

This behavior is in contrast to that found in most magnetic systems where a lower volume typically causes a decrease in the magnetic moment (see, e.g. Refs. [7,11]). In order to shed a light on our findings, we have evaluated pressure-induced changes in the magnitudes of local magnetic moments and we present them in **Figure 2(d)**. Importantly, all the local magnetic moments of Mn atoms behave as is typical for magnetic systems, i.e. the magnitude of their local magnetic moments decreases with increasing pressure (decreasing volume). But their pressure dependences differ. The swapped Mn atoms on the Ni sublattice are in different atomic environment than the Mn atoms on the Mn sublattice. As a consequence, not only that the magnitude of local magnetic moments of the swapped Mn atoms is lower and their orientation is antiparallel to that of the total magnetic moment, as discussed above, but also their pressure dependence is different. In particular, a slope of their decrease as a function of the pressure is steeper than the slope of Mn atoms on the Mn sublattice (that also have higher magnitude of their local magnetic moments, see **Figure 2(d)**). The swapped atoms are more sensitive to the applied hydrostatic pressures. The anomalous trend of the magnitude of the total magnetic moment is, therefore, partly caused by the fact that this magnitude is a sum of pressure-dependent local magnetic moments with some of them contributing with an opposite sign due to the ferromagnetic nature of this state. A major part of the total magnetic moment of the whole supercell is the difference between the magnitudes of two Mn atoms on the Mn sublattice with parallel orientation to that of the total magnetic moment (positive contributions) and the magnitudes of the two swapped Mn atoms on the Ni sublattice with antiparallel orientation to the total magnetic moment (negative contributions).

As far as pressure-dependences of local magnetic moments of Ni atoms are concerned, their magnitudes typically decrease or stay constant. Some of the Ni atoms have the magnitude of their local magnetic moments very slightly increased by the applied pressure but the increase is within an expected error bar of our calculations, e.g. an increase from 0.08 μ_B to 0.11 μ_B in the case of the application of 9.6 GPa of hydrostatic pressure, and do not represent any important physical phenomenon. It is worth noting that the atomic distribution of two Mn-Ni swaps per 16-atom supercell of the stoichiometric Ni_2MnSn reduces the symmetry from a cubic one in the case of a defect-free Heusler structure (see **Figure 1**) to a tetragonal one. Therefore, there are two lattice parameters, see them defined in **Figure 2(a)**, and each of them responds differently to the applied pressures. Consequently, their ratio changes with the pressure and the trend is provided in **Figure 3(a)**. Interestingly, the zero-pressure state has the *c/a* ratio lower than one but it increases with pressure, reaches the value of 1 for about 7.6 GPa, and is higher than one for yet higher pressures. For the sake of completeness, we also present the changes in the overall volume, see **Figure 3(b)**. Last, but not least, it should be emphasized that the studied ferrimagnetic state (with the anomalous pressure-dependence of the total magnetic moment) is an excited state with respect to thermodynamically more stable ferromagnetic state that has the energy lower by about 18 meV/atom.

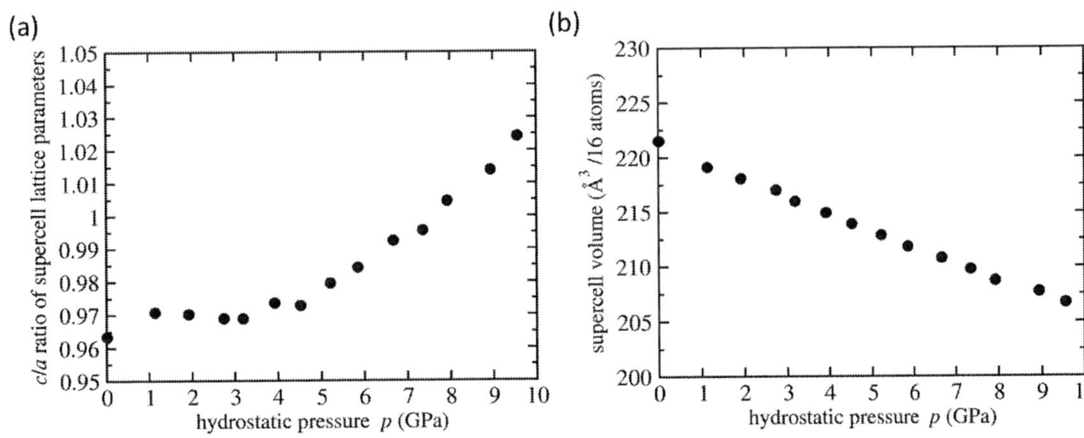

Figure 3 Computed pressure dependences of the ratio of lattice parameters *c/a* (a) and the volume (b) of our computational 16-atom supercell of the stoichiometric Ni_2MnSn with two Mn-Ni swaps.

Next we present our results obtained for a highly off-stoichiometric Ni_2MnSn state with the composition $Ni_7Mn_8Sn_1$ that has a cubic symmetry, i.e. an austenite. The 16-atom supercell is schematically shown in **Figure 4(a)**. There is one Mn atom on the Ni sublattice, the replaced Ni atom is completely missing and there are three extra Mn atoms replacing three Sn atoms within the 16-atom supercell. The zero-pressure ferrimagnetic state is shown in **Figure 4(b)**. The Mn atom on the Ni sublattice has a lower magnitude of its magnetic moment and this magnetic moment is antiparallel with respect to the orientation of the total magnetic moment (similarly as discussed above for Mn atoms on the Ni sublattice in the case of Ni_2MnSn with two Mn-Ni swaps). Magnetic moments of all Mn atoms on the Sn sublattice are also antiparallel but their magnitudes are similar to those of Mn atoms on the Mn sublattice. The pressure dependence of the total magnetic moment is again anomalous but the trend is different, convex, see **Figure 4(c)**, when compared with the concave trend in the case of Ni_2MnSn with two Mn-Ni swaps, see **Figure 2(c)**. The anomaly can be again tracked down to the behavior of magnetic moments of individual Mn atoms. The magnitudes of their magnetic moments decrease with increasing pressure, i.e. they do not exhibit any anomalous pressure dependence. But their responses to the applied pressures are differ for different Mn atoms. Also this ferrimagnetic state is an excited one with respect to a state that exhibits Mn atoms on the Mn and Sn sublattices having parallel orientation of their magnetic moments (the energy is then lower by 23 meV/atom).

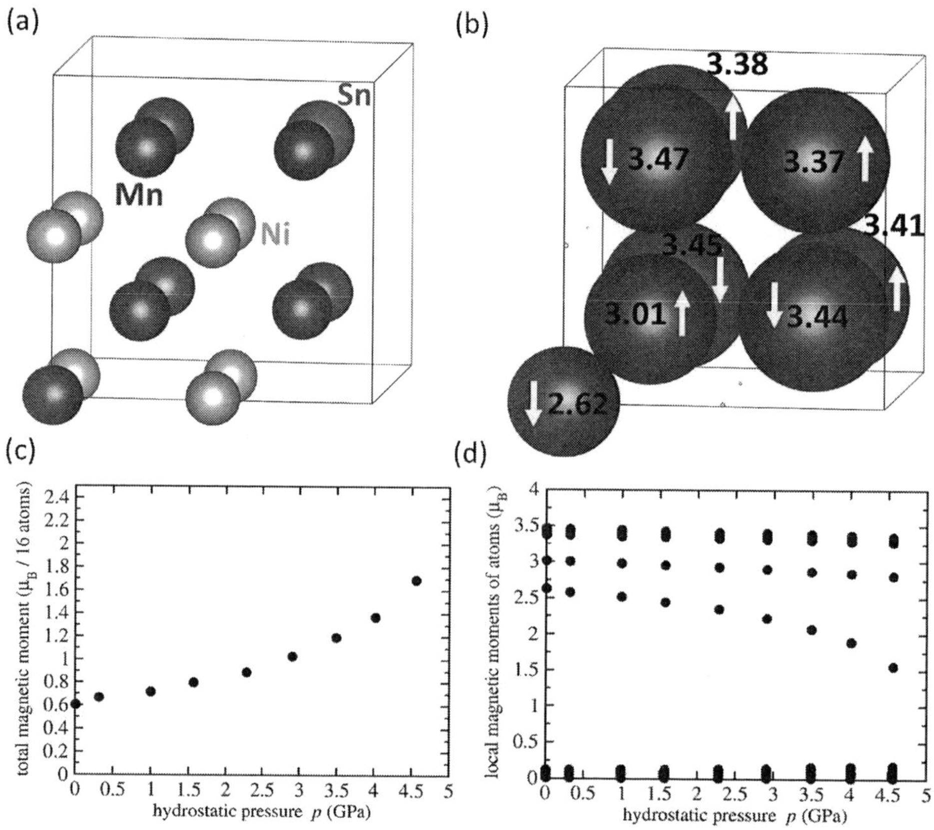

Figure 4 Schematic visualization of our 16-atom computational supercell of the studied austenite-structure off-stoichiometric Ni-Mn-Sn alloy with the composition $Ni_7Mn_8Sn_1$ (a). Part (b) contains the magnitudes of local magnetic moments of Mn and Ni atoms (in Bohr magnetons) as indicated by the diameter of spheres representing the atoms and their orientation is shown by the arrows. The calculated pressure dependence of the total magnetic moment of the 16-atom supercell is shown sub-figure (c) and those of the local magnetic moments of constituent atoms are exhibited in sub-figure (d). The local magnetic moments lower than 0.5 μ_B are those of Ni atoms while values above 1.5 μ_B represent those of Mn atoms. Magnetic moments of Sn atoms are lower than 0.1 μ_B and are not shown.

4. CONCLUSION

We have performed an *ab initio* study of two ferrimagnetic states, a Ni_2MnSn state with Mn-Ni swaps and an off-stoichiometric $Ni_7Mn_8Sn_1$ alloy, modelled by 16-atom supercells with the atomic positions related to those in the full Heusler structure. We have determined thermodynamic, magnetic and structural properties of the both pressure-free states as well as those corresponding to hydrostatic pressures of a few GPa. Both studied states exhibit anomalous pressure-induced changes in the total magnetic moment when the total magnetic moments of the supercells increase with increasing hydrostatic pressure. Despite of this peculiar trend in the total magnetic moment, the magnitudes of local magnetic moments of individual Mn atoms (that play a decisive role) decrease with increasing hydrostatic pressure. The identified anomalous phenomena may be related to an interplay of a few atomic-level mechanisms. First, the magnetic moments of Mn atoms, that are either parallel or antiparallel to the orientation of the total magnetic moment of the supercells, nearly compensate each other due to the ferrimagnetic nature of the studied magnetic states. Second, the swapped and off-stoichiometric atoms lead to different local atomic environments of Mn atoms and, consequently, to different local magnetic moments of these Mn atoms as well as their different response to hydrostatic pressures. Importantly, the local magnetic moments of Mn atoms, that are antiparallel to the orientation of the total magnetic moment, are more sensitive to the applied pressures. Regarding thermodynamic properties, the studied states are excited ones. As the next step we plan to test the mechanical stability of the studied states by evaluating their elastic constants and phonon spectra similarly as done in Ref. [19]

ACKNOWLEDGEMENTS

The authors acknowledge the Czech Science Foundation for the financial support received under the Project No. 20-16130S. Computational resources were provided by the Ministry of Education, Youth and Sports of the Czech Republic under the Projects e-INFRA CZ (ID:90140) at the IT4Innovations National Supercomputing Center and e-Infrastructure CZ (e-INFRA LM2018140) at the MetaCentrum as well as CERIT Scientific Cloud, all provided within the program Projects of Large Research, Development and Innovations Infrastructures. Sub-figures 2(a,b) and 4(a,b) were visualized using the VESTA software [20].

REFERENCES

[1] HEUSLER, F.; STARCK, W.; HAUPT, E. Über die Synthese ferromagnetischer Manganlegierungen. *Verhandlungen der Deutschen Physikalischen Gesellschaft.* 1903, vol. 5, p. 219.

[2] GRAF, T.; FELSER, C.; PARKIN, S.S.P. Simple rules for the understanding of Heusler compounds. *Progress in Solid State Chemistry.* 2011, vol. 39, p. 1.

[3] RECARTE, V.; *et al.* Direct evidence of the magnetoelastic interaction in Ni_2MnGa magnetic shape memory system. *Applied Physics Letters.* 2013, vol. 102, Art. No. 201906.

[4] LEE, S. J.; *et al.* Magnetic, magneto-optical, and transport properties of ferromagnetic shape-memory Ni_2MnGa alloy. *Journal of Applied Physics.* 2003, vol. 93, pp. 6975-6977.

[5] BROWN, P. J.; *et al.* The crystal structure and phase transitions of the magnetic shape memory compound Ni_2MnGa. *Journal of Physics: Condensed Matter.* 2002, vol. 14, pp. 10159-10171.

[6] ENTEL, P.; *et al.* Complex magnetic ordering as a driving mechanism of multifunctional properties of Heusler alloys from first principles. *The European Physics Journal B.* 2013, vol. 86, no. 65.

[7] KAŠTIL, J.; *et al.* Effect of pressure and high magnetic field on phase transitions and magnetic properties of $Ni_{1.92}Mn_{1.56}Sn_{0.52}$ and Ni_2MnSn Heusler compounds. *Journal of Alloys and Compounds.* 2015, vol. 650, pp. 248-255.

[8] SUTOU, Y.; *et al.* Magnetic and martensitic transformations of NiMnX(X=In,Sn,Sb) ferromagnetic shape memory alloys. *Applied Physics Letters.* 2004, vol. 85, pp. 4358-4360.

[9] BROWN, P.J.; *et al.* The magnetic and structural properties of the magnetic shape memory compound $Ni_2Mn_{1.44}Sn_{0.56}$. *Journal of Physics: Condensed Matter.* 2006, vol. 18, pp. 2249-2259.

[10] ÇAKIR, A.; *et al.* Intermartensitic transitions and phase stability in $Ni_{50}Mn_{50-x}Sn_x$ Heusler alloys. *Acta Materialia.* 2015, vol. 99, pp. 140-149.

[11] FRIÁK, M.; *et al.* An ab initio study of pressure-induced changes of magnetism in austenitic stoichiometric Ni_2MnSn. *Materials.* 2021, vol. 14, no. 523.

[12] KRESSE, G., HAFNER, J. Ab initio molecular dynamics for liquid metals. *Physical Review B.* 1996, vol. 47, pp. 558-561.

[13] KRESSE, G., FURTHMŰLLER, J. Efficient iterative schemes for ab initio total-energy calculations using a plane-wave basis set. *Physical Review B.* 1996, vol. 54, pp. 11169-11186.

[14] HOHENBERG, P., KOHN, W. Inhomogeneous electron gas. *Physical Review B.* 1964, vol. 136, pp. B864-B871.

[15] KOHN, W., SHAM, L. J. Self-consistent equations including exchange and correlation effects. *Physical Review A.* 1965, vol. 140, pp. A1133-A1138.

[16] BLŐCHL, P. E. Projector augmented-wave method. *Physical Review B.* 1994, vol. 50, pp. 17953-17979.

[17] KRESSE, G., JOUBERT, D. From ultrasoft pseudopotentials to the projector augmented-wave method. *Physical Review B.* 1999, vol. 59, pp. 1758-1775.

[18] PERDEW, J. P., BURKE, K., ERNZERHOF, M. Generalized gradient approximation made simple. *Physical Review Letters.* 1996, vol. 77, pp. 3865-3868.

[19] FRIÁK, M., *et al.* Pressure-induced increase of the total magnetic moment in ferrimagnetic $Ni_{1.9375}Mn_{1.5625}Sn_{0.5}$ martensite: A quantum-mechanical study. *Accepted for publication in Materials Transactions* (2022). Available from: https://doi.org/10.2320/matertrans.MT-MA2022006.

[20] MOMMA, K., IZUMI, F. VESTA 3 for three-dimensional visualization of crystal, volumetric and morphology data. *Journal of Applied Crystalography.* 2011, vol. 44, pp. 1272-1276.

GENERAL OVERVIEW OF GAN DEVICES AND TRANSPORT PROPERTIES OF ALGAN/GAN HEMT STRUCTURES - IMPACT OF DISLOCATION DENSITY AND IMPROVED DESIGN

[1]Eduard HULICIUS, [1,2]František HÁJEK, [1]Alice HOSPODKOVÁ, [1]Pavel HUBÍK, [1]Zuzana GEDEONOVÁ, [1]Tomáš HUBÁČEK, [1]Jiří PANGRÁC, [1]Karla KULDOVÁ

[1]*Institute of Physics CAS, v.v.i., Prague, Czech Republic, EU, hulicius@fzu.cz*

[2]*Faculty of Nuclear Sciences and Physical Engineering, CTU in Prague, Czech Republic, EU*

https://doi.org/10.37904/nanocon.2021.4309

Abstract

GaN-based nanostructures are used for many present semiconductor devices. The main topics are structures for blue LEDs and LDs, but there are also other interesting and important GaN devices namely for power electronics, scintillators and detectors as well as High Electron Mobility Transistors (HEMT). Several of our previous NANOCON contributions were devoted to the scintillator and detector structures; one was a description of optimization of HEMT structures and their growth parameters.

Keywords: GaN devices, HEMT, MOVPE epitaxy, dislocation

1. INTRODUCTION

This contribution is devoted to a general review of GaN power electronic devices, a comparison with other materials (mainly Si and SiC), a description of some fields of application and a future forecast.

The second part will describe our recent results - the influence of dislocation density on transport properties of HEMT structures is reported. Experimental results, obtained on HEMT structures prepared in the same growth run on templates with different dislocation densities, are compared. We show experimentally that lowering the dislocation density considerably increases the electron mobility in 2DEG. It is shown that the dislocation density has also a significant influence on the Fermi level position in GaN. Presented results suggest that reduction of the dislocation density can improve the performance of high frequency GaN HEMTs applications.

High frequency power electronic devices have many application fields, as it is shown in **Figures 1a** and **b**.

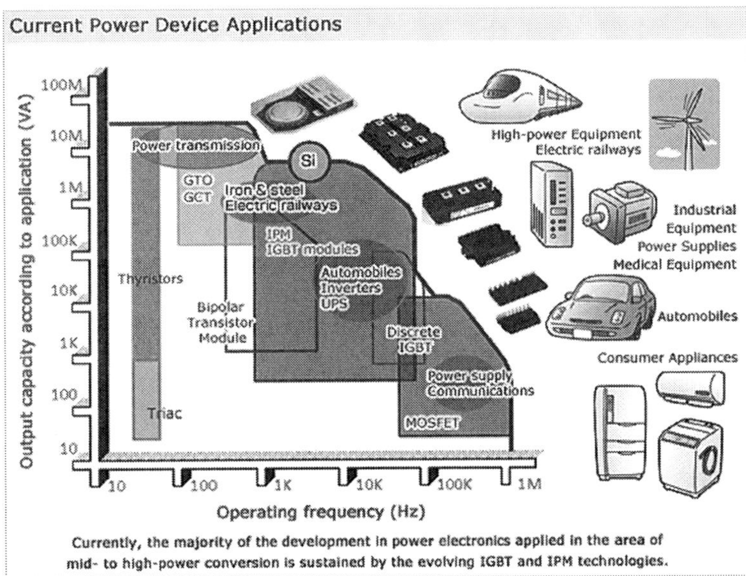

a) Output power (W) versus operating frequencies (Hz) for some current and future devices. The majority of the development in power electronic applied in the area of mid- to high-power conversion is sustained by the evolving IGBT and IPM technologies.

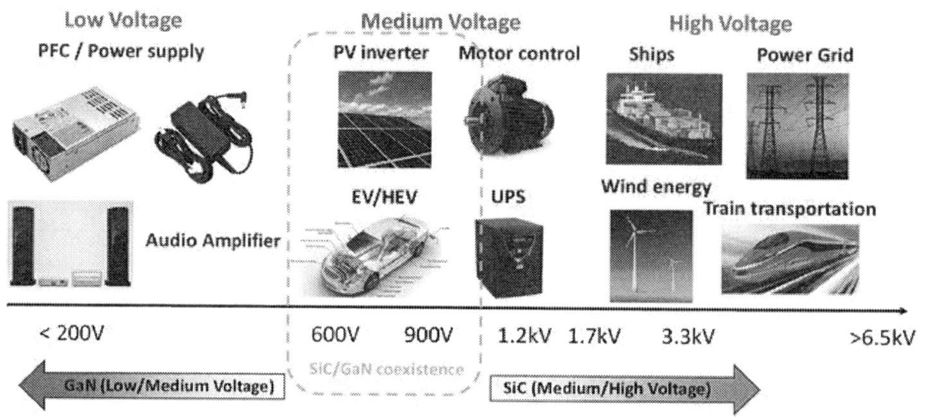

b) Different devices (current and future) according to the maximum applied "Voltages".

Figures 1a, b Application fields for three the most important semiconductor materials from the output power versus operating frequencies and voltage point of view.

There are three most important and promising materials: Si, SiC and GaN. Devices are based on different structures - see **Figure 2**, but the material properties are determinative (**Figure 2b**). Of course, there exist many other interesting materials, as Ga_2O_3, GaAs, InP or old school vacuum tubes - see **Figure 2a**, at the edge of electric device parameters. However, the main industrial players are only three. They also meet the requirements on temperature and radiation resistance as well as sufficient lifetime.

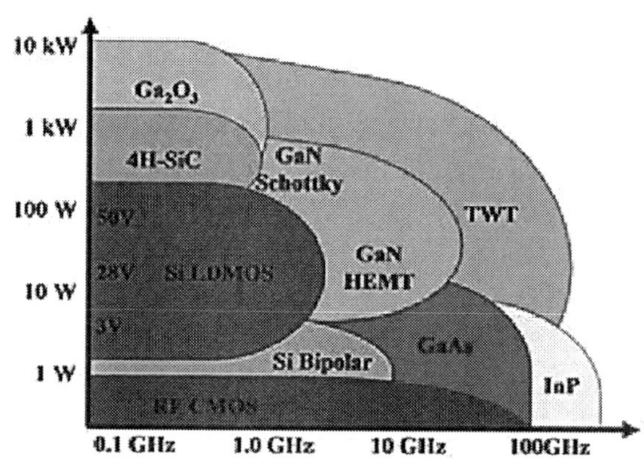

(a)

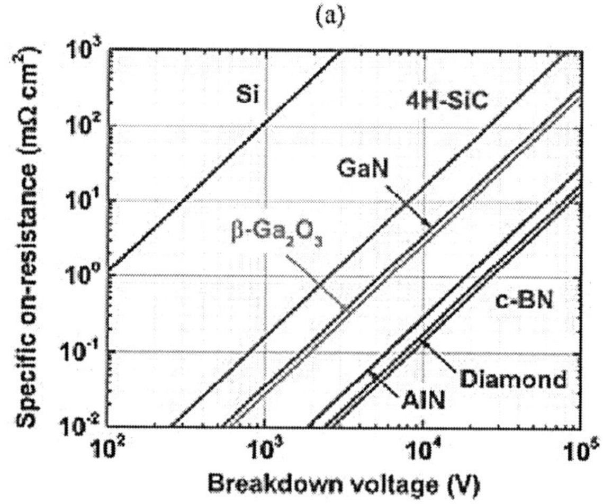

(b)

Figure 2 Comparison of some semiconductor materials and structures for different Output power versus Operating frequencies - upper part (a), and key physical parameters, which are responsible for device and structure limits - lower part (b). (***TWT*** *is* ***Traveling-wave tube amplifier,*** *a specialized vacuum tube that is used in electronics to amplify radio frequency (RF) signals in the microwave range.*)

We prepare and study InGaN/GaN and AlGaN/GaN scintillator and HEMT structures not only for scientific research, but also for industrial applications for the last 8 years, thus we will put here stress on GaN. In **Figure 3**, there are shown development and predictions of GaN power devices market.

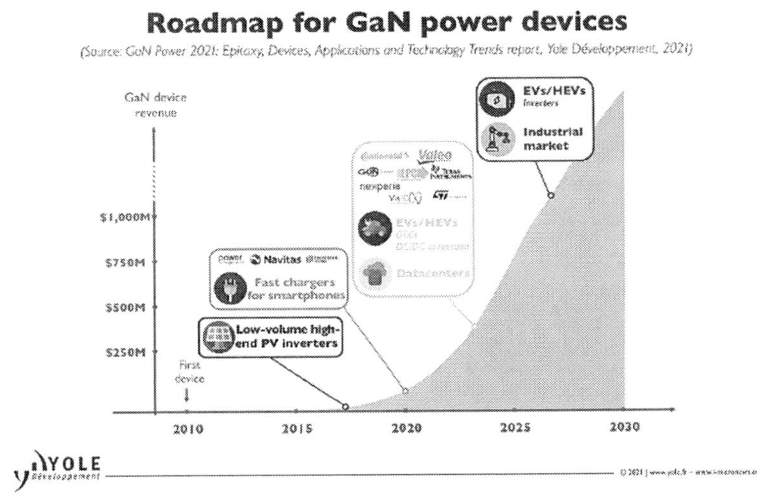

a) Rather optimistic YOLE prediction (2021), *starting from the "Low-volume high-end PV inverters (2017)" over the "Fast chargers for smartphones (2020)", "E-cars, and Datacenters (2023)" to the "EVs/HVs Inverters and Industrial market (2027)"*

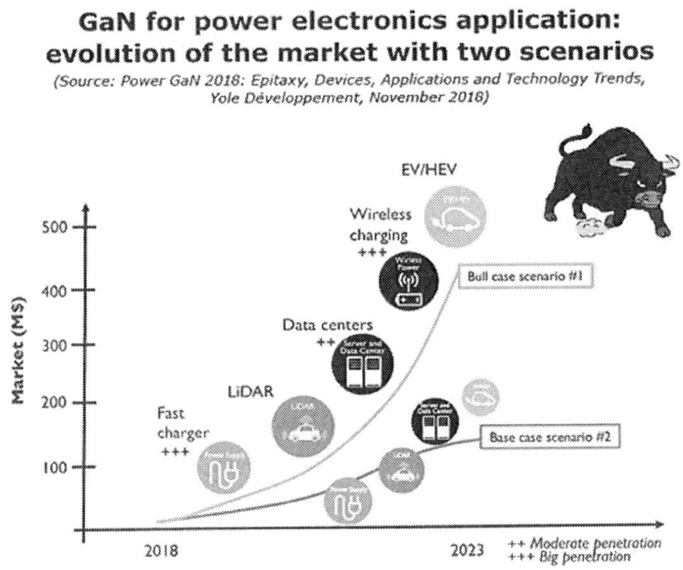

b) GaN power devices market forecasts. ***Bull case scenario*** = *higher or moderate penetration and earlier realizations and as new: "Wireless charging (2022)"* It is about ¼ of all GaN possible applications

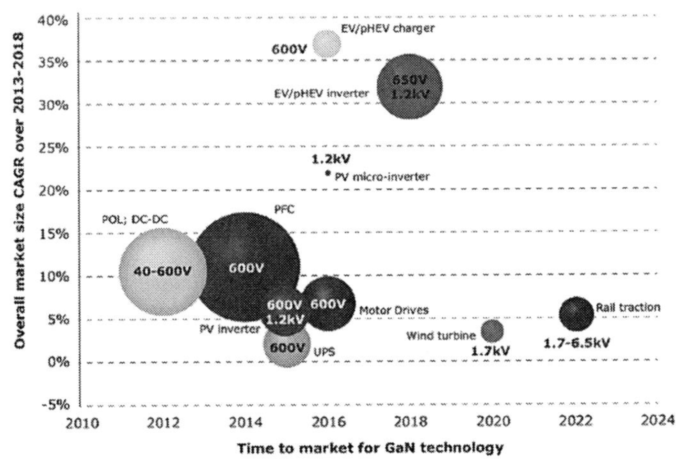

c) GaN devices development versus voltage. ***Dark red 600 V*** *Power Factor Correction (PFC) single-phase AC/DC;* ***Green 40-600 V*** *DC/DC converters and UPS;* ***Red 650 - 1 200 V*** *- EV/pHEV inverter;* ***Violet 600 - 1 200 V*** *- PV inverter, Motor drivers and Rail traction*

Figures 3a, b, c GaN power devices market forecast for different application fields and producer companies

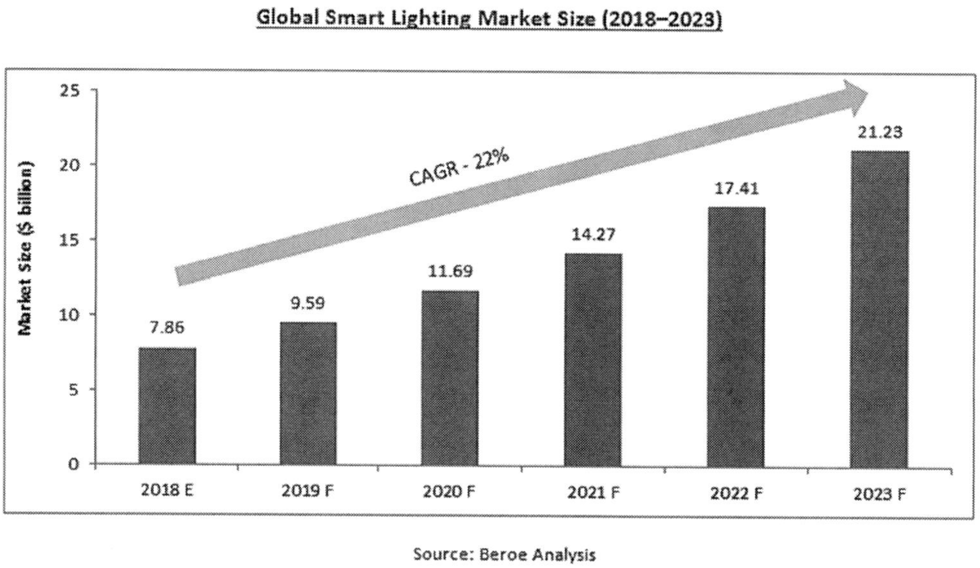

Figure 4 GaN LED market predictions are more than order of magnitude higher than GaN power devices

In comparison with GaN blue/white LED huge market - see **Figure 4**, the GaN power devices market is only 10 % of the LED market, but absolute value of the GaN power devices market is currently about 2 billion USD with estimated strong growth, as it is shown in **Figure 3**.

2. RESULTS OF OUR RESEARCH ON ALGAN/GAN HEMT STRUCTURE

2.1. Regular AlGaN/GaN HEMT structure (without AlGaN back barrier)

In our laboratory were prepared different HEMT structures [1]. The scheme are depicted in **Figure 5**. Structures were prepared on three types of GaN templates differing in dislocation density: GaN substrate with a very low dislocation density of approximately 10^6 cm^{-2}, and two templates on sapphire differing by the GaN buffer technology leading to lower (LDD) and higher dislocation density (HDD). For comparison of transport properties of these structures, see **Table 1**.

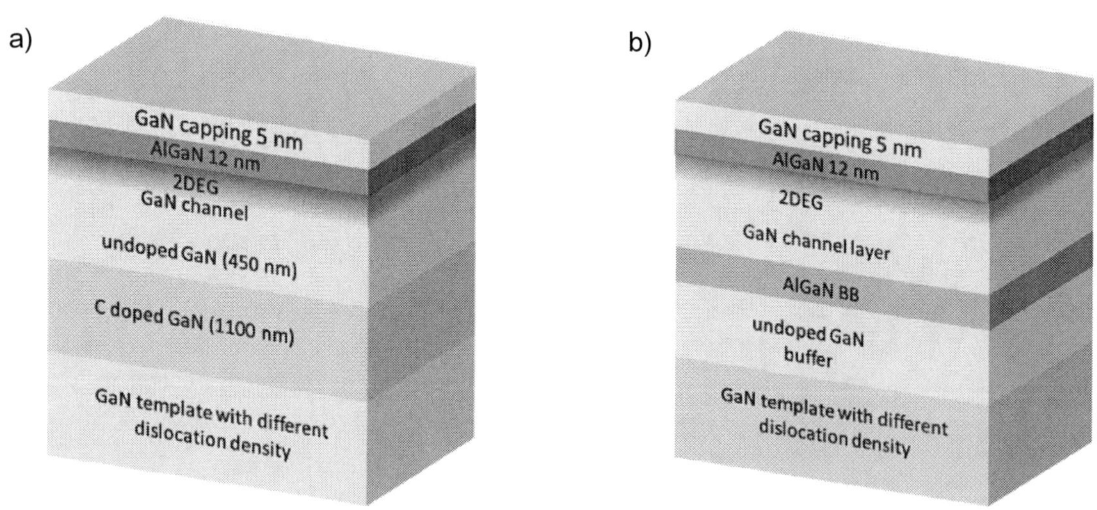

Figure 5 Scheme of **(a)** conventional HEMT structure and **(b)** HEMT structure with AlGaN back-barrier

Table 1 Dislocation density according to XRD, concentration and mobility of samples on different templates

Sample	**Type of template and substrate**	**Dislocation densities** (cm^{-2})	**Apparent 2DEG concentration from CV** (cm^{-2})	**Sheet Hall concentration** (cm^{2})	**Mobility** (cm^2/Vs)
HDD template	HDD on sapphire	2.79×10^{10}	1.20×10^{13}	8.77×10^{12}	1187
LDD template	LDD on sapphire	9.36×10^{8}	1.18×10^{13}	1.56×10^{13}	1360
PLDD	LDD on patterned sapphire	6.81×10^{8}	1.17×10^{13}	1.09×10^{14}	539
GaN substrate	GaN substrate	3×10^{6}	9.78×10^{12}	3.18×10^{15}	798

Surprisingly, poor mobility results were obtained on samples prepared on patterned sapphire or GaN substrates with lower dislocation densities. According to Hall measurements, these samples also revealed very high sheet carrier concentrations. Capacity measurements giving carrier profiles revealed that this unexpectedly high carrier concentration is not caused by carriers in the 2DEG channel, but by carriers from a deeper layer in the heterostructure. In structures prepared on LDD substrate, deeper layers have low resistivity, and conductive channels parallel to 2DEG are dominant in Hall measurement. Parallel currents in nitride structures grown on sapphire substrates were also previously reported [2]. These parasitic channels not only influence Hall measurement results, but they are also detrimental for HEMT applications. In the case of HDD structures, undoped GaN and C-doped GaN regions become highly resistive and they insulate the channel region from the deeper layer of the structure. Thus, only for the HDD structures was the 2DEG active region measured by Hall measurement. For the same reason, it is not a problem to grow industrial HEMT structures on Si substrates, where a high density of dislocations is formed.

In order to study the influence of dislocation density on transport properties of HEMT structures a method for improved localization of electrons in the channel region is required. Such a solution could also be useful for the application of HEMT structures with a low dislocation density. We have found that the widely used carbon doping of buffer layers (see conventional HEMT structure in **Figure 5a** is not reliable and does not provide sufficient isolation, especially for structures with a much lower dislocation density. The lower resistance of C-doped layers causes this, when the dislocation density is low. Fe doping creating a semi-insulating buffer may help to solve this problem [3]. AlN/GaN superlattice can also be used [4]. Another possible solution, which has been recently published [5,6], and which we have decided to follow and investigate in more detail, is the introduction of a so called "back barrier" (BB) layer below the channel region, see **Figure 5b**.

2.2. HEMT structures with AlGaN back barrier - experimental results

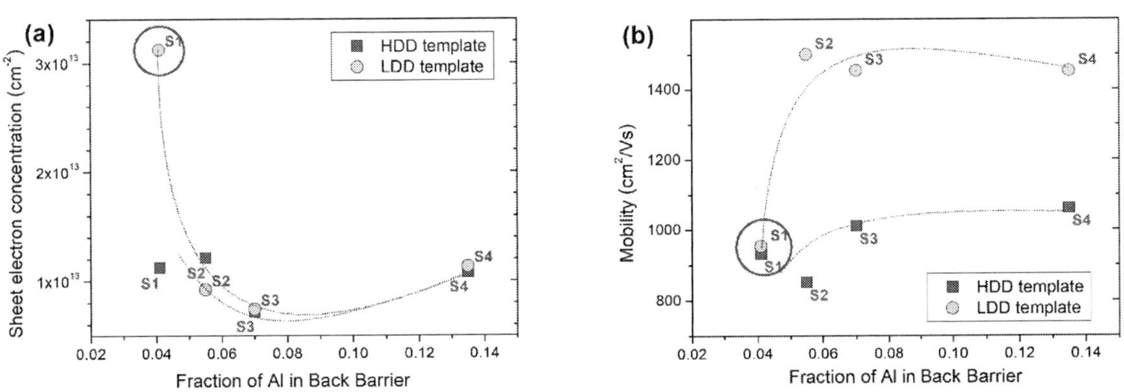

Figure 6 (a) Measured sheet carrier concentration and **(b)** mobility as a function of Al concentration in AlGaN BB for HDD (red squares) and LDD (green circles) samples. The results marked by circle suggest leaking AlGaN BB for LDD sample S1, [1].

Three sets of structures were prepared to identify optimal parameters for a structure with an AlGaN BB and to compare experimental and simulated results. Particular sets of structures used to study the influence of BB composition, thickness of BB and GaN channel layer thickness are color-coded in the table. All structures were prepared on LDD and HDD templates during the same procedure to study the influence of dislocation density on transport properties in 2DEG.

Results of transport measurements are shown in **Figure 6**. The most obvious result is that, in the case of structures prepared on HDD templates, structure parameters do not significantly influence electron mobility in the 2DEG channel.

3. CONCLUSION

We have shown that reduction of dislocation density considerably increases electron mobility in 2DEG. All presented results support our expectation that a suitably designed AlGaN back barrier can help to prevent this phenomenon. The best results were obtained for an $Al_xGa_{1-x}N$ BB composition, with x in the range 0.07-0.15, a BB thickness of 17 nm and a GaN channel layer thickness > 140 nm. For structures prepared on HDD buffer, parallel currents through deeper parts of the heterostructure were not detected, even in the case where no BB or other isolation layer was used. We did not observe any influence of BB parameters on transport properties of samples prepared on HDD templates.

The exact mechanism of how the dislocation density significantly influences 2DEG transport will be the subject of further investigation. Since scattering on dislocations was excluded as a significant scattering mechanism in 2DEG, other mechanisms, such as scattering on ionized impurities surrounding dislocations or the influence of the interface morphology, may be responsible for lowering the electron mobility in structures with a high dislocation density. Finally, the presented results suggest that reduction of dislocation density increases electron mobility in 2DEG, which is beneficial for GaN HEMTs high frequency applications.

ACKNOWLEDGEMENTS

The authors acknowledge support from MEYS project LTAIN19163; from CzechNanoLab project LM2018110 at LNSM Res. Infrastructure and from the Operational Programme RDE financed by EU Structural and Invest. Funds and by MEYS (Project No. SOLID21 CZ.02.1.01/0.0/0.0/16_019/0000760).

REFERENCES

[1] HAJEK, F. et al. Transport properties of AlGaN/GaN HEMT structures with back barrier: impact of dislocation density and improved design. *Sem. Sci. and Technol.* 2021, vol. 36, p. 075016.

[2] LOOK, D. C., MOLNAR, R. J. Degenerate layer at GaN/sapphire interface. *Appl. Phys. Lett.* 1997, vol. 70, pp. 3377-3379.

[3] BOUGRIOUA, Z., et al. Some benefits of Fe doped less dislocated GaN templates for AlGaN/GaN HEMTs grown by MOVPE. *Phys. Status Solidi A.* 2005, vol. 202, pp. 536-544.

[4] SHEN, X. Q., et al. High-quality GaN film and AlGaN/GaN HEMT grown on 4-inch Si (110) substrates by MOCVD using an ultra-thin AlN/GaN superlattice interlayer. *Phys. Status Solidi B.* 2015, vol. 252, pp. 1075-1078.

[5] GU, Y., CHANG, D., SUN, H., ZHAO, J., YANG, G., DAI, Z., DING, Y. Theoretical Study of InAlN/GaN High Electron Mobility Transistor (HEMT) with a Polarization-Graded AlGaN Back-Barrier Layer. *Electronics.* 2019, vol. 8, p. 885.

[6] LIU, X., et al. Analysis of the back-barrier effect in AlGaN/GaN high electron mobility transistor on free-standing GaN substrates. *J. Alloys Compd.* 2020, vol. 814, p. 152293.

MICROSTRUCTURE AND OPTO-ELECTRONIC EFFECTS IN MXENES SPINCOATED FROM POLAR APROTIC SOLVENTS ON ITO

[1]Kezia SASITHARAN, [1]Jaroslav KULICEK, [2]Yaryna SOYKA, [2]Michal PROCHAZKA, [2]Maria OMASTOVA, [1]Bohuslav REZEK

[1]*Centre for Advanced Photovoltaics, Faculty of Electrical Engineering, Czech Technical University in Prague, Prague, Czech Republic, EU*

[2]*Polymer Institute SAS, Bratislava, Slovakia, EU*

https://doi.org/10.37904/nanocon.2021.4307

Abstract

MXenes have drawn considerable attention in the past decade, thanks to their attractive properties such as metallic conductivity and surface hydrophilicity. While MXenes form highly stable dispersions in water, it can act as a limitation for certain applications such as photoactive layer in photovoltaic devices. In this work, delaminated MXenes of aqueous solution type Ti_3C_2 were prepared first, and then we have used a solvent-exchange technique to prepare suspensions of MXenes in three polar aprotic solvents namely, DMSO, DMF and NMP. Upon spin-coating under the same conditions, each solvent variation yields a different thin film morphology - in terms of particle size and surface coverage, as evidenced from AFM investigations. While the MXenes in DMSO yielded large aggregated particles with µm-sized islands in the film, MXenes in DMF and NMP were found to form films with well-dispersed MXene sheets in the size range 250 nm-50 nm and 80 nm-10 nm, respectively. This study also provides additional insights into the microstructure and opto-electronic properties of the MXene thin films using correlative Raman microscopy and photoluminescence spectroscopy. The information provided by this study on the variation in the properties depending on the solvent used to process and spin-cast the films are important for evaluating MXenes in thin film device applications.

Keywords: MXenes, spin-coating, thin-film microstructure, atomic force microscopy, photoluminescence, polar aprotic solvents

1. INTRODUCTION

Ever since the successful exfoliation of a single layer graphene from graphite, there has been an increased interest in two-dimensional materials.[1] In recent years, two dimensional materials such as boron nitride, transition metal dichalcogenides, phosphorenes, co-ordination polymers such as metal-organic framework nanosheets (MONs) and coordination organic frameworks (COFs), have been explored and were found to possess exceptional opto-electronic and mechanical properties.[2] MXenes are a novel class of two-dimensional materials that can be referred to with the general formula $M_{n+1} X_n T_x$ where M is a transition metal, X is carbon or nitrogen, and T represents the surface functional groups (-OH, -O and/or -F).[3] MXenes are derived from inorganic ternary carbides or nitrides (referred to as the MAX phase, where A represents an A-group element). The M-A bonds are more chemically reactive, and therefore the A-layer can be selectively etched upon reaction with hydrofluoric acid (HF).[3] This is followed by a de-lamination step in which molecules such as water, dimethyl sulfoxide (DMSO), tetrabutylammonium hydroxide (TBAOH), etc. are intercalated into the interlayer spacing thereby forcing the layers apart and generating single-flake MXene dispersions.[4] The quality of the produced sheets, their surface chemistry and structural defects depend on the etching conditions.[5]

MXenes are highly conductive and therefore exhibit superior electrical and mechanical properties as compared to other low dimensional materials.[6] The functional groups on the surface make the MXenes hydrophilic and stable when stored as highly concentrated suspensions in water.[7] This enables the creation of MXene inks which can be conveniently stored, transported, and coated onto substrates. These properties have enabled the exploration of MXenes as conductive supports for photocatalysis, as photocatalysts for H_2 production, as electrodes in Li-ion batteries and as micro-supercapacitors.[8] The numerous data available in the literature on the colloidal properties of MXenes in water are useful for these applications. While a lot of new research is focused on using antioxidants as additives to enhance the shelf-life and stability of MXenes in water, very limited studies exist about the evaluation of MXene thin-films cast from different solvents.[9] The lack of these studies limit the possibility of exploring MXenes in applications where water exposure is detrimental, such as in organic or perovskite solar cells.

In this work, we report a solvent-exchange technique replacing water with polar aprotic solvents - dimethyl sulfoxide (DMSO), dimethyl formamide (DMF), and N-methyl pyrrolidone (NMP) - and we explore the microstructure and optoelectronic properties of the thin films spin-coated from each of these solvents. We study the thin film morphology using atomic-force microscopy (AFM) and evaluate the topography and phase images to understand the distribution of the material on the substrate after spin-coating. We have used Raman spectroscopy as a tool for evaluating the structural composition of the films in combination with microstructure mapping which gives additional information about the film morphology. We also report photoluminescence spectra as an important tool in determining the opto-electronics of the films. We anticipate that this preliminary investigation into the thin-film morphology of MXene films when spin-coated from polar aprotic solvents can act as a guideline for applications where MXenes being stored in water can be a limiting factor.

2. EXPERIMENTAL METHODS

2.1. Delaminated MXenes

The starting material for the preparation of MXenes was a MXene paste obtained from Drexel University (internal batch labelling MX_201123_Drexel_paste). This paste was prepared from Ti_3AlC_2 MAX phase using MILD etching method (HCl/LiF) previously reported in literature.[3] Upon etching, an expanded MXene form is obtained, from which the single layers have to be separated. For delamination, the expanded MXenes were mixed with LiF followed by centrifugation to remove and collect single layered nanoparticles. For getting the required concentration, a vacuum concentrator (SpeedVac) was used to remove water and increase the concentration. This resulted in the 0.9 mg/mL dispersion of MXene in water used in this study.

2.2. Solvent -exchange with polar aprotic solvents

For solvent exchange, a known volume of the MXene-water sample was taken and centrifuged at 13000 rpm for 30 minutes and the supernatant above the sediment was pipetted out. The sedimented MXenes were re-dispersed in the polar aprotic solvents - DMSO, DMF and NMP respectively and centrifuged at 13000 rpm for 5 minutes and the supernatant above the sediment was pipetted out. This process of addition of fresh solvent followed by centrifugation and pipetting out the supernatant liquid was performed for 3 cycles, to ensure the complete removal of any residual water. This repeated washing cycle can result in aggregation of the particles; therefore, the final sediment was re-dispersed in fresh solvent and sonicated at 37kHz for 30 minutes and stored for further analysis.

2.3. Spin-coated films

ITO-coated glass substrates obtained from Ossila Ltd were used for the film preparation. The films were prepared using a dynamic spin-coating technique where 50 µl of the MXene dispersion was dropped onto the substrate, followed by a spin-coating at 1000 rpm for 60 seconds, with acceleration of 200 rpm/s. The substrates were then placed on a pre-heated hotplate at 100 °C for 30 seconds for quick drying of any residual solvent. The substrates were then used for further characterization.

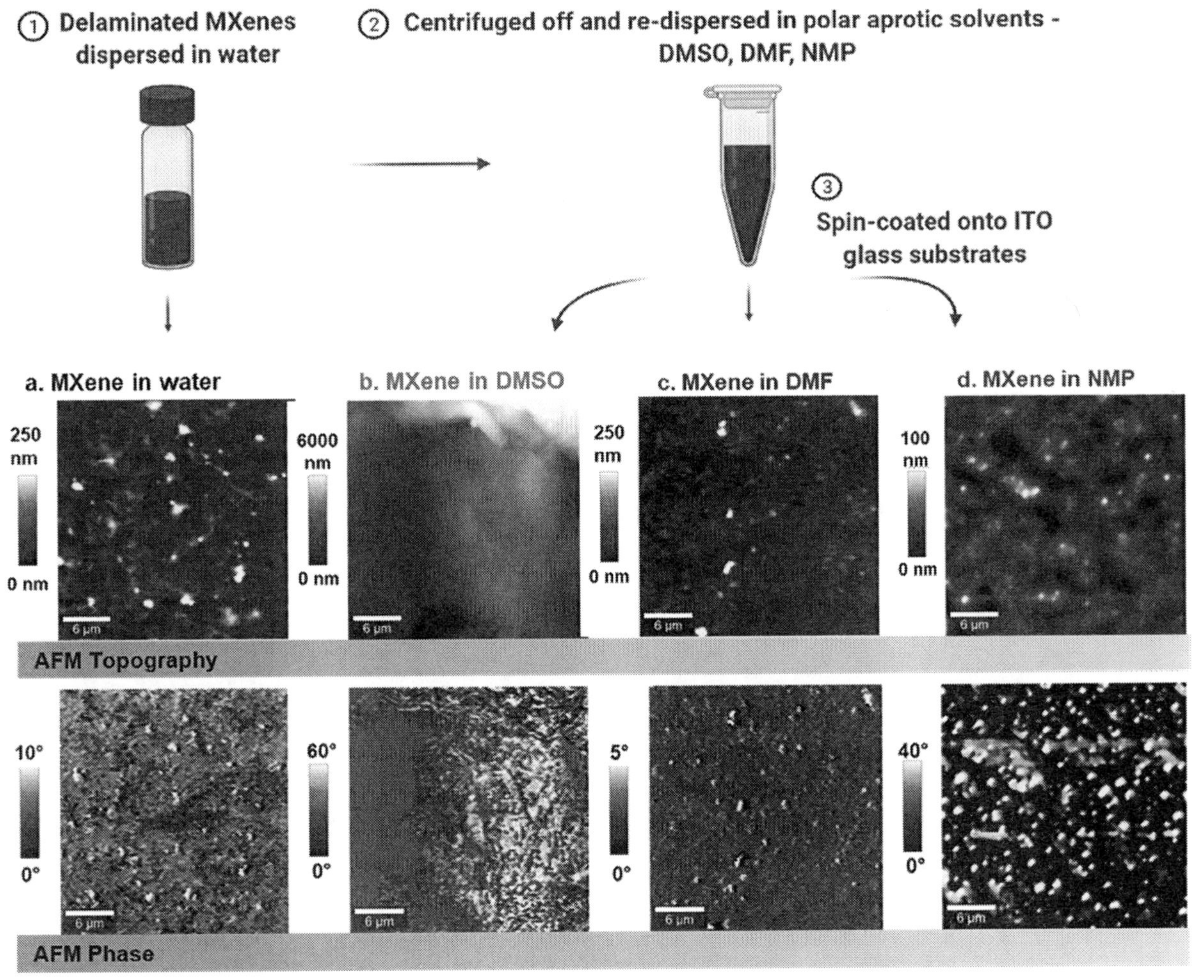

Figure 1 Schematic representation of the processes used for sample preparation in this work and AFM topography and phase images of a representative 30 × 30 μm² area of the samples: (a) MXene-Water, (b) MXene-DMSO, (c) MXene-DMF, and (d) MXene-NMP

3. RESULTS AND DISCUSSION

3.1. Atomic Force Microscopy

Atomic Force Microscopy (AFM) was used as an easy technique to analyze the MXene morphology in the films spin-coated from different solvents. The MXene-water film (**Figure 1a**) was found to be uniform with the sheets appearing in rolled or wrinkled morphology, interspersed with some larger particles of ~200 nm height. In the film cast from MXene-DMSO (**Figure 1b**), there is a stark difference in increased aggregation of MXene particles ending up in μm-sized islands on the surface. In the film cast from MXene-DMF and MXene-NMP (**Figures 1 c,d**) we did not observe such aggregation effects as in DMSO. The particles appear more uniformly distributed. The parameters evaluated from the AFM topography and phase data are summarized in **Table 1**. The average size of the particles with standard deviation in nm calculated from the 30 × 30 μm² scan area shows that the size of particles decreases in the order MXene-DMSO > MXene-Water > MXene-DMF >

MXene-NMP. When we compare the peak to peak roughness, which is the vertical difference between the highest and lowest points on the sample surface, we find that MXene-DMSO gives the highest value at ~3.7 µm, followed by MXene-water at ~603 nm, MXene-NMP at ~133 nm and MXene-DMF at ~72 nm. These results show that MXene in DMF is an ideal solvent medium, resulting in more uniformly sized particle distribution when spin-coated, even better than in commonly used water. On the other hand, MXene in NMP results in a relatively higher number of smaller particles and high AFM phase contrast. Each of these samples have a different phase value than the MXene-water sample, which indicates that there is a variation in the surface properties (including chemistry) depending on the solvent from which the material was spin-coated.

Table 1 Comparison of the particle size, peak to peak roughness and RMS phase values between MXene-water, MXene-DMSO, MXene-DMF and MXene-NMP calculated from Figure 1

Values from AFM	MXene-water	MXene-DMSO	MXene-DMF	MXene-NMP
Average size of particles (nm)	75 ± 49	1153 ± 934	42 ± 14	36 ± 15
Roughness peak to peak (nm)	603	3771	72	133
Phase peak to peak (°)	33	76	24	51

3.2. RAMAN MICROSTRUCTURE AND SPECTRUM

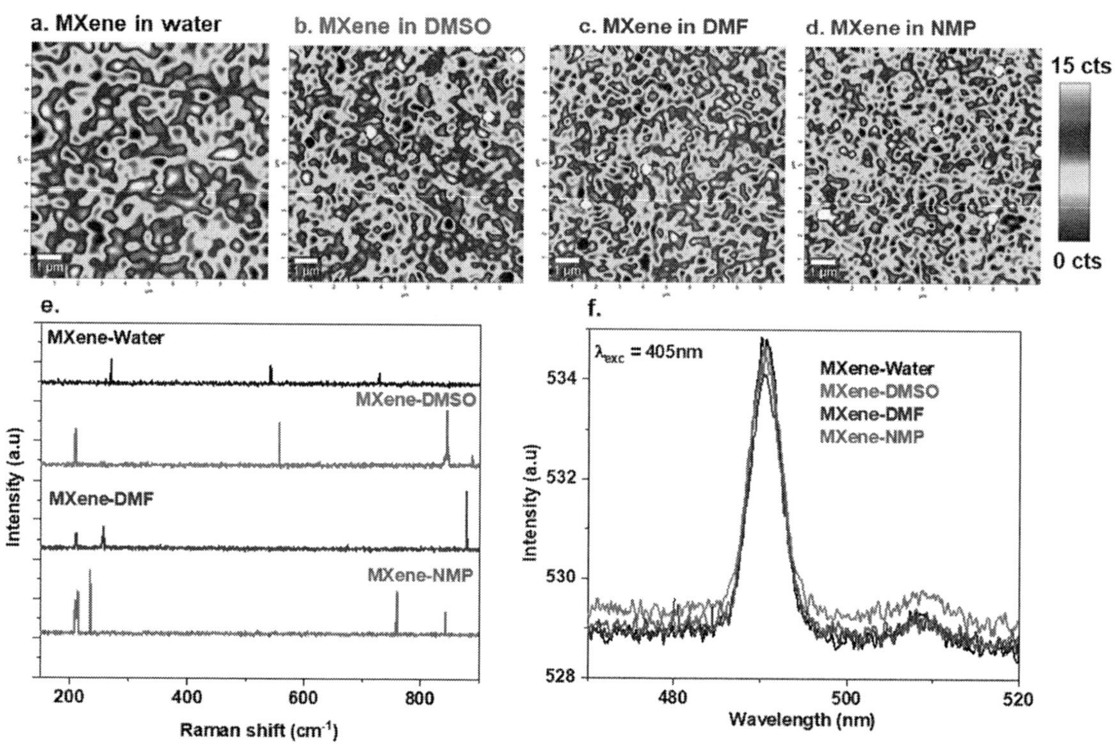

Figure 2 Raman microstructure of the -(a) MXene- Water; (b) MXene-DMSO; (c) MXene-DMF and (d) MXene-NMP films; (e) Acquired Raman single spectrum of the films (f) Photoluminescence spectrum of the samples

Raman spectroscopy has been used as a technique for MXene films to measure the lattice vibrations and provide information about bonding in the structure. When surface groups are added, it is expected that the number of atoms in the unit cell increases and therefore the number of vibrations increases accordingly resulting in an increased number of bands in the Raman spectra. Unit cell distortion and distribution of surface groups result in peak shifting and peak broadening as well. The peaks have been assigned based on the date from previously reported literature.[10] In the spectra reported below in **Figure 2**, the peak at ~200 cm^{-1} corresponds to the A1g vibrations of the MXene flake region. The region between 230 and 470 cm^{-1} correspond to the in-plane E1g vibrations of the surface groups attached to the Ti atoms. We observe the appearance of new peaks in this region in the MXene-DMF and MXene-NMP. The region between 580 and 730 cm^{-1} corresponds to carbon vibrations. One of the issues with MXenes is that they degrade over time upon exposure to oxygen and moisture, resulting into degradation and formation of TiO_2. It is worth noting that the typical Raman peaks of TiO_2 (Eg at 150 cm^{-1}; B1g at 400 cm^{-1}, A1g at 600cm^{-1}) are missing from these films, probably due to its relatively low content in the films.[11] **Figure 2** shows the Raman mapping of the film microstructure corresponding to the A1g vibrations in the MXene flake region. The size of the flakes or particles corresponding to the Raman vibrations are well correlated with the lateral sizes of the protrusions observed in the AFM topography of the MXene films. This confirms that the protrusions or dot-like structures in the AFM images are indeed small-sized MXenes and not TiO_2 dots, and the varying phase contrast is due to the different surface chemistry introduced by the different solvents. The Raman analysis therefore shows the distinction in the molecular fingerprints of the differently processed films, confirms the absence of degradation products and gives a more detailed information about the MXene film microstructure.

3.3. PHOTOLUMINESCENCE

Figure 2f shows that our MXene films exhibit photoluminescence. While MXenes photoluminescence is in contradiction to their metallic band structure, such light emission has previously been attributed to a band gap enlargement effect induced by quantum confinement, applicable to small scale MXene quantum dots.[12] Larger flakes have also previously been reported to show photoluminescence.[13] Another possible mechanism for photoluminescence in MXenes is based on the theory that TiOx can form around Ti vacancies on the surface of $Ti_3C_2T_x$ MXene.[3],[4] While the Raman spectra do not show the typical peaks of TiO_2, it is only an indication of the absence of TiO_2 in large quantities, and is not enough evidence to rule out the possibility of TiO_2 interspersed within the $Ti_3C_2T_x$ films. Investigating these films using X-ray diffraction combined with photoluminescence mapping of the films to gather more insights into the photoluminescence mechanism will be part of our future studies.

4. CONCLUSIONS

Delaminated MXene sheets in aqueous media and MXenes processed in polar aprotic solvents (DMSO, DMF, NMP) were spin-coated onto ITO coated glass substrates and comparatively evaluated using AFM, Raman mapping and photoluminescence. MXenes spin-coated from water, DMF, and NMP formed uniform films, with particle sizes being even smaller and more uniform in the case of DMF and NMP as compared to water. On the other hand, MXenes spin-coated from DMSO gave rise to agglomerated particles with µm-sized features. The Raman single spectra show the presence of all the characteristic Raman peaks of MXenes with distinctive differences in the number of bands and peak shifts. These differences may be attributed to the surface modifications induced by the different solvent environments. Raman mapping showed that the microstructure correlates with the topography observed in AFM. The films exhibit strong photoluminescence upon excitation under UV/Visible light, which can be attributed to either strong confinement effects induced by the small sized particles, or the presence of TiO_2 defects around Ti vacancies in the MXene structure. Further characterization using XPS, FT-IR, XRD, PL, photo-Kelvin probe and APS will be our future work to determine the type of surface modification induced by the different solvents, its correlation with the film microstructure and the resulting effects on the photovoltage. We anticipate that these results will be useful for evaluating MXenes as electrodes, buffer layers or surface passivation layers in devices where use of water-based solvents is detrimental and better control of MXenes dispersions and chemistry may be beneficial.

ACKNOWLEDGEMENTS

The work has been supported by the Centre for Advanced Photovoltaics (CZ.02.1.01/0.0/0.0/15_003/0000464) and from the European Union's Horizon 2020 project Nano2Day, grant agreement No. 777810.

REFERENCES

[1] J. C. MEYER, A. K. GEIM, M. I. KATSNELSON, K. S. NOVOSELOV, T. J. BOOTH and S. ROTH. The structure of suspended graphene sheets. *Nat.* [online]. 2006, vol. 446, no. 7131, pp. 60-63. Available from: https://doi.org/10.1038/nature05545.

[2] K. SASITHARAN *et al.* Metal-organic framework nanosheets for enhanced performance of organic photovoltaic cells. *J. Mater. Chem. A.* [online]. Mar. 2020, vol. 8, no. 12, pp. 6067-6075. Available from: https://doi.org/10.1039/c9ta12313j.

[3] M. ALHABEB *et al.* Guidelines for Synthesis and Processing of Two-Dimensional Titanium Carbide (Ti3C2Tx MXene). *Chem. Mater.* [online]. Sep. 2017, vol. 29, no. 18, pp. 7633-7644. Available from: https://doi.org/10.1021/ACS.CHEMMATER.7B02847.

[4] L. H. KARLSSON, J. BIRCH, J. HALIM, M. W. BARSOUM, and P. O. Å. PERSSON. Atomically Resolved Structural and Chemical Investigation of Single MXene Sheets. *Nano Lett.* [online]. Aug. 2015, vol. 15, no. 8, pp. 4955-4960. Available from: https://doi.org/10.1021/ACS.NANOLETT.5B00737.

[5] X. SANG *et al.* Atomic Defects in Monolayer Titanium Carbide (Ti3C2Tx) Mxene. *ACS Nano.* [online]. Oct. 2016, vol. 10, no. 10, pp. 9193-9200. Available from: https://doi.org/10.1021/ACSNANO.6B05240.

[6] H. TANG, R. WANG, L. SHI, E. SHEREMET, R. D. RODRIGUEZ and J. SUN. Post-processing strategies for improving the electrical and mechanical properties of Mxenes. *Chem. Eng. J.* [online]. Dec. 2021, vol. 425, p. 131472. Available from: https://doi.org/10.1016/J.CEJ.2021.131472.

[7] B. AKUZUM *et al.* Rheological Characteristics of 2D Titanium Carbide (MXene) Dispersions: A Guide for Processing Mxenes. *ACS Nano.* [online]. Mar. 2018, vol. 12, no. 3, pp. 2685-2694. Available from: https://doi.org/10.1021/ACSNANO.7B08889.

[8] A. BHAT, S. ANWER, K. S. BHAT, M. I. H. MOHIDEEN, K. LIAO and A. QURASHI. Prospects challenges and stability of 2D MXenes for clean energy conversion and storage applications. *npj 2D Mater. Appl.* [online]. Jun. 2021, vol. 5, no. 1, pp. 1-21. Available from: https://doi.org/10.1038/s41699-021-00239-8.

[9] X. ZHAO *et al.* Antioxidants Unlock Shelf-Stable Ti3C2Tx (MXene) Nanosheet Dispersions. *Matter.* [online]. Aug. 2019, vol. 1, no. 2, pp. 513-526. Available from: https://doi.org/10.1016/J.MATT.2019.05.020.

[10] M. SHEKHIREV, C. E. SHUCK, A. SARYCHEVA, and Y. GOGOTSI. Characterization of MXenes at every step, from their precursors to single flakes and assembled films. *Prog. Mater. Sci.* [online]. Jul. 2021, vol. 120, p. 100757. Available from: https://doi.org/10.1016/J.PMATSCI.2020.100757.

[11] A. GRZEGÓRSKA *et al.* Enhanced photocatalytic activity of accordion-like layered Ti3C2 (MXene) coupled with Fe-modified decahedral anatase particles exposing {1 0 1} and {0 0 1} facets. *Chem. Eng. J.* [online]. Dec. 2021, vol. 426, p. 130801. Available from: https://doi.org/10.1016/J.CEJ.2021.130801.

[12] Q. XUE *et al.* Photoluminescent Ti3C2 MXene Quantum Dots for Multicolor Cellular Imaging. *Adv. Mater.* [online]. Apr. 2017, vol. 29, no. 15, p. 1604847. Available from: https://doi.org/10.1002/adma.201604847.

[13] L. ZHANG *et al.* Tuning the photoluminescence of large Ti3C2Tx MXene flakes. *Ceram. Int.* [online]. Jun. 2019, vol. 45, no. 9, pp. 11468-11474. Available from: https://doi.org/10.1016/J.CERAMINT.2019.03.014.

GROWTH MECHANISM AND FUNCTIONAL PROPERTIES OF GAS-PHASE SYNTHESIZED FEW-LAYER GRAPHENE

[1]Ondřej JAŠEK, [1]Jozef TOMAN, [1]Miroslav ŠNÍRER, [1]Vít KUDRLE, [1]Jana JURMANOVÁ, [1]Dalibor VŠIANSKÝ

[1]*Faculty of Science - Masaryk University, Brno, Czech Republic, EU, jasek@physics.muni.cz*

https://doi.org/10.37904/nanocon.2021.4305

Abstract

Microwave plasma-based gas-phase synthesis of graphene and its derivatives represents a simple, environmentally friendly, and easily scalable production method enabling emerging large-scale applications. Microwave plasma decomposition of organic precursors, hydrocarbons, and alcohols, forms a high-temperature environment, where the process of catalyst-free dehydrogenation and consequent formation C atoms and C_2 molecules leads to the nucleation and growth of high-quality few-layer graphene nanosheets. Structure of the synthesized carbon material changed from graphitic nanoparticles to graphene nanosheets with increasing H/C atomic ratio in the precursors, for C_2H_2 and CH_4, respectively. Using alcohols, ethanol - C_2H_5OH and diethylether - $(C_2H_5)_2O$, the main hydrocarbon fragments generated during the decomposition of C_xH_yO structure resulted in the gas composition with high enough H/C atomic ratio to growth nanosheets with low intensity Raman D band and high integrated intensity ratio of 2D/G Raman band, 1.5 and higher. However, the yield of the synthesis strongly depended on the arrangement of atoms in the precursor molecule. Prepared material was analyzed by electron microscopy, Raman and X-ray photoelectron spectroscopy and thermogravimetry. High temperature oxidation resistance and electrical conductivity of prepared samples was investigated.

Keywords: Graphene, microwave plasma, growth mechanism

1. INTRODUCTION

Gas-phase synthesis of graphene represents a simple, environmentally friendly, and easily scalable production method [1] enabling deployment of carbon nanomaterials in the wide range of emerging large-scale applications such as energy storage, internet-of-things sensors and flexible electronics [2].

Gas-phase growth of graphene from hydrocarbon and alcohol precursors proceeds through consecutive dehydrogenation reactions forming carbon gas composed from C atoms and C_2 molecules which condensate into graphene nuclei in the supersaturation conditions. However, this synthesis process is realized only at a high-enough temperature environment, above 4500 K, and some of the gas-products stay in the form of simple hydrocarbons (CH_4, C_2H_2, C_2H_4 etc.) [3,4]. Depending on the plasma source and stability of the environment [5], the precursor can be partially decomposed by a so-called low temperature channel, where higher hydrocarbons and aromatic compounds containing large amounts of hydrogen are produced [6].

Such reactions can also produce solid state material in the form of soot - carbon black. Frenklach and Wang [7] suggested so called hydrogen abstraction acetylene addition (HACA) mechanism, where benzene molecule, formed by conversion from acetylene, further grows into more complex aromatic molecules, such as naphthalene, by addition of another acetylene molecule. As proposed in the HACA mechanism, the addition of acetylene is followed by consecutive hydrogen abstraction, which removes the H atoms during the growth phase from the graphene scaffold, leaving the lattice composed of carbon atoms. In case of alcohol based precursors, the presence of oxygen modifies the reaction pathway leading to the formation of C atoms and C_2

molecules from hydrocarbon based products of initial decomposition reaction while the other part of the precursor, containing oxygen atom, is converted into CO or CO_2 byproduct depending on the high or low temperature channel of the reaction pathway [4].

In this work we study the growth of graphene nanosheets from precursors with different H/C ratio at atmospheric pressure microwave plasma torch and investigate the high temperature oxidation resistance and electrical conductivity of such prepared material.

2. EXPERIMENTAL

Graphene nanosheets were synthesized by decomposition of hydrocarbons (methane CH_4, acetylene C_2H_2 (19 - 38 sccm)) or alcohols (ethanol C_2H_5OH (99.8 % p.a. Penta), diethylether $(C_2H_5)_2O$ (99 % p.a. Penta) (10 - 100 sccm)) mixed with argon carrier gas using a dual-channel microwave plasma torch at atmospheric pressure. The discharge was ignited in a reactor chamber formed by a quartz tube (8 cm in diameter and 20 cm length enclosed by aluminium flanges) in argon working gas, Q_{ArC} = 500 sccm, flowing through the stainless-steel nozzle central channel, 0.8 mm diameter (**Figure 1**), applying microwave power of 270 W. The secondary channel (annulus with outer radius 8.4 mm and inner radius 7.7 mm) was used for the introduction of carbon precursor and argon carrier gas (Q_{ArS} - 100 - 700 sccm). The deposition time was 5 - 10 minutes and the synthesized nanopowder was scraped by brush from the reactor walls after cooling down in argon. More details about the experimental procedure can be found in our recent publications [5,8].

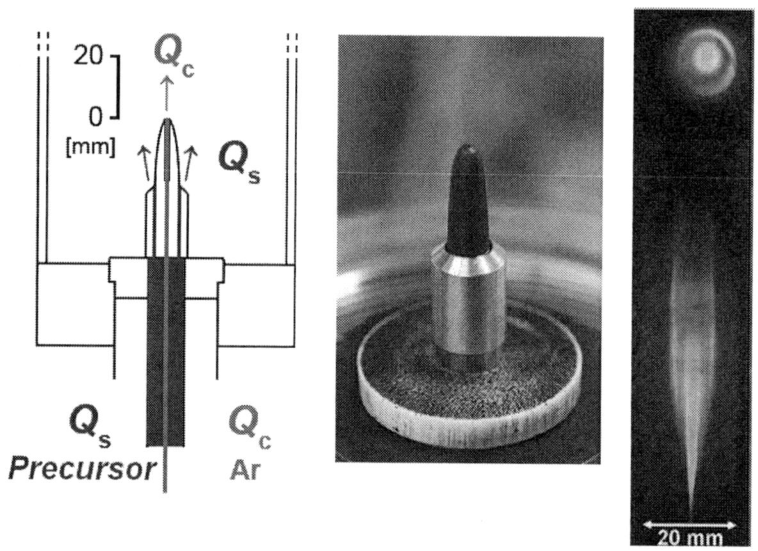

Figure 1 Detail of dual-channel nozzle scheme (left) and its photograph (center). Photograph of the microwave plasma torch discharge in argon with ethanol admixture (right), including top view.

Samples were imaged with MIRA3 (Tescan, Czech Republic) scanning electron microscope (SEM) with Schottky field emission electron gun and equipped with a secondary electron (SE) and back-scattered electron (BSE) detectors as well as Oxford Instruments EDX analyzer (Oxford, England). XPS analysis was performed using a Kratos Analytical Axis Supra (KRATOS-XPS, Kratos Analytical Ltd., UK) spectrometer with a monochromatic Al source using a pass energy of 80 eV for wide spectra and of 20 eV for high-resolution analysis of C1s and O1s peaks. The analyzed regions were 300 × 700 μm in size. The deconvolution of mentioned peaks was carried out in the CasaXPS 2.3.22PR 1.0 software. Raman spectroscopy was carried out using the HORIBA LabRAM HR Evolution system with 532 nm laser, using 100x objective, and 5% ND filter (500 mW maximum power), 600 gr/mm grating, and 30 s acquisition time in the range from 500 to 3350 cm^{-1}.

Thermogravimetric analysis (TGA) was carried out using a Setaram Setsys Evoluton 1750 instrument. The analyses were conducted in a dynamic air atmosphere (20 $sccm^{-1}$) with a constant heating rate of 10 $°C.min^{-1}$ in the temperature range from 40 to 1000 °C. The obtained data were processed using the Setaram Processing software. Electrical conductivity of mechanically pressed carbon nanomaterial tablets (10 mm in diameter, 10 mg of nanopowder) were measured by the Ossila 4-probe point system.

3. GRAPHENE NANOSHEETS SYNTHESIS BY DECOMPOSITION OF HYDROCARBONS

The decomposition of hydrocarbons by atmospheric pressure microwave discharges leads to formation of a high temperature environment, with gas temperatures above 4500 K. At such conditions, the reaction pathway of the high temperature decomposition of hydrocarbons preferred the dehydrogenation reaction leading to the formation of C_2H molecules and consequently C atoms and C_2 molecules. However, the equilibrium between C_xH_y products and graphene nanosheets nucleation and growth depends on the amount of carbon and hydrogen atoms in the reaction [9].

To study this process, we carried out synthesis using CH_4 (38 sccm) and C_2H_2 (19 sccm) precursors with different H/C ratio, using the same amount of carbon in the deposition mixture. SEM analysis of the samples synthesized under these conditions can be seen in **Figure 2a**. In the case of CH_4, we expect the formation of C_2H_x molecules and subsequent formation of C_2H_2 and C_2 molecules leading to the formation of graphene nanosheets. Indeed, we can observe objects with nanosheet structure with lateral dimensions of hundreds of nm synthesized using the CH_4 precursor. Such structures exhibited a low number of defects as determined by Raman spectroscopy analysis (**Figure 3a**). Low D peak, at 1340 cm^{-1}, intensity was accompanied with high intensity G peak, at 1575 cm^{-1}, and high intensity 2D peak at 2682 cm^{-1}. The full-width half maximum (FWHM) of the 2D peak was 55 cm^{-1}, consistent with a few-layer graphene structure [10]. The integrated intensity ratio of ID/IG and 2D/IG was 0.54 and 1.39, respectively. Low intensity D', G* and D+G peaks at 1614, 2450 and 2928 cm^{-1}, respectively, could be observed as well.

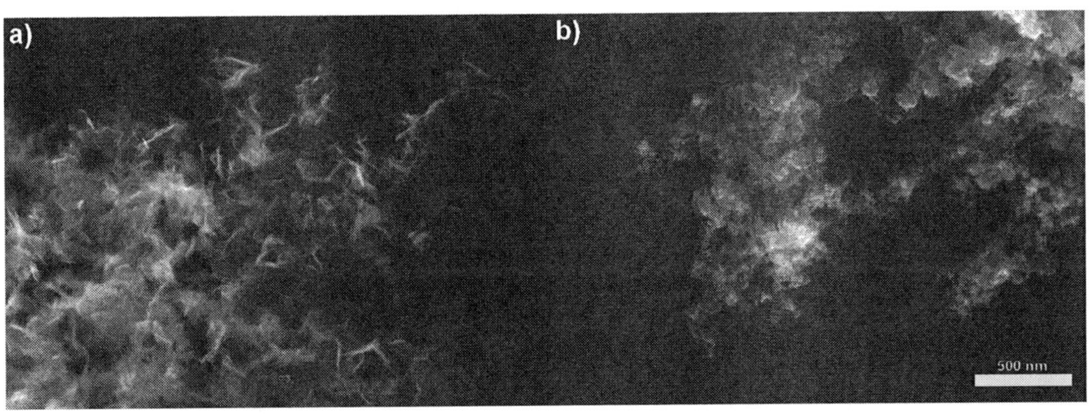

Figure 2 SEM analysis of graphene nanosheets prepared by decomposition of a) CH_4 (38 sccm), H/C = 0.25 and b) C_2H_2 (19 sccm), H/C = 1. Q_{ArC} = 500 sccm, P = 270 W

On the other hand, synthesis using acetylene (C_2H_2) as a precursor led to the formation of spherical carbon nanoparticles (**Figure 2b**) with diameters ranging from 10 to 100 nm. The high degree of disorder in such a structure resulted in the change of character of its Raman spectra (**Figure 3a**), where high intensity D peak, at 1336 cm^{-1}, as well D*, D** and D' peaks at 1210, 1450 and 1610 cm^{-1} could be observed. Intensity of the 2D peak was very small in comparison to the CH_4 sample. This result was consistent with structures obtained by Choi et al. [11], using high-temperature laser irradiation of acetylene. They suggested that a process like the HACA mechanism or direct decomposition of acetylene into C_2 and H_2 on the surface of graphene nuclei at high temperatures led to the defect formation and enclosing of the structure.

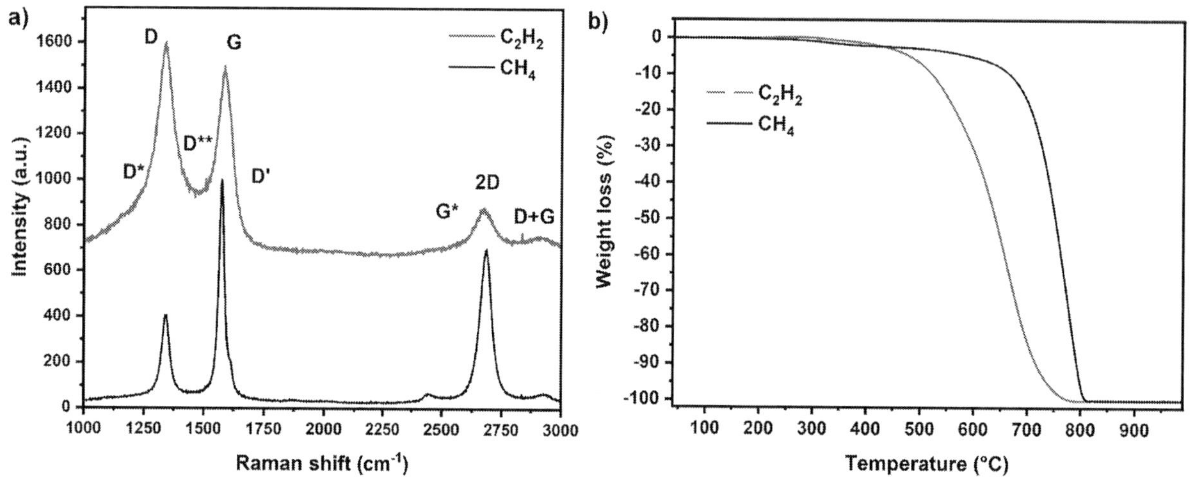

Figure 3 a) Raman spectroscopy and b) TGA of samples prepared by decomposition of CH_4 and C_2H_2. Q_{CH4} = 38 sccm, H/C = 0.25, Q_{C2H2} = 19 sccm, H/C = 1, Q_{ArC} = 500 sccm, P = 270 W.

As expected, XPS analysis of both samples showed a high content of carbon, over 99 %, the rest being oxygen. The presence of oxygen can be explained by a small amount of air impurities in the reaction chamber during the experiment as well as the surface contamination of the sample after being exposed to the air atmosphere. The sp^2/sp^3 ratio determined by fitting of C1s peak profile was 9.9 and 6.8 for CH_4 and C_2H_2 samples, respectively.

Amount of disorder in the material substantially influenced its high temperature oxidation resistance measured by TGA as can be seen in **Figure 3b**. Carbon nanoparticles prepared by decomposition of C_2H_2, started to combust, around 500 °C, and the exhibited total weight loss around 780 °C, while the graphene nanosheets prepared from CH_4 were stable up to 650 °C with total weight loss temperature of 815 °C.

4. GRAPHENE NANOSHEETS SYNTHESIS BY DECOMPOSITION OF ALCOHOLS

In recent years, the synthesis of graphene nanosheets using high temperature plasma discharges was focused, except for methane, to ethanol as a precursor. Dato and Frenklach [12] suggested that ethanol possessed an ideal H/C/O ratio for graphene nanosheet synthesis. This idea was suggested after successful synthesis of graphene from ethanol and its isomer dimethylether CH_3OCH_3. Diethylether structure with two CH_3CH_2 groups connected by an oxygen atom represents an extension of the ethanol molecular arrangement C_2H_5OH. However, in this case the ratio between the number of H, C and O atoms changes substantially.

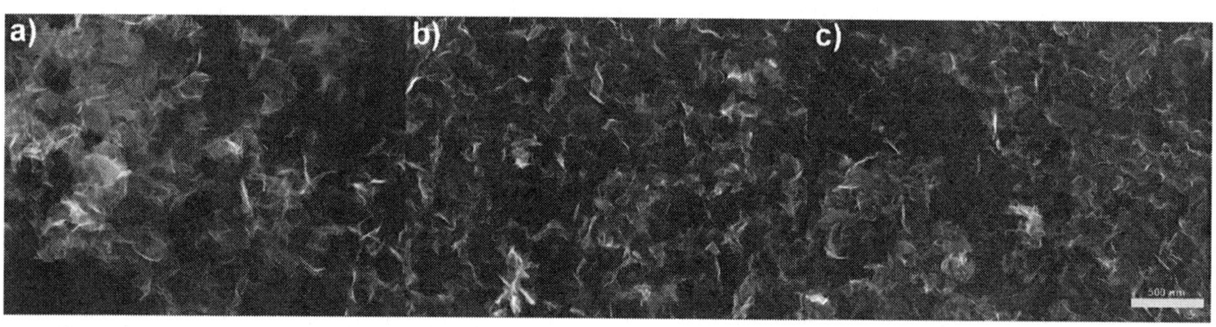

Figure 4 SEM analysis of graphene nanosheets prepared by decomposition of a) ethanol (Q_{eth} = 50 sccm), H/C = 3, b) diethylether+H_2 (Q_{dieth}/Q_{H2} = 100/50 sccm), H/C = 2.75 and c) diethylether (Q_{dieth} = 100 sccm), H/C = 2.5. Q_{ArC} = 500 sccm, P = 270 W

We carried out synthesis using $(C_2H_5)_2O$ with and without H_2 admixture to investigate the influence of various H/C atomic ratio on the synthesis process and compared the results with material prepared using ethanol.

As can be seen on SEM micrographs in **Figure 4**, the synthesis of graphene nanosheets was successful using all precursors. Raman spectroscopy analysis of prepared structures can be seen in **Figure 5a**. All samples exhibited a low amount of structural disorder, ID/IG Raman band ratio 0.5 - 0.6, and high ratio of I2D/IG, of Raman bands, 1.6 - 1.7, leading to the conclusion that the structural properties did not change even with change of H/C ratio from 3 to 2.5. Also, the change of O/C ratio from 0.5to 0.25 did not influence the quality of prepared nanostructures. Therefore, the conversion of part of the precursor molecule into CO during the synthesis had no effect on the quality of the material. Additional hydrogen had negligible influence on the synthesis process. At the same time, we could observe around 4 times lower yield of the synthesis, by weighting the produced nanopowder, using diethylether precursor. This result suggests precursor molecular arrangement plays an important role in the efficiency of the synthesis process.

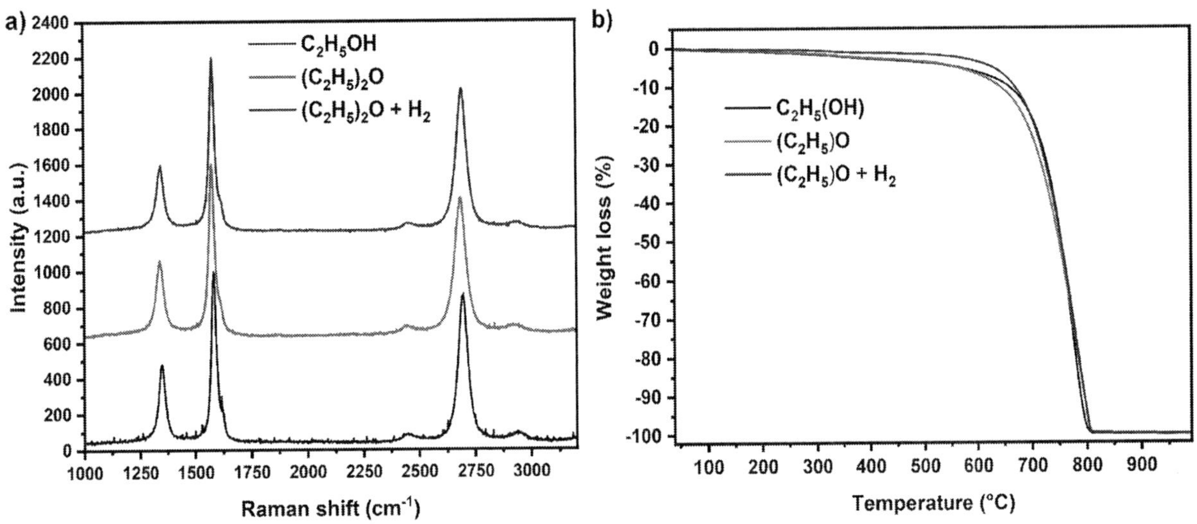

Figure 5 a) Raman spectroscopy and b) TGA of samples prepared by decomposition of ethanol and diethylether precursors. Q_{eth} = 50 sccm, Q_{dieth} = 100 sccm, Q_{H2} = 50 sccm, Q_{ArC} = 500 sccm, P = 270 W

XPS analysis of all samples showed similarly to samples prepared from hydrocarbons a high, but slightly lower content of carbon, over 98 %, the rest being oxygen. The sp^2/sp^3 ratio determined by fitting C1s peak profile was 19 and 21 for C_2H_5OH and $(C_2H_5)_2O$ samples, respectively.

Given the low amount of disorder and high $sp^{2/}sp^3$ ratio of all the samples prepared by decomposition of alcohol precursors exhibited excellent oxidation resistance as can be seen in **Figure 5b**. The measured TG curves in air had almost identical profiles with total mass loss around 810 °C, like the CH_4 sample, and in good agreement with our previously published study [8].

5. MEASUREMENT OF ELECTRICAL CONDUCTIVITY OF GRAPHENE NANOPOWDER TABLETS

From an application point of view the good thermal stability under oxidation atmosphere combined with a good electrical conductivity can be of an advantage. Different carbon nanomaterials create conductive pathways depending on their geometry and a form of layers created from them. To minimize such an effect, we measured conductivity of tablets with the same diameter formed from the same amount of nanopowder (**Table 1**). As we can see, all the graphene nanosheet samples showed similar value of conductivity ~ 1000 $S.m^{-1}$, while the conductivity of the tablet formed from carbon nanoparticles synthesized using C_2H_2 precursor was much lower 60 $S.m^{-1}$. The electrical conductivity was higher for samples with a lower amount of disorder in the material, lower ID/IG ratio, but this comparison was not valid for thermally expanded graphite. This sample exhibited a high ID/G ratio, but the conductivity was comparable to graphene nanosheets. This can be caused by sample structure, nanoflakes of pure graphite, which exhibit very high conductivity and form a very good conductive path.

Table 1 Electrical conductivity of carbon nanomaterial pellets prepared by decomposition of various carbon precursors. Conductivity measurement uncertainty ± 4 %

Precursor	CH_4	C_2H_2	C_2H_5OH	$(C_2H_5)_2O$	$(C_2H_5)_2O$+ H_2	expanded graphite
Conductivity ($S.m^{-1}$)	980	60	1140	1090	1150	1180
Raman ID/IG ratio	0.54	1.60	0.54	0.61	0.54	2.48

6. CONCLUSIONS

We investigated the synthesis of graphene nanosheets using microwave plasma torch at atmospheric pressure by decomposition of hydrocarbon and alcohol precursors with various H/C/O composition ratio. We showed that the use of all precursors with H/C ratio higher than 2, led to the synthesis of graphene nanosheets with low ID/IG and high I2D/IG Raman band integrated intensity ratio and that presence of the oxygen did not substantially influence the nanosheet's structure. However, the arrangement of atoms of the precursor molecule influenced the decomposition reaction pathway and consequently the yield of the synthesis. This result was consistent with the dehydrogenation reaction growth mechanism of graphene nanosheets. Higher quality material exhibited substantially higher electrical conductivity and high temperature oxidation resistance.

ACKNOWLEDGEMENTS

This work was supported by The Czech Science Foundation under project 18-08520S and in part by project LM2018097 funded by the Ministry of Education, Youth and Sports of the Czech Republic. CzechNanoLab project LM2018110 funded by MEYS CR is gratefully acknowledged for the financial support of the XPS measurements by David Pavliňák at CEITEC Nano Research Infrastructure.

REFERENCES

[1] DATO, A. Graphene synthesized in atmospheric plasmas-A review. *J. Mater. Res.* 2019, vol. 34, no. 1, pp. 214-230.

[2] YANG,J., HU, P. and YU, G. Perspective of graphene-based electronic devices: Graphene synthesis and diverse applications. *APL Mater.* 2019, vol. 7, no. 2, p. 020901.

[3] SHAVELKINA, M. B., FILIMONOVA, E. A. and AMIROV, R. K. Effect of helium/propane-butane atmosphere on the synthesis of graphene in plasma jet system. *Plasma Sources Sci. Technol.* 2020, vol. 29, no. 2, p. 025024.

[4] RINCON, R. et al. Experimental research on ethanol-chemistry decomposition routes in a microwave plasma torch for hydrogen production. *Chem. Eng. J.* 2016, vol. 284, pp. 1117-1126.

[5] TOMAN, J., JASEK, O., SNIRER, M., KUDRLE, V., and JURMANOVA, J. On the interplay between plasma discharge instability and formation of free-standing graphene nanosheets in a dual-channel microwave plasma torch at atmospheric pressure. *Journal of Physics D: Applied Physics.* 2019, vol. 52, no. 26. p. 265205.

[6] HARRIS, S. J. and WEINER, A. M. Chemical kinetics of soot particle growth. *Annu. Rev. Phys. Chem.* 1985, vol. 36, no. 1, pp. 31-52.

[7] FRENKLACH, M. and WANG, H. Detailed modeling of soot particle nucleation and growth. *Symp. Combust.* 1991, vol. 23, no. 1, pp. 1559-1566.

[8] JAŠEK, O. et al. Controlled high temperature stability of microwave plasma synthesized graphene nanosheets. *J. Phys. D Appl. Phys.* 2021, vol. 54, no. 16, p. 165201.

[9] TSYGANOV, D.et al. On the plasma-based growth of 'flowing' graphene sheets at atmospheric pressure conditions. *Plasma Sources Sci. Technol.* 2016, vol. 25, no. 1, p. 015013.

[10] FERRARI, A. C. et al. Raman spectrum of graphene and graphene layers. *Phys. Rev. Lett.* 2006, vol. 97, no. 18, p. 187401.

[11] CHOI, M. et al. Formation of shell-shaped carbon nanoparticles above a critical laser power in irradiated acetylene. *Adv. Mater.* 2004, vol. 16, no. 19, pp. 1721-1725.

[12] DATO, A. and FRENKLACH, M. Substrate-free microwave synthesis of graphene: experimental conditions and hydrocarbon precursors. *New J. Phys.* 2010, vol. 12, no. 12, p. 125013.

ACTIVATED CARBON FIBERS DERIVED FROM SEA BALLS FOR THE USE AS SUPERCAPACITOR ELECTRODES

[1,2]Stefan BREITENBACH, [1]Christoph UNTERWEGER, [2]Achim Walter HASSEL, [1]Christian FÜRST

[1]*Wood K plus - Kompetenzzentrum Holz GmbH, Linz, Austria, EU,* s.breitenbach@wood-kplus.at

[2]*Institute of Chemical Technology of Inorganic Materials (TIM), Johannes Kepler University Linz, Linz, Austria, EU*

https://doi.org/10.37904/nanocon.2021.4308

Abstract

Sea balls (pillae marinae) were impregnated with diammonium hydrogen phosphate (DAHP), carbonized, and activated using carbon dioxide or steam. Thus, activated carbon fibers (ACFs) with unique morphology and specific surfaces of 243 $m^2 g^{-1}$ were produced. These ACFs were studied as electrode materials for supercapacitors that achieved specific capacities of 38 F g^{-1} in an organic electrolyte, resulting in specific energy for the supercapacitor of 9.7 W h kg^{-1}.

Keywords: Activated carbon, electrode materials, supercapacitor, energy storage, sea balls, pillae marinae

1. INTRODUCTION

Energy storage is a key technology in the ongoing efforts to combat climate change. Supercapacitors are highly regarded energy storage devices due to their ability to absorb and release large amounts of energy very quickly. They are also characterized by their long service life of over 1 million charging cycles and low maintenance costs [1-3]. In this type of energy storage, the mostly carbon-based electrode materials are of central interest.

A promising material for supercapacitor electrodes is activated carbon fibers (ACFs). These ACFs have a high specific surface area, a well-suited pore structure, and possess excellent conductivity in the electrode due to their fiber shape, as we were able to find out in previous work [4-6]. When ACFs are used in electrode materials, high-purity substances such as viscose fibers, pitch fibers, or polyacrylonitrile fibers are usually used. These precursors have to be synthesized in advance in a complex process.

In this study, sea balls will be used as a starting material for the production of ACFs. In contrast to other starting materials for ACFs, no complex manufacturing process is required here; the sea balls are a naturally occurring, biological raw material. They are produced by the wave action in the sea by tearing out and felting dead parts of the seagrass (often neptune grass). When they are washed up on beaches they are removed by heavy machinery. Apart from its use as an insulating material, there are no known industrial applications for this raw material.

2. MATERIALS AND METHODS

The sea balls were collected on the beach of Talamone, Italy. They were washed in distilled water in a drum hoop mixer for 1 h and dried at 80 °C in a drying oven. Depending on the sample, treatment with a 10 wt.% DAHP solution was performed (see **Table 1**). For this purpose, the samples were soaked in the 10 wt.% DAHP solution for 15 min and then dried at 80 °C overnight. Carbonization was carried out in a chamber furnace (HTK8, Gero, Germany) with a heating rate of 10 °C min^{-1} to 850 °C under nitrogen atmosphere. Before cooling, the samples were held isothermally at the target temperature for 30 min. Activation was carried out in a rotary kiln (RSR-B 120/500/11, Nabertherm GmbH, Germany). Prior to use, N_2 was used to purge the apparatus at a flow rate of 100 l h^{-1}. Samples were then heated from room temperature to the desired activation temperature T under an N_2 flow rate of 50 l h^{-1}. To ensure that the temperature was uniform throughout the reaction chamber, the sample was kept isothermal under the nitrogen flow for 30 min. Then, the N_2 flow was terminated and replaced by CO_2 at the desired flow rate Q (80 l h^{-1}). The completion of the activation process was marked by stopping the CO_2 flow and restarting the N_2 flow until the oven cooled down to room temperature.

Table 1 Overview of the prepared samples

Sample	Sample treatment	Activation conditions
SB_165	Washing, drying; carbonization	T = 870 °C; Q = 80 l h^{-1}; t = 165 min
SB_DAHP_165	Washing, drying; impregnation with a 10 wt.% DAHP soultion; drying; carbonization	T = 870 °C; Q = 80 l h^{-1}; t = 165 min
SB_DAHP_285	Washing, drying; impregnation with a 10 wt.% DAHP soultion; drying; carbonization	T = 870 °C; Q = 80 l h^{-1}; t = 285 min

The morphology of the samples was investigated using Scanning Electron Microscopy (SEM) with Phenom ProX (Thermo Fisher Scientific, USA). The chemical composition was determined by using energy dispersive X-ray spectroscopy (EDX) by using the same instrument. The acceleration voltage was 15kV and a backscattered electron detector (BSD) was used.

The specific surface area of the activated sea balls was determined using physical adsorption of nitrogen at -196 °C on an automatic volumetric sorption analyzer AutosorbiQ MP-XR (Quantachrome Instruments, USA) using the Brunauer-Emmet-Teller (BET) method. The adsorbed gas volume was measured at a relative pressure range of 0.1 - 0.3 using 40 - 50 mg of sample. Prior to the measurements, the samples were outgassed at 300 °C for 4 h under vacuum.

The electrode production can be found elsewhere [6,7]. The cells were assembled using a Swagelok®-type 2-electrode test cell with two symmetrical ACF electrodes as a counter and working electrode soaked with a 1 M triethylmethylammonium tetrafluoroborate (TEMA BF4, >98.0%, TCI Deutschland GmbH, Germany) solution in propylene carbonate (PC, 99.5%, Acros Organics N.V., Belgium) as an electrolyte with a separator (Celgard® 3401, Celgard, USA) between them. A C-coated aluminum foil (z-flo 2651, Coveris Management GmbH, Austria) was used as a current collector. All cells were assembled under nitrogen atmosphere avoiding moisture and oxygen. Cyclic voltammetry (CV) and galvanostatic discharge curve (GDC) measurements were performed using a potentiostat (Vertex.One, Ivium Technologies BV, The Netherlands). The gravimetric capacitance C_S was determined by CV and GDC using the following equations (1) and (2):

$$C_S = \frac{\int_{V_1}^{V_2} i\, dV}{2\, m_E\, v} \tag{1}$$

$$C_S = \frac{2\, I\, t}{m_E\, \Delta V} \tag{2}$$

The specific capacitance C_S of one electrode is determined by the integral $\int_{V_1}^{V_2} i\, dV$ of the positive part of the 5th CV curve, limited by the V_2 and V_1 of the cell voltage ΔV, scan rate v and the mass of the active material m_E of one electrode. For the GDC measurements, C_S is calculated using the current I, the discharge time t, m_E and ΔV. Energy density E_S and power density P_S for the device were calculated using equations (3) and (4):

$$E_S = \frac{1}{8}\, C_{S,GDC}\, (\Delta V)^2 \tag{3}$$

$$P_S = \frac{E_S}{t} \tag{4}$$

3. RESULTS AND DISCUSSION

The sea balls are 3 - 7 cm in diameter before carbonization and consist of a dense mesh of fibers (**Figure 1**). This structure is retained even after carbonization.

Figure 1 Impregnated sea ball before and after carbonization

The morphology of the sea balls after carbonization and after activation (SB_DAHP_165) was studied by SEM (**Figure 2**). In the case of the impregnated samples, the fiber structure is preserved both after carbonization and after activation by carbon dioxide. Without the DAHP treatment, the sea ball disintegrates during activation, leaving only ash and no carbon structure. The fibers differ strongly in thickness and length (see **Figure 2a**). At the fracture points of the fibers, it can be seen that the fibers possess a unique honeycomb-like structure.

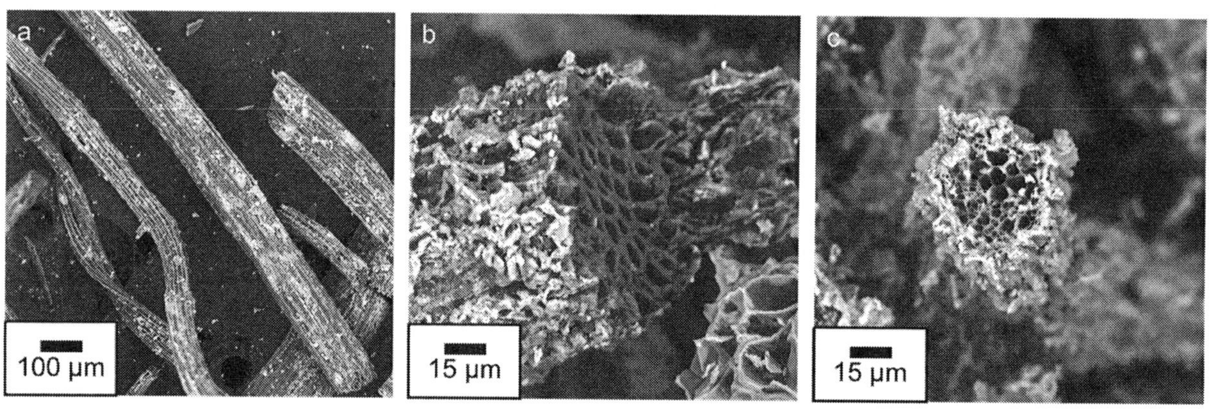

Figure 2 SEM micrograph of the carbonized (a and b) and the activated sea ball (b)

After activation, only the samples treated with DAHP form a carbon structure. The yield after 165 min activation time is 68.4 %, the specific surface area of the sample is 200 $m^2\ g^{-1}$. With the longer activation time, the activation yield decreases to 33.9 %, but the specific surface area increases to 243 $m^2\ g^{-1}$.

Table 2 Yield of the activation step, specific surface areas, and total pore volumes of the activated samples

Sample	Yield of the acticvation / %	Specific surface area / $m^2\ g^{-1}$	Total pore volume / $cm^3\ g^{-1}$
SB_165	-	-	-
SB_DAHP_165	68.4	200	0.13
SB_DAHP_285	33.9	243	0.15

For further characterization, test cells were prepared from the ACFs and examined by CV and GDC. The sample activated for 165 min reached a maximum specific capacitance of 17 F g^{-1} at CV at the lowest scan rate of 2 mV s^{-1}. The sample activated for 285 min showed a maximum specific capacitance of 28 F g^{-1}. With increasing scan rate the specific capacitance decreased to 13 F g^{-1} at 300 mV s^{-1}. For the specific capacitances calculated from the galvanostatic measurements (**Figure 3b**), the sample SB_DAHP_165 reaches a maximum value of 33 F g^{-1}, while the sample reaches a value of 38 F g^{-1}. The longer activated DAHP-impregnated sea ball also reaches the highest specific energies, calculated per test cell, of 9.7 W h kg^{-1} and the highest specific power of 3800 W kg^{-1} (**Figure 3c**).

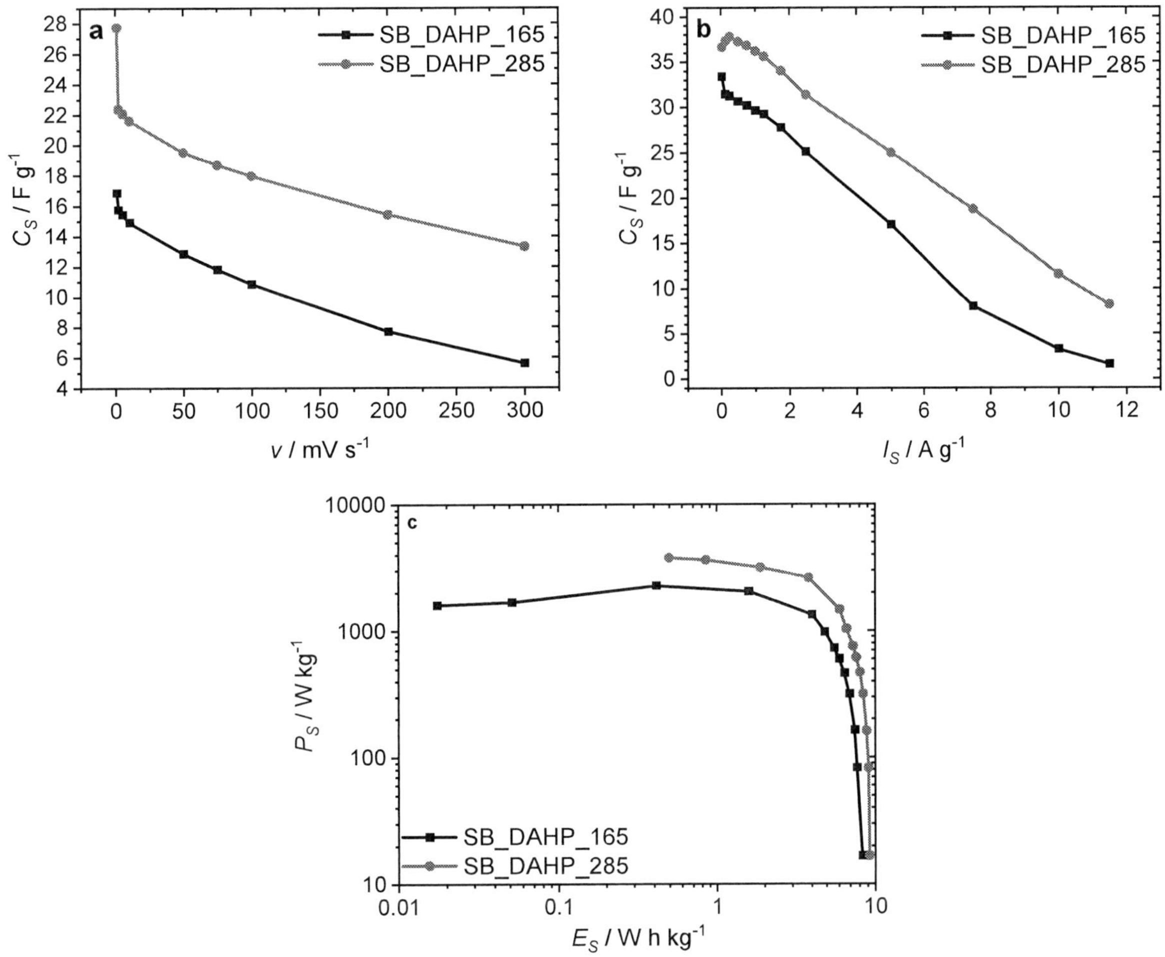

Figure 3 Specific capacitance per electrode Cs derived from CV (a) and from GDC (b) as well as specific power Ps and specific energy Es derived by GDC (c)

4. CONCLUSION

Sea balls could be successfully used to produce ACFs that can be used as electrode material for supercapacitors. It was shown that the use of an impregnate is essential, otherwise no ACFs could be obtained. In the prepared AFCs, the unique honeycomb-like porous macroporous structure from the precursor remains intact. Overall, it can be easily seen that the higher specific surface area achieved by the longer activation results in higher specific capacities, powers, and energies. The achieved capacities per electrode of 38 F g^{-1} are relatively low compared to other precursors, but the achieved specific energy density of 17 W h kg^{-1} per test cell is promising.

ACKNOWLEDGEMENTS

The authors wish to thank the European Regional Development Fund (EFRE) and the province of Upper Austria for financial support of this study through the program IWB 2014-2020 (project BioCarb-K). The authors would also like to thank Sarah Neustl for her assistance with the practical work.

REFERENCES

[1] ENOCK, T.K., KING'ONDU, C.K., POGREBNOI, A., JANDE, Y.A.C. Status of Biomass Derived Carbon Materials for Supercapacitor Application. *Int. J. Electrochem.* 2017, pp. 1-14.

[2] HAO, L., LI, X., ZHI, L. Carbonaceous electrode materials for supercapacitors. *Adv. Mater.* 2013, vol. 25, pp. 3899-3904.

[3] MURRAY, D.B., HAYES, J.G. Cycle Testing of Supercapacitors for Long-Life Robust Applications. *IEEE Trans. Power Syst.* 2015, vol. 30, pp. 2505-2516.

[4] PANDOLFO, A.G., HOLLENKAMP, A.F. Carbon properties and their role in supercapacitors. *J. Power Sources.* 2006, vol. 157, pp. 11-27.

[5] BEGUIN, F., FRACKOWIAK, E. *Supercapacitors: Materials, systems, and applications.* Weinheim: Wiley, 2013.

[6] BREITENBACH, S., LUMETZBERGER, A., HOBISCH, M.A., UNTERWEGER, C., SPIRK, S., STIFTER, D., FÜRST, C., HASSEL, A.W. Supercapacitor Electrodes from Viscose-Based Activated Carbon Fibers: Significant Yield and Performance Improvement Using Diammonium Hydrogen Phosphate as Impregnating Agent. *C.* 2020, vol. 6, pp. 17-29.

[7] BREITENBACH, S., GAVRILOV, N., PASTI, I., UNTERWEGER, C., DUCHOSLAV, J., STIFTER, D., HASSEL, A.W., FÜRST, C. Biomass-derived carbons as versatile materials for energy-related applications - capacitive properties vs. oxygen reduction reaction catalysis. *C.* 2021, vol. 7, pp. 55-70.

CONTROLLABLE MODIFICATION OF THE BATIO3 NANOPARTICLES USING SI-ATRP APPROACH AND IMPACT ON THE VIBRATION SENSING CAPABILITIES OF THEIR PVDF-BASED COMPOSITES

[1*]Miroslav MRLIK, [1]Přemysl FAJKUS, [1]Danila GORGOL, [1]Martin CVEK, [1]Josef OSICKA

[1]Centre of Polymer Systems, Tomas Bata University in Zlín, Zlín, Czech Republic, EU,
**mrlik@utb.cz*

https://doi.org/10.37904/nanocon.2021.4322

Abstract

In this contribution the ceramic nanoparticles, $BaTiO_3$ (BT), were successfully and controllably coated with two monomers, poly(butyl acrylate) (PBA) and poly(methyl methacrylate) (PMMA) using SI-ATRP approach. The presence of the polymers on the surface was investigated using FTIR and TGA techniques. The molecular weight and polydispersity index were calculated from the gel permeation chromatography, while monomer conversion was confirmed using nuclear magnetic resonance. Both methods provide clear statement that controlled radical polymerization was successful. The surface of the neat BT, BT-PBA and BT-PMMA nanoparticles was investigated using BET analysis and using contact angle. The nanoparticles were mixed with poly(vinylidene fluoride) (PVDF) using micro compounder. The compatibility of the nanoparticles and PVDF matrix was evaluated investigation of the viscoelastic properties using dynamic mechanical analyzer. The contribution of the enhanced compatibility between the nanofiller and matrix was elucidated by electrical response to vibration testing. Finally, it will be presented how the SI-ATRP modification of BT nanoparticles with PBA and PMMA influence the capability for vibration sensing applications.

Keywords: Poly(vinylidene fluoride), $BaTiO_3$, nanoparticles, SI-ATRP, poly(butyl acrylate), poly(methyl methacrylate), vibration sensing

1. INTRODUCTION

The poly(vinylidene fluoride) (PVDF) is a material with unique properties showing simple processing, mechanical and thermal stability, tunable electrical and also structural properties [1-4]. Such capabilities can be very effectively utilized in various applications i.e. controllable actuation, sensing or energy harvesting devices [5-7]. Generally, the PVDF performance strongly depends on the electrically-active β-phase [8-10]. However, without any treatment PVDF forms the α-crystalline phase which is useless in a view of mentioned applications. There are various methods to enhance the β-phase content in PVDF material even for already prepared films or non-processed granules. In the case of PVDF films, the useful methods include film stretching [11,12] or poling the films by the application of high external electric field (several kV) [13]. The combination of both methods provides an improved β-phase content in comparison to the individual single method [14]. Another approach for β-phase development is the fabrication of PVDF films via electrospinning [15], or their PVDF co-blending with other polymers such as poly(methyl methacrylate) [16] or the introducing of various fillers forming the PVDF composites. These fillers can be organics (graphite, graphene oxide, reduced graphene) [1,17,18], or inorganic Barium titanate ($BaTiO_3$) or (BT) [16]. Therefore, this study is aimed on the fabrication of the BT hybrid particles those surface is coated with short polymer brushes. Neat PVDF sheets were compared to the composites containing either neat BT or BT hybrid particles and their dynamical mechanical properties were elucidated. Finally, vibration sensing performance of was analyzed in order to confirm the enhancing effects of BT hybrid filler.

2. EXPERIMENTAL

2.1. Materials

Barium titanate (BT) (99%), 3-aminopropyl triethoxy silane (APTES) (97%), touluene (p.a.), α-bromoisobutyryl bromide (BiBB, 98%) served as an initiator was linked onto BT surface. Initiator bonding was performed in the presence of proton scavenger, triethyleneamine (TEA, ≥99%). Methyl methacrylate (MMA, 99%), *n*-butyl acrylate (BMA, 99%), ethyl α-bromoisobutyrate (EBiB, 98%), *N,N,N′,N′′,N′′*-pentamethyldiethylenetriamine (PMDETA, ≥99%), copper bromide (CuBr, ≥99%) and anisole (99%) were used as a monomers, sacrificial initiator, ligand, catalyst and solvent, respectively. Poly(vinylidene fluoride) (PVDF; molecular weight of 275 000 $g \cdot mol^{-1}$) were purchased from Sigma-Aldrich (USA). All chemicals were of analytical grade and used as received. Demineralized (DEMI) water (conductivity of 25 $\mu S \cdot m^{-1}$) was used for purifying procedure after ATRP initiator bonding.

2.2. Synthesis of $BaTiO_3$ hybrid particles

Firstly, the neat BT particles were functionalized using silane coupling agent APTES, when for 2g of BT particles 4 ml of APTES was applied. The particles were put into the 3-neck flask and dispersed in toluene. The APTES was added to the mixture dropwise and after 30 minutes, the mixture was heated to 110 °C and refluxed for 8 hours. The particles collected using centrifuge (8000 rpm) and were washed with toluene and acetone three times. Centrifugation was repeated with every single washing cycle.

Then the functionalized BT particles were placed to Schlenk flask and evacuated and backfilled with argon. Then the freshly distilled dry THF was transferred to the flask under argon atmosphere and cooled down to 0° C. The proton scavenger TEA was added and solution was mixed for 30 minutes. Later, the BiBB initiator was added to the system dropwise (1ml/hr). Reaction was carried out overnight and then refluxed at 70 °C for additional 2 hours.

The BT particles modified with ATRP initiator (0.75 g) were transferred into a Schlenk flask equipped with a gas inlet/outlet and a septum. The flask was evacuated and replaced with argon for several times. BA (17.5 mL, 110mmol) or MMA (11.8 mL, 110 mmol), EBiB (0.162 mL, 1.1 mmol), PMDETA (0.92 mL, 4.4 mmol), and anisole (15 mL) were then added to the flask and the flask was degassed using several freeze-pump-thaw cycles. The CuBr catalyst (0.156 g, 1.1 mmol) was quickly added under argon flow into the frozen system. Anisole was used as a solvent in the amount of 50 vol.%. The polymerization mixture was stirred at 60 °C for two hours and then the polymerization was stopped by exposure to air. The product (hybrid pyarticles) was filtered, washed by DMF (2x200 mL), acetone (2x200 mL) and diethyl ether (2x100 mL). Particles were dried at 30 °C overnight under reduced pressure (100 mbarr)

2.3. Preparation of PVDF/BT-based composites

In order to prepare the composites containing neat BT and hybrid particles, the polymer matrix PVDF and BT particles with 1 wt. % content were mixed using microextrusion machine (Xplore, USA) at 200 °C. Homogenization of the prepared composites carried out 5 minutes. Then the mixed systems were compressed to the sheets with thickness of 100 μm again at 200 °C. The compression takes place for 5 minutes, while cooling procedure was ensured using cooling device providing 15 °C constant cooling for 10 minutes.

2.4. Characterization

Proton nuclear magnetic resonance (1H NMR) spectra were recorded at 25 °C using an instrument (400 MHz VNMRS Varian, Japan) and deuterated chloroform (CDCl3) as a solvent. The molar mass (Mn) and dispersity (Đ) of polymer chains were investigated by gel permeation chromatography (GPC) using a GPC instrument (PL-GPC220, Agilent, Japan) equipped with GPC columns (Waters 515 pump, two PPS SDV 5 μm columns (diameter of 8 mm, length of 300 mm, 500 Å + 105 Å)) and a Waters 410 differential refractive index detector tempered to 30 °C. The samples for NMR spectroscopy and GPC analysis were prepared by their dilution with CDCl3 and THF, respectively, followed by the purification process, in which they were passed through a neutral alumina column. The contact angle (CA) measurements were evaluated from the static sessile drop method on the pellets carried out on a Surface Energy Evaluation system equipped with a CCD camera (Advex Instruments, Czech Republic). A droplet (5 µL) of distilled water was carefully dripped onto surface and the CA value was recorded. The presented CA results are the average values from 10 independent measurements. The viscoelastic properties of both the nanocomposite and pure polymer matrix were studied by dynamic mechanical analysis (DMA) in tensile mode. All measurements were performed at linear viscoelastic region. The measurement was done at 1 Hz in the temperature range from -150 °C to 150 °C. The dielectric spectroscopy in temperature range from -150 °C to 100 °C and in frequency range from 10-1 to 107 Hz was employed to investigate the polymer chains dynamics. Investigation of the d33 coefficient to prove the vibration sensing capabilities was performed in the similar way as was published elsewhere [19].

3. RESULTS AND DISSCUSSION

The successfully functionalized BT particles with immobilized ATRP initiator on the surface can be clearly seen in the **Figure 1**. The SI-ATRP polymerization from the surface is showed as well and conditions for the polymerization is clearly visible. The resulting hybrids as the final product of the demonstrated reaction are summarized in the **Table 1**. It can be seen that controlled conditions were confirmed by low value of PDI while monomer conversion calculated from ^{1}HNMR well correlates with results of Mn from GPC measurements.

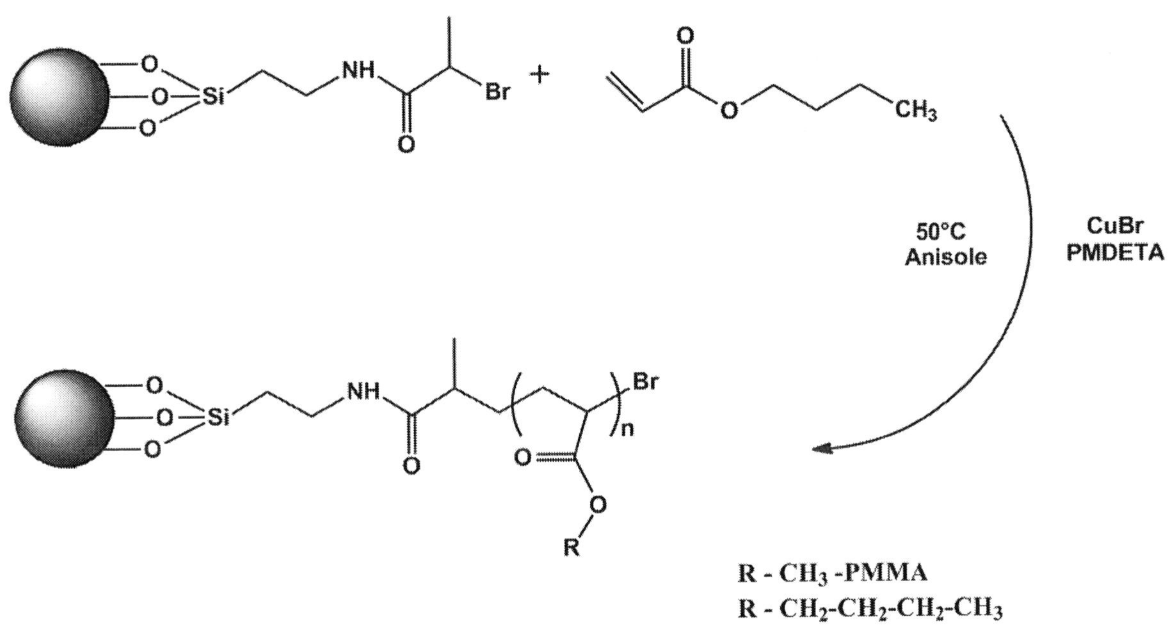

Figure 1 Schematic visualization of the SI-ATRP modification of BT particles with various monomers

Table 1 Summarized results from GPC and ^{1}HNMR measurements

Sample ID	Neat BT	BT-PMMA	BT-PBA
Mn (kDa)	N/A	3.3	4.7
PDI (-)	N/A	1.07	1.22
Conversion (%)	N/A	35	37

The clear confirmation of the BT particles surface modification the specific surface area and contact angle was investigated. The results from this measurements are visible in the **Table 2**. It can be seen that specific surface area of the modified particle significantly decreased from 17 m^2/g to 12 m^2/g and 11 m^2/g, for neat BT, BT-PMMA and BT-PBA, respectively. On the other hand the water contact angle increased nearly twice indicating the improved surface properties and thus better compatibility of such particles with PVDF can be expected.

Table 2 Summarized values of the surface properties measurements

Sample ID	Neat BT	BT-PMMA	BT-PBA
Specific surface area (m^2/g)	17	12	11
Water contact angle (°	64	91	118

Since the monitoring of the vibrations is dynamic process and mechanical properties play crucial role, the dynamic mechanical characterization was performed (**Figure 2**). In case of the storage modulus, the lower values is representing by neat PVDF matrix while both BT-modified samples showed enhanced capability to store the mechanical energy and effectively release. On the other hand, the investigation of the Tan delta, clearly showed that short polymer chains polymerized from the BT surface acting as the plasticizers and shifted the values of Tg towards lower temperatures. In this respect can be concluded, that SI-ATRP modification provide system with more flexible polymer chains which can positively contribute to the enhanced vibration sensing performance.

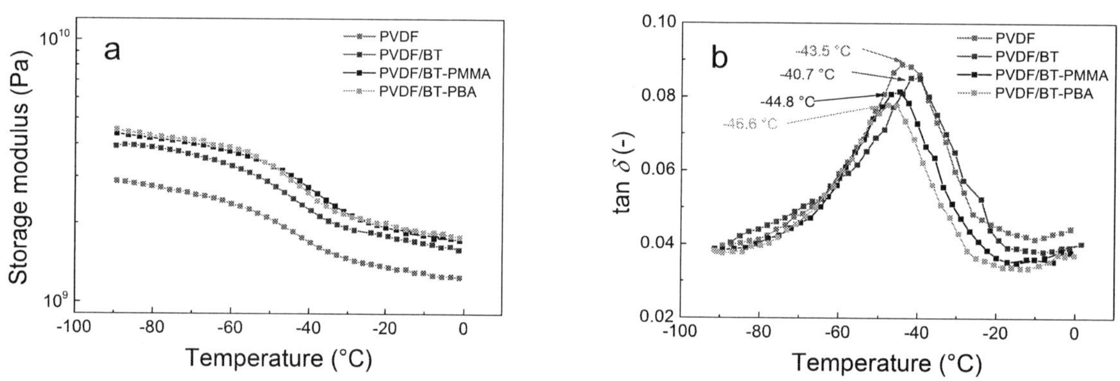

Figure 2 Temperature dependence of the Storage (a) and Tan delta (b) for various composite systems

Finally, in order to investigate the capabilities of the fabricated PVDF-based composite systems in their potential application, d33 coefficient was measured (**Table 3**). In this case, the presence of the short polymer chains positively influencing the vibration sensing and such samples provide better response on the vibrations than neat PVDF or PVDF/BT composite. Therefore, only small addition of the BT particles positively influencing their vibration sensing capabitilies.

Table 3 The d33 values for the neat PVDF and PVDF/BT-based composites

Sample ID	Neat PVDF	PVDF/BT	PVDF-BT-PMMA	PVDF/BT-PBA
10 kΩ	12.7±0.7	13.0±1.3	14.2±1.1	15.8±0.7
30 kΩ	12.9±0.9	13.1±1.4	14.5±1.7	16.0±0.4
50 kΩ	13.1±0.5	13.2±1.1	14.8±1.4	16.2±0.5
average	12.9±0.4	13.1±1.3	14.5±1.4	16.0±0.5

4. CONCLUSION

In this paper, the successful modification of the BT particles with PMMA and PBA short polymer chains using ATRP approach is presented. The investigations of the polymerization products clearly proved that polymerizations were controlled and provided low PDIs. Furthermore, the surface properties of the BT particles were significantly changed after modification and contact angle increased from 64°for neat BT to 91° and 118° for BT-PMMA and BT-PBA, respectively. Moreover, the dynamic mechanical investigations showed that Tg for modified BT particles was shifted towards lower temperatures indicating plasticizing effect of the short polymer brushes. Therefore, also vibration sensing capabilities were proved to be best for hybrid-based PVDF composites in comparison to neat PVDF and PVDF/BT ones.

ACKNOWLEDGEMENTS

The authors gratefully acknowledge the Czech Science Foundation grant no. 19-17457S for the financial support. This work was supported by the Ministry of Education, Youth and Sports of the Czech Republic - DKRVO (RP/CPS/2020/003).

REFERENCES

[1] HUANG, LY., LU, CX., WANG, F., WANG, L. Preparation of PVDF/grahene ferroelectric composite films by in situ reduction with hydrobromic acids and their properties. *RSC Advances*. 2014, vol. 4, pp. 45220-45229.

[2] CHANG, CE., TRAN, VH., WANG, JB., FUH, YK., LIN, LW. Direct-Write Piezoelectric Polymeric Nanogenerator with High Energy Conversion Efficiency. *Nano Letters*. 2010, vol. 10, pp. 726-731.

[3] ISSA, A. A., Al-MAADEED, M. A.S.A., MRLIK, M. and LUYT, A. S. Electrospun PVDF graphene oxide composite fibre mats with tuanble physical properties. *Journal of Polymer Research*. 2016, vol. 23, p. 232

[4] RAMADAN, K.S., SAMEOTO, D., EVOY, S. A review of piezoelectric polymers as functional materials for electromechanical transducers. *Smart Materials and Structures*. 2014, vol. 23, 033001.

[5] LEE, C., TARBUTTON, JA. Electric poling-assisted additive manufacturing process for PVDF polymer-based piezoelectric device applications. *Smart Materials and Structures*. 2014, vol. 23, 095044.

[6] SARAVANKUMAR, B., SOVOON, S., KIM, SJ. Self-Powered pH Sensor Based on a Flexible Organic-Inorganic Hybrid Composite Nanogenerator. *ACS Applied Materials and Interfaces*. 2014, vol. 6, pp. 13716-13723.

[7] FLORCZAK, S., LORSON, T., ZHENG, T., MRLIK, M., HUTMACHER, D.W. HIGGINS, M.J. LUXENHOFER, R., DALTON, P. D. Melt electrowriting of electroacitve poly(vinylidene fluoride) fibers. Polymer International. 2019, vol. 68, pp. 735-745.

[8] MARTINS, P., LOPES, AC., LANCEROS MENDEZ, S. Electroactive phases of poly(vinylidene fluoride): Determination, processing and applications. *Progress in Polymer Science*. 2014, vol. 39, pp. 683-706.

[9] MAJÍ, S., SARKAR, PK., AGGARWALI, L., GHOSH, SK., MANDALI, D., SHEET, G., ACHARYA, S. Self-oriented beta-crystalline phase in the polyvinylidene fluoride ferroelectric and piezo-sensitive ultrahin Langmuir-Schaefer film, *Physical Chemistry Chemical Physics*, 2015, 17, 8159-8165.

[10] KANIK, M., AKTAS, O., SEN, HS., DURGUN, E., BAYINIR, M. Spontaneous High Piezoelectricity in Poly(vinylidene fluoride) Nanoribbons Produces by Iterative Thermal Size Reduction Technique. *ACS Nano*. 2014, vol. 8, pp. 9311-9323.

[11] LI, L., ZHANG, MQ., RONG, MZ., RUAN, WH. Studies on the transformation process of PVDF from alpha to beta phase by stretching. *RSC Advances*. 2014, vol. 4, pp. 3938-3943.

[12] LIU, GM., SCHNEIDRER, K., ZHENG, LC., ZHANG, XQ., LI, CC., STAMM, M., WANG, DJ. Stretching induced phase separation in poly(vinylidene fluoride)/poly(butylene succinate) blends studies in-situ X-ray Scattering, *Polymer*. 2014, vol. 55, pp. 2588-2596.

[13] SHARMA, M., MADRAS, G., BOSE, S. Process induced electroactive beta-polymorph in PVDF: effect on dielectric and ferroelectric properties. *Physical Chemistry Chemical Physics*. 2014, vol. 16, pp. 14792-14799.

[14] MRLIK, M., OSICKA, J., CVEK, M., ILCIKOVA, M., SRNEC, P., GORGOL, D., TOFEL, P. Comparative study of PVDF sheets and their sensitivity to mechanical vibrations: The role of dimensions, molecular weight, stretching and poling. *Nanomaterials*. 2021, vol. 11, 1637.

[15] FANG, J., NIU, HT., WANG, HX., WANG, XG., LIN, T. Enhanced mechanical energy harvesting sing needleless electrospun poly(vinylidene fluoride) nanofibre webs. *Energy and Enviromental Science*. 2013, vol. 6, pp. 2196-2202.

[16] MOKOFENG, TG., LUYT, AS., PAVLOVIC, VP., PAVLOVIC, VB., DUDIC, D., VLAHOVIC, B., DJOKOVIC, V. Ferroelectric nanocomposites of polyvinylidene fluoride/ polymethyl methacrylate blend and BaTiO3 particles: Fabrication of beta-crystal polymorph rich matrix through mechanical activation of the filler. *Journal of Applied Physics*. 2014, vol. 115, p. 084109.

[17] ZHANG, YY., JIANG, SL., YU, Y., ZENG, YK., ZHANG, GZ., ZHANG, QF., HE, JG. Crystallization behavior and phase-tranformation mechanism with the use of graphite nanosheets in poly(vinylidene fluoride) nanocomposites. *Journal of Applied Polymer Science*. 2012, vol. 125, pp. E314-E319.

[18] THANGAVEL, E., RAMASUNDARAM, S., PITCHAIMUTHU, S., HONG, SW., LEE, SY., YOO, SS., KIM, DE., ITO, E., KARIG, YS. Structural and tribological characteristics of poly(vinylidene fluoride)/functionalized graphene oxide nanocomposite thin films. *Composite Science and Technology*. 2014, vol. 90, pp. 187-192.

[19] CHAMAKH, M.M., MRLIK, M., LEADENHAM, S., BAZANT, P., OSICKA, J., ALMAADEED, M.A., ERTURK, A.; KURITKA, I. Vibration sensing systems based on poly(vinylidene fluoride) and microwave-assisted synthesized ZnO star-like particles with controllable structural and physical properties. *Nanomaterials*. 2020, vol. 10, p. 2345.

CHARACTERIZATION OF THE VERY LOW CONTACT RESISTANCE ON HEAVILY BORON DOPED (113) CVD DIAMOND

[1]Jan VOVES, [1]Alexandr LAPOSA, [2]Zbyněk ŠOBÁŇ, [1]Pavel HAZDRA, [1]Vojtěch POVOLNÝ, [2,3]Vincent MORTET, [1,2]Nicolas LAMBERT, [2]Marina DAVYDOVA

[1]*Faculty of Electrical Engineering, Czech Technical University in Prague, Prague, Czech Republic, EU,* voves@fel.cvut.cz

[2]*Institute of Physics of the Czech Academy of Sciences, Prague, Czech Republic, EU,* mortet@fzu.cz

[3]*Faculty of Biomedical Engineering, Czech Technical University in Prague, Kladno, Czech Republic, EU*

https://doi.org/10.37904/nanocon.2021.4320

Abstract

The low resistance of ohmic contacts on diamond layers is important for the fabrication of diamond power electronic devices with fast switching capabilities for future high voltage applications. The low barrier height between the metal and diamond, high level of boron doping and annealing at elevated temperatures are the most critical parameters to reach the lowest contact resistivity. In this work, we report on titanium/gold ohmic contacts prepared on the heavily boron-doped (113) epitaxial diamond layers. The contact resistance has been characterized by the Circular Transmission Line Model (cTLM) structures. We used the analytical model of field enhanced emission, tunneling and the image force influence including Fermi level position dependence on the boron concentration for theoretical Ti/Au contact analysis and the Silvaco TCAD 2D simulation to estimate the measurement error associated with the nonzero metal resistance. We show that the resulting simulation values are consistent with the experimental results.

Keywords: Diamond, ohmic contacts, power devices, TCAD simulation

1. INTRODUCTION

CVD diamond layers with (113) surface orientation are studied as a new platform for preparation of high-performance diamond power devices [1,2]. They offer better surface morphology and lower surface roughness in comparison with (100) and (111) oriented substrates. High-power diamond devices require reliable and low-resistance ohmic contacts to minimize their total ON–state resistance [3]. In this work, we compare the experimental electrical properties of Ti/Au ohmic contacts on the epitaxial (113) highly boron-doped diamond (BDD) with theoretical calculations based on the cTLM analytical method and on the physical TCAD simulations. This analysis is necessary for the accurate determination of very low values of contact resistance on highly conductive diamond layers, the sheet resistance of which is comparable to the resistance of metallization.

2. EXPERIMENTAL

BDD layers were grown on (113) oriented substrates in a commercial microwave plasma enhanced chemical vapor deposition (MWPECVD) reactor AX5010 from Seki Diamond Systems. The microwave power was 700 W, pressure 100 mbar, methane concentration 1%, and different B/C ratios in the gas phase (100 to 2000 ppm) provided boron concentrations in the epilayer between 10^{19} to 10^{21} cm^{-3}. The surface morphology and the thickness of deposited layers were determined by atomic force (AFM) and cross-section scanning electron microscopy (X-SEM). Their resistivity, carrier concentration, and carrier mobility at room temperature were determined by Hall effect measurement. To characterize contact properties, circular Transmission Line Model (cTLM) patterns using a 75 µm radius of the inner circle (L) and 10 to 75 µm electrode gap spacing (d) to the outer electrode were prepared using a laser lithography system (Microwriter ML) and wet chemical etching (shown in **Figure 1**). The 10 nm thick Ti contact layers covered by a 100 nm Au capping layer were evaporated using a Covap PVD system from Angstrom Engineering with a base pressure below 3 × 10^{-7} mbar on an ozone treated (113) surface and annealed at temperature up to 700 °C. Titanium (Ti) is one of the most commonly used metals to fabricate ohmic contacts on boron-doped diamond. Ti ohmic contacts covered by the gold capping layer have a good thermal stability. The Kelvin method, Cascade Microtech M150 probing system and an Agilent B4156A semiconductor parameter analyser were used for the measurement of current-voltage (I-V) characteristics. Contact resistances R_{Csp}, transfer lengths L_T and sheet resistances R_{sh} were obtained by the linear least square fitting of the dependence of the cTLM structure resistance R_T with d, which is valid only when the d is much smaller than L [4].

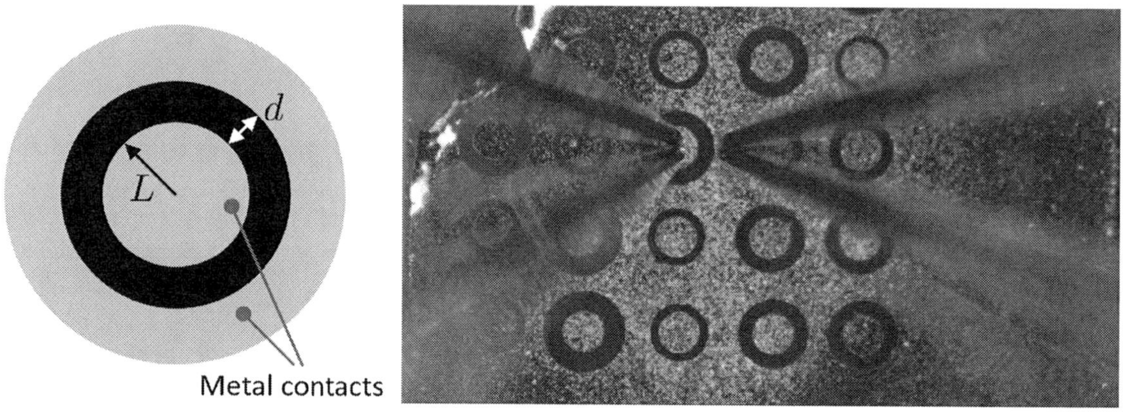

Figure 1 The cTLM structure used for measurement of the contact resistance on the BDD layers. The schematic picture (left) and the photo of four-probe measurement of the real structure (right)

2.1. The correction factor

In the case when $d \geq 4L$ the measured resistance should be multiplied by the correction factor C during the R_c determination. The total measured C-TLM resistance is given by [4]:

$$R_T = \frac{R_{sh}}{2\pi L}(d + 2L_T)C \quad (1)$$

where R_{sh} is the diamond sheet resistance, L_T is the transfer length where the current flows from metal to semiconductor layer and the correction factor is defined by:

$$C = \frac{L}{d} ln(1 + \frac{d}{L}) \quad (2)$$

The corrected resistance value R_{Tcorr} is then linearly dependent on d according to:

$$R_{T_{corr}} = \frac{R_T}{C} = \frac{R_{sh}}{2\pi L}(d + 2L_T) \quad (3)$$

As shown in **Figure 2**, where the open squares are the measured values and full squares are the corrected values, we can see the perfect linear behaviour in respect to the gap spacing d. The gradient multiplied by $2\pi L$ gives the sheet resistance of the BDD layer and the specific contact resistance is calculated as:

$$R_{C_{SP}} = R_{sh}\, 2L_T^2 \quad (4)$$

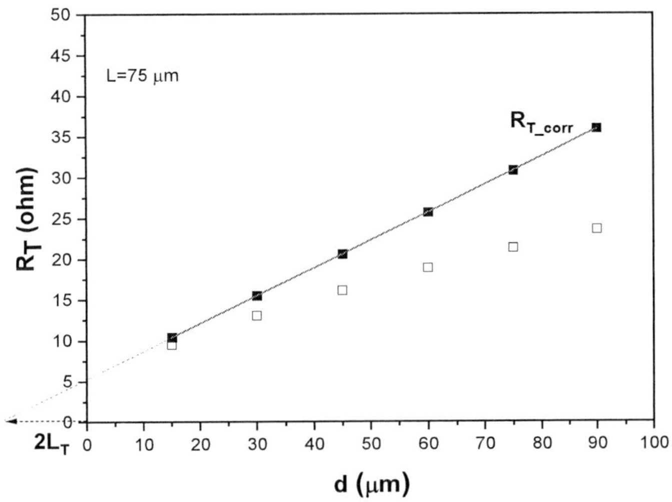

Figure 2 The total resistance of the cTLM structure in the dependence on the gap spacing. Experimental results are the open squares and corrected values are the full squares with the linear fit.

3. MODELS OF THE CONTACT RESISTANCE

3.1. The analytical model

To determine the barrier height Φ_b between the Ti/Au and p-type diamond we used the analytical model of the contact resistance of the Schottky diode [5 - 7]. The built-in voltage V_{bi} is defined by:

$$V_{bi} = \Phi_b - \frac{1}{q}(E_F - E_V), \tag{5}$$

Where Φ_b is the barrier height and $E_F - E_V$ is the distance between Fermi level and the valence band energy:

$$E_F - E_V = k_B \mathrm{T}.\ln\left(\frac{N_V}{N}\right) \tag{6}$$

N_V is the valence band effective density of states equal to 10^{19} cm^{-3} for diamond at room temperature [8] T is the absolute temperature, k_B is the Boltzmann constant and N is the doping concentration. We included the Schottky barrier lowering $\Delta\varphi_b$ due to image force:

$$\Phi'_b = \Phi_b - \Delta\Phi_b = \Phi_b - \left(\frac{q^3 N\left(V_{bi} - \frac{k_B T}{q}\right)}{8\pi^2(\varepsilon_r \varepsilon_0)^3}\right)^{\frac{1}{4}} \tag{7}$$

where ε_r is the diamond relative permittivity (5.5) in [9]. For the thermionic field emission model of the charge transfer the contact resistance $\rho_{C(TFE)}$ is given by:

$$\rho_{C(TFE)} = c_{TFE}\frac{k_B}{qA^*T}\exp\left(\frac{q\Phi'_b}{E_0}\right) \tag{8}$$

where A^* is the modified Richardson constant [10] and the coefficient c_{TFE} is calculated by the formula:

$$c_{TFE} = \frac{k_B T \cosh\left(\frac{E_{00}}{k_B T}\right)\sqrt{\coth\left(\frac{E_{00}}{k_B T}\right)}}{\sqrt{\pi(q\Phi'_b + (E_F - E_V))E_{00}}}\exp\left(\frac{E_F - E_V}{E_0} - \frac{E_F - E_V}{k_B T}\right) \tag{9}$$

where the exponential term can be approximated by 1 for higher doping level and

$$E_0 = E_{00}.\coth(\frac{E_{00}}{k_B T}) \tag{10}$$

with

$$E_{00} = \frac{q\hbar}{2}\sqrt{\frac{N}{m_t^* \varepsilon_r \varepsilon_0}} \tag{11}$$

m_t^* is hole effective mass (0.908 m_0 for diamond) [8]. At high diamond doping levels, the field emission (tunnelling) is the most important transport mechanism. The contact resistance is given by:

$$\rho_{C(FE)} = c_{FE}\frac{k_B}{qA^*T}\exp\left(\frac{q\Phi'_b}{E_{00}}\right) \tag{12}$$

where

$$c_{FE} = \frac{q}{E_{00}} \quad . \tag{13}$$

The resulting contact resistance is determined from the formula for parallel combination of both:

$$\rho_{C(FE,TFE)} = \left(\frac{1}{\rho_{C(FE)}} + \frac{1}{\rho_{C(TFE)}}\right)^{-1} \tag{14}$$

3.2. The TCAD model

The influence of the finite resistivity of the Ti/Au layer was inspected by the Silvaco TCAD two dimensional simulation. In the ideal case, we expect that the isotropic current flows between the outer and inner electrode during cTLM measurement. In the real case, the finite resistivity of the metal layer creates nonisotropic potential distribution at the circular structure. For the simulation, we have chosen the silicon model with the concentration 10^{19} cm^{-3}, thickness 1 µm and the artificial mobility of 18199 $cm^2/V.s$ which gives the same resistivity as our Ti/Au layer (the real case). For the simulation of an „ideal contact metal", we increased the carrier mobility of Ti/Au layer100 times. The difference in resistance between the real and ideal case is larger with the lower ratio between the sheet resistance of the diamond and the metal layers as shown in the **Figure 3** for the L=75 µm and d=20 and 75 µm with the probe distance 200 µm. The simulated distributions of current densities in the C-TLM structure corresponding to the real case are shown in **Figure 4**.

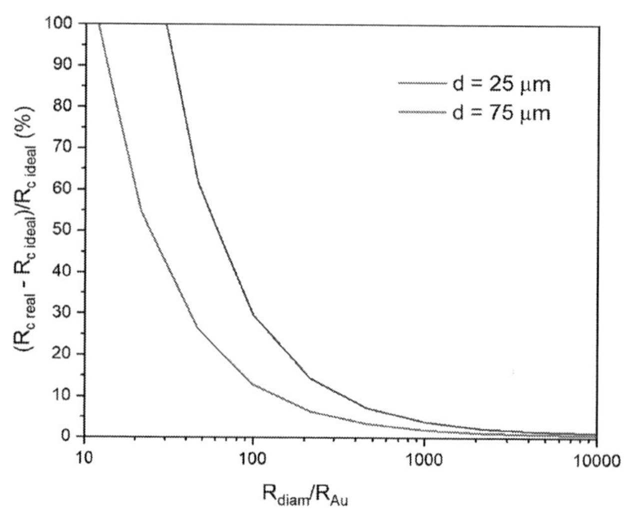

Figure 3 The relative difference between the ideal and real cTLM structure in the dependence on the diamond/metal sheet resistance ratio

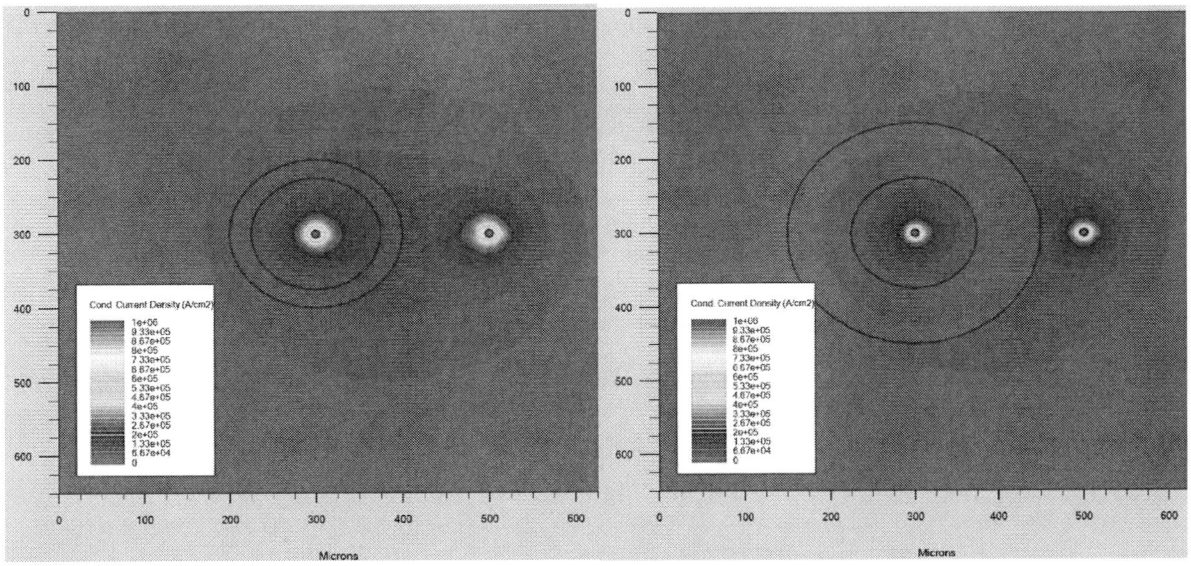

Figure 4 The simulated two-dimensional plot of the current density with d = 20 and 75 µm and the sheet resistance of the metal layer only 10 times lower than for the diamond layer.

4. RESULTS

We compared the result of analytical model with the experimental values of cTLM method. The result is shown in the **Figure 5**.

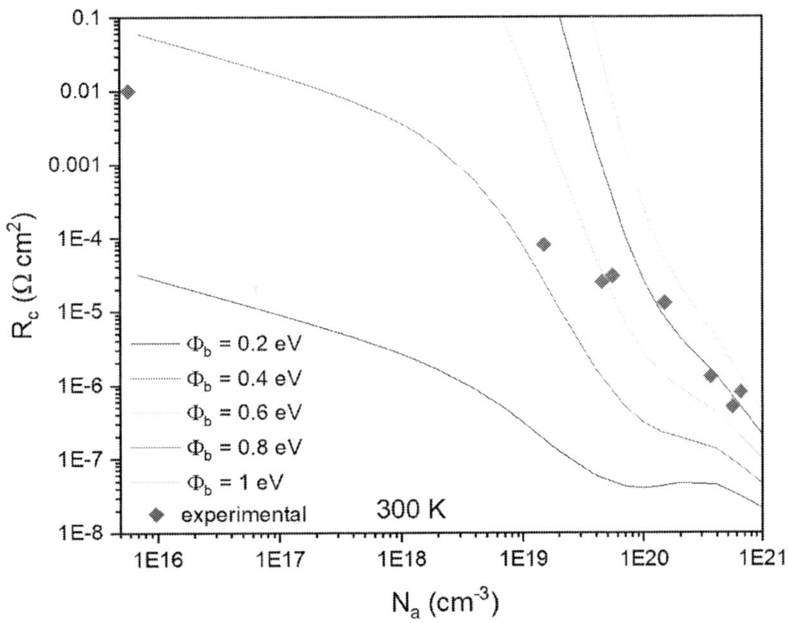

Figure 5 The calculated contact resistances in the dependence on the diamond boron doping levels for different contact barrier heights together with the experimental results.

The lowest values of the specific contact resistance reached was around 10^{-6} Ω.cm^2, which is the resolution limit of the cTLM method in our configuration. At lower diamond boron doping levels the experimental data are close to the curve with barrier height 0.4 eV, which is consistent with the published values [11]. We found out a significant decrease of the contact resistance compared to the values, which we obtained on epilayers with (100) and (111) orientation [3]. Moreover, Ti/Au contacts showed a stable ohmic behaviour in the whole range of annealing temperatures.

5. CONCLUSION

This study confirms the high suitability of (113) oriented diamond epitaxial layers in the combination with Ti/Au metal layers for the fabrication of diamond power electronic devices with low resistance ohmic contacts. The cTLM method could be used for the specific contact resistance characterization. The correction of the results due to low L/d ratio are important to include and the error connected with the low diamond/metal sheet resistance ratio should be taken into account.

ACKNOWLEDGEMENTS

This work was supported by the Czech Science Foundation grant No. 20-11140S by the Czech Technical University Student Grant No. SGS20/176/OHK3/3T/13 and by the Ministry of Education, Youth and Sports of the Czech Republic project LM2018110.

REFERENCES

[1] PINAULD-THAURY, M.-A. et al. Phosphorus-doped (113) CVD diamond: A breakthrough towards bipolar diamond devices. *Appl. Phys. Lett.* 2019, vol. 114, p. 112106.

[2] MORTET, M. et al. Properties of boron-doped (113) oriented homoepitaxial diamond layers*, Diam. Relat. Mater.* 2021, vol. 111, p. 108223.

[3] KOIZUMI, S. et al. *Power Electronics Device Application of Diamond Semiconuctors*. Woodhead Publishing, 2018.

[4] KLOOTWIJK, J.H., TIMMERING, J.H. Merits and Limitations of Circular TLM structures for contact resistance determination for novel 111-V HBTs, In *Proc. of IEEE 2004 Int. Conference on Microelectronic Test Structures*, Avaji, IEEE. 2004, pp. 247-252.

[5] LOHMÜLLER, E., WERNER, S., HOENIG, GREULICH, R.- J., CLEMENT, F. Impact of boron doping profiles on the specific contact resistance of screen printed Ag-Al contacts on silicon, *Solar Energy Materials & Solar Cells* 2015, vol. 142, pp. 2-11.

[6] VARAHRAMYAN, K., VERRET, E. J., A Model for Specific Contact Resistance Applicable for Titanium Silicide-Silicon Contacts. *Solid-Slate Electronics*. 1996, vol. 39, pp. 1601-1607.

[7] YU, A. Y. C. Electron Tunneling and Contact Resistance Metal-Silicon Contact Barriers. *Solid-State Electronics*. 1970, vol. 13, pp. 239-247.

[8] FONTAINE, F. Holes in boron-doped diamond: comparison between experiment and an improved model, *Diamond and Related Materials*. 2000, vol. 9, pp. 1076-1080.

[9] DONATO, N. et al., Diamond power devices: state of the art, modelling, figures of merit and future perspective, *J. Phys. D: Appl. Phys*. 2020, vol. 53, p. 093001.

[10] CROWELL, C. R. The Richardson Constant for Thermionic Emission in Schottky Barrier Diodes, *Solid-State Electronics*. 1965, vol. 8, pp. 395-399.

[11] LI, F. N. Barrier heights of Au, Pt, Pd, Ir, Cu on nitrogen terminated (1 0 0) diamond determined by X-ray photoelectron spectroscopy. *Appl. Surf. Sci.* 2018, vol. 456, pp. 532-537.

QUALITY CONTROL OF GRAPHENE OXIDE PRODUCTION WITH LARGE AREA SHEETS

[1,2]Pavel SVEC, [1,2]Zuzana BYTESNIKOVA, [1,2]Vojtech PRIKRYL, [2]Jana PEKARKOVA, [1]Lenka VANKOVA, [2]Sanam GAREHBAGHI, [1,2]Lukas RICHTERA, [1,2]Vojtech ADAM

[1]Department of Chemistry and Biochemistry, Mendel University in Brno, Brno, Czech Republic, EU, svecpavel@centrum.cz

[2]Central European Institute of Technology, Brno University of Technology, Brno, Czech Republic, EU, svecpavel@centrum.cz

https://doi.org/10.37904/nanocon.2021.4319

Abstract

There are many processes and modifications for the production of graphene oxide (GO) that are based on the original synthesis by W.S.Hummers and R.E.Offeman (1957), B.C.Brodie (1859), L.Staudenmaier (1898). Not all modifications produce high-quality material. The GO production process we use is a modified production according to Mr. W.S.Hummers and Mr. R.E.Offeman. Because the synthesis itself consists of several individual steps each step must be controlled in order to achieve the best result. Quality control at every step helps us to reduce the content of impurities and increase the yield of the final product. Impurities can be toxic or electrochemically active and thus have a direct effect on the application of the produced material. The ideal result is that the final product will contain only GO with individual large sheets and will not contain any other impurities and byproducts from synthesis such as potassium, chlorine, sulfur, manganese salts, or non-exfoliated graphite oxide. At the same time, we must reach sufficient production and small losses. We use Scanning Electron Microscopy with Energy Dispersive X-ray Analysis (SEM-EDX), conductometer, and pH meter to control the individual steps. The result of our production is single layer GO with large area sheets.

Keywords: Graphene oxide, large area sheets, impurities, nanomaterials, synthesis

1. INTRODUCTION

Graphene oxide (GO) was first synthesized by Brodie in 1855 [1]. Later, Hummers and Offeman [2] improved its production. GO is a promising material applied in many industries, such as the food, electrical, health, and agricultural sectors. The GO is involved in packaging material due to its modifying agents in the food industry. It can increase the surface of the electrodes, while in electrochemistry devices such as biosensors and a protective layer while developing anodes in the electrical industry. It can form composites with other nanomaterials having adhesive and antibacterial properties, suitable for healthcare instruments. The antibacterial and antifungal properties of GO have encouraged to be applied in the agricultural sector also [3,4,5].

The reasons have encouraged us to start our production of GO due to low quality and increasing prices of commercially available GOs from different manufacturers. These commercial GOs contained many impurities such as fibers, residues of nano-oxidized graphite, nano-exfoliated graphite oxide, unwashed salts, and other undefined contaminants. Despite the proclaimed sizes of individual GO sheets, the sheets were not homogenous, having significant differences in size and quality.

2. MATERIALS AND METHODS

2.1. Chemicals

Chemicals used in this study (graphite flakes, H_2SO_4, HCl, $KMnO_4$, H_2O_2) were purchased from Sigma-Aldrich (St. Louis, MO, USA) in ACS purity. Deionized water was purified using a Milli-Q Direct QUV apparatus equipped with a UV lamp (Merck Millipore, Darmstadt, Germany). The resistance was 18.2 MΩ [5].

2.2. pH and conductivity

The pH meter inoLab pH 720 (WTW, Wellheim, Germany) was used for the pH measurement with a standard pH electrode. A conductometer inoLab Cond7110 (Xylem Analytics, Wellheim, Germany) was used to measure conductivity with a common electrode where the cell volume was 1 cm^3 and temperature was 21 °C. The conductivity and pH measurements were done while washing the GO with HCl. A shallow pH value was observed during synthesis.

2.3. Scanning electron microscopy

The well-dispersed GO sample solution was prepared by dilution. The dilution ratio was 1:200 with demineralized water, and then the sample was drop cast on a silicon wafer and allowed to dry at laboratory temperature (22 °C). This wafer was adhered by carbon tape to the stub and later inserted into the SEM. The morphology and composition of the GO were examined by scanning emission microscopy with a Tescan MAIA 3 equipped with a field emission gun (Tescan Ltd., Brno, Czech Republic). The best images were obtained using the In-Lens SE detector at a working distance between 3 - 5 mm and 5 kV acceleration voltage. A pixel image of 768×858 dimensions was obtained at 5,000 - 50,000 fold magnification. In UH Resolution mode, a full-frame was captured, and the accumulation of image with image shift correction was enabled, and it took about 1.5 minutes with the ~1 μs/pixel dwell time. The 2.4 nm spot size was set.

2.4. Synthesis of GO

The GO was synthesized by following the procedure reported earlier in Richtera et al. 2015 [5]. Each synthesizing steps of GO were monitored by SEM, including the graphite that entered into the reaction.

2.4.1. Start of production (green phase)

The graphite flakes were mixed with H_2SO_4, and then $KMnO_4$ was added. Graphite was placed in water for SEM image only to ensure the same condition (liquid medium). Graphite entered the production process in dry form. The resulting reaction mixture was green. At each addition, a sample was taken and characterized in SEM (**Figure 1**).

2.4.2. Oxidation at time and termination (brown and pink phase)

The green reaction mixture was stirred for four days until the green color changed to brown. After four days, the mixture was allowed to stand for another five days until it turned into pink color. H_2O_2 was used to terminate the reactions (**Figure 2**).

2.4.3. Washing with HCl and water

A color change manifested the completed reactions from pink to yellow. This was followed by a gradual and intensive washing with HCl (**Figure 3**). Eventually, water started to be added to separate the GO sheets because the water intercalated into the layers of graphite oxide (**Figure 4**).

3. RESULTS

3.1. SEM observation

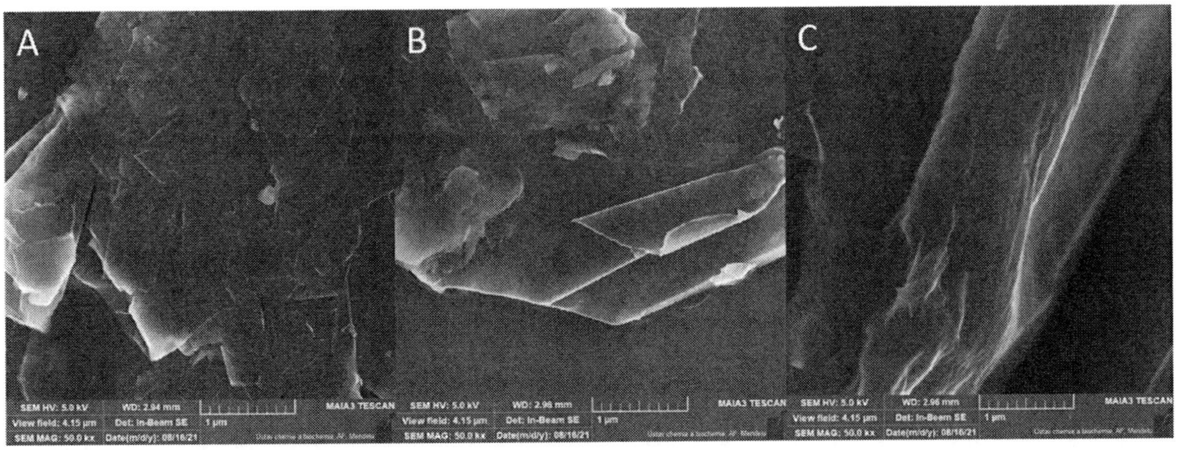

Figure 1 Graphite flakes in green phase. A - graphite flakes, B - graphite flakes with H_2SO_4, C - graphite flakes with H_2SO_4 and $KMnO_4$

In **Figure 1A,** we can see the structure of graphite entering the reaction. A large block of graphite flakes consists of smaller and thinner sheets stacked on top of each other. The surface is smooth and flat, without wrinkles or other deformations. **Figure 1B**, after the addition of H_2SO_4, shows the edge of one graphite flakes, where the sandwich structure is visible. By sandwich, the system is meant the composition of many 2D sheets of graphite. After the addition of $KMnO_4$, the individual graphite layers oxidized (**Figure 1C**). It is most noticeable on the surface. The original smooth surface of graphite is wrinkled. This wrinkling is due to the local separation of the individual graphite layers from each other during oxidation. Oxidation proceeds most rapidly at the edges and outer layers and progresses to the depth of the flakes [6].

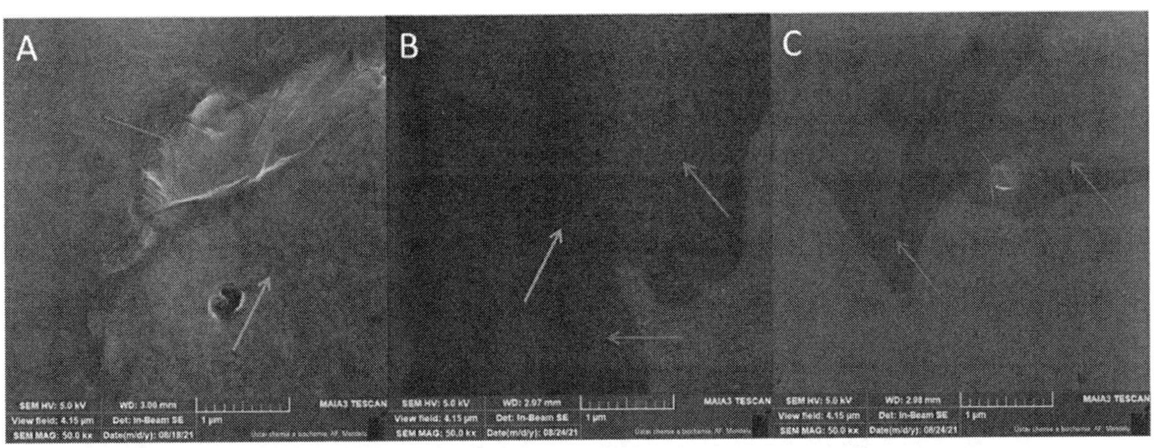

Figure 2 Oxidation phase of graphite. A - graphite flakes after four days of oxidation, B - graphite oxide sheets before termination, C - graphite oxide after termination

Four days of oxidation of graphite with constant stirring helped break down the sandwich structure into thinner layers. In **Figure 2A** (red arrow), we see a smaller piece of graphite. This piece has no sharp edges and is wrinkled. The individual thinner layers of graphite oxide slip from the original graphite on the upper and left sides. The blue arrow indicates where the graphite oxide layer covered the salts formed in the reaction. This layer of graphite oxide prevented the formation of larger salt crystals. The reaction mixture stood for the next five days. During this time, the graphite flakes oxidized, and the block of sheets was separated. **Figure 2B** shows the block of sheets (red arrows) with a large area and not torn. The blue arrow indicates the residues of the reaction mixture between the two graphite oxide sheets. In **Figure 2C** we see graphite oxide after termination. Upon termination with hydrogen peroxide, the oxidation number changed from Mn^{+VII} to Mn^{+II}. This reaction altered the color of the suspension from a pink to a yellow, clear liquid containing thin layers of graphite oxide [6].

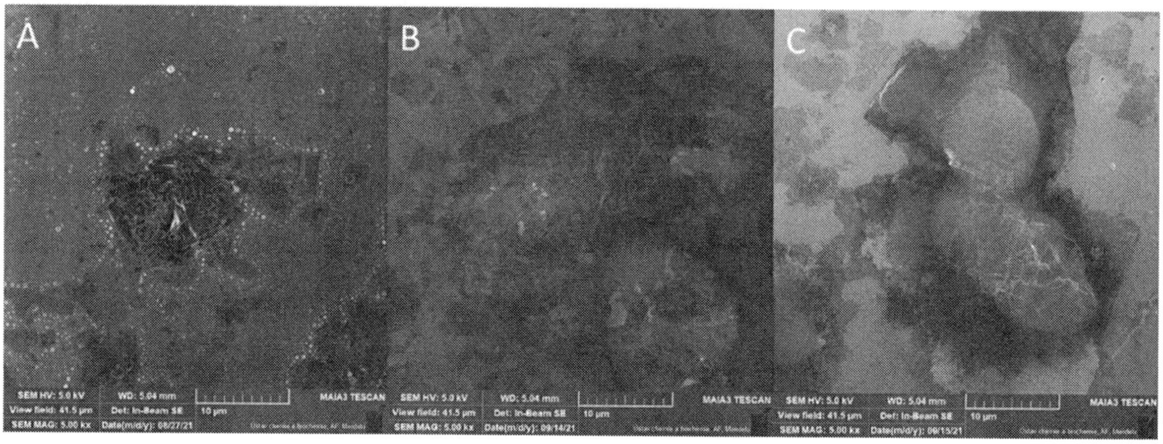

Figure 3 Washing graphite oxide with HCl. A - graphite oxide with impurities, B - the mount of impurities decreases, C - state of graphite oxide at the end of washing with HCl

As we can see in **Figure 3A**, the sample contains many salts in the form of crystals. These are $MnCl_2$ and $MnSO_4$. Gradual washing with HCl reduces the amount of Mn^{+II} bound to the graphite oxide. The gradual decrease of Mn^{+II} can be seen in **Figure 3B**, where almost no crystals of the formed salts are visible. However, we can see areas where acid residues under the graphite oxide layer are. These residues are well visible in **Figure 3C**, where they have an ellipse shape and are covered with graphite oxide [6].

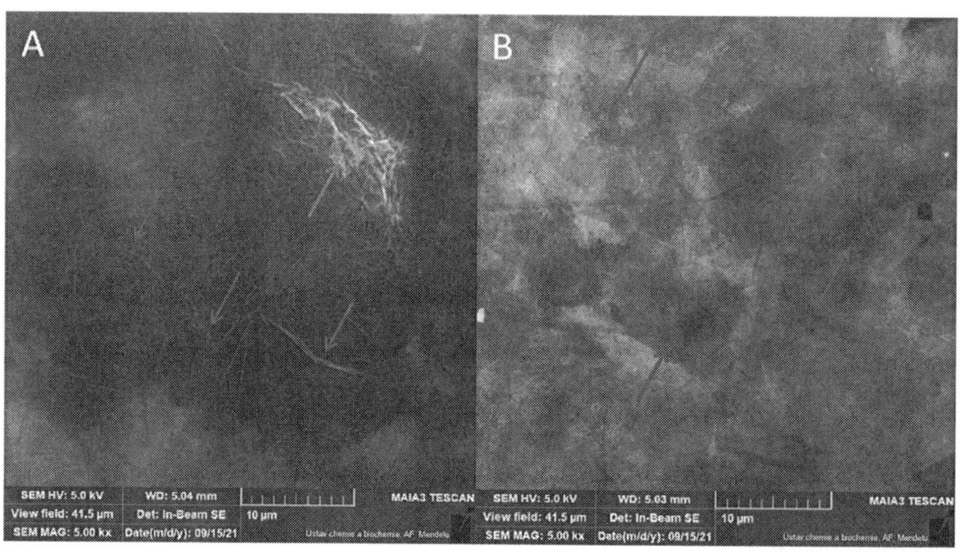

Figure 4 Separating individual GO sheets. A - after 8 L of water, B - after 16 L of water. The red arrows show the wrinkles that occur when the individual leaves of the GO are separating. The green arrow indicates where the acid residues are.

At the beginning of the washing with water (first wash), acid residues were still present in the sample, as seen in **Figure 4A** (green arrow). The red arrows show where the wrinkles of GO are visible. There are also many layers of GO, which lie loosely on top of each other and no longer hold in the block as previously with graphite flakes. After further washing with water, we no longer see any acid residues, and the exfoliation of the individual GO layers continues, as shown in **Figure 4B**. Red arrows indicate a fine wrinkle of the GO [6].

3.2. pH and conductivity

Following individual washing phases, pH and conductivity values of GO were measured (**Table 1**). The HCl fractions were diluted 10-fold before measuring the pH because these fractions were too acidic.

Table 1 Values of pH and conductivity

Sample	pH	Conductivity (mS/cm)
1. fraction of HCl	0.52	379.00
2. fraction of HCl	0.64	360.00
3. fraction of HCl	0.75	320.00
1. fraction of water	0.52	83.60
2. fraction of water	1.41	13.13

Increasing pH values and decreasing conductivity values show that the number of ions released in previous synthesis steps (impurities) decreases, indicating a successful cleaning. These changes were visible in the **Figures 3A, B, C,** and **4A, B**. In HCl fractions (1 - 3), the pH increase very slowly. In these 3 fractions, the conductivity slowly decreases because only Mn ions are removed, but the anions from acids remain in excess. In the individual washing steps with water, we noticed that the conductivity continued to decrease sharply due to continuously removing anions through washing following a constant increase in the pH.

4. CONCLUSION

The modified Hummers method is very effective in the production of GO. The individual synthesis steps and gradual improvement allowed us to produce GO sheets with a large surface. The synthesized GO has low impurities after washing. The decreasing conductivity and simultaneous increasing pH during each washing step confirm the removal of impurities. SEM monitored each step of GO synthesis and cleaning. Corresponding changes in SEM images of GO were observed during individual cleaning steps and the following measurement in conductometer and pH value. The quality of the synthesis was continuously monitored from graphite to GO along with the course of oxidation. Continual quality check with SEM helped clarify the oxidation of the outer layers of graphite, the separation of individual blocks of graphite oxide, and the effect of washing after the oxidation to the GO.

ACKNOWLEDGEMENTS

This research was carried out under the project CEITEC 2020 (LQ1601) with financial support from the Ministry of Education, Youth and Sports of the Czech Republic under the National Sustainability Programme II.

The work was supported from ERDF " Multidisciplinary research to increase application potential of nanomaterials in agricultural practice " (No. CZ.02.1.01/0.0/0.0/16_025/0007314).

REFERENCES

[1] BRODIE, B.C., XII. On the atomic weight of graphite. *Philosophical transactions of the royal society of London.* 1859, vol. 149, pp. 249-259.

[2] HUMMERS, W.S., OFFEMAN, R.E. Preparation of graphitic oxide. *Journal of the American Chemical Society*. 1958, vol. 80, pp. 1339.

[3] ZHU, Y., JAMES, D.K., TOUR, J.M. New routes to graphene, graphene oxide and their potential applications. *Advanced Materials*. 2012, vol. 24, pp. 4924-4955.

[4] KIM, J., COTE, L.J., HUANG, J., Two dimensional soft materials: New faces of graphene oxide. *Accounts of chemical research*. 2012, vol. 45, pp. 1356-1364.

[5] RICHTERA, L., et. al. The composites of graphene oxide with metal or semimetal nanoparticles and their effect on pathogenic microorganisms. *Materials*. 2015, vol. 8, no. 6, pp. 2994-3011.

[6] DIMIEV, A.M., TOUR, J.M. Mechanism of graphene oxide formation. *ACSNANO*. 2014, vol. 8, pp. 3060-3068.

LIGHTWEIGHT, FLEXIBLE AND HIGH-PERFORMANCE NANOCOMPOSITES BASED ON REDUCED GRAPHENE OXIDE AND SPINEL FERRITE ($ZnFe_2O_4$ / $CoFe_2O_4$) NANOPARTICLES IN THERMOPLASTIC POLYURETHANE MATRIX FOR ELECTROMAGNETIC INTERFERENCE SHIELDING APPLICATIONS

[1]ANJU*, [1]Raghvendra SINGH YADAV, [1]Ivo KURITKA, [1]Jarmila VILCAKOVA, [1]David SKODA, [1]Pavel URBANEK, [1]Michal MACHOVSKY, [1]Milan MASAR, [1]Michal URBANEK

[1]Centre of Polymer Systems, University Institute, Tomas Bata University in Zlín, Czech Republic, EU, deswal@utb.cz

https://doi.org/10.37904/nanocon.2021.4323

Abstract

Lightweight and flexible material with enhanced electromagnetic interference shielding effectiveness is highly in demand in the electronics and communication industry. Herein, we synthesized $CoFe_2O_4$ and $ZnFe_2O_4$ nanoparticles using the sonochemical method, and further, these nanoparticles were embedded inside thermoplastic polyurethane (TPU) along with reduced graphene oxide (rGO) by a melt-mixing approach using a microcompounder. $CoFe_2O_4$ or $ZnFe_2O_4$ nanocomposites in the TPU matrix with rGO showed outstanding electromagnetic shielding performance having a constant thickness of 0.80 mm, only. The maximum total shielding effectiveness (SE_T) was 48.3 dB for $ZnFe_2O_4$ nanocomposite and 50.76 dB for $CoFe_2O_4$ nanocomposite. These results indicate that the developed nanocomposite can be potentially utilized for electromagnetic shielding applications.

Keywords: Nanocomposites, Electromagnetic Interference Shielding, Graphene, Spinel ferrites, Thermoplastic Polyurethane

1. INTRODUCTION

Owing to the demand for electronic devices and telecommunication technology, electromagnetic interference (EMI) has become a serious and severe concern in various areas such as aerospace, defense, intelligence, and electronic appliances [1]. Electromagnetic interference shielding is a mechanism of reflection and absorption of electromagnetic radiation by a material that prevents the penetration of harmful electromagnetic waves into electronic devices. For commercial applications, an EMI shielding value greater than 20 dB corresponding to 99.9 % of the attenuation of the electromagnetic radiation power is suitable [2].

With the broad investigation in the many previous years, polymer nanocomposites have gained a lot of attention due to their lightweight, corrosion resistance, flexibility, high conductivity, and ease of processability [3]. A homogeneous dispersion of fillers in the polymer matrix is required to determine the final properties of fabricated nanocomposites [4]. Thermoplastic polyurethane (TPU) has been broadly used in electronic applications due to its outstanding high extensibility, flexibility, low glass transition temperature, good mechanical properties, and chemical resistance [5]. Therefore, TPU can be considered as an ideal choice as a polymer matrix for fillers. However, the EMI shielding performance of a polymer nanocomposite is based on the electrical and magnetic properties of the fillers. Ferrites have been extensively utilized as magnetic material because of their low cost, ease of preparation, and large magnetic momentum. A combination of polymer-magnetic ferrite composites is a potential candidate as shielding material due to a hybrid structural and functional combination between the organic and inorganic materials [6]. Despite several attempts to design a highly conductive network in a polymer matrix, there are still gaps in utilizing spinel ferrite nanoparticles with reduced graphene oxide in the TPU matrix for EMI shielding in the literature. From this standpoint, we have utilized cobalt ferrite and zinc ferrite nanoparticles along with reduced graphene oxide as a filler in the TPU polymer matrix for electromagnetic interference shielding applications in the X-band frequency region.

2. MATERIALS AND METHODS

2.1. Materials

Cobalt nitrate $Co(NO_3).6H_2O$, zinc nitrate hexahydrate $Zn(NO_3).6H_2O$, and iron nitrate $Fe(NO_3)_3.9H_2O$ were procured from AlfaAesar GmbH & Co. KG (Germany). Sodium nitrate ($NaNO_3$) was purchased from Lach-Ner, Czech Republic. Potassium permanganate ($KMnO_4$) powder and graphite flakes were purchased from Sigma-Aldrich, Germany. Vitamin C (Livsane) was sourced from Dr. Kleine Pharma GmbH, Germany.

2.2. Methods

$CoFe_2O_4$ and $ZnFe_2O_4$ nanoparticles were synthesized using the sonochemical method reported in the previous literature [7] using an ultrasonic homogenizer (UZ SONOPULS HD 2070) with a frequency of 20 kHz and power of 70 W for 60 min. Graphene oxide (GO) was synthesized by the modified hummer's method [8]. Reduced graphene oxide (rGO) was obtained by chemical reduction of synthesized graphene oxide (GO) using vitamin C as a reducing agent. For the preparation of TPU based nanocomposites, 20 wt% nanofillers (spinel ferrites nanoparticles and RGO in 9:1 ratio) were mixed with TPU (C80A10) in A DSM MICRO 5 micro-compounder (Research B.V., The Netherlands) with a capacity of 5 cm^3. TPU based nanocomposites were prepared by melt-mixing method at 200 °C for 7 min at 150 rpm, and further, these were hot-pressed to create a rectangular-shaped sheet of a 22.86 × 10.16 × 0.8 mm^3 dimension. The developed nanocomposites with 0.80 mm thickness were designated as $ZnFe_2O_4$-rGO-TPU and $CoFe_2O_4$-rGO-TPU nanocomposites.

2.3. Characterization

X-ray diffraction pattern of nanoparticles and nanocomposites was recorded on RigakuMiniFlex 600 X-ray Spectrometer. The microstructures study was done by using a scanning electron microscope Nova NanoSEM450 (FEI company). An Agilent N5230A vector network analyzer in the frequency range 8.2 -12.4 GHz was utilized to evaluate the electromagnetic interference shielding efficiency of prepared nanocomposites. Magnetic characteristics were assessed using a vibrating sample magnetometer (VSM 7407, Lake Shore).

3. RESULTS AND DISCUSSION

3.1. Structural study

Figure 1 (a) depicts the XRD pattern of $ZnFe_2O_4$ and $CoFe_2O_4$ nanoparticles. The distinctive XRD peaks assigned to crystalline planes (220), (311), (400), (422), (511), (440), and (533) which confirmed the cubic spinel structure with the Fd3m space group symmetry for both $ZnFe_2O_4$ and $CoFe_2O_4$ nanoparticles [9]. Moreover, the crystallite size was calculated by debye Scherrer's formula [7], which was found to be 3.45 nm and 11.13 nm for $ZnFe_2O_4$ and $CoFe_2O_4$ nanoparticles, respectively. Further, The XRD investigation of $ZnFe_2O_4$ -rGO-TPU and $CoFe_2O_4$ -rGO-TPU nanocomposites displayed in **Figure 1 (b),** which reveals the presence of the diffraction planes (220), (311), (400), (422), (511), and (440) associated with the face-centered cubic structure of spinel ferrites. It signifies the presence of the spinel ferrite nanoparticles inside the TPU matrix. Moreover, TPU exhibits broad diffraction peaks ranging from 10° to 28° associated with an ordered structure of the hard phase and disordered structure of the amorphous phase [10], which were present in the XRD pattern of the nanocomposites as well (see **Figure 1 (b)**) and thus remarked the existence of TPU in nanocomposites.

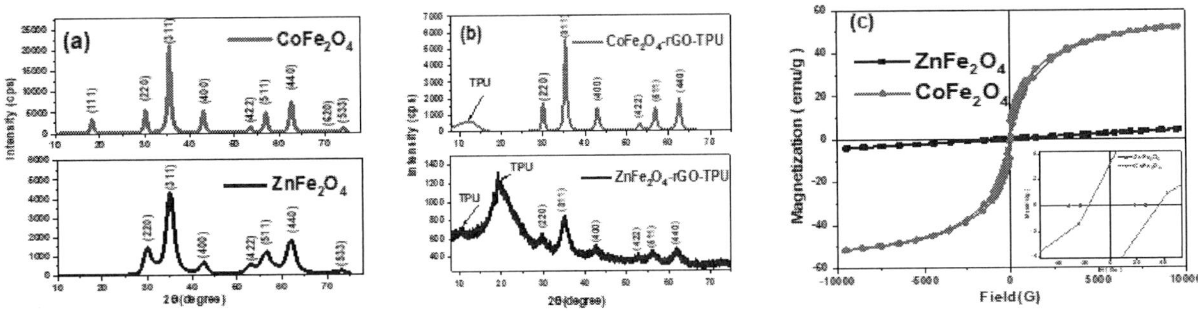

Figure 1 (a) XRD of $ZnFe_2O_4$ and $CoFe_2O_4$ nanoparticles, **(b)** XRD of $ZnFe_2O_4$-rGO-TPU and $CoFe_2O_4$-rGO-TPU nanocomposites, **(c)** Magnetic hysteresis curves of $ZnFe_2O_4$ and $CoFe_2O_4$ nanoparticles

3.2. Magnetic Study

The magnetic properties of synthesized $ZnFe_2O_4$ and $CoFe_2O_4$ nanoparticles were investigated using a vibrating sample magnetometer (VSM) at room temperature with an applied magnetic field up to 10 kOe, as displayed in **Figure 1(c).** The saturation magnetization value is 4.07 emu/g and 51.98 emu/g for $ZnFe_2O_4$ and $CoFe_2O_4$ nanoparticles, respectively. The low value of saturation magnetization of spinel ferrite nanoparticles compared with bulk is due to its small particle size [11]. Further, the value of coercivity $H_c \approx 36.17$ Oe and remanent magnetization M_r 3.37emu/g, for $CoFe_2O_4$ nanoparticles revealing its ferromagnetic behavior. However, the value of coercivity $H_c \approx 0$ Oe and remanent magnetization $M_r \approx 0$ emu/g for $ZnFe_2O_4$ nanoparticles, proves superparamagnetic characteristics [7].

3.3. TEM and SEM study

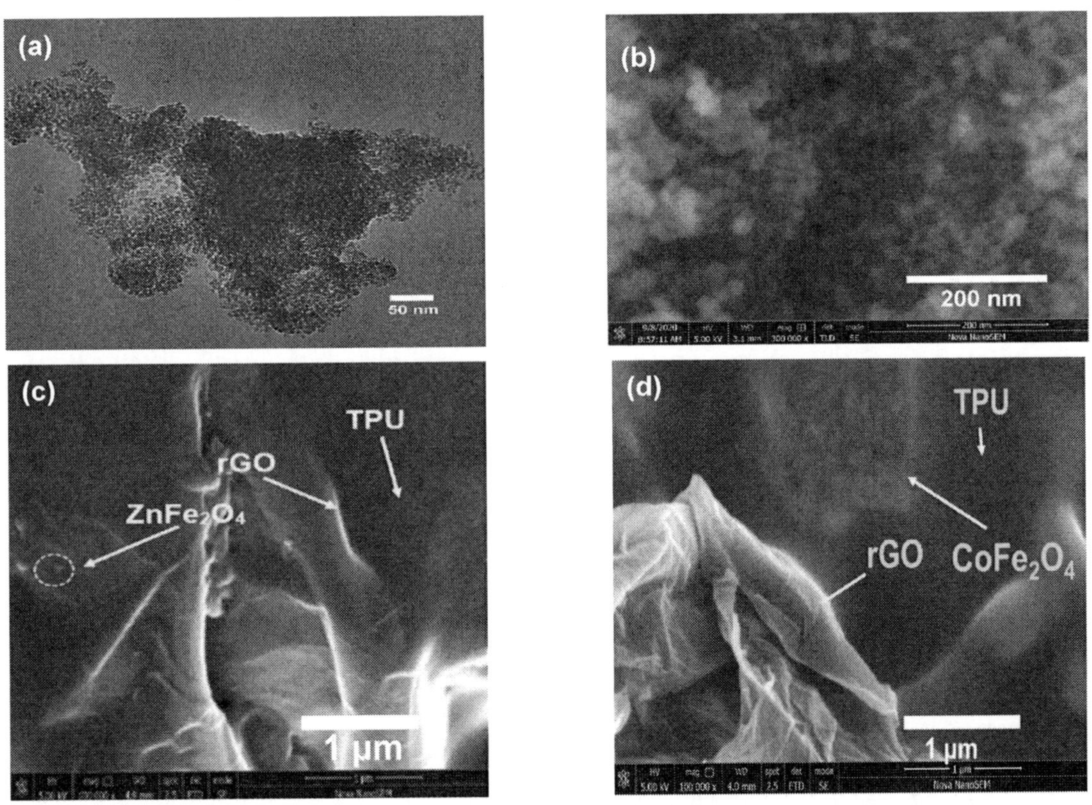

Figure 2 (a) TEM image of the $ZnFe_2O_4$ nanoparticles **(b)** FE-SEM image of $CoFe_2O_4$ nanoparticles, **(c)** FE-SEM image of the $ZnFe_2O_4$-rGO-TPU, **(d)** FE-SEM image of $CoFe_2O_4$-rGO-TPU

Figure 2(a) displays the TEM image of the $ZnFe_2O_4$ nanoparticles revealing the spherical particle of size 3-5 nm. A typical FE-SEM image of $CoFe_2O_4$ nanoparticles is shown in **Figure 2(b),** which shows spherical particles of the size 9-15 nm. **Figures 2(c-d)** depicts the FE-SEM image showing the cross-sectional view of the developed $ZnFe_2O_4$ -rGO-TPU and $CoFe_2O_4$ -rGO-TPU nanocomposites, which confirm the presence of nanoparticles, rGO embedded in the TPU matrix.

3.4. Electromagnetic Interference Shielding Effectiveness

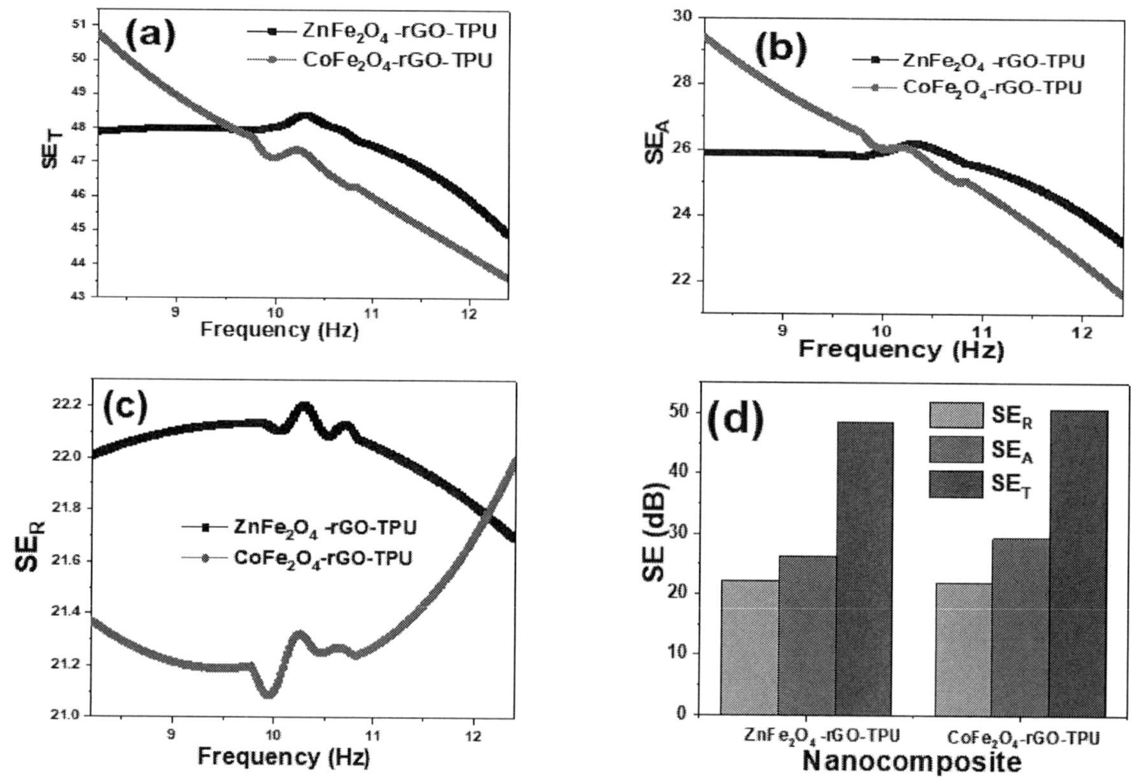

Figure 3 Frequency response of $ZnFe_2O_4$ -rGO-TPU and $CoFe_2O_4$ -rGO-TPU nanocomposites with **(a)** SE_T, **(b)** SE_A , **(c)** SE_R **(d)** comparison chart

Figures 3(a-c) displays the SE_T, SE_A, and SE_R of $ZnFe_2O_4$ -rGO-TPU and $CoFe_2O_4$ -rGO-TPU nanocomposites with a thickness of 0.80 mm in the frequency range of 8.2 to 12.4 GHz. The value of SE_T, which is a sum of the magnitude of SE_A and SE_R, is found to be 48.39 dB for $ZnFe_2O_4$ -rGO-TPU and 50.76 dB for $CoFe_2O_4$ -rGO-TPU nanocomposites [7]. It can be noticed from **Figure 3(b)** that the maximum value of SE_A for $ZnFe_2O_4$ -rGO-TPU is 26.19 dB. On the other hand, the maximum value of SE_A for $CoFe_2O_4$ -rGO-TPU is 29.39 dB. However, the value of SE_R is 22.20 dB for $ZnFe_2O_4$ -rGO-TPU and 21.99 dB for $CoFe_2O_4$ -rGO-TPU, as shown in **Figure 3(c)**. It signifies that absorption is the major contributor to the shielding performance of $ZnFe_2O_4$ -rGO-TPU and $CoFe_2O_4$ -rGO-TPU nanocomposites [11]. The high shielding effectiveness due to absorption is attributed to the interaction of electric and magnetic dipoles. Moreover, the conducting network of the rGO sheet with nanoparticles enhanced interfacial polarization, which might have created a good balance between impedance matching and attenuation constant [7,12]. A comparison chart displayed in **Figure 3(d)** reveals the better performance of $CoFe_2O_4$ -rGO-TPU nanocomposite compared with $ZnFe_2O_4$ -rGO-TPU nanocomposite. Since lightweight is a crucial factor for choosing a shielding material, especially in the aircraft industry, on taking this into account, the specific shielding effectiveness (SSE), which is the ratio of SE_T and density of the material, was calculated [13]. The density of the nanocomposites is in the range of 0.93 to 1.06 cm^3/g, and the thickness of the nanocomposites is 0.80 mm. The calculated value of SSE is found to be in the range of 47 dB cm^2/g to 51 dB cm^2/g [13]. However, for practical application, the SSE alone is not sufficient for analyzing the lightweight shielding performance of a material. Considering this, the absolute shielding efficiency (SSE/t), defined as the ratio of specific shielding efficiency to the thickness of the material, is also evaluated. The magnitude of SSE/t is in the range of 597 dB cm^3/g to 648 dB cm^3/g for developed nanocomposites [13].

4. CONCLUSION

Herein, we developed lightweight and flexible nanocomposites based on TPU as matrix and spinel ferrite nanoparticles ($CoFe_2O_4$ and $ZnFe_2O_4$) and rGO as nano-fillers. The developed nanocomposites exhibit total shielding effectiveness of 48.39 dB and 50.76 dB for $ZnFe_2O_4$ -rGO-TPU and $CoFe_2O_4$ -rGO-TPU nanocomposites, respectively, in the X-band frequency range with a thickness of only 0.80 mm. Also, both the nanocomposites are suitable for commercial applications as they possess a shielding effectiveness value greater than 20 dB. Therefore, lightweight and flexible $ZnFe_2O_4$ -rGO-TPU and $CoFe_2O_4$ -rGO-TPU nanocomposites can be a promising candidates for electromagnetic interference shielding applications.

ACKNOWLEDGEMENTS

We thank the financial support of the Czech Science Foundation (Ga19-23647s) project at the Centre of Polymer Systems, Tomas Bata University in Zlin, Czech Republic. R.S.Y. thanks the Leibniz Institute of polymer research Dresden, Germany, for hosting as a guest scientist in its laboratories.

REFERENCES

[1] MA, Z., KANG, S., MA, J., SHAO, L., ZHANG, Y., LIU, C., WEI, A., XIANG, X., WEI, L., And GU. J., Ultraflexible and Mechanically Strong Double-Layered Aramid Nanofiber-Ti3C2Tx MXene/Silver Nanowire Nanocomposite Papers for High-Performance Electromagnetic Interference Shielding. *ACS Nano.* [online]. 2020, vol. 14, pp. 8368–8382.

[2] SONG, W.-L., GONG, C., LI, H., CHENG, X.-D., CHEN, M., YUAN, X., CHEN, H., YANG, Y., FANG, D. Graphene-Based Sandwich Structures for Frequency Selectable Electromagnetic Shielding. *ACS Appl. Mater. Interfaces.* [online]. 2017, vol. 9, no. 41, pp. 36119-36129. Available from: https://doi.org/10.1021/acsami.7b08229.

[3] JIANG, D., MURUGADOSS, V., WANG, Y., LIN, J., DING, T., WANG, Z., SHAO, Q., WANG, C., LIU, H., LU, N., WEI, R., SUBRAMANIA, A., GUO, Z. Electromagnetic Interference Shielding Polymers and Nanocomposites - A Review. *Polymer Reviews.* [online].2019, vol. 59, no. 2, pp. 280-337. Available from: https://doi.org/10.1080/15583724.2018.1546737.

[4] SONG, P., QIU, H., WANG, L., LIU, X., ZHANG, Y., ZHANG, J., KONG, J., GU, J. Honeycomb Structural RGO-MXene/Epoxy Nanocomposites for Superior Electromagnetic Interference Shielding Performance. *Sustainable Materials and Technologies.* [online]. 2020, vol. 24, p. e00153. Available from: https://doi.org/10.1016/j.susmat.2020.e00153.

[5] SOBHA, A.P., SREEKALA, P.S., NARAYANANKUTTY, SUNIL. K. Electrical, Thermal, Mechanical and Electromagnetic Interference Shielding Properties of PANI/FMWCNT/TPU Composites. *Progress in Organic Coatings.* [online]. 2017, vol. 113, pp. 168-174. Available from: https://doi.org/10.1016/j.porgcoat.2017.09.001.

[6] ISMAIL, M. M., RAFEEQ, S. N., SULAIMAN, J. M. A., MANDAL, A. Electromagnetic Interference Shielding and Microwave Absorption Properties of Cobalt Ferrite CoFe2O4/Polyaniline Composite. *Appl. Phys. A,* [online]. 2018, vol. 124, no. 5, p. 380. Available from: https://doi.org/10.1007/s00339-018-1808-x.

[7] YADAV, R. S., ANJU, JAMATIA, T., KUŘITKA, I., VILČÁKOVÁ, J., ŠKODA, D., URBÁNEK, P., MACHOVSKÝ, M., MASAŘ, M., URBÁNEK, M., KALINA, L., HAVLICA, J. Superparamagnetic $ZnFe_2O_4$ Nanoparticles-Reduced Graphene Oxide-Polyurethane Resin Based Nanocomposites for Electromagnetic Interference Shielding Application. *Nanomaterials.* [online]. 2021, vol. 11, no. 5, p. 1112. Available from: https://doi.org/10.3390/nano11051112.

[8] BAI, Y., RAKHI, R. B., CHEN, W., ALSHAREEF, H. N. Effect of PH-Induced Chemical Modification of Hydrothermally Reduced Graphene Oxide on Supercapacitor Performance. *Journal of Power Sources.* [online]. 2013, vol. 233, pp. 313-319. Available from: https://doi.org/10.1016/j.jpowsour.2013.01.122.

[9] ATIF, M., ASGHAR, M. W., NADEEM, M., KHALID, W., ALI, Z., BADSHAH, S. Synthesis and Investigation of Structural, Magnetic and Dielectric Properties of Zinc Substituted Cobalt Ferrites. *Journal of Physics and Chemistry of Solids.* [online]. 2018, vol. 123, pp. 36-42. Available from: https://doi.org/10.1016/j.jpcs.2018.07.010.

[10] BARICK, A.K., TRIPATHY, D.K. Preparation, Characterization and Properties of Acid Functionalized Multi-Walled Carbon Nanotube Reinforced Thermoplastic Polyurethane Nanocomposites. *Materials Science and Engineering: B.* [online]. 2011, vol.176, pp. 1435-1447.

[11] YADAV, R. S., ANJU, JAMATIA, T., KUŘITKA, I., VILČÁKOVÁ, J., ŠKODA, D., URBÁNEK, P., MACHOVSKÝ, M., MASAŘ, M., URBÁNEK, M., KALINA, L., HAVLICA, J. Excellent, Lightweight and Flexible Electromagnetic Interference Shielding Nanocomposites Based on Polypropylene with $MnFe_2O_4$ Spinel Ferrite Nanoparticles and Reduced Graphene Oxide. *Nanomaterials.* [online]. 2020, vol. 10, pp. 2481.

[12] RAJU, P., SHANKAR, J., ANJAIAH, J., MURTHY, S. R. Shielding Effectiveness Studies of NiCuZn Ferrite-Polyaniline Nanocomposites for EMI Suppression Applications. In: *Proceedings of the International Conference on Advanced Materials.* Kerala: AIP. [online]. 2019, p. 020027. Available from: https://doi.org/10.1063/1.5130237.

[13] YADAV, R. S., KUŘITKA, I., VILČÁKOVÁ, J., MACHOVSKÝ, M., ŠKODA, D., URBÁNEK, P., MASAŘ, M., JURCA, M., URBÁNEK, M., KALINA, L., HAVLICA, J. $NiFe_2O_4$ Nanoparticles Synthesized by Dextrin from Corn-Mediated Sol−Gel Combustion Method and Its Polypropylene Nanocomposites Engineered with Reduced Graphene Oxide for the Reduction of Electromagnetic Pollution. *ACS Omega.* [online]. 2019, vol. 4, pp. 22069−22081.

STUDY OF CRACKS FORMATION IN HIGHLY - LOW BORON-DOPED EPITAXIAL (113) DIAMOND BILAYERS

[1,2]Vincent MORTET, [1]Ladislav KLIMSA, [1,3]Nicolas LAMBERT, [1]Marina DAVYDOVA, [1]Jaromir KOPECEK, [1]Silvia IGNACOVA SEDLAKOVA

[1]*FZU - Institute of Physics of the Czech Academy of Sciences, Prague, Czech Republic, EU,* mortetv@fzu.cz
[2]*Faculty of Biomedical Engineering, Czech Technical University in Prague, Kladno, Czech Republic, EU*
[3]*Faculty of Electrical Engineering, Czech Technical University in Prague, Prague, Czech Republic, EU*
lambenic@fel.cvut.cz

https://doi.org/10.37904/nanocon.2021.4326

Abstract

In this work, we present the study of the formation of cracks in high and low boron-doped diamond epitaxial bilayers necessary in the fabrication process of Schottky diodes. Epitaxial diamond layers were grown on (113) oriented diamond substrates by Microwave Plasma Enhanced Chemical Vapor Deposition. The effect of the thickness and the methane concentration during the growth of the undoped diamond layer on the crack formation have been studied using optical and scanning electron microscopy (SEM). We experimentally observed a critical thickness of ca. 3.5 µm above which all undoped layers are cracked. The formation of these cracks is attributed to the relaxation of the elastic energy stored in the epitaxial undoped layer due to the significant lattice mismatch (ca. 0.8 %) between the undoped and highly boron-doped diamond layers with a boron concentration of 10^{21} cm^{-3} as determined by Raman spectroscopy analysis.

Keywords: Diamond, boron-doping, epitaxy, cracks

1. INTRODUCTION

Many acknowledge diamond as the best wide-bandgap semiconductor for fabricating high-performance, high power, and high-frequency electronic devices due to its outstanding properties. Yet many scientific and technical issues must be resolved before the realization of reliable diamond electronic devices. Recent studies have demonstrated the interest in (113) epitaxial diamond layers in the fabrication of diamond electronic devices as this crystalline orientation offers better surface morphology, higher growth rate, high boron incorporation efficiency, lower surface roughness and defect-free surface in comparison with (100) and (111) orientations [1-3]. The fabrication process of semi-vertical diamond Schottky diodes requires the growth of epitaxial doped diamond bilayer with low and high dopant concentrations, which is the second necessary step in fabricating diamond electronic devices like Schottky diodes. However, the fabrication of boron-doped diamond bilayer structure is not straightforward due to cracks in the diamond epitaxial layer preventing the final realization of Schottky diodes. This work studies the formation of cracks in non-intentionally doped diamond epitaxial bilayers deposited on a heavily boron-doped (113) oriented diamond layer.

2. EXPERIMENTAL

Epitaxial boron-doped diamond bilayers were grown on (113) oriented high-pressure high-temperature substrates in a commercial AX5010 microwave plasma-enhanced chemical vapor deposition (MWPECVD) reactor from Seki Diamond Systems using the deposition conditions reported in **Table 1**. A ca. 5 µm thick highly boron-doped epitaxial diamond layer is first deposited on the (113) oriented diamond substrate with a boron to carbon ratio (B/C) in the gas phase of 2000 ppm for a deposition time of 2 hours. Immediately after the flow of the boron precursor (trimethylborane) is stopped, the methane concentration is adjusted to the desired concentration. The methane concentration during this second step has been varied from 0.3 % to 1 %. The deposition time of this second layer is variable. The boron concentration of the highly boron-doped diamond layer has been evaluated from the analysis of its Raman spectrum between 1000 and 1500 cm^{-1} measured at room temperature using a Renishaw InVia Raman Microscope with a 488 nm excitation laser. The diamond layer thickness is estimated using known deposition rates. The deposition rates were determined from a multilayer boron-doped epitaxial diamond sample characterized by secondary ion mass spectroscopy (SIMS). The formation of cracks on fabricated samples was determined by optical and scanning electron microscope (SEM) using a Zeiss Imager Z1m light optical microscope and a TESCAN FERA3 GM SEM.

Table 1 Deposition conditions of epitaxial boron-doped diamond layers

	Pressure (mbar)	Microwave power (W)	Total gas flow (sccm)	[CH4] (%)	B/C (ppm)	Deposition time (h)
1st step	100	700	700	1	2000	2
2nd step	100	700	700	variable	0	variable

3. RESULTS AND DISCUSSION

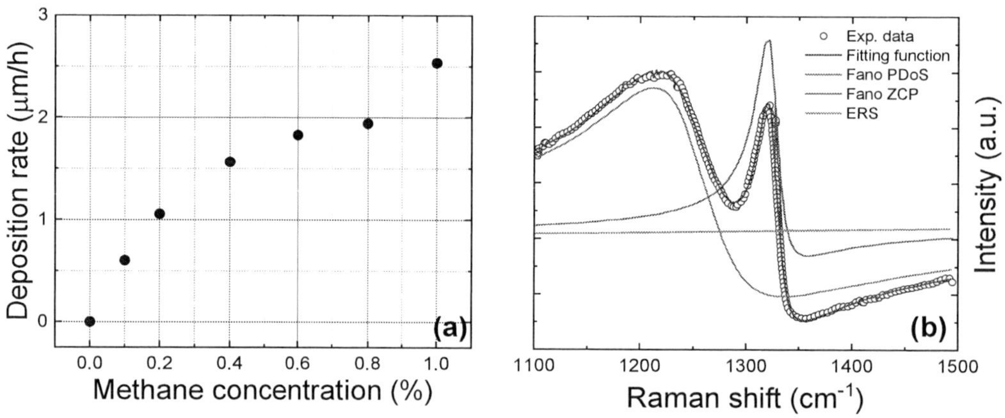

Figure 1 (a) Deposition rate of epitaxial boron-doped diamond as a function of the methane concentration and a boron to carbon ratio of 100 ppm in the gas phase in an AX5010 MWPECVD reactor operating at 100 mbar and 700 Watts **(b)** Experimental Raman spectrum (Exp. Data) of the epitaxial boron-doped diamond layer grown at a boron to carbon ratio of 2000 ppm and its decoupled double Fano-function fit (fitting function) including electronic Raman scattering (ERS), the Fano shaped zone center phonon diamond line (Fano ZCP) and the Fano shaped maximum of the phonon density of states peak (Fano PDoS).

Figure 1a shows the deposition rate of epitaxial diamond grown on (113) oriented substrates as a function of the methane concentration between 0 and 1 % for a boron to carbon ratio in the gas phase of 100 ppm. The growth rate grows sub-linearly with a value of ca. 2.5 µm at a methane concentration of 1 %. The deposition rate is independent of the boron to carbon ratio up to 2000 ppm, as demonstrated by SIMS analysis of epitaxial boron-doped diamond layer prepared with different boron to carbon ratios for the two methane concentrations 0.1 and 1 %. Therefore, the thickness of the different diamond layers prepared in this work has been estimated from the growth rate reported in **Figure 1a**. **Figure 1b** shows the Raman spectrum of the heavily boron-doped diamond underlayer grown at B/C = 2000 ppm between 1100 and 1500 cm^{-1} with a laser excitation wavelength of 488 nm and its decoupled double Fano-function fit using the online fitting tool at https://ofm.fzu.cz/cs/raman-tool to determine the boron concentration of ca. 10^{21} cm^{-3} from the width (12.32 cm^{-1}) of the undisturbed zone center phonon line of diamond [4]. This highly boron-doped diamond layer has a low resistivity of 2 mΩ.cm, allowing the formation of ohmic contact with a specific contact resistance below 10^{-6} Ω.cm^{-2} suitable for the fabrication of low ON resistance Schottky diodes [1].

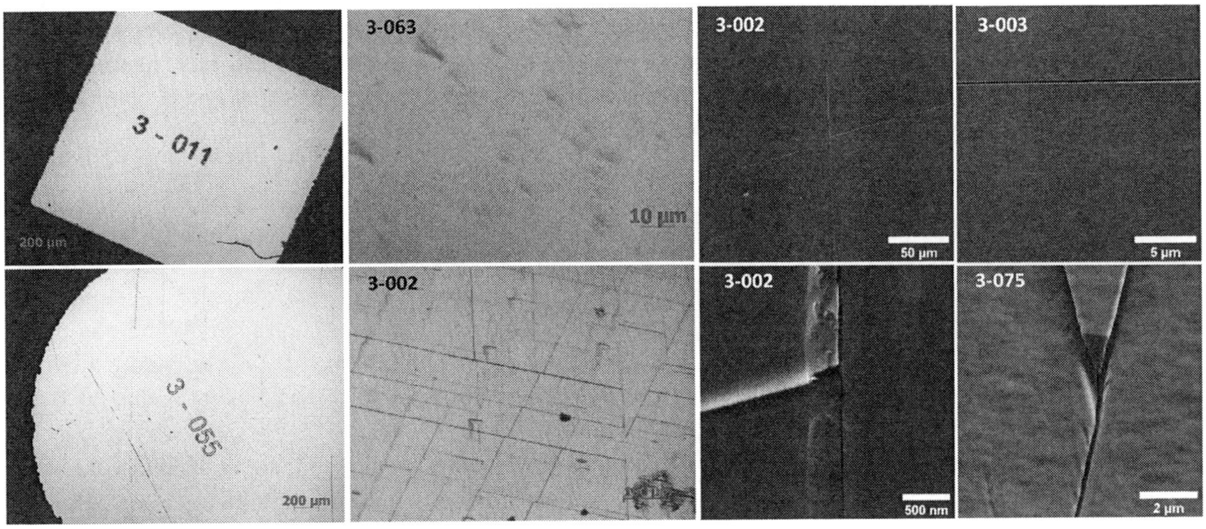

Figure 2 Optical (left) and scanning electron microscopy images (right) of the highly boron-doped/undoped epitaxial diamond bilayers of samples with different thicknesses (t) of the undoped layer ($t_{3\text{-}011}$ < 3 µm, $t_{3\text{-}055}$ = 1.4 µm, $t_{3\text{-}002}$ = 10.1 µm; $t_{3\text{-}063}$ = 5.1 µm, $t_{3\text{-}003}$ = 7 µm, $t_{3\text{-}075}$ = 7 µm). Examples of characteristic cracks in higher magnification are shown on samples 3-002 and 3-075 (bottom-right images).

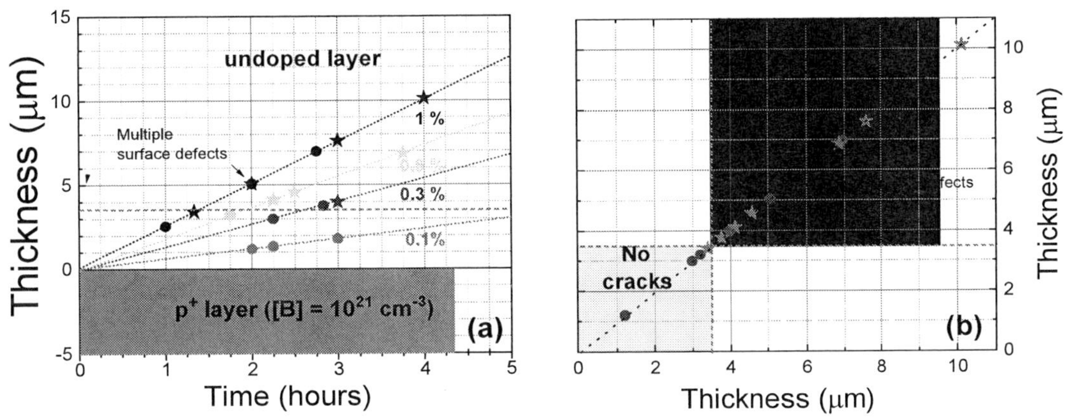

Figure 3 (a) Thickness of undoped diamond layer and observation of cracks (star symbols) or not (dotted symbols) on the sample surface as a function of the deposition time for undoped diamond layers grown with different methane concentrations **(b)** Graphical representation of the formation of cracks (star symbols) or not (dot symbols) on the surface of the sample as a function of the thickness of the undoped diamond layer independently of the methane concentration during the layer deposition.

Figure 2 (left) shows optical microscopy images of epitaxial diamond bilayers with different undoped diamond layer thicknesses and methane concentrations. Samples with a thin undoped diamond layer (e.g. sample 3-011 and 3-055) are smooth and without the defects observed on conventional (100) or (111) epitaxial layers [5]. They also do not exhibit visible cracks. On the contrary, most thick samples exhibit cracks (sample 3-002) or a high density of surface defects (sample 3-063). **Figure 2 (right)** shows SEM images with detailed morphology of cracks filled by polycrystalline-like diamonds immediately after their formation during the deposition process. **Figure 3a** shows the estimated thickness from the deposition time of undoped layer deposited at different methane concentrations on the 5 µm thick heavily boron-doped epitaxial layer and if the sample is cracked (star symbol) or not (dot symbol). Apart from the sample with a 7 µm thick undoped layer and the 5.1 µm thick 3-063 sample, all samples with a thick undoped layer exhibit cracks on their surface. These results are well illustrated in **Figure 3b,** which indicates the existence of a critical thickness between 3.5 and 4 µm above which the samples are cracked independently on the growth conditions. The formation of these cracks is attributed to the relaxation of the mechanical energy in the undoped diamond layer due to a significant lattice mismatch between the two doped diamond layers.

4. CONCLUSION

In this work, highly boron-doped / undoped epitaxial diamond bilayers have been grown by microwave plasma-enhanced chemical vapor deposition technique on (113) oriented single crystal diamond substrates with different thicknesses of the undoped diamond layer and different methane concentrations in the gas phase. The boron concentration is ca. 1021 cm^{-3} in the 5 µm thick highly doped diamond layer as determined from Raman analysis. Optical and scanning electron microscopy investigation shows the formation of cracks unwanted for the fabrication process of Schottky diodes on fabricated bilayers for samples with an undoped diamond layer thickness larger than ca. 3.5 µm independently on the methane concentration in the gas phase during the plasma-enhanced chemical vapor deposition. The formation of cracks is attributed to the relaxation of the mechanical energy in the undoped diamond layer due to a significant lattice mismatch between the two doped diamond layers.

ACKNOWLEDGEMENTS

The Czech Science Foundation primarily supported this work grant No. 20-11140S, the Czech Technical University Student Grant No. SGS20/176/OHK3/3T/13 and Education financed by European Structural and Investment Funds and the Czech Ministry of Education, Youth and Sports (Project No. SOLID21 - CZ.02.1.01/0.0/0.0/16_019/0000760). CzechNanoLab project LM2018110 funded by MEYS CR is also gratefully acknowledged for the financial support of the measurements at LNSM Research Infrastructure.

REFERENCES

[1] MORTET, V., TAYLOR, A., LAMBERT, N., GEDEONOVÁ, Z., FEKETE, L., LORINČIK, J., KLIMŠA, L., KOPEČEK, J., HUBÍK, P., ŠOBÁŇ, Z., LAPOSA, A., DAVYDOVA, M., VOVES, J., POŠTA, A., POVOLNÝ, V., HAZDRA, P. Properties of boron-doped (113) oriented homoepitaxial diamond layers. *Diamond and Related Materials.* 2021, vol. 111, p. 108223.

[2] PINAULT-THAURY, M.-A., TEMGOUA, S., GILLET, R., BENSALAH, H., STENGER, I., JOMARD, F., ISSAOUI, R., BARJON, J. Phosphorus-doped (113) CVD diamond: A breakthrough towards bipolar diamond devices. *Applied Physics Letters.* 2019, vol. 114, p.112106.

[3] TALLAIRE, A., VALENTIN, A., MILLE, V., WILLIAM, L., PINAULT-THAURY, MA, JOMARD, F., BARJON, J., ACHARD, J. Growth of thick and heavily boron-doped (113)-oriented CVD diamond films. *Diamond and Related Materials.* 2016, vol. 66, pp. 61-66.

[4] MORTET, V., ŽIVCOVÁ, ZV, TAYLOR, A., DAVYDOVÁ, M., FRANK, O., HUBÍK, P., LORINCIK, J., ALESHIN, M. Determination of atomic boron concentration in heavily boron-doped diamond by Raman spectroscopy. *Diamond and Related Materials.* 2019, vol. 93, pp. 54-58.

[5] TALLAIRE, A., KASU, M., UEDA, K., MAKIMOTO, T. Origin of growth defects in CVD diamond epitaxial films, *Diamond and Related Materials.* 2008, vol. 17, pp. 60-65.

Al_2O_3-Ta_2O_5 MULTILAYER THIN FILMS DEPOSITED BY PULSED DIRECT CURRENT MAGNETRON SPUTTERING FOR DIELECTRIC APPLICATIONS

[1]Richard DREVET, [1]Pavel SOUČEK, [2]Pavel MAREŠ, [2]Martin DUBAU, [1]Petr VAŠINA

[1]*Department of Physical Electronics, Masaryk University, Brno, Czech Republic, EU,* drevet@mail.muni.cz

[2]*HVM Plasma, spol s ro, Prague, Czech Republic, EU,* pavel.mares@HVM.cz

https://doi.org/10.37904/nanocon.2021.4324

Abstract

This research aims at synthesizing multilayer oxide thin films made of Al_2O_3 and Ta_2O_5 for dielectric applications. Multilayer thin films made of two, four, or eight oxide layers are synthesized by physical vapor deposition, specifically the mid-frequency pulsed direct current magnetron sputtering. The thin films are made of stoichiometric Al_2O_3 and Ta_2O_5 layers having a specific morphology observed from cross-section images obtained by scanning electron microscopy (SEM). The Al_2O_3 layers have a columnar structure, whereas the Ta_2O_5 layers are uniformly dense. X-ray diffraction (XRD) characterizations show that these oxide layers have very limited crystallinity due to the experimental conditions used during the magnetron sputtering process, particularly the low temperature of the substrate.

The dielectric behavior of the multilayer oxide thin films is assessed by measuring their dielectric breakdown potential. The two-layer and four-layer systems have intermediate values compared to the dielectric breakdown potentials measured for a monolayer of Al_2O_3 and a monolayer of Ta_2O_5 produced under the same experimental conditions. In the case of the eight-layer system, the dielectric breakdown potential value is the highest one, even higher than that measured for a monolayer of Ta_2O_5.

Keywords: Reactive magnetron sputtering, aluminum oxide, tantalum oxide, multilayer thin films, dielectric breakdown

1. INTRODUCTION

The thin oxide coatings are widely studied in the microelectronics industry for their insulation properties characterized by high dielectric constants [1]. They are used for various applications such as CMOS transistors, flash memory, and surface passivation of solar cells [2]. The magnetron sputtered aluminum oxide (Al_2O_3) coatings are among the most common ones, mainly due to the low-cost production. On the other hand, the coating process is hard to control in order to achieve high reproducibility and the achieved dielectric strength is far from the ideal values measured for bulk Al_2O_3. Tantalum oxide (Ta_2O_5) coatings could be a valuable alternative since they have better dielectric properties and the reactive process is easier to control, but they are also more expensive [3,4].

This work aims to improve the dielectric properties of surface oxide layers and, at the same time, to maintain a reasonable production cost. A relevant solution is the combination of both oxides in a multilayer system that alternates Al_2O_3 layers and Ta_2O_5 layers. According to Martinez-Perdiguero *et al.*, the interface between two oxide layers is advantageous to improve the dielectric properties of surface oxide thin films [5]. They have observed a significant improvement in the dielectric breakdown potential of a multilayer thin film made of Al_2O_3 and SiO_2 compared to the average value of a single monolayer of each oxide. They explain that the interface between two oxide layers acts as a barrier able to slow down the electron flows inside the insulating material. The resulting dielectric breakdown potential is higher than the average value of both oxide layers [6].

Multilayer oxide thin films made of Al_2O_3 and Ta_2O_5 are barely studied in the literature. The present work explores a solution to synthesis this multilayer system at low temperature by pulsed direct current magnetron reactive sputtering. Multilayer thin films made of two, four, and eight layers are studied. The experimental conditions are described, and the synthesized thin films are characterized by scanning electron microscopy associated with energy-dispersive X-ray spectroscopy (SEM-EDS) and X-ray diffraction (XRD). Finally, the dielectric properties of the thin films are assessed by measuring their dielectric breakdown potential.

2. MATERIALS AND METHODS

The Al_2O_3 and Ta_2O_5 layers are deposited by pulsed direct current magnetron sputtering using the HVM Flexilab thin film deposition system (HVM Plasma spol s ro). The aluminum and tantalum targets are powered by a Pinnacle Plus + power supply (Advanced Energy) at 250 W. The Si (100) and steel substrates were ultrasonically cleaned for 3 min in acetone and 3 min in isopropyl alcohol. Then, they are placed inside the chamber of the deposition system. After pumping for two hours, a background pressure value of 10^{-3} Pa is reached inside the chamber. Then, argon and oxygen are continuously injected inside the chamber. The flow of Ar is 60 sccm. The flow of O_2 is 4 sccm in the case of the Al_2O_3 layers and 19.2 sccm in the case of the Ta_2O_5 layers. The corresponding pressures are 1.0 Pa and 1.2 Pa, respectively. The temperature of the substrates is maintained at 180 °C during deposition. The direct current is pulsed at a frequency of 350 kHz and a pulse reversal time of 0.6 μs.

The morphology and the chemical composition of the obtained thin films are studied from cross-sections of the samples with a scanning electron microscope (SEM, Tescan MIRA3) equipped with an energy dispersive X-ray spectrometer (EDS). The phases of the oxide layers are characterized by X-ray diffraction (XRD) with a Rigaku SmartLab Type F diffractometer using monochromatic $Cu_{K\alpha}$ radiation (λ = 0.15406 nm). The phases are identified from the diffraction files provided by the International Centre for Diffraction Data.

The dielectric breakdown potentials are measured from the multilayer oxide thin films deposited on the steel substrates. Each side of one sample is connected to one output of a DC power supply, progressively increasing the voltage between the two electrodes. Because the multilayer oxide thin film is an electrical insulator, there is no current flow through the material until dielectric breakdown occurs. The corresponding potential value is the dielectric breakdown potential of the multilayer oxide thin film. The measurement is repeated at ten different positions for each sample to obtain an average value.

3. RESULTS AND DISCUSSION

The cross-section SEM image of **Figure 1** shows the morphology of the oxide thin film made of two layers deposited on a silicon substrate. According to the EDS characterizations, the lower part of the thin film is stoichiometric aluminum oxide, and the upper part of the thin film is stoichiometric tantalum oxide.

The continuity at the interface is good, with no separation between the two oxide layers. The Al_2O_3 layer has a columnar structure. This type of growth is commonly observed for thin films deposited by magnetron sputtering [7].

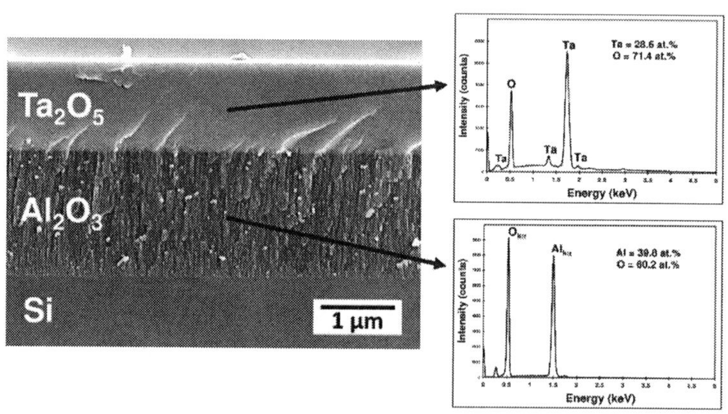

Figure 1 Cross-section SEM image and pointed EDS spectra of the bilayer oxide thin film deposited on a silicon substrate by pulsed direct current magnetron sputtering

This morphology indicates that the experimental conditions (target power, partial pressure, and substrate temperature) provide enough energy to the aluminum and oxygen atoms to direct the columnar growth of the Al_2O_3 layer. The morphology of the Ta_2O_5 layer looks different, being dense and uniform with no specific growth direction observable at this scale.

The four-layer and the eight-layer thin films are observed in the cross-sectional SEM images of **Figure 2**. The characteristic morphologies of the oxide layers are also observable, i.e., the columnar structure of the Al_2O_3 layers and the dense and uniform morphology of the Ta_2O_5 layers.

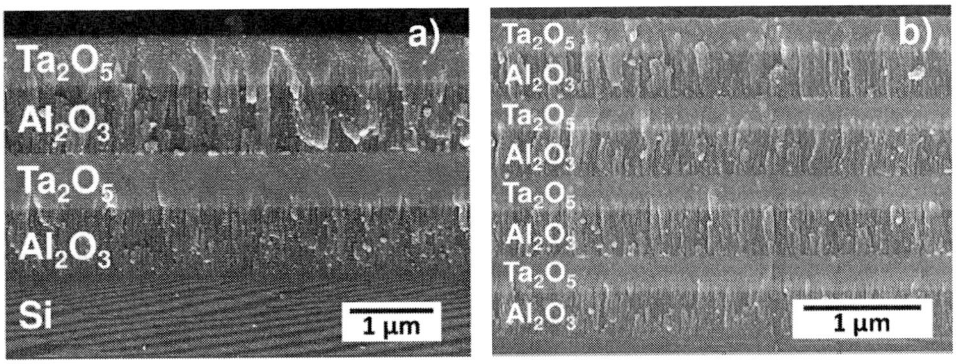

Figure 2 Cross-section SEM images of (a) the four-layers and (b) the eight-layers Al_2O_3-Ta_2O_5 thin films

The interfaces between the oxide layers are good for both samples, inducing continuity within the thin films without separation. All of these interfaces show that the tantalum oxide layers firstly grow by mimicking the columnar morphology of the aluminum oxide layer below. After a few nanometers, the Ta_2O_5 layer becomes more uniform with no specific growth direction as previously described for the bilayer oxide film (**Figure 1**).

The XRD patterns of the three multilayer oxide thin films are shown in **Figure 3**. Whatever the number of oxide layers, the XRD patterns show broad peaks corresponding to a nanocrystalline material. The low temperature used during the magnetron sputtering process induces a very low crystallization of the synthesized oxide layers [8]. However, a characteristic signal corresponding to Ta_2O_5 (pdf file # 00-025-0922) is observable for the three multilayer thin films corresponding to the upper oxide layer of each sample.

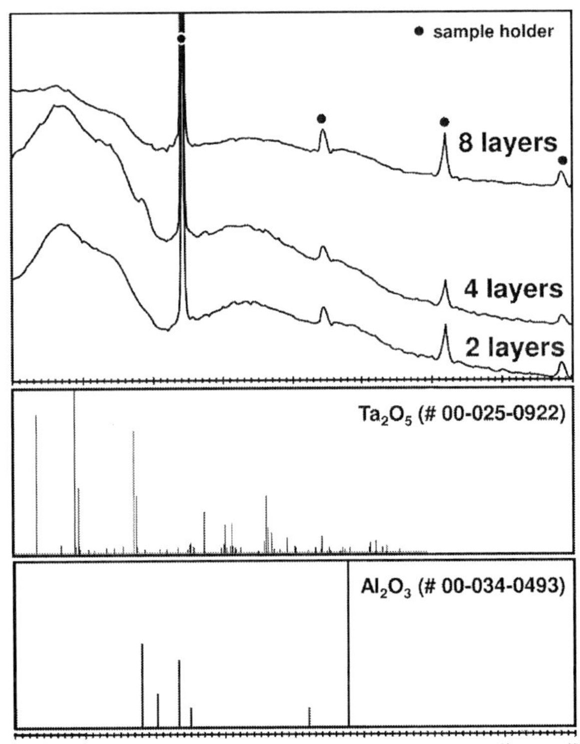

Figure 3 XRD patterns of the multilayer Al_2O_3-Ta_2O_5 thin films

The dielectric breakdown potentials of the three multilayer oxide thin films are shown in **Figure 4** and compared to the values obtained for a monolayer of Al_2O_3 and a monolayer of Ta_2O_5 produced under the same experimental conditions.

As expected, the dielectric breakdown potential of the Al_2O_3 monolayer (100 V μm^{-1}) is lower than that of the Ta_2O_5 monolayer (171 V μm^{-1}). The two-layer thin films (152 V μm^{-1}) and the four-layer thin films (105 V μm^{-1})

have intermediate values compared to both oxide monolayers. In the case of the value of the eight-layer thin film, the dielectric breakdown potential value is the highest (184 V µm^{-1}), even higher than that measured for the Ta_2O_5 monolayer.

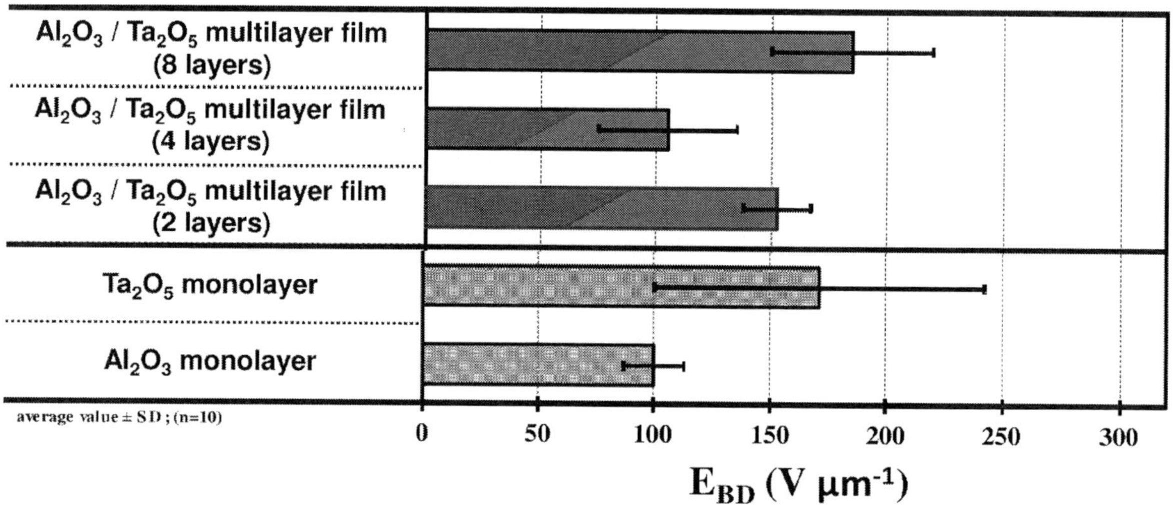

Figure 4 Dielectric breakdown potential (E_{BD}) of multilayer Al_2O_3-Ta_2O_5 thin films

The result of the four-layer oxide thin film is surprising in comparison with that of the other two multilayer thin films. The obtained dielectric breakdown potential value is barely higher than that of the Al_2O_3 monolayer despite the presence of Ta_2O_5 layers in the thin film in a multilayer system. This result may be due to some defects in the layers generated during deposition. This phenomenon is well known in reactive magnetron sputtering of oxide thin films [9]. The presence of oxygen in the atmosphere of the chamber contributes to the poisoning of the target, changing the surface of the metallic target to a dielectric oxide. The electric charges involved in the process are trapped in the dielectric at the surface of the target, and they accumulate until being released by arcing. This abrupt discharge results in an instability of the process, inducing uncontrollable damage to the deposited thin film [10]. However, the reason why this phenomenon occurs more intensively for the sample with four layers remains unanswered.

On the other hand, the two-layer and eight-layer thin films have higher dielectric breakdown potentials than the average value of the monolayers of Al_2O_3 and Ta_2O_5. This improvement is due to the impact of the interfaces in the multilayer thin films, which slow the propagation of the electric charges during an electrical breakdown. Gefle *et al.* have studied the mechanisms implied in this barrier effect. They describe the influence of several factors that prevent charge propagation at the interface between two solid media [11]. The main factors involved are local mechanical strain, space charge, permittivity gradient, and conductivity gradient. These factors are specific to the interfaces and increase the breakdown voltage and the breakdown time [12]. As a result, the dielectric strength of the multilayer oxide thin films is improved.

4. CONCLUSION

Thin films made of two, four, or eight alternate layers of Al_2O_3 and Ta_2O_5 were synthesized by pulsed direct current magnetron reactive sputtering. The process produced stoichiometric oxide layers of very low crystallinity with a characteristic morphology for each oxide layer. Thanks to the interfaces of the multilayer systems, the two-layer and eight-layer thin films show improved dielectric breakdown potentials compared to the average value measured for a monolayer of Al_2O_3 and a monolayer of Ta_2O_5. The value of the four-layer thin film is lower, certainly because of some defects in the layers generated during the magnetron reactive sputtering deposition.

REFERENCES

[1] ROBERTSON, J. High dielectric constant oxides. *The European Physical Journal Applied Physics*. 2004, vol. 28, pp. 265-291.

[2] The International Technology Roadmap for Semiconductors - emerging research materials. [online]. 2013. [viewed: 2021-10-01]. Available from: https://irds.ieee.org/.

[3] MAREŠ, P., KADLEC, S., DUBAU, M., MAREK, A., VYSKOČIL, J. Long-term stability and disappearing anode effects during reactive DC and pulsed bipolar magnetron sputtering of Al_2O_3. *Vacuum*. 2020, vol. 173, 109161.

[4] CHANELIERE, C., AUTRAN, J.L., DEVINE, R.A.B., BALLAND, B. Tantalum pentoxide (Ta_2O_5) thin films for advanced dielectric applications. *Materials Science and Engineering R*. 1998, vol. 22, pp. 269-322.

[5] MARTINEZ-PERDIGUERO, J., MENDIZABAL, L., MORANT-MIÑANA, M.C., CASTRO-HURTADO, I., JUARROS, A., ORTIZ, R., RODRIGUEZ, A. Electrical insulation and breakdown properties of SiO_2 and Al_2O_3 thin multilayer films deposited on stainless steel by physical vapor deposition. *Thin Solid Films*. 2015, vol. 595, pp. 171-175.

[6] HANBY, B.V.T., STUART, B.W., GIMENO-FABRA, M., MOFFAT, J., GERADA, C., GRANT, D.M. Layered Al_2O_3-SiO_2 and Al_2O_3-Ta_2O_5 thin-film composites for high dielectric strength, deposited by pulsed direct current and radio frequency magnetron sputtering. *Applied Surface Science*. 2019, vol. 492, pp. 328-336.

[7] MIRZAEI, S., ALISHAHI, M., SOUČEK, P., BURŠÍKOVÁ, V., ZÁBRANSKÝ, L. GRÖNER, L., BURMEISTER, F., BLUG, B., DAUM, P., MIKŠOVÁ, R., VAŠINA, P. Effect of substrate bias voltage on the composition, microstructure and mechanical properties of W-B-C coatings. *Applied Surface Science*. 2020, vol. 528, 146966.

[8] FRUNZA, R.C., KMET, B., JANKOVEC, M., TOPIC, M., MALIC, B. Ta_2O_5-based high-K dielectric thin films from solution processed at low temperatures. *Materials Research Bulletin*. 2014, vol. 50, pp. 323-328.

[9] MEISSNER, M., TOLG, T., BAROCH, P., MUSIL J. Elimination of Arcing in Reactive Sputtering of Al_2O_3 Thin Films Prepared by DC Pulse Single Magnetron. *Plasma Processes and Polymers*. 2011, vol. 8, pp. 500-504.

[10] WIATROWSKI, A., PATELA, S., KUNICKI, P., POSADOWSKI, W. Effective reactive pulsed magnetron sputtering of aluminum oxide - Properties of films deposited utilizing automated process stabilizer. *Vacuum*. 2016, vol. 134, pp. 54-62.

[11] GEFLE, O.S., LEBEDEV, S.M., USCHAKOV, V.Y. The mechanism of the barrier effect in solid dielectrics. *Journal of Physics D: Applied Physics*. 1997, vol. 30, pp. 3267-3273.

[12] LEBEDEV, S.M., GEFLE, O.S., POKHOLKOV. Y.P. The Barrier Effect in Dielectrics: The Role of Interfaces in the Breakdown of Inhomogeneous Dielectrics. *IEEE Transactions on Dielectrics and Electrical Insulation*. 2005, vol. 12, pp. 537-555.

TRANSFORMATION OF ZNO-BASED STRUCTURES UNDER HEAVY MO DOPING: DEFECT STATES AND LUMINESCENCE

[1]Maksym BURYI, [1]Zdeněk REMEŠ, [1,2]Kateřina DĚCKÁ, [3]Júlia MIČOVÁ, [1,4]Lucie LANDOVÁ

[1]Institute of Physics CAS in Prague, Prague, Czech Republic, EU, buryi@fzu.cz

[2]Czech Technical University in Prague, Faculty of Nuclear Sciences and Physical Engineering, Prague, Czech Republic, EU, katerina.decka@fjfi.cvut.cz

[3]Institute of Chemistry SAS, Bratislava, Slovakia, chemjumi@savba.sk

[4]Czech Technical University in Prague, Faculty of Electrical Engineering, Prague, Czech Republic, EU, abelova@fzu.cz

https://doi.org/10.37904/nanocon.2021.4321

Abstract

Hydrothermally grown ZnO-based structures were heavily doped with Mo (various doping levels from 2 to 25 %). Mo was found to strongly affect the structure and morphology of ZnO, resulting in a complex mixture of zinc oxide and molybdenum. Moreover, the transformation of material phases upon the increased Mo content was observed. ZnO phase was observed only at low Mo doping level (2 and 5 %). This correlated very well to the changes in the luminescence and electron paramagnetic resonance signals.

Keywords: ZnO, Mo doping, material phase transformation, luminescence, electron paramagnetic resonance

1. INTRODUCTION

ZnO is a smart material which finds application in various fields of human activities: as scintillator, including application in time-of-flight positron emission tomography [1,2]; as electrode material in supercapacitors; in photocatalysis, energy harvesting and storage systems [3,4].

In majority of cases luminescence of ZnO nanostructures is composed of fast exciton luminescence peaks having maximum in near ultraviolet region at about 380 nm with decay time in the order of hundreds of ps and a relatively slow defects-related bands decay time in the order of tens of μs. For example, the experimentally observed broad red bands ranging from 1 to 2 eV were related to zinc (V_{Zn}) or oxygen (V_O) vacancies [5]. The red emission bands (1.8-2 eV) related to V_{Zn} [5,6] were recently also observed in the hydrothermally grown ZnO nano- and microrods [7,8]. It is known that low-level Mo doping (below 1 %) can positively affect the luminescence properties of ZnO (both defect-related and exciton luminescence) [7,8], whereas doping with 10 and 30% of Mo deteriorates them. Moreover, Mo has the tendency to couple with ZnO creating some new phase other than ZnO itself.

ZnO structure can also be characterized by the typical electron paramagnetic resonance (EPR) single-line signal observed at the *g* factor $g \approx 1.96$ ascribed to the shallow Ga, Al, H donors (SD) [9-12], the Zn^+ + D complex, D = Ga, Al, H [8]. The influence of Mo on this shallow donor was studied earlier [6,8] in the ZnO:Mo(0.05, 0.25, 1%) and ZnO:Mo(10, 30%).

However, there is no systematic investigation of the influence of Mo at the doping level within 2-25% on ZnO structure, morphology and creation of zinc-molybdenum complex oxides, luminescence and magnetic properties. Therefore, this is the aim of the present paper.

2. EXPERIMENTAL

2.1. Samples preparation

All reagents were analytical grade and were used as-received without further purification. Zinc nitrate hexahydrate ($Zn(NO_3)_2 \cdot 6H_2O$) and hexamethylenetetramine (HMTA, $C_6H_{12}N_4$) were purchased from Slavus and ammonium heptamolybdate tetrahydrate $(NH_4)_6Mo_7O_{24} \cdot 4H_2O$ (NHMO) was obtained from Sigma-Aldrich. Deionized water was purified with a system So~Safe Water Technologies, having a conductivity 0.20 $\mu S \cdot cm^{-1}$ at 25 °C.

Undoped and Mo-doped ZnO nanorods with the Mo content 2-25% were prepared by the hydrothermal growth method. First, the corresponding stoichiometric amounts of $Zn(NO_3)_2 \cdot 6H_2O$ and $(NH_4)_6Mo_7O_{24} \cdot 4H_2O$, were dissolved in 200 ml and 50 ml of deionized water, respectively, with vigorous stirring by a magnetic stirrer. At the same time, a 25 mM aqueous solution of HMTA was prepared. All the solutions were filtered through a Whatman 2 filter. Subsequently, the filtered solutions were stirred vigorously for 15 min. The nominal concentration of the ZnO:Mo(2-25%) in the final suspensions was 25 mM. The solution of HMTA was added to the mixture. Stirring of the reaction mixture was followed by hydrothermal growth of nanorods by heating at 90 °C for 3 hours. The grown nanorods were isolated and then purified (removal of residual salts), washing 5 times with deionized water and subsequent centrifugation at 18,000 rpm (RCF: 23,542 x g) for 15 minutes. Finally, the samples were lyophilized.

The amount of powder samples was approximately the same in all experiments.

2.2. Experimental techniques

XRD on powder samples was performed using RigakuMiniFlex 600 (Ni-filtered Cu-$K_{\alpha1,2}$ radiation) equipped with the NaI:Tl scintillation detector and XRD patterns were compared to the relevant records in the ICDD PDF-2 database (version 2013). The angular range was 10°-80°, with a step of 0.02° and a scanning speed of 2°/min.

The size and morphology of ZnO NRs has been checked by scanning electron microscopy (SEM) using MAIA3, TESCAN electron microscope with the in-beam SE detector placed in objective lens and the electron beam energy 5 keV.

The steady-state photoluminescence (PL) spectra in the 350-800 nm spectral range were excited at frequency 333 Hz by pulsed 1 mW UV LED optically filtered by narrow band-pass filter centered at the wavelength 340 nm and recorded with 2 nm spectral resolution using the spectrally calibrated f/4 double gratings monochromator, long-pass filters, a red sensitive photomultiplier, current preamplifier (10 μA/V) and a lock-in amplifier referenced to the UV LED frequency. All PL spectra were spectrally calibrated, normalized at the wavelength 355 nm on the same value (dominated by optical scattering of excitation light) and converted from wavelength to energy scale taking into account the Jacobian correction [13]. PL spectra were measured with 5 mg ZnO powder pressed in a Suprasil glass tube with the inner (outer) diameter 2 mm (3 mm) featuring a low fluorescence background.

EPR measurements were performed with a commercial Bruker EMXplus spectrometer in the X-band (9.4 GHz) within the 4-296 K temperature range using Oxford Instruments ESR900 cryostat. The spectrometer sensitivity is about 10^{12} spins/mT.

3. RESULTS AND DISCUSSION

3.1. Phase purity and morphology

The XRD patterns of ZnO:Mo(2-25%) samples are demonstrated in **Figure 1**.

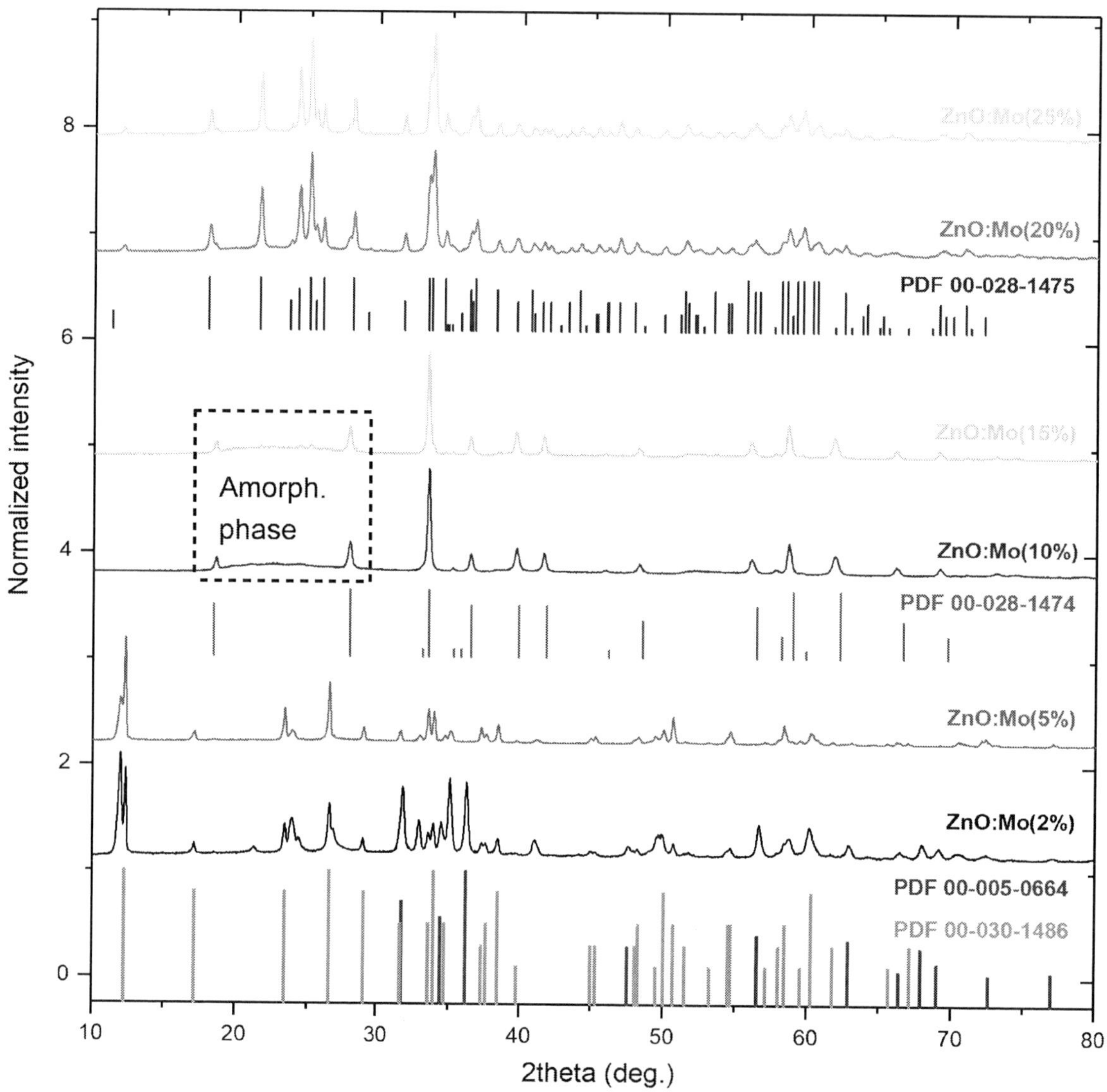

Figure 1 XRD patterns of the ZnO:Mo(2-25%). The positions of the XRD reflections are indicated by vertical solid lines distinguished by color according to the corresponding record in the ICDD PDF-2 database: PDF 00-005-0664 (hexagonal ZnO), PDF 00-030-1486 ($Zn_5Mo_2O_{11}{\cdot}5H_2O$), PDF 00-028-1474 ($MoO_3{\cdot}2ZnO{\cdot}H_2O$), PDF 00-028-1475 ($2MoO_3{\cdot}3ZnO{\cdot}H_2O$).

The XRD patterns of ZnO:Mo(2-25%) are very complex, especially at the largest Mo content, composed of different material phases (they were tentatively ascribed to the complex zinc-molybdenum oxides). The negligibly small presence of ZnO phase could be detected only in the ZnO:Mo(2, 5%) dominated by some other material phase. This phase was tentatively ascribed to $Zn_5Mo_2O_{11}{\cdot}5H_2O$. Further increase of the Mo content to 10 and 15% resulted in the disappearance of the ZnO phase and transformation of the $Zn_5Mo_2O_{11}{\cdot}5H_2O$ to the $MoO_3{\cdot}2ZnO{\cdot}H_2O$. Moreover, some unknown amorphous phase appeared. Its origin is unknown. Further increase of the Mo content to 20 and 25% results in the $MoO_3{\cdot}2ZnO{\cdot}H_2O$ transformation into the $2MoO_3{\cdot}3ZnO{\cdot}H_2O$ while amorphous phase disappeared.

SEM provides an insight into the morphology of the ZnO:Mo(2-25%) samples. The corresponding images are shown in **Figure 2**. The ZnO:Mo(2%) sample is the only consisting of hexagonal prismoid microrods, in average, of the size 1-2 μm in circumference and over 3 μm long (**Figure 2A**). However, the microrods are outnumbered by the irregular flat slab-like structures with the averaged area of about 1-10 μm^2. Interestingly, the ZnO:Mo(5%) sample is composed of hexagons, varying in "diameter" within the 5-10 μm range and about 0.5 μm thick (**Figure 2B**).

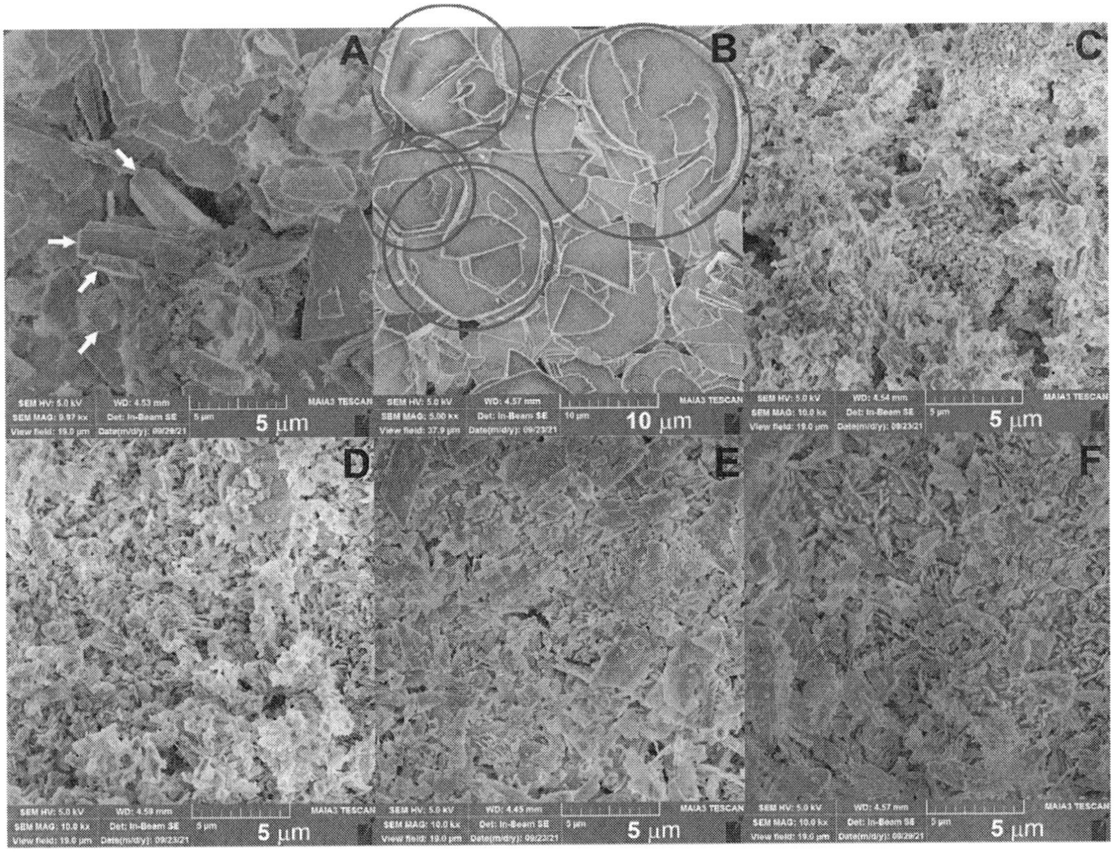

Figure 2 SEM images of the ZnO:Mo(2-25%) samples with different magnification as indicated in Figures: A - ZnO:Mo(2%); B - ZnO:Mo(5%); C - ZnO:Mo(10%); D - ZnO:Mo(15%); E - ZnO:Mo(20%); F - ZnO:Mo(25%). ZnO microrods are stressed by white arrows. Hexagons are marked by red circles.

The ZnO:Mo(10-25%) samples consist of flake-like sheared structures with the averaged area about 0.25 μm^2 (**Figures 2C-F**). In the ZnO:Mo(20, 25%) samples the flakes are getting mixed with the larger slabs of the averaged area of about 5 μm^2. This is consistent with the previous study of the ZnO:Mo(30%) [7].

Based on the XRD and SEM analyses one may conclude that the increased Mo doping level results in the phase transformation and changes in morphology of the ZnO:Mo(2-25%) samples.

3.2. PL and EPR characterization

PL spectra have been measured in all the ZnO:Mo(2-25%) samples at room temperature as can be seen in **Figure 3A**. The spectrum of ZnO:Mo(2%) is composed of two very broad overlapped bands (B1 (~2.14 eV) and B3(~2.80 eV)) - covering the 1.4-3.4 eV energy region. Surprisingly, the PL spectrum of the ZnO:Mo(5%) is single-band peaking at 2.64 eV (B2 band with 1 eV full width at half maximum (FWHM)). Furthermore, the B2 band is getting weaker upon the increased Mo content to 10 and 15% and the new clearly visible very narrow bands appear (B4-7 peaking at 2.96, 3.16, 3.24 and 3.34 eV, respectively). The B4-7 bands may be ascribed to exciton-like emission observed in ZnO [14], especially considering the existence of ZnO part in the complex zinc-molybdenum oxide creating ZnO:Mo(10, 15%) samples as discussed in the subsection above (see also **Figure 1**). The B4-7 bands are weaker in the ZnO:Mo(15%) sample. The new single-band spectra, B3' (~2.91 eV), of the same amplitude are observed in the ZnO:Mo(20, 25%) samples. All of this confirms the gradual phase transformation upon the increased Mo content in the ZnO:Mo(2-25%) samples.

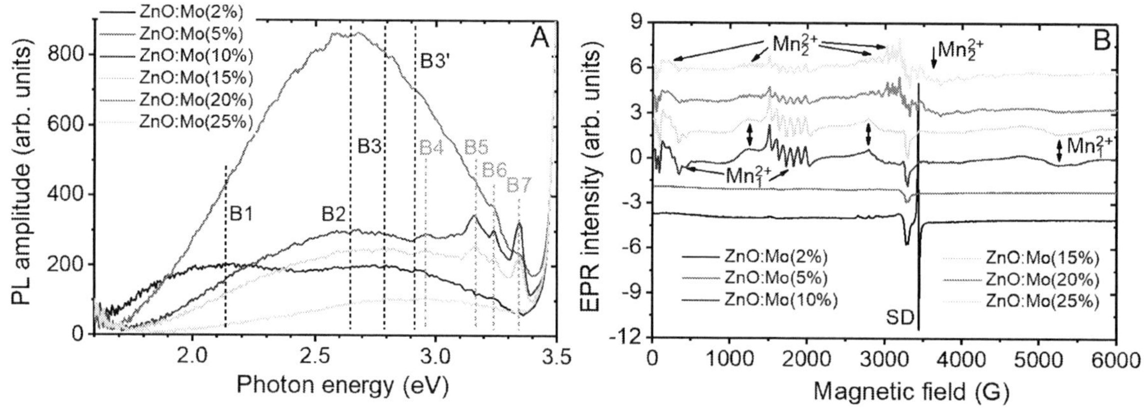

Figure 3 A - PL spectra measured in the ZnO:Mo(2-25%). B1-3,3′,4-7 indicate specific emission bands. The signal above 3.4 eV is scattered light. **B** - EPR spectra measured in the ZnO:Mo(2-25%). "SD" stands for "shallow donors". $Mn^{2+}_{1,2}$ and vertical double arrows indicate the spectra of two different Mn^{2+} centers.

In order to find out the presence of paramagnetic centers and their evolution upon the increased Mo content, EPR spectra were measured in all the ZnO:Mo(2-25%) samples and shown in **Figure 3B**. Only the spectrum of the ZnO:Mo(2%) was attributed with the SD signal at the *g* factor $g = 1.954$. The rest of visible signals were not identified. This signal is absent in the spectrum of the ZnO:Mo(5%). Based on this fact and considering also the XRD results above, one may conclude that only the tiniest amount of ZnO phase exists in the ZnO:Mo(5%) sample. The SD signal is absent in the EPR spectra of the ZnO:Mo(10, 15%). However, there are some resonances which were identified as belonging to Mn^{2+} (electron spin $S = 5/2$ and the ^{55}Mn nucleus with 100% abundance and nuclear spin $I = 5/2$), Mn^{2+}_1 - there are five components due to $S = 5/2$ with the one exhibiting the most pronounced hyperfine structure (sextet of lines because of $I = 5/2$ with the center at 1738 G) [15]. The Mn^{2+}_1 spectrum is getting weaker in the ZnO:Mo(20, 25%) samples, and the new Mn^{2+} spectrum appears (Mn^{2+}_2). It was identified using similar consideration as for the Mn^{2+}_1 identification above. The Mn^{2+} is an accidental impurity but serves as a good indicator of the gradual material phase transformation. The SD signal is absent in the EPR spectra of the ZnO:Mo(20, 25%) as well. This correlates well to the XRD and PL measurements above.

4. CONCLUSION

It has been found that heavily Mo-doped hydrothermally grown ZnO-based structures appear a mix of different zinc-molybdenum complex oxides upon the increased Mo content. Their morphology is changing upon the Mo doping level from microrods to microhexagons, then to micro- and nanoflakes and microslabs as well. This is confirmed by the observed transformations of PL and EPR spectra.

ACKNOWLEDGEMENTS

This work was supported by the Czech Science Foundation by the project No. 20-05497Y, CAS Mobility Plus project SAV-AV CR-21-09 and by the Scientific Grant Agency of Ministry of Education, Science, Research and Sport of Slovak Republic and Slovak Academy of Sciences (VEGA 2/0157/20).

REFERENCES

[1] LEITER, F., ZHOU, H., HENECKER, F., HOFSTAETTER, A., HOFMANN, D. M., MEYER, B. K. Magnetic resonance experiments on the green emission in undoped ZnO crystals, *Physica B.* 2001, vol. 308-310, pp. 908-911.

[2] LECOQ, P. Pushing the Limits in Time-of-Flight PET Imaging, *IEEE Trans. Radiat. Plasma Med. Sci.* 2017, vol. 1, pp. 473-485.

[3] NAJIB, S., BAKAN, F., ABDULLAYEVA, N., BAHARIQUSHCHI, R., KASAP, S., FRANZÒ, G., SANKIR, M., SANKIR, N. D., MIRABELLAD, S., ERDEM, E. Tailoring morphology to control defect structures in ZnO electrodes for high-performance supercapacitor devices, *Nanoscale.* 2020, vol. 12, pp. 16162-16172.

[4] TOUFANI, M., KASAP, S., TUFANI, A., BAKAN, F., WEBER, S., ERDEM, E. Synergy of nano-ZnO and 3D-graphene foam electrodes for asymmetric supercapacitor devices, *Nanoscale.* 2020, vol. 12, pp. 12790-12800.

[5] FRODASON, Y. K., JOHANSEN, K. M., BJØRHEIM, T. S., SVENSSON, B. G., ALKAUSKAS, A. Zn vacancy as a polaronic hole trap in ZnO. *Phys. Rev. B.* 2017, vol. 95, 094105.

[6] BURYI, M., REMEŠ, Z., BABIN, V., ARTEMENKO, A., VANĚČEK, V., AUBRECHTOVÁ DRAGOUNOVÁ, K., LANDOVÁ, L., KUČERKOVÁ, R., MIČOVÁ, J. Transformation of free-standing ZnO nanorods upon Er doping, *Applied Surface Science.* 2021, vol. 562, 150217.

[7] BURYI, M., REMEŠ, Z., BABIN, V., VANĚČEK, V., AUBRECHTOVÁ DRAGOUNOVÁ, K., MIČOVÁ, J., LANDOVÁ, L., KUČERKOVÁ, R. ZnO nanorods alloyed with Mo/Er. The effect of postdeposition treatment on defect states and luminescence. *IOP Conf. Ser.: Mater. Sci. Eng.* 2021, vol. 1050, 012002.

[8] BURYI, M., REMEŠ, Z., BABIN, V., NOVOTNÝ, M., VANĚČEK, V., AUBRECHTOVÁ DRAGOUNOVÁ, K., MIČOVÁ, J., LANDOVÁ, L., KUČERKOVÁ, R., MORE-CHEVALIER, J., CHERTOPALOV, S., FITL, P., KMJEČ, T. Influence of Mo doping on the luminescence properties and defect states in ZnO nanorods. Comparison with ZnO:Mo thin films, Applied Surface Science. 2021, vol. 555, 149679.

[9] JANOTTI, A., VAN DE WALLE, CH. G. Native point defects in ZnO. *Phys. Rev. B.* 2007, vol. 76, 165202.

[10] JANOTTI, A., VAN DE WALLE, CH. G. Fundamentals of zinc oxide as a semiconductor. *Rep. Prog. Phys.* 2009, vol. 72, 126501.

[11] GONZALES, C., BLOCK, D., COX, R. T., HERVÉ, A. Magnetic resonance studies of shallow donors in zinc oxide. *J. Crystal Growth* 1982, vol. 59, 357.

[12] BLOCK, D., HERVÉ, A., COX, R. T. Optically detected magnetic resonance and optically detected ENDOR of shallow indium donors in ZnQ. *Phys. Rev. B* 1982, vol. 25, 6049.

[13] MOONEY, J., KAMBHAMPATI, P. Get the Basics Right: Jacobian Conversion of Wavelength and Energy Scales for Quantitative Analysis of Emission Spectra. *J. Phys. Chem. Lett.* 2013, vol. 4, pp. 3316-3318.

[14] ÖZGÜR, Ü., ALIVOV, YA. I., LIU, C., TEKE, A., RESHCHIKOV, M. A., DOĞAN, S., AVRUTIN, V., CHO, S.-J., MORKOÇ, H. A comprehensive review of ZnO materials and devices. *J. Appl. Phys.* 2005, vol. 98, 041301.

[15] ABRAGAM, *A.,* BLEANEY, B. *Electron Paramagnetic Resonance of Transition Ions.* Oxford: Clarendon Press, 1970, vol. 1.

HYDROPHILIZATION OF POLYVINYLIDENE FLUORIDE (PVDF) NANOFIBERS BY PLASMA TREATMENT AND ITS EFFECT ON NANOPARTICLE DEPOSITION

[1]Veronika KAŽIKOVÁ, [1]Július VIDA, [2]Petr RYŠÁNEK, [2]Martin KORMUNDA, [2]Pavla ČAPKOVÁ, [1]Tomáš HOMOLA

[1]*MUNI - Masaryk University, Brno, Czech Republic, EU, 517378@mail.muni.cz*

[2]*J. E. Purkyne University, Ústí nad Labem, Czech Republic, EU*

https://doi.org/10.37904/nanocon.2021.4325

Abstract

Polyvinylidene fluoride (PVDF) polymeric nanofibers deposited by roller electrospinning on polypropylene non-woven fabric were treated with atmospheric pressure plasma to hydrophilize its surface. Diffuse coplanar surface barrier discharge (DCSBD) was employed to generate low-temperature atmospheric plasma in ambient air. Changes in wettability were determined by evaluating water contact angles before and after plasma treatment for various treatment times. The effect of aging of plasma-treated surfaces stored in laboratory conditions was studied in the course of storage time. The chemical composition of the surfaces was determined by X-ray photoelectron spectroscopy (XPS) which showed a decrease in carbon and an increase in oxygen atomic concentrations. Subsequently, significant differences in loading of iron oxide nanoparticles between treated and untreated PVDF samples were observed by scanning electron microscopy.

Keywords: DCSBD plasma source, PVDF, wettability, water contact angle, iron oxide nanoparticles

1. INTRODUCTION

Polyvinylidene fluoride (PVDF) nanofibers are a promising type of polymeric nanofibers thanks to their large surface area-to-volume ratio, flexibility, fine pore texture and more advantageous properties used in many nano- or biotechnological applications in medicine, cosmetics, electronics manufacturing, metallurgy or for designing nanocomposite membranes. Such membranes can utilize the adsorption of porous fibres and loaded nanoadsorbers for water filtration, which is one of the most significant concerns in many ever-growing industries, polluting Earth's water resources [1,2]. As nanoadsorbers, iron oxide nanoparticles, attracted interest, due to many desirable properties of reactivity, adsorption, catalytic properties or low cost [1]. Especially magnetite nanoparticles represent a promising method of adsorbing heavy metal pollutants from water, with an ability to easily manipulate them with a magnetic field [3].

In such processes, the wettability of used porous material is crucial and low wettability is a disadvantage of using PVDF in contrast to its superior chemical, thermal and mechanical properties. Thus, enhancing the wettability of PVDF is desired not only in water filtration, but for example also in the preparation of stimuli-responsive membranes [4,1]. Traditionally, chemical hydrophilization techniques such as coating with hydrophilic polymer solutions are widely applied [1]. Creating a PVDF mixed matrix by blending PVDF with the deformed rebar-like Fe_3O_4-palygorskite nanocomposites has also proved to lower the water contact angle by around 20° [5]. In this work, we present an alternative low-cost method for hydrophilization, which is more environmentally friendly than often used chemical methods, by using low-temperature plasma surface treatment. Plasma treatment is a very effective method for changing surface properties and enhancing wettability [6]. Especially for treatments of polymers, plasma sources with low working temperatures are desirable. This condition is met with diffuse coplanar surface barrier discharge (DCSBD) with a working temperature of around 70 °C, the ability to use in ambient air, so no vacuum setup is needed and configuration ideal for the treatment of larger surfaces.

2. EXPERIMENT

PVDF polymeric nanofibers were deposited on polypropylene (PP) non-woven fabric by electrospinning. Samples were cut into approximately 1 cm x 1 cm pieces, placed onto a glass substrate and its surface was treated with plasma. DCSBD, commercialized by MUNI spin-off Roplass s.r.o., Czech Republic was used as a source of low-temperature atmospheric pressure plasma. It is a type of dielectric barrier discharge consisting of parallel electrodes embedded in a ceramic plate. As working an ambient air was used. To ensure the samples are treated with a diffuse part of a plasma, samples were moved during treatment back and forth on the discharge plane. The samples were during treatment in direct contact with a discharge plate to ensure plasma was present in the space between the nanofibers inside the sample. The treatment time was varied between 2 s and 16 s. Aging of the samples was determined after 7, 14 and 30 days of storage.

Wettability of untreated and treated samples was evaluated by dropping 1 µm water droplet with a pipette on the surface and contact angle was analysed with See system (Advex Instruments, Czech Republic).

The chemical composition of the surface and bonding states were analysed from X-ray photoelectron spectroscopy (XPS) spectra, obtained by AXIS Supra spectrometer (Kratos Analytical Ltd., United Kingdom). Data were calibrated to C-H peak at 285.0 eV in C1s narrow region. Background U Poly Tougaard was used, line shapes for fitting were mixed Gaussian-Lorentzian.

Fe_2O_3 magnetite nanoparticle powder (Sigma-Aldrich, <100 nm) was dispersed in dowanol PM (1:4) and milled in a glass vial with glass balls at 900 rpm for 10 days. After milling, the suspension was mixed with polysiloxane binder and dissolved in ethanol in ratio 6:2. The concertation of polysiloxane in ethanol was 20% and the process is described in detail in [7]. Then the suspension was further diluted in dowanol PM in a 1:10 ratio and dropped with a 2 µm volume pipette on the untreated and plasma treated PVDF. After drying, the samples with deposited nanoparticles were sputtered with 10 nm thick Au + Pd film and imaged with MIRA3 scanning electron microscope (Tescan, Czech Republic) with Schottky field emission electron gun. Images were obtained by secondary electron detector, while the accelerating voltage was set to 8 kV and working distance to 15 mm.

3. RESULTS AND DISCUSSION

3.1. Wettability

Changes in water contact angle (WCA) were observed after plasma treatment in ambient air. The untreated PVDF surface showed WCA of 95° ± 6° and ambient air plasma treatment for 2 s was enough to increase the wettability of PVDF to the extend it was not possible to measure the static WCA, as droplets were immediately absorbed by nanofibers PVDF surface. The aging behaviour of the PDFV after the plasma treatment for 8 s was evaluated after 7, 14 and 30 days. The results showed that the ambient air plasma treated sample retained its increased wettability even after a month. The plasma treatment was repeated with setting the distance between PVDF sample and plasma plane to 0.3 mm, as usually used to treat surfaces with DCSBD to ensure the best interaction with plasma. However, this led to no change in wettability whatsoever, suggesting that PVDF nanofibers must be in direct contact with the DCSDB ceramics in order to plasma sufficiently enter the space between nanofibers.

3.2. Chemical analysis of PVDF after plasma treatment

In **Table 1** can be seen individual atomic concentration changes obtained from high resolution X-ray photoelectron spectroscopy (XPS) regions and the ratio of C:O concentrations. The ratio decrease can be clearly seen after plasma treatment. This suggests surface cleaning processes of plasma from carbon contaminations. Even 2 seconds of plasma treatment was enough to decrease the carbon concentration and increase the concentration of oxygen and fluorine. Longer plasma treatments do not change the chemical composition of the surface significantly. Traces of potassium, phosphorus and sodium were detected on the surface, decreasing slightly after plasma treatment. Moreover, some amount of nitrogen was incorporated into the surface during treatment.

Table 1 Atomic concentrations of untreated and ambient air plasma-treated PVDF polymer, obtained from high-resolution XPS regions

Treatment time	C [%]	F [%]	O [%]	K [%]	P [%]	Na [%]	N [%]	C:O ratio
0 s	56.3 ± 0.4	23 ± 1	12.3 ± 0.3	2.7 ± 0.2	3.5 ± 0.2	1.7 ± 0.1	0 ± 0	4.6 ± 0.1
2 s	48.6 ± 0.4	29.90 ± 0.01	15.0 ± 0.3	2.44 ± 0.01	2.6 ± 0.2	0.68 ± 0.01	0.8 ± 0.1	3.25 ± 0.08
4 s	49.0 ± 0.1	30.26 ± 0.08	14.7 ± 0.4	2.13 ± 0.08	2.28 ± 0.07	0.58 ± 0.07	1.0 ± 0.1	3.32 ± 0.08
8 s	48.58 ± 0.06	29.5 ± 0.7	16.1 ± 0.5	2.0 ± 0.2	2.3 ± 0.2	0.47 ± 0.02	1.1 ± 0.3	3.0 ± 0.1
16 s	50.93 ± 0.07	29.7 ± 0.5	14.1 ± 0.9	1.5 ± 0.2	2.1 ± 0.3	0.36 ± 0.06	1.3 ± 0.3	3.6 ± 0.2

Figure 1 shows intensities of XPS C 1s and O 1s peaks individual components and their evolution with plasma treatment time. In **Figure 2**, shapes of C 1s XPS spectra and deconvoluted components can be seen for reference and 2 s plasma treated sample in ambient air. C 1s spectrum was deconvoluted into five components: C-H/C-C (285.0 eV), C-OH/C-O-C (286.3 eV), C=O (288.0 eV), O-C=O (289.0 eV) and C-F2 (290.7 eV). Because K 2p region overlaps with the C 1s region, the intensities of K 2p peaks was set to zero. After ambient air plasma the intensity of C-C/C-H component decreased, indicating surface cleaning from hydrocarbon contamination. Moreover, a significant increase in O-C=O component is observed, caused by oxygen species in the plasma.

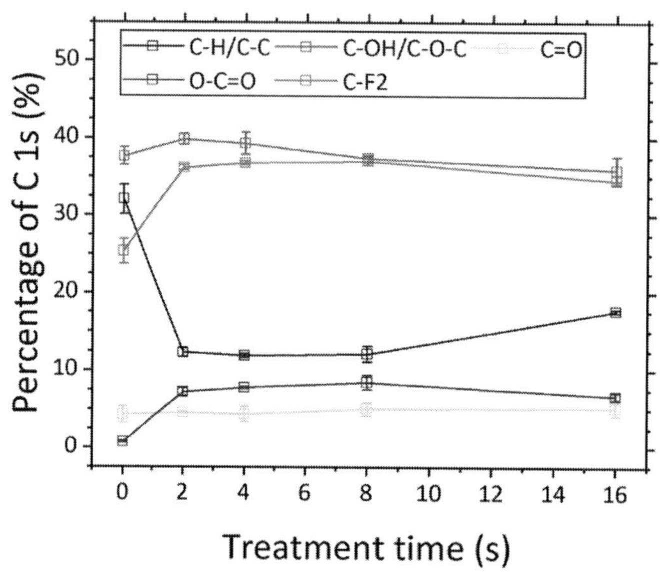

Figure 1 Changes in the intensities of individual components of the C 1s peak in XPS spectrum

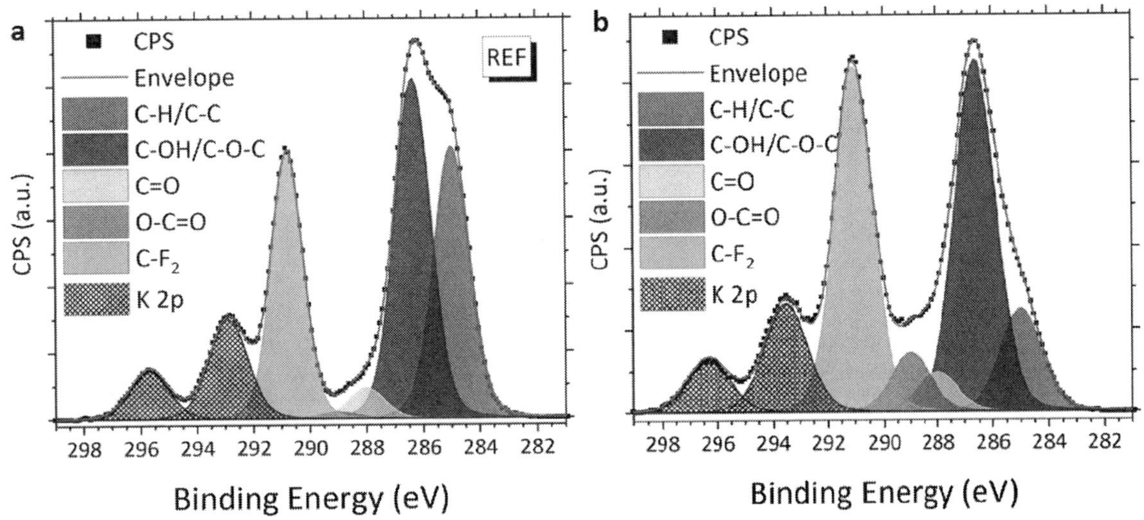

Figure 2 Changes in the shape and components of C 1s narrow regions in XPS spectra due to ambient air plasma treatment for 2 s (b) compared with reference sample (a).

3.3. Deposition of nanoparticles on plasma treated PVDF

The untreated and plasma treated PVDF nanofibers with deposited magnetite particles, are shown in **Figure 3**. Due to the wettability results, only an ambient air plasma was utilized in this experiment and treatment time was set to 2 seconds, as longer treatment did not produce significantly better results. Images revealed that fibres on the sample without plasma treatment were wrapped in dried solution. However, on the sample with plasma treatment, solution wetted the fibres to the higher extend, getting deeper into the sample and leaving more nanoparticles on the surface of fibres.

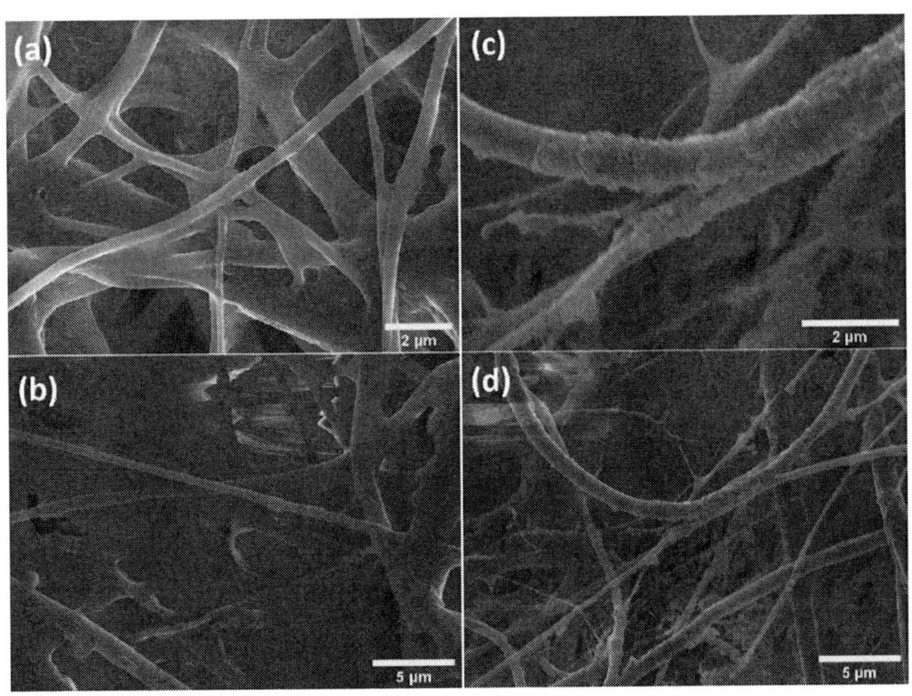

Figure 3 SEM images of nanoparticle solution deposited on untreated PVDF nanofibers (**a**, and **b**,) and 2 s ambient air plasma treated PVDF nanofibers (**c**, and **d**,).

4. CONCLUSION

We achieved a change of the water contact angle of PVDF nanofibers increasing its wettability with a low temperature DCSBD plasma source in ambient air. Chemical composition analysis revealed a possible cleaning effect of plasma from hydrocarbon contamination and functionalization of its surface by polar oxygen-containing groups. The PVDF surface with enhanced wettability showed an increased ability to adsorb more magnetite nanoparticles on nanofibers surface.

ACKNOWLEDGEMENTS

This research was supported by GACR projects 19-14770Y and projects LM2018097, LM2018110 funded by the Ministry of Education, Youth and Sports of the Czech Republic.

REFERENCES

[1] ABDALI, N., MARJANI, A., HEIDARY, F., ADIMI, M. Fabrication of PVA coated PES/PVDF nanocomposite membranes embedded with in situ formed magnetite nanoparticles for removal of metal ions from aqueous solutions. *New j. Chem.* 2017, vol. 41, pp. 6405-6414.

[2] ASMATULU, R., MUPPALLA, H., VEISI, Z., KHAN, W.S., ASADUZZAMAN, A., NURAJE, N. Study of hydrophilic electrospun nanofiber membranes for filtration of micro and nanosize suspended particles. *Membranes.* 2013, vol. 3, pp. 375-388.

[3] PADMAVATHY, K.S., MADHU, G., HASEENA, P.V. A study on effects of pH, adsorbent dosage, time, initial concentration and adsorption isotherm study for the removal of hexavalent chromium (Cr(VI)) from wastewater by magnetite nanoparticles. *Procedia Technology.* 2016, vol. 24, pp. 585-594.

[4] HSU, C.C., WU, C.S., LIU, Y.L. Multiple stimuli-responsive poly(vinylidene fluoride) (PVDF) membrane exhibiting high efficiency of membrane clean in protein separation. *Journal of Membrane Science.* 2014, vol. 450, pp. 257-264.

[5] ZHU, J., ZHOU, S., LI, M., XUE, A., ZHAO, Y., PENG, W., XING, W. PVDF mixed matrix ultrafiltration membrane incorporated with deformed rebar-like Fe_3O_4-palygorskite nanocomposites to enhance strength and antifouling properties. *Journal of Membrane Science.* 2020, vol. 612, pp. 118467.

[6] LAI, J., SUNDERLAND, B., XUE, J., YAN, S., ZHAO, W., FOLKARD, M., MICHAEL, B.D., WANG, Y., Study on hydrophilicity of polymer surfaces improved by plasma treatment. *Applied Surface Science.* 2006, vol. 252, pp. 3375-3379.

[7] HOMOLA, T., DZIK, P., VESELY, M., KELAR, J., CERNAK, M., WEITER, M. Fast and low-temperature (70 °C) mineralization of inkjet printed mesoporous TiO_2 photoanodes using ambient air plasma. *ACS Appl. Mater. Interfaces.* 2016, vol. 8, pp. 33562-33571.

SESSION B
INDUSTRIAL AND ENVIRONMENTAL APPLICATIONS OF NANOMATERIALS

Chairmen

Prof. Dr. Radek ZBORIL	Palacky University in Olomouc, CATRIN-RCPTM, Czech Republic, EU
Prof. Dr. Miroslav CERNIK	Technical University of Liberec, Czech Republic, EU
Prof. Dr. Miloslav PEKAR	Brno University of Technology, Czech Republic, EU

COMPOSITE COATINGS CONTAINING NANOMETRIC CONSTITUENTS PREPARED BY PLASMA SPRAYING WITH LIQUID FEEDSTOCKS

Radek MUSALEK, Tomas TESAR, Jan MEDRICKY, Jan CIZEK, Frantisek LUKAC

Institute of Plasma Physics of the Czech Academy of Sciences, Prague, Czech Republic, EU,
musalek@ipp.cas.cz

https://doi.org/10.37904/nanocon.2021.4310

Abstract

Plasma spraying of liquid feedstocks allows deposition of materials with novel microstructures. Due to the combination of a liquid carrier and ultrafine powders (in case of suspensions) or in-situ formation of the final material in the plasma jet (in case of solutions), preparation of ultrafine composites is possible, typically in a form of coatings on the substrates. Moreover, both suspension and solution spraying routes can be easily combined, either together or even with conventional plasma spraying of coarse dry powders, thus merging benefits of "nanometric" and "micrometric" constituents in the final composite microstructure. As a consequence, liquid feedstock plasma spraying represents an industrially-relevant deposition process with wide variability, providing coatings with properties tailored for various applications ranging from compact wear-resistant layers, porous thermal barrier coatings, bio-compatible medical coatings, catalyst carriers, fuel-cells, etc. The paper introduces several examples of such ultrafine coatings recently deposited at IPP CAS using hybrid water-stabilized plasma torch.

Keywords: Plasma spraying, liquid feedstock, coatings, functionally graded materials (FGMs)

1. INTRODUCTION

Conventional plasma spraying is based on spraying of relatively coarse dry powders, with typical size in tens of micrometers [1]. Decreasing particle size is needed to achieve finer coatings microstructures. However, too small particles do not possess high enough momentum to penetrate the plasma jet and get heat-treated properly and thus provide poor deposition efficiency. Small dry particles may even get entirely evaporated in the plasma jet. These problems may be solved by plasma spraying of liquid feedstocks allowing deposition of coatings from suspensions and solutions [2]. Due to the combination of a liquid carrier and ultrafine powders (suspensions route) or in-situ formation of the final material in the plasma jet (solutions route), preparation of ultrafine composites is possible, as the liquid carrier provides the needed momentum and the injected material can thus penetrate the plasma jet. However, as the primary stream of the liquid gets fragmented in the plasma due to the immense shear forces, each created droplet contains multiple primary nanoparticles, which agglomerate after the liquid carrier evaporation. It has therefore turned out, that introduction of nanometric features into the microstructure may be quite challenging as most of the primary particles agglomerate during the in-flight stage [3]. However, the deposited droplets are still significantly smaller (particularly in terms of volume) than in the case of conventional spraying of dry powders, enabling formation of features with size of several tens of nanometers. Also, nanometric features may be present in the coating in the form of ultrafine porosity, untreated primary particles (e.g., particles which were deliberately not entirely melted), condensates (i.e. material which was evaporated but condensed again on the coated part-sample) or from solution.

This paper introduces examples of several coatings prepared recently at the Institute of Plasma Physics of the Czech Academy of Sciences having submicron-sized features and deposited with different geometries, including relatively homogeneous single-layers, heterogeneous multilayers, intermixed coatings and functionally graded coatings (FGM).

2. EXPERIMENTALS

All coatings shown in this study were deposited by hybrid water/argon stabilized plasma torch WSP-H 500 (ProjectSoft HK a.s., Czech Republic) with typical operation power 150 kW. The deposition took place in the open air atmosphere onto the substrates attached to a revolving carousel. The substrates were first grit-blasted by alumina, residues of which may be found on the substrate-coating interface. Injection of the liquid feedstocks into plasma was radial with typical injection distance from the plasma nozzle in the range of 20 to 30 mm. When spraying liquid feedstocks, typical stand-off distance of the plasma torch from the coated substrates was 100 mm, which guaranteed proper heat treatment of the fragmented particles along their whole trajectory in the plasma jet (typical jet length was also ~100 mm) without losing too much velocity, ensuring high deposition efficiency. Due to the immense heat transfer from the plasma towards the samples, these were cooled by compressed air. The plasma torch was mounted on programmable robotic arm allowing deposition by consecutive strokes of the torch in front of the carousel separated by cooling periods. Liquid feedstocks were injected into plasma with very high feed rates, typically around 100 ml/min. More details on the deposition with WSP-H are provided e.g. in [4,5] or papers referenced for each coating example.

Shadowgraphy of the feedstock injection into plasma jet was carried out by SprayCam imaging system (Control Vision Inc., USA). Metallographic cross-sections and free-surfaces of the deposited coatings were investigated by EVO MA 15 (Carl Zeiss, Germany) scanning electron microscope in back-scattered electron detection mode.

3. RESULTS

3.1. Coating Formation

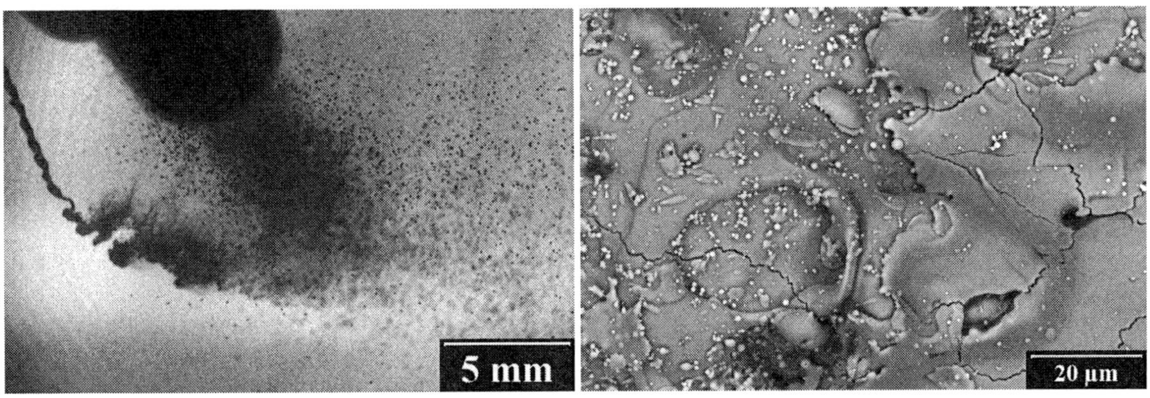

Figure 1 Left: Shadowgraphy of concurrent injection of liquid stream and coarse dry powder into the plasma jet during so-called hybrid plasma spraying. **Right:** Illustration of the hybrid YSZ coating surface incorporating both conventional large splats from coarse dry powder and miniature splats deposited from suspension.

Both main setups of thermal spraying using conventional dry powder or liquid feedstocks may be illustrated in **Figure 1**. In the left figure, shadowgraphy of the injection point is depicted. In this so-called hybrid spraying setup [6], continuous liquid stream is fragmented in the plasma jet into smaller droplets which are further downstream accompanied by coarser dry particles. Subsequently, all droplets are dragged towards the coated part while the liquid carrier is evaporated and the solid phase(s) melted. After the impact of molten droplets on the substrate, material is deposited in form of flattened lamellae - splats (**Figure 1 - right**). These are relatively big when originated from the coarse dry powders, having characteristic diameter of several tens of micrometers. In case of ceramics, their cracking is common, due to the evolution of quenching stresses exceeding their relatively low local strength and limited ability to deform plastically [7]. Nevertheless, these cracks may be welcome as they improve for example thermal shock resistance or strain tolerance of the deposited ceramic layers. On the other hand, splats originating from liquids are much smaller which may provide other benefits if needed, such as suppressed formation of the intrasplat cracks. Also, due to their rather lower momentum at the time of impact, they often tend to retain their round shape even when deposited (**Figure 2 - right**). In case of plasma spraying with liquid feedstocks only, the deposited coatings may consist solely of such submicron-sized particles. Naturally, also the porosity within the coating becomes much finer when compared to the conventional coatings sprayed from coarse powders.

3.2. Coatings with Ultrafine Features

It is a great benefit of thermal spraying that different layers may be easily combined e.g. by alternating feedstocks in the feedline. Example of such layered coating is shown in **Figure 2**, where a triple layer consisting of metallic bond-coat was overlaid by yttria stabilized zirconia (YSZ) deposited first from dry coarse powder and consequently from ethanol-based YSZ suspension [8]. A magnified detail of interface between ceramic sublayers (**Figure 2-right**) illustrates a substantial difference in the size of achievable microstructural features, including sub-micron sized lamellae and pores.

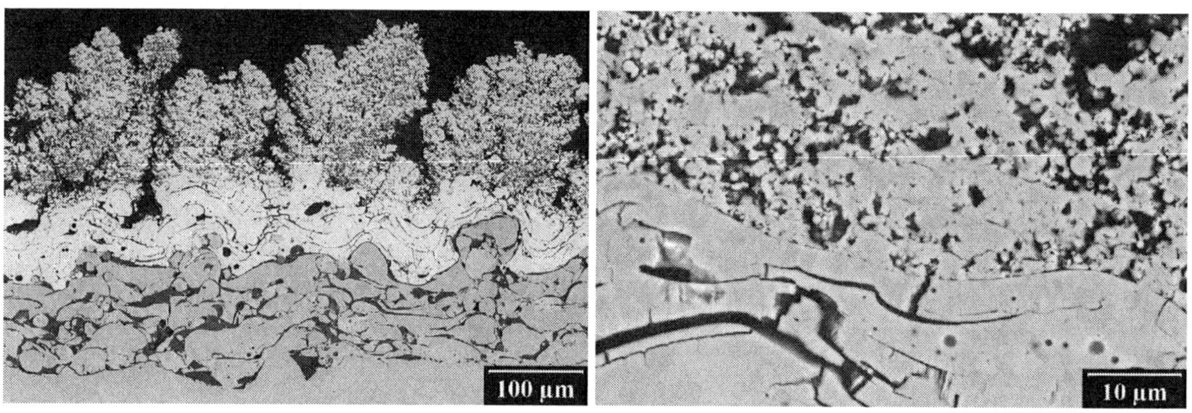

Figure 2 Layered coating consisting of NiCrAlY bond-coat (at the bottom) oversprayed by coarse YSZ from powder and columnar ultrafine YSZ deposited from suspension

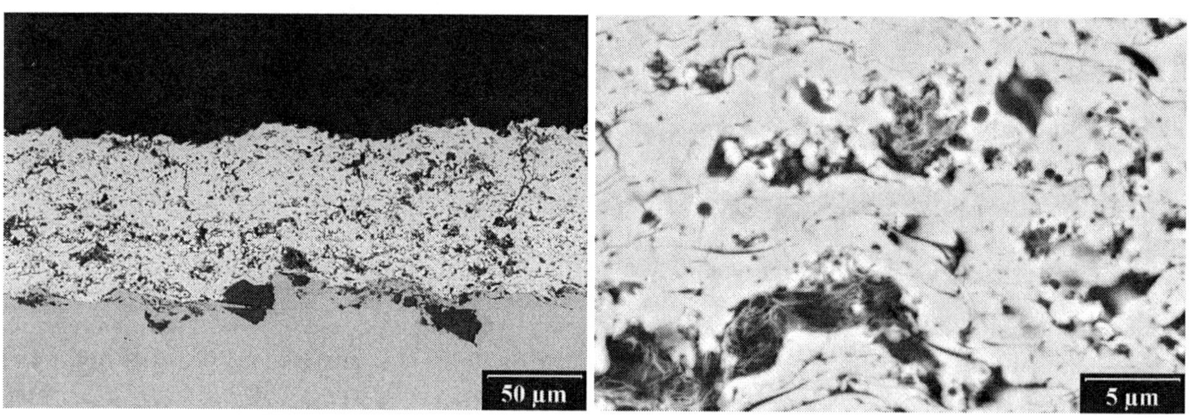

Figure 3 Vertically cracked YSZ ceramic deposited from solution

Aside from the size of the features, also their overall morphology may be altered. **Figure 3** shows a typical microstructure of a vertically cracked YSZ ceramics deposited from solution (mixture of yttrium nitrate hexahydrate + zirconium acetate in dilute acetic acid) [4]. Areas of rather dense coating intermingled with highly porous thin-walled "foamy" areas may be observed. A much more porous alumina coating was deposited from aqueous solution of aluminium nitrate nonahydrate (**Figure 4**) [9]. Extremely porous ultrafine coating, which was nevertheless quite fragile and thin (**Figure 5**), was deposited from solution of titanium isopropoxide in ethanol [10]. In this case, a "tree-like" morphology of the coating suggests that the coating was predominantly formed by condensation of a vaporized feedstock rather than conventional deposition of the molten droplets.

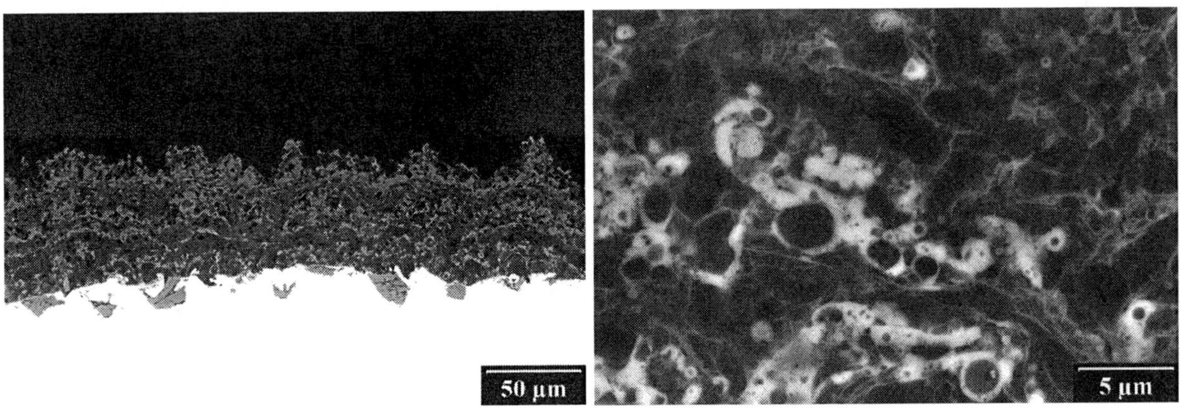

Figure 4 Porous alumina ceramic deposited from solution

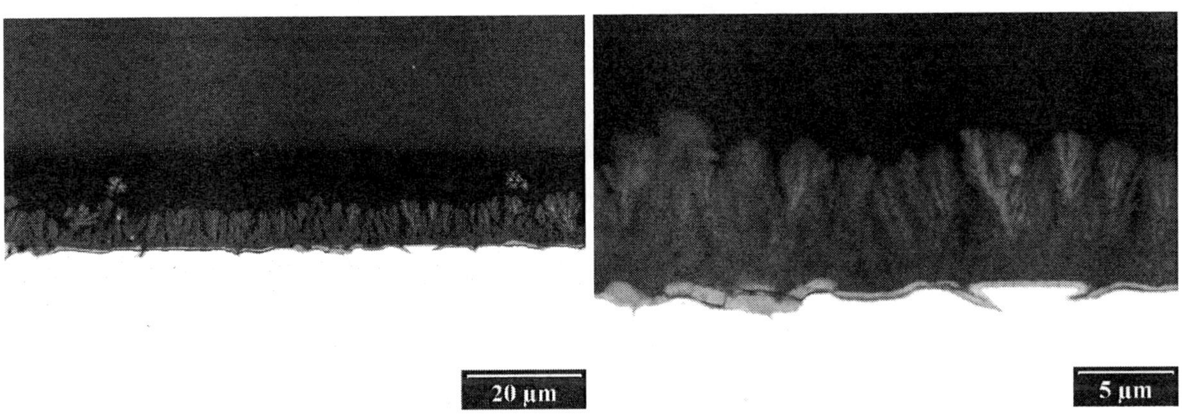

Figure 5 Ultra-fine titania condensed during spraying of solution of titanium isopropoxide

3.3. Intermixed Coatings

Another possibility to achieve novel ultrafine microstructures by thermal spraying is deposition of composite coatings. **Figure 6** illustrates such coating deposited from aqueous hydroxyapatite suspension [11]. Two types of features may be observed within the coating; rather dense flattened splats with characteristic diameter around 10 micrometers and much finer droplets originating from the feedstock that was not thoroughly processed, i.e. where the liquid carrier was evaporated but the primary particles were deposited without considerable agglomeration. In the following experiments (**Figure 7**), a composite coating was deposited also by using aqueous hydroxyapatite (HAp) suspension as feedstock. However, this time, with addition of aqueous silver nitrate solution, which resulted into incorporation of homogeneously dispersed Ag nanoparticles within the coating. Integration of such submicron-features may be beneficial e.g. for improvement of antibacterial properties of the HAp coatings (currently under survey).

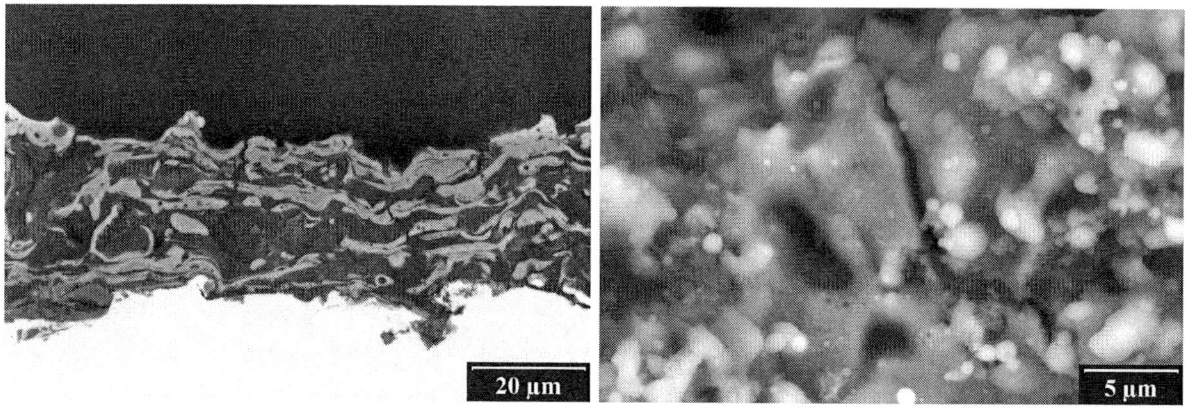

Figure 6 Hydroxyapatite coating. Cross-section (left) and free-surface (right) showing submicron features.

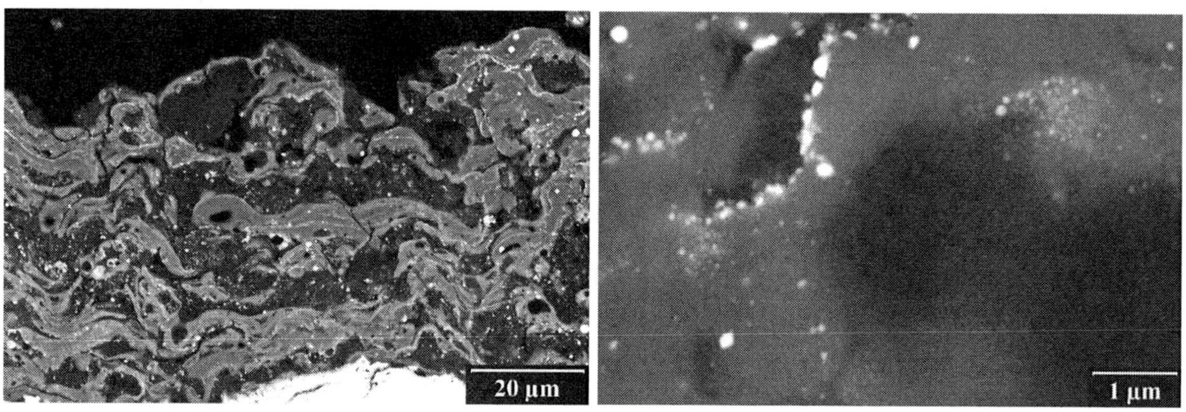

Figure 7 Hydroxyapatite coating with homogenously dispersed antibacterial Ag nanoparticles (brightest phase)

3.4. GRADIENT COATINGS

Recent modifications to the liquid injection unit enabled continuous change of the feedstock composition, for example, from suspension A to suspension B. In this way, a novel coating with gradual change from alumina to YSZ was deposited (**Figure 8**). Such microstructure may potentially provide substantial functional benefits such as mitigation of stresses evolving during thermal loading on the sharp interfaces of conventional layered coatings.

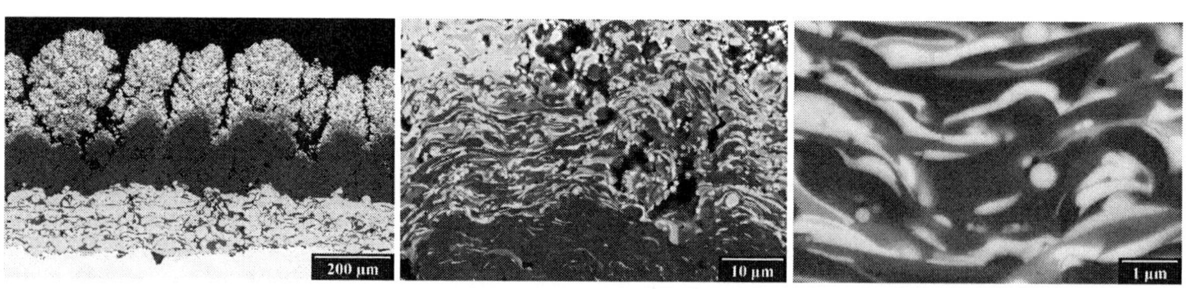

Figure 8 Microstructure of multilayer coating with functionally graded interface between alumina (dark) and yttria-stabilized zirconia (bright). Ceramic coatings were deposited over NiCrAlY bond-coat.

4. CONCLUSION

As illustrated in the selected examples, recent advances in plasma spraying with liquid feedstocks enabled deposition of novel coatings with interesting sub-micron/nanometric sized features. These may be reflected in improved functional properties of the deposits when compared to conventional coatings or may be used in novel applications such as catalysis, SOFCs, etc., where ultrafine composite coatings/layers are needed. Due to the high feed rates achievable with WSP-H technology, plasma spraying with liquid feedstocks may be also both cost- and time-efficient. For example, deposition of application-relevant coating thickness (typically several tens to several hundreds of micrometers) onto 20 test coupons (20 × 30 mm) usually takes just several minutes. The ability of the developed method to coat large areas in the open-air atmosphere (i.e., without costly and space-constrained chamber providing vacuum or protective gas atmosphere) and high efficiency of the deposition process makes it an ideal candidate for deposition of such coatings even at industrial scale.

ACKNOWLEDGEMENTS

Financial support through project 19-10246S "Deposition mechanism and properties of multiphase plasma sprayed coatings prepared with liquid feedstocks" funded by Czech Science Foundation is gratefully acknowledged.

REFERENCES

[1] DAVIS, J.R. *Handbook of Thermal Spray Technology*. Davis J.R., ed. Materials Park, OH, USA: ASM International, 2004.

[2] FAUCHAIS, P., JOULIA, A., GOUTIER, S. et al. Suspension and solution plasma spraying. *Journal of Physics D: Applied Physics*. 2013, vol. 46, pp. 224015.

[3] FAUCHAIS, P., VARDELLE, M., GOUTIER, S., VARDELLE, A.. Key Challenges and Opportunities in Suspension and Solution Plasma Spraying. *Plasma Chemistry and Plasma Processing*. 2014, vol. 35, pp. 511-25.

[4] MUSALEK, R., MEDRICKY, J., TESAR, T. et al. Controlling Microstructure of Yttria-Stabilized Zirconia Prepared from Suspensions and Solutions by Plasma Spraying with High Feed Rates. *Journal of Thermal Spray Technology*. 2017, vol. 26, pp. 1787-803.

[5] MUSALEK, R., MEDRICKY, J., TESAR, T. et al. Suspensions Plasma Spraying of Ceramics with Hybrid Water-Stabilized Plasma Technology. *Journal of Thermal Spray Technology*. 2017, vol. 26, pp. 37-46.

[6] JOSHI, S.V., SIVAKUMAR, G. Hybrid Processing with Powders and Solutions: A Novel Approach to Deposit Composite Coatings. *Journal of Thermal Spray Technology*. 2015 vol. 24, pp. 1166-86.

[7] KURODA, S., CLYNE, T.W. The quenching stress in thermally sprayed coatings. *Thin Solid Films*. 1991, vol. 200, pp. 49-66.

[8] TESAR, T., MUSALEK, R., MEDRICKY, J., CIZEK, J., LUKAC, F., CHRASKA, T. Deposition of multiphase coatings from liquid feedstock using hybrid water-stabilized plasma torch. In: *ITSC 2018 - Proceedings of the International Thermal Spray Conference*. Orlando: ASM International, 2018, pp. 456-62.

[9] TESAR, T., MUSALEK, R., LUKAC, F. et al. Solution Precursor Plasma Spraying of Cr-Doped Al_2O_3 Thermochromic Coatings. *Journal of Thermal Spray Technology*. 2019, vol. 29, pp. 199-211.

[10] MUSALEK, R., MEDRICKY, J., TESAR, T., KOTLAN, J., LUKAC, F. Deposition of titania from solution by hybrid water-stabilized plasma torch. In: *METAL 2016 - Conference Proceedings: 25th Anniversary International Conference on Metallurgy and Materials*. Brno: Tanger, 2016, pp. 1121-6.

[11] CIZEK, J., BROZEK, V., CHRASKA, T. et al. Silver-Doped Hydroxyapatite Coatings Deposited by Suspension Plasma Spraying. *Journal of Thermal Spray Technology*. 2018, vol. 27, pp. 1333-43.

DIFFUSE COPLANAR SURFACE BARRIER DISCHARGE PLASMA TREATMENT AS A PART OF TECHNOLOGY FOR MANUFACTURING OF NANOFIBER-BASED FILTERS

[1]Oleksandr GALMIZ, [1]Michal FLEISCHER, [1]David PAVLIŇÁK, [1]Jakub KELAR, [1]Ali Jamaati KENARI, [2]Jana SVOBODOVÁ, [1]Mirko ČERNÁK, [1]Dušan KOVÁČIK

[1]*Department of Physical Electronics, CEPLANT, Masaryk University, Brno, Czech Republic, EU,* o.galmiz@mail.muni.cz

[2]*NAFIGATE Corporation a.s., Prague, Czech Republic, EU*

https://doi.org/10.37904/nanocon.2021.4311

Abstract

Nanofiber membranes are made of synthetic polymers mainly by electrospinning technology. The key point for creating a functional nanofiber membrane for water and air filters is to meet basic key properties such as filtration efficiency, mechanical resistance, and resistance to fouling and chemicals. Design and manufacturing of the advanced nanofiber-based filters urgently require new environment-friendly and cost-effective surface treatments without the use of organic solvents and caustic solutions. To address this need, as an alternative, the atmospheric-pressure plasma treatment offers to be used for surface activation of polymer textile materials serving as a substrate for electrospun nanofiber. Nanofiber carriers represented by polypropylene non-woven were pre-treated by dielectric barrier discharge in continuous mode to improve the adhesion between the produced nanofibers and substrate. The increased adhesive forces to carrier substrate were confirmed by two peeling tests. The fact that the robust and effective atmospheric-pressure diffuse coplanar surface barrier discharge technology, primarily developed and optimized for the plasma treatment of textile and fibrous material, can be easily implemented in the industrial production lines predetermines this technology for in-line a large throughput manufacturing of advanced nanofiber-based filters.

Keywords: Nanofibers, plasma, adhesion, spunbond non-woven polypropylene

1. INTRODUCTION

Currently, the requirements to achieve filtration efficiency in conjunction with the requirement for the highest possible breathability of filter material force the manufacturer to look for new material combinations and technological processes for the production of filter media. Filter media containing a nanofiber filter membrane have proven to be an advantageous and economical option for capturing submicron particles. Such media can therefore be advantageously used for the production of filters and protective aids with FPP3 filtration efficiency, including protection against viral particles. These nanofiber membranes are made of synthetic polymers mainly by electrospinning technology. This fabrication method which is based on high electrostatic forces to elongation and thinning of flowed polymer fiber is very popular for its simplicity, efficiency, and usability of most industrially expanded polymers.

Such nanofiber properties as great porosity and high surface area make this nanomaterial interesting for applications focusing on filtration, fibrous reinforcements in synthetic or biological composites, electronic devices, or tissue engineering [1]. The nanofibers themselves are tender material and are very sensitive to mechanical manipulation. Moreover, the low adhesion forces of the nanofibers to the untreated carrier make them very complicated material to manipulate with, and a high degree of the nanofibers damage during their processing. The filtration efficiency and mechanical resistance are also determined by such parameters as good adhesion of the membrane to the substrate.

At present, many efforts are being addressed to modify the surface properties of polymer fibers to improve the performance of these materials and to pave the way to new technological applications. Nowadays a plasma treatment is a frequently used method for physical and chemical modification of polymer surfaces [2]. Under the bombardment of active species generated by the plasma, the polymer surface can be modified by removing surface contamination, introducing new chemical functional groups, or deposition of thin coating [3]. The diffusivity of the diffuse coplanar surface barrier discharge (DCSBD) plasma and the high efficiency showed to be one of the most efficient in such cases.

In this work, the DCSBD plasma sources operated in the air at atmospheric pressure were used to improve the adhesion properties of the substrate represented by spunbond non-woven polypropylene (SB NWPP) after plasma treatment and the nanofibers' adhesion to the substrate.

2. EXPERIMENTAL SETUP

2.1. Nanofibres deposition

The spinning of nanofiber mesh was carried out on the NanospiderTM NSLAB 500 machine (Elmarco, Czech Republic). Conditions of electrospinning were as follows: 55 kV applied HV, 140 mm electrode to collector distance, 6 rpm rotating speed of the working electrode. As the working electrode was used the 10 cm length rotating wire electrode. The nanofiber mats were collected on spunbond non-woven polypropylene of 17 gsm substrate with a 25 mm/min shift speed. The rewinding itself was provided by a self-standing rewinding device, ensuring the correct fabric downforce to the electrodes, the displacement of the carrier during electrospinning, and the speed and feed direction.

Polyvinylidene fluoride (PVDF) was chosen as a starting polymer with well-defined properties for the study of the adhesion of nanofibers to the substrate and their modification in plasma. Electrospinning solution was prepared by dissolving the PVDF powder (purchased by Nafigate a.s., Czech Republic) in a solution (6.75/50 g/ml) of N,N-Dimethylacetamide (anhydrous, 99.8 %, Sigma Aldrich). The 1 % of tetraethylammonium bromide (Sigma Aldrich) was added to the final solution to increase the electrical conductivity. After pellets agitation, the solution was tempered with continuous stirring for 1h at 40 °C and later 23h at 25 °C.

2.2. Plasma treatment

The processing of the carrier fabric was performed using a planar or curved DCSBD (CDCSBD) electrode. Plasma was generated in the form of a surface dielectric barrier discharge in air at atmospheric pressure at a supplied sinusoidal voltage of 15-30 kHz frequency and amplitude of 10 kV. The tested power of the DCSBD was in the range 200 - 600 W.

The main advantage of a concavely curved DCSBD electrode allowing the processing of sensitive materials, as well as various foils and soft fabrics. The process parameters were the same as for the planar electrode. Photos in **Figures 1 (a, c)** shows the DCSBD electrodes prepared for the plasma treatment process.

2.3. Strike-through time measurements

The test to determine the liquid penetration time (urine imitation) for nonwovens was performed according to European standard ISO 9073-8. The materials used for this test consist of infiltrating reference pads (filter papers) and urine imitation according to ISO 3696 (a solution with 9 g/l chloride sodium in the water of purity 3). An electronic time meter, with the possibility of reading the time up to hundreds of seconds, was used to measure the strike-through time for 5 ml of testing solution.

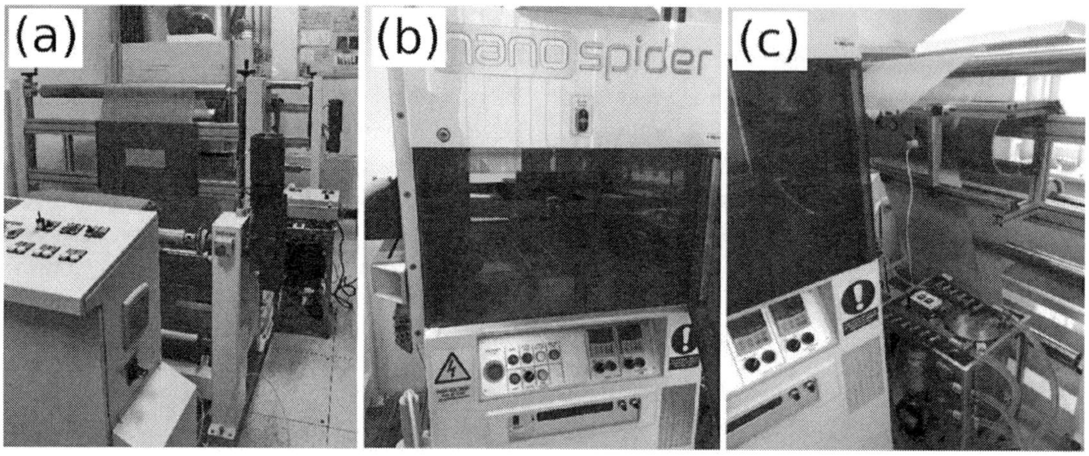

Figure 1 The line for the carrier plasma pre-treatment and the nanofibers production: a) planar DCSBD electrode, b) the Nanospider NSLAB 500 device, c) concavely curved DCSBD electrode

2.4. Adhesion measurements - Peel test

Adhesion measurements were performed on a TA.XTplusC Texture Analyzer with a 50 N load cell at a head speed of 1 mm/s from Stable Micro Systems. The force required to detach the calibrated adhesive tape, under 90 degree angle, from the test specimens (peel test) was measured. Reference samples and samples that underwent plasma treatment on DCSBD and CDCSBD systems were tested. A peel test was done immediately after the plasma treatment. Testing was performed on more than 10 samples and the resulting values were obtained from the arithmetic mean of the measured data. Exponent Connect software was used to process the data.

2.5. Adhesion measurements - Loop test

Adhesion tests were performed by measuring the force required to tear the deposited nanofiber layer from an area of 25 mm × 25 mm. A TA.XTplusC Texture Analyzer (Stable Micro Systems, UK) with a 5 N load cell was used for the measurement. A method that was used for these measurements called "loop test" is a modified method based on the method used for adhesion measurement of the adhesive tape.

Figure 2 Measurement of the adhesive force between the SB NWPP substrate and the PVDF nanofibers by the loop test method

The sample with cut 25 mm × 25 mm sections, was glued to a plastic holder with double-sided adhesive tape on the side of the substrate, which was then fixed in the lower jaws of the measuring device. A test adhesive tape with sufficiently high adhesion, 10 cm long, was placed in a controlled manner in the upper movable jaws so as to form a loop (see **Figure 2**). Keeping the same length (10 cm) and the shape of the loop a reproducible pressure force of the test adhesive tape was achieved. The tape was not pressed to the sample in any other manner. Subsequently, the jaws were moved away, and the magnitude of the applied force vs. jaw position was recorded. The speed of movement of the jaw head was 5 mm/s. The trigger force was 1 g (i.e. 10 mN). As the parameter for comparison, the area under the obtained curve, corresponding to the mechanical work necessary to tear off the layer was calculated and compared. The maximum force and the mean force were then determined from the digital recording of the measurement. The maximum force was always reached just before tearing the nanofiber layer from the textile substrate. In a general minimum of 10 samples were tested for every condition.

3. RESULTS

3.1. Strike-through time measurements

It was observed that the urine imitation was able to penetrate through the reference SB NWPP substrate sample. However, non-uniform behavior was recorded. On some samples of the non-woven, strike-through time values exceeded 600 seconds. Mainly the average time of penetration was around 16 seconds with a standard deviation of 10 seconds. Even such a short treatment time as 1 second by DCSBD plasma was enough to reduce this time to 2 seconds. The effect of plasma treatment using both DCSBD configurations was comparable and did not exceed 3 seconds even with error ranges.

3.2. Adhesion measurements - Peel test

For this experimental setup, both planar and curved configurations of DCSBD were tested. While treating the carrier material both configurations of the electrodes proved to have a desirable effect in adhesion increase for a chosen carrier material. In the case of porous carrier textiles treated in direct contact with the thin plasma layer generated on the ceramic plate of the DCSBD electrode, the surface free energy increases, which also indicates the increase of the adhesion forces. Plasma treatment applying the 400 W power increases the adhesion force of the SB NWPP material significantly (see **Figure 3**). The application of the planar electrode configuration resulted in the increase of the adhesion force by 450 %. With a low treatment time (up to 3 seconds) the measurements behave in the same manner. No significant tearing and therefore no major force deviations were observed, see **Figure 4**. Measuring the samples after plasma exposure for 5 seconds and longer, a different behavior was recorded. In most cases, the material still showed results close to the previous and did not have any significant influence on the measured force values. But, in some cases, a complete tearing of the used fabric was observed (**Figure 4c**). This difference could be explained by the inhomogeneity of the material that also corresponds to the differences in the strike-through time measurements. Some samples were plasma treated at 600 W of the supplied power. Samples were torn in the same way as after the plasma exposure at 400 W. With the increase of the exposure time more and more fabrics stayed glued to the adhesion tape. The change occurs at an exposure time of 10 s when tearing becomes more pronounced and the material damage could take place shortly after the beginning of the test.

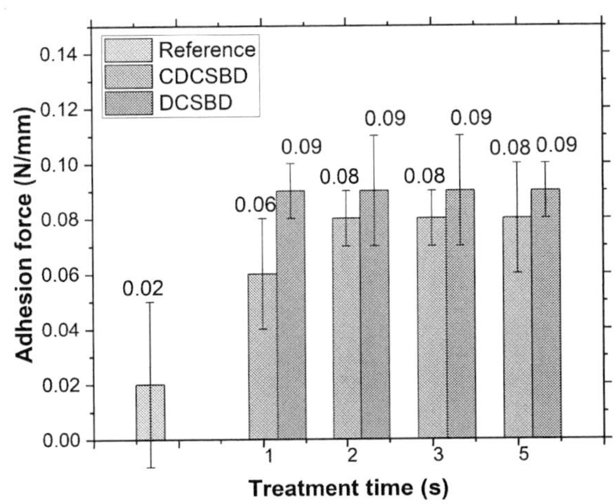

Figure 3 Adhesion force measurements on SB NWPP 17gsm

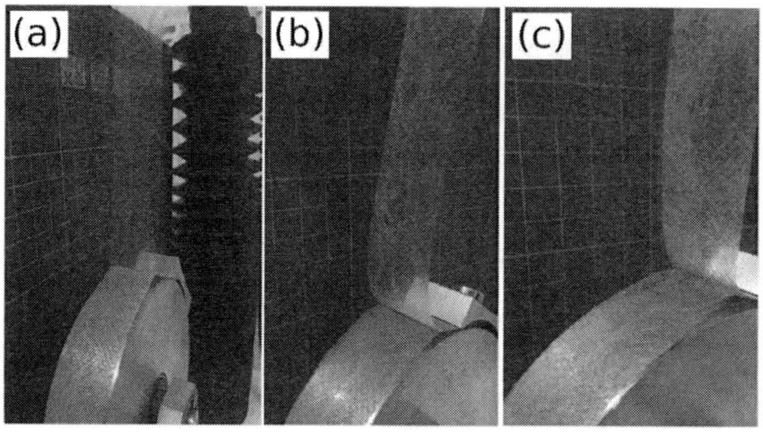

Figure 4 Photo of the a) reference sample SB NWPP after the peel test, samples that have undergone plasma treatment using DCSBD with an exposure time of b) 2 s, c) 5 s

3.3. Adhesion measurements - Loop test

When measuring reference samples, the applied nanofiber layer was always completely separated from the textile substrate. The values of adhesion force in most cases were so small that even a very sensitive strain gauge could not observe them. In some cases, a force of 0.01 N was observed. However, the plasma treatment often increased the adhesion to the substrate. The samples of the carrier substrate were plasma treated with the exposure time of 1, 2, 5, and 10 seconds at the supplied power of 400 and 600 Watts. The PVDF nanofibers were deposited on the carrier right after the plasma treatment. The results obtained after the loop test measurements are summarized in **Table 1** and **Table 2**.

Increasing the treatment time as well as supplied power results in the increased value of the work needed for the detachment of fibers from the carrier. Even one second plasma treatment of the carrier gives the opportunity for further material processing. Increasing the exposure time to 10 seconds results in more than 550% increase of the adhesion force in comparison to the 1 second plasma pre-treatment.

Table 1 The results of the loop test after 400 W plasma treatment

Treatment time (s)	Max (N)	Mean (N)	Area (N*mm)
1	0.07 ± 0.01	0.022 ± 0.006	0.37 ± 0.06
2	0.08 ± 0.02	0.05 ± 0.01	0.45 ± 0.016
5	0.13 ± 0.02	0.05 ± 0.02	0.72 ± 0.21
10	0.25 ± 0.04	0.09 ± 0.01	2.27 ± 0.58

Table 2 The results of the loop test after 600 W plasma treatment

Treatment time (s)	Max (N)	Mean (N)	Area (N*mm)
1	0.09 ± 0.02	0.02 ± 0.01	0.46 ± 0.08
2	0.12 ± 0.04	0.05 ± 0.01	0.78 ± 0.18
5	0.14 ± 0.05	0.05 ± 0.02	0.89 ± 0.46
10	0.28 ± 0.04	0.11 ± 0.02	2.56 ± 0.61

From the results above it is obvious that the adhesion forces of the carrier substrate, as well as the adhesion force between the carrier and the nanofibers, increased significantly after the plasma treatment.

4. CONCLUSION

It was shown that the DCSBD plasma treatment has an outstanding effect on the adhesion of the SB NWPP textile. The strike-through time test showed that even after seconds of plasma treatment the permeability of urine imitation (according to ISO 9073-8) increased significantly. The material became hydrophilic what is a good indicator for the adhesion enhancement. The results of the peel force test indicate that the plasma treatment promotes the adhesion force increase at least 3 times. The dependence of this increase on the exposure time is not reflected in the results of the measured strain gauges, but with the help of photographs, we observe an increased tearing of the fabric with increasing time. A special method for measuring nanofibers' adhesion to the substrate was invented. With the help of this method the samples used for this study were tested. The results show a weak holding of nanofibers on the reference carrier substrate. The manipulation with the untreated carrier of nanofibers is a problem now and requires further processing. Samples that were plasma treated show dependence in increasing adhesion with increasing the exposure time. It was shown that the work needed to detach the nanofibers from the substrate increased significantly.

By testing two configurations of the DCSBD electrode, two applied powers, and several exposure times, it was concluded that the best parameters for treating the SB NWPP substrate without damaging it are planar electrode, 400 W of supplied power, and ten seconds of exposure time. Applying these parameters, homogeneous plasma treatment was achieved with the reproducible results of increased adhesion force.

ACKNOWLEDGEMENTS

This work was supported by project In-line plasma technology for a large throughput manufacturing of advanced nanofiber-based filter fabrics (No. E01-14) funded by the European Light Industries Innovation and Technology initiative, which is funded by the European Union COSME Program for Competitiveness of Companies and Small and Medium Enterprises.

REFERENCES

[1] ŠVACHOVÁ, V., VOJTOVÁ, L., PAVLINÁK, D., VOJTEK, L., SEDLÁKOVÁ, V., HYRŠL, P., ALBERTI, M., JAROŠ, J., HAMPL, A., JANČAŘ, J. Novel electrospun gelatin/oxycellulose nanofibers as a suitable platform for lung disease modeling. *Mater. Sci. Eng. C.* [online]. 2016, vol. 67, pp. 493-501. Available from: https://doi.org/10.1016/j.msec.2016.05.059.

[2] FRIEDRICH, J. *The Plasma Chemistry of Polymer Surfaces.* Wiley-VCH Verlag GmbH & Co. KGaA: Weinheim, Germany, 2012. ISBN 9783527648009.

[3] KEDRONOVÁ, E., ZAJÍČKOVÁ, L., HEGEMENN, D., KLÍMA, M., MICHLÍČEK, M., MANAKHOV, A. Plasma Enhanced CVD of Organosilicon Thin Films on Electrospun Polymer Nanofibers. *Plasma Process. Polym.* [online]. 2015, vol. 12, pp. 1231-1243. Available from: https://doi.org/10.1002/ppap.201400235.

MAGNETRON SPUTTERING DEPOSITION OF HIGH ENTROPY NITRIDES FROM CHROMIUM-HAFNIUM-MOLYBDENUM-TANTALUM-WOLFRAM SYSTEM

Tomasz STASIAK, Pavel SOUČEK, Vilma BURŠÍKOVÁ, Petr VAŠINA

Department of Physical Electronics, Faculty of Science, Masaryk University, Brno, Czech Republic, EU,
tomasz.stasiak@mail.muni.cz

https://doi.org/10.37904/nanocon.2021.4312

Abstract

High entropy alloys (HEAs) are multicomponent materials with at least five elements close to the equiatomic ratio. The multielement composition leads to many effects that stabilize solid solution phases instead of intermetallic compounds. It results in promising mechanical properties, high thermal stability, good corrosion resistance, etc. After the discovery of HEAs in 2004, the concept only concerned metallic materials. Later, the idea was extended to high entropy ceramics (carbides, nitrides, oxides). In this study, the high entropy nitrides (HENs) from the Cr-Hf-Mo-Ta-W system were deposited by magnetron sputtering from elemental segmented targets. Two series of samples were deposited under different nitrogen flow from 0 sccm to 20 sccm at ambient temperature and 700 °C. The coatings deposited at low nitrogen flow consist predominantly of metallic and amorphous phases. At higher nitrogen flow (above 10 sccm), the coatings mainly consist of multielement face-centered cubic nitride with nanometric grain size, which lattice parameter increases with increasing nitrogen content. The chemical composition of films varies as a function of nitrogen flow and temperature. With increasing nitrogen flow and temperature, the deposition rate decreases from 67 nm min^{-1} at ambient temperature with no nitrogen to 38 nm min^{-1} at 700 °C under 20 sccm nitrogen flow. The coatings revealed high hardness up to 26.6 GPa and Young's modulus up to 531 GPa.

Keywords: High entropy alloys, high entropy ceramics, nitrides, magnetron sputtering deposition, thin films

1. INTRODUCTION

High entropy alloys (HEAs) are multicomponent alloys that contain at least five elements in a near equiatomic ratio [1-3]. The concept was introduced independently by two research groups led by Cantor [1] and Yeh [2] in 2004. Since then, many articles describing different systems from almost infinite possible compositions of HEAs have been published. HEAs reveal many promising properties, such as high mechanical resistance at high and low temperatures, creep, oxidation, corrosion, and hydrogen embrittlement resistance, that can compete with many traditional alloys [4]. The reason for these properties could be explained by four core effects. The effects, such as high entropy, lattice distortion, sluggish diffusion, and cocktail effect, present due to the multielement composition of HEAs, can partially explain the exceptional properties of this group of materials [3]. Among HEAs, the alloys based on refractory elements were given significant attention in recent years. It is mainly thanks to exceptional mechanical properties in bulk materials [5] and thin films [6].

The concept of metallic HEAs was quickly expanded to high entropy ceramics (HECs), i.e., carbides, nitrides, oxides. The high entropy nitrides (HENs) deposited as thin films seem to be studied the most frequently among HECs. The first publication dealing with a reactive magnetron sputtering deposition of nitrides from the Al-Cr-Ta-Ti-Zr system already showed high hardness and elastic modulus, 32 and 368 GPa, respectively [7]. Later studies confirmed the very high mechanical resistance of other HENs, often reaching hardness values above 25 GPa [8-10]. Most HENs consist of a single phase of multielement NaCl-type face-centered cubic structure [11]. The formation of a single solid solution phase confirms the presence of a high entropy effect in this group of materials. Moreover, HENs reveal other promising properties such as good corrosion resistance [8,10], good oxidation resistance [12], promising tribological properties (low friction coefficient and high wear resistance) [9]. The properties mentioned above make HENs very good candidates for high resistant protective coatings in many industrial applications. In this paper, the investigation focus on the deposition of the HEAs and HENs coatings from the Cr-Hf-Mo-Ta-W system by reactive magnetron sputtering. The microstructure and mechanical properties were investigated as a function of nitrogen flow and deposition temperature.

2. EXPERIMENTAL METHODS

The coatings were deposited on silicon substrate by an HVM Flexilab magnetron sputtering device (HVM Plasma spol s.r.o., Czech Republic) from elemental segmented targets. Each circular target of a diameter of 50.8 mm and thickness of 3 mm was cut into four equal pieces. Based on the sputtering yield [13] and preliminary experimental data, the segments of the targets were divided into three cathodes to obtain a near-equiatomic composition of coatings. Cathodes contained four parts of Cr (Cathode 1), two pieces of Mo and two pieces of Hf (Cathode 2), two segments of Ta and two segments of W (Cathode 3). The power delivered to the cathodes was set as 80, 150, and 150 W for cathode 1, 2, and 3, respectively. Si (100) was used as the substrate for the films. The argon flow was fixed at 80 sccm (standard cubic centimeter per minute), resulting in a pressure of 1.20 Pa. Two series of depositions were performed under various nitrogen flows between 0-20 sccm at room temperature (RT) and high temperature (HT, 700 °C). The base pressure before a deposition was about $2 \cdot 10^{-4}$ Pa. The deposition time was 30 minutes. Rotation of the substrate holder was set to 5 rpm to ensure homogenous distribution of elements. The structure of the deposited coatings was investigated by X-Ray Diffraction (XRD) using a Rigaku SmartLab diffractometer (Rigaku, Japan) with a copper Kα radiation source (λ = 1.54056 Å). The microstructure, chemical composition, and deposition rate were studied using a TESCAN MIRA 3 scanning electron microscope (SEM) (TESCAN, Czech Republic) coupled with energy-dispersive X-ray spectroscopy (EDX) detector X-MAX50 (Oxford Instruments, UK). The hardness and effective Young's modulus were measured using a Hysitron 950 (Bruker, Germany/USA) using the XPM mode under a load of 20 mN. At least 100 indentations were acquired for each sample to obtain reliable results.

3. RESULTS AND DISSCUSSION

3.1. Chemical composition

The SEM-EDX measurements, shown in **Table 1**, revealed significant changes of chemical composition.

Table 1 Chemical composition (at. %) of coatings deposited at RT and HT under different flows of nitrogen

Nitrogen fl./ depostion temp.	Cr (at. %)		Hf (at. %)		Mo (at. %)		Ta (at. %)		W (at. %)		N (at. %)	
	RT	HT	RT	HT	RT	HT	RT	HT	RT	HT	RT	HT
0 sccm	22.7	23.1	16.7	14.9	22.8	23.3	19.6	19.4	18.2	19.3	0.0	0.0
5 sccm	14.8	14.8	10.4	9.9	16.5	19.7	14.1	13.2	16.2	16.5	28.0	25.9
10 sccm	13.4	13.9	6.0	6.4	14.7	17.9	10.4	11.4	15.0	16.7	40.5	33.7
15 sccm	12.7	13.3	4.2	5.0	14.3	17.6	9.1	9.7	15.2	15.7	44.5	38.7
20 sccm	12.0	12.8	3.6	4.5	14.5	18.3	8.7	9.5	15.3	17.3	45.9	37.6

As expected, with increasing N_2 flow, the nitrogen content in the coatings increases. It is the highest for the sample deposited at RT under 20 sccm N_2 flow. It should be noted that the rise of nitrogen content with increasing its flow is not linear. The increase is fast for a small N_2 flow; however, the difference between 15 and 20 sccm is minimal due to saturation. The content of the metallic elements decreases with increasing N_2 flow, especially the drop of Hf and Ta is higher than that of other elements. It can be explained by more significant poisoning of these metals, i.e., the formation of the N-containing layer on the target surface, which decreases sputtering yield. The temperature during deposition also affects the chemical composition. The effect of temperature on the metallic elements is relatively small; however, the impact on the nitrogen content is significant.

3.2. Crystallographic structure

The XRD patterns, shown in **Figure 1,** showed important differences between samples deposited under different nitrogen flows. The coatings prepared only under Ar flow at RT and HT reveal a crystallographic structure composed of three body-centered cubic (bcc) phases with the lattice parameters close to pure Ta, Mo, and W. The temperature during deposition has an impact on the structure of the metallic coatings. The texture shift is observed from a (110) and (211) at RT to a (200) at HT. The N_2 flow of 5 sccm leads to the formation of an amorphous structure at RT, while the samples prepared at HT present mainly bcc phase (a = 3.092 Å) and probably some fraction of the amorphous phase too. The amorphization of the structure under low nitrogen flow could be related to the excess interstitial nitrogen in the bcc phases and the simultaneous lack of sufficient nitrogen content to form multielement nitride. The higher N_2 flow results in multielement face-centered cubic (fcc) NaCl-type nitride formation. The formation of nitride solid-solution could be attributed to the high entropy effect, which by increasing solubility, limits the formation of intermetallic phases between nitrogen and metallic elements [9]. Similar nitride phases were found in other HEN coatings [8,9,14]. Besides the main phase, some small peaks of fcc phases with slightly different lattice parameters are also present. The lattice parameter and crystallite size of multielement fcc nitride change as a function of N_2 flow and temperature, which is shown in **Table 2**. The deposition at the HT leads to a smaller lattice parameter and bigger crystallite size for a given N_2 flow than the deposition at RT. The HT enhances diffusivity and mobility, and therefore it results in bigger crystallite size. The increase of nitrogen flow results in a bigger lattice parameter and a smaller crystallite size.

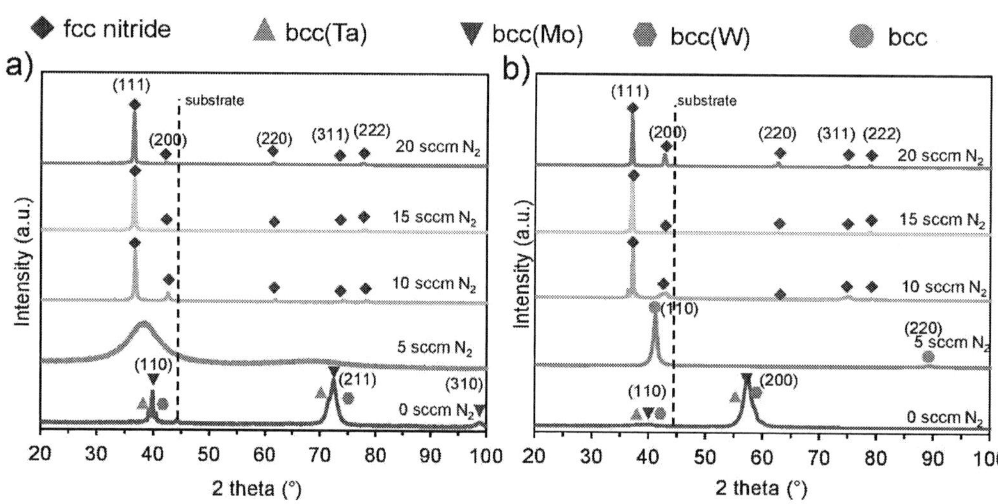

Figure 1 X-ray diffraction patterns of coatings deposited at RT (a) and HT (b) under flow of nitrogen

Table 2 Lattice parameter and crystallite size of fcc multielement nitrides

Nitrogen Flow (sccm)	Lattice parameter (Å)		Crystallite size (nm)	
	RT	HT	RT	HT
10	4.236	4.193	17.2	33.0
15	4.252	4.202	16.3	32.4
20	4.268	4.220	15.4	23.7

3.3. Microstructure and deposition rate

The SEM observations of the cross-section of coatings, shown in **Figure 2,** reveal significant differences in microstructure between conditions. The sample prepared at RT without nitrogen flow presents a columnar microstructure, while the coating deposited at HT shows a more dense microstructure without easily distinguishable columns. Both samples (RT and HT) prepared under 5 sccm nitrogen flow present a very dense cross-section. The coatings deposited under higher nitrogen flow (10-20 sccm) show a typical columnar structure. The columns are slightly wider for the HT series than for the RT series, and their size decreases with increasing nitrogen flow. It seems that the coatings deposited at high nitrogen flow show more porosity.

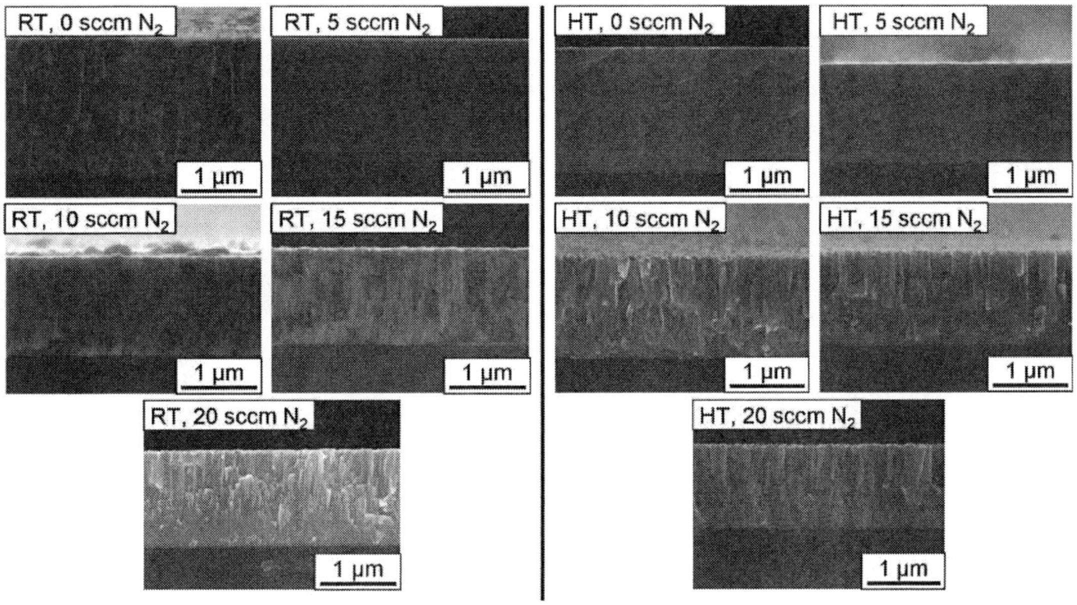

Figure 2 SEM images of the cross section of coatings deposited at RT and HT under flow of nitrogen

The deposition rate, shown in **Figure 3**, is relatively high for all prepared coatings. Nevertheless, it is decreasing with increasing nitrogen flow. It can be a result of more severe partial poisoning of the targets [14]. The temperature seems to play a less significant role, mainly by slightly reducing the deposition rate of samples prepared without nitrogen. However, the HT could lead to densification of coatings, and therefore slightly decreasing the thickness and consequently the deposition rate.

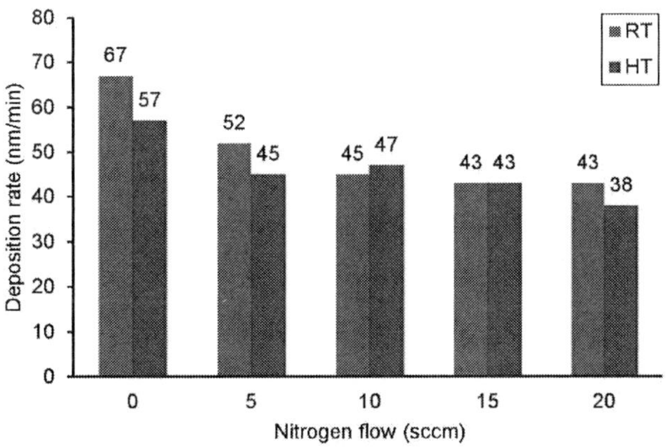

Figure 3 Deposition rate of coatings prepared under different flows of nitrogen at RT and HT

3.4. Mechanical properties

The mechanical properties evaluated based on the microindentation test, shown in **Figure 4**, are very promising. The pure metallic samples reveal a good hardness level, with the sample deposited at HT showing higher hardness due to lower porosity. The highest hardness (26.6 GPa) among tested samples was measured for the coating deposited at HT under 5 sccm N_2 flow. This sample consists mainly of the bcc phase and some fraction of the amorphous phase. The cross-section images revealed a very high density. The completely amorphous sample deposited at 5 sccm N_2 flow present also high hardness. The coatings with the fcc multielement nitride deposited at higher nitrogen flow show lower hardness. The HT deposition series between 10 and 20 sccm reveal hardness of about 14.0 GPa. It is a quite low value compared to metallic and amorphous coatings due to significantly higher porosity. The RT deposition series show decreasing hardness with increasing N_2 flow. It could be related to a slight increase of porosity. Nevertheless, the deposition at RT leads to a slightly higher hardness than at HT. It could be associated with a smaller crystallite size.

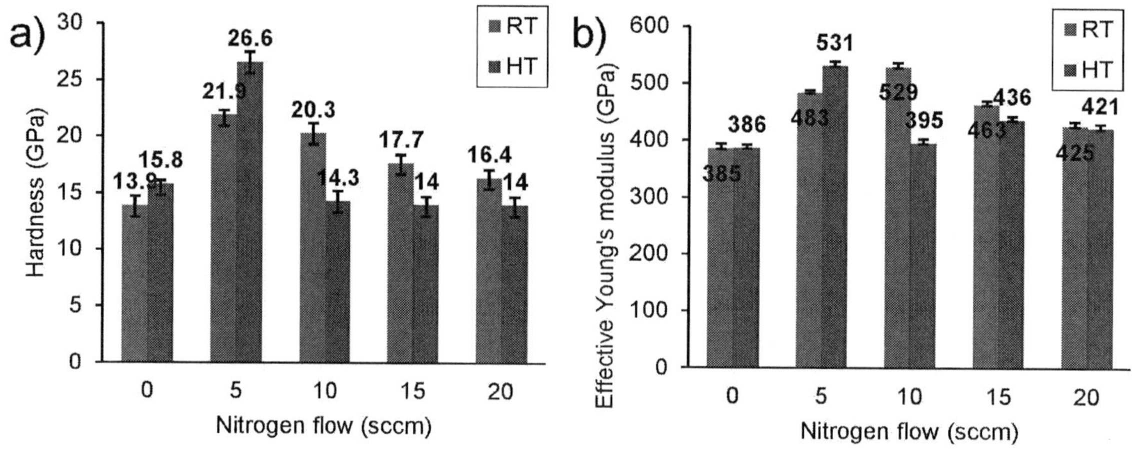

Figure 4 Hardness (a) and effective Young's modulus (b) of coatings deposited at RT and HT

The effective Young's modulus (E) of metallic coatings is about 385 GPa for samples deposited at RT and HT. Similar to the hardness, the effective Young's modulus is the highest for the samples deposited at 5 sccm, especially for that prepared at HT (531 GPa). For the coatings which consist of multielement nitride, the E decreases with increasing nitrogen flow for the RT series. For the HT series, the effective Young's modulus is somewhat similar between samples prepared between 10 and 20 sccm nitrogen flow. The decrease of the effective Young's modulus could be associated with the formation of defects such as pores.

4. CONCLUSION

Coatings of HEAs and HENs from the Cr-Hf-Mo-Ta-W family were successfully deposited by reactive magnetron sputtering from segmented elemental metallic targets. The paper describes the effect of the deposition conditions (temperature, nitrogen flow) on the microstructure and mechanical properties. The results show a decrease of metallic element content and an increase of nitrogen with the increasing nitrogen flow. Coatings deposited at low N_2 flows consist either of metallic or amorphous phases. The higher nitrogen flow leads to the formation of a major multielement fcc nitride phase with the lattice parameter slightly increasing with increasing nitrogen content and smaller for high-temperature series. The multielement nitride consists of nanometric grains between 15.4 and 33.0 nm, which decrease with increasing nitrogen flow. The deposition rate drops with increasing nitrogen flow, and it is slightly lower for the HT series. Cross-section observations reveal columnar structure in samples consisting of the fcc nitride and dense microstructure deposited at low nitrogen flow. The deposited coating present very high hardness and effective Young's modulus. Especially high hardness up to 26.6 GPa was found in dense coating deposited at low nitrogen flow. The promising results presented in this paper, which are the first experimental study of the HENs from the Cr-Hf-Mo-Ta-W system, encourage further, more detailed investigations.

ACKNOWLEDGEMENTS

This research was supported by project LM2018097 funded by the Ministry of Education, Youth and Sports of the Czech Republic.

REFERENCES

[1] CANTOR, B., CHANG, I.T.H., KNIGHT, P., VINVENT, A.J.B. Microstructural development in equiatomic multicomponent alloys. *Mater. Sci. Eng. A.* 2004., vol. 375-377, pp. 213-218.

[2] YEH, J.-W., CHEN, S.-K., LIN, S.-J., GAN, J.-Y., CHIN, T.-S., SHUN, T.-T., TSAU, C.-H., CHANG, S.-Y. Nanostructured High-Entropy Alloys with Multiple Principal Elements: Novel Alloy Design Concepts and Outcomes. *Adv. Eng. Mater.* 2004, vol. 6, pp. 299-303.

[3] YEH, J.-W. Recent Progress in High Entropy Alloys. *Ann. Chim. - Sci. Des Matériaux*. 2006, vol. 31, pp. 633-648.

[4] GEORGE, E.P., RAABE, D., RITCHIE, R.O. High-entropy alloys. *Nat. Rev. Mater.* 2019, vol. 4, pp. 515-534.

[5] SENKOV, O.N., WILKS, G.B., SCOTT, J.M., MIRACLE, D.M. Mechanical properties of Nb25Mo25Ta25W25 and V20Nb20Mo20Ta20W20 refractory high entropy alloys. *Intermetallics*. 2011, vol. 19, pp. 698-706.

[6] KIM, H., NAM, S., ROH, A., SON, M., HAM, M.H., KIM, J.H., CHOI, H. Mechanical and electrical properties of NbMoTaW refractory high-entropy alloy thin films. *Int. J. Refract. Met. Hard Mater.* 2019, vol. 80, pp. 286-291.

[7] LAI, C.H., LIN, S.J., YEH, J.W., CHANG, S.Y. Preparation and characterization of AlCrTaTiZr multi-element nitride coatings. *Surf. Coatings Technol.* 2006, vol. 201, pp. 3275-3280.

[8] VON FIEANDT, K., PASCHALIDOU, E.M., SRINATH, A., SOUCEK, P., RIEKEHR, L., NYHOLM, L., LEWIN, E. Multi-component (Al,Cr,Nb,Y,Zr)N thin films by reactive magnetron sputter deposition for increased hardness and corrosion resistance. *Thin Solid Films*. 2020, vol. 693, 137685.

[9] CUI, P., LI, W., LIU, P., ZHANG K., MA, F., CHEN, X., FENG, R., LIAW, P.K. Effects of nitrogen content on microstructures and mechanical properties of (AlCrTiZrHf)N high-entropy alloy nitride films. *J. Alloys Compd.* 2020, vol. 834, 155063.

[10] SHU, R. PASCHALIDOU, E.M., RAO, S.G., LU, J., GRECZYNSKI, G., LEWIN, E., NYHOLM, L., LE FEBVRIER, A., EKLUND, P. Microstructure and mechanical, electrical, and electrochemical properties of sputter-deposited multicomponent (TiNbZrTa)Nx coatings. *Surf. Coatings Technol.* 2020, vol. 389, 125651.

[11] OSES, C., TOHER, C., CURTAROLO, S., High-entropy ceramics. *Nat. Rev. Mater.* 2020, vol. 5, pp. 295-309.

[12] KHAN, N.A., AKHAVAN, B., ZHOU, C., ZHOU, H., CHANG, L., WANG, Y., LIU, Y., BILEK, M.M., LIU, Z. High entropy nitride (HEN) thin films of AlCoCrCu0.5FeNi deposited by reactive magnetron sputtering. *Surf. Coatings Technol.* 2020, vol. 402, 126327.

[13] YAMAMURA Y., TAWARA H. Energy dependence of ion-induced sputtering yields from monatomic solids at normal incidence, *Atomic Data and Nuclear Data Tables* 1996, vol. 62, pp. 149-253.

[14] EL GARAH, M., ACHACHE, S., MICHAU, A., SCHUSTER F., SANCHETTE, F., Structural and mechanical properties of AlTiTaZr(-N) medium entropy films (MEF) obtained by DC magnetron sputtering in dynamic mode, *Surf. Coatings Technol.* 2020, vol. 396, 125941.

SILICA SUPPORTED WO_3/CU_2O HETEROSTRUCTURED NANOPARTICLES FOR PHOTOCATALYTIC DEGRADATION OF HORMONES

Hassan ALI, Milan MASAR, Muhammad YASIR, Tomas SOPIK, Pavel URBANEK, Michal MACHOVSKY, Ivo KURITKA

Centre of Polymer Systems, Tomas Bata University in Zlin, Zlin, Czech Republic, EU, hali@utb.cz

https://doi.org/10.37904/nanocon.2021.4343

Abstract

In this study, a WO_3/Cu_2O based heterojunction was synthesized via a facile solvothermal method. The as-prepared nanocomposite was characterized by XRD, SEM, and UV-vis spectroscopy. The photocatalytic activity of the as-prepared heterojunction was evaluated by the means of estradiol (E2) and 17α-ethinylestradiol (EE2) hormone degradation under UV LED illumination wavelength at λ_{max}~ 365 nm. The degradation results showed that the prepared photocatalyst was able to achieve considerable photoactivity owing to the intrinsic heterojunction charge transfer mechanism. Photocatalytic degradation rates of 27 and 35 % were achieved for the E2 and EE2 hormones, respectively.

Keywords: WO_3/Cu_2O, heterojunction, degradation, photocatalysis, estrogens

1. INTRODUCTION

Over the last few decades, extensive use of chemicals and its production on industrial scale have resulted in serious environmental and ecological issues, mainly due to the contamination of aqueous resources vital for sustaining the life cycle [1]. Among the commonly discharged chemical affluents, pharmaceuticals like antibiotics and hormonally active agents, also classified as endocrine-disrupting chemicals (EDC) are of particular importance due to relatively high bio toxicity, even at minuscule concentrations [2]. The elimination of these chemical toxicants including EDC is imperative since the exposure of such chemicals are associated with several chronic diseases such as cancer, ADHD, learning disability, brain development and retarded sexual development. Therefore, it is imperative to develop sustainable water treatment systems capable of eliminating these toxic chemicals with low cost and energy requirements. So far, various strategies have been implemented to address this issue, including adsorption, biological degradation, electrochemical oxidation (EO), membrane technology, sonolysis, and photocatalysis, each having specific advantages and disadvantages [3,4]. For example, frequent replacement of electrodes and membranes due to fouling increases the operational costs of EO and membrane filtration techniques, respectively [5]. Photocatalysis, in this regard, is currently considered the most sustainable method and has been employed in various applications such as pollutants removal, fuel generation, pathogen elimination, and nitrogen fixation [6-8]. A typical photocatalysis process requires only a suitable semiconductor photocatalyst, reaction medium and solar light irradiation as an energy input. The process is initiated by the absorption of photons having energy equal to or greater than the bandgap of the photocatalyst. This results in the excitation of the electrons in the valence band (VB) to the conduction band (CB) and subsequent generation of electron-hole pairs. These excited electrons and holes then participate in the reduction and oxidation of the water molecules and the adsorbed chemical species on the surface of the photocatalyst. However, currently developed photocatalysts suffer from several critical issues which limit their photoactivity, i.e., the wide bandgap of commonly used photocatalysts results in poor visible light harvesting and futile recombination of the generated photoexcited charge carriers [8]. To address these fundamental drawbacks, several modification strategies have been implemented so far to-date such as surface engineering, sensitization, defect engineering, doping, and heterojunction [9,10]. Out of these, the construction of heterojunction has compelling advantages of addressing the previously mentioned issues associated with the single component photocatalysts, namely, the suppression of charge recombination and extension of light absorption to visible range. Numerous heterojunctions systems have been investigated so far with promising photocatalytic activities [11]. Among several dual photocatalytic systems, construction of WO_3 and Cu_2O based heterojunction presents an attractive approach primarily due to favorable to band position.

2. EXPERIMENTAL

2.1. Materials

Hydrogen peroxide (H_2O_2) (Lachner), copper(II) sulfate ($CuSO_4$) (Lachner), glucose ($C_6H_{12}O_6$) (Lachner), and sodium hydroxide (NaOH) (Lachner) were used for the preparations of the sample. The tungsten powder was obtained from ISS Nippon Kayaku group. Estradiol (E2) and 17α-ethinylestradiol (EE2) were purchased from Sigma-Aldrich, Germany. All chemicals were of analytic grade with p.a. ≥ 99.5 % and were used as received without further purification. All aqueous solutions were prepared by demineralized water.

2.2. Synthesis of WO_3/Cu_2O nanocrystals

In the first step, WO_3 was prepared from starting solution of peroxotungstic acid ($[WO_2(O_2)H_2O]\cdot nH_2O$, PTA) which was obtained by dissolving 1 g of tungsten powder in 20 mL of H_2O_2. Tungsten powder dissolved quickly under intense stirring and at elevated temperature (55-60 °C). Next, 5 g of silica was added into the prepared starting solution, and by a Teflon stick thoroughly mixed to the resulting paste. This paste was dried at 45 °C for 24 hours. The as-prepared powder was crushed in a mortar with a pestle. Subsequently, this powder was annealed for 2 hours at 550 °C in a muffle furnace. The obtained WO_3 powder prepared by this synthesis method was used in the next step for Cu_2O decoration as follows: 300 mg of previously prepared WO_3 powder and 8 mL of ammonia solution with 75 mg of $CuSO_4$ were mixed with a magnetic stirrer. After a short time, 150 mg of glucose was added with continuous stirring, followed by adjusting pH of the mixture with NaOH.

2.3. Characterization methods

Phase structure of all samples was investigated by XRD diffractometer MiniFlex600 (Japan, RIGAKU) with a Co-Kα X-ray source (λ = 1.7903 Å) in the diffraction angle range of 5-90° 2θ. Crystallite sizes were estimated according to the Scherrer's equation as follows:

$$d = K\lambda/\beta cos\theta \quad (1)$$

where *d* is the diameter and the shape factor *K* is 0.89, since no preferential orientation was observed, *λ* is the wavelength (CoKα 1,2) = 0.179 nm, the angle θ is the full width half maximum FWHM of the corresponding highest diffraction peak, i.e., (002) and (111) line for WO_3 and Cu_2O, respectively, and β is the line broadening at half the maximum intensity corrected for the instrumental response. Sample analysis was performed by NovaNanoSEM 450 microscope (The Netherland, FEI company). Microscopic images were taken using an ETD (topographic contrast) and CBS (material contrast) detector accelerated at 5 kV and 15 kV voltages.

2.4. Photocatalytic experiment

The photocatalytic adsorption and degradation experiment was performed using WO_3/Cu_2O as a photocatalyst in an aqueous mixture of hormone solution containing E2 and EE2 hormones. Each photocatalytic test was conducted by transferring 5 mg of the powdered catalyst into a beaker containing 10 mL of hormone solution with each hormone at a concentration of 0.2 mg/L and a total solution concentration of 0.8 mg/L. Separate beakers were used to evaluate the hormones adsorption rate under constant magnetic stirring at 450 rpm. Additionally, a separate beaker containing solution only was kept as a reference to calculate the removal percentage from the initial concentration. Results were obtained from high-performance liquid chromatography (HPLC) using calibration after running each sample twice.

2.5. HPLC Method

HPLC analysis was performed on a Dionex UltiMate 3000 Series equipped with a diode array detector (Thermo Fisher Scientific, Germany) according to the previously reported technique [12]. The concentration of hormones was calculated from the results of the 200 nm performed test (concentration of calibration standards 0.20 - 0.02 mg/L; Data were recorded and processed in Chromeleon 7.2 software (Thermo Fisher Scientific) [13].

3. RESULTS AND DISCUSSION

The crystalline phase structures of the WO_3/Cu_2O heterojunction are shown in **Figure 1**. For reference, peroxotungstic acid (PTA) and silica-supported WO_3 diffractograms are also given. The XRD pattern of the PTA exhibits intense peaks in the range of small Braggs angles, typically associated with layered compounds. The as-prepared silica supported WO_3 exhibits several characteristic peaks, indexed to (002), (020), (200), (112), (202), (220), (222), (004), (040), (223), and (402) planes as reported in PDF Card No. 01-071-4310. From **Figure 1c**, it can be seen that the characteristic peaks of Cu_2O, corresponding to the (111), (200), (220), (311) planes appeared as reported in PDF card No. 01-071-0305, indicating successful formation of the heterojunction. The peaks are broader and of low intensity due to the presence of silica. Moreover, the decrease of signal intensity for WO_3 is observed which can be attributed due to possible shielding of newly formed cupric crystalline phase. No other crystalline phases such as $CuWO_4$ were detected in the samples. The average crystallite sizes of the WO_3 and Cu_2O were calculated to be 57.64 and 50.24 nm, respectively, according to Scherrer's formula.

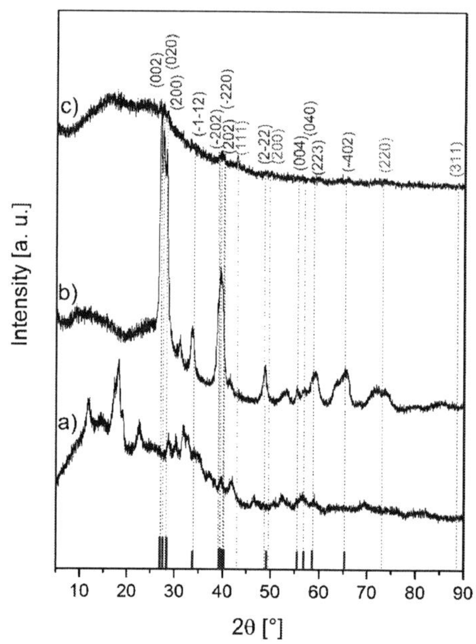

Figure 1 XRD patterns for silica treated by peroxotungstic acid precursor (a), silica-supported WO_3 obtained by calcining precursor at 550 °C (b) and silica-supported WO_3/Cu_2O heterostructure.

The morphology of the as-prepared samples was investigated using SEM as shown in **Figure 2**. The SEM images of silica-supported PTA precursor exhibits course morphology with small grain size, which is consistent with poor crystallinity as evidenced by the XRD results (**Figures 2A-B**). For the annealed samples, WO_3 and

WO_3/Cu_2O, an increase in the temperature led to smaller grain size and a large number of Cu_2O particles were observed on the surface of the WO_3 embedded with silica. No agglomeration of Cu_2O particles was noticed.

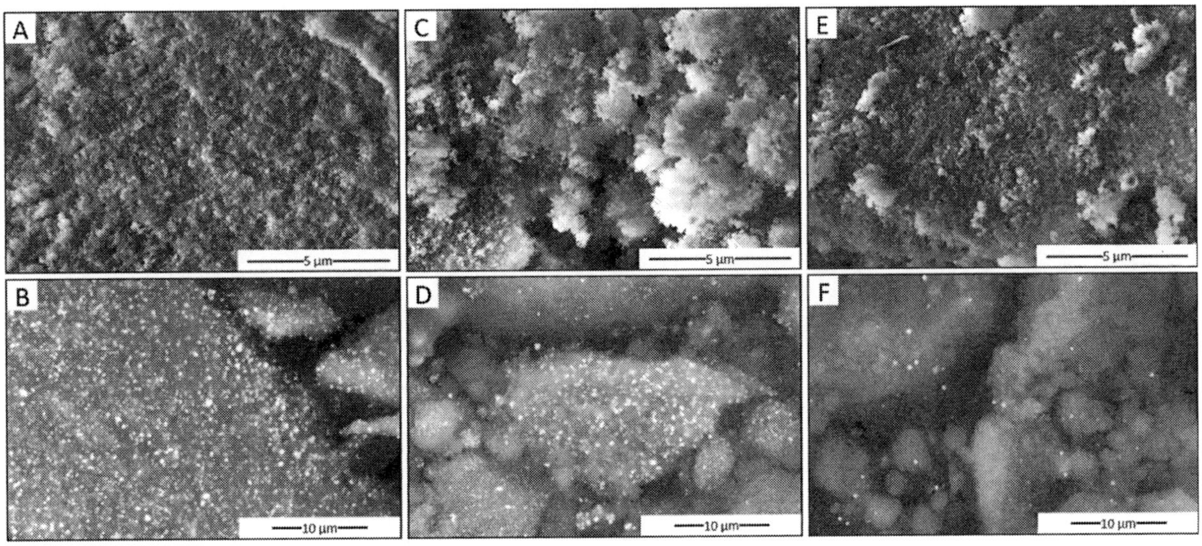

Figure 2 SEM images of peroxotungstic acid precursor (A-B), silica-supported WO_3 (C-D), and silica-supported WO_3/Cu_2O heterojunction (E-F). Top and bottom images were taken by ETD and CBS detector, respectively.

The electronic structure of semiconductor which plays a crucial role in the photoactivity is closely related to its bandgap and its alignment. The as-prepared WO_3/Cu_2O sample was also characterized by UV-vis analysis to evaluate the optical bandgap value via Kubelka-Munk equation, as previously reported [14].

$$[\alpha hv]^p = A(hv - Eg) \tag{2}$$

where α is the optical absorption coefficient, hv stands for quantized photon energy, A is the constant of proportionality, and the value p indicates the electronic transition type. The value of the exponent, p, was determined by plotting a $[F(R)\cdot hv]^{1/2}$ vs hv graph and calculating the best fit, which turned out to be ½, implying an indirect allowed transition. Finally, by plotting a $[F(R)\cdot hv]^{1/2}$ vs hv graph and extrapolation of the graph slope to $F(R) \rightarrow 0$, the optical bandgap values were obtained as shown in **Figure 3**.

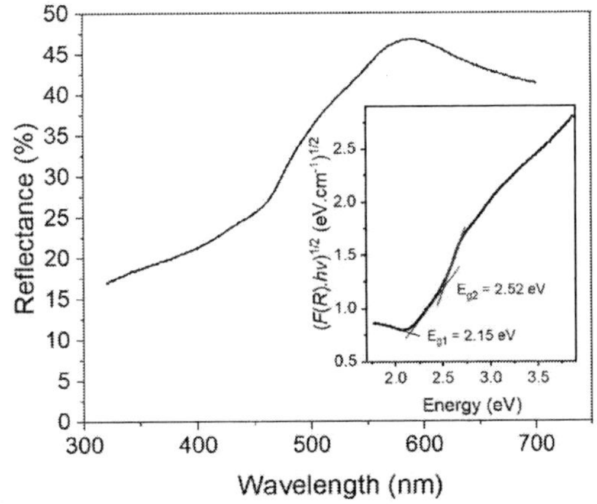

Figure 3 Reflectance spectrum and the calculated bandgap energies (inset) of the silica-supported WO_3/Cu_2O composite

The photocatalytic activity of the as-prepared nanocomposite was evaluated via E2 and EE2 hormones degradation having an initial concentration of 0.2 mg/L under UV light irradiation (Roithner LaserTechnik, UVLUX 340-HL-3) with a maxima wavelength of 343 nm (corresponding to $E_g \sim 3.6$ eV). From **Figure 4**, it is obvious that the photocatalytic activity gradually increased with increasing time, reaching 26 and 35 % after 1 h duration, for E2 and EE2, respectively. However, it should be noted that the reference sample without light irradiation also recorded some hormonal removal due to adsorption, with contribution reaching as high as 5.4 (E2) and 7.8 % (EE2), within 1 h duration. This high degree of hormonal adsorption could be ascribed due to the presence of a large number of silica particles, possessing highly porous structure and large surface area.

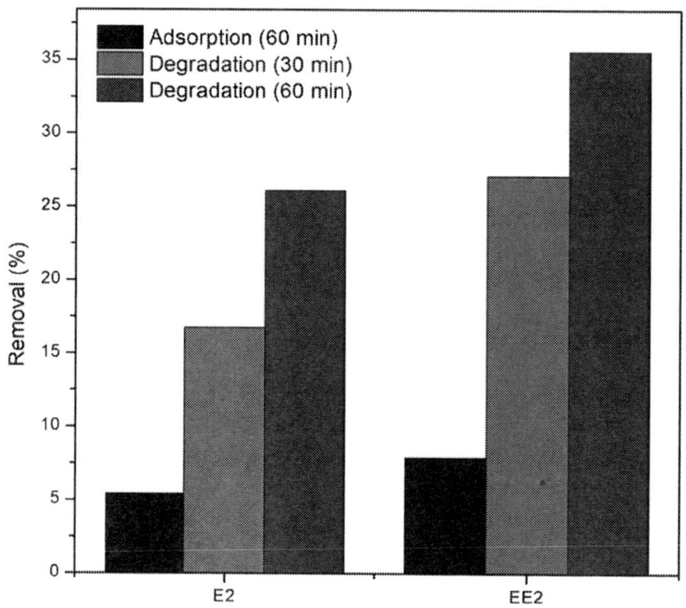

Figure 4 Photocatalytic and adsorption activity of WO_3/Cu_2O heterojunction

4. CONCLUSION

A silica supported WO_3/Cu_2O heterojunction was prepared via a facile solvothermal method employing peroxotungstic acid and copper sulfate as precursors. XRD and SEM analysis showed the formation of a suitable heterojunction was successful. The as-prepared WO_3/Cu_2O showed significant photocatalytic activity towards estrogenic hormonal degradation under UV light irradiation. The enhanced photoactivity was attributed due to the advantageous charge transfer route of heterojunction. This work may provide insights into developing new highly efficient materials for environmental pollutants degradation.

ACKNOWLEDGEMENTS

This research obtained funding from the Ministry of Education, Youth and Sports of the Czech Republic in the frame of the project LTT20010. M.Y. and T.S. acknowledge the support of the Technological Agency of the Czech Republic (grant no. FW01010588). M.Y. also appreciate the financial support from the Internal Grant Agency of Tomas Bata University in Zlin; contract grant number: IGA/CPS/2021/002.

REFERENCES

[1] ARYAL, N., WOOD, J., RIJAL, I., DENG, D., JHA, M. K., OFORI-BOADU, A. Fate of environmental pollutants: A review. *Water Environment Research*. 2020, vol. 92, no. 10, pp. 1587-94. Available from: https://doi.org/doi.org/10.1002/wer.1404.

[2] LA MERRILL, M. A., VANDENBERG, L. N., SMITH, M. T., GOODSON, W., BROWNE, P., PATISAUL, H. B., GUYTON, K. Z., KORTENKAMP, A., COGLIANO, V. J., WOODRUFF, T. J., RIESWIJK, L., SONE, H., KORACH, K. S., GORE, A. C., ZEISE, L., ZOELLER, R. T. Consensus on the key characteristics of endocrine-disrupting chemicals as a basis for hazard identification. *Nature Reviews Endocrinology*. 2020, vol. 16, no. 1, pp. 45-57. Available from: https://doi.org/10.1038/s41574-019-0273-8.

[3] KUMAR, S., AHLAWAT, W., BHANJANA, G., HEYDARIFARD, S., NAZHAD, M. M., DILBAGHI, N. Nanotechnology-based water treatment strategies. *Journal of Nanoscience and Nanotechnology*. 2014, vol. 14, no. 2, pp. 1838-1858. Available from: https://doi.org/10.1166/jnn.2014.9050.

[4] UDDIN, Z., AHMAD, F., ULLAN, T., NAWAB, Y., AHMAD, S., AZAM, F., RASHEED, A., ZAFAR, M. S. Recent trends in water purification using electrospun nanofibrous membranes. *International Journal of Environmental Science and Technology*. 2021. Available from: https://doi.org/10.1007/s13762-021-03603-9.

[5] JAFARI, M., VANOPPEN, M., VAN AGTMAAL, J. M. C., CORNELISSEN, E. R., VROUWENVELDER, J. S., VERLIEFDE, A., VAN LOOSDRECHT, M. C. M., PICIOREANU, C. Cost of fouling in full-scale reverse osmosis and nanofiltration installations in the Netherlands. *Desalination*. 2021, vol. 500, article no. 114865. Available from: https://doi.org/10.1016/j.desal.2020.114865.

[6] GAO, Y., DUAN, J., ZHAI, X., GUAN, F., WANG, X., ZHANG, J., HOU, B. Photocatalytic Degradation and Antibacterial Properties of Fe3+-Doped Alkalized Carbon Nitride. *Nanomaterials*. 2020, vol. 10, no. 9, pp. 1751. Available from: https://doi.org/10.3390/nano10091751.

[7] ALI, H., MASAR, M., GÜLER, A. C. C., URBÁNEK, M., MACHOVSKY, M., KUŘITKA, I. Heterojunction-based photocatalytic nitrogen fixation: Principles and current progress. *Nanoscale Advances*. 2021, vol. 3, no. 22, pp. 6358-6372. Available from: https://doi.org/10.1039/D1NA00565K.

[8] DJURIŠIĆ, A. B., HE, Y., NG, A. M. C. Visible-light photocatalysts: Prospects and challenges. *APL Materials*. 2020, vol. 8, no. 3, pp. 030903. Available from: https://doi.org/10.1063/1.5140497.

[9] XIONG, L., TANG, J. Strategies and Challenges on Selectivity of Photocatalytic Oxidation of Organic Substances. *Advanced Energy Materials*. 2021, vol. 11, no. 8, pp. 2003216. Available from: https://doi.org/10.1002/aenm.202003216.

[10] SAPUTERA, W. H., RIZKIANA, J., WULANDARI, W., SASONGKO, D. Role of defects on TiO2/SiO2composites for boosting photocatalytic water splitting. *RSC Advances*. 2020, vol. 10, no. 46, pp. 27713-27719. Available from: https://doi.org/10.1039/d0ra05745b.

[11] NADEEM, M. A., KHAN, M. A., ZIANI, A. A., IDRISS, H. An overview of the photocatalytic water splitting over suspended particles. *Catalysts*. 2021, vol. 11, no. 1, pp. 60. Available from: https://doi.org/10.3390/catal11010060.

[12] YASIR, M., ŠOPÍK, T., KIMMER, D., SEDLAŘÍK, V. Facile HPLC technique for simultaneous detection of estrogentic hormones in wastewater In: *NANOCON 2020 Conference Proceedings*. Brno: Nanocon, 2020, pp. 272-276. Available from: https://doi.org/10.37904/nanocon.2020.3710.

[13] YASIR, M., ŠOPÍK, T., LOVECKÁ, L., KIMMER, D., SEDLAŘÍK, V. The adsorption, kinetics, and interaction mechanisms of various types of estrogen on electrospun polymeric nanofiber membranes. *Nanotechnology*. 2022, vol. 33, no. 7, pp. 075702. Available from: https://doi.org/10.1088/1361-6528/ac357b.

[14] ALI, H., GULER, A. C., MASAR, M., URBANEK, P., URBANEK, M., SKODA, D., SULY, P., MACHOVSKY, M., GALUSEK, D., KURITKA, I. Solid-State Synthesis of Direct Z-Scheme Cu_2O/WO_3 Nanocomposites with Enhanced Visible-Light Photocatalytic Performance. *Catalysts*. 2021, vol. 11, no. 2, pp. 1-26. Available from: https://doi.org/10.3390/catal11020293.

PHOTOCATALYTIC HYDROGENATION OF CARBON-CARBON TRIPLE BOND

[1]Anna ZABELINA, [1]Denis ZABELIN, [1]Vaclav ŠVORČÍK, [1,2]Oleksiy LYUTAKOV

[1]*VŠCHT - University of Chemistry and Technology, Prague, Czech Republic, EU, lyutakoo@vscht.cz*

[2]*Tomsk Polytechnic University, Research School of Chemistry and Applied Biomedical Sciences, Tomsk, Russian Federation*

https://doi.org/10.37904/nanocon.2021.4334

Abstract

Hydrogenation reactions are one of the main reactions in the chemical industry. In this work, we have demonstrated a plasmon assisted hydrogenation, performed on bimetallic (Au@Pt) nanoparticles. Using phenylacetylene as an example, it was shown that plasmon accelerates the hydrogenation process and makes it possible at mold conditions with high conversion and reaction yield.

Keywords: Photocatalysis, hydrogenation, core-shell nanoparticles, selectivity, alkyne

1. INTRODUCTION

Selective hydrogenation of unsaturated organic compounds such as alkynes (C≡C) and aldehydes, ketones and esters (C = O) is a fundamental transformation in organic chemistry, commonly used in synthetic, petroleum and natural products chemistry [1,2]. Hydrogenation is usually carried out at elevated temperatures and with an external source of hydrogen at high pressure [3-5].

Selective hydrogenation, which excludes undesirable processes of hyperhydrogenation of organic compounds is especially technically important. Currently, there are two approaches to carrying out selective semihydrogenation - utilization of specific metal catalysts in combination with molecular hydrogen [6-8]. At the same time, hydrogen gas, which is usually used for hydrogenation at high pressures and temperatures, is a source of high danger, so it is more advisable to use another sources of hydrogen.

Therefore, from safety and selectivity points of view, the utilization of alternative hydride reagents such as lithium aluminum hydride ($LiAlH_4$), triethylsilane (Et_3SiH) and sodium borohydride ($NaBH_4$) is suitable [9-13]. However, these compounds do not show sufficient activity in the hydrogenation of some organic groups and molecules. Here we have demonstrated plasmon-assisted photocatalysis, performed at mild conditions under visible light illumination. Generally, similar approach, i.e. plasmon assisted photocatalysis has attracted a lot of attention because the excitation of localized surface plasmon resonance (LSPR) accelerates wide range of relevant chemical transformations [14-17].

In this work, a highly selective photocatalytic hydrogenation of the C≡C triple bond with bimetallic Au@Pt particles was carried out using sodium borohydride as a hydrogen source.

2. EXPERIMENTAL

2.1. Materials

Deionized water (EMD MILLIPORE), trisodium citrate, gold(III) chloride, L-Ascorbic acid (99%), chloroplatinic acid solution (8 wt. % in H_2O), Phenylacetylene, Sodium borohydride (99%), Methanol (≥99.9%) were ordered from Sigma-Aldrich and used without additional purification.

2.2. Sample preparation

Gold nanoparticles were obtained by the Turkevich method [18]. After the preparation of the seed, Au was introduced into a round-bottom flask with a solution of platinum hydrochloric acid (8 wt% B H_2O). The mixture was stirred for 20 minutes. Then L-ascorbic acid was added very slowly dropwise until the colour of the solution changed from red to black. Then the resulting bimetallic nanoparticles were washed three times with methanol.

The solution of Au@Pt in methanol and deionized water, was sonicated for 3 min after the addition of 0.13mM of phenylacetylene. The hydrogenation reaction of phenylacetylene was performed in thermal chamber under careful temperature control. Immediately before the experiment, 1.5 equivalents of $NaBH_4$ were added to the reaction mixture.

The mixture was stirred and irradiated illuminated with LED (595 nm central emission wavelength). At the end of the experiment, nanoparticles were removed from the reaction mixture and reaction progress was monitored using gas chromatography.

2.3. Measurement Techniques

Ultraviolet-visible spectroscopy (UV-Vis) spectra were measured using Spectrometer Lambda 25 (Perkin-Elmer) in 300-1100 nm wavelength range. The nano-morphology of catalyst and the average particle size were analysed by high-resolution transmission electron microscopy HRTEM with Jeol 2200 FS microscope (Jeol, Japan). The conversion of reagents and products was analysed by gas chromatograph Agilent 8860 series GC system (Agilent Technologies, Inc., Wilmington, DE).

3. RESULTS AND DISCUSSION

Successful preparation of gold and bimetallic nanoparticles were examined using UV-Vis (**Figure 1**) and HRTEM-EDX (**Figure 2**) measurements. Gold nanoparticles show a peak of plasmon resonance located near 530 nm. The addition of Pt led to the significant broadening of the absorption band and its shift to 586 nm. The HRTEM image of the AuNPs (**Figure 2A**) shows a typical spherical shape with an average size about 30 nm. In the case of Au@Pt NPs HRTEM result (**Figure 2B**) reveal a uniform dendritic shell (Pt) and spherical core (Au). Elemental mappings of Au and Pt also show the corresponding distribution of Au and Pt metals across the nanoparticle. For further experiments, an LED with emission wavelength of 595 nm was chosen, which corresponds to the Au@Pt peak in the UV-visible spectrum.

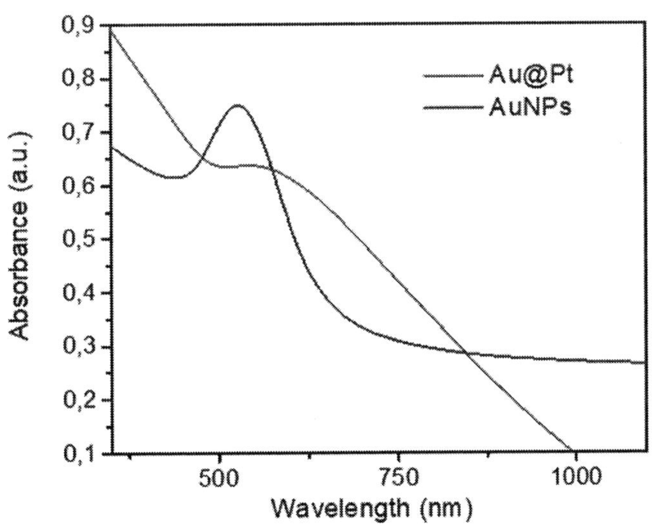

Figure 1 The UV-Vis absorption spectra of: AuNPs and Au@Pt NPs

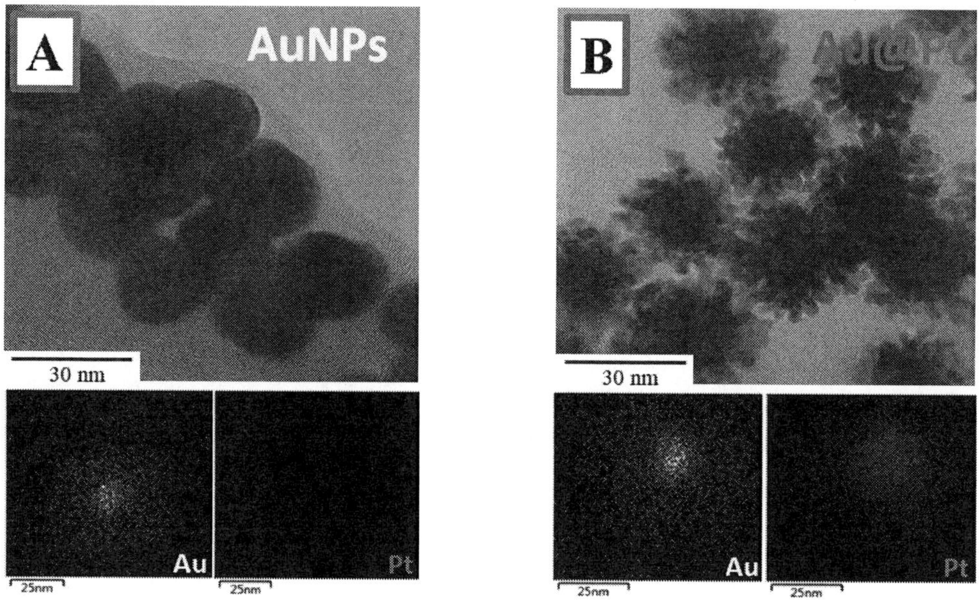

Figure 2 HRTEM-EDX images of (A) - AuNPs and (B) - Au@Pt nanoparticles

The conversion and selectivity results were monitored by a gas chromatograph. **Figure 3** shows the curves of the dependence of the conversion of phenylacetylene on the reaction time at room temperature in the dark and under irradiation with LEDs (595 nm). It is well visible that reaction progress is influenced by the illumination. In particular, in the absence of illumination, the reaction proceeds with a maximum conversion of 60 % after 8 h. and about 100 % in the case of plasmon triggering.

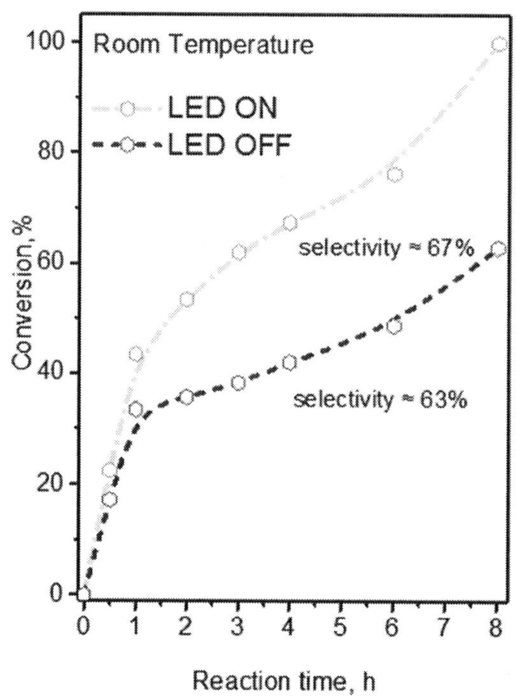

Figure 3 Dependence of Phenylacetylene conversion on the reaction time at RT in dark and under LED irradiation (595 nm)

4. CONCLUSION

In this work, we have successfully carried out the hydrogenation of phenylacetylene under plasmon assistance. Reaction was performed with utilization of $NaBH_4$ as a hydrogen source. Surface plasmon was excited on the bimetallic metal nanoparticles, which consist of plasmon active gold core and catalytically active Pt shell. The reaction conversion, closed to 100 % was reached.

ACKNOWLEDGEMENTS

This work was supported by the project Internal Grant Agency UCT No A2_FCHT_2021_086

REFERENCES

[1] DE VRIES, J.G, ELSEVIER, C.J. *The handbook of homogeneous hydrogenation*. WeinheimWiley-VCH, 2007.

[2] LIST, G.R., JACKSON, M., ELLER, F., ADLOF, R.O. Low trans spread and shortening oils via hydrogenation of soybean oil. *Journal of the American Oil Chemists' Society.* 2007, vol. 84, no.6, pp. 609-612.

[3] HUANG, Y., LIU, Z., GAO, G., XIAO, Q., MARTENS, W., DU, A., ZHU, H. Visible light-driven selective hydrogenation of unsaturated aromatics in an aqueous solution by direct photocatalysis of Au nanoparticles *Catalysis Science & Technology*. 2018, vol. 8, no. 3, pp. 726-734.

[4] IMAMURA, K., OKUBO, Y., ITO, T., TANAKA, A., HASHIMOTO, K., KOMINAMI, H. Photocatalytic hydrogenation of alkenes to alkanes in alcoholic suspensions of palladium-loaded titanium (iv) oxide without the use of hydrogen gas. *RSC Advances*. 2014, vol. 4, no. 38, pp. 19883-19886.

[5] QUIROZ, J. ,BARBOSA, E. C., ARAUJO, T. P., FIORIO, J. L., WANG, Y. C., ZOU, Y. C., HAIGH, S. J. Controlling reaction selectivity over hybrid plasmonic nanocatalysts. *Nano letters*. 2018, vol. 18, no. 11, pp. 7289-7297.

[6] LU, Y., FENG, X., TAKALE, B. S., YAMAMOTO, Y., ZHANG, W., BAO, M. Highly selective semihydrogenation of alkynes to alkenes by using an unsupported nanoporous palladium catalyst: No leaching of palladium into the reaction mixture. *ACS Catalysis*. 2017, vol. *7*, no. 12, pp. 8296-8303.

[7] ZHANG, X. B. Preparation of 1-phenylethanol by selective hydrogenation of acetophenone over alumina-supported Co catalysts. *Reaction Kinetics, Mechanisms and Catalysis*. 2011, vol. 102, no. 2, pp. 417-424.

[8] SANDOVAL, C. A., OHKUMA, T., MUÑIZ, K., NOYORI, R. Mechanism of asymmetric hydrogenation of ketones catalyzed by BINAP/1, 2-diamine− ruthenium (II) complexes. *Journal of the American Chemical Society*. 2003, vol. 125, no. 44, pp. 13490-13503.

[9] ELSEN, H., LANGER, J., BALLMANN, G., WIESINGER, M., HARDER, S. Alkaline Earth Metal Aluminates as Catalysts for Imine Hydrogenation. *Organometallics*. 2020, vol. 39, no. 23, pp. 4238-4246.

[10] TITOVA, Y.Y., SUKHOV, B.G., SCHMIDT, F.K. Nano-size bimetallic ternary hydrogenation catalysts based on nickel and copper complexes. *Journal of Organometallic Chemistry*. 2020, vol. 928, p. 121485.

[11] WÖLKE, C., DANILIUC, C. G., KEHR, G., ERKER, G. Diels-Alder route to norbornane derived vicinal phosphane/borane frustrated Lewis pairs for the metal-free catalytic hydrogenation of α, β-unsaturated ketones. *Journal of Organometallic Chemistry*. 2019, vol. 899, p. 120879.

[12] JIA, W., XI, Q., LIU, M. T., YANG, CHEN, Y., YIN, D., WANG, X. One-pot synthesis of O-heterocycles or aryl ketones using an $InCl_3/Et_3SiH$ system by switching the solvent. *The Journal of organic chemistry*. 2019, vol. 84, no. 9, pp. 5141-5149.

[13] WEN, X., SHI, X., QIAO, X., WU, Z., BAI, G. Ligand-free nickel-catalyzed semihydrogenation of alkynes with sodium borohydride: a highly efficient and selective process for cis-alkenes under ambient conditions. *Chemical Communications*. 2017, vol. 53, no 39, pp. 5372-5375.

[14] KALE, M. J., AVANESIAN, T., XIN, H., YAN, J., CHRISTOPHER, P. Controlling catalytic selectivity on metal nanoparticles by direct photoexcitation of adsorbate-metal bonds. *Nano letters*. 2014, vol. 14, no. 9, pp. 5405-5412.

[15] LINIC, S., ASLAM, U., BOERIGTER, C., MORABITO, M. Photochemical transformations on plasmonic metal nanoparticles. *Nature materials*. 2015, vol. 14, no. 6, pp. 567-576.

[16] ZABELINA, A., ZABELIN, D., MILIUTINA, E., LANCOK, J., SVORCIK, V., CHERTOPALOV, S., LYUTAKOV, O. Surface plasmon-polariton triggering of $Ti_3C_2T_x$ MXene catalytic activity for hydrogen evolution reaction enhancement. *Journal of Materials Chemistry A*. 2021, vol. 9, no. 33, pp. 17770-17779.

[17] GUSELNIKOVA, O., POSTNIKOV, P., KOSINA, J., KOLSKA, Z., TRELIN, A., SVORCIK, V., LYUTAKOV, O. A breath of fresh air for atmospheric CO_2 utilisation: a plasmon-assisted preparation of cyclic carbonate at ambient conditions. *Journal of Materials Chemistry A*. 2021, vol. 9, no. 13, pp. 8462-8469.

[18] KIMLING, J., MAIER, M., OKENVE, B., KOTAIDIS, V., BALLOT, H. AND PLECH, A. Turkevich method for gold nanoparticle synthesis revisited. *The Journal of Physical Chemistry B*. 2006, vol. 110, no. 32, pp.15700-15707.

STABLE DEEP EUTECTIC SOLVENT DOPED WITH METAL NANOPARTICLES

[1,2]Anastasiia TULUPOVA, [1]Vasilii BURTSEV, [1]Vaclav ŠVORČÍK, [1,2]Oleksiy LYUTAKOV

[1]*VŠCHT - University of Chemistry and Technology, Prague, Czech Republic, EU,* lyutakoo@vscht.cz

[2]*Tomsk Polytechnic University, Research School of Chemistry and Applied Biomedical Sciences, Tomsk, Russian Federation*

https://doi.org/10.37904/nanocon.2021.4340

Abstract

In this work, we prepared MeNPs doped DES systems. Gold nanoparticles were prepared on aluminum oxide microparticles surface. DES was prepared from choline chloride and urea, and embedded with the nano/micro particles. Created system was characterized using UV-Vis and FTIR spectroscopy, as well as TEM technique. As the result, we demonstrate that stable deep eutectic solvents that can be used for a range of applications.

Keywords: Deep eutectic solvents, doped, metal nanoparticles

1. INTRODUCTION

Metal nanoparticles (MeNPs) are of high interest since their unique optical, magnetic, electrical and physiological properties [1]. MeNPs can be used in different application areas, such as biomedicine [2], electrochemical catalysis [3], chemical sensing [4] and many others. MeNPs are commonly used as a colloidal suspension in water or traditional organic solvents. In turn, Deep Eutectic Solvents (DES) are recently proposed new alternative media for applications as a "green" reaction medium in organic synthesis. In contrast to traditional organic solvents, DES is non-toxic and non-flammable. Finally, deeply eutectic solvents do not produce toxic metabolites and are biodegradable [5]. The main uses of DES technology include the synthesis of nanostructures with a controlled shape [6], metal electrodeposition [7], CO2 adsorption [8] and others. But recently, attention has been paid to the study of DES as alternative solvents for organic synthesis and photochemistry. In this work we propose a creation of MeNPs doped DES for a range of potential applications.

2. EXPERIMENTAL

2.1. Materials

Gold(III) chloride trihydrate (HAuCl4, ≥99.9%), sodium citrate tribasic dihydrate (≥99.5%), choline chloride (≥98.0%), urea (≥99.0%), aluminum oxide, 4-aminothiophenol (≥97.0%), p-Toluenesulfonic acid monohydrate (≥98.5%), tert-Butyl nitrite (≥90.0%), methanol, acetonitrile and deionized water were purchased from Sigma-Aldrich and used without further purification.

2.2. Samples preparation

Preparation of Deep eutectic solvent

At first stage, choline chloride is dried under vacuum at 90 °C and then mixed with powdered urea in a ratio of 1:2. The two components were heated with constant stirring until a homogenous mixture was obtained (around 5 hours).

Preparation of AuNPs

Gold nanoparticles on aluminum oxide microparticles surface was prepared according to following methodology: Al_2O_3 powder first was modified with functional group -SH using a diazonium salt. Modified 0.2 g

Al_2O_3 was added to deionized water and boil. After the water boiled was added 10 mM aqueous $HAuCl_4$ solution to it, then 40 mM sodium citrate tribasic dihydrate water solution was added dropwise within 30-40 seconds. The resulting solution was boiled for 10 min, then was transferred to a cold hotplate and continue stirring for 12 hours. The resulting particles were washed by water and methanol by centrifugation at 7000 rpm for 5 min and dried in oven at 60 °C. **Figure 1** schematically shows the preparation route Al_2O_3-AuNPs.

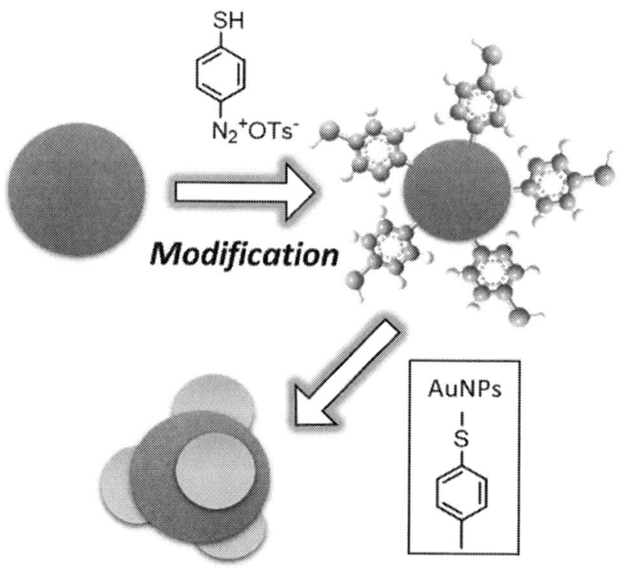

Figure 1 Schematic representation of the preparation of gold nanoparticles on aluminum oxide microparticles surface

2.3. Measurement Techniques

UV-Vis absorbance spectra of the samples were measured using a HR2000 (Ocean Optics) spectrometer in 200-800 nm wavelengths range using the AvaLight-DHS light source (Avantes).

Fourier transform infrared (FTIR) attenuated total reflection (ATR) spectra were recorded using a Nicolet 6700 spectrometer (Thermo Scientific, France) with a Smart ATR accessory device (1000 scans and 4 cm^{-1} resolution).

Transmission electron microscopy (TEM) images were obtained with a JEOL JEM-1010 instrument (JEOL Ltd., Japan), with a SIS MegaView III digital camera (Soft Imaging Systems, acceleration voltage 80 kV) and analysis was performed by using AnalySIS Software 2.0.

3. RESULTS AND DISCUSSION

Deep eutectic solvent including choline chloride (ChCl) as the hydrogen bond acceptor urea as a hydrogen bond donor were prepared. The synthesized DES had the following physicochemical properties: a homogenous and transparent liquid of high viscosity, without any evident solid particles and stable at room temperature. To confirm the formation of hydrogen bonding, infrared spectra of choline chloride, urea and deep eutectic solvents were obtained. The results are presented in **Figure 2**. Absorption bands of urea at 3425 cm^{-1} and 3333 cm^{-1} correspond to the stretching mode of $-NH_2$. On the contrary, these absorption bands moved towards the lower values of wavenumber, and specifically 3402 cm^{-1} and 3315 cm^{-1}. This shift could be affected by the formation of hydrogen bonds between urea and ChCl. Also, the bands associated to ChCl and urea appeared in the spectrum of DES. Their corresponding wave number of the vibrational modes and associated terminal groups are listed in **Table 1**.

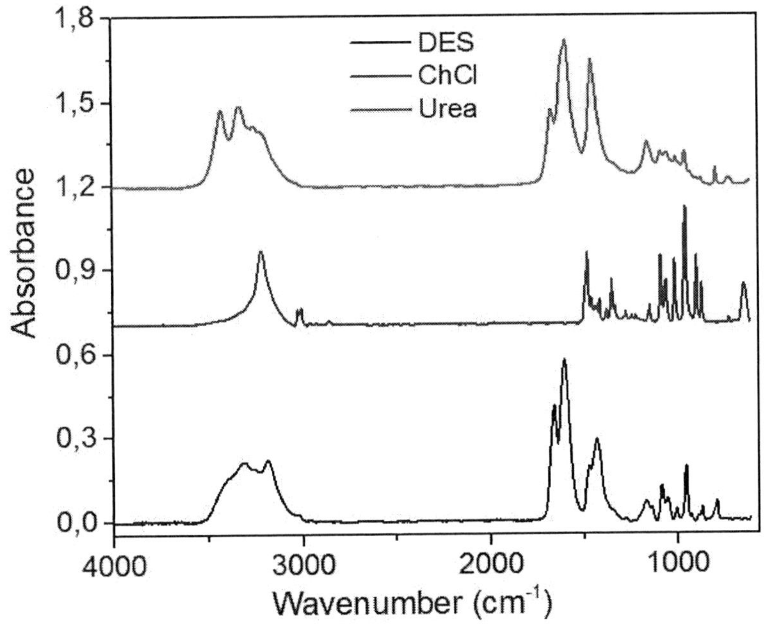

Figure 2 FTIR spectra of choline chloride, urea and synthesized DES

Table 1 Wave numbers and their assignments obtained from FTIR spectra

Observed frequencies (cm^{-1})			**Assignments**
Urea	**Choline Chloride**	**DES**	
3425		3402	ν_{as} NH_2
3333		3315	ν_s NH_2
3256			ν_{as} NH_2
	3219	3182	ν_{as} OH
	3016	3026	δ_{as} OH
1674		1665	δ_a NH_2
1592		1601	δ_{as} NH_2
	1482	1472	ρ CH_3
1454		1430	ρ_s CH_3
1151		1169	ν_{as} CN
	1086	1082	ρ CH_2
	1008	1008	ν C-O
	953	953	ν_{as} CCO
	889	870	ν_{as} N-CH_3
788		783	ω C=O

In the next step, prepared material was characterized using TEM technique. **Figure 3A** shows the TEM image of microparticle of Al_2O_3 before modification. **Figure 3B** indicates the apparent presence of gold nanoparticles located over the surface of aluminum oxide. In the results, we confirmed the success of synthesis of gold nanoparticles on Al_2O_3 microparticles surface.

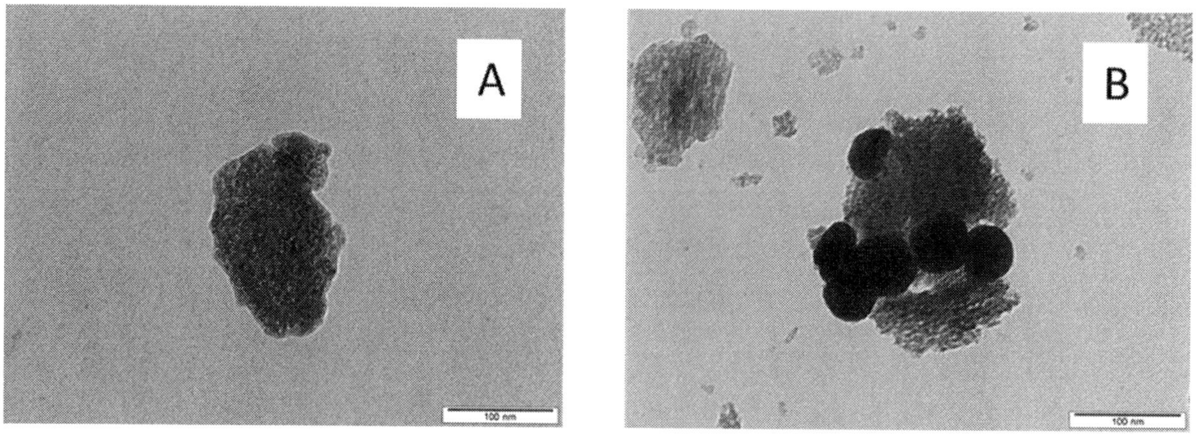

Figure 3 TEM images of Al_2O_3 (A), Al_2O_3-AuNPs (B).

The optical characterization of the created system was performed by UV-Vis measurement (**Figure 4**). The characteristic for pristine Al_2O_3 absorption band is located at 260 nm. Synthesis of AuNPs on Al_2O_3 surface leads to the appearance of peak with the maximum at 540 nm, which is typical absorption band, characteristic for gold nanoparticles.

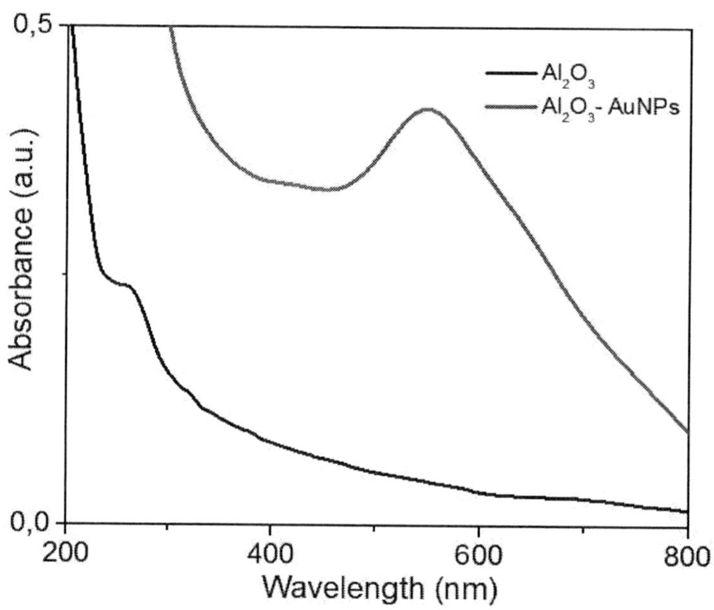

Figure 4 UV-VIS spectrum of Al_2O_3 and gold nanoparticles on Al_2O_3

4. CONCLUSION

In this work, deep eutectic solvent was prepared and doped with metal nanoparticles, previously immobilized on Al_2O_3 powder surface. Created system was characterized by FTIR, TEM and UV-Vis method. Such system can be further used for a range of attractive application, with especial attention on green chemistry field.

ACKNOWLEDGEMENTS

This work was supported by under the project IGA 126 88 2109.

REFERENCES

[1] MCCONNELL, W. P., NOVAK, J. P., BROUSSEAU III, L. C., FUIERER, R. R., TENENT, R. C. FELDHEIM, D. L. Electronic and Optical Properties of Chemically Modified Metal Nanoparticles and Molecularly Bridged Nanoparticle Arrays. *The Journal of Physical Chemistry B.* 2000, vol. 104, pp. 8925-8930.

[2] MITCHELL, M. J., BILLINGSLEY, M. M., HALEY, R. M. Engineering precision nanoparticles for drug delivery. *Nature Reviews Drug Discovery.* 2021, vol. 20, pp. 101-124.

[3] ZABELINA, A., ZABELIN, D., MILIUTINA, E., LANCOK, J., SVORCIK, V., CHERTOPALOV, S., LYUTAKOV, O. Surface plasmon-polariton triggering of $Ti_3C_2T_x$ MXene catalytic activity for hydrogen evolution reaction enhancement. *Journal of Materials Chemistry A.* 2021, vol. 9, pp. 17770-17779.

[4] OLSHTREM, A., GUSELNIKOVA, O., POSTNIKOV, P., TRELIN, A., YUSUBOV, M., KALACHYOVA, Y., LAPCAK, L., CIESLAR, M., ULBRICH, P., SVORCIK, V., LYUTAKOV, O. Plasmon-assisted grafting of anisotropic nanoparticles - spatially selective surface modification and the creation of amphiphilic SERS nanoprobes. *Nanoscale.* 2020, vol. 12, pp. 14581-14588.

[5] ZHANG, Q., VIGIER, K., ROYERA, S., JEROME, F. Deep eutectic solvents: syntheses, properties and applications. *Chemical Society Reviews.* 2012, vol. 41, pp. 7108-7146.

[6] LIAO, H., JIANG, Y., ZHOU, Z., CHEN, S., SUN, S. Shape-Controlled Synthesis of Gold Nanoparticles in Deep Eutectic Solvents for Studies of Structure-Functionality Relationships in Electrocatalysis. *Angewandte Chemie.* 2008, vol. 120, pp. 9240-9243.

[7] ABBOTT, A. P., TTAIB, K. E., FRISCH, G., RYDERA, K. S., WESTON, D. The electrodeposition of silver composites using deep eutectic solvents. *Physical Chemistry Chemical Physics.* 2012, vol. 14, pp. 2443-2449.

[8] ZHU, A., JIANG, T., HAN, B., ZHANG, J., XIEA, Y., MAA, X. Supported choline chloride/urea as a heterogeneous catalyst for chemical fixation of carbon dioxide to cyclic carbonates. *Green Chemistry.* 2007, vol. 9, pp. 169-172.

PLASMA ENHANCED ATOMIC LAYER DEPOSITION OF SOLID-STATE ELECTROLYTE Li-Ta-O FOR SOLID-STATE BATTERIES

[1]Pavel FEDOROV, [1,2]Denis OLKHOVSKII, [1,3]Denis NAZAROV, [1]Vladislav CHERNYAVSKY, [1,4]Yury KOSHTYAL, [4]Aleksander RUMYANTSEV, [1]Anatoly POPOVICH, [1*]Maxim Yu. MAXIMOV

[1]Peter the Great Saint-Petersburg Polytechnic University, Saint Petersburg, Russia, fedorovpavel99@yandex.ru, Odl-ne@yandex.ru, dennazar1@ya.ru, vlad.djxdr@gmail.com, yury.koshtyal@gmail.com, rumyantsev.amr@gmail.com, director@immet.spbstu.ru,

[1]Correspondence: maximspbstu@mail.ru*

[2]Saint-Petersburg State Institute of Technology, Saint Petersburg, Russia

[3]Saint Petersburg State University, Saint Petersburg, Russia

[4]Ioffe Institute, Saint Petersburg, Russia

https://doi.org/10.37904/nanocon.2021.4336

Abstract

Solid-state batteries (SSBs) are regarded as the next step in energy storage technology. One of the main problems in developing such batteries is developing and synthesizing solid-state electrolytes (SSE). The main factor that stops the introduction of SSBs into everyday life is the low ionic conductivity of Li ions in modern SSEs. Therefore, the primary purposes of the work are to establish the reasons for the manifestation of this inhibiting factor and to assess the possibility of using new materials for SSLIB. Solution these problems ultimately can give an impetus to the development and promotion of such type batteries. In this work, SSE was obtained by the atomic layer deposition (ALD) method, allowing the formation of homogeneous coatings with precision control of the thickness. ALD of LixTaOy thin films on silicon and stainless steel substrates using Ta(OEt)5 and LiOtBu was studied. The synthesis temperature was 300 °C, which is based on the previous research. Samples were synthesized with different ratios of Li:Ta = 1:2, 1:3, 1:7, and depending on the ratio of metals, the growth rate per supercycle varied from 0.21 to 0.46 nm. The film thickness was determined by spectral ellipsometry. A scanning electron microscopy was used to determine the conformity and morphology of the coatings. X-ray Photoelectron Spectroscopy was used to determine the elemental composition on the surface and in the bulk of SSE. According to X-ray diffraction, thin films are amorphous. Cyclic voltammetry has been used to study how thin films will respond to different voltages. The cathodic region cycling at various discharge currents (from 20 to 80 μA/cm2) presents a low capacity.

Keywords: Atomic layer deposition, lithiated tantalum oxide, solid-state electrolyte, thin films, solid-state batteries

1. INTRODUCTION

Lithium-ion batteries (LIBs), due to their unique characteristics (specific energy, specific power, wide operating temperature range, recharge resource, etc.), have already taken root in many areas of human activity, from industrial technology to household and portable appliances [1]. But due to the use of a liquid organic electrolyte in LIBs, they are characterized by the following disadvantages: first of all, we need to introduce additional process - formation, which complicates the operation of their production, and the second drawback is the decrease in output parameters and maximum capacity with use due to the formation of solid-electrolyte interphase (SEI) [2,3]. Therefore, today in lithium-ion energy, quite a lot of research is devoted to solid-state batteries (SSBs). This is because they use a solid inorganic compound instead of the usual liquid electrolyte [4]. Just because of this, the main factor that distinguishes SSBs from ordinary LIBs has increased levels of fire and explosion safety due to a decrease in the proportion of combustible material. So it becomes possible to introduce lithium-ion energy in new promising areas, such as electric vehicles.

Compounds similar in structure to LISICON (Li Super Ion CONductors) [5] are chosen as a solid electrolyte: $Li_{1+x}Ti_{2-x}M_x(PO_4)_3$ (M = Ge, Ti, Sn, Zn, Hf [5]), Li_2S-P_2S_5, and others [6]. As specific examples, we can cite studies of structures: $Li_xLa_3Zr_2O_{12}$, $Li_{1+x}Al_xGe_{2-x}(PO_4)_3$, $Li_{10}GeP_2S_{12}$, etc. [1,7,8,9]. The standard method for obtaining these structures is the sol-gel method [5, 6]. Still, the growing popularity of synthesis using atomic layer deposition (ALD) can be considered a progressive and better method for obtaining solid electrolytes. The most important advantages of ALD over the sol-gel method, in this case, can be called: obtaining conformal coatings, obtaining the required thickness with a minimum error, and the possibility of covering complex geometric shapes [1,10]. Less important, but still advantages can be called: the repeatability of the process, its self-limiting, and synthesis at relatively low temperatures [10,11,12].

This article describes the results of studies of obtained LiTaO thin films on steel and silicon substrates by the ALD method. We studied samples with the ratio Li:Ta = 1:2, 1:3, and 1:7, their growth rate on the substrate, and the elemental composition in-depth. Based on which the optimal Li-Ta-O films were selected for subsequent syntheses on the cathode material [13].

2. MATERIALS AND METHODS

The deposition of Li-O and Ta-O for Li-Ta-O films was carried out by ALD with Picosun R-150 setup at 300 °C at a base pressure of 8-12 hPa. Lithium tert-butoxide (LiOtBu, 97 %, Sigma Aldrich) and tantalum(V) ethoxide ($Ta(OEt)_5$, 99.98 % Sigma-Aldrich), were used as metal-containing reagents. Remote oxygen plasma (O_2-p) was applied as a counter-reagent.

Precursors' containers were heated to 190 °C for $Ta(OEt)_5$ and 220 °C for LiO^tBu. All depositions were performed as follows. After pulsing of metal-containing reagent ($Ta(OEt)_5$ - 1.5s or LiO^tBu - 3.0s) the excess of the precursor was purged with nitrogen (99.999 %) within 10s. Pulse times for O_2 plasma was 10s. For deposition of Li-Ta-O thin films we used super cycles consisting of one LiO^tBu/O_2-p pulses and two, three or seven $Ta(OEt)_5/O_2$-p pulses: [LiO^tBu/O_2-p] + [$Ta(OEt)_5/O_2$-p]*2 = LiTaO-1/2, [LiO^tBu/O_2-p] + [$Ta(OEt)_5/O_2$-p]*3 = LiTaO-1/3, [LiO^tBu/O_2-p] + [$Ta(OEt)_5/O_2$-p]*7 = LiTaO-1/7. To obtain the required films, 100-150 synthesis supercycles were performed. Estimated growth rates for Ta-O was 0.05 nm/cycle, for Li-O - 0.11 nm/cycle.

Monocrystalline silicon wafers (surface orientation 100, the size 4x4 cm, Telecom-STV Co., LTD, Zelenograd, Moscow, Russia) and stainless-steel plates (316SS, Tob New Energy Technology Co., LTD, diameter 15.8 mm) were used as substrates.

The thicknesses of the films were measured by spectral ellipsometry (wavelength range 350-1000 nm) using Ellips-1891 SAG ellipsometer (CNT, Novosibirsk, Russia). The morphology of the films was studied by scanning electron micrographs of flat and cross-sections were obtained by a Supra 55 VP scanning electron microscope (SEM, Zeiss, Oberkochen, Germany). X-ray diffraction (XRD) studies were performed using a Bruker D8 ADVANCE (Cu-Kα). The obtained results were processed by the Rietveld method using TOPAS 5 software (XRD). Finally, the chemical composition of the films was studied by X-ray photoelectron spectra (XPS) and was obtained on an Escalab 250Xi spectrometer (Thermo Fisher Scientific, Waltham, MA, USA).

Electrochemical studies were performed in CR2032 coin cells, the samples of Li-Ta-O were deposited on the 316SS surface. Lithium foil, polyolefin porous film 2325 (Celgard, Charlotte, NC, USA), and TC- E918 (Tinci, Guangzhou, China) solution were used as the counter electrode, separator, and electrolyte, respectively. The coin cells (CR2032) were assembled in OMNI-LAB (VAC) glove box under argon atmosphere. Cyclic voltammetry (CV) was performed using a potentiostat PGSTAT302N+ (Autolab, Utrecht, the Netherlands) in the range of 0.1 - 4.3 V with a scan rate of 0.5 mV/s. Cyclic charge/discharge was performed using the battery testing system CT-3008W-5V10mA (Neware, Shenzhen, China) at room temperature in the range of voltage 3.0-4.3 V, at current 20-80 μA/cm^2.

3. RESULTS AND DISCUSSIONS

3.1. Synthesis of Li-Ta-O structures

Previous experiments with the synthesis of tantalum structures as thin films [1,14] made it possible to select the conditions for obtaining the Li-Ta-O system; the corresponding synthesis parameters are presented in the Materials and Methods section.

A series of experiments were carried out to obtain Li-Ta-O with the ratios of the pulse $LiO^tBu/Ta(OEt)_5$ in the supercycle: 1:2, 1:3, and 1:7. The ratios were selected based on research results published elsewhere [15]. As a result, the investigated growth rate of LiTaO-1/2 corresponds to the expected - 0.21 nm/supercycle (**Table 1**). According to spectral ellipsometry, the samples LiTaO-1/3 and LiTaO-1/7 have a higher growth rate than expected, possibly caused by the stimulation of growth with an increase in the number of pulse $Ta(OEt)_5$.

Table 1 The thicknesses of the Li-Ta-O films measured using SEM and ellipsometry and the growth parameters of the films according to the ellipsometry data

Samples - Li/Ta ratio	SEM average thickness, nm	Ellipsometry average thickness, nm	Number of supercycles	Growth rate, nm/supercycle	Estimated growth rate, nm/supercycle
LiTaO-1/2	45.05	30.9	150	0.21	0.21
LiTaO-1/3	39.44	37.9	120	0.32	0.26
LiTaO-1/7	58.75	62.61	100	0.63	0.46

The thicknesses measured by SEM (**Table 1**) for the LiTaO-1/3 and LiTaO-1/7 samples are close to the data obtained by ellipsometry. For LiTaO-1/2, there is a discrepancy between the two measurement methods. This may be due to the film thickness gradient and edge effects when measuring thin films on SEM. According to the SEM micrograph (**Figure 1**), the surface of LiTaO-1/7 is homogeneous without visible defects.

Figure 1 SEM image demonstrates a LiTaO-1/7 coating sample on the silicon substrate

For a detailed study of the chemical composition of the Li-Ta-O system, studies were carried out by the XPS method (LiTaO-1/7 - **Table 2**). As a result, the composition of the samples' surface and the bulk of the coating were investigated before and after etching the surface layer with argon ions (90 and 300 sec, 500 eV). As a result, the following elements were found on the sample's surface: lithium, tantalum, oxygen, and carbon (**Table 2**). Also, by the etching proceeds, it can be seen the number of lithium decreases and the proportion of tantalum increases. This may be due to the presence of a gradient of elements in the bulk of the film. Thus, the surface is enriched with lithium, while tantalum is found mainly in the bulk of the film.

According to the CIL peaks of carbon (**Figure 2**), before etching, C is present on the surface in the composition of compounds, including the C-OH, C-C, C-H groups, and C=O bonds (peak shoulder at 284.8 eV) and COOH (289 eV peak). After 90 sec etching of the surface layer, the carbon signal has dropped noticeably, which indicates its absence in bulk (concentration in the analysed sample is below the detection threshold).

The spectra of Li1s (**Figure 2**), obtained without preliminary etching and after 90 and 300 sec etching, exhibit a broad maximum in the region of manifestation of lithium compounds. In the region of the location of the maximum, bands appear that characterize the presence of lithium in various compounds: Li_2O, Li_2CO_3, LiOH. The low intensity and signal/interference ratio of the Li1s spectra do not allow one to reveal the preferred state. After etching, the intensity of the peak drops by almost a factor of two.

The position of the maxima in the Ta4f spectrum (**Figure 2**) before etching corresponds to the TaO_2 compound, which is inconsistent with the data obtained from the O1s (Ta_2O_5) spectrum. This discrepancy can be caused by the high described difficulties and errors in the calibration of the position of the peaks (charge compensation). Nevertheless, in the spectra after 90 sec etching, the appearance of new peaks (components) in the region of lower energies is noticeable, these peaks correspond to the Ta^0 and Ta^{2+} states of tantalum. Therefore, an unambiguous conclusion can be made about reducing tantalum (possibly even to metallic Ta) during etching.

Table 2 Element ratios before and after etching a Li-Ta-O-1/7 film

Element	Before etching (at. %)	After etching 90 s (at. %)	After etching 300 s (at. %)
O	47.65	55.29	3.15
Ta	12.45	31.94	1.15
Li	20.96	12.78	-
C	18.95	-	0.97
Si	-	-	94.74

After etching for 300 s, most of the signals from the silicon substrate and weak signals of pure tantalum are present in the spectra, so after 300 s, the film is completely etched away, so the etching rate is about 0.21 nm/s.

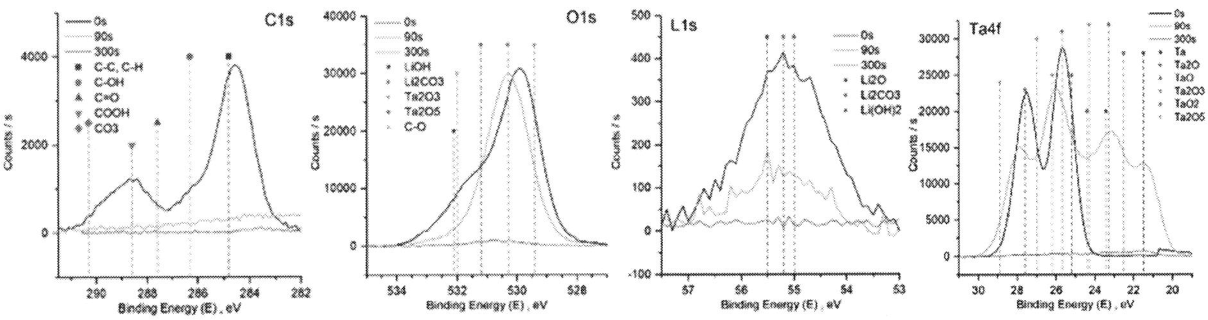

Figure 2 XPS spectra of the Li-Ta-O-1/7 film before and after etching

The X-ray phase analysis pattern (**Figure 3**) shows two relatively intense peaks characteristic of the Ta structures in the region 27° (TaO_2) and 33° (TaO), as well as one less intense peak in area 62 (TaO_2). An ultra-intense peak of 33° is also observed, which can be associated with a defect in the silicon substrate. Comparing the obtained diffractogram with cards from the database of diffractograms PDF, we can conclude that the obtained LiTaO-1/7 film is X-ray amorphous.

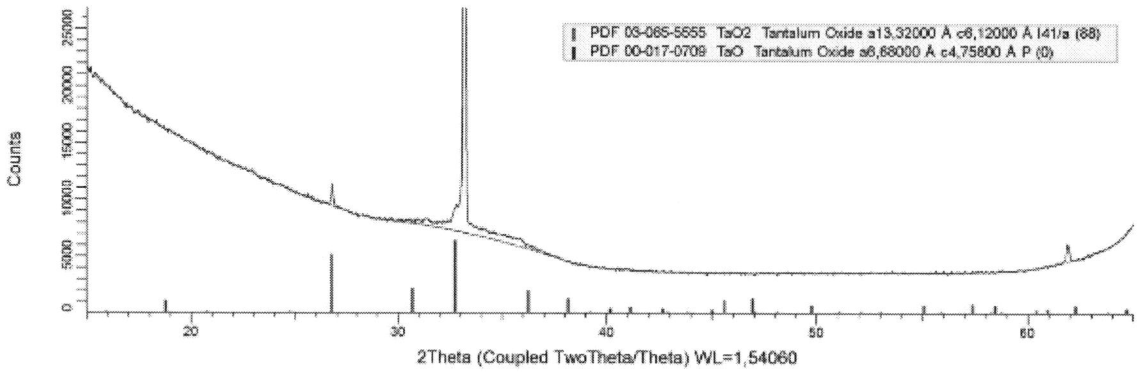

Figure 3 Diffraction pattern of the LiTaO-1/7 sample (Si substrate)

3.2. Electrochemical tests of Li-Ta-O structures

The LiTaO-1/2 samples on steel (SS_LiTaO-1/2) with a calculated ratio Li/Ta = 1/1 were selected to study the electrochemical characteristics. This ratio was selected based on research results published elsewhere [15], which showed that with an increase of lithium content (from x = 0.32 to 0.98 in Li_xTaO_y), the lithium-ion conductivity is improved by two orders of magnitude. Based on the results of work [15] and the obtained XPS graphs (**Figure 3**), we can conclude that with an increase in the Li concentration (x = 1.73), it tends to form harmful bonds with carbon (Li_2CO_3), hydroxide ($Li(OH)_2$) and oxygen (Li_2O), instead of being in a free state in the $[TaO_6]^-$ crystal lattice. As a result of the formation of such bonds, we lose lithium ions, thereby lowering the overall conductivity of the material.

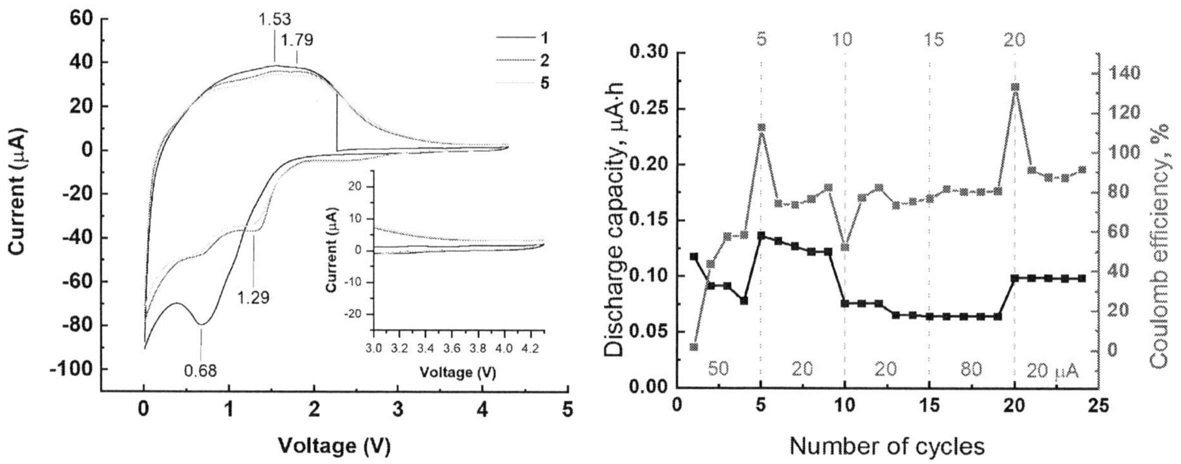

Figure 4 Cyclic voltammetry (CV) curves and results of cyclic tests at different discharge currents

The CV curves of a series of SS_LiTaO-1/2 samples (**Figure 4**) and coatings of the Ta-O system have a general trend: in the anodic region (3.0-0.1 V), there is current amplification, which is most likely associated with the formation of SEI film [16], and no current increases are observed in the potential cathode range (4.3-3.0 V - inset in **Figure 4**). Further cycling at currents from 20 to 80 μA/cm2 in the cathode potential range (4.3-3.0 V) showed the presence of a minimum capacity of about 2.0-1.5 μA•h•μm-1•cm-2 (0.14-0.07 μA•h), for example in our previous work on the lithium-nickel-silicon oxide cathode study [1], the capacity of the cathode material was 30-20 μA•h•μm-1•cm-2, so we can consider that this thin film of LiTaO-1/2 does not affect on the final capacity. Thus, the investigated coatings of the LiTaO-1/2 system do not significantly contribute to the electrochemical capacity in the potential cathode range. Therefore, this thin film can be studied like cathode-solid electrolyte systems in the future.

4. CONCLUSION

First, thin films carried out at different synthesis ratios showed different growth rates for LiTaO-1/2, LiTaO-1/3, and LiTaO-1/7: 0.21, 0.32, and 0.63 nm/supercycle respectively. Secondly, the physicochemical analysis showed: thin films are amorphous; lithium diffuses from the bulk closer to the surface, while tantalum is concentrated in bulk or bottom of coating; traces of carbon (which are absent in bulk) were also found on the surface, which may indicate its interaction with lithium or the formation of organic acids/bases/substances. Finally, electrochemical tests have shown no current amplification in the cathode region (4.3-3.0 V); also, Li-Ta-O films do not mainly introduce additional capacity, at currents equal 2.0-1.5 $\mu A \cdot h \cdot \mu m^{-1} \cdot cm^{-2}$.

ACKNOWLEDGEMENTS

The research was supported by the Ministry of Science and Higher Education of the Russian Federation (State Assignment for basic research 0784-2020-0022).

REFERENCES

[1] MAXIMOV, M. et al. Atomic layer deposition of lithium-nickel-silicon oxide cathode material for thin-film lithium-ion batteries. *Energies*. 2020, vol. 13, no. 9.

[2] CHARLES-BLIN, Y. et al. Surface atomic layer fluorination of Li4Ti5O12: Investigation of the surface electrode reactivity and the outgassing behavior in LiBs. *Appl. Surf. Sci. Elsevier*. 2020, vol. 527, p. 146834.

[3] LI, X. et al. Atomic layer deposition of solid-state electrolyte coated cathode materials with superior high-voltage cycling behavior for lithium ion battery application. *Energy Environ. Sci.* 2014, vol. 7, no. 2, pp. 768-778.

[4] TALIN, A.A. et al. Fabrication, Testing, and Simulation of All-Solid-State Three-Dimensional Li-Ion Batteries. *ACS Appl. Mater. Interfaces*. 2016, vol. 8, № 47, pp. 32385-32391.

[5] PERSHINA, S. V. et al. Glass-ceramics in Li1+xAlxGe2-x(PO4)3 system: The effect of Al2O3 addition on microstructure, structure and electrical properties. *J. Alloys Compd. Elsevier B.V.* 2020, vol. 835, p. 155281.

[6] RAJAGOPA,L R., RYU, K.S. A novel Na substituted Li3-xNaxPS4 mixed ions solid electrolyte for all solid-state lithium batteries. *Applied Surface Science. Elsevier B.V.* 2020, vol. 527, p. 146739.

[7] KIM, G.H. et al. Preparation and Characterization of Sol-Gel-Driven Li x La 3 Zr 2 O 12 Solid Electrolytes and LiCoO 2 Cathodes for All-Solid-State Lithium-Ion Batteries. *J. Nanosci. Nanotechnol.* 2020, vol. 20, № 11, pp. 7002-7009.

[8] KIM, Y.C. et al. Improving the ionic conductivity of Li1+xAlxGe2-x(PO4)3 solid electrolyte for all-solid-state batteries using microstructural modifiers. *Ceram. Int. Techna Group S.r.l.* 2020, vol. 46, № 14, pp. 23200-23207.

[9] HAYASHI, D. et al. Synthesis of Li10GeP2S12-type lithium superionic conductors under Ar gas flow. *J. Power Sources. Elsevier B.V.* 2020, vol. 473, p. 228524.

[10] CUEVAS, A.L. et al. Chemical, optical and transport characterization of ALD modified nanoporous alumina based structures. *J. Ind. Eng. Chem.* 2020, vol. 91, pp. 139-148.

[11] NGUYEN, V.H. et al. Impact of precursor exposure on process efficiency and film properties in spatial atomic layer deposition. *Chem. Eng. J.* 2021, vol. 403.

[12] POPOVICH, A.A. et al. Improvement of the cycle life of LiCoO2-based electrodes used in lithium-ion batteries. *Russ. J. Appl. Chem.* 2015, vol. 88, № 5, pp. 898-899.

[13] KOLCHANOV, D.S. et al. Inkjet Printing of Li-Rich Cathode Material for Thin-Film Lithium-Ion Microbatteries. *Energy Technol.* 2020, vol. 8, № 3, pp. 1-11.

[14] FEDOROV, P., NAZAROV, D., MEDVEDEV, O. et al. Plasma enhanced atomic layer deposition of tantalum (V) oxide for solid-state batteries. *Coatings*. 2021.

[15] HU, Y. et al. Ionic conductivity in LixTaOy thin films grown by atomic layer deposition. *Electrochim. Acta. Elsevier Ltd.* 2020, vol. 361.

[16] KOSHTYAL, Y. et al. Atomic layer deposition of ni-co-o thin-film electrodes for solid-state libs and the influence of chemical composition on overcapacity. *Nanomaterials*. 2021, vol. 11, № 4.

PLASMA OXIDATION OF PRINTED POLYSILOXANE LAYERS

[1]Július VIDA, [2]Petr DZIK, [1]Tomáš HOMOLA

[1]*Department of Physical Electronics, Faculty of Science, Masaryk University, Brno, Czech Republic, EU, jvida@mail.muni.cz*

[2]*Faculty of Chemistry, Brno University of Technology, Brno, Czech Republic, EU*

https://doi.org/10.37904/nanocon.2021.4339

Abstract

Organosilica binders are a promising way to interconnect various nanoparticles in printed coatings for photovoltaic cells and other applications. Post-deposition treatment of printed nanoparticle films with organosilica binders is required to remove the organic moieties and thus achieve the optimal optoelectronic properties of the resulting film. As a result, the polysiloxane binder is converted to almost fully amorphous silica and this process is called mineralization. Atmospheric pressure plasma operating in open air has proven to be the most promising method, as the operating temperature can be kept below 70 °C and the overall treatment time required is in order of minutes. These are significant advantages compared to alternative approaches like thermal sintering, where temperatures required are around 450 °C, or chemical or UV light treatment, where the treatment times extend into the order of hours. To better understand the underlying chemistry in the interaction of the ambient air plasma and the organosilica binder we performed an X-ray photoelectron spectroscopy (XPS) study on the plasma-treated films of silica binder, regularly used with titania nanoparticles. The detailed analysis of core-level spectra of C 1s, O 1s and Si 2p were used to observe the removal of methyl groups from the film and gradual transformation into amorphous SiO_2. The scanning electron microscope revealed significant patterning of the surface by interaction with plasma after exposures longer than 16 seconds.

Keywords: Nanoparticle binder, organosilica, plasma mineralization, DCSBD, atmospheric pressure plasma

1. INTRODUCTION

Sol-gel processes are widely used to prepare various silica gels for different applications including protective hydrophobic coatings [1]. Recently a process involving methyltriethoxysilane (MTEOS) precursor was developed and used to prepare a matrix in which titania nanoparticles were dispersed to obtain porous and flexible photocatalytically active coating [2]. The organosilica matrix acted as a binder interconnecting the nanoparticles. This has substituted the connections between nanoparticles that form during traditional high-temperature sintering and are necessary for desired properties of the film. It has been shown that these new coatings can be used as photoanodes in photovoltaic devices such as perovskite solar cells, which is the most interesting application of the coating so far [4].

As-deposited coatings without any curing, however, don't display satisfactory performance. The methyl groups in the organosilica binder hinder the electrical properties of the films and cause poor wettability if any subsequent overprinting is desired for more complicated devices. Three approaches for post-printing curing were compared: thermal sintering, UV light treatment, and air plasma treatment [6]. It was shown that all three methods cause minerallization of the binder resulting in fully insoluble and stable coating with improved mechanical properties of the film. Out of the three, the plasma approach is extremely attractive as the treatment times necessary can be reduced to one minute, the temperature can be kept below 70 °C providing compatibility with the temperature-sensitive flexible substrates, and the process is readily compatible with roll-to-roll processing for preparing large area coatings.

The mechanism of minerallization of the organosilica binder is believed to involve the cleavage of methyl groups bonded to silicon and transformation into amorphous SiO_x. The energetic nitrogen and oxygen species in plasma, specifically oxygen radicals, react with $Si-CH_3$ groups on the surface resulting in surface active sites [8]. We chose to study in the detail the chemical changes happening on the surface of polysiloxane coating. Up to now, all the studies reported the effect of curing on coatings of binder together with the nanoparticles. Elimination of the nanoparticles and analysing the polysiloxane coating by itself can give more detailed information about the minerallization process.

2. EXPERIMENTAL PROCEDURE

The preparation of recently developed organosilica binder, a hybrid silica sol based on MTEOS [2] was adopted without modification. The binder was diluted with isobutanol in ratio 1:20. The mixture was spin-coated onto a crystalline silicon wafer at 800 RPM for 10 s and then 3000 RPM for 30 s.

Plasma was generated in ambient air by diffuse coplanar surface barrier discharge (DCSBD) - a dielectric barrier discharge with a coplanar arrangement of electrodes; commercialized by Roplass s.r.o. (Czech Republic) [3]. DCSBD generates plasma of high uniformity and power density up to 100 W/cm^3, while maintaining temperature below 70 °C. The plasma region of the DCSBD unit was around 0.3 mm thick. To ensure a good contact of the sample surface with the plasma region the sample holder was distanced from the ceramics of the DCSBD unit by 0.3 mm of Kapton tape. The exposure time of the treated sample surface was varied from 0 s to 128 s.

The surface chemistry of the films was determined from X-ray photoelectron spectra (XPS) obtained by AXIS Supra spectrometer from *Kratos Analytical Ltd.* (United Kingdom). The instrument uses Al Kα spectral line (1486.6 eV photon energy) and an electron flood gun was utilized for charge compensation. The acquisition of narrow regions of the spectrum was performed with the pass energy 20 eV. For analysis of the spectra CasaXPS software was used. The spectra were calibrated to C-C/C-H peak of C 1s at 285 eV binding energy.

Mira3 scanning electron microscope (SEM) from *Tescan* (Czech Republic) was employed for the analysis of the morphology of the coatings and the effect of plasma. Prior to analysis the films were coated with 10 nm of Au/Pd conductive film to reduce the charging of the surface by incident electrons. The images were captured at with 2000x and 5000x magnification, 10 kV accelerating voltage and at a working distance of 10 mm.

3. RESULTS AND DISCUSSION

Table 1 Atomic concentrations on the surface of polysiloxane film before and after plasma exposure, obtained from XPS

Treatment time	Silicon [%]	Oxygen [%]	Carbon [%]	Nitrogen [%]
0 s	23.7 ± 0.1	44.8 ± 0.6	31.5 ± 0.6	0
1 s	25.5 ± 0.6	61 ± 1	13.0 ± 0.7	0.51 ± 0.01
8 s	25.4 ± 0.1	62 ± 1	12 ± 1	0.6 ± 0.3
16 s	24.9 ± 0.2	62.8 ± 0.1	11.6 ± 0.1	0.8 ± 0.2
32 s	25.2 ± 0.1	62.1 ± 0.7	12.0 ± 0.3	0.7 ± 0.5
64 s	25.7 ± 0.1	61.6 ± 0.1	12.1 ± 0.4	0.7 ± 0.2
128 s	25.8 ± 0.1	62.1 ± 0.2	11.7 ± 0.1	0.4 ± 0.1

We used X-ray photoelectron spectroscopy (XPS) was used to analyze the changes in the surface chemistry on the surface of the polysiloxane coatings. In **Table 1** the atomic concentrations of the elements detected on the surface are shown. Immediately after the exposure to plasma for just one second, we observe a significant increase in the oxygen concentration from the original 45 % to 61 % and a decrease of the carbon concentration from 32 % to 13 %. A small amount of nitrogen was introduced to the surface (below 1 %) during the ambient air plasma treatment. Prolonged plasma exposure did not cause any further changes in the surface elemental composition. After 128 seconds the surface contained 62 % of oxygen and 12 % of carbon. The silicon content remained relatively stable during the plasma treatment, 24 % on the original coating and 26% after 128 seconds of treatment. The removal of carbon and replacement by oxygen agrees with the results of previous studies [4].

The high-resolution core-level XPS spectra of the films are shown in **Figure 1**. **Figure 1a** compares the C 1s region of the untreated and plasma-treated film. The main peak at low binding energy is made up of two components corresponding to Si-CH_3 environment and the carbon in C-C/C-H hydrocarbon from some organic contamination on the surface at binding energies 284.5 eV and 285.0 eV, respectively. Furthermore, we observe contribution from C-OH at 286.5 eV and after plasma treatment contributions from O-C=O at 288 eV appear. This is a result of oxidation by plasma. We also observe significant shift of the main peak from 284.5 eV to 285.0 eV over the increasing plasma treatment times. This is a direct observation of the removal of methyl groups that are bound to silicon in the binder.

Figure 1b shows the high-resolution O 1s XPS spectrum. The reference sample of untreated displayed a single peak at 532.4 eV corresponding to a oxygen in the CH_3-Si-$O_{3/2}$ [2, 5]. After the exposure to ambient air plasma for just one second a shift towards higher binding energies is observed. This shift is believed to be caused by a transformation of the surface of the coating into amorphous silica under the influence of oxygen species in plasma. After exposure to plasma the position of the peak is 532.9 eV corresponding to the oxygen in SiO_2. The Si 2p region of the XPS spectra is shown in **Figure 1c**. The contribution from the silicon atoms in CH_3-Si-$O_{3/2}$ and SiO_2 are located at 102.5 eV and 103.7 eV [7]. Since the difference in binding energy is larger than in the O 1s spectrum, the changes in the chemistry can be observed more easily. After the plasma treatment we can see a shift from the original position 102.5 eV to higher binding energies. After short treatments the resulting peak is broader, indicating a combination of contribution from both environments. With prolonged treatment time the peak gets narrower and moves to higher binding energies, suggesting a complete transition to amorphous silica and an increased depth in the film where the plasma has an effect. This ultra-fast minerallization of the polysiloxane film offers the same result as a much slower minerallization under exposure to UV light presented in previous studies [2].

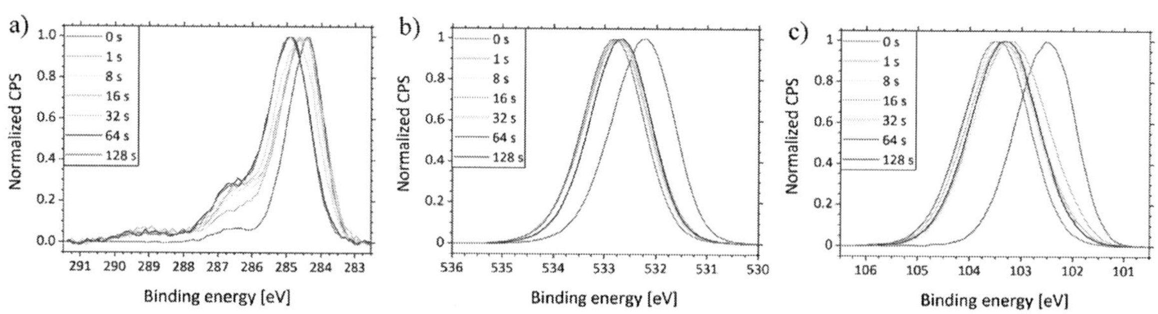

Figure 1 Comparison of a) C 1s, b) O 1s and c) Si 2p high-resolution core-level X-ray photoelectron spectra of untreated and plasma treated films

Figure 2 shows SEM images of untreated and plasma-treated coatings. The 16 s plasma-treated sample in **Figure 2c** begins to show patterning of the surface because of the interactions with the plasma. This patterning becomes more pronounced as the plasma exposure is prolonged. **Figure 2f** shows the sample treated for 128 s. The SEM image shows very sharp features of hills and valleys presumably caused by the etching of some areas on the surface of the film by plasma.

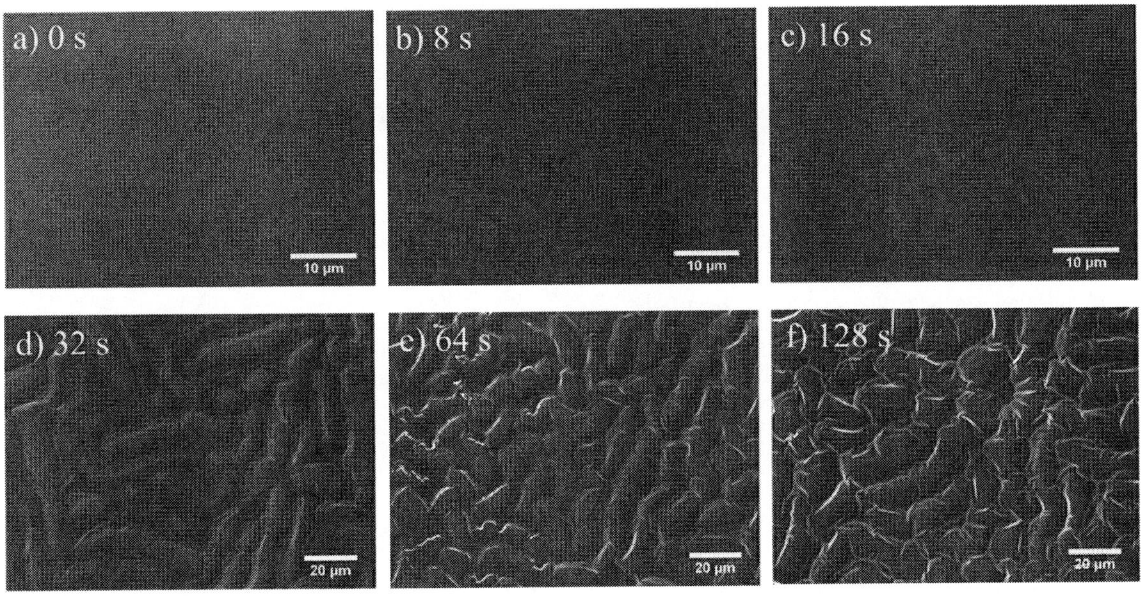

Figure 2 Images from scanning electron microscope of a) untreated and b)-f) plasma treated for 8-128 s. The a) untreated, b) 8 s and c) 16 s treated samples are magnified 5000 times. The d) 32 s, e) 64 s and f) 128 s treated samples are shown magnified 2000 times.

4. CONCLUSION

In this paper we report a detailed XPS study of an ultra-fast minerallization of polysiloxane coating by atmospheric pressure ambient air plasma. Energetic species from plasma react with the $Si\text{-}CH_3$ bonds on the surface transforming them into amorphous silica. The whole surface is converted after just one second of treatment. Prolonged exposure to plasma causes the minerallization to occur deeper in the film. It also causes etching resulting in patterning of the surface.

ACKNOWLEDGEMENTS

This research has been supported by Czech Science Foundation project 19-14770Y and projects LM2018097 and LM2018110 funded by Ministry of Education, Youth and Sports of Czech Republic.

REFERENCES

[1] DARMAWAN, A., HANDAYANI, D.L., SAPUTRA, R. E. Hydrophobic silica films derived from methyltriethoxysilane (MTES): Effect of pH and calcination temperature. *Applied Physics A*. [online]. 2021, vol. 127, no. 9, pp. 1-8. Available from: https://doi.org/10.1007/S00339-021-04794-1.

[2] GRÉGORI, D., BENCHENAA, I., CHAPUT, F., THÉRIAS, S., GARDETTE, J. L., LÉONARD, D., GUILLARD, S., PAROLA, S.. Mechanically stable and photocatalytically active TiO_2/SiO_2 hybrid films on flexible organic substrates. *Journal of Materials Chemistry A*. [online]. 2014, vol. 2, no. 47, pp. 20096-20104. Available from: https://doi.org/10.1039/c4ta03826f.

[3] ČERNÁK, M., ČERNÁKOVÁ, L., HUDEC, I., KOVÁČIK, D., ZAHORANOVÁ, A.. Diffuse coplanar surface barrier discharge and its applications for in-line processing of low-added-value materials. *EPJ Applied Physics*. [online]. 2009, vol. 47, no. 2, Available from: https://doi.org/10.1051/epjap/2009131.

[4] HOMOLA, T., POSPISIL, J., SHEKARGOFTAR, M., SVOBODA, T., HVOJNIK, M., GEMEINER, P., WEITER, M., DZIK, P.. Perovskite Solar Cells with Low-Cost TiO_2 Mesoporous Photoanodes Prepared by Rapid Low-Temperature (70 °C) Plasma Processing. *ACS Applied Energy Materials*. [online]. 2020, vol. 3, no. 12, pp. 12009-12018. Available from: https://doi.org/10.1021/acsaem.0c02144.

[5] O'HARE, L. A., HYNES, A., ALEXANDER, M. R.. A methodology for curve-fitting of the XPS Si 2p core level from thin siloxane coatings. *Surface and Interface Analysis*. [online]. 2007, vol. 39, no. 12-13, pp. 926-936. Available from: https://doi.org/10.1002/sia.2634.

[6] DZIK, P., VESELÝ, M., PACHOVSKÁ, M., NEUMANN-SPALLART, M., BURŠÍKOVÁ, V., HOMOLA, T.. The influence of curing methods on the physico-chemical properties of printed mesoporous titania patterns reinforced by methylsilica binder. *Catalysis Today*. [online]. 2018, vol. 313, pp. 26-32. Available from: https://doi.org/10.1016/J.CATTOD.2018.01.012.

[7] O'HARE, L. A., PARBHOO, B., LEADLEY, S. R.. Development of a methodology for XPS curve-fitting of the Si 2p core level of siloxane materials. *Surface and Interface Analysis*. [online]. 2004, vol. 36, no. 10, pp. 1427-1434. Available from: https://doi.org/10.1002/sia.1917.

[8] BAKLANOV, M. R., MARNEFFE, J. F., SHAMIRYAN, D., URBANOWICZ, A. M., SHI, H., RAKHIMOVA, T. V., HUANG, H., HO, P.S.. Plasma processing of low-k dielectrics. *Journal of Applied Physics*. [online]. 2013, vol. 113, no. 4, p. 041101. Available from: https://doi.org/10.1063/1.4765297.

PVDF/PVDF-TRFE BLENDS LOADED WITH $BATIO_3$: FROM PROCESSING TO PERFORMANCE TESTING

[1]Martin CVEK, [1]Miroslav MRLÍK, [1]Josef OSIČKA, [1]Danila GORGOL, [2]Pavel TOFEL

[1]Tomas Bata University in Zlín, University Institute, Centre of Polymer Systems, Zlín, Czech Republic, EU, cvek@utb.cz

[2]Brno University of Technology, Department of Physics, Faculty of Electrical Engineering and Communication, Brno, Czech Republic, EU

https://doi.org/10.37904/nanocon.2021.4327

Abstract

Concerns surrounding the limited supply of fossil fuels have been the subject of much debate. As of promising solutions, polymers like poly(vinylidene fluoride) (PVDF) have gained attention due to their ability to generate electrical energy from the waste mechanical vibrations. The energy harvesting and vibration sensing potential of PVDF is however limited due to its low content of electroactive β-phase, which has to be increased by indirect post-processing. Recently, a synergistic effect was found in PVDF directly blended with its trifluoroethylene copolymer (PVDF-TrFE) due to strong interfacial polarization. In this study, we aim to further increment the piezoelectric performance of PVDF/PVDF-TrFE blends by incorporating a small amount of $BaTiO_3$ nanocrystals via a facile and scalable processing route. The β-phase content was monitored using FTIR and XRD. Melt rheology experiments showed that co-blending of PVDF-TrFE as well as the addition of $BaTiO_3$ slightly increased melt viscosity and complex modulus. Despite that, rheological data suggested that developed formulations can be processed by conventional techniques intended for a large-scale production. More importantly, PVDF/PVDF-TrFE binary blends supplemented with $BaTiO_3$ are expected to exhibit superior d33 compared to conventional neat blends, which could make them highly promising for modern energy harvesting and sensor-related applications.

Keywords: Nanohybrid, vibration sensing, pvdf, blending, d33, piezoelectricity

1. INTRODUCTION

The increasing demand for the electrical energy and depletion of fossil fuels are highly concerning facts of the present time [1]. The energy harvesting systems represent an alternative concept to address this challenge by converting the waste mechanical energy, such as vibrations, air/fluid movement, body motion etc., into the useful electrical form [2]. For this purpose, piezoelectric materials are well-suited since they spontaneously polarize when subjected to a mechanical stress [1,3]. Although the energy nanogenerators based on various inorganic materials ($BaTiO_3$, ZnO, InN, GaN, CdS) or ceramics ($NaNbO_3$, $KNbO_3$) have shown high piezoelectric coefficients, they are brittle, heavy and difficult to process [4]. These drawbacks can be eliminated/reduced by using ferroelectric polymers, mainly poly(vinylidene fluoride) (PVDF) and its copolymers with trifluoroethylene (PVDF-TrFE) [5] or hexafluoropropylene (PVDF-HFP) [6].

PVDF is a semi-crystalline polymer existing in five phases (α-, β-, γ-, δ- and ε-) depending on the processing routes and post-treatment. The α- and ε-phases are easily accessible directly from polymer melt [7], they are however non-polar due to *trans-gauche* conformation self-canceling the dipoles [5]. The γ- and δ-polymorphs are polar to some extent, and hence piezo- and ferroelectric [8]. While γ-phase PVDF is hardly accessible from α-phase using spatially confined systems; the α- to δ-phase changeover can be realized by so-called electroforming process under electric fields (~150 kV/mm) [5,8]. The most desirable phase for the construction of energy harvesting and sensing devices, i.e. β-phase, can be obtained from α-phase by stretching (uniaxial or biaxial) at the elevated temperatures [9,10], and/or by polling of PVDF films under strong electric fields (above 500 kV/mm) [5, 11].

Contrary to PVDF, the copolymer PVDF-TrFE crystalizes directly into the electroactive β-phase regardless the processing conditions or post-treatment. Besides, PVDF-TrFE shows an increased chain mobility enabling to reach crystallinity of up to 90 %, compared to PVDF that attains the maximal crystallinity of ~50 % [5]. At present, the use of PVDF-TrFE is limited due to its complicated synthesis, which is reflected in high price, and relatively narrow working temperature range [12]. From these reasons, co-blending strategy was adopted to synergistically combine properties of the individual components. PVDF/PVDF-TrFE blends were found to be immiscible in the crystalline phase [13], however, their miscibility on the lamellar level was later confirmed due to co-existing mixed amorphous phase [14]. Meng et al. [5] found that minor amounts of PVDF-TrFE significantly increased β-phase content of PVDF, even beyond the rule-of-mixtures. This synergism with origins in the interfacial polarization resulted to a larger dielectric constant of PVDF/PVDF-TrFE when compared to pure components. Other strategy for enhancing the piezoelectric activity of PVDF is based on the incorporation of small amounts of various nanofillers [15,16] to maintain flexibility of the PVDF sensors, and even enhance their thermal stability [6].

Despite significant progress, a little attention has been paid to co-blended PVDF-based copolymers with inclusions of piezoelectric nanocrystals. Yang et al. [6] found that PVDF/PVDF-HFP blends loaded with dopamine-modified $BaTiO_3$ exhibited superior polarization level ($d33 = 40$ pc/N) compared to neat PVDF/PVDF-HFP blend ($d33 = 23$ pc/N), and obviously, neat PVDF ($d33 = 18$ pc/N). In this work, we aim to further increase the piezoelectric performance of PVDF/PVDF-TrFE blends by embedding the piezoelectric $BaTiO_3$ nanocrystals. The blending process was performed on a compounding device that allows the large-scale production. From the processing perspective, the rheological behavior of the binary blends as well as their $BaTiO_3$-loaded analogues was studied. Finally, the effects of PVDF-TrFE and $BaTiO_3$ on the formation of the electroactive β-phase content, and consequential d33 coefficient were investigated, implying a great potential of the developed nanocomposites.

2. EXPERIMENTAL PART

2.1. Materials

PVDF beads (CAS number: 24937-79-9, density of 1.78 g/cm^3) having the average molecular weight (M_W) of 107 000 g/mol, PVDF-TrFE (Solvene®300/P300, CAS number: 28960-88-5, density of 1.90 g/cm^3) with the composition of VDF and TrFE of 70 and 30 mol%, respectively, and $BaTiO_3$ nanopowder (CAS number: 12047-27-7, density of 6.08 g/cm^3) were purchased from Sigma-Aldrich (USA).

2.2. Fabrication of nanocomposites

The desired amounts of PVDF and PVDF-TrFE were melt-blended using a micro-compounder DSM Xplore MC15 (Xplore Instruments, The Netherlands) at 240 °C to produce binary blends with the weight ratios of 100/0 and 80/20. In parallel, analogous PVDF/PVDF-TrFE blends were fabricated containing 5 wt% of $BaTiO_3$ nanopowder. In a dosing sequence, the feedstock was gradually introduced into the compounder within 2 minutes at 25 rpm, followed by 10 minute blending at 40 rpm. A constant torque indicated homogeneity of the blends (and dispersion of the $BaTiO_3$). The product was collected in a form of filament.

Each filament was cut into pieces that were compression molded to produce films with a thickness of 0.5 mm. The calculated amounts of granules were placed into metallic molds, pre-heated for 5 minutes, and compressed with a pressure of 10 MPa for the additional 5 minutes, while temperature was set to 210 °C. Afterwards, the mold was cooled down in a controlled manner to ensure repeatability of the process.

2.3. General characterizations

The dimensions and morphology of $BaTiO_3$ was studied by transmission electron microscopy (TEM) on a JEM-2100Plus (JEOL, Japan) device equipped with a LaB6 cathode operating under the accelerating voltage of 200 kV. Prior to the analysis, the nanopowder was dispersed in acetone and dripped onto a carbon-coated TEM grid (300 mesh, Agar Scientific, UK). The TEM images were analyzed using ImageJ software (National Institutes of Health, USA).

The surface properties of $BaTiO_3$ were investigated through nitrogen adsorption/desorption isotherm collected on a volumetric gas adsorption analyzer (BELsorp Mini II, BEL, Japan) at 77K. The sample was degassed at 60 °C for 5 hours before starting the measurement. The specific surface area was determined by Brunauer-Emmett-Teller (BET) method using data points of relative pressure range from 0.05 to 0.30.

The Fourier transform infrared spectroscopy (FTIR) was performed on Nicolet 6700 (Thermo-Scientific, USA) spectrometer equipped with ATR accessory using a germanium crystal. The spectra were acquired in a wavenumber range of 4000-500 cm^{-1} with a spectral increment of 2 cm^{-1} at laboratory conditions.

The crystallographic structure of the samples was examined via X-Ray diffractions (XRD) using Miniflex 600 (Rigaku, Japan) diffractometer with a Co-Kα radiation source (λ = 1.789 Å) operating within 2θ range of 10-95° with a scan speed of 3°/min.

Rheological behavior of the molten-state samples was studied on a Physica MCR502 (Anton Paar, Austria) rheometer equipped with the CTD600 heating chamber and TC30 temperature control unit. The complex viscosity was recorded during the frequency sweep from 0.1 to 100 Hz with a constant amplitude strain of 0.05 % at the temperature of 230 °C.

2.4. Vibration sensing assembly

Vibration sensing capability of the PVDF and PVDF-TrFE films and their $BaTiO_3$-loaded analogues was investigated on circular samples of 30 mm in a diameter and a thickness of 0.5 mm. Prior to the measurement, the poling of samples was performed at 100 °C using a field of 20 kV/mm. The samples were pressed using a single-point bar; the applied mass equaled 400 grams and frequency of its oscillation was set to 1 Hz. The output signal was collected using NI-4331 resistance substitution box (National Instruments, USA) and the final d33 coefficients were calculated according to the equations described elsewhere [4,10].

3. RESULTS AND DISCUSSION

Although $BaTiO_3$ nanopowder is a commercial product, its relevant properties were studied to understand its implications on the structural and electromechanical characteristics of PVDF-based blends. **Figure 1a** shows its XRD pattern with a number of sharp peaks at 25.9° (001), 36.9° (101), 45.6° (111), 53.1° (002), 60.0° (102), 66.2° (112), 78.6° (202) and 84.3° (212) proving purity and crystalline character of the material (PDF Card No. 01-075-2121). The inset displays the size and morphology; the primary particles possessed rounded-shape with the average diameter of ~48±8 nm as determined by image analysis. The recorded physisorption isotherm curve was classified as reversible Type-II isotherm [17], and specific surface area calculated using BET method equaled 17 m^2/g.

The FTIR spectra of the PVDF and PVDF/PVDF-TrFE films are displayed in **Figure 1b**. As seen, both spectra were almost identical and demonstrated the coexistence of β- and α-phase [18]. In more details, the data showed that the addition of the PVDF-TrFE incremented β-phase content, which was reflected in a higher intensity of β-peak around 841 cm^{-1}. At the same time, the intensity of α-peaks around 763 cm^{-1} and 1210 cm^{-1} diminished confirming the desirable phase transformation [5]. Therefore, binary PVDF/PVDF-TrFE blend was expected to achieve a superior piezoelectric performance. Such trend as not revealed in the FTIR spectra of $BaTiO_3$-loaded samples due to overlapping intensity coming from the filler.

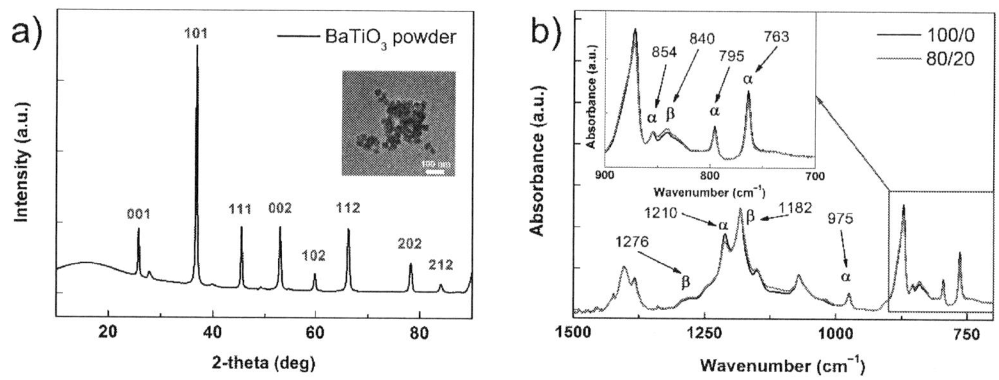

Figure 1 The XRD pattern (a) for $BaTiO_3$ nanopowder with inset TEM figure, and the FTIR spectra (b) for the neat PVDF and PVDF/PVDF-TrFE binary mixture.

The XRD analysis was used to determine the evolution of the crystalline phase in the samples. As seen in **Figure 2a**, the diffractogram ($K\alpha_1$, λ = 1.790 Å) of neat PVDF exhibited 2-theta peaks at 21.9°, 40.7° and 43.9° corresponding to β-phase, and a small peak around 42.1° related to α-phase. After the addition of PVDF-TrFE, the intensity of the β-peak signal at 21.9° dramatically increased, which demonstrated the ability of TrFE to spontaneously crystalize into this phase, regardless the post-treatment [5,10]. Moreover, the binary blend exhibited a significant development of α-phase corresponding to a double-peak at 19.4° and 20.1°; the latter is sometimes associated also to γ-phase [18]. Introducing the $BaTiO_3$ nanopowder modified the XRD spectra (**Figure 2b**) and significantly affected the intensities of the PVDF-related peaks. By comparing the spectra, it appears that the crystalline phase was developed to a lower extent, when compared to the neat analogues. This phenomenon most likely occurred due to nano-confinement effects on polymer crystallization, since $BaTiO_3$ nanoparticles represent steric obstacles restricting the mobility of the PVDF chains [19]. Despite this effect, the PVDF/PVDF-TrFE blend loaded with $BaTiO_3$ exhibited remarkably higher β-peak intensity than its counterpart based on PVDF. The rest of the XRD spectra (2-theta above 28°) resembled the pattern for $BaTiO_3$ nanopowder.

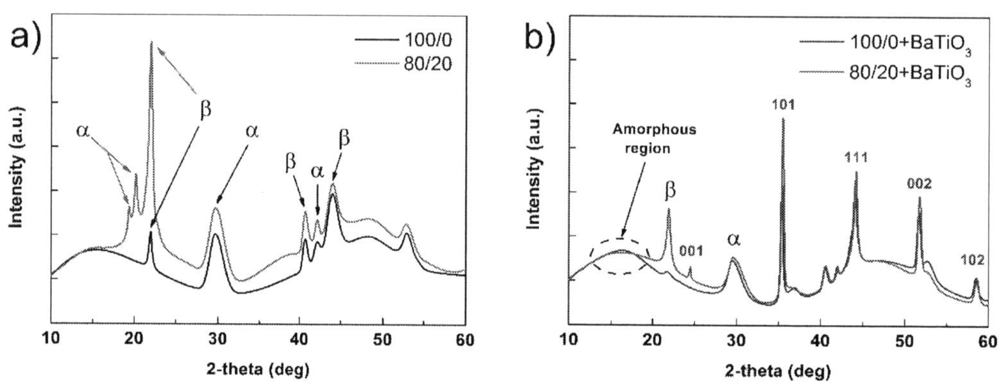

Figure 2 XRD patterns for PVDF and PVDF/PVDF-TrFE blends (a) and their $BaTiO_3$-loaded analogues (b)

Rheological study was employed to gain a useful insight to the processing ability of the PVDF-based systems and the results are shown in **Figure 3**. It was found that storage modulus, *G'*, and complex viscosity, η^*, values of PVDF/PVDF-TrFE blend were higher than those of neat PVDF, which was attributed to a higher polarity of the former providing stronger interactions, and thus more resistance to flow [20]. The addition of $BaTiO_3$ nanocrystals further increased the *G'* and η^* values due to well-known stiffening effects of rigid inorganic particles when dispersed in polymer matrix.

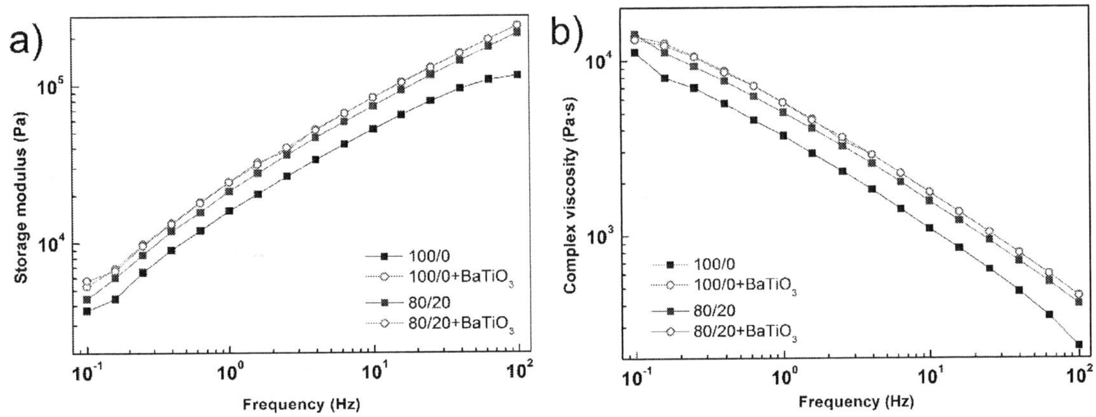

Figure 3 The melt rheology data for the PVDF and PVDF/PVDF-TrFE blends (solid symbols) and their $BaTiO_3$-loaded analogues (open symbols) represented as storage modulus (a) and complex viscosity (b).

The electromechanical capability of the PVDF-based elements can be predetermined by the β-phase content [11] and the presence of nanofillers [6], besides the other relevant factors [10]. **Table 1** displays the recorded d33 values for the investigated samples. As seen, the PVDF/PVDF-TrFE blend exhibited a superior d33 coefficient when compared to the neat PVDF due to the ability of PVDF-TrFE to crystallize directly into the β-phase [5], which correlates well with the XRD pattern (**Figure 2a**). The addition of $BaTiO_3$ nanopowder further enhanced d33 values (by 23.6 % and 41.7 %, respectively, compared to neat analogues), which stems from its inherent piezoelectric character. From these reasons, the PVDF/PVDF-TrFE with $BaTiO_3$ inclusions exhibited the highest d33, and thus, the best vibration sensing and energy harvesting capability among the samples. It is presumed that further improvements could be achieved by increasing the concentration of $BaTiO_3$ nanocrystals.

Table 1 The d33 values for the PVDF and PVDF/PVDF-TrFE films and their $BaTiO_3$-loaded analogues

Sample ID	**100/0**	**80/20**	**100/0 + $BaTiO_3$**	**80/20 + $BaTiO_3$**
d33 (pC/N)	14.8±1.7	15.1±1.4	18.3±1.1	21.4±0.7

4. CONCLUSION

In this work, the processing and energy harvesting capabilities of the neat PVDF, PVDF/PVDF-TrFE blend, and their $BaTiO_3$-loaded analogues were investigated. The materials were fabricated using industrially scalable techniques, such as compounding and compression molding. The FTIR showed that binary blend exhibited a higher content of the electroactive β-phase, without any additional interventions or post-processing. The XRD analysis confirmed the presence of β-phase via characteristic signal at 21.9°, and also, a minor content of α-phase at 19.4° and 20.1°. The intensity of these peaks was attenuated in $BaTiO_3$-loaded analogues, most probably, due to nano-confinement effects on polymer crystallization. The co-blending of PVDF-TrFE and the presence of $BaTiO_3$ nanocrystals resulted to increased G' and η^* values, but these factors synergistically enhanced d33 coefficient up to 21.4±0.7 pC/N, which is highly relevant for intended applications.

ACKNOWLEDGEMENTS

The authors gratefully acknowledge the Czech Science Foundation (grant no. 19-17457S) for the financial support. This work was also supported by the Ministry of Education, Youth and Sports of the Czech Republic - DKRVO (RP/CPS/2020/003) and (RP/CPS/2020/006).

REFERENCES

[1] WANKHADE, S.H., TIWARI, S., GAUR, A., MAITI, P. PVDF-PZT nanohybrid based nanogenerator for energy harvesting applications. *Energy Reports*. 2020, vol. 6, pp. 358-364.

[2] GAUR, A., KUMAR, C., TIWARI, S., MAITI, P. Efficient energy harvesting using processed poly(vinylidene fluoride) nanogenerator. *ACS Applied Energy Materials*. 2018, vol. 1, pp. 3019-3024.

[3] RAMADAN, K.S., SAMEOTO, D., EVOY, S. A review of piezoelectric polymers as functional materials for electromechanical transducers. *Smart Materials and Structures*. 2014, vol. 23, 033001.

[4] CHAMAKH, M.M., MRLIK, M., LEADENHAM, S., BAZANT, P., OSICKA, J., ALMAADEED, M.A., ERTURK, A.; KURITKA, I. Vibration sensing systems based on poly(vinylidene fluoride) and microwave-assisted synthesized ZnO star-like particles with controllable structural and physical properties. *Nanomaterials*. 2020, vol. 10, 2345.

[5] MENG, N., ZHU, X.J., MAO, R., REECE, M.J., BILOTTI, E. Nanoscale interfacial electroactivity in PVDF/PVDF-TrFE blended films with enhanced dielectric and ferroelectric properties. *Journal of Materials Chemistry C*. 2017, vol. 5, pp. 3296-3305.

[6] YANG, L., ZHAO, Q.Y., HOU, Y., HONG, L.Y., JI, H.L., XU, L, ZHU, K., SHEN, M., et al. Flexible polyvinylidene fluoride based nanocomposites with high and stable piezoelectric performance over a wide temperature range utilizing the strong multi-interface effect. *Composites Science and Technology*. 2019, vol. 174, pp. 33-41.

[7] LI, L., ZHANG, M.Q., RONG, M.Z., RUAN, W.H. Studies on the transformation process of PVDF from alpha to beta phase by stretching. *RSC Advances*. 2014, vol. 4, pp. 3938-3943.

[8] MARTIN, J., ZHAO, D., LENZ, T., KATSOURAS, I., DE LEEUW, D.M., STINGELIN, N. Solid-state-processing of delta-PVDF. *Materials Horizons*. 2017, vol. 4, pp. 408-414.

[9] MOHAMMADI, B., YOUSEFI, A.A., BELLAH, S.M. Effect of tensile strain rate and elongation on crystalline structure and piezoelectric properties of PVDF thin films. *Polymer Testing*. 2007, vol. 26, pp. 42-50.

[10] MRLIK, M., OSICKA, J., CVEK, M., ILCIKOVA, M., SRNEC, P., GORGOL, D., TOFEL, P. Comparative study of PVDF sheets and their sensitivity to mechanical vibrations: The role of dimensions, molecular weight, stretching and poling. *Nanomaterials*. 2021, vol. 11, 1637.

[11] SHARMA, M., MADRAS, G., BOSE, S. Process induced electroactive beta-polymorph in PVDF: Effect on dielectric and ferroelectric properties. *Physical Chemistry Chemical Physics*. 2014, vol. 16, pp. 14792-14799.

[12] FURUKAWA, T. Structure and functional properties of ferroelectric polymers. *Advances in Colloid and Interface Science*. 1997, vol. 71-72, pp. 183-208.

[13] TANAKA, H., LOVINGER, A.J., DAVIS, D.D. Miscibility and isomorphic cocrystallization in blends of ferroelectric copolymers of vinylidene fluoride and trifluoroethylene. *Journal of Polymer Science Part B-Polymer Physics*. 1990, vol. 28, pp. 2183-2198.

[14] GREGORIO, R., CHAUD, M.R., DOS SANTOS, W.N., BALDO, J.B. Miscibility and morphology of poly(vinylidene fluoride)/poly (vinylidene fluoride)-ran-trifluorethylene blends. *Journal of Applied Polymer Science*. 2002, vol. 85, pp. 1362-1369.

[15] ISSA, A.A., AL-MAADEED, M., LUYT, A.S., MRLIK, M., HASSAN, M.K. Investigation of the physico-mechanical properties of electrospun PVDF/cellulose (nano)fibers. *Journal of Applied Polymer Science*. 2016, vol. 133, 43594.

[16] SINGH, H.H., SINGH, S., KHARE, N. Design of flexible PVDF/NaNbO3/RGO nanogenerator and understanding the role of nanofillers in the output voltage signal. *Composites Science and Technology*. 2017, vol. 149, pp. 127-133.

[17] CVEK, M., KOLLAR, J., MRLIK, M., MASAR, M., SULY, P., URBANEK, M., MOSNACEK, J. Surface-initiated mechano-ATRP as a convenient tool for tuning of bidisperse magnetorheological suspensions toward extreme kinetic stability. *Polymer Chemistry*. 2021, vol. 12, pp. 5093-5105.

[18] CAI, X.M., LEI, T.P., SUN, D.H., LIN, L.W. A critical analysis of the alpha, beta and gamma phases in poly(vinylidene fluoride) using FTIR. *RSC Advances*. 2017, vol. 7, pp. 15382-15389.

[19] PAPANANOU, H., PERIVOLARI, E., CHRISSOPOULOU, K., ANASTASIADIS, S.H. Tuning polymer crystallinity via the appropriate selection of inorganic nanoadditives. *Polymer*. 2018, vol. 157, pp. 111-121.

[20] CARBONE, C., BENWADIH, M., D'AMBROGIO, G., LE, M.Q., CAPSAL, J.F., COTTINET, P.J. Influence of matrix and surfactant on piezoelectric and dielectric properties of screen-printed BaTiO3/PVDF composites. *Polymers*. 2021, vol. 13, 2166.

STUDY OF THE EFFECT OF PLASMA ON THE EFFICIENCY OF NANOFIBER FILTERS

[1]David PAVLIŇÁK, [1]Oleksandr GALMIZ, [1]Jakub KELAR, [1]Dušan KOVÁČIK, [1]Mirko ČERNÁK, [1]Michal FLEISCHNER, [1]Ali Jamaati KENARI, [1]Pavel KOSÍK, [2]Jana SVOBODOVÁ, [3]Jakub ONDRÁČEK

[1]CEPLANT, Department of Physical Electronics, Masaryk University, Brno, Czech Republic, EU, d.pavlinak@mail.muni.cz

[2]NAFIGATE Corporation a.s., Prague, Czech Republic, EU

[3]ICPF CAS, Department of Aerosol Chemistry and Physics, Prague, Czech Republic, EU

https://doi.org/10.37904/nanocon.2021.4335

Abstract

The aim of this work was to determine the effect of plasma on the substrate and on the nanofibers themselves. The PA6 nanofibrous material and the PP carrier were exposed to the effects of dielectric barrier plasma generated in air at atmospheric pressure. Changes in chemical composition, the effect on adhesions to the support substrate and especially the plasma effect on the filtration efficiency of the nanofiber material were studied.

Keywords: Nanofibers, plasma, adhesion, filtration efficiency

1. INTRODUCTION

Although nanofibers have been known to the scientific community for almost a hundred years, they have long remained on the brink of interest. It was not until the 1990s that there was a renaissance of nanofiber materials research. Thanks to their unique properties, high filtration efficiency, great free surface combined with good breathability, nanofiber materials are directly destined for filtration applications and come to the forefront of industrial interest.[1] At the beginning of the millennium, the last obstacle was solved - the efficient industrial production, and since then nothing has hindered mass commercial production. Thus today, it is common to encounter a nanofiber membrane as a component sportswear or separation filter in gas filters.[2] The recent COVID-19 pandemic, which globally started the mass production of cheap protective devices capable of trapping virus particles, also played a significant role in commercial spreading.[3] The nanofibers fabricated in a thin layer are a very fine material and cannot be easily handled. In practice, this is solved by their application on a carrier substrate, protecting them from damage. Unfortunately, the nanofiber membrane adheres very reluctantly to these substrates, so it must be properly fixed. In industrial production, this is usually solved by a lamination process, which means that the nanofibers are sophisticatedly glued to the substrate with an adhesive and covered with a protective mesh. However, this technique brings several complications and limitations to the useful properties of the resulting material. In addition, the lamination process significantly increases the cost of the production process.[4] Thus, for the production of disposable protective devices, such as respirators or protective suits, alternative possibilities of attaching nanofibers to the carrier are being sought. A relatively elegant method is the application of plasma to a support substrate, on which nanofibers are subsequently applied. If an atmospheric plasma source is used, this technology can be directly implemented in the in-line production process. It is also possible to apply plasma directly to the nanofiber network, which allows us to modify their chemical and physical properties. In this way, we obtain a functional, inexpensive nanomaterial that we can use for products with high added value. [5,6]

In this work, we focused on the use of dielectric barrier discharges in a coplanar arrangement. These plasma sources are able to generate low temperature energetic plasma over a relatively large area. Therefore, due to their construction, they are particularly suitable for processing textile materials. However, barrier discharges generated at atmospheric pressure also have their disadvantages. From a microscopic point of view, the energy in the plasma is focused into tiny channels called streamers. There is a risk that these energy channels can damage the material during treatment and create small holes inside it. In the case of minor damage to the carrier material, this is not yet a significant problem. However, if we wanted to modify the nanofibers themselves, the perforated nanofiber mesh would of course lose its unique filtering ability. In this work, we mainly focused on the preparation of nanomaterial for air filtration, and we will monitor the positive or adverse effect of plasma on filtration efficiency.

2. EXPERIMENTAL PART

2.1. Materials and methods

Nanofibers were prepared by electrospinning on a laboratory instrument Nanospider NS LAB 500 at spinning parameters: substrate feed rate 50 mm / min, distance between electrodes 140 mm, applied voltage 55 kV. Spinning was performed on a 10 cm long rotating wire electrode. A 9 % solution of polyamide 6 (PA6), (Ultramide B24, BASF) was used as spinning solution. The polymer granulate was dissolved in a mixture of acetic acid (glacial A.G., 99.8 %, PENTA s.r.o.) and formic acid (p.a. 99 %, PENTA s.r.o.) at room temperature. The support carrier material was a nonwoven polypropylene fabric (PP) weighing 17 and 80 g / m^2 (Pegatex S, PF Nonwovens Czech s.r.o.). Plasma treatment of materials was performed using a barrier plasma discharge (diffuse coplanar surface barrier discharge, DCSBD) with a coplanar arrangement of electrodes in a corundum dielectric. The operating frequency at 400 W of electrical input was approximately 15 kHz at a sinusoidal applied voltage of 20 kV (peak / peak). Due to the active cooling of the DCSBD with synthetic mineral oil, the surface temperature of the electrode did not exceed 80 °C during the experiment. The treated materials were subsequently characterized by electron microscopy (SEM, MIRA 3, Tescan Brno s.r.o.), Infrared spectroscopy (ATR-FTIR, Bruker Vertex 80V, Bruker Optics) and photoelectron spectroscopy (XPS, AXIS Supra TM, Kratos). Adhesion tests were performed by the Loop test method, top-hung adhesive tape wound into a loop (TA.XTplusC, Stable Microsystem) with a load cell of 10 N and a peel speed of 5 mm / min. The size-resolved penetration of particles through the tested materials was measured by a filter testing system developed at ICP CAS (Czech Republic). Samples were tested for size-resolved filtration efficiency at a flow rate of 8.7 l/min (for a test sample with a diameter of 47 mm) guaranteeing a forward speed of 10.6 cm/s. This frontal velocity corresponds to a flow rate of 95 l/min (with an average respirator surface of 150 cm^2) - according to EN 149 standards.

2.2. Influence of plasma treatment on the support substrate and increase of adhesive energy

First of all, we studied the effect of plasma on the nonwoven PP fabric itself. The fabric was treated at 400 W for 10 and 60 s in the open air. Due to plasmachemical reactions, we expected an increase in the surface of the PP fabric of new polar functional groups (especially oxygen-rich groups such as -OH, C=O, COO). The presence of carbonyl and hydroxyl groups was confirmed by infrared spectroscopy, where we found small increase of vibrations in the region of 1630 and 1725 cm^{-1}.[7,8] These vibrations do not occur in the reference material. The intensity of these vibrations increases with the time of treatment. Similarly, we observed an increase in oxygen content using EDX and XPS spectroscopy. The values are given in **Table 1**. However, as the treatment time increases, so does the damage to the PP material itself. This damage is observed by SEM and manifests itself in the formation of small craters and bulges on the surface of the PP fiber (see **Figure 1**). For this reason, we chose the mean golden way for further analyzes, i.e. a treatment time of 10 s. A more detailed C 1s XPS analysis showed the presence of carbonyl, hydroxyl and carboxyl groups after plasma treatment of the material and is consistent with the results from ATR-FTIR. We obtain similar results even after plasma processing the nanofibers themselves. Here, strong oxidative effects and morphological changes occur. The values are again given in **Table 1**.

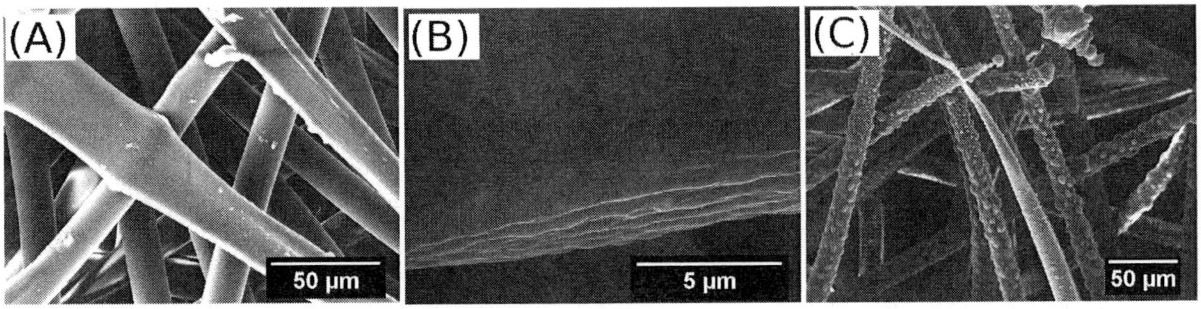

Figure 1 SEM Image of plasma treated PP substrate. (A) Reference PP - magnification 1 500x; (B) PP plasma treatment 10 s - detail on fiber surface, magnification 50 000x; (C) PP plasma treatment 60 s - magnification 1000x

Table 1 Elemental composition of reference a plasma treated substrate and nanofibers measured by EDX and XPS spectroscopy

Sample	EDX (wt. %)			XPS (at.%)		
	C	N	O	C	N	O
PP ref	97.4	0.0	2.6	98.6	0.0	1.4
PP plasma 10s	95.4	0.0	4.6	83.6	2.2	12.5
PA6 NF ref	54.4	30.3	15.4	81.6	9.0	9.4
PA6 NF plasma 60 s	52.4	28.3	19.3	77.6	9.2	13.2

Of interest is the analysis of adhesion energy performed using a peel test. This test is normally performed using a rotating wheel to which the test sample is attached. A standardized adhesive tape is then adhered to the sample. The adhesive tape is then gradually mechanically torn off the sample. **Figure 2** shows the result of the peel test. While the gradual tear-off of the untreated substrate is measurable, after the plasma treatment the adhesive force is too great, and the sample is damaged. The fibers of the PP substrate remain firmly attached directly to the surface of the tape, and in this way, we measure the cohesive forces inside the PP substrate itself (**Figure 2**). This method is also not suitable for determining the adhesion forces of nanofibers to a substrate. For this reason, we have developed a new type of method for determining the so-called loop test. The innovation of the method consists in omitting the rotating wheel, the samples are attached to a horizontal base, and the twisting of the standardized tape into a loop. The loop is then gradually lowered from above towards the sample, it is carefully attached to it and then it starts to tear away from it in the opposite movement. The device then tears the layers of the sample apart at a constant speed and at the same time measures the dependence of the adhesion force at a distance from the zero position of the instrument. Finally, it integrates this dependency to get the mechanical work done $W_{adh.}$ (N·mm). The adhesive energy is determined here using the following formula:

$$E_{adh.} = \frac{W_{adh.}}{S} \; ; (J \cdot m^{-2})$$

where the S is sample area (in our case square 2x2 cm). The method of determining the adhesive energy is inspired by the work of Rombaldoni et al.[9]

The resulting data for treated and untreated PP substrate are shown in **Table 2**. These results are of good order in correlation with our previous publication [5], where a different type of test was performed to determine the adhesive energy. Specifically, in this work the tearing test was proved of 180° sample attachment in mechanical jaws. Because the results of both methods are mutually comparable, we can say that the determination of adhesion energy using the innovative Loop test is valid and applicable for the determination of adhesion energy. This method can determine both the increase in the adhesion of the treated substrate to the tape and the adhesion of the nanofibers to the support substrate.

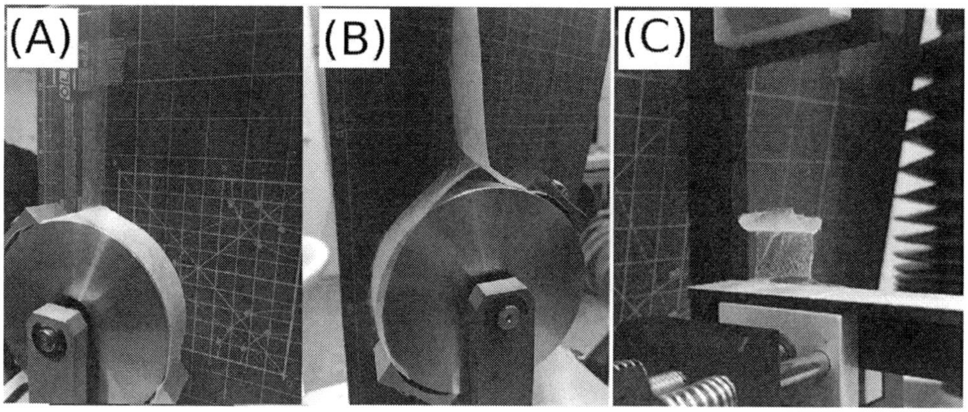

Figure 2 Peel test of plasma untreated (A) or treated 10 s PP substrate (B) and the Loop test of plasma treated 10 s PP substrate with deposited PA6 nanofibers (C)

Table 2 Determination of the average work required to tear nanofiber layer from the plasma treated, or untreated polypropylene substrate

Sample	Loop test		
	$W_{adh.}$ (N·mm)	Maximal force of adhesion (mN)	$E_{adh.}$ (J·m^{-2})
NF PA6 on PP ref	0.344	40	0.860
NF PA6 on plasma PP 10s	0.959	154	2.397

For creating high-quality adhesive joints (joint between textile fiber and other material) the material composition of the polymer fibers of the substrate itself has a great influence. The following sequence of the quality of the adhesive joint of basic types of textile fibers to various ones is investigated types of binders was experimental given for the most common textile materials [10].:

Viscose > polyamide > cotton > polyester > wool > polypropylene

Thus, the strongest adhesive joints are to viscose, the weakest for polypropylene substrate. This can be clearly seen also in the measured values of adhesion energy, when polyamide nanofibers have a very low adhesion to the polypropylene substrate. The adhesion force is influenced, among other things, by the morphology and chemical composition of the fibers surface. In practice, therefore, the fabrics are in industrial processes treated with chemical binders. [11,12] The effect of plasma on adhesion is very significant here. At almost all investigated values, we measured approximately threefold improvement in the adhesion of nanofibers to the substrate.

The exact causes of the increase in adhesion after exposure to plasma are not yet known. However, two probable hypotheses can be established from the above-mentioned analyses.

- Plasma causes unevenness on the surface of the substrate fibers - nanofibers are better mechanically captured to substrate

- Oxidizing effects of plasma - adhesion energy could be increased due to the presence of reactive polar groups, which can stimulate the formation of new chemical bonds or physical interactions between materials (e.g. hydrogen bonds, electrostatic forces)

2.3. Influence of plasma treated materials on filtration efficiency

Table 3 shows the values of filtration efficiency depending on the treatment of the material in the plasma. As can be seen here, the PP substrate alone does not achieve filtration efficiency or the FFP1 category. In contrast, the nanofiber membrane achieves a filtration efficiency of 97 % and therefore clearly meets the conditions for filtration category FPP1. from the measured values it is clear that the plasma does not have a negative effect on the filtration of the material and that both in the samples where only the carrier PP substrate was treated and sample where the nanofibers were also plasma treated.

Table 3 Filtration efficiency test

Sample		Filtration efficiency test	
		Filtration efficiency (%)	**Filtration class (EN 149)**
1	PP ref	<7	<FFP1
2	PP plasma (10 s)	<7	<FFP1
3	PP ref + PA6 NF	97	FFP3
4	PP plasma + PA6 NF	97	FFP3
5	PP ref + PA6 NF (plasma)	91	FFP3(2)
6	PP plasma + PA6 NF (plasma)	96	FFP3

During the analysis, we noticed that some samples where the carrier substrate was not treated with plasma abruptly decreased in filtration efficiency (see sample 5). This phenomenon appears to be random in this set of samples. We explain it that due to the poor adhesion of used materials. It is possible that between the nanofiber layer and substrate, the small air pockets can be formed through the manipulation with sample. It can cause that the testing gas can penetrate through the material without effective filtration which results in overall reduction in filtration efficiency up to limit of FFP2.

3. CONCLUSION

In conclusion, the application of plasma to the support substrate leads to improved adhesion properties to the nanofiber membrane. Plasma can also be applied to the nanofibers themselves, which can modify their morphology and their chemical composition. No negative effect of plasma on the filtration efficiency of the material was observed.

ACKNOWLEDGEMENTS

This work was supported by project In-line plasma technology for a large throughput manufacturing of advanced nanofiber-based filter fabrics (No. E01-14) funded by the European Light Industries Innovation and Technology initiative, which is funded by the European Union COSME Program for Competitiveness of Companies and Small and Medium Enterprises.

REFERENCES

[1] BARHOUM, A., RASOULI R., YOUSEFZADEH, M., RAHIER, H., BECHELANY, M. *Nanofiber Technology: History and Developments.* In: Barhoum A., Bechelany M., Makhlouf A. (eds) Handbook of Nanofibers, Cham, Springer, 2018

[2] TEO, W-E, INAI, R., RAMAKRISHNA, S. Technological advances in electrospinning of nanofibers. *Science and technology of advanced materials*. [online]. 2011, vol.12, no. 1. Available from: https://doi.org/10.1088/1468-6996/12/1/013002

[3] ESSA, W. K., et al. Nanofiber-Based Face Masks and Respirators as COVID-19 Protection: A Review. *Membranes*. [online]. 2021, vol. 11, no. 4, p. 250. Available from: https://doi.org/10.3390/membranes11040250.

[4] YALCINKAYA, F.; HRUZA, J. Effect of laminating pressure on polymeric multilayer nanofibrous membranes for liquid filtration. *Nanomaterials*. [online]. 2018, vol. 8, no. 5, p. 272. Available from: https://doi.org/10.3390/nano8050272.

[5] PAVLIŇÁK, D. et al. Application of dielectric barrier plasma treatment in the nanofiber processing. *Materials Today Communications*. [online]. 2018, vol. 16, pp. 330-338. Available from: https://doi.org/10.1016/j.mtcomm.2018.07.010

[6] PAVLIŇÁK, D. et al. Functionalisation and pore size control of electrospun PA6 nanofibres using a microwave jet plasma. *Polymer degradation and stability*. [online]. 2014, vol. 108, pp. 48-55. Available from: https://doi.org/10.1016/j.polymdegradstab.2014.05.027

[7] KUPSTOV, A.H., ZHIZIN G.N. Handbook of Fourier transform Raman and infrared spectra of polymers. Amsterdam: Elsevier, 1998

[8] BARBES, L., C. RADULESCU, STIHI, C. ATR-FTIR spectrometry characterisation of polymeric materials. *Romanian Reports in Physics*. 2014, vol. 66, no. 3, pp. 765-777.

[9] ROMBALDONI F. et al. Adhesion enhancement of electrospun nanofiber mats to polypropylene nonwoven fabric by low-temperature oxygen plasma treatment. *Surface and Coatings Technology*. [online]. 2013, vol. 216, pp. 178-84. Available from: https://doi.org/10.1016/j.surfcoat.2012.11.056.

[10] KRCMA R. *Teorie netkaných textilií*. Liberec: Vysoká škola strojní a textilní, 1985.

[11] RUSSELL, S. J. Handbook of nonwovens. Cambridge: Woodhead publishing limited, 2006.

[12] SHERWOOD, N. Binders for nonwoven fabrics. *Industrial & Engineering Chemistry*. [online]. 1959, vol. 51, no. 8, pp. 907-910. Available from: https://doi.org/10.1021/ie50596a026.

ALTERNATING CURRENT ELECTROSPINNING OF PA 6 USING ADDITIVES IN FORM OF OXOACIDS

*Pavel HOLEC, Tomas KALOUS, Pavel POKORNY, Ondrej BATKA, Josef SKRIVANEK

*Technical University of Liberec, Liberec, Czech Republic, EU, *pavel.holec@tul.cz*

https://doi.org/10.37904/nanocon.2021.4329

Abstract

Nanofibrous materials are currently in transition between laboratory and industrial-scale production. The main factor that hinders transition speed is relatively low productivity compared to widely used technologies such as melt-blown or force-spinning. One possible solution is to use a different technology than a typical direct current electrospinning (DC). It is possible to electrospin solutions using another high-voltage source - providing alternating current energy (AC). The potential of AC technology lies in increased productivity of the process and the possibility of electrospinning solutions containing a high dose of additives (15-30 % wt.) such as hydroxyapatite, graphene, active carbon, ceramics, clay particle, aluminium or metallic oxides. When using AC technology, some problems can arise. The main one is the transition of a working solution that spins well in small laboratory DC systems or industrial-scale Nanospider™ machines driven by direct current. It is usually required to modify the solution to improve the spinnability of the solution in the AC system. In our case, we experienced the problem in the spinnability of PA 6 when using the so-called "overflow electrode" that enables the spinning of polymer for a more extended time (multiple hours). The initial solution of PA 6 did not produce a nanofibrous layer but technologically unprocessable small flakes. We show that minor modification of solutions using oxoacids can lead to significant improvement of the nanofibrous product. The results are presented in the form of SEM images and histograms showing the change in diameters of nanofibers.

Keywords: Electrospinning, polyamide 6, nanofibers, oxoacids

1. INTRODUCTION

Materials with high added value are currently increasing in demand. Currently, there is a wide range of technologies that can provide such materials. In general, we can divide the technologies as mentioned above into two primary groups. The first one contains systems initially designed to produce fibers with diameters in the range of micrometers, such as melt-blown [1], which are currently modified to provide fibers with decreased diameter, leading to approximately 500 nm [2]. The second group is represented by the processes that are directly aimed at the production of nanofibers. The most well-known approach for the production of nanofibers is electrospinning [3]. Since then, most experiments/systems have been driven by direct current power sources [4]. Recently, the so-called alternating current (AC) electrospinning has been discovered [5]. Initially, experiments with AC electrospinning contained electrically active collectors similar to the DC systems [6]. The modification of the AC system was done by our group, removing the electrically active collector [7]. This seemingly small change brings a whole new set of possibilities for the concept of electrospinning. By its nature, AC electrospinning produces a nanofibrous structure that spontaneously creates a tube-like structure that contains almost no charge due to a constant change in polarity, leading to self-discharging. This structure is subsequently blown away from the electrode (to a specific place where the electrode is aimed at) by so-called ionic wind [8]. Without the necessity to use the electrically active collector (for the neutralization of charge), fibers can be used to coat any object of desired shape [7]. It is necessary to mention that the AC variant of electrospinning also brings a new challenge that can be represented by the required (often) modification of electrospun solution. The solution that works well on the DC system can manifest difficulties during the transfer to the AC electrospinning due to its nonstatic nature. The rapid change of polarity and constant change of the voltage gives only a limited time [9] for actual electrospinning, and polymeric solution needs to be able to offset the spinning process in this limited window. This paper presents an example of such modification based on the spinning of Polyamide 6 (PA 6). We show that adding multiple oxoacids can lead to a major improvement of the spinning process and provide nanofibrous materials of good quality. The level of quality is dependant on the used oxoacid.

2. METHODS

The technologically sustainable long-term AC electrospinning of Polyamide 6 can be arranged using a pump with an overflowing effect. This method provides more solutions than can be electrospun at a given time, so the whole body of the electrode is always coated with the solution. The new solution flows down - back to the reservoir. Without this overflow effect, the dry parts of the electrode are slowly covered by a growing mass of the dried polymer. This leads to diminishing quality of the electrospinning and subsequent end of the stable process.

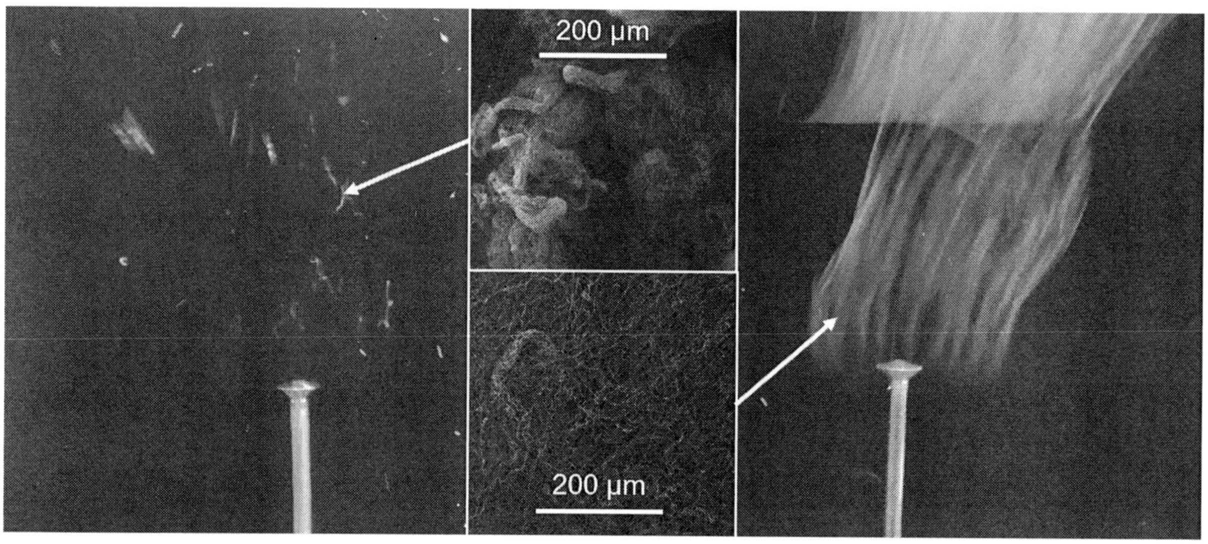

Figure 1 AC electrospinning of basic PA 6 solution (left) and one enhanced by an oxoacid (right) with corresponding details of the nanofibres

Materials and preparation of solutions

The experiments used the basic solution of PA 6 - Ultramid B27 (M_w 66360 g/mol, D=3.2) provided by the BASF company. The polymer was dissolved in a mixture of two organic acids, namely formic acid (Penta Chemicals) and acetic acid (Penta Chemicals), mixed in weight ratio 1:1. The basic solution contained 10 % (wt.) PA 6. For the following experiments with additives, the set of oxoacids that expanded our previous research [10] were chosen: nitric acid (Penta Chemicals), phosphoric acid (Penta Chemicals), and p-toluenesulfonic acid monohydrate (Sigma Aldrich).

Table 1 List of additives successfully improving the quality of AC electrospinning of PA 6

Acid	Formula	Concentration [mol/l]
nitric	HNO_3	0.441
phosphoric	H_3PO_4	0.863
p-toluenesulfonic	$C_6H_4(CH_3)SO_3H$	0.298

Table 1 shows the various additives and their respective concentration in the basic solution of PA 6. It is worth mentioning that the dose of additives was always very small. Only a few drops are required to modify the relatively high volume of a basic solution (for example, 300 ml). The values in **Table 1** represent the optimal dose of the additive. A smaller dose does not sufficiently affect the spinning quality. Exceeding the optimal value also depresses the production of quality materials.

Electrospinning system

The quality of electrospinning was ensured by the use of a custom-made screw pump [7]. The fresh polymer was constantly overflown from the top of the electrode and over the edge, where maximal electrical intensity can be found. The signal (type and frequency) was provided by the Owon AG 1022 and transformed to high voltage values by TREK 50/20 high-voltage amplifier. All experiments were performed using the same 42 kV and 50 Hz settings.

3. RESULTS

The solvents were electrospun using the screw-pump mentioned above and high voltage power source TREK. Each sample was collected using a PMMA (polymethyl methacrylate) plate. Samples were examined using an SEM microscope.

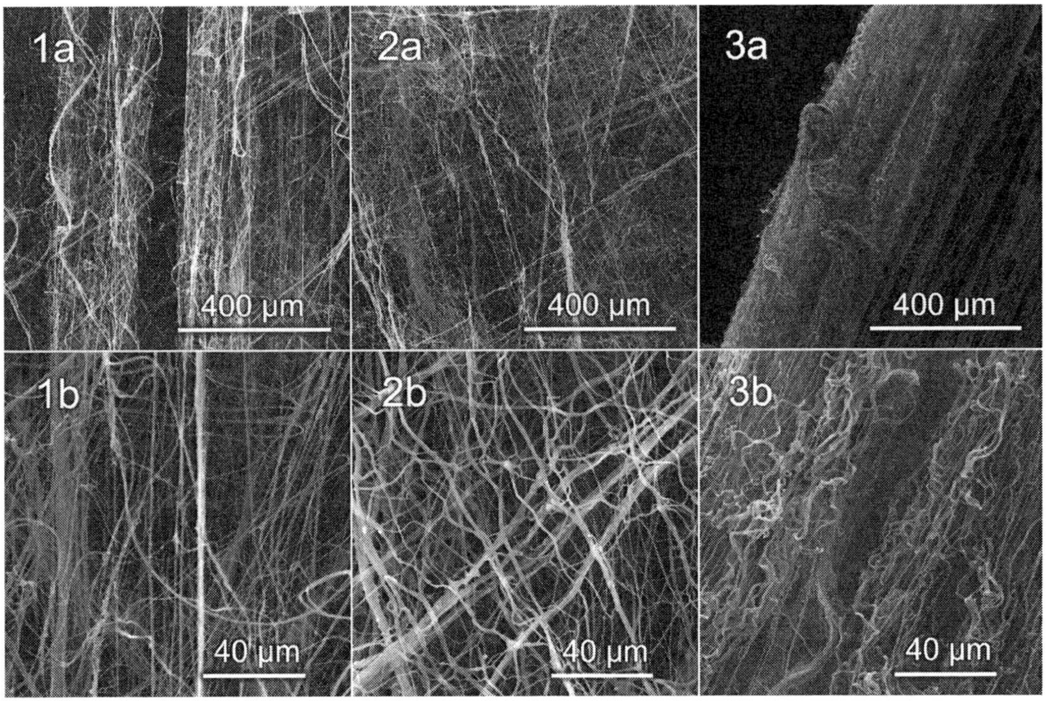

Figure 2 Nanofibrous products provided by basic solution of PA 6 enhanced by nitric (1a, 1b), phosphoric (2a, 2b) and p-toluenesulfonic acid (3a, 3b)

The materials produced using enhanced solutions (optimal dose) produced the process-able fibrous layers in all cases (**Figure 2**). The morphology and diameters of the fibers differ, and they are shown in histograms (**Figure 3**). Each oxoacid created a different structure of the fibers. Compared to previously tested oxoacids (sulfuric and methanesulfonic) published in [10], the quality of the fibrous product was decreased. The relatively significant differences in the diameters of nanofibers were caused by the nature of AC electrospinning that does not provide constant energy through the whole half-wave of the signal.

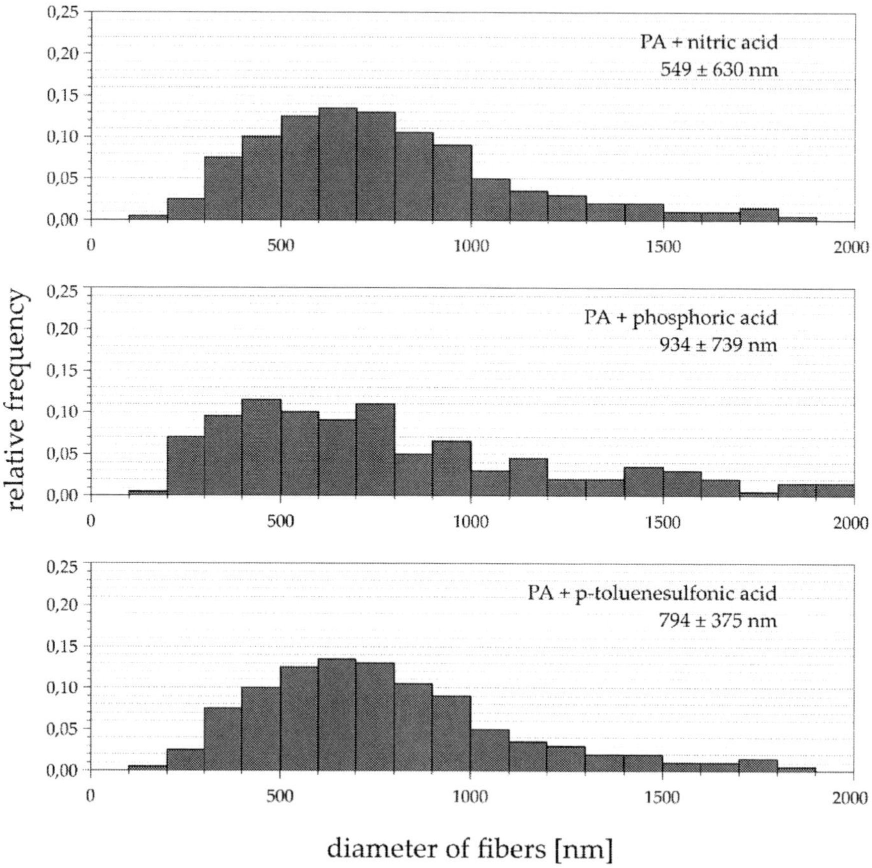

Figure 3 Histograms of fibrous diameters for each modified PA6 solution. Figures represent the solution enhanced by nitric, phosphoric and p-toluenesulfonic acid with a corresponding average diameter and standard deviation.

The positive influence on the spinnability of Polyamide 6 with additive in the form of oxoacids might be explained by the physical bonding of the carboxyl groups of the acids to the amide groups of the polymer. Each oxoacid can dissociate in the basic solution and create ions in the solution. Dissociated acid particles are probably able to interact with the electrically imbalanced CONH group in amides. Although no covalent bond is created, the dissociated particles can influence the properties of polyamide macromolecules through relatively weak but numerous interactions between the amide group and dissociated ions. This might lead to a virtual increase in molecular weight, number of entanglements or improve macromolecule stability in changing alternating current electric fields. This behavior was not observed when using different chemicals as additives that change the physical properties of solutions, such as viscosity or conductivity (salts or surfactants).

4. CONCLUSION

The presented work shows that oxoacids can improve the spinnability of PA 6 solutions for AC electrospinning systems and can lead to a fibrous material with acceptable diameters and structures. We have shown that acids that can dissociate can improve AC electrospinning when using an overflowing system (improving the sustainability of the process). The theory explaining this behavior was also mentioned. Further research could improve the spinnability of various polymers using AC field and optimization of fibrous diameters and morphology of produced nanofibrous layers and bulky structures.

ACKNOWLEDGEMENTS

I want to acknowledge the support given by the Technical University of Liberec. Especially the Department of Textile Machine Design and Department of Nonwovens and Nanofibrous Materials for the financial support of the research

REFERENCES

[1] DRABEK, J., ZATLOUKAL, M. Meltblown technology for production of polymeric microfibers/nanofibers: A review. *Physics of Fluids.* 2019, vol. 31, 091301.

[2] ELLISON, J., PHATAK, A., GILES, D., MACOSKO, CH., BATES, F. Melt blown nanofibers: Fiber diameter distribution and onset of fiber breakup. *Polymer.* 2007, vol. 48, pp. 3306-3316.

[3] RAMAKRISHNA, S., FUJIHARA, K., TEO, W., LIM, T., MA, Z. *An Introduction to Electrospinning and Nanofibers.* Singapore: World Scientific Publishing Co, 2005. ISBN 981-256-415-2.

[4] DOSHI, J., RENEKKER D.; Electrospinning process and application of electrospun fibers. *Journal of electrostatics* 1995, vol. 35, pp. 151-160.

[5] KESSICK, R., FENN, J., TEPPER, G. The Use of AC Potentials in Electrospraying and Electrospinning Processes. *Polymer.* 2004, vol. 45, pp. 2981-2984.

[6] MAHESHWARI, S., CHANG, H.-C. Assembly of Multi-Stranded Nanofiber Threads through AC Electrospinning. *Adv. Mater.* 2009, vol. 21, pp. 349-354.

[7] POKORNY, P., KOSTAKOVA, E., SANETRNIK, F., MIKES, P., CHVOJKA, J., KALOUS, T., BILEK, M., PEJCHAR, K., VALTERA, J., LUKAS, D. Effective AC Needleless and Collectorless Electrospinning for Yarn Production. *Phys. Chem. Chem. Phys.* 2014, vol. 16, pp. 26816-26822.

[8] WANG, J., ZHU, T., CAI, Y., ZHANG, J., WANG, J. Review on the recent development of corona wind and its application in heat transfer enhancement. *International Journal of Heat and Mass Transfer.* 2020, vol. 152.

[9] FARKAS, B., BALOGH, A., CSELKÓ, R., MOLNÁR, K., FARKAS, A., BORBÁR, E., MAROSI, G., NAGY, Z. Corona alternating current electrospinning: A combined approach for increasing the productivity of electrospinning. *International Journal of Pharmaceutics*. 2019, vol. 561, pp. 219-227.

[10] KALOUS, T., HOLEC, P., JIRKOVEC, R., LUKAS, D., CHVOJKA, J. Improved spinnability of PA 6 solutions using AC electrospinning. *Materials Letters.* 2021, vol. 283.

ROLL-TO-ROLL CLEANING AND ACTIVATION OF ULTRA-THIN FLEXIBLE GLASS USING ATMOSPHERIC-PRESSURE PLASMA IN AMBIENT AIR

Slavomír SIHELNÍK, Richard KRUMPOLEC, Monika STUPAVSKÁ, Jianyu FENG, Eva KOSOVÁ, Jakub KELAR, Mirko ČERNÁK, Dušan KOVÁČIK

Masaryk University, Faculty of Science, Department of Physical Electronics, CEPLANT, Brno, Czech Republic, EU, sihelnik@mail.muni.cz

https://doi.org/10.37904/nanocon.2021.4344

Abstract

Flexible electronics are nowadays manufactured on substrates made particularly of plastics and paper. In demand for better optical and chemical properties, glass is also used in its ultra-thin flexible form (UTFG). Digital ink-jet printing is an attractive method for deposition of conductive precursors on flexible substrates, mainly due to its ability to be incorporated into roll-to-roll systems for massive production. Plasma cleaning and activation is proposed here as a dry, efficient, and controllable method for pre-treatment of glass substrate before deposition of conductive inks. Tested plasma sources were chosen with respect to the specific thermomechanical properties of UTFG and requirements of large area printing on flexible substrates. Both diffuse coplanar surface barrier discharge (DCSBD) and industrial corona achieved a significant wetting improvement of UTFG, which results from modification of properties governed by surface nanolayer. Water contact angle (WCA) measurement and X-ray photoelectron spectroscopy revealed strong activation of UTFG surface due to incorporation of oxygen-based polar groups. The ageing of plasma-treated glass, monitored for one week, showed better stability of the DCSBD plasma-treated UTFG surface. We observed the WCA recovery from 7 % to 38 % in respect to the initial value measured on the untreated UTFG (68°). The impact of plasma on the morphology of the glass surface was analysed using atomic force microscopy.

Keywords: Flexible glass, roll-to-roll, cleaning, plasma activation, DCSBD, industrial corona

1. INTRODUCTION

Surface modification of glass is a widely used approach of tailoring its properties when interference with its bulk is not desirable or necessary. Tools for glass surface modifications could consist of any state of matter; therefore even plasma treatment has been used over decades, mainly in laboratory environments. Application of plasma treatment could be especially advisable in processes when desired properties are governed by a surface nano-layer [1].

The contemporary flexible electronics are usually made on plastic substrates using coating and printing methods. Paper and flexible glass are other substrate materials researched and applied in this sphere, connected to fields like photovoltaics, biophysics and optoelectronics [2]. Ultra-thin flexible glass (UTFG) is a novel material, and so far, only a few studies reported the application of plasma treatment to its surface. UTFG's potential and its advantageous properties were shown in the studies of indoor photovoltaics, wireless antennas and organic light-emitting diodes (OLED) [3]. In microelectronics and printed flexible electronics manufactured on UTFG, the bonding of thin structures deposited using printed conductive inks is extremely important, and it can be improved by cleaning and chemical activation of the glass surface before deposition.

The exposure of glass surface to ambient air conditions, which causes adsorbing of air pollutants to it, plays an essential role on the scale of molecular size. The naturally adsorbed layer consisting of water and organic contamination with a thickness of several nanometers could hinder processes based on adding material to the glass surface. Exposure of glass surface to active species included in plasma seems to be a potential method for fusion of cleaning and activation into one processing step, which can be realized immediately before the deposition of coating, thin film or layered structure. Plasma cleaning of the glass surface is provided through free electrons with energy high enough to break hydrogen bonds of adhered layer contaminating surface [4]. Active species produced by plasma interaction with surrounding gas can be incorporated onto the plasma-treated surface during the exposure. Activation of the UTFG surface by promoting functional groups generated in plasma could facilitate the coupling of precursor molecules in ink-jet printing processes.

The choice of a useful plasma system that can be applied on materials like UTFG is limited by its thermal and mechanical sensitivity. Different non-thermal plasma sources need to be tested and compared to obtain desirable modifications that could be suitable for improving concrete applications on UTFG. The impact of parameters like discharge regime, exposure time or distance are tested to achieve better efficiency, uniformity and stability of the effect, which can be analyzed through wetting measurements and XPS analyses of chemical composition. The influence of plasma interaction to morphology of treated glass surface is analyzed with AFM measurements of roughness.

2. EXPERIMENT

UTFG AF32 sheets with a thickness of 30 μm and a size of 101.6 × 101.6 mm were purchased from Schott AG. AF32 samples of this original waffer size were used only in the experiments in planar configuration. Other UTFG samples were been cut to form of bands 12 × 80 mm in the order to prevent conveying errors on R2R systems and thus improve the achieved effect. A part of experiments focused on impact of plasma to surface morphology was realised on common float soda-lime glass (SLG) from Polartherm-Flachglas GmbH with 2.8 mm thickness and a size of 20 × 80 mm. All of the experiments were carried on the side of glass without tin content. UTFG samples were only blown over to eliminate the presence of residuals from laser cutting. Samples of SLG were contamined by wet cutting process, so they were precleaned in ultrasonic cleaner with acetone, isopropyl alcohol and deinoised (DI) water.

Table 1 Setup of applied devices using plasma treatment in ambient air at atmospheric pressure

Device	Geometry	Dimensions (mm)	Power (W)	Areal power density (W/cm^2)	Volume power density (W/cm^3)
DCSBD "Air-pillow"	planar	80 x 195	400	2.57	85.47
Roll-To-Roll DCSBD	curved	80 x 195	400	2.57	85.47
Roll-To-Roll VDBD	curved	75 x 210	400	2.54	25.40

Two basic configurations of dielectric barrier discharge for generating atmospheric pressure plasma in air were been compared. Variations of the used devices employing diffuse coplanar surface barrier discharge (DCSBD) and volume dielectric barrier discharge (VDBD) units and their parameters are summarised in **Table 1.** Devices using DCSBD plasma units provided by ROPLASS, s. r. o. were tesed in two different geometries, indeed both are respecting demand of contactless treatment on a large area. Experiments realised on small samples with planar DCSBD unit were carried with holder with spacers ensuring distance 0.3 mm between samples and ceramic that is advisable for effective treatment. DCSBD "Air-pillow" is equipped with pneumatic bearings that enables contactless leading of discharge unit over smooth planes of unlimitted size [1]. Closed-chamber Roll-To-Roll DCSBD reactor (TYPE III) employs concavely curved discharge units and is designed for treatment of flexible materials using various working gas [5]. Compared commercial device manufactured by Ahlbrandt System GmbH is using so-called industrial corona generating VDBD plasma of comparable areal power density as DCSBD [6].

The static WCA measurement using 1 μl of DI water was realised with See System by Advex Instruments, s. r. o. The XPS analysis was done with ESCALAB 250 Xi analytic platform by Thermo Fisher

Scientific, using Al Kα source and Rowland circle monochromator. Spectra calibration, processing and fitting routines were accomplished in Avantage software. AFM measurements were carried on Ntegra Prima by NT-MDT in semi-contact mode using silicone tips HA_NC with curvature radius under 10 nm. Surface roughness was evaluated from 5 measurements on 5 × 5 µm areas that were scanned with 1 Hz frequency.

3. RESULTS AND DISCUSSION

The comparision of water contact angles (WCA) measured repeatedly during one week on UTFG treated 1 and 3 s with tested plasma sources is presented in **Figure 1.** The graph illustrates ageing study of samples stored in ambient atmosphere, which was particulary carried out on the first day of experiment-after 0, 2, and 8 hours. The rest of WCAs were measured 1, 3, and 7 days later. The WCA (68 ± 2)° is the initial value of untreated 30 µm thick AF32 obtained independently on the analysed side. The most significant wettability improvement of UTFG surface immediately after treatments was achieved even after the shortest applied exposure time 1 s using both roll-to-roll plasma systems. The lowest WCA drop (4.6 ± 0.7)° acquired after 1 s exposure with the industrial corona was matched with DCSBD "Air-pillow" when 3 s exposure was applied. More significant and uniform effect on DCSBD plasma treated UTFG surface was observed with 1 s exposure using curved unit, when the lowest achieved error affected the measured values resulting in (5.0 ± 0.5)°. The treatment time of 3 s with R2R DCSBD reactor proved less effective, which was caused by charging of the UTFG sample from plasma and consequent impact on its leading under the curved DCSBD ceramic, which was electricaly charged, too.

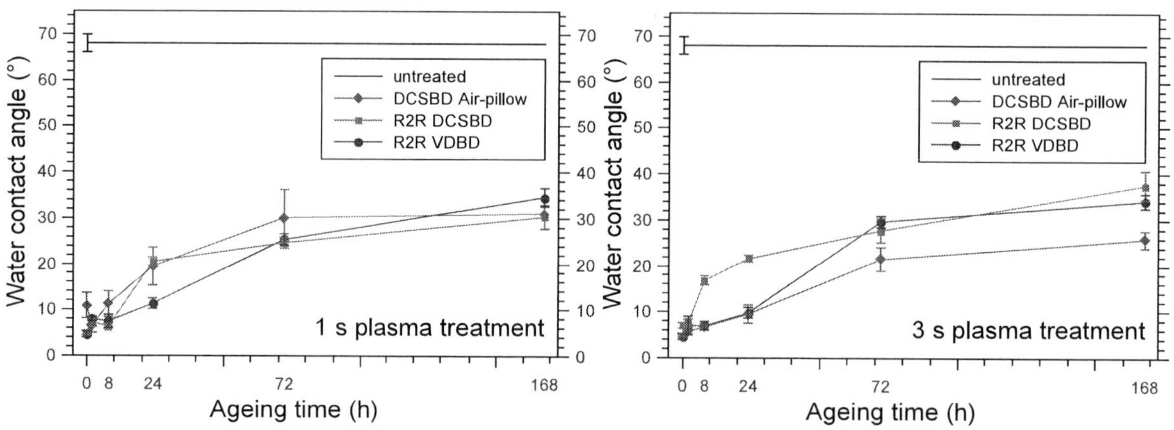

Figure 1 Wettability of ultra-thin flexible glass (UTFG) surface treated with plasma

Hydrophilic effect achieved with roll-to-roll systems was relatively stable in the first 8 hours after exposures of 1 s. The results of experiments applying 3 s treatment time followed similar tendency, with an exception of values from the R2R DCSBD. Faster recovery of WCAs measured on UTFG treated 3 s with R2R DCSBD corresponds to the lowering of its efficiency by prolonging treatment time, which was reported immediately after treatment. Stability in the first 8 hours for shorter treatment time is a satisfying result taking into account the duration of one work shift in the industry. Longer storage of plasma-treated UTFG demonstrated gradually recovering tendency, which was more significant in the first 3 days. Afterwards, in the range of 7 days, WCAs have reached a value of only 50 % of the untreated UTFG value for all configurations.

The present study using roll-to-roll plasma treatement with focus on improved conveying of UTFG samples resulted in optimalization of treatment time with curved DCSBD to 1 s, which is closer to industrial requirements, compared to 5 s from our preliminary study aimed at contactless large area treatment [1]. The reported saturation of WCAs measured after 3 days of ageing is in good accordance with the studies using DCSBD on common float glass with exposure times 3 s [7]. However, the results presented here vary in the initial value of untreated surface, which is much higher, hence changes and stability are relatively more markable on UTFG. Interesting is that minimum and saturation levels are similar independently on the used glass substrate and its initial WCA value. The essence of wettability change of glass surface induced by plasma modification seems to be independent from its initial surface morphology or its eventual alteration by interaction with plasma.

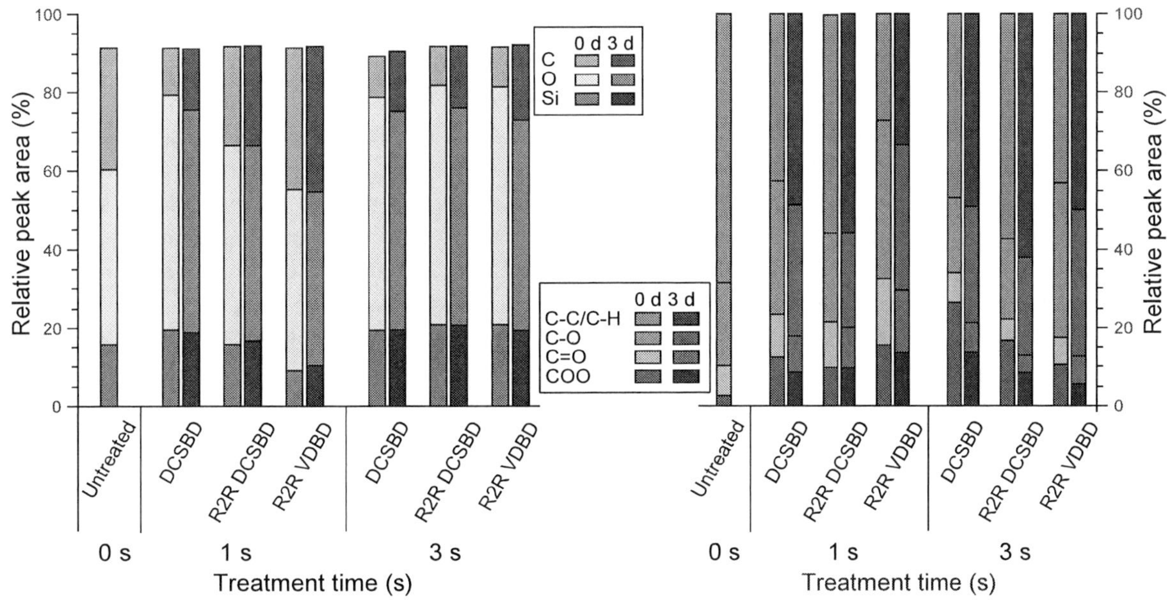

Figure 2 XPS analysis of a) elemental composition and b) carbon bonds partition on UTFG treated with plasma carried before and after 3 days of storage on ambient atmosphere.

The XPS analysis of UTFG treated 1 and 3 s with all of the tested plasma systems revealed a dominant presence of carbon, oxygen and silicon (C1s, O1s and Si2p) in the survey spectra. Atomic concentrations of C, O and Si are presented in **Figure 2**, their values on untreated sample were 31 %, 45 % and 16 %, respectively. Carbon could be present in organic pollution which is contained in the water film adsorbed to UTFG surface from air humidity. Another possible source of carbon on the plasma-treated UTFG surface are carbonyl and carboxyl functional groups incorporated into it. All the applied plasma treatments induced reduction of atomic carbon concentration, with an exception of 1 s treatment with VDBD. Storage of 3 days induced alterations of relative carbon content only in the case of longer exposure times 3 s, when it increased from 10 % to 15 % for both DCSBDs and to 19 % for VDBD. The presence of oxygen is related to bulk material, water film and also to the occurrence of chemically active functional groups. After all of the tested plasma treatments, oxygen content increased, while the most significant growth to 61 % was achieved by 3 s treatment with R2R DCSBD reactor. After 3 days of storage, oxygen concentration is reduced, which was more pronounced for the treatments of 3 s. Silicon concentration represents the bulk material, and its concentration depends on its presence in the analyzable depth profile (~3 nm). In the carried out experiments, the Si content did not vary so significantly as it for C and O, which can be explained by their rearrangement on the surface without a markable change in thickness of the surface nano-layer. The achieved results confirmed the expected activation by the rise of oxygen-based groups present after air plasma treatment of even a short duration for both tested plasma sources. However, 1 s treatment time with VDBD was not enough to decrease atomic carbon relative content, hence its cleaning effect was not achieved, contrarily to the DCSBD plasmas.

High-resolution spectra of C1s peak revealed the principle behind the rearrangement of carbon and oxygen induced by plasma interaction with UTFG surface. Breaking of C—C and C—H bonds is related to an oxidation process resulting in the bonding of highly reactive oxygen to the surface, which was confirmed by a significant increase of relative peak areas of C—O, C=O and O—C=O bonds. The acquired data showed an increase of oxygen-based carbon bonds, which was the most significant after 1 s treatment with VDBD plasma. The fact that atomic carbon content on the UTFG surface increases only after 1 s VDBD, together with a simultaneous increase of atomic oxygen content implies the assumption that for all other tested conditions the rest of oxygen was bonded directly to the silicone-based structure of UTFGs bulk. Direct oxidation of the silicone structure cleaned from the residuals from organic pollution seems to be more suitable for the improvement of uniformity and stability in time, which is in accordance with the results of WCAs after treatment with R2R VDBD reactor and its uncertainties measured immediately. After 3 days of storage on ambient air, the highest alterations in the C1s bonds concentration were shown in C—C and C—H bonds that increased and O—C=O bonds that decreased approximately by the same relative area. C—O and C=O bonds changed only slightly, which is in accordance with the higher reactivity of carboxyl groups that were dissolving faster in humid air. Recombination of oxygen-based polar groups in air humidity during storage is in accordance with the decrease of relative peak areas of oxygen-containing bonds and hindering of wettability, which supports the interpretation that the impact of adsorbed water is more pronounced on the sample with a highly oxidised surface. This phenomenon limits the applicability of plasma surface modification, while prolonging exposure does not provide any additional hydrophilic improvement or better stability of the chemical composition of the UTFG surface. The present results of experiments carried out on the R2R DCSBD reactor showed the best match in cleaning rate and activation efficiency for reaching the most desirable effect using treatment with duration of 1 s. Better uniformity and stability of the effect achieved with DCSBD plasmas could be explained with higher local concentrations of active species generated in its thin active plasma region compared to 1 mm wide gap of VDBD plasma.

Table 2 RMS roughness and arithmetical mean roughness of glass surface treated with DCSBD plasma

3-LIQUID CLEANED	Sq (nm)	u(Sq) (nm)	Sa (nm)	u(Sa) (nm)
Untreated sample	1.4	0.4	0.5	0.1
DCSBD 3 s	1.5	0.2	1.1	0.2
DCSBD 10 s	1.7	1.3	0.7	0.5

Additionally, the DCSBD as the optimal plasma source has been tested for its plasma impact to the morphology of glass surface. Results of AFM measurement on the chemically precleaned samples of common soda-lime glass are summarized in **Table 2**. Evaluation of root mean square roughness (Sq) and arithmetical mean roughness (Sa) showed significant growth of uncertainties of values obtained in measurements carried out on the sample treated with the longest tested time of 10 s. There was no significant change in Sq of the sample treated for 3 s with DCSBD, although the resulting Sa values doubled, the uncertainties are significantly lower in comparison to the values measured after 10 s treatment. The reason of higher uncertainties reported on the samples treated with longer exposure time lies in deterioration of glass surface. The resulting Sq values measured on samples treated with shorter exposure time approved DCSBD plasma as a gentle and uniform instrument for fast surface modification of glass, which seems to be harmless to the bulk glass structure. The achieved results are in accordance with the results reported in former studies, when using DCSBD plasma at shorter exposure times than 10 s did not cause any significant deterioration of the glass surface [7].

4. CONCLUSION

The tested plasma sources achieved a significant wetting improvement of 30 μm thick UTFG using optimal treatment time 1 s with the curved DCSBD unit. WCA and XPS revealed fast pollution removal and strong activation of UTFG surface due to incorporation of oxygen-based polar groups. The ageing of plasma-treated UTFG, monitored for one week, showed the best stability of the effect on the DCSBD plasma-treated UTFG surface. Plasma seems to be an efficient and controllable cleaning method that respects the specific thermomechanical properties of UTFG and the requirements of contactless large area printing on fragile flexible substrates processed on roll-to-roll devices.

ACKNOWLEDGEMENTS

This research has been supported by the projects LM2018097 funded by the Ministry of Education, Youth and Sports (CEPLANT centre) and project FV40114 funded by the Ministry of Industry and Trade (program TRIO).

REFERENCES

[1] SIHELNÍK, S., KRUMPOLEC, R., STUPAVSKÁ, M., FENG, J., KOSOVÁ, E., KELAR, J., ZEMÁNEK, M., KOVÁČIK, D., ČERNÁK, M. (2020). Large area cleaning and activation of flexible ultra-thin glass by non-thermal atmospheric-pressure plasma. In: *Nanocon 2020 Conference Proeedings.* [online]. Brno: Tanger, 2020, pp. 305-310. Available from: https://doi.org/10.37904/nanocon.2020.3717.

[2] PARK, S., KIM, H., KIM, J. H., & YEO, W. H. Advanced nanomaterials, printing processes, and applications for flexible hybrid electronics. *Materials.* [online]. 2020, vol. 13., iss. 16, pp. 1-34. Available from: https://doi.org/10.3390/MA13163587.

[3] CASTRO-HERMOSA, S., LUCARELLI, G., TOP, M., FAHLAND, M., FAHLTEICH, J., BROWN, T.M. Perovskite Photovoltaics on Roll-To-Roll Coated Ultra-thin Glass as Flexible High-Efficiency Indoor Power Generators. *Cell Reports Physical Science.* [online]. 2020, vol. 1, iss. 5, pp. 100045 (17). Available from: https://doi.org/10.1016/j.xcrp.2020.100045.

[4] DEY, T., NAUGHTON, D. Cleaning and anti-reflective (AR) hydrophobic coating of glass surface: a review from materials science perspective. *Journal of Sol-Gel Science and Technology.* [online]. 2016, vol. *77*, iss. 1, pp. 1-27. Available from: https://doi.org/10.1007/s10971-015-3879-x.

[5] R&D centre for plasma and nanotechnology surface modifications. *DCSBD plasma for the large area roll-to-roll applications (Hydrogen plasma).* [online]. 2021 [viewed: 2021-11-01]. Available from: Available from: https://ceplant.cz/equipment-and-services/low-cost-plasma.

[6] Ahlbrandt System GmbH. *Corona Pure.* [online]. 2021 [viewed: 2021-12-02]. Available from: Available from: https://www.ahlbrandt.com/corona-pure.

[7] BUČEK, Andrej, HOMOLA, Tomáš, ARANYOSIOVÁ, Monika, VELIČ, Dušan, PLECENIK, Tomáš, HAVEL, Josef, SŤAHEL, Pavel, and ZAHORANOVÁ, Anna. Atmospheric pressure non-equilibrium plasma treatment of glass surface. *Chemické Listy.* 2008, vol. 102, pp. 1459-1462.

THE INFLUENCE OF THE SIZE OF HEMATITE PARTICLES ON THE PROPERTIES OF POLYETHYLENE/HEMATITE COMPOSITES

[1]Ana PERŠIĆ, [2]Nina POPOV, [1]Elza ŠTEFANOVIĆ, [2]Stjepko KREHULA, [1]Emi GOVORČIN BAJSIĆ, [1]Mirela LESKOVAC, [1]Ljerka KRATOFIL KREHULA

[1]Faculty of Chemical Engineering and Technology, University of Zagreb, Zagreb, Croatia, EU, apersic@fkit.hr, estefanov@fkit.hr, egovor@fkit.hr, mlesko@fkit.hr, krehula@fkit.hr

[2]Division of Materials Chemistry, Ruđer Bošković Institute, Zagreb, Croatia, EU, Nina.Popov@irb.hr, krehul@irb.hr

https://doi.org/10.37904/nanocon.2021.4330

Abstract

The incorporation of metal oxide particles into polymer matrix in many cases leads to improved characteristics of the material: thermal stability, mechanical strength, light absorbing or antibacterial properties etc.

This study aims to develop polyethylene/hematite composites prepared with hematite (α-Fe_2O_3) particles of different size. Such types of composites have not been yet thoroughly studied but have a potential to reveal improved properties in comparison to the pure polymer. Polyethylene is a material with a broad field of application, primarily as packaging material for food and other products. Hematite is non-toxic, thermally and chemically stable low-cost metal oxide. There is also a demanding task to prepare composites with non-aggregated hematite particles and their fine dispersion in polyethylene matrix.

In this work two types of hematite with well-defined shape and uniform size were used for preparation of the composites: hematite HC1 with cubic particles of average size of 2 µm and HS2 with spherical particles sized of about 100 nm. The mass fraction of hematite in the composites was 0.25, 0.5 and 1 %. Prepared polyethylene/hematite composites were characterized by thermogravimetric analysis (TGA), differential scanning calorimetry (DSC) and diffuse reflectance UV-Vis-NIR spectroscopy. The mechanical properties were also studied. The results show that such composites have improved properties in comparison to a pure polyethylene, especially the composites with hematite HS2 due to its particles of lower size and large surface. Obtained results reveal that such composites may be a promising material for wide range of applications.

Keywords: Composites, polyethylene, hematite, thermal stability, UV blocking

1. INTRODUCTION

An increase in the use of polymer materials recently demands their new or improved types and roles in many different fields. Polymer composites represent an outstanding choice for development of materials because they unite the properties of polymer matrix and filler which will ultimately result with completely new or improved material types. Fillers are usually used to improve polymer's properties, especially mechanical properties, or to replace the part of polymer matrix with less expensive substance. Furthermore, fillers are nowadays also used to develop some functional properties of polymer composites: UV blocking, water absorbtion, flavour releasing, oxygen scavenging, antimicrobial activity etc. The choice of additives must especially take care of health and safety, i.e. it demands the use of nontoxic components [1]. Hematite (α-Fe_2O_3) represents good option for application in preparation of polymer composites due to its non-toxicity, thermal and chemical stability, low cost and ageing resistance. Because of such suitable properties, the polymer/hematite composites may possess improved thermal, mechanical and magnetic properties as well as catalytic characteristics [2]. Development of advanced functional polymer materials includes sometimes additives that contribute to improved UV blocking property, which is very important in prevention of material's degradation and in prolonging its lifetime [3]. To obtain a composite material with good and balanced properties, it is very important to use the filler with the particles of uniform shape and size as well as to avoid their aggregation in the polymer matrix. This study aims to develop polyethylene/hematite composites with the hematite particles of different size and to analyze their influence on the overall composite properties. The reason for this research was also that polyethylene/hematite composites have not been yet thoroughly studied but their development may be a promising choice for creation of polymer materials with advanced applications.

2. EXPERIMENTAL PART

2.1. Materials

For the preparation of the polyethylene/hematite composites LDPE polymer granulate (Dow Chemical) was used. Hematite particles HC1 and HS2 were synthesized in Laboratory for Synthesis of New Materials, Division of Materials Chemistry, Ruđer Bošković Institute. Uniform hematite particles were prepared according to slightly modified synthesis procedure reported by Sugimoto et al. [4]. For preparation of cubic hematite particles (sample HC1) 2.5 ml of H_2O and 47.5 ml of 6 M NaOH aqueous solution were slowly added to 50 ml of 2 M $FeCl_3$ aqueous solution in a polypropylene (PP) bottle under the strong magnetic stirring. Obtained suspension was heated at 100 °C in laboratory oven for 8 days. For preparation of spherical hematite particles (sample HS2) 2.2 ml of H_2O, 1.5 ml of 1M Na_2SO_4 aqueous solution and 46.3 ml of 6 M NaOH aqueous solution were slowly added to 50 ml of 2 M $FeCl_3$ aqueous solution in a PP bottle under the strong magnetic stirring. Obtained suspension was heated at 100 °C in laboratory oven for 6 days. Precipitated hematite particles were washed (using milli-Q water and centrifuge) and dried at 60 °C in air.

SEM images of the synthesized hematite samples HC1 and HS2 were recorded using a field emission scanning electron microscope (FE-SEM) model JSM-7000F manufactured by Jeol Ltd. They are presented in **Figures 1** and **2**. Hematite HC1 has cubic particles of average size of 2 µm while HS2 has spherical particles sized of about 100 nm.

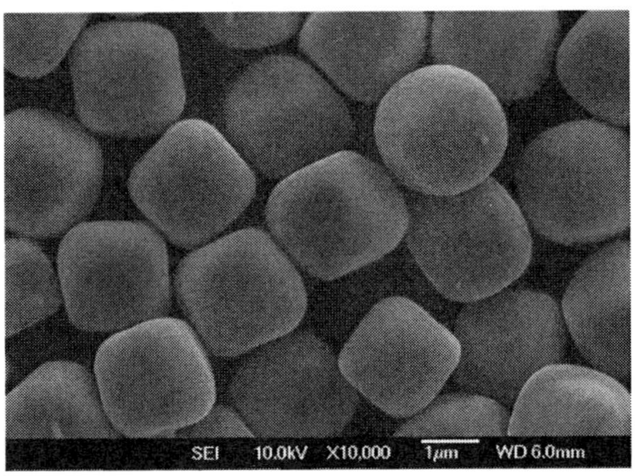

Figure 1 SEM image of hematite sample HC1

Figure 2 SEM image of hematite sample HS2

2.2. Preparation of the composites

Polyethylene/hematite composites denoted as LDPE/HC1 and LDPE/HS2 are prepared by mixing in a Brabender kneader, at 180 °C over a period of 3 minutes, 45 rpm. The content of hematite in the samples was 0.25, 0.5 and 1 %. The obtained composite materials were prepared for further characterization by pressing into the foils and plates. The pressing was carried out by hydraulic press Dake, model 44-226 at a temperature of 190 °C.

2.3. Characterization

UV-Vis-NIR spectroscopy. Diffuse reflectance UV-Vis-NIR spectra of polymer composite films were recorded at 20 °C using a Shimadzu UV-3600 UV-Vis-NIR spectrophotometer with integrating sphere. Barium sulfate was used as a reference material.

Thermogravimetric analysis (TGA). The thermal stability of the obtained polymer composites was determined by the thermogravimetric analyzer TA Instruments Q500. Mass specimens of 10 mg were analyzed in the nitrogen stream at a heating rate of 10 °C/min in the temperature range of 25 to 800 °C.

Differential scanning calorimetry (DSC). The thermal properties were determined on Mettler Toledo DSC822e device. Samples were heated from 25 °C to 180 °C and then they were cooled down to -150 °C (heating/cooling rate was 10 °C/min), two heating and cooling cycles were performed.

Determination of mechanical properties. Mehanical properties were determined on Zwick 1445 universal device. Samples were 100 mm long and 10 mm wide (~ 1mm thick). The stretching speed was 50 mm/min.

3. RESULTS

3.1. UV/Vis spectroscopy

Ultraviolet (UV) radiation induces harmful influences to polymer materials. It may significantly reduce the properties and lifetime of polymers. In order to prevent such degradation effects, different types of UV stabilizers must be used. Some of them include metal oxides like titanium or zinc oxide. There is also a potential of hematite to act as an UV stabilizer i.e. UV blocking agent due to its UV absorption capability [5]. From UV-Vis-NIR spectra of polyethylene/hematite, **Figures 3** and **4**, it can be seen that some differences are observed in comparison with the pure polyethylene. The composites show absorption in ultraviolet region and in the visible light region. The particles of lower size (HS2), due to their large surface, enhance the UV absorbance capacity of the composites. For that reason, such prepared polyethylene/hematite composites may be considered as suitable materials for making different packaging types with good UV blocking properties.

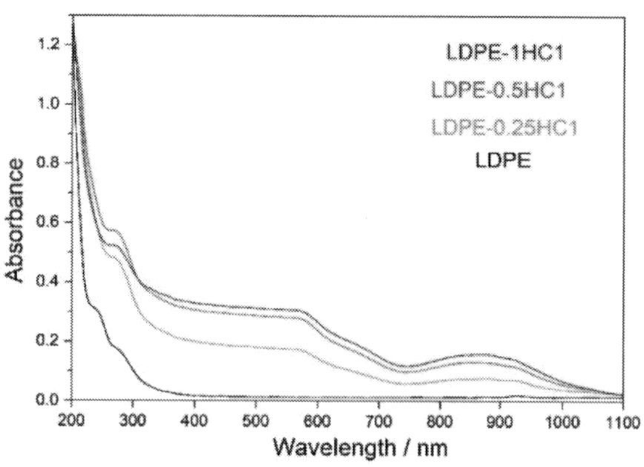

Figure 3 UV/Vis spectra of LDPE and LDPE/HC1 composites

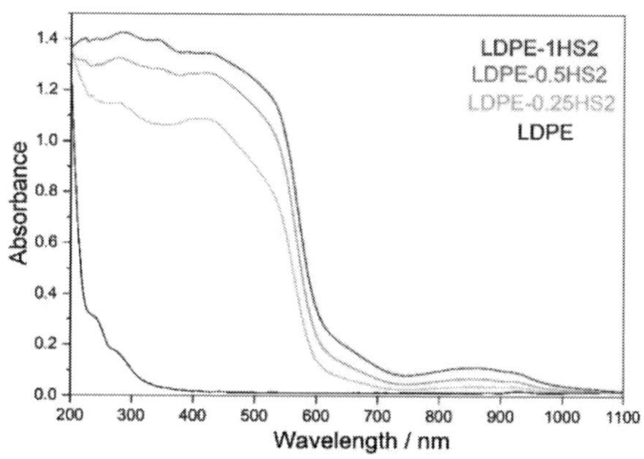

Figure 4 UV/Vis spectra of LDPE and LDPE/HS2 composites

3.2. Thermogravimetric analysis

It is known that hematite may improve the thermal stability of some types of polymers e. g. polystyrene [6] but there is a lack of information regarding use of hematite in polyethylene matrix. For that reason, this study brings the results of the thermogravimetric analysis of polyethylene/hematite composites, presented in **Table 1**.

Table 1 Initial decomposition temperature of the samples ($T_{95\%}$) and temperature of maximum rate of decomposition (T_{max}).

Sample	$T_{95\%}$ (°C)	Tmax (°C)
LDPE	422.60	467.07
LDPE + 0.25 % HC1	421.88	466.35
LDPE + 0.5 % HC1	419.01	467.07
LDPE + 1 % HC1	426.16	468.67
LDPE + 0.25 % HS2	438.39	487.35
LDPE + 0.5 % HS2	407.53	483.57
LDPE + 1 % HS2	436.23	491.46

The results show that LDPE is degraded in one step. Samples of LDPE/hematite composites also degrade in one step but on higher temperatures, what proves improvement of their thermal stability in comparison to the pure LDPE. LPDE composites with HS2 (hematite particles of smaller size, average size of about 100 nm) showed a significant increase of $T_{95\%}$ and T_{max}. $T_{95\%}$, which is the temperature at which 5 mas. % of the sample is decomposed (initial decomposition temperature), is improved even up to 16 °C. The temperature of maximum rate of decomposition, T_{max}, for that composites, is also improved (even up to 24 °C) compared to the pure LDPE.

3.3. Differential scanning calorimetry (DSC)

The results of DSC analysis, presented in **Table 2**, show that LDPE is a crystalline polymer with a glass transition of -129.75 °C. The addition of hematite to the LDPE matrix leads to a slight increase in the degree of crystallinity. The increase in the degree of crystallinity can be attributed to the action of filler as a nucleating agent [7].

Table 2 The results of the analysis by differential scanning calorimetry

Sample	T_g (°C)	T_m (°C)	ΔH_m (Jg^{-1})	T_c (°C)	χ_c (%)
LDPE	-129.75	114.41	-86.91	96.08	29.66
LDPE + 0.25 % HC1	-123.07	111.84	-111.46	95.24	38.14
LDPE + 0.5 % HC1	-130.55	112.00	-91.83	95.31	31.50
LDPE + 1 % HC1	-131.86	112.44	-97.85	94.88	33.73
LDPE + 0.25 % HS2	-128.02	112.50	-115.87	95.22	39.64
LDPE + 0.5 % HS2	-126.72	111.88	-109.76	95.12	37.65
LDPE + 1 % HS2	-125.84	111.99	-111.91	95.59	38.58

3.4. Determination of mechanical properties

Hematite may significantly improve the mechanical properties of some types of polymer matrix [8]. The results of the mechanical properties of LDPE and the composite samples, expressed as dimensionless tensile strength and dimensionless elongation at break, are presented in **Figures 5** and **6**. Mechanical properties show the improvement of the tensile strength for all composite samples in comparison to the pure LDPE**.** The elongation at break is improved for the samples LDPE + 1% HC1 and LDPE + 0.25% HS2.

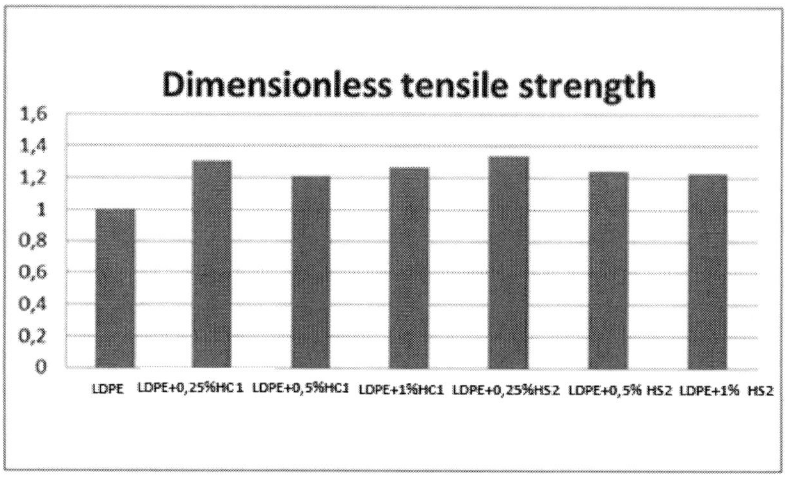

Figure 5 Tensile strength of the studied samples

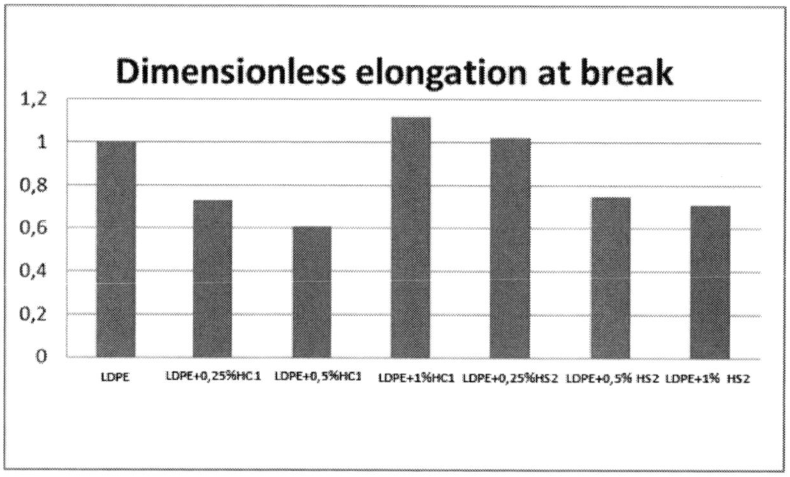

Figure 6 Elongation at break of the studied samples

4. CONCLUSION

The results of UV-Vis-NIR absorption measurements show that all prepared composites (LDPE/HC1 and LDPE/HS2) are suitable for making different packaging materials with good UV blocking properties. The LDPE + 0.25% HS2 composite can be considered as the most thermally stable sample because it shows high temperature stability despite the low filler content.

The results show that LDPE/hematite composites have improved UV blocking, thermal and mechanical properties in comparison to a pure polyethylene, especially the composites with hematite HS2 due to its particles of lower size and large surface. Such composites may be a promising material for wide range of applications.

ACKNOWLEDGEMENTS

The part of this research (the synthesis of hematite particles) was supported by the Croatian Science Foundation, project number IP-2016-06-8254.

REFERENCES

[1] LLORENS, A., LLORET, E., PICOUET, P.A., TRBOJEVICH, R., FERNANDEZ, A. Metallic-based micro and nanocomposites in food contact materials and active food packaging. *Trends in Food Science & Technology.* 2012, vol. 24, pp. 19-29.

[2] KAUSAR, A. Polymeric materials filled with hematite nanoparticle: current state and prospective application. *Polymer-Plastics Technology and Materials.* 2020, vol. 59, no. 3, pp. 1-16.

[3] HRNJAK-MURGIĆ, Z., REŠČEK, A. PTIČEK SIROČIĆ, A., KRATOFIL KREHULA, Lj., KATANČIĆ, Z. *Nanoparticles in Active Polymer Food Packaging.* Surrey: Smithers Pira, 2015.

[4] SUGIMOTO, T., SAKATA, K. Preparation of monodisperse pseudocubic α-Fe_2O_3 particles from condensed ferric hydroxide gel. *Journal of Colloid and Interface Science.* 1992, vol. 152, no. 2, pp. 587-590.

[5] TRUFFAULT, L., CHOQUENET, B., KONSTANTINOV, K., DEVERS, T., COUTEAU, C., COIFFARD, L. J. M. Synthesis of Nano-Hematite for Possible Use in Sunscreens. *Journal of Nanoscience and Nanotechnology.* 2011, vol. 11, pp. 2413-2420.

[6] KULJANIN, J., MARINOVIĆ CINCOVIĆ, M., ZEC, S., ČOMOR, M. I., NEDELJKOVIĆ, J. M. Influence of Fe_2O_3 - filler on the Thermal Properties of Polystyrene. *Journal of Materials Science Letters.* 2003, vol. 22, pp. 235-237.

[7] ZALDUA, N., MAIZ, J., de la CALLE. A., GARCÍA-ARRIETA, S., ELIZETXEA, C., HARISMENDY, I., TERCJAK, A., MÜLLER, A. J. Nucleation and Crystallization of PA6 Composites. *Polymers.* 2019, vol. 11, pp. 1680-1699.

[8] DJOKOVIĆ, V., NEDELJKOVIĆ, J. M. Stress Relaxation in Hematite Nanoparticles Polystyrene Composites. *Macromolecular Rapid Communications.* 2000, vol. 21, pp. 994-997.

THE ROLE OF HUMIC ACID IN THE CORROSION OF AZ31 MAGNESIUM ALLOY

Leoš DOSKOČIL, Jiří HONČ, Matěj BŘEZINA, Jaromír WASSERBAUER

Brno University of Technology, Faculty of Chemistry, Brno, Czech Republic, EU,
doskocil@fch.vut.cz

https://doi.org/10.37904/nanocon.2021.4331

Abstract

Magnesium alloys used as sacrificial anodes can interact with humic acids (HAs) in soil and water environments. Therefore, these natural organic molecules can have an impact on the corrosion behaviour of magnesium materials. In this work, the aim was to investigate the effect of lignitic HA on the corrosion of AZ31 magnesium alloy in 3.5 % NaCl at different pH (5, 7 and 9). The results showed that HA increased the corrosion rate about four times at pH 7, while it weakly decreased the corrosion rate at pH 5 and 9. The acceleration of corrosion is probably related to the depletion of Mg^{2+} ions by forming complexes with HA instead of producing Mg(OH)2. The increase in corrosion resistance may be related to the reduced stability of the complexes and adsorption on the AZ31 alloy surface at pH 5 and 9. Surface analysis suggested that the presence of HA promotes the formation of $Mg(OH)_2$ nanoparticles.

Keywords: Corrosion, AZ31 magnesium alloy, humic acid

1. INTRODUCTION

Magnesium is a lightweight metal with superior strength to weight ratio, high specific stiffness, electromagnetic shielding property and high thermal conductivity [1]. These features make it a great choice for various applications in the automotive industry, aeronautics, electronics, biodegradable medical implants etc. [1,2]. Unfortunately, a poor corrosion resistance is the biggest limiting factor for its wider use. This disadvantage can be overcome to some extent by the preparation of magnesium alloys, which most often contain alloying elements such as aluminium, zinc, manganese. For example, the popular magnesium alloy AZ31 is made up of about 3 % aluminium and about 1 % zinc [1].

High electrochemical activity and high susceptibility to corrosion in many environments predetermine magnesium and its alloys to be used as a sacrificial anode in cathodic protection of steel structures such as of oil and gas pipelines, oil drilling platforms cables, heat exchangers, aircraft, ships (ship hulls) and bridges [1].

Magnesium anodes are of two generic types, Mg-Mn and Mg-Al-Zn. Both alloy systems have a high driving voltage a therefore find application in high resistivity environments such as soil and fresh, or brackish waters. Because magnesium is nontoxic, its use is permissible in potable water systems wherein the conductivity is low [3].

As far as the authors know, very little is known about the corrosion of magnesium and its alloys in the soil. The literature commonly states that magnesium is quite resistant to corrosion in soil. The presence of calcium and magnesium salts in most soils tends to inhibit corrosion, but the corrosivity is different in different kinds of soil [4].

Humic substances are an important component of soils, which can have an impact on the corrosion behavior of materials. Previous publications have shown that humic substances have an impact on the corrosion process of steel, galvanized steel and brass [5,6]. However, the effect of humic substances on magnesium corrosion is not known in soil or water environment.

Humic substances are the heterogeneous mixture of organic molecules which are traditionally divided based on their alkaline and acid solubility into three classes: humic acid (HA), fulvic acid, and humin [7]. HAs are characterized by high molecular weight and a high content of O-functional groups such as carboxylic and hydroxyl groups. HA occurs not only in water and soil but also in peat, coal and leonardite, whereas their chemical-physical properties may differ from each other depending on their origin. An increase in the content of HAs (generally organic matter) in soil and water can be expected due to the growing interest in green agriculture and other interventions of human to the natural environment [7]. In connection with this fact, the interactions of organic substances with magnesium anode (generally metal materials) will become increasingly important. Therefore, the investigation of changes in the corrosion resistance/mechanisms of magnesium materials in the presence of HAs may be a desirable challenge for corrosion research.

Corrosion of magnesium and its alloys in the presence of HAs can also have an impact on the presence of inorganic ions in the corrosive environment. Corrosion of the magnesium sacrificial anode results in the release of predominantly magnesium ions as well as aluminum or zinc ions due to oxidation of the anode.

It is also conceivable that HAs could be involved in the formation of magnesium hydroxide nanoparticles, which form on the surface of the magnesium anode during corrosion.

Pankratov et al. [8] recently discovered that magnetic nanoparticles of iron oxides can be stabilized by HAs and suggested that the nanoparticles can be formed also during corrosion of iron in contact with soil.

The aim of this work is to determine the effect of lignite HA on the corrosion of AZ31 alloy in the environment of 3.5 % NaCl at different pH (5, 7 and 9).

2. EXPERMENTAL

HA was isolated from South Moravian lignite, which was obtained from the locality Mikulčice, Czech Republic. Lignitic HA was extracted using a modified procedure recommended by the International Humic Substances Society. Briefly, air-dried lignite was demineralized with 0.1 mol/l hydrochloric acid and the suspension was then washed with deionized water until an almost neutral pH was achieved. Suspension was shaken overnight with 0.1 mol/l NaOH and 0.1 mol/l $Na_4P_2O_7$. After centrifugation, HA was precipitated from the supernatant by means of acidification to pH < 2 with 6 mol/l HCl, and further separated by centrifugation. Precipitated HA was purified and dialysed using a Spectra/Por cellulose membrane (1 kDa) until the free chloride ions were washed out and the purified lignite HA was finally freeze-dried.

A stock solution of 100 mg/l HA was prepared by dissolving in 0.1 mol/l NaOH containing 3.5 % NaCl. Three fractions of HAs were obtained from the stock solution by lowering the pH to 9, 7 and 5 by sequential addition of HCl with stirring. The precipitated HA was centrifuged, and the supernatant was filtered through a 0.45 µm filter. The obtained pH-fractionated HAs were stored in a refrigerator and used for corrosion tests within one week.

AZ31 magnesium alloy samples (20 × 20 × 6 mm) were used for corrosion tests. Their chemical composition was determined using Glow-Discharge Optical Emission Spectroscopy (GDOES); 3.60 % Al, 1.34 % Zn, 0.28 Mn, 0.03 Si, 0.002 Fe and 0.01 Sn. The samples of AZ31 alloy were ground using 320 and 1200 SiC grit abrasive papers, washed in distilled water and isopropyl alcohol and then dried by hot air.

Total organic carbon (TOC) was determined by accredited testing laboratory LABTECH using a Shimadzu TOC 5000A analyzer according to Czech technical standard ČSN EN 1484.

The UV/Vis absorption spectra of HA and pH-fractionated HAs at a concentration of 5 mg/l C were recorded on Hitachi U-3900H double beam UV/Vis spectrometer (Hitachi, Tokyo, Japan) in the spectral range between 200 and 800 nm. Sodium borate buffer (1 mol/l H_3BO_3, 0.25 mol/l NaOH) was used as a solvent of the fractions and as well as a blank solution.

The Attenuated Total Reflectance (ATR-FTIR) technique was used for a structural characterization of organic substances. Fourier transform infrared (FTIR) spectra of the substances were recorded on Nicolet iS50 spectrometer (Thermo Fisher Scientific, Waltham, MA, USA) using the Attenuated Total Reflectance (ATR) measuring technique (single reflection built in diamond ATR crystal). All measurements were taken in the spectral range 4000-400 cm^{-1} at 4 cm^{-1} resolutions as an average of 256 scans. A background spectrum was collected from the clean dry surface of the ATR crystal in an ambient atmosphere.

The short-time potentiodynamic measurements were performed in 3.5 % NaCl solution with and without pH-fractionated HAs at laboratory temperature using the Bio-Logic VSP-300 potentiostat (BioLogic, Seyssinet-Pariset, France). The corrosion solutions contained 0 and 15 mg/l C of pH-fractionated HAs. Measured samples with exposed area of 1 cm^2 served as a working electrode. Saturated calomel electrode (SCE) was used as a reference electrode and Pt wire as a counter-electrode. Open circuit potential (OCP) variations with time were recorded up to 10 min of exposure. Afterward, Tafel plot was obtained by carrying out partial potentiodynamic polarization in the potential range from −200 mV to 250 mV from OCP at the potential scan rate of 1 mV/s.

Surface analysis of the samples after corrosion was performed using scanning electron microscopy (SEM; JEOL JSM-7600F, Tokyo, Japan). Energy-dispersive X-ray spectroscopy (EDS, model OXFORD INSTRUMENTS X-MAX 80 mm^2) was used to examine the surface of the samples after electrochemical testing to specify the character of the corrosion products. Process of Au sputtering was used to prevent unfavorable sample conductivity.

3. RESULTS AND DISCUSSION

TOC analysis showed that all three fractions of HA contained 30 mg/l of carbon, whereas the starting HA before fractionation contained 50 mg/l of carbon. HA and its fractions soluble at pH 5, 7 and 9 were characterized using UV-Vis spectrometry and FTIR spectrometry. The UV-Vis absorbance spectra (not shown) were featureless and without the presence of any indication of a maximum and/or a minimum. The absorbance values decreased exponentially with increasing acquisition wavelength. Absorbance ratios $E_{ET/Bz}$ and $E_{2/4}$ were calculated. The $E_{ET/Bz}$ ratio (the ratio of absorbance at 253 nm to that 220 nm) is higher for compounds in which the aromatic structural units are predominantly substituted with oxygen-containing functional groups, whereas lower values are associated with aliphatic substituents. The $E_{2/4}$ ratio (the ratio of absorbance at 265 nm to that 465 nm) indicates the molecular size of the molecules and its aromaticity; the higher $E_{2/4}$ ratio, the smaller size molecular size and the lower aromaticity. The pH-fractionated HAs were very similar to each other ($E_{ET/Bz}$ and $E_{2/4}$ were about 0.90 and 7.81, resp.), but different from the original HA ($E_{ET/Bz}$ and $E_{2/4}$ were 0.87 and 6.24, resp.). The results suggest that the fractionated HAs are composed of smaller molecules with lower aromaticity, and richer in oxygen functional groups.

FTIR spectra of HA and its fractions were very similar (not shown). They contained aliphatic groups (2905 and 2840 cm^{-1}), including methyl and methylene groups. These groups were also confirmed by the deformation band (1457 cm^{-1}). The bands at 1564 and 1369 cm^{-1} can be attributed to aromatic substances and methyl groups bound to aromatics, respectively. These bands can also be related to carboxylates. The presence of carboxyl groups was indicated by a band at 1700 cm^{-1}. Spectra also contained aryl ether compounds (1262 cm^{-1}), phenols and ethers (1198 and 1154 cm^{-1}), alcohols (1094 and 1037 cm^{-1}). Out-of-plane C-H deformation bands were found in the 900-700 cm^{-1} region.

Potentiodynamic polarization was performed in 3.5 % NaCl medium with 15 mg/l carbon of pH-fractionated HAs and in the absence of these HAs. All corrosion solutions were tested at three different pHs (5, 7 and 9). The potentiodynamic curves are shown in **Figure 1** and the quantified values (corrosion current density i_{cor} and corrosion potential E_{cor}) are given in **Table 1**. The determined values should be considered rather as indicative, because the obtained polarization curves do not exhibit conditions for Tafel analysis. As a result,

Tafel fitting of data was only performed from the cathodic branch. However, a comparison of the position and shape of the polarization curves with the determined parameters (E_{cor} and i_{cor}) shows that there are no mutual contradictions. The corrosion current density of AZ31 alloy in 3.5 % NaCl decreased with increasing pH. This is in line with Pourbaix diagram and the literature [9,10]. The i_{cor} value is a measure of the corrosion rate; the higher the i_{cor}, the higher the corrosion rate. **Table 1** and **Figure 1** show that AZ31 alloy had the lowest corrosion rate in 3.5 % NaCl at pH 9, whereas the highest corrosion rate was in 3.5 % NaCl at pH 5. The corrosion current density decreased approximately sixfold by increasing the pH from 5 to 9. The surface of the AZ31 alloy was slightly protected by the formation of a layer of $Mg(OH)_2$ in an alkaline environment; the protective function of the layer decreased with lower pH. In addition, the $Mg(OH)_2$ layer is disrupted by the attack of chloride ions, which dissolve $Mg(OH)_2$ into soluble $MgCl_2$ [9,10]. The corrosion potential of E_{cor} shifted slightly to a more positive value with increasing pH, indicating better corrosion resistance of AZ31 alloy in the given environment.

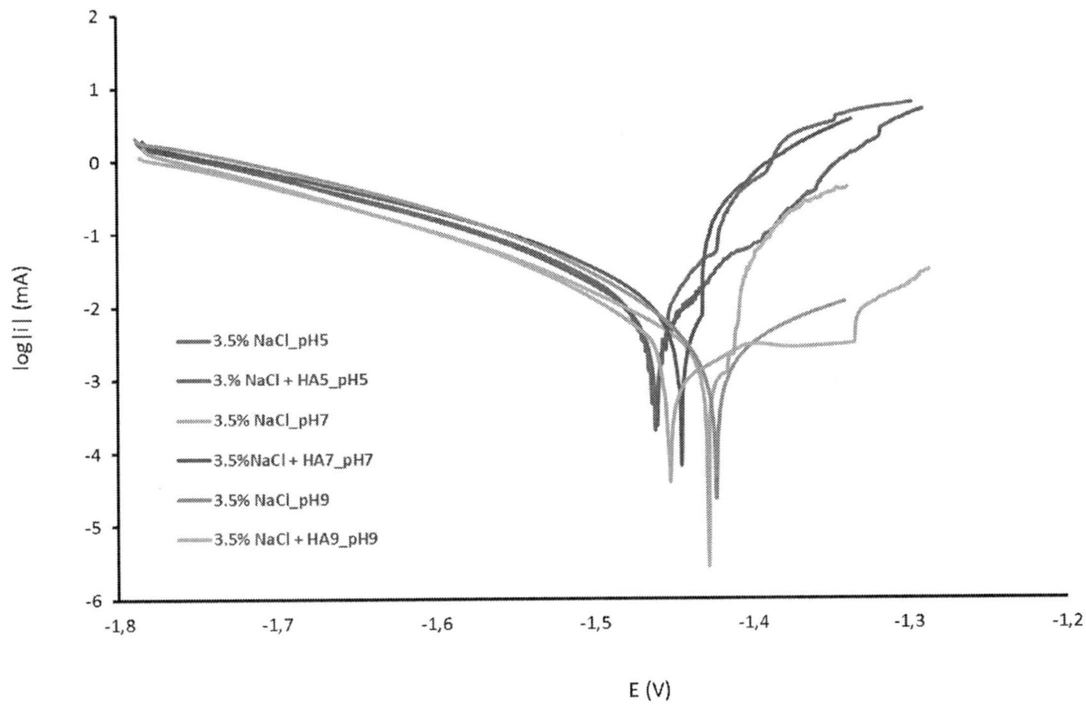

Figure 1 Potentiodynamic polarization curves of the AZ31 alloy in 3.5% NaCl with and without HAs

The addition of HA to the 3.5 % NaCl solution led to a change in the corrosion resistance of the AZ31 alloy, as can be seen from the potentiodynamic curves. A significant change in corrosion resistance was found in the case of a corrosive environment with pH 7. The value of the corrosion current density increased four times after the addition of HA. This increase in corrosion rate may be related to the interaction of released magnesium ions with HAs to form complex compounds. This depletion of Mg^{2+} ions can lead to the formation of a poor layer of $Mg(OH)_2$ on the surface. In contrast, the addition of HA to 3.5 % NaCl at pH 5 and 9 caused a slight decrease in the corrosion rate of the AZ31 alloy. The decrease in corrosion current density in alkaline media may be related to the a less willing interactions between Mg^{2+} ions and HA molecules to form stable complex compounds. In addition, a larger proportion of HAs can be adsorbed on the surface of $Mg(OH)_2$, which is easily formed mainly in a strongly alkaline environment. The decrease in the corrosion rate in an acidic environment can be explained by the lower solubility of magnesium humates, which can be deposited on the surface of the AZ31 alloy. The layer formed can limit the access of aggressive ions to the surface and, as a result, slightly increase the corrosion resistance of the material. These observations are valid at least for short-term measurements. A more detailed analysis of the potentiodynamic curves shows that the change in pH and the addition of HAs to 3.5 % NaCl mainly have an impact on the shape of the anodic branch compared to the cathodic branch. A break in the anodic branches of the potentiodynamic curve indicates pitting, which is characteristic of corrosion in the presence of chlorides. The shape of the potentiodynamic curves suggests that the presence of HAs and pH do not affect the corrosion mechanism of AZ31 in 3.5 % NaCl. The results also indicate that the corrosion behavior of AZ31 alloy in 3.5% NaCl is primarily controlled by pH and physicochemical properties of HAs. It should be noted that the chemical composition of the individual fractions of HAs is without significant differences. The more detailed role of HAs in corrosion will have to be studied in the future.

Table 1 Electrochemical parameters of potentiodynamic polarization.

Corrosion solution		E_{cor} (V vs. SCE)	i_{cor} (μA/cm^2)
pH 5	3.5 % NaCl	-1.46	18.4
	3.5 % NaCl + HA5	-1.46	17.7
pH 7	3.5 % NaCl	-1.43	4.5
	3.5 % NaCl + HA7	-1.45	18.1
pH 9	3.5 % NaCl	-1.42	3.4
	3.5 % NaCl + HA9	-1.45	2.8

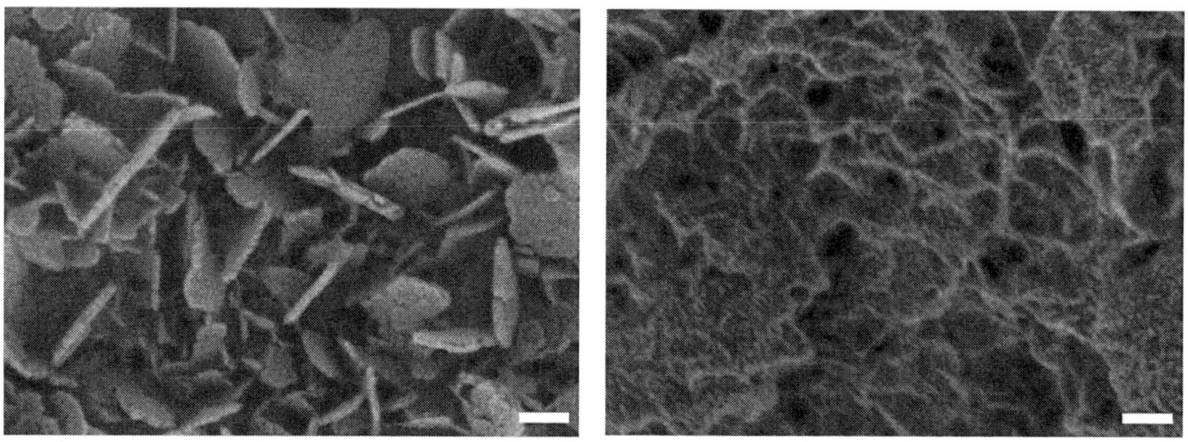

Figure 2 SEM images of AZ31 alloy surface after corrosion at pH 7 in 3.5% NaCl (left) and 3.5% NaCl containing HA (right). Bar represents 100 nm

Surface analysis showed that the surface of the AZ31 alloy after corrosion is highly heterogeneous regardless of the presence of HAs and pH. **Figure 2** shows the characteristic surface of the samples after seven days of corrosion in 3.5 % NaCl medium with and without HAs at pH 7. The other samples are not presented as they did not differ significantly from each other. Particular emphasis was placed on detecting the presence of nanoparticles and/or nanoshape. Surface and elemental analysis revealed that the surface of the AZ31 alloy in 3.5 % NaCl was formed by $Mg(OH)_2$ flat pieces with a thickness of about 40 nm. The elemental composition of the corrosion products included approximately 69 at.% O and 26 at.% Mg and a minor occurrence of about 3 at.% aluminum (AZ31 alloy signal) and 2 at.% Cl. The surface of the samples, which

corroded in the environment containing HA molecules, was characterized by a network of flat pieces consisting probably of $Mg(OH)_2$ and nanometer-sized spheres. Unfortunately, high quality images for samples corroded in HA environments could not be obtained due to difficulties associated with sample preparation for SEM analysis and during SEM analysis. As a result, it cannot be ruled out that the observed nanoparticles originate in gold that was deposited by the sputtering device. A typical elemental composition of corrosion products included about 61 at.% O, 19 at.% Mg and 18 at.% C, and about 2 at.% aluminum (AZ31 alloy signal). A significant proportion of carbon indicates the adsorption of HAs on the surface of $Mg(OH)_2$ during corrosion.

4. CONCLUSION

The results indicate that HA can affect the corrosion resistance of AZ31 alloy in 3.5 % NaCl depending on pH. HA increased the corrosion rate about four times at pH 7, while it weakly decreased the corrosion rate at pH 5 and 9. The acceleration of corrosion is probably related to the depletion of Mg^{2+} ions by forming complexes with HA instead of producing $Mg(OH)_2$. This facilitates the attack of aggressive chloride ions with the magnesium alloy surface. The slight increase in corrosion resistance may be related to the reduced stability of the complexes and adsorption (mainly of HA molecules) on the AZ31 alloy surface at pH 5 and 9. The effect of HA on the corrosion potential values was negligible. The shape of the potentiodynamic curves indicated that HA have no effect on the corrosion mechanism of AZ31 alloy in 3.5 % NaCl at different pH. Surface analysis of AZ31 alloy after corrosion suggested that the presence of HA promotes the formation of $Mg(OH)_2$ nanoparticles. Further research is needed to gain a deeper understanding of the role of HAs in the corrosion behavior of AZ31 alloys.

ACKNOWLEDGEMENTS

This work was supported by Specific University Research at FCH BUT, Project Nr. FCH-S-21-7553, Ministry of Education, Youth and Sports of the Czech Republic.

REFERENCES

[1] PATHAK, S., MENDON, S., BLANTON, M., RAWKINS, J.W. Magnesium-based sacrificial anode cathodic protection coatings (Mg-rich primers) for aluminium alloys. *Metals.* 2012, vol. 2, pp. 353-376.

[2] WASSERBAUER, J., BUCHTÍK, M., TKACZ, J., FINTOVÁ, S., MINDA, J., DOSKOČIL, L. Improvement of AZ91 Alloy Corrosion Properties by Duplex NI-P Coating Deposition. *Materials*, 2020, vol. 13, p. 1-17.

[3] CRUNDWELL, R.F. Sacrificial anodes. In: Shreir's Corrosion. Elsevier BV, 2010, pp. 2773.

[4] SONG, G.L., ATRENS, A. Corrosion Mechanisms of Magnesium Alloys. *Advanced Engineering Materials.* 1999, vol. 1, pp. 11-33.

[5] DICK, L.F.P., RODRIGUES, L.M. Influence of Humic Substances on the Corrosion of the API 5LX65 Steel. *Corrosion*. 2006, vol. 62, pp. 35-45.

[6] SORIANO, C., ALFANTAZI, A. Corrosion behavior of galvanized steel due to typical soil organics. *Construction and Building Materials.* 2016, vol. 102, 904-912.

[7] DOSKOCIL, L., BURDIKOVÁ-SZEWIECZKOVA, J., ENEV. V., KALINA, L., WASSERBAUER, J. Spectral characterization and comparison of humic acids isolated from some European lignites. *Fuel.* 2018, vol. 213, pp. 123-132.

[8] PANKRATOV, D.A., ANUCHINA, M.M. Nature-inspired synthesis of magnetic non-stoichiometric Fe_3O_4 nanoparticles by oxidative in situ method in a humic medium. *Materials Chemistry and Physics*. 2019, vol. 231, pp. 216-224.

[9] MINDA, J., FINTOVÁ, S. HADZIMA, B., DOLEŽAL, P., HASOŇOVÁ, M., DOSKOČIL., WASSERBAUER, J. Electrochemical corrosion behavior of pure Mg processed by powder metallurgy. *Coatings.* 2021, vol. 11, pp. 986.

[10] THIRUMALAIKUMARASAMY, D., SHAMMUGAM, K., BALASUBRAMANIAN, V. Comparison of the corrosion behavior of AZ31B magnesium alloy under immersion test and potentiodynamic polarization test in NaCl solution. *Journal of Magnesium and Alloys*. 2014, vol. 2, pp. 36-49.

SCREEN-PRINTED MOLYBDENUM DISULFIDE ELECTRODES FOR ELECTROCHEMICAL SENSING OF DOPAMINE

[1]Michaela PAVLIČKOVÁ, [2]Lenka LORENCOVÁ, [1]Michal HATALA, [2]Ján TKÁČ, [1]Pavol GEMEINER

[1]*Faculty of Chemical and Food Technology, Slovak University of Technology in Bratislava, Slovakia, EU*
michaela.pavlickova@stuba.sk

[2]*Institute of Chemistry, Slovak Academy of Sciences, Slovakia, EU*

https://doi.org/10.37904/nanocon.2021.4332

Abstract

Two-dimensional (2D) molybdenum disulphide (MoS_2) belongs to a class of materials called transition metal dichalcogenides (TMDCs). Due to its large surface area, high biocompatibility, non-zero band-gap, structural versatility and mechanical flexibility, MoS_2 can be used for next-generation sensing devices. In this work, printable high viscosity inks containing MoS_2 in various ratios and ethyl cellulose (EC) as a binder were prepared. The effect of different MoS_2 concentrations was investigated by measuring inks rheological properties. Structural and chemical properties of electrodes were studied by optical microscopy and SEM/EDX analysis. In addition, prepared MoS_2/EC inks were screen-printed onto fluorine-doped tin oxide (FTO) coated glass substrates to develop working electrodes for electrochemical sensing of dopamine. Finally, differential pulse voltammetry measurements by applying FTO/MoS_2 sensors showed that dopamine's limit of detection (LOD) was below the micromolar level. This work marks to the potential for fast, simple and mass printing production of MoS_2 sensors to detect further target analytes.

Keywords: Molybdenum disulfide, screen-printing, printed electronics, printable inks, electrochemical sensors

1. INTRODUCTION

Molybdenum disulfide (MoS_2) is a typical layered transition metal sulfide. Its structure consists of three stacked atom layers (S-Mo-S) held together by weak van der Waals forces [1]. MoS_2 has unique electronic, optical properties, and due to biocompatibility, high electrochemical catalytic activity, electron transfer ability and specific surface area is suitable for sensing and biosensing applications [2]. Therefore, MoS_2 can be used for electrochemical detection of dopamine, glucose, tyrosine and other target analyses [2-4].

Dopamine (DA) is a neurotransmitter regulating the function of the central nervous and cardiovascular systems in the human brain. Deficiency of DA can lead to various neurological diseases, e.g. schizophrenia, Alzheimer´s and Parkinson´s disease. Hence, the determination of the DA concentrations is beneficial for disease diagnosis [5]. Standard sensors for detecting various analyses are prepared by modifying commercial carbon screen-printed electrodes (SPE). This step is usually performed by the drop-casting method [2,4,6]. A drop-casting method is defined as the process of dispersing an electrocatalytic material in a suitable solvent followed by the pipetting of the dispersion onto a working carbon electrode. The solvent evaporates, and electrocatalytic material is immobilized on the electrode surface. Although drop-casting is a simple method, it has several disadvantages, e.g. low reproducibility or uncontrollable distribution of the deposited material [7]. In contrast, printing techniques (screen-printing) have high reproducibility and the possibility to prepare various fine structures for a wide range of layer thicknesses. Rowley-Neale et al. prepared screen-printable carbon/MoS_2 inks by the modification of commercial carbon-graphite ink with MoS_2 particles. In their work, screen-printed carbon/MoS_2 working electrodes were used for electrochemical evaluation of oxygen reduction reaction (ORR) [7]. This work aims at the fast and simple preparation of printable high viscosity MoS_2 inks, which can be used to fabricate screen-printed FTO/MoS_2 sensors for dopamine detection at submicromolar level.

2. EXPERIMENTAL SECTION

Materials

Molybdenum disulfide (MoS_2) with particle size ~ 6 µm, ethylcellulose (viscosity 22 cP 5 % in toluene/ethanol 80:20), terpineol (≥ 96 %, bp 213 °C) and FTO/glass substrates (fluorine-doped tin oxide, 7Ω/sq) were purchased from Sigma Aldrich. Dopamine hydrochloride (DA, $C_8H_{11}NO_2$.HCl) and phosphate buffer (PB) components (KH_2PO_4 and K_2HPO_4, pH 7.0) used for electrochemical analysis were also ordered from Sigma Aldrich. All solutions were freshly prepared in ultrapure deionized water (0.055 µS/cm).

MoS_2 working electrode preparation

The first step was the preparation of screen-printable MoS_2 inks. The polymeric binder was prepared by dissolution 8 wt% of ethylcellulose in terpineol. Subsequently, different concentrations of MoS_2 powder (25, 45 and 60 wt%; samples Mo-25, 45 and 60) were homogenized with a polymeric binder in a hand-held mixing unit. The second step was the screen-printing process, where prepared MoS_2 inks were printed on 15×20 mm FTO substrates and used as working electrodes (**Figure 1**). Before printing, FTO substrates were gradually cleaned in an ultrasonic bath using detergent, acetone and isopropyl alcohol (IPA) for 20 minutes in every solvent. Substrates were also UV-C pretreated for 20 minutes. The yellow high modulus polyester yarn mesh, with mesh count 71 cm^{-1} and 48 µm thread diameter (PME 71-48 Y, SEFAR), was used in the screen-printing process. The deposition was performed on the manual screen-printing machine (Drucktech). After printing, MoS_2 layers were left for 10 minutes for levelling and then dried in a laboratory oven at 120 °C for 30 minutes.

Figure 1 Image of screen-printed MoS_2 working electrodes on FTO substrates for electrochemical sensing

Characterization methods

The rheological behaviour of MoS_2 inks was characterized with a rheometer (HAAKE, MARS iQ) using a 35 mm diameter parallel plate geometry with a gap set at 0.4 mm. The temperature was 25 °C. Dynamic viscosity [η (Pa.s)] and shear stress [τ (Pa)] were measured at a shear rate [$\dot{\gamma}$ (s^{-1})] from 0 to 2000 s^{-1}. The level of thixotropy was characterized by time-dependent structural regeneration after shearing. Structural and chemical properties of MoS_2 printed layers were analysed by optical microscopy (LEICA DM 2700 M) and scanning electron microscopy with energy-dispersive X-ray spectroscopy (SEM/EDX, JEOL JSM-IT500HR). Differential pulse voltammetry (DPV) was used for the detection of dopamine. MoS_2 layers on FTO substrates were used as working electrodes; the platinum wire was used as the counter-electrode and a argentochloride electrode (Ag/AgCl/3 M KCl) as the reference electrode.

3. RESULTS AND DISCUSSION

In the first step, we analyzed the rheological behavior of prepared viscous and screen-printable MoS_2 inks with different concentrations of MoS_2. **Figure 2** shows flow curve (a) and viscosity curve for typical pseudoplastic behaviour. This behaviour is characterized by decreasing dynamic viscosity with increasing shear rates and is crucial for the screen-printing process. It is also obvious that the structural breakdown of more concentrated MoS_2 ink (Mo-60) occurs at a lower shear rate (~140 s^{-1}) compared to ink with 25 wt% of MoS_2 (~1170 s^{-1}). Dependence of shear stress on shear rate was measured from 0 to 1200 s^{-1}, which corresponds to the printing process when ink is in the rest (low $\dot{\gamma}$) and when is pressed through the screen mesh (high $\dot{\gamma}$). Values of dynamic viscosity for selected shear rates: 10, 100 and 400 s^{-1} are listed in **Table1**.

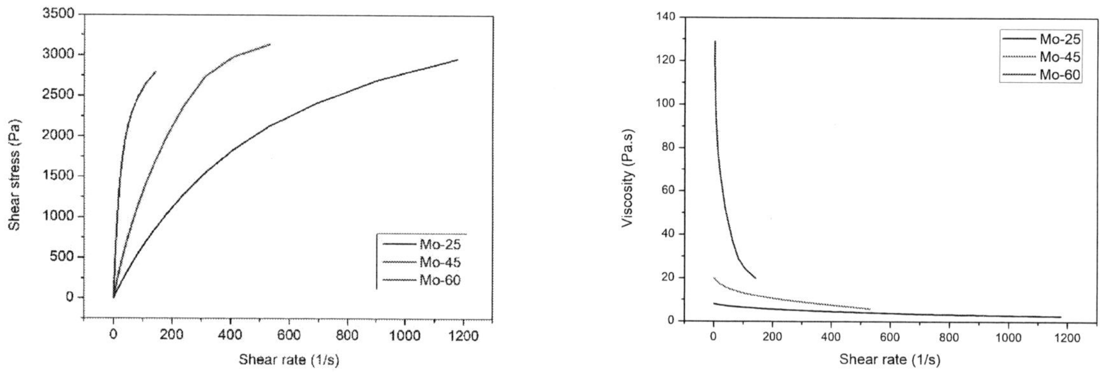

Figure 2 Flow curve (left) and viscosity curve (right) for printable inks with 25, 45 and 60 wt% of MoS_2 (Mo-25, 45 and 60)

Table 1 Dynamic viscosity of screen-printable MoS_2 inks at different shear rates

	10	**100**	**400**
Mo-25	8	6	4
Mo-45	19	13	7
Mo-60	85	24	-

To investigate inks behaviour during the screen-printing process, time-dependent flow curves were measured (**Figure 3**). The first interval of the curve simulates behaviour at rest at a preset low shear rate (0.5 s^{-1}). The second interval simulates the structural breakdown of the ink during the screen-printing process at a preset high shear rate (2000 s^{-1}). The third interval simulates structural regeneration at rest after application using the same preset low shear rate as in the first interval. The recovery rates of samples Mo-25 and Mo-45 were 74 and 90 %, respectively. In the case of the sample Mo-45, viscosity regeneration was rapid from the beginning and then increased linearly (**Figures 3 (a), (b)**). **Figure 3 (c)** shows that structure of sample Mo-60 was totally destroyed after applying high shear rates, and thus viscosity regeneration could not be evaluated.

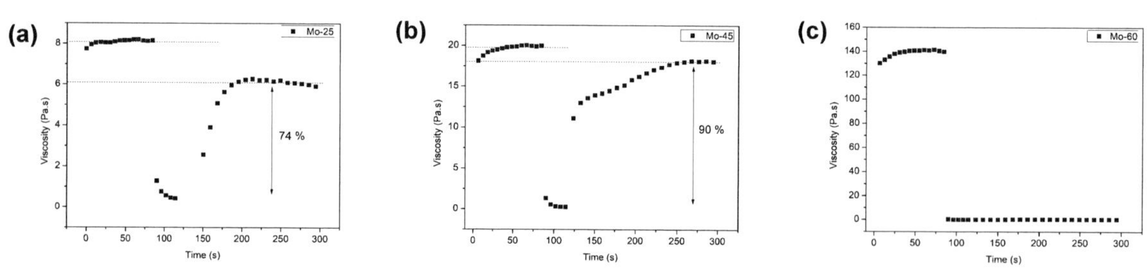

Figure 3 Comparison of the structural regeneration of inks with different MoS_2 concentrations using time-dependent viscosity curves; (a) 25, (b) 45 and (c) 60 wt% MoS_2

The structural and chemical properties of screen-printed MoS_2 layers were analyzed. The thickness of screen-printed MoS_2 layers was measured using an optical microscope, by a 3D image sequential recording method. The recorded values confirmed the influence of particle concentration. The thickness of layers increased with the amount of MoS_2 particles (**Table 2**).

Table 2 Thickness of screen-printed MoS_2 layers

Sample	Mo-25	Mo-45	Mo-60
Thickness (µm)	4 ± 1	11 ± 2	18 ± 1

Figure 4 shows results of SEM/EDX analysis where it can be clearly seen that the layer of sample Mo-25 is not as homogenous as thicker layers of samples with higher MoS_2 content. The atomic representation of elements (atom %) was further investigated through EDX and confirmed the presence of Sn, which is a component of FTO substrates. The atom% of C, which belongs to the binder, decrease with increasing MoS_2 concentration (**Table 3**). **Table 3** also shows that the atomic ratios of S/Mo are 2.4, which corresponds to MoS_2 structure.

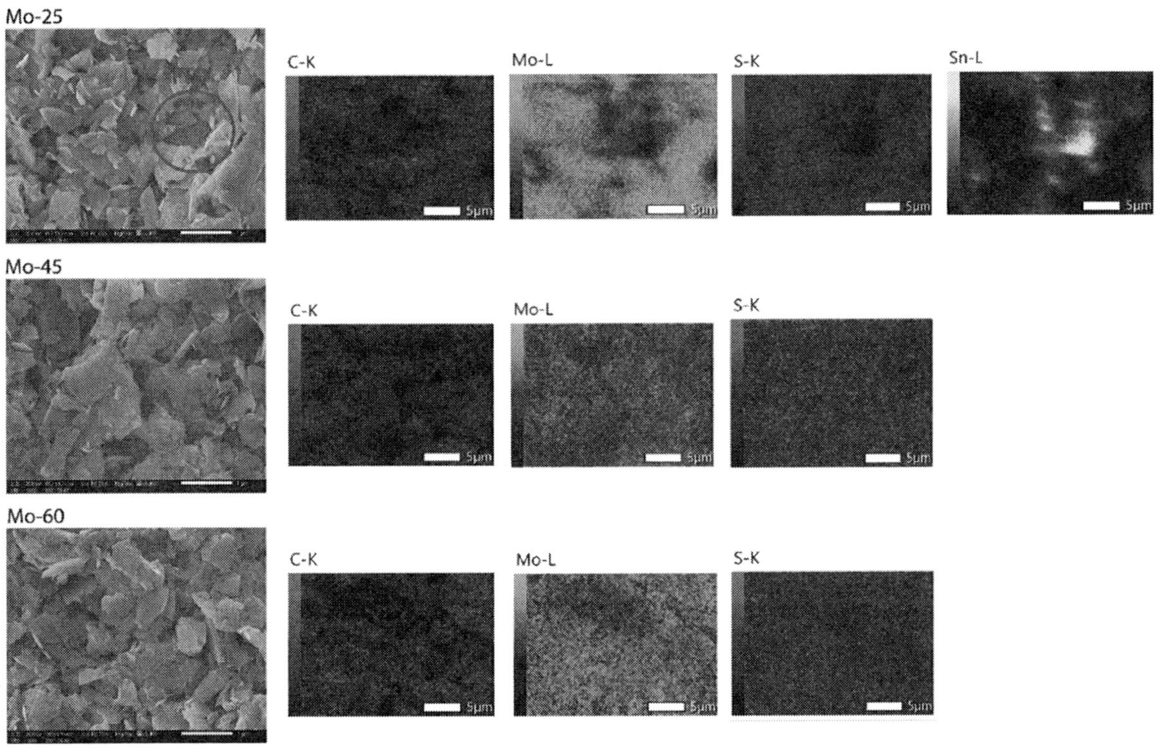

Figure 4 SEM/EDX analysis of printed Mo-25, Mo-45 and Mo-60 layers

Table 3 Atomic representation of elements in MoS_2 printed layers obtained by EDX analysis

	Atom%		
	Mo-25	**Mo-45**	**Mo-60**
C	48.21 ± 0.06	41.35 ± 0.06	39.55 ± 0.06
S	33.07 ± 0.02	41.53 ± 0.02	42.75 ± 0.02
Mo	14.02 ± 0.02	17.13 ± 0.03	17.70 ± 0.02
Sn	4.70 ± 0.01	-	-

The electrochemical behaviour of screen-printed FTO/MoS_2 electrodes was assessed by DPV experiment. The DPV curves showing electrochemical oxidation of dopamine in 0.1 M phosphate buffer (pH 7.0) as a supporting electrolyte at above-mentioned manufactured FTO/MoS_2 electrodes can be clearly seen in **Figure 5**. The linear dependence of i_p (oxidation peak current) *vs.* c_{DA} (concentration of dopamine) was investigated in the range of 10 - 300 µM for the measured analyte. The limit of detection (LOD) was calculated according to a signal-to-noise ratio (S/N) = 3. The best LOD of 0.342 µM (R^2 = 0.995) was evaluated for Mo-45 sample. LODs of 2.1 µM (R^2 = 0.996) and 1.2 µM (R^2 = 0.982) were achieved for Mo-25 and Mo-60 samples, respectively.

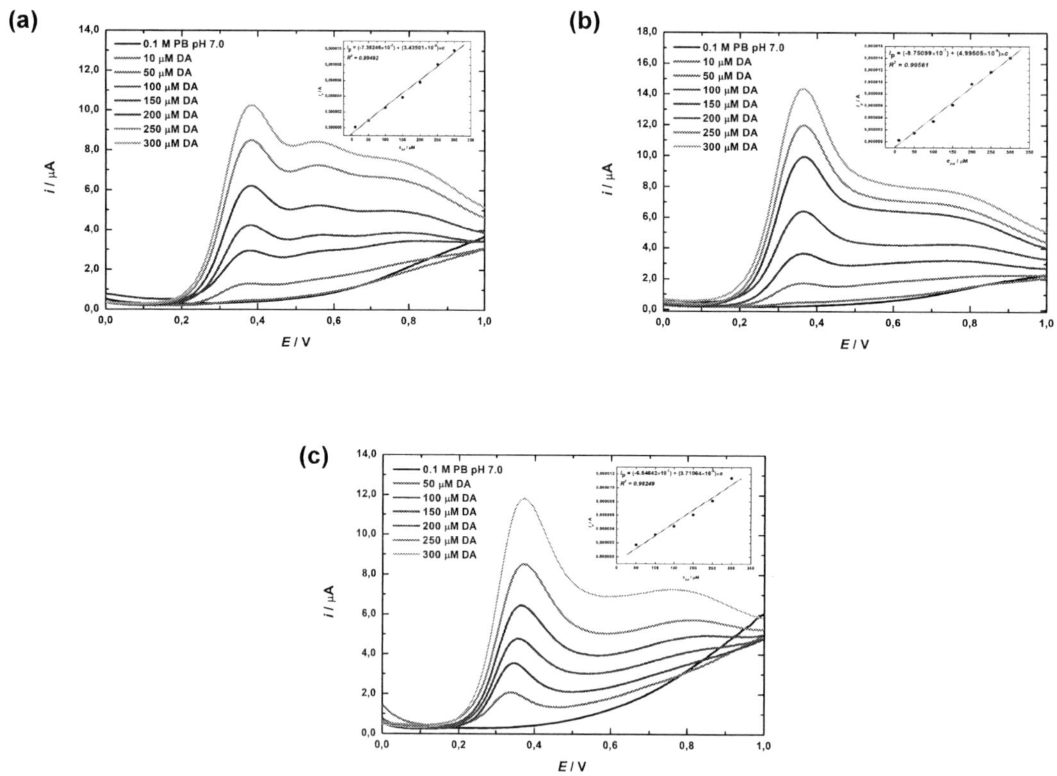

Figure 5 DPVs recorded for different concentrations of DA at Mo-25 (a), Mo-45 (b) and Mo-60 (c) in 0.1 M PB pH 7.0

4. CONCLUSION

In this work, printable inks were prepared containing different concentrations of MoS_2 and ethylcellulose as a binder. After rheological behaviour analysis, MoS_2 inks were screen-printed on substrates with a conductive FTO layer and were used as a working electrode for the electrochemical determination of dopamine. Prepared MoS_2 inks exhibit pseudoplastic flow behaviour. The printable ink containing 45 wt% MoS_2 showed the ideal rheological properties because 90 % of viscosity recovered, which is important in the screen-printing process. SEM/EDX analysis confirmed that the layer of sample Mo-45 is homogenous, and the Sn from FTO substrate is not present. Finally, FTO/MoS_2 working electrodes for the detection of dopamine were successfully screen-printed. The differential pulse voltammetry measurements showed that the best limit of detection of 0.342 µM was evaluated for the working electrode printed from ink containing 45 wt% MoS_2.This work confirmed that it is possible to prepare MoS_2 sensors by simple screen-printing technique and that the sensors show great potential towards detection of further target analytes.

ACKNOWLEDGEMENTS

This research was supported by the Slovak Grant Agency (VEGA 1/0488/19), the STU Grant Scheme for Support of Young Researches "GRAPHENEPRINT" and APVV-17-0300.

REFERENCES

[1] CHENG, M., ZHANG, X.,WANG, M., HUANG, H., MA, J. A facile electrochemical sensor based on well-dispersed graphene-molybdenum disulfide modified electrode for highly sensitive detection of dopamine.*J. Electroanal. Chem.* [online]. 2017, vol. 786, pp. 1-7. Available from: https://doi.org/10.1016/J.JELECHEM.2017.01.012.

[2] VAN TUAN, D., THUY NGAN, D.T., THUY, N.T., LAN, H., NGUYET, N.T., VAN THU, V., HUNG, V., TAM, P.H. Effect of nanostructured MoS_2 morphology on the glucose sensing of electrochemical biosensors. *Curr. Appl. Phys.* [online]. 2020, vol. 20, no. 9, pp. 1090-1096. Available from: https://doi.org/10.1016/j.cap.2020.06.027.

[3] SOOKHAKIAN, M., BASIRUN, W.J., GOH, B.T., WOI, P.M., ALIAS, Y.Molybdenum disulfide nanosheet decorated with silver nanoparticles for selective detection of dopamine. *Colloids Surfaces B Biointerfaces.* [online]. 2019, vol. 176, pp. 80-86. Available from: https://doi.org/10.1016/J.COLSURFB.2018.12.058.

[4] ZRIBI, R., MAALEJ, R., MESSINA, E., GILLIBERT, R., DONATO, M.G., MARAGÓ, O.M., GUCCIARDI, P.G., LEONARDI, D.G., NERI, G. Exfoliated 2D-MoS_2 nanosheets on carbon and gold screen printed electrodes for enzyme-free electrochemical sensing of tyrosine. *Sensors Actuators, B Chem.* [online]. 2020, vol. 303, p. 127229. Available from: https://doi.org/10.1016/j.snb.2019.127229.

[5] YIN, A., WEI, X., CAO, Y., Li, H. High-quality molybdenum disulfide nanosheets with 3D structure for electrochemical sensing.*Appl. Surf. Sci.* [online]. 2016, vol. 385, pp. 63-71. Available from: https://doi.org/10.1016/J.APSUSC.2016.05.066.

[6] PARRA-ALFAMBRA, A.M., CASERO, E., VÁZQUEZ, L., QUINTANA, C., DEL POZO, M.,PETIT-DOMÍNGUEZ, M.D. MoS_2 nanosheets for improving analytical performance of lactate biosensors. *Sensors Actuators, B Chem.* [online]. 2018, vol. 274, pp. 310-317. Available from: https://doi.org/10.1016/j.snb.2018.07.124.

[7] ROWLEY-NEALE, S.J., SMITH, G.C., BANKS, C.E.Mass-Producible 2D-MoS_2-Impregnated Screen-Printed Electrodes That Demonstrate Efficient Electrocatalysis toward the Oxygen Reduction Reaction. *ACS Appl. Mater. Interfaces.* [online]. 2017, vol. 9, no. 27, pp. 22539-22548. Available from: https://doi.org/10.1021/ACSAMI.7B05104.

PLASMON-MEDIATED ELECTROCATALYTIC ACTIVITY OF TIB2 TOWARDS WATER SPLITTING

[1]Vladislav BURAVETS, [1]Anna ZABELINA, [1]Elena MILIUTINA, [2]Daniela POPELKOVÁ, [3]Jiří HENYCH, [2]Martin ŠŤASTNÝ, [1]Václav ŠVORČÍK, [1]Oleksiy LYUTAKOV

[1]*University of Chemistry and Technology, Prague, Czech Republic, EU, info@vscht.cz*

[2]*Institute of Inorganic Chemistry of the Czech Academy of Sciences, Husinec-Řež, Czech Republic, EU, sekretar@iic.cas.cz*

[3]*Jan Evangelista Purkyně University, Ústí nad Labem, Czech Republic, EU, incoming@ujep.cz*

https://doi.org/10.37904/nanocon.2021.4341

Abstract

Environmental problems related to the economy based on fossil fuels are of paramount importance. However, transition to renewable energy sources is restrained by the availability of storage technologies. Electrochemistry is a widely recognized prominent tool to achieve this goal by converting renewable energy into the form of chemical bonds accessible further as fuels, such as hydrogen produced by water splitting. Crucial losses in such process are caused by the high overpotentials, required for water splitting as a hydrogen source. To achieve required efficiency of water splitting appropriate catalysts have to be found with the suitable combination of activity, stability and cost. Nano-structured, two-dimensional materials (2D) are attractive candidates due possessing many of the desired properties and highly tunable characteristics. Employing light allows additional degree of freedom to boost conventional photo-electrocatalysis, in particular plasmon-mediated electrocatalysis. Enhancement of the catalytic activity can be increased even further if the catalytic system absorbs wide range of light spectrum. In this study, we combine plasmon-active Au grating with 2D flakes of TiB_2, to perform plasmon-mediated water splitting half-reaction - hydrogen evolution.

Keywords: Hydrogen evolution, photo-electrocatalysis, plasmonics, 2D materials

1. INTRODUCTION

Nowadays society consumes on the range of 16 TW*hours of primary energy per year, with 84.3 % of it coming from fossil fuels [1]. This causes a significant impact on the environment being one of the major causes of climate crisis. In order to switch for renewable energy sources to mitigate impact on the environment, a method for efficient storage and use on demand of this energy has to be found. Hydrogen is a prominent candidate for such a purpose due to high energy density and water as the only by-product during combustion [2]. Efficient water electrolysis is an important technology for energy conversion into H_2 for the sustainable society [3,4]. A targeted catalytic activity is proposed to require overpotential lower than 0.45V combined for both half-reactions - oxygen evolution reaction (OER) and hydrogen evolution reaction (HER) [5]. Thus, different opportunities to enhance both half-reactions are required. A promising approach to enhance catalytic activity is to employ light and plasmonic materials to enhance the light absorption and widen range of useful light spectrum [6]. The plasmonic resonant wavelength depends on the material, size and shape of the catalyst and can be fine-tuned by manipulating this variables [7].

In this work, we propose a model to enhance catalytic activity towards HER by combining plasmon-active materials and 2D flakes (TiB_2). Proposed approach allows us to significantly increase the HER efficiency, under the plasmon assistance in photo-electrochemical process.

2. EXPERIMENTAL PART

2.1. Materials

Au target for metal deposition (4N purity) was purchased from Safina (Czech Republic). Sulfuric acid, titanium boride (TiB_2), chloroauric acid ($HAuCl_4$), trisodium citrate ($Na_3C_6H_5O_7$) were purchased from Sigma-Aldrich.

2.2. Sample preparation

For AuNPs synthesis Turkevich method was used [8]. AuNPs were synthesized by reduction of gold from the $HAuCl_4$ solution initialized by sodium tris-citrate. For preparation of Au-grating a commercially available polymer grating were used as template and covered by thin gold layer. TiB_2 2D flakes were obtained by exfoliation in water/alcohol mixture under the high-power ultrasound treatment.

2.3. Characterization methods

Ultraviolet-visible spectra were measured using Spectrometer Lambda 25 (Perkin-Elmer). Crystal structure of the TiB_2 was investigated by XRD technique (PANalytical X'Pert PRO). The micro-morphology and elemental analysis of the samples were performed by high-resolution transmission electron microscopy EFTEM Jeol 2200 FS microscope (Jeol, Japan). Linear sweep voltammetry measurements were carried out with PalmSense4 potentiostat, controlled by PSTrace5 software. Experiments were performed in a 15 ml quartz cuvette, using a three-electrode configuration at room temperature, with platinum and Ag/AgCl electrodes as counter and reference, respectively. The LEDs light sources (Thorlabs) with different wavelengths were placed at a distance of ~1 cm from sample surface and the power on sample surface was adjusted to 0.4 W/cm^2.

3. RESULTS AND DISCUSSION

First, delaminated TiB_2 flakes were subjected to XRD measurements. Results are presented in **Figure 1** and indicate that obtained XRD pattern correspond well with previously published (reference number 03-065-1073).

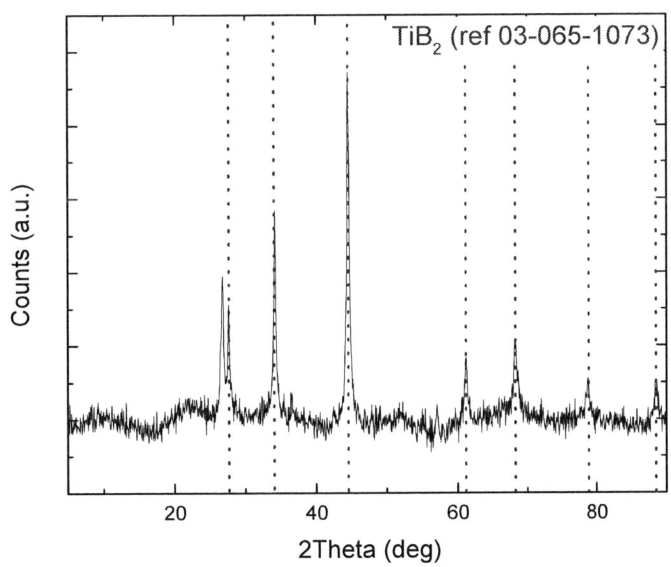

Figure 1 XRD diffraction pattern of ultrasonicated TiB_2

The absorption of all used materials (i.e. 2D flakes, gold nanoparticles and gold grating) are presented in **Figure 2**. It can be seen, that main absorption peaks are spectrally separated, which should ensure the excellent overlapping with sunlight spectrum and allow to utilize sunlight spectrum in plasmon*-assisted photo-electrochemical water splitting.

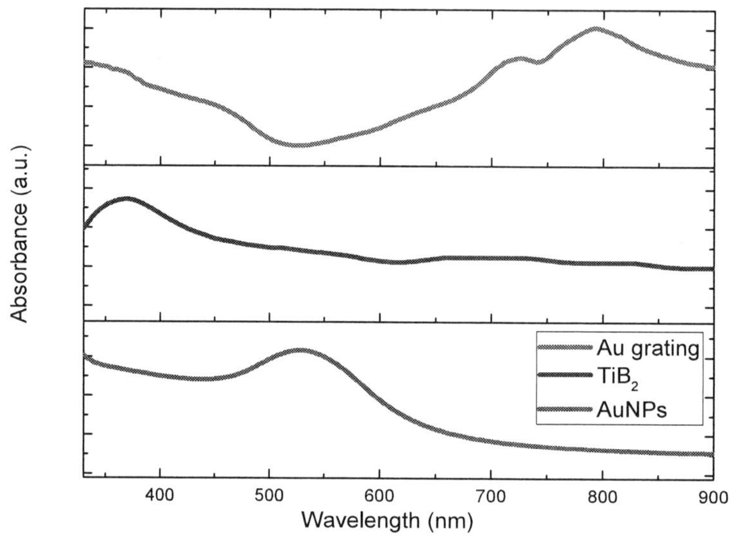

Figure 2 UV-Vis spectra of the Au grating (red), ultrasonicated TiB2 (blue) and Au nanoparticles (green)

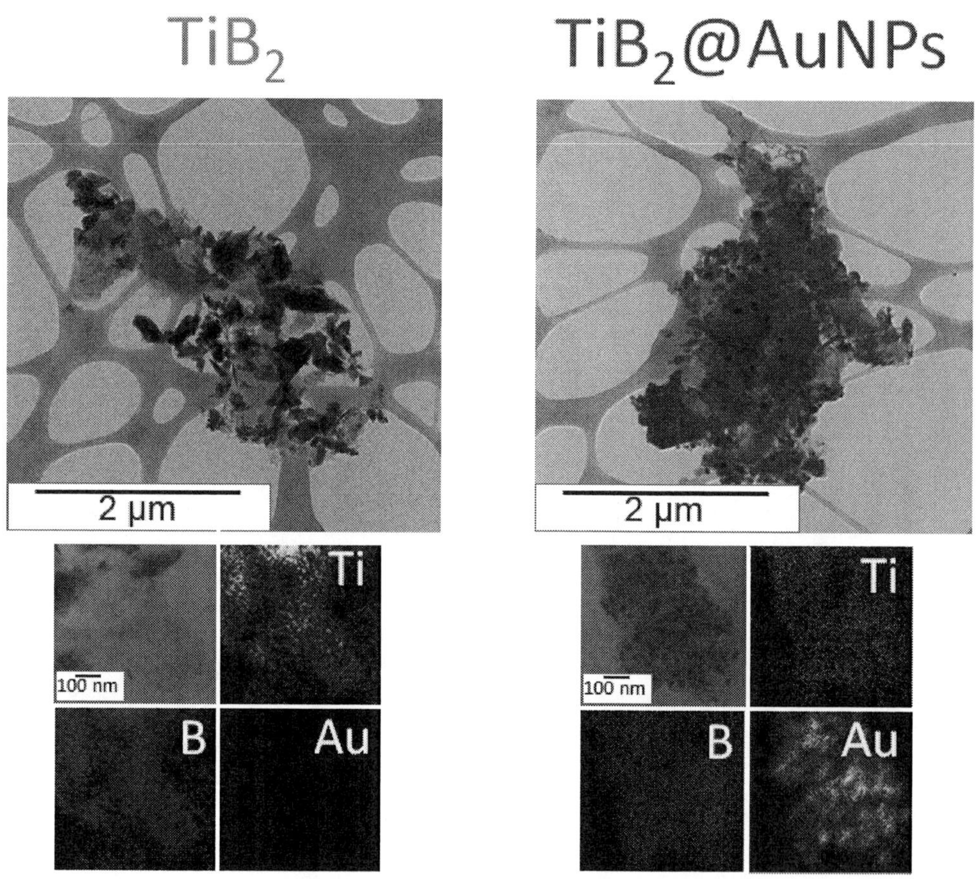

Figure 3 TEM images (top) and TEM/elemental mapping merged images (bottom) of TiB2 paticle (left) and TiB2 particle populated with AuNPs (right)

In **Figure 3** TEM images and TEM-EDX elemental mapping are presented. It can be seen that TiB_2 particles are of the similar size after modification with AuNPs. In turn, TEM-EDX mapping indicates that after surface-assisted grafting of nanoparticles, the TiB_2 flakes surface are uniformly covered by AuNPs.

Linear sweep voltammetry (LSV) experiments, performed in dark or under illumination are shown in **Figure 1**. It is evident that Illumination of Au grating with red light (**Figure 4A**) resulted in the increase of current, hence, rate of the hydrogen evolution reaction. However, illuminating by the blue or green light did not result in the further enhancement of the catalytic activity. When TiB_2 was deposited on the Au grating (**Figure 4B**) it was possible to harvest light of wider spectral range which also caused increase of the HER rate. Illuminating such system as TiB_2/Au grating with red light resulted in similar current as obtained for the solely Au grating, while blue and green lights caused further increase in the current. Recorded currents under the green and blue lights were very similar on the TiB_2/Au grating. Finally, to harvest the light of green spectrum, AuNPs were seeded on TiB_2. Current response of such system is shown in (**Figure 4C**). It can be seen, that a further increase of materials HER catalytic activity under illumination by green light was achieved. So, we are showing, that different combination of plasmon active and catalytic-active 2D materials with different absorption wavelengths allows to increase the light harvesting with corresponded enhancement of photo-electrochemical catalytic activity towards hydrogen evolution reaction.

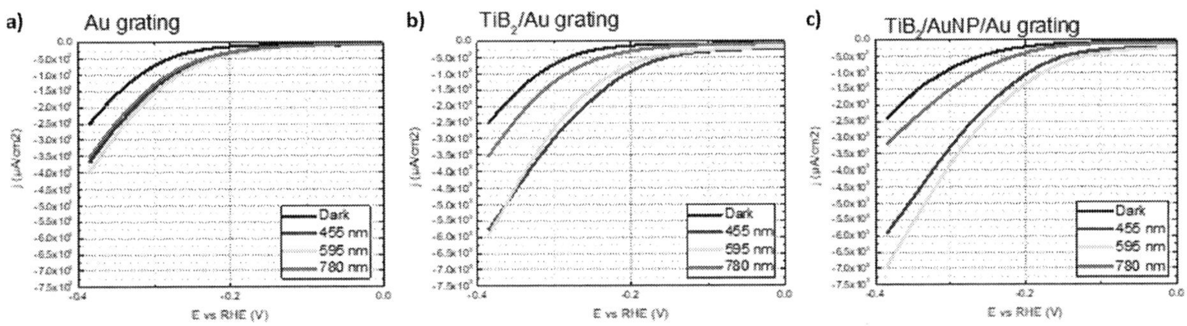

Figure 4 Linear sweep voltammetry (LSV) of Au grating (a), TiB2/Au grating (b) and TiB2/AuNPs/Au grating (c) in dark and under LED illumination with wavelengths 455 nm (blue), 595 nm (green) and 780 nm (red)

4. CONCLUSION

A system to harvest different range of (sun)light spectrum was proposed based on the different combination of plasmon active materials (Au grating and AuNPs) and catalytically-active 2D flakes (TiB_2). Each of three components has shown its own peak of absorbance in the visible and near UV (365 nm) wavelength range. Created hybrid systems were tested for its photo-electrocatalytic activity towards HER. The samples structure and materials coupling was characterized/confirmed using XRD, UV-Vis, and TEM-EDX techniques. The functionality of created materials was checked in water splitting half-reaction - hydrogen evolution using the photo-electrochemical measurements in LSV mode (performed in dark or under illumination). A significant increase in current was achieved under illumination by light with different wavelengths allowing potential enhancing HER rate by wide range of light spectrum. In particular it was shown, that plasmon triggering and corresponded increase in catalytic activity TiB_2 can be significantly reached under the application of wavelengths, corresponded to plasmon absorption band.

ACKNOWLEDGEMENTS

This work was supported by the MV, VS IMPAKT 1, grant number VJ01010065.

REFERENCES

[1] bp global. *Statistical Review of World Energy.* [online]. 2021. [viewed 2021-11-07] Available from: https://www.bp.com/en/global/corporate/energy-economics/statistical-review-of-world-energy.html.

[2] TURNER JA. Sustainable Hydrogen Production. *Science.* [online]. 2004, no. 305, pp. 972-974. Available from: https://doi.org/10.1126/science.1103197.

[3] GUSELNIKOVA, O., TRELIN, A., MILIUTINA, E., ELASHNIKOV, R., SAJDL, P., POSTNIKOV, P., KOLSKA, Z., SVORCIK, V., LYUTAKOV, O. Plasmon-induced water splitting-through flexible hybrid 2D architecture up to hydrogen from seawater under NIR Light. *ACS Appl Mater Interfac.* [online]. 2020, no.12, pp. 28110-28119. Available from: https://doi.org/10.1021/acsami.0c04029.

[4] ZABELINA, A., ZABELINA, D., MILIUTINA, E., LANCOK, J., SVORCIK, V., CHERTOPALOV S., & LYUTAKOV, O. Surface plasmon-polariton triggering of Ti 3 C 2 T x MXene catalytic activity for hydrogen evolution reaction enhancement. *J Mater Chem A.* [online]. 2021, no. 9, pp.17770-17779. Available from: https://doi.org/10.1039/D1TA04505A.

[5] MCRORY CCL, JUNG S, FERRER IM, CHATMAN SM, PETERS JC, JARAMILLO TF. Benchmarking Hydrogen Evolving Reaction and Oxygen Evolving Reaction Electrocatalysts for Solar Water Splitting Devices. *J Am Chem Soc.* [online]. 2015, no. 137, pp.4347-4357. Available from: https://doi.org/10.1021/ja510442p.

[6] CHOI CH, CHUNG K, NGUYEN T-TH, KIM DH. Plasmon-Mediated Electrocatalysis for Sustainable Energy: From Electrochemical Conversion of Different Feedstocks to Fuel Cell Reactions. *ACS Energy Lett.* [online]. 2018, no. 3, pp. 1415-1433. Available from: https://doi.org/10.1021/acsenergylett.8b00461.

[7] LINIC S, CHRISTOPHER P, INGRAM DB. Plasmonic-metal nanostructures for efficient conversion of solar to chemical energy. *Nature Mater.* [online]. 2011, no. 10, pp. 911-921. Available from: https://doi.org/10.1038/nmat3151.

[8] KIMLING J, MAIER M, OKENVE B, KOTAIDIS V, BALLOT H, PLECH A. Turkevich Method for Gold Nanoparticle Synthesis Revisited. *J Phys Chem.* [online]. 2006, no. 110, pp. 15700-15707. Available from: https://doi.org/10.1021/jp061667w.

SUPERABSORBENTS ENRICHED BY ANORGANIC (NPK) AND ORGANIC (LIGNOHUMATE) NUTRIENTS IN ENVIRONMENTAL SYSTEMS

Petra ZÁVODSKÁ, Martina KLUČÁKOVÁ

Brno University of technology, Faculty of Chemistry, Brno, Czech Republic, EU, Petra.Zavodska@vut.cz

https://doi.org/10.37904/nanocon.2021.4337

Abstract

This contribution is focused on superabsorbent materials based on polymerization of partially neutralized acrylic acid. Several types of superabsorbents were prepared. They differed by content of inorganic (NPK fertilizer) and organic (lignohumate) nutrients for their release in soil systems. Superabsorbents were characterized from point of view of their swelling capacity in order to evaluate their ability to control water management soil. Gradual release of incorporated nutrient into water was monitored in time and complemented by measurements of pH and conductivity, UV-VIS, ICP-OES and Ion chromatography. It was found that the nutrients were released from superabsorbents for the duration of several days in the dependence on the type of prepared superabsorbent. In conclusion, simple pot experiments with corn were realized. Corn was grown in presence of each type of superabsorbent and even without it to get differences not only between samples but between samples and corn itself. It was confirmed that the application of superabsorbents can improve soil properties, water management, nutrient uptake, and growth of plants.

Keywords: Superabsorbents, NPK, lignohumate, nutrients, soil system

1. INTRODUCTION

Superabsorbents (SAPs) are three-dimensionally crosslinked hydrophilic polymers that are capable of swelling and retaining large volumes of water in swollen state [1]. They're not only able to absorb large amount of water but they can do it during short period of time. Also, they can retain absorbed water even under the pressure. Due to this fact they are used widely in traditional water-absorbing materials, such as nappies, hygienic napkins and so on [2,3]. For their ability to absorb and contain water and ability to carry nutrients essential for plants, they are used in agriculture as source of water and nutrients. SAPs are used in agriculture not only for this purpose but also for their help to reduce irrigation water consumption, reduce the death rate of plants, improve fertilizer retention in soil, and increase plant growth rate [4]. SAPs find usage in many industries thanks to their ability to absorb water. For example, they are used in diapers or pads. Except for personal care industry they can be find in the construction industry as an addition to concrete in places where is a chance of leaking water, for example they were used construction of English Channel tunnel [5].

Nutrients are essential for plants to be able to grow, develop and produce at their best. Plants need approx. 17 essential nutrients that were grouped into three groups by the amount that plants need. The largest amount of nutrients that plants needs are called primary nutrients or macronutrients. Primary nutrients contain carbon, hydrogen, nitrogen, oxygen, phosphorus, and potassium. Secondary nutrients are needed in moderate amounts compared to primary nutrients and contain calcium, magnesium, and sulphur. Last are micronutrients that are needed in tiny amounts and contain boron, chlorine, copper, iron, manganese, molybdenum, and zinc [6]. As mentioned before plants need nutrients for their growth. Each type of plant has its own optimum nutrient range as well as minimum required level. If the nutrient amount doesn't reach the minimum level plants start to show deficiency symptoms specific to each nutrient. Excessive uptake can also cause improper growth because higher concentration of nutrients can be toxic for the plant [7]. Therefore, motivation for this work was to determinate superabsorbents' ability to absorb and hold water and nutrients and to use it in pot experiment that simulated natural conditions in fields. Results from this work could be very helpful to use SAPs in field and especially in the places where it's normally difficult to grow plants.

2. EXPERIMENTS

Six different SAPs were prepared (**Table 1**) using acrylic acid (AA) as monomer for polymerization. Acrylic acid was partially neutralized by adding potassium hydroxide (KOH). Solution of acrylic acid was placed on magnetic stirrer and while stirring constantly potassium hydroxide was added by drops. After partial neutralization cross-linker was added, in this case it was methylene bisacrylamide (MBA). Potassium persulfate (KPS) was added after and was used as an initiating agent. As inorganic nutrients we used NPK and as organic nutrients we used lignohumate (LH) which were added to the mixture and heated up to range of temperatures from 90 to 100 °C based on the type of SAP. Every type of SAP was labelled by a letter, specifically letters A, B, C, D, H and I and their exact recipe is shown in **Table 1**. Prepared SAPs were dried in dryer before they were used for swelling capacity measurements.

Table 1 Recipe of used SAPs [8]

SAP	A	B	C	D	H	I
AA (g)	14.25	14.25	14.25	14.25	14.25	14.25
KOH (g)	4.75	4.75	4.75	4.75	4.75	4.75
MBA (g)	0.016	0.016	0.016	0.016	0.016	0.016
KPS (g)	0.5	0.5	0.5	0.5	0.5	0.5
NPK (g)	0.6602	6.602	0.6602	6.602	-	-
LH (g)	-	-	1	1	1	-

Prepared SAPs were characterized by point of view of their swelling capacity. Approximately 50 mg of each SAP was mixed with 100 ml of distilled water and at 24 hours interval for four days was swelling capacity measured by weighing swelled SAPs. SAPs were taken out after the time interval and extracts were used for measuring pH and conductivity, UV-VIS, ICP-OES and ion chromatography (IC) to monitor release of nutrients from SAPs.

For the study of SAPs influence on plants growth pot experiments with corn were performed. Firstly, corn was sprouted by immersing in the water for 24 hours and put on wet cotton wool then. Sprouted corns were planted to pots with approximately 250 g of soil and 1 g of each SAPs. Control samples were also prepared, so corn was planted to pots with only approx. 250 g of soil. Experiment was performed without water-stress, so every week the plants were watering with approx. 50 ml of water. Three pots were made for each sample and each time interval (46, 60 and 74 days). In one pot were three sprouted corns set. For three time periods it makes 63 pots and 189 sprouted corns total.

3. RESULTS AND DISCUSSION

Six types of SAPs were prepared following the recipe shown in **Table 1**. Each one of them was characterized by swelling capacity that shows that sample H absorbed the largest amount of water comparing to other types. For example, after 72 hours was swelling coefficient equal to 19.4. On the contrary, the smallest amount of absorbed water is seen in sample D, where after 72 hours was swelling coefficient only 5.5. Chyba! Nenalezen zdroj odkazů. shows dependency of swelling capacity on time intervals for each SAPs. It's apparent that sample H (after 72 hours equals to 19.4) and sample C (after 72 hours equals to 15.2) have the largest swelling capacity and sample D (after 72 hours equals 5.5 which was the highest swelling coefficient for sample D) and

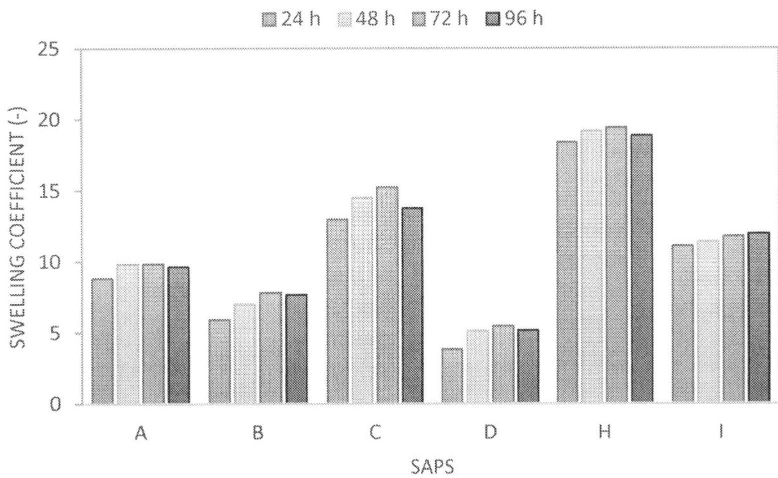

Figure 1 Swelling capacity of each type of SAPs

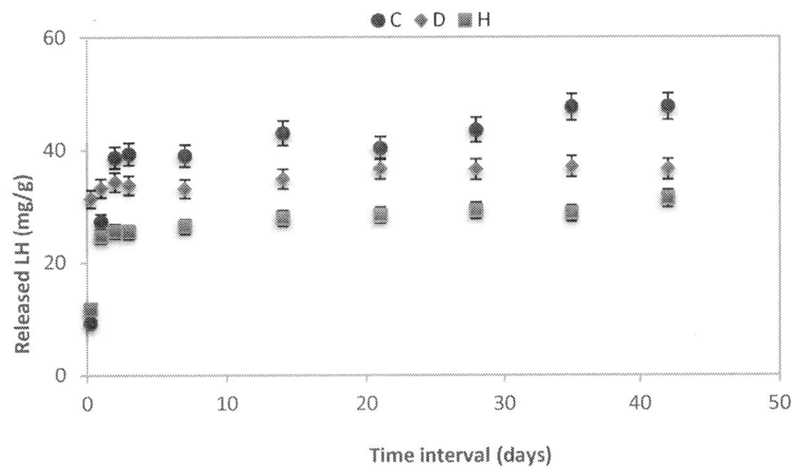

Figure 2 Results from UV-VIS measurement to monitor release of LH

sample B (after 72 equals to 7.8 which was again the highest swelling coefficient for sample B) has the smallest. According to our results shown in Chyba! Nenalezen zdroj odkazů. it is apparent that presence of lignohumate helps absorbing water whereas presence of higher amount of NPK reduces the possible amount of absorbed water. Results from pH and conductivity measurement (**Table 2**) shown that incorporated nutrients in SAPs were released and, in every type of SAP, the longer the time was the more nutrients were released. For example sample D, after 6 hours (0.25 day in **Table 2)** was pH equals to 5.37 and after 42 days was equals to 5.5. Conductivity for sample D after 6 hours was 255.7 µS and after 42 days was conductivity equal to 275.4 µS. Contrary to measurement of swelling capacity, NPK helps releasing nutrients. In addition, UV-VIS spectrum was measured for samples C, D and H to monitor release of organic nutrient lignohumate (Chyba! Nenalezen zdroj odkazů.). Results shows that sample C has the highest amount of released lignohumate comparing to other two samples. Specifically, after 42 was 47.67 mg/g from sample C released while in sample D was released 36.58 mg/g and in sample H only 31.42 mg/g. Results from ICP-OES (**Table 3**) and IC (**Table 4**) shown that in every measured nutrient were highest amounts measured by sample B and D. In **Table 2** and **Table 3** it's apparent that sample D released the highest amounts of nutrients. For example, release of potassium. In **Table 2** and **Table 3** we can see that in sample A released amount of potassium is equal to 30.87 % wt., in sample B 52.29 % wt., in sample C 35.78 % wt., in sample D 56.32 % wt., in sample H 33.94 % wt. and in sample I 28.96 % wt. As mentioned before sample B and D have the highest amounts of released nutrients. But between these two samples, sample D has higher amounts when releasing potassium, phosphorus and NO_3^-. It's apparent that presence of lignohumate helps release of nutrients.

Also, it's apparent that presence of NPK helps releasing lignohumate as an organic nutrient. At the same time the results shown that higher amount of NPK incorporated (in our work it's 6.602 g) slows releasing lignohumate. So, higher amount of NPK incorporated (6.602 g) has negative influence on swelling capacity and release of lignohumate. The only positive influence was in releasing inorganic nutrients where more elements could and were released.

Table 2 pH and conductivity measurement

Days	pH						Conductivity (μS)					
	A	B	C	D	H	I	A	B	C	D	H	I
0.25	5.65	5.37	5.61	5.38	5.74	5.75	351.5	387.2	149.4	255.7	108.3	78.8
1	5.59	5.24	5.88	5.45	5.96	6.55	197.6	290.3	155.7	283.9	123.6	85.7
2	6.11	5.67	5.84	5.47	5.98	6.35	130.5	314.9	164.2	285.6	125.9	74.0
3	6.01	5.52	5.76	5.40	5.70	5.51	119.1	310.7	146.0	318.4	118.2	79.4
7	6.36	6.10	5.62	5.53	5.82	5.59	133.7	276.0	141.0	268.4	107.2	82.5
14	5.93	5.64	5.87	5.49	5.70	5.68	124.7	271.6	147.6	256.2	114.6	112.7
21	6.53	6.17	5.94	6.08	6.23	5.93	953	281.0	154.0	302.4	107.1	83.3
28	6.01	5.89	6.07	5.73	6.18	5.27	114.6	282.1	141.9	267.1	103.3	85.0
35	6.63	6.16	6.33	6.09	6.30	6.18	103.5	287.8	139.6	285.9	113.3	95.0
42	6.34	5.77	6.18	5.50	5.90	5.85	75.1	270.6	124.8	275.4	130.6	34.3

Table 3 ICP-OES measurement of potassium and phosphorus

Days	Potassium [% wt.]						Phosphorus [% wt.]			
	A	B	C	D	H	I	A	B	C	D
0.25	37.63	59.66	42.03	54.18	30.98	21.93	24.55	84.58	16.20	94.37
1	32.63	63.22	46.52	59.83	34.62	23.74	26.52	95.76	57.68	89.78
2	36.57	62.75	48.03	57.63	34.72	27.07	34.79	86.70	59.91	99.66
3	34.96	57.16	42.76	66.48	34.29	23.95	33.10	88.31	61.52	93.14
7	43.50	62.60	43.81	56.48	31.22	24.03	36.06	89.45	71.87	95.71
14	33.22	55.19	44.56	54.45	32.69	33.39	39.64	85.92	91.77	95.76
21	29.53	59.45	44.66	62.23	31.35	34.17	33.13	89.76	100.00	100.00
28	34.88	58.35	43.16	54.37	30.23	23.61	37.17	100.00	100.00	100.00
35	29.99	59.22	44.19	57.45	30.43	26.56	22.58	100.00	100.00	100.00
42	30.87	52.29	35.78	56.32	33.94	28.96	33.00	100.00	100.00	100.00

Table 4 IC measurement of NO_3^- and PO_4^{3-}

Days	NO3- [% wt.]				PO43- [% wt.]			
	A	B	C	D	A	B	C	D
0.25	47.51	57.61	38.95	75.80	26.00	60.03	43.58	66.45
1	28.99	89.36	46.76	97.16	48.82	80.82	72.00	84.77
2	78.53	82.11	81.09	100.00	58.35	88.53	89.82	93.11
3	63.68	90.65	81.01	95.03	81.10	99.13	100.00	83.93
7	65.26	82.43	86.63	94.28	89.72	100.00	83.56	99.17
14	72.93	98.10	66.80	95.81	72.32	100.00	59.93	99.10
21	78.96	89.68	79.15	98.85	66.61	98.55	57.29	91.16
28	56.88	100.00	77.00	100.00	73.20	100.00	50.46	96.56
35	66.67	100.00	85.57	100.00	74.58	100.00	37.77	100.00
42	78.92	94.76	74.13	100.00	68.73	100.00	62.55	97.79

As we can see in **Table 3** and **Table 4** some results are showing 100 % wt. of released nutrients. These results were our intentions because SAPs were used only as a carrier of nutrients. Therefore, we used them in pot experiment because results showed no interaction between SAPs and nutrients.

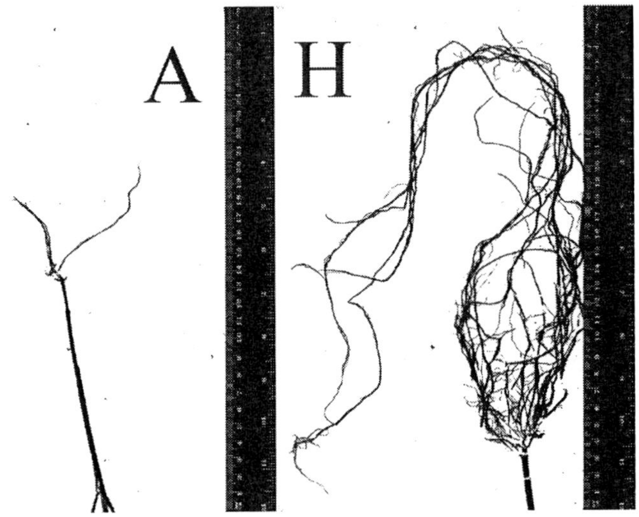

Figure 3 Scans of roots A and H after 74 days

Pot experiments were realized in controlled environment with UV lamp simulated day and night. After each cultivation corn plants were weighed, roots were measured and scanned. Scans were analysed in program Harmonic and Fractal analyser (HARFA) to characterize branching of root systems. Result showed positive correlation between roots length and weigh of the plant which means the heavier the plants were the longer the roots were. In addition, the heavier plants had more complex root systems than the lighter ones. Scans of roots showed only a small difference after the first cultivation time (46 days) therefore there is apparent that almost all plants had their optimal range of nutrients and water. After second cultivation (60 days) there were difference in roots and even in plants' weigh. The largest roots were measured by plants with sample A and B. And after the last cultivation time (74 days, results shown in **Figure 4**) some plants died already, for example by sample B all plants died, but plants by samples with lignohumate were still viable and in good condition. **Figure 3** shows root of sample A and H after 74 days (last cultivation time). After the pot experiment it was apparent that presence of lignohumate and the right amount of NPK (0.6602 g in our work) incorporated in superabsorbents has positive influence on plant growth.

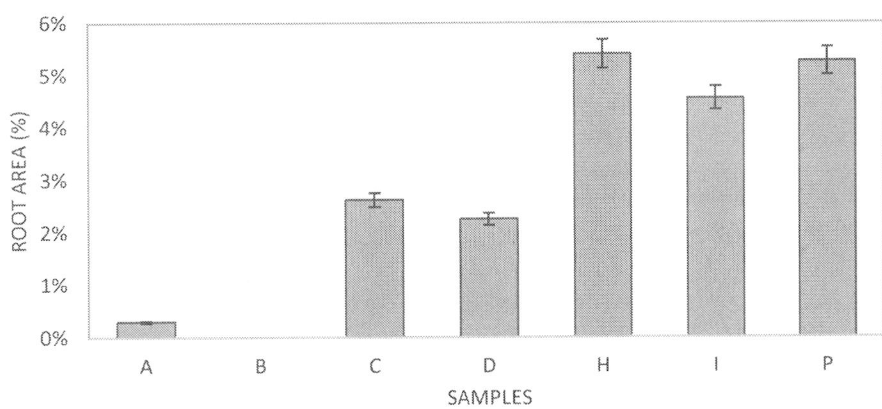

Figure 4 Results from HARFA after last cultivation (74 days)

4. CONCLUSION

Superabsorbents were prepared, characterized, and used in pot experiment. Measurement of swelling capacity showed the highest amount of absorbed water by sample H after 72 hours that equals to 19.4 and the lowest by sample D after 72 hours that equals to 5.5. Both samples contain lignohumate and sample D has higher amount of NPK (6.602 g) on top of it. It's apparent that higher amount of NPK reduce the possible amount of absorbed water. On the contrary presence of lignohumate increase the possible amount of absorbed water. Conductivity and pH measurements confirmed releasing of nutrients into distilled water withing samples. As mentioned before pH of sample D after 6 hours was equal to 5.38 and after 42 days was equal to 5.5. ICP-OES and IC also confirmed releasing of nutrients. In **Table 2** and **Table 3** is shown that in sample A released amount of potassium is equal to 30.87 % wt., in sample B 52.29 % wt., in sample C 35.78 % wt., in sample D 56.32 % wt., in sample H 33.94 % wt. and in sample I 28.96 % wt. Therefore, it's apparent that lignohumate incorporated in superabsorbent helps releasing of nutrients. As well as inorganic nutrients the lignohumate can be released. The highest amount of released lignohumate was by sample C, that has lower amount of NPK than sample D. Specifically, after 42 was 47.67 mg/g from sample C released while in sample D was released 36.58 mg/g and in sample H only 31.42 mg/g. As mentioned before the higher amount of NPK incorporated (6.602 g) in SAPs has negative influence on swelling capacity and release of lignohumate.

Pot experiment was made to use prepared SAPs in real condition. Corns were growth with and without presence of SAPs in soil and measured after cultivation time. After the first cultivation time there were barely any differences between plants with each SAP and even the soil itself. After the second cultivation time plants in soil without SAPs started to die while plants growth in presence of SAPs were still viable and in good condition. After the last cultivation time plants started to die and wither even in presence of SAPs. But still there were plants that stayed viable and in good condition.

Usage of SAPs has its advantages but also some disadvantages, for example they remain in soil even after end of cultivation. Even so, the advantages outweigh the disadvantages and usage of SAPs has a bright future in agriculture, because they are able to absorb and contain water and ability to carry nutrients essential for plants. Also, they can reduce irrigation water consumption, reduce the death rate of plants, improve fertilizer retention in soil, and increase plant growth rate.

REFERENCES

[1] BEHERA, Sabyasachi and Prakash A. MAHANWAR. Superabsorbent polymers in agriculture and other applications: a review. *Polymer-Plastics Technology and Materials*. 2020, vol. 59, no. 4, pp. 341-356.

[2] DADHANIYA, P., PATEL, M. & PATEL, R. Swelling and dye adsorption study of novel superswelling [Acrylamide/N-vinylpyrrolidone/3(2-hydroxyethyl carbamoyl) acrylic acid] hydrogels. *Polym. Bull.* 2006, vol. 57, pp. 21-31. Available from: https://doi.org/10.1007/s00289-006-0531-5

[3] LIANG, Rui, Mingzhu LIU and Lan WU. Controlled release NPK compound fertilizer with the function of water retention. *Reactive and Functional Polymers*. 2007, vol. 67, no. 9, pp. 769-779.

[4] MOHANA RAJU, K and M PADMANABHA RAJU. Synthesis of novel superabsorbing copolymers for agricultural and horticultural applications. *Polymer International*. 2001, vol. 50, no. 8, pp. 946-951.

[5] BUCHHOLZ, Fredric L. Superabsorbent Polymers: An Idea Whose Time Has Come. *Journal of Chemical Education*. 1996, vol. 73, no. 6, pp. 512-515.

[6] RILEY, LEE E.; DUMROESE, R. KASTEN; LANDIS, THOMAS D. National Proceedings: Forest and Conservation Nursery Associations-2003. Proc. RMRS-P-33. Fort Collins, CO: U.S. Department of Agriculture, Forest Service, Rocky Mountain Research Station. 2004, p. 156.

[7] SILVA, J. A., UCHIDA, R. S. (ed.). *Plant Nutrient Management in Hawaii's Soils*. 1st ed. Honolulu: University of Hawaii at Manoa, 2000. ISBN 1-929235-08-8.

[8] KRATOCHVILOVA, Romana, SEDLACEK, Petr, PORIZKA, Jaromir and KLUCAKOVA, Martina. 2021. Composite materials for controlled release of mineral nutrients and humic substances for agricultural application. *Soil Use and Management*. 2021, vol. 37, no. 3, pp. 460-467.

SURFACE MODIFICATION OF NANOFIBROUS MEMBRANES FOR WATER PURIFICATION

[1]Izabela Joanna GALLUS, [1]Evren BOYRAZ, [1]Fatma YALCINKAYA

[1]*TUL - Technical University of Liberec, Faculty of Mechatronics, Informatics and Interdisciplinary Studies, Liberec, Czech Republic, EU, izabela.gallus@tul.cz*

https://doi.org/10.37904/nanocon.2021.4328

Abstract

Around 1.2 billion people have limited access to drinkable water and millions succumb every year to sicknesses caused due to the consumption of impure water or lack of water. In the coming years, this problem will only progress at alarming rates due to the aggressive growth of population, urbanization and droughts. It is essential to investigate new approaches for water treatment with low energy costs and economic benefits. The biggest problem in membrane filtrations is fouling, which causes membrane pore blockages and creates a cake layer on the membrane surface. The aim of the project is to obtain nanofibrous membranes that will have high fouling resistance and antibacterial effects. Nanoparticles of Ag, ZnO, TiO2 and CuO are known for their antibacterial properties. In order to attach nanoparticles to the membrane surface, appropriate functional groups are required. The first step of the project involves the attachment of functional groups capable of binding nanoparticles to the surface of the nanofiber membrane. For this alkaline surface modification using NaOH, KOH, DETA under different conditions was performed. After modification, changes in hydrophilicity and structure of membranes, pore size and appearance of nanofibers were checked. Defluorination reaction conditions with NaOH and KOH proved to be insufficient to obtain the desired functional groups on the membrane surface. In the case of the reaction with DETA, amine groups were obtained, which will be used in the next step to attach the nanoparticles.

Keywords: PVDF, nanofibers, surface, alkaline, modification

1. INTRODUCTION

Various methods have been already investigated in the purification of water like distillation, sand filtration, reverse osmosis, membrane filtration. Membrane filtration is a relatively new approach and has numerous advantages when compared to other methods. Nanofiltration allows the removal of minerals, salts, pathogens (bacteria, viruses, fungus), total dissolved solids (TDS), cations and anions from the water, requiring low power and low operational temperatures [1]. Till date, numerous nanomaterials have been developed for water treatment. Nanomaterials can be divided into various groups based on their physicochemical properties and surface characteristics. The most specified materials included carbon nanotubes (CNTs), nanofibers, polymer-based nanoadsorbents, nanoclays (natural and modified), nanoparticles (Au, Ag, Pt NPs) and metal-based nanoadsorbents (Fe2O3, ZnO, CuO, TiO2 NPs) and mixed oxide nanoparticles. Each of these materials have specific properties and should be used accordingly in the purification of, e.g., carbon nanotubes are recommended for removal organic contaminations, whereas bioactive nanoparticles are applied to remove fungi and bacteria [2]. Thin-Film Nanocomposite (TFC) membranes are prepared by interfacial polymerization and are other promising method for water treatment. Nanoparticles are integrated with a thin polymer film layer to provide extra membrane properties. In 2010/2011, based on Kurth and Lind's research, zeolite-polyamide TFN membranes for reverse osmosis are commercially available. They achieved higher salt rejection than regular reverse osmosis membranes. According to LG NanoH2O, Inc. research, TFN membranes show 99.4 % salt rejection and two times higher water flux than polyamide membranes. Incorporation of Ag nanoparticles into thin film layer reduces biofouling, increases water permeability and shows antibacterial effects [3]. TFC membranes can also remove heavy metals from water, their application for water treatment is wide and surface properties can be changed and functionalized according to the requirements [4]. Nanofibrous membranes might be used for water desalination as well. Feng et al fabricated a PVDF nanofibrous membrane by electrospinning. According to their research, they obtained potable water (NaCl concentration was lower than 280 ppm), where saline water consisted 6 wt % NaCl [5]. The goal of the project is to obtain nanofiber membranes for water purification that can provide high fouling resistance and antimicrobial properties. To obtain antibacterial properties, nanoparticles (Ag, ZnO, CuO), which are well known for their antibacterial activities, will be attached to the membrane surface. To achieve the desired effect, functional groups must first be located on the membrane surface that are capable of binding the nanoparticles. The defluorination reaction developed by Ross et al. allows fluorine to be eliminated from the PVDF chain and replaced with more hydrophilic groups. This reaction can be performed using alkaline reagents such as KOH, NaOH [6,7]. As a result the hyydrophobicity of the membranes can increased. Groups were introduced to bind nanoparticles, in this project using the PVDF degradation reaction.

2. EXPERIMENTAL

The hydrophobic PVDF nanofibers were prepared before by electrospinning at the Technical University of Liberec (Czech Republic). Nanofibrous membranes were prepared by the lamination of 3 layers: nonwoven support (PET - 100 GSM), adhesive web and PVDF nanofibers. As an adhesive web used either polyurethane (PUR), polyamide (PA) or polyethersulfone (PES). As an alkaline agent potassium hydroxide and sodium hydroxide in pellet form (Penta Chemicals UNLIMITED) and diethylenetriamine (Sigma Aldrich) were used. Isopropanol and ethanol supplied by Penta Chemicals UNLIMITED were used to wash membranes and dissolve hydroxides. Deionized water was used to prepare solution and rinse membranes surfaces.

2.1. Alkaline treatment with NaOH and KOH

Membranes were washed in isopropanol and rinsed with deionized water before modification. The membranes were then dried at 40 °C. 2M NaOH solution was prepared in ethanol while 2M KOH solution was prepared in isopropanol. The reaction was run for 0.5 and 2 hours. After treatement the membranes were washed with deionized water and dried at 40 °C. The modification was conducted at a temperature of 25 °C.

2.2. Alkaline treatment with dietylenetriamine (DETA)

Membranes were washed in isopropanol and rinsed with deionized water before modification. The membranes were then dried at 40 °C. Membranes were modified in 4M aqueous DETA solution for 1, 3 and 5 hours at 60 °C. After treatment the membranes were washed with deionized water and dried at 40 °C.

2.3. Membrane characterization

The scanning electron microscopy (SEM) and Fourier transfer infrared spectroscopy (FTIR) were used to estaminate changes in nanofibers structure and surface content. The Vega 3SB scanning electron microscope (SEM; TESCAN VEGA, Brno, Czech Republic) was used to observe nanofibers surface. The samples were covered with 7 nm gold layer before the analysis. The diameter of nanofibers was measured by ImageJ program (National Institute of Health). The samples were analyzed by Nicolet iZ10 Fourier transform infrared spectroscope (FTIR, Thermo Scientific, Prague, Czech Republic). The water contact angle was determined by Drop Shape Analyzer DS4 (Krüss GmbH). Membrane pore size was determined by bubble point method using Porometer 3G through a pore size analyzer (Quantachrome Instruments, Anton Paar GmbH).

3. RESULTS

In this research work it was noticed that the layers of nanofibers were separated from the membrane support when treated for long period of time or at high concentration levels of alkaline agent. The adhesive webs used did not have sufficient chemical resistance against the chemicals used resulting in damage during modification. The smallest chemical resistance was shown by PES- adhesive web membranes. The nanofibers increased their volume after the modification, independently from the chemical agent used. The diameter of pristine PVDF nanofibers was 198 nm while after modification it could reach up to 300 nm. The mean pore size of pristine membranes was cca. 0.7 µm whereas after alkaline treatment it became smaller and stood at 0.3-0.4 µm. The differences are visible on SEM images shown in **Figure 1**.

Figure 1 The PVDF nanofibers before and after modification

The measurements of contact angle show that the hydrophilicity of the modified membranes significantly increased after modification. The contact angle of pristine membranes was 97.1°, 81.9°, 93° for PA-, PES-, PUR-based membranes, respectively. The contact angle decreased gradually with the time when treated with KOH, whereas after others modifications reached 0°. Below on **Figure 2** showed selected results.

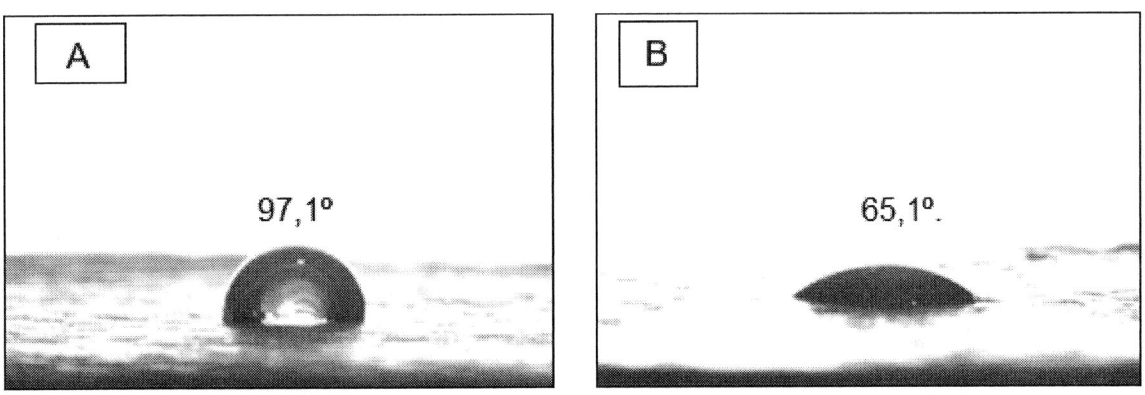

Figure 2 The contact angle for pristine PVDF membrane and after 1 hour of modification in KOH

The increase in membrane hydrophilicity may indicate that some of the fluorine has been removed from the membrane and replaced with more hydrophilic groups. The FTIR results have shown changes in the membrane structure and may confirm the above assumption.

The presence of -OH, -CF, -C=C and C=O groups is essential to estaminate the changes in PVDF membranes sturcture. In the range of 1050-1250 cm^{-1} two sharp peaks appear from the vibrations of bihalide. The -OH groups shown near 3400 cm^{-1}, the radical forms of -OH groups may occurs either in dissociation (3580-3650 cm^{-1}) or in association (3200-3400 cm^{-1}). In the range of 1600-1750 cm^{-1} peaks appear from C=O, C=C bonds, their presence in samples may suggest that reaction of defluorization has not been completed. In the **Figure 3** gives the FTIR results after KOH modification.

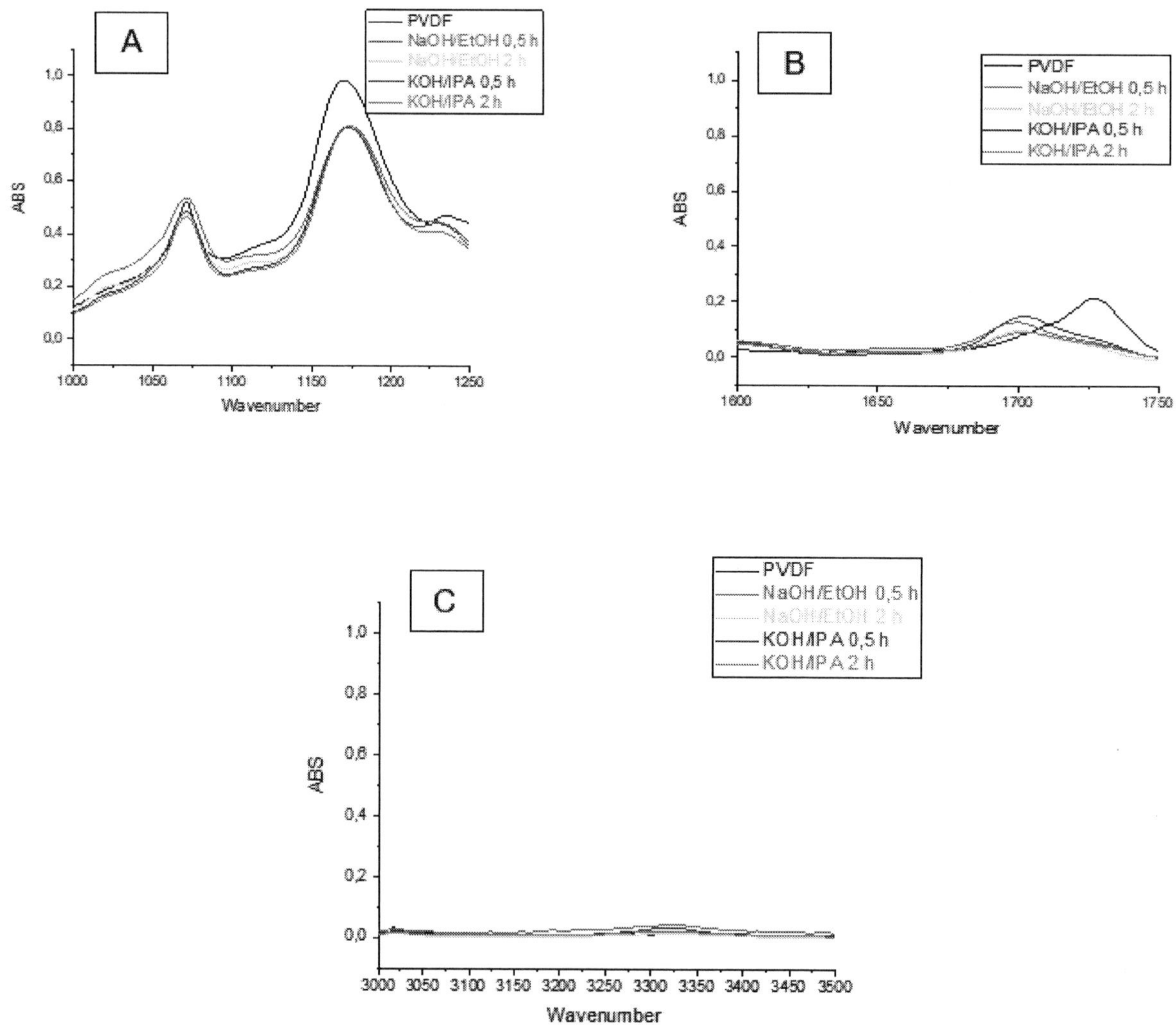

Figure 3 The FTIR results in ranges A) 1000-1250 cm-1, B) 1600-1740 cm-1, C) 3000-3500 cm-1 for PVDF membranes after treatment in 2M KOH or 2M NaOH after 0.5 and 2 hours

The results show that the amount of C-F groups is reduced after alkaline treatment. Determination of the presence of -OH groups was not possible based on the results obtained. However, the intensivity of peaks for -C=O, -C=C may suggest, that the reaction of defluorization was initiated, but not completed.

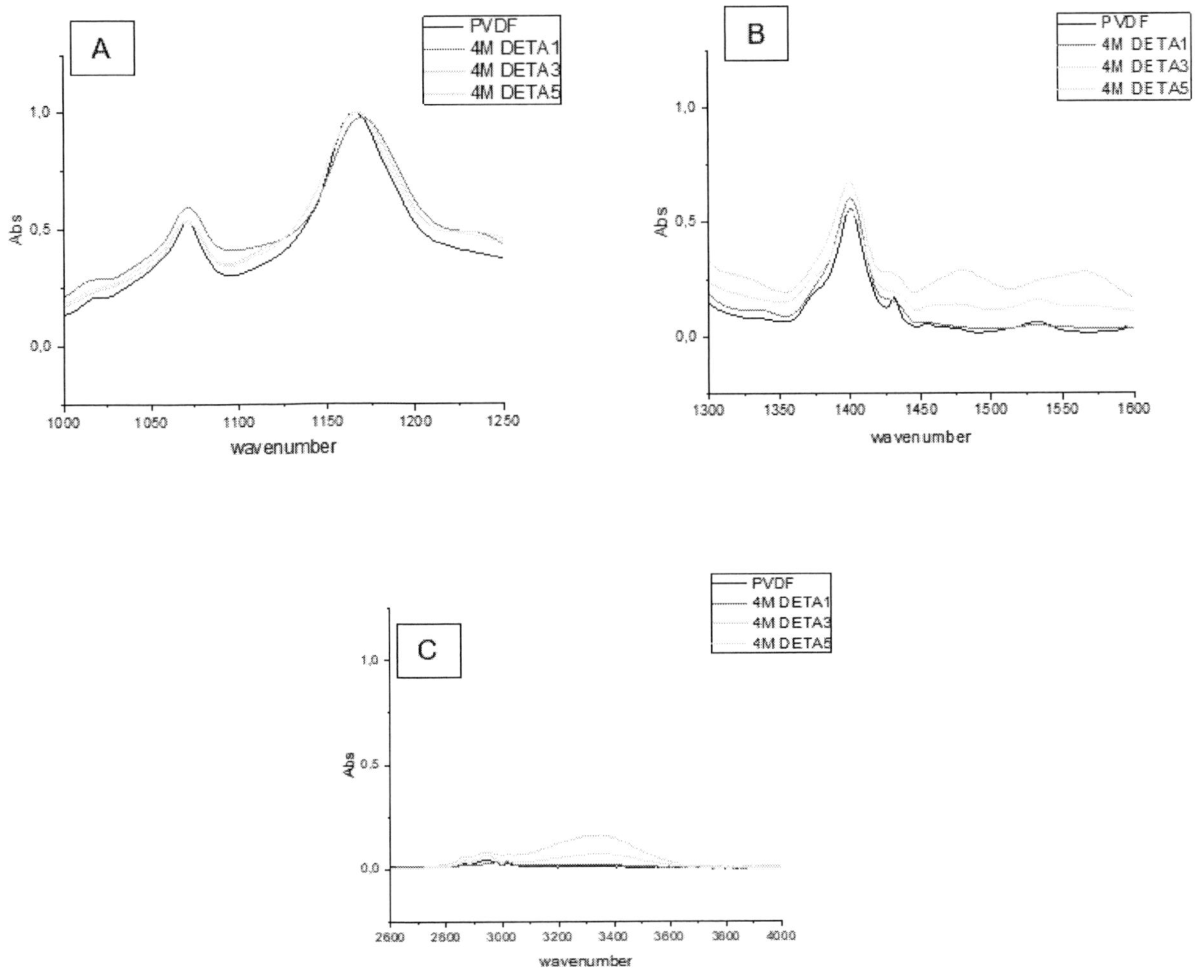

Figure 4 The FTIR results in ranges A) 1000-1250 cm-1, B) 1300-1600 cm-1, C) 2600-4000 cm-1 for PVDF membranes after treatment in 4M DETA after 1,3 and 5 hours

It can be seen in **Figure 4** that clear deformation vibrations of N-H bonds from amine (approx. 1490 + 1580 cm-1) appeared after modification of membranes in 4M DETA solution. The stretching vibrations of N-H bonds might be obscured by the stretching vibrations of O-H bonds from residual moisture. However, it can be concluded that $-NH_2$ groups appeared on the membrane surface.

4. CONCLUSION

According to the results obtained, alkaline treatment by KOH and NaOH does not provide sufficient amount of -OH groups into surface of nano PVDF membranes. The high concentration of alkaline agent causes membrane disintegration and leads to disassociation of layers. The small concentration of modified agents initiates a reaction of defluorization, but is not completed. However, in the literature, where fluoride was successfully removed from the polymer chain and replaced by -OH groups. Reducing the concentration of the alkaline agent and increasing the reaction time may be the key to success. Modification with DETA yielded -NH2 groups on the membrane surface that could be used to attach nanoparticles in the future.

ACKNOWLEDGEMENTS

The research presented in this article was partly supported by the Internal Grant of the Technical University of Liberec, Czech Republic (Project No. SGS-2021-3027). The authors acknowledge the assistance provided by the Research Infrastructures NanoEnviCz (Project No. LM2018124), which is supported by the Ministry of Education, Youth and Sports of the Czech Republic.

REFERENCES

[1] SINGH, R., BHADOURIA, R., et al. Nanofiltration technology for removal of pathogens present in drinking water. *Waterborne Pathogens*. 2020, p.463-489.

[2] PATANJALI, Pooja, et al. Nanotechnology for water treatment: A green approach. In: *Green Synthesis, Characterization and Applications of Nanoparticles. Elsevier*. 2019, p. 485-512.

[3] REDDY K., et al. Nanotechnology for water purification: applications of nanotechnology methods in wastewater treatment. In: *Water Purification. Academic Press*. 2017, p. 33-74.

[4] MADHURA, Lavanya; SINGH, Shalini. A Review on the Advancements of Nanomembranes for Water Treatment. *Nanotechnology in Environmental Science*. 2018, pp. 391-412.

[5] WANG, Zhuqing, et al. Recent advances in nanoporous membranes for water purification. *Nanomaterials*. 2018, vol. 8, no. 2, p. 65.

[6] ROSS, G. J., et al. Surface modification of poly (vinylidene fluoride) by alkaline treatment1. The degradation mechanism. *Polymer*. 2000, vol. 41, no. 5, pp. 1685-1696.

[7] ROSS, G. J., et al. Surface modification of poly (vinylidene fluoride) by alkaline treatment Part 2. Process modification by the use of phase transfer catalysts. *Polymer*. 2001, vol. 42, no. 2, pp. 403-413.

GREEN SYNTHESIS OF TITANIUM AND ZINC OXIDE NANOPARTICLES FOR SIMULTANEOUS PHOTOCATALYTIC REMOVAL OF ESTROGENS IN WASTEWATER

Muhammad YASIR, Tomáš ŠOPÍK, Hassan ALI, Dušan KIMMER, Vladimír SEDLAŘÍK

Centre of Polymer Systems, University Institute, Tomas Bata University in Zlín, Zlín, Czech Republic, EU, yasir@utb.cz

https://doi.org/10.37904/nanocon.2021.4333

Abstract

This study reports on the green synthesis of titanium and zinc oxide nanoparticles (ZnO-NPs) using lemon juice and peel extract, zinc acetate, ethylene glycol, and titanium IV isopropoxide as precursors. The prepared TiO_2 and ZnO-NPs were characterized via X-ray Diffraction analysis, UV-vis spectroscopy, dynamic light scattering, scanning electron microscopy, and EDAX. The as-prepared samples, TiO_2 and ZnO-NPs, were further subjected for the photocatalytic degradation of estrogenic hormones (Estrone, Estradiol, Ethinylestradiol, and Estriol) under UV light irradiation at 365 nm, which resulted in promising photocatalytic activity. All four hormones were significantly degraded owing to the photocatalytic activity combined with a slight contribution (4-11 %) from the hormonal adsorption onto the surface of the photocatalysts. Overall hormonal degradation rates in the range of 84-93 % and approximately 99 % were achieved in 60 minutes under UV light irradiation by ZnO and TiO_2, respectively.

Keywords: Wastewater treatment, effluents, contraceptive, estrogen degradation, nanoparticles, green synthesis, liquid chromatography, HPLC

1. INTRODUCTION

Over the past several decades, the scarcity of clean water has increased with the rising demand for treated wastewater due to the ever-growing human population. Water pollution due to toxic chemical effluents such as estrone (E1), β-estradiol (E2), 17α-ethinylestradiol (EE2), and estriol (E3) steroid hormones, also known as endocrine-disrupting chemicals, is a serious threat to marine life, humans, and animals present in the environment [1]. Significantly high potencies of these hormones have been reported in water streams [2]. Such levels are being monitored to control the toxicity tolerance of these effluents at wastewater treatment plants [3,4]. In the Czech Republic, steroid hormones in the concentration range of 3.4-41 ng/L have been detected in wetlands [5]. Synthetic hormone such as EE2 is most commonly used in oral contraceptive pills, although it is also used for the treatment of menstrual problems and prostate cancer. Due to the hormonal metabolites interaction in nature and their subsequent de-conjugation after the partial breakdown at sewage facilities, they become active again, and their remnants are also of significant contamination source in the environment [6,7]. These medicines have long-lasting adverse effects on living creatures when released in water sources [8]. Fertility disorder, rapid fish femininity, and testicles weight loss in Japanese quails are some of its consequences [9]. Similar impacts, such as the drop in male sperm count, higher risk of breast and ovarian cancer, have also been observed in humans [10]. Therefore, in this regard, immediate remediation of such toxic pollutants and their disposal is considered an environmental urgency. Conventional water treatment techniques are either less efficient for total toxicants removal or require other sophisticated methods for purification, usually high in energy and cost. In this context, photocatalytic degradation is the most promising strategy for environmental pollutants remediation due to simplicity, ease of operation, cost, and environmental sustainability. Photocatalysis has been successfully used in various other applications such as fuel production, antimicrobial activity, hydrogen production, and nitrogen fixation [11,12,13]. Two of the most commonly used photocatalysts to date are ZnO and TiO_2, each possessing superior chemical stability and photocatalytic activity over the UV light range. However, it is essential to adopt a green synthesis approach with minimal environmental impact in regard to their synthesis. Lemon peels contain citric acid, which can act as a reducing agent in the green synthesis of TiO_2-NPs. Whereas esterification occurs when citric acid in lemon extract mixes with ethylene glycol at 90 °C. A stable and soluble chelate complex of Zn is formed in the solution. When the solution is evaporated, Zn^{2+} ions uniformly bind to the active available binding sites of precursor resin, and this precursor breaks when calcinated to form ZnO-NPs. Therefore, it can be a useful green approach for making such NPs.

This study reports on the green synthesis and characterization of ZnO and TiO_2-NPs, which are highly efficient in the photocatalytic degradation removal of estrogenic hormones, simultaneously in a one-step using high-performance liquid chromatography (HPLC).

2. MATERIALS AND METHODS

2.1. Materials and reagents

The estrogenic hormones: E1 ≥99 %, E2 ≥98 %, E3 ≥97 %, and EE2 ≥98 % were purchased from Sigma Aldrich Chemie GmbH, Germany. The HPLC grade solutions: acetonitrile, and ethanol were purchased from Honeywell and VWR, Czech Republic, respectively. Deionized water (pH 7.3, 18.3 MΩ/cm) was sourced from Milli-Q ultra-pure water purification system, Biopak® Polisher, Merck, USA. Glass microfiber (GMF) filters (Whatman, Czech Republic) with pore size 0.45 µm and 25 mm diameter were used prior to HPLC. The primary precursor of ZnO is zinc acetate dihydrate ≥99.5 %, purchased from Fisher Scientific, U.K, and Sigma Aldrich's titanium IV isopropoxide (TTIP) for TiO_2, which has a purity of >99.5 %. Ethylene glycol was purchased from VWR, Czech Republic. For the bio-mediated synthesis, fresh and clean lemon peels (Citrus limon) were sliced into roughly 2 cm^2 area pieces. All reagents were used without further purification.

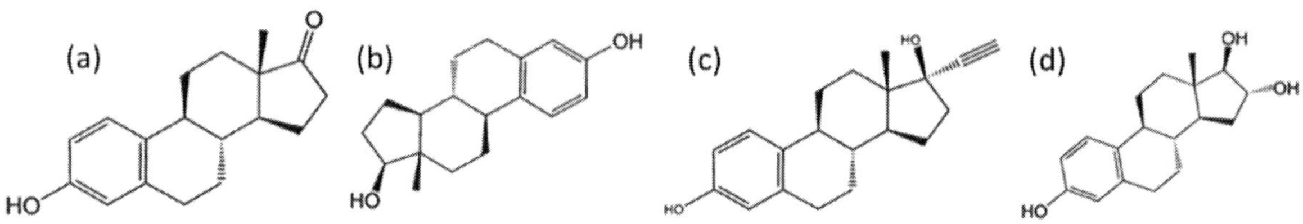

Figure 1 Steroid estrogenic hormones a) E1, b) E2, c) EE2, d) E3

2.2. Synthesis of ZnO and TiO_2 NPs

For ZnO-NPs synthesis, 20 mL of ethylene glycol was taken, and 2.5 g of zinc acetate was uniformly dissolved with constant stirring. Lemons were purchased from the market and thoroughly cleaned before peeling. They were hand-squeezed for juice while solid particles and seeds were filtered off through common lab filter paper. The precursor solution was vigorously stirred at 1200 rpm, and the clean lemon extracted juice was drop-wise added with the help of a micropipette over a 1 h period. The solution was further stirred at 90 °C for 3 h using a preheated oil bath to obtain a yellowish gel. It was dried in a vacuum for 2 h at 110 °C. Afterward, the obtained product was calcinated for 2 h at 700 °C in the muffle furnace to obtain ZnO nanoparticles which were crushed to have uniform fine particles without agglomeration.

TiO_2-NPs were also produced by green synthesis. Initially, lemon peel extract was obtained by boiling 50 g of lemon peels in 150 ml of distilled water at 90 ± 3 °C for 2.5 h while swirling magnetically at 400 rpm. The extract was combined with a 1.5 N TTIP solution in the following proportions: A mixture of 68 mL TTIP solution and 55 mL extract was used to prepare the sample. The lemon extract had a pH of 4.5 and TTIP of 5.5; both were brought to a pH of 5 before mixing. The sample was agitated at 700 rpm for 3 h, then centrifuged to separate the aqueous phase. The precipitated sample was then dried at 100 °C for 8 h and calcined at 700 °C for 2 h in a muffle furnace to obtain TiO_2-NPs powder.

Figure 2 a) Collection of lemons from the market, b) waste lemon peels, c) hand-squeezed extracted juice from lemon, d) filtered juice, e) peels cut into small pieces, f) peels boiled and continuously stirred, g) extract collected from boiled peels

2.3. Characterization

Different analytical techniques were deployed to characterize the samples. Morphological analysis was carried out using Nova 450 scanning electron microscope (SEM) (FEI, Thermo Fisher Scientific, USA) via a through-the-lens detector (TLD) at 5-10 kV applied potential and additionally equipped with Octane plus energy dispersive X-ray (EDX) spectroscopy (EDAX, Ametek Inc., PA, USA). X-ray diffractogram (XRD) of NPs was recorded using Miniflex^{TM} 600 X-ray diffractometer (Rigaku, Japan), with CoK_{β} (λ=1.79 Å) as a source. The angle 2θ was in the range from 3-90° with operating current, step size, step time, and operational voltage set to 15 mA, 0.02°, 10°/s, and 40 kV, respectively. The obtained diffractograms using Co source were converted to Cu using PowerDLL software converter 2.93 to compare data in the literature. Zetasizer Nano ZS (Malvern Instruments, Malvern, UK) was used to determine the particle size distribution with diiodomethane as a dispersant at ambient temperature and viscosity of 2.6 cP. The UV-vis absorption of the as-prepared NPs was conducted with the UV-vis spectrometer LAMBDA 1050+ (PerkinElmer, UK). The estrogenic hormones concentration was simultaneously quantified and recorded on HPLC DionexUltiMate 3000 Series (Thermo Fisher Scientific, Germany), using the method devised by Yasir *et al.* The limit of detection and quantification were found to be 0.560 and 1.867, 1.189 and 3.963, 0.920 and 3.067, and 1.883 and 6.280 µg/L for E3, E2, EE2, and E1, respectively [14].

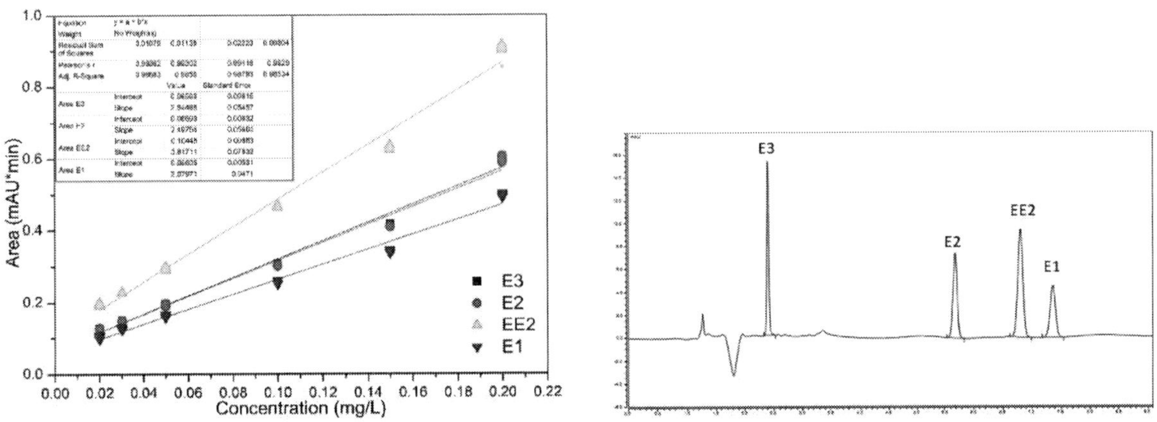

Figure 3 Calibration curve and chromatogram for estrogenic hormones (E1, E2, EE2, E3)

2.4. Photocatalytic degradation

The photocatalytic degradation was conducted with the synthesized ZnO and TiO_2 nanogranules (5 mg) and 10 mL hormone solution (0.8 mg/L) under UV lamp irradiation at 365 nm with continuous stirring at 450 rpm. After 60 min, the samples were drawn from the solution, the catalyst was removed by centrifuging at 14000 rpm, and the remaining concentration of hormones in the solution was monitored using HPLC. A separate beaker without NPs was used as a control. The degradation efficiency is defined by equation (1).

$$Degradation\ removal\ (\%) = \frac{C_i - C_t}{C_i} \times 100 \qquad (1)$$

Where C_i is the initial concentration (mg/L) and C_t is the solution concentration at time t (mg/L).

3. RESULTS AND DISCUSSION

3.1. Characterization of adsorbents

Figure 4 below shows the appearance, surface morphology, and elemental analysis of NPs.

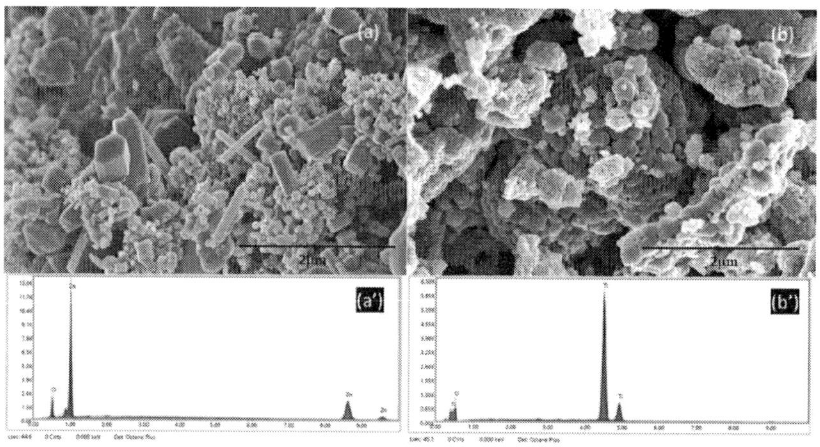

Figure 4 SEM micrographs and EDAX spectrum of, (a and a´) ZnO-NPs, (b and b´) TiO_2-NPs, respectively

Figure 4a displays two distinguished types of ZnO-NPs, the cubic shaped with diameter ~50 nm, and hexagonal rods ~100-250 nm, length ~900-1100 nm. In **Figure 4a´** the peak at 0.55 keV for O (wt.% 6.51 and at.% 22.15), and Zn at 1.00, and 8.60 keV represent wt.% 93.49 and at.% 77.85. **Figure 4b** demonstrates TiO_2-NPs; the NPs varied in size less than 100 nm, and also some NPs on agglomerates can be seen. **Figure 4b´** characteristic peaks at 0.4, 4.5, and 4.9 keV for Ti (wt.% 77.71 and at.% 53.80), while at 0.5 keV for O (wt.% 22.29 and at.% 46.20). The elemental analysis proves that no traces of any other element was found, and the synthesized NPs were pure.

Figure 5a displays the diffractogram of ZnO and TiO_2-NPs. The sharp intensity peaks at 2θ with miller indices were 31.7 (100), 34.4 (002), 36.2 (101), 47.4 (102), 56.5 (110), 62.8 (103), 66.2 (200), 67.8 (112), 68.9 (201), and 72.5 (004). These characteristic peaks represent the hexagonal structure of ZnO matching with DB card: 01-080-0075. For TiO_2, characteristic peaks at 25.3 (101), 36.9 (103), 37.7 (004), 38.5 (112), 48.0 (200), 53.8 (105), 55.0 (211), 62.1 (213), 62.6 (204), 68.7 (116), and 70.2 (220) represent anatase phase (DB card: 01-071-1166) while peaks at 27.4 (110), 36.0 (101), 39.1 (200), 41.2 (111), 44.0 (210), 54.3 (211), 56.6 (220), 62.6 (002), 64.0 (310), 68.9 (301), and 69.8 (112) show rutile phase matching DB card: 01-086-0148. The calculated average particle size values for ZnO and TiO_2 from the well-known Debye-Scherrer formula are 33.32 and 45.85 nm, respectively. **Figure 5b** shows ZnO-NPs size in the range of 900-1100 nm, which can be the length of hexagonal rods that are in good compliance with the results from SEM and **Figure 5c** presents TiO_2-NPs size density peak at ~60 nm; similar particle size is also evident from SEM micrograph.

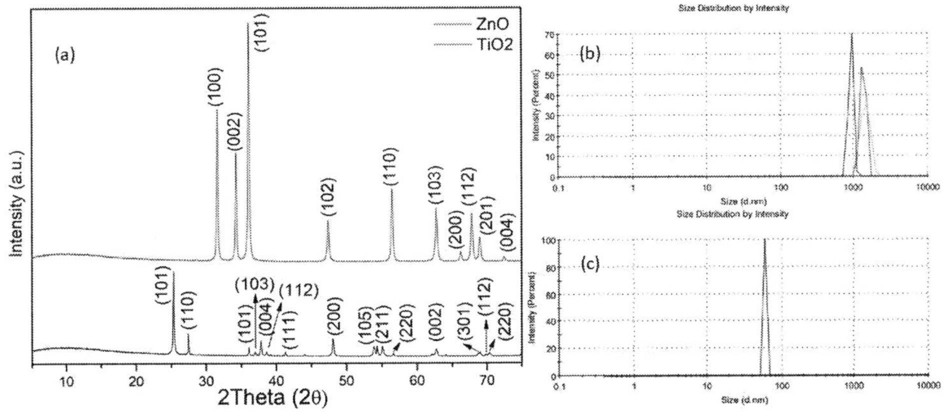

Figure 5 (a) X-ray diffraction pattern of green synthesized ZnO and TiO_2-NPs, the particle size distribution of (b) ZnO-NPs, and (c) TiO_2-NPs

3.2. UV-vis diffuse reflectance spectroscopy and degradation mechanism

The factor that plays a crucial role in photoactivity and optical properties is the electronic structure associated with its bandgap. The as-prepared ZnO and TiO_2 samples were characterized by UV-vis spectroscopy to determine the optical bandgaps using the Kubelka-Munk equation (2).

$$[F(R).hv]^{1/2} = [hv - Eg] \qquad (2)$$

By plotting the line of best fit on *[F(R)·hv]1/2* vs. *hv* graph gives the optical bandgap energy values using the extrapolation of its slope to *F(R)* → 0, as shown in **Figure 6**.

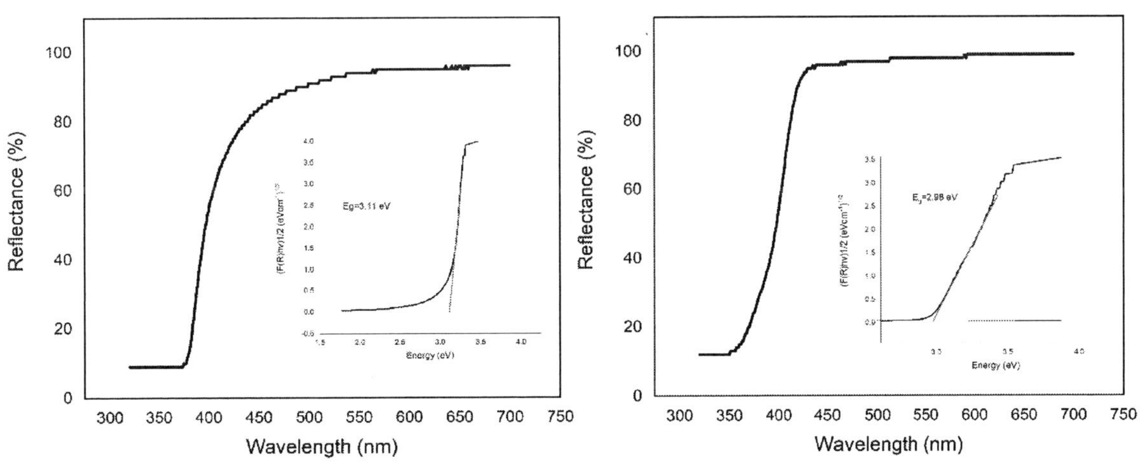

Figure 6 [F(R)·hv]1/2 vs. hv graph for ZnO (left) and TiO_2 (right)

Optical bandgap values of 2.98 and 3.11 eV were obtained for the TiO_2 and ZnO, respectively. The obtained values are similar to the bandgap values reported in the literature. However, for the TiO_2 sample, a blueshift in the wavelength is observed, due to which the bandgap decreased from the typically reported value of 3.2 to 2.98 eV. This indicates that the as-prepared TiO_2 sample possesses the ability to excite electrons even in the visible light range.

The as-prepared TiO_2 and ZnO samples exhibited superior photocatalytic performance towards hormones degradation. A schematic illustration of the proposed photocatalytic mechanism is given in **Figure 7**. Since TiO_2 and ZnO are two of the most commonly investigated photocatalytic semiconducting materials, their degradation mechanism is well established in the literature. A near-total degradation was achieved by the following process, i.e., (i) electronic excitation to the conduction after absorbing photons with energy higher than the bandgap (ii) formation of holes in the valence band (iii) participation of the excited electrons and holes in water to produce highly reactive radicals. The produced reactive species such as •OH, •O^{2-} possess the strong oxidizing ability and were able to degrade the organic hormones adsorbed onto the surface of the photocatalysts.

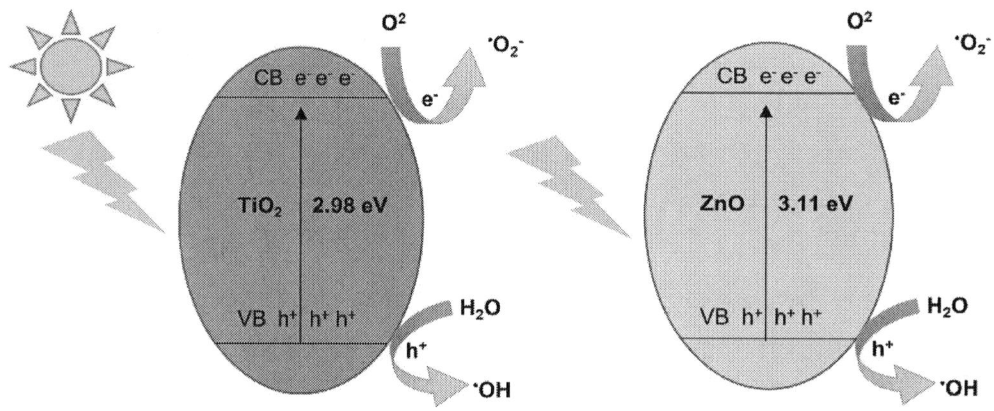

Figure 7 Degradation mechanism of TiO_2 and ZnO-NPs

Simultaneous adsorption and photocatalytic degradation removal of four estrogenic hormones were determined, as shown in **Figure 8**.

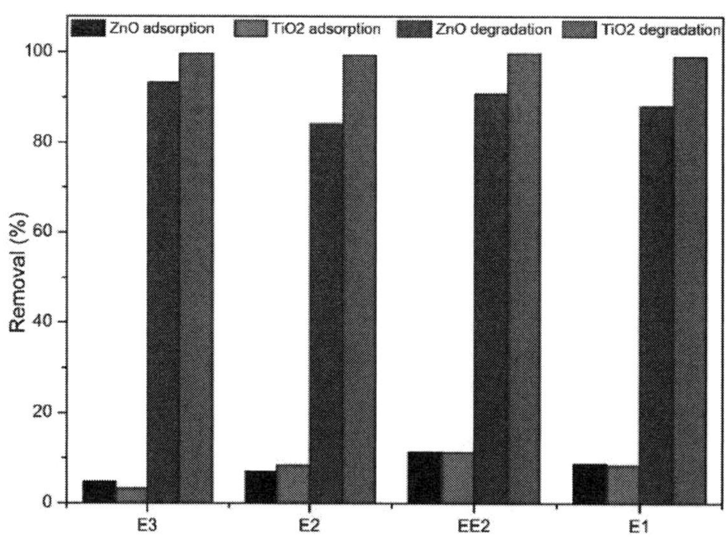

Figure 8 Adsorption and UV-Photocatalytic degradation of steroid estrogen hormones in 1 h time

It can be seen that the adsorption of each hormone is almost similar and nearly equal on both types of NPs. However, the difference is evident in degradation. TiO_2 is more efficient than ZnO-NPs for each hormone reaching approximately 99% degradation in 60 min. Also, ZnO has the highest degradation for E3 (93 %) with the least adsorption (4 %), while EE2 has the highest adsorption (11%) and photodegradation with TiO_2.

4. CONCLUSION

The green synthesized ZnO and TiO_2-NPs were successfully prepared from lemon extract and peels. The X-ray diffractograms determined the phases, particle size, and EDAX elemental analysis confirmed the purity of NPs. The NPs achieved a rapid photodegradation tendency with overall hormonal degradation in the range of 84-93% by ZnO and approximately 99% by TiO_2 in 60 min under UV light irradiation. In the future, these NPs can be used as additives in nanofiber modification for enhanced adsorption and photodegradation of dyes, other steroid hormones, and micro-pollutants in wastewater.

ACKNOWLEDGEMENTS

The authors gratefully acknowledge the financial support by the Ministry of Education, Youth, and Sports of the Czech Republic (grant no. RP/CPS/2020/002) and the Internal Grant Agency, IGACPS2021002 from the Centre of Polymer System (CPS), Tomas Bata University in Zlin, Czech Republic.

REFERENCES

[1] SCHÄFER, A.I., STELZL, K., FAGHIH, M., SEN GUPTA, S., KRISHNADAS, K. R., HEIßLER, S., PRADEEP, T. Poly(ether sulfone) Nanofibers Impregnated with β-Cyclodextrin for Increased Micropollutant Removal from Water. *ACS Sustainable Chemistry and Engineering.* [online]. 2018, vol. 6, no. 3, pp. 2942-2953. Available from: https://doi.org/10.1021/acssuschemeng.7b02214.

[2] CARTINELLA, J. L., CATH, T. Y., FLYNN, M. T., MILLER, G. C., HUNTER, K. W., CHILDRESS, A. E. Removal of natural steroid hormones from wastewater using membrane contactor processes. *Environmental Science and Technology.* [online]. 2006, vol. 40, no. 23, pp. 7381-7386. Available from: https://doi.org/10.1021/es060550i.

[3] JOHNSON, A. C., AERNI, H. R., GERRITSEN, A., GIBERT, M., GIGER, W., HYLLAND, K., JÜRGENS, M., NAKARI, T., PICKERING, A., SUTER, M. J. F., SVENSON, A., WETTSTEIN, F. E. Comparing steroid estrogen, and nonylphenol content across a range of European sewage plants with different treatment and management practices. *Water Research.* [online]. 2005, vol. 39, no. 1, pp. 47-58. Available from: https://doi.org/10.1016/j.watres.2004.07.025.

[4] SARMAH, A. K., NORTHCOTT, G. L., LEUSCH, F. D. L., TREMBLAY, L. A. A survey of endocrine disrupting chemicals (EDCs) in municipal sewage and animal waste effluents in the Waikato region of New Zealand. *Science of the Total Environment.* [online]. 2006, vol. 355, no. 1-3, pp. 135-144. Available from: https://doi.org/10.1016/j.scitotenv.2005.02.027.

[5] VYMAZAL, J., BŘEZINOVÁ, T., KOŽELUH, M. Occurrence and removal of estrogens, progesterone and testosterone in three constructed wetlands treating municipal sewage in the Czech Republic. *Science of the Total Environment.* [online]. 2015, vol. 536, pp. 625-631. Available from: https://doi.org/10.1016/j.scitotenv.2015.07.077.

[6] HAN, J., QIU, W., CAO, Z., HU, J., GAO, W. Adsorption of ethinylestradiol (EE2) on polyamide 612: Molecular modeling and effects of water chemistry. *Water Research.* [online]. 2013, vol. 47, no. 7, pp. 2273-2284. Available from: https://doi.org/10.1016/j.watres.2013.01.046.

[7] LIMPIYAKORN, T., HOMKLIN, S., ONG, S. K. Fate of estrogens and estrogenic potentials in sewerage systems. *Critical Reviews in Environmental Science and Technology.* [online]. 2011, vol. 41, no. 13, pp. 1231-1270. Available from: https://doi.org/10.1080/10643380903488680.

[8] CHEN, Y., ZHANG, Y., LUO, L., SHI, Y., WANG, S., LI, L., LONG, Y., JIANG, F. A novel templated synthesis of C/N-doped β-Bi2O3 nanosheets for synergistic rapid removal of 17α-ethynylestradiol by adsorption and photocatalytic degradation. *Ceramics International.* [online]. 2018, vol. 44, no. 2, pp. 2178-2185. Available from: https://doi.org/10.1016/j.ceramint.2017.10.173.

[9] NGHIEM, L. D., SCHÄFER, A. I. Adsorption and transport of trace contaminant estrone in NF/RO membranes. *Environmental Engineering Science.* [online]. 2002, vol. 19, no. 6, pp. 441-451. Available from: https://doi.org/10.1089/109287502320963427.

[10] SOLOMON, G. M., SCHETTLER, T. Environment and health: 6. Endocrine disruption and potential human health implications. *Cmaj*. 2000, vol. 163, no. 11, pp. 1471-1476.

[11] ALI, H., GULER, A. C., MASAR, M., URBANEK, P., URBANEK, M., SKODA, D., SULY, P., MACHOVSKY, M., GALUSEK, D., KURITKA, I. Solid-state synthesis of direct z-scheme cu2o/wo3 nanocomposites with enhanced visible-light photocatalytic performance. *Catalysts*. [online]. 2021, vol. 11, pp. 1-26. Available from: https://doi.org/10.3390/catal11020293.

[12] DJURIŠIĆ, A. B., HE, Y., NG, A. M. C. Visible-light photocatalysts: Prospects and challenges. *APL Materials*. [online]. 2020, vol. 8, no. 3. Available from: https://doi.org/10.1063/1.5140497.

[13] ALI, H., MASAR, M., GÜLER, A. C., URBÁNEK, M., MACHOVSKY, M., KUŘITKA, I. Heterojunction-based photocatalytic nitrogen fixation: Principles and current progress. *Nanoscale Advances*. [online]. 2021. Available from: https://doi.org/10.1039/D1NA00565K.

[14] YASIR, M., ŠOPÍK, T., KIMMER, D., SEDLAŘÍK, V. Facile hplc technique for simultaneous detection of estrogentic hormones in wastewater In: *NANOCON 2020 Conference Proceedings*. [online]. Brno: Nanocon, 2020, pp. 272-276. Available from: https://doi.org/10.37904/nanocon.2020.3710.

SILVER/ALUMINA CORE/SHELL NANOPATTERNED FILMS FOR IN-FIELD DETECTION OF SOIL AND WATER PHOSPHORUS USING SURFACE ENHANCED RAMAN SPECTROSCOPY

[1,2]Jan PRÁŠEK, [1,2]Radim HRDÝ, [3]Dušan HEMZAL, [1,2]Jana DRBOHLAVOVÁ, [4]Radovan KOPP

[1]*CEITEC, Brno University of Technology, Brno, Czech Republic, EU,*
prasek@vutbr.cz, jana.drbohlavova@ceitec.vutbr.cz

[2]*Faculty of Electrical Engineering, Brno University of Technology, Brno, Czech Republic, EU,* hrdy@vutbr.cz,

[3]*Faculty of Science, Masaryk University, Brno, Czech Republic, EU,* hemzal@physics.muni.cz

[4]*Faculty of Agronomy, Mendel University, Brno, Czech Republic, EU,* radovan.kopp@mendelu.cz

https://doi.org/10.37904/nanocon.2021.4338

Abstract

In this paper, results of our running work on a sensitive and efficient detection of phosphates are presented. We develop a testing kit for rapid chemical detection based on surface-enhanced Raman scattering (SERS). We combine highly SERS active silver layer with a thin overlayer of alumina that is able to adsorb phosphorus compounds, in the form of silver/alumina core/shell nanopatterned film. In the first step, the silver layer was prepared by physical vapor deposition, followed by thermal annealing to form silver nanodots. Subsequently, alumina film was deposited over the silver nanodots using atomic layer deposition to form the silver/alumina core/shell nanopatterned surface. The prepared Ag/Al_2O_3 films can vary in overall thickness from 4 to 10 nm by changing the deposition parameters. The produced Ag/Al_2O_3 nanodots exhibit high SERS activity and significant adsorption of phosphate anions on these nanofilms has been confirmed under 532 nm excitation, which produced characteristic phosphate bands in the Raman spectra. The method will be further applied for quantitative detection of inorganic phosphates as well as organophosphates in real pond water samples.

Keywords: Phosphorus, phosphate, detection, core/shell, nanoparticles, Raman spectroscopy, SERS

1. INTRODUCTION

Water quality is likely to be one of the main limiting factors for the Central European pond industry in the future [1]. Understanding and timely assessment of chemical balance in the composition of aquatic ecosystems, namely in ponds and lakes, can prevent heavy economic losses in fish production. The phosphorus concentration plays a key role in the development of the phytoplankton community as it is influencing the basic physicochemical parameters in water. A sufficient concentration of dissolved oxygen in the water is a crucial parameter for the survival of fish in ponds. Rapid detection of changes in the phosphate compounds concentration in water, organic sediments, and pond soil, is therefore very important.

Phosphorus is a biogenic element that has a major influence on the primary production of green plants. Phytoplankton can utilise not only phosphorus from orthophosphate but also orthophosphate absorbed on the surface of suspended solids. Sorption and desorption processes of phosphate in sediments thus play an important role in changes of phosphorus concentration [2]. On contrary to lakes and reservoirs, not only the residence time of water in ponds but also the trophic level, sediment quality, and the intensity of pond management are crucial for phosphorus cycling [3]. Estimating nutrient retention in ponds using simple models is therefore not realistic and highly accurate field data measured in an appropriate way are urgently needed.

One of the techniques suitable for in-field analysis of phosphorus compounds in water and water sediment (soil) in real environmental conditions is Raman spectroscopy, as it offers a measurement using a portable device [4]. The advantage of this analytic tool is that water does not generally interfere with Raman spectral analysis. However, the technique needs to be sensitive enough to qualitatively as well as quantitatively distinguish among various types of phosphates. Surface-enhanced Raman scattering (SERS) might provide such an improvement of sensitivity, but so far, there is only minimum scientific evidence for this application in phosphate analysis [5].

In this paper, we report on the design and development of testing kit based on SERS for rapid and precise detection of phosphate compounds dissolved in pond water. The testing kit is developed as a core-shell nanopatterned surface formed by silver nanodots to enhance the Raman signals and ultra-thin alumina layer to bind the free phosphate from the tested sample. While silver is well known as one of the excellent candidates for enhancing the signal of Raman scattering, alumina in combination with silver was used for SERS so far only rarely [6]. For instance, the deposition of alumina on Ag to create a hybrid SERS platform has been previously reported for spectroscopic detection of water splitting reaction intermediates, where a thin layer of alumina helped to prevent charge transfer between Ag and the reaction system [7]. In another case, nanoporous alumina served as a template for the subsequent deposition of silver nanoparticles to be used in SERS [8]. Oliveira et al. found that Al_2O_3 seems to be the most suitable material for removing phosphate from aquatic systems with eutrophication problems [9]. The reason is that alumina has a very high surface area, is insoluble in water and is selective for P in the presence of chloride and sulphate ions.

According to the literature, our Ag/Al_2O_3 nanopatterned system, where alumina facilitates phosphate scavenging [10] and simultaneously enables oxidative protection of core silver, is the first one to be used for the direct phosphate SERS analysis in a pond aquatic environment. The objective of our study was to verify a suitability of chosen materials and the methods for our samples' fabrication for their subsequent application in SERS analysis of phosphate.

2. EXPERIMENTAL

2.1. Testing kit fabrication

Two types of testing kit were developed: the first type was fabricated on a silicon substrate: 8 nm Ag dots covered with 2 nm of Al_2O_3 and pure 8 nm Ag dots for reference. In particular, the 15 × 15 mm substrates were cut from 5" p-type Si wafers (Siegert Wafer GmbH, Germany) of crystallographic orientation (100), thickness ≈ 525 μm, and specific resistivity < 5 mΩ·cm. The silver (99.99% pure, Kurt J. Lesker, USA) film was deposited using e-beam thermal evaporation (Bestec, Germany), while its thickness was monitored during the deposition using calibrated quartz crystal microbalance. Subsequently, the 2 nm thick alumina layer was deposited using Atomic layer deposition (ALD) from trimethylaluminum precursor (99.999+% pure, PURATREM-High Purity Inorganic, Strem, USA) using Fiji 200 ALD deposition system (Ultratech/CambridgeNanoTech/Veeco, USA). Before the ALD process, the substrate with Ag film was inserted to the process chamber and annealed at 300 °C for 15 minutes to create the nanodot structure and to remove the water residues from the surface. The 2 nm thick Al_2O_3 film was deposited in 20 cycles with the thickness of 1 Å per cycle at 300 °C. The O_2 plasma was used to enhance the deposition.

The second type of testing kit was fabricated on a sapphire substrate. The sapphire was chosen just as a comparative material to verify if silicon does not interfere with phosphates in Raman measurement via false positive signals. The 15 × 15 mm substrates were cut from C-plane orientation (0001) sapphire wafers (Siegert Wafer GmbH, Germany) with diameter of ≈ 100 mm, thickness of ≈ 650 μm, and 99.998% purity. The silver film of 10 nm thickness was deposited using e-beam thermal evaporation; the Ag thickness was monitored by calibrated quartz crystal microbalance method during the deposition process. The 2 nm thick alumina layer was deposited using the same ALD process as described above, with annealing and deposition temperatures set to 300 °C.

2.2. Morphology analysis of Ag/Al_2O_3 core/shell nanopatterned film

The surface morphology and homogeneity of the fabricated samples were investigated with field scanning electronic microscope (MIRA-II LMU, Tescan, Czech Republic) operated at 1-30 keV in high vacuum mode. Images were captured under 15 kV using InBeam or standard SE detector.

2.3. Raman measurements

To check the basic SERS activity of the prepared kits, the Si/Ag and $Si/Ag/Al_2O_3$ nanopatterned substrates have been used in combination with an organic dye; providing several additional orders of resonant enhancement, the SERRS spectra of organic dyes allow for sensitive quantification of SERS activity of the substrates.

In particular, 3 μL droplets of 1 μM aqueous solution of (brilliant) cresyl blue (Sigma-Aldrich) were cast on the substrates and let dry. The subsequent Raman characterization has been accomplished using 532 nm diode-pumped solid-state (DPSS) laser with nominal power of 10 mW and x20 objective with NA=0.4. By stretching the laser spot into a line, the StreamLine™ modality of the used Renishaw inVia spectrometer allows to combine low sample heating with shortened acquisition time thanks to parallelized accumulation scheme. In our case, the resulting power in the spot of measurement was about 250 μW and the Raman maps with resolution around 100x100 pixels over an area of about 3x3 mm with 1 s accumulation at each pixel were measured in about 6 min overall accumulation time. The obtained Raman maps overcome single-point measurements by detailed information on SERS activity distribution over the sample and allow to safely evaluate the mean SERS enhancement of the substrates.

The ability of sapphire-based Ag/Al_2O_3 nanopatterned film to detect the phosphate was tested with 10 mM K_3PO_4 (Sigma-Aldrich) aqueous solution prepared from high purity (>18 MΩ.cm) demineralized water (DI). The kits were immersed in the phosphate solution for 30 min and let dry on air. Consequently, Raman maps were taken under conditions similar to Si-based substrates measurement described above.

3. RESULTS AND DISCUSSION

3.1. SEM analysis of silver/alumina core/shell nanodots

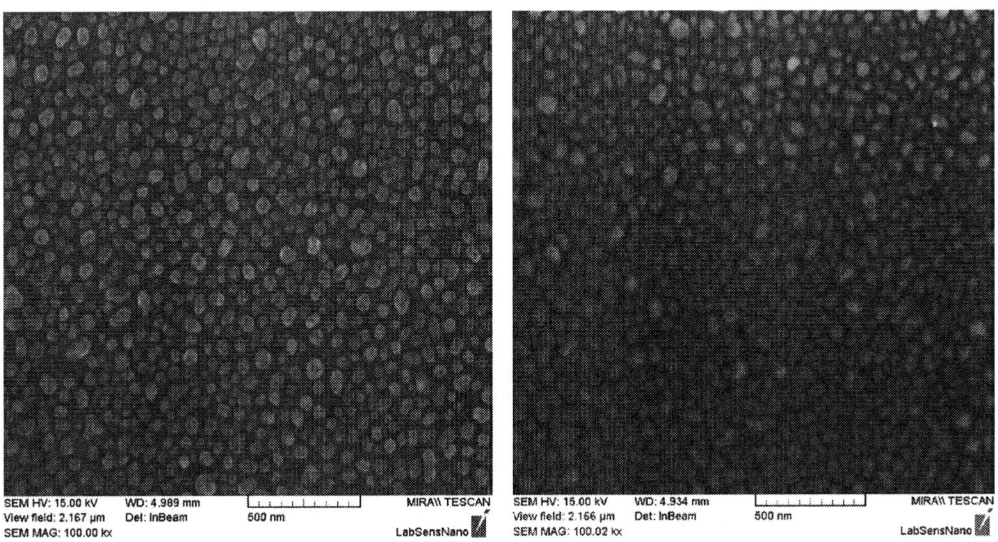

Figure 1 The SEM comparison of 8 nm thick silver nanodots without (left) and with (right) alumina shell prepared on silicon substrate using ALD fabrication process at 300 °C

The testing kits with core/shell structures were successfully fabricated using the thermal e-beam evaporation and ALD techniques on the silicon substrate as mentioned in chapter 2 and investigated using scanning electron microscopy (SEM). The comparison of 8 nm thick silver nanodots without (left) and with (right) alumina shell prepared on the silicon substrate using ALD fabrication process at 300 °C are shown in **Figure 1**. From the comparison is clear that there is no significant change in the structure after the alumina shell deposition. The success deposition can be deducted from worse focus which is given by worse charge transfer from the sample covered with non-conductive alumina shell.

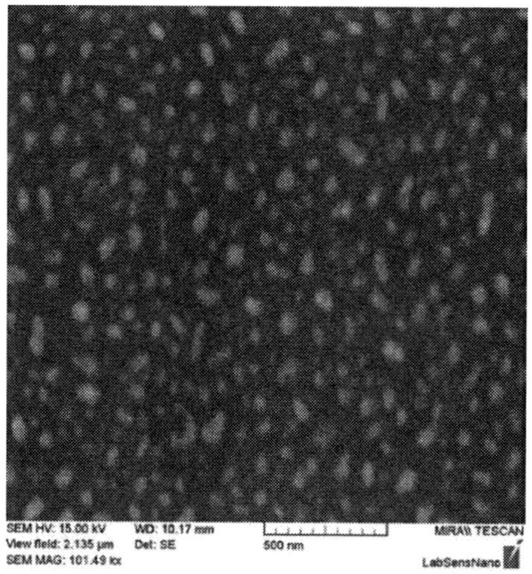

Figure 2 The 10 nm thick silver/alumina core/shell nanodots prepared on sapphire substrate by ALD fabrication process at 300 °C.

Figure 2 shows the 10 nm thick silver/alumina core/shell nanodots prepared on sapphire substrate by ALD fabrication process at 300 °C. In comparison with the samples fabricated on the silicon substrate the structure on sapphire substrate resulted in less dense structure which is probably given by the initial silver layer thickness.

3.2. SERS activity of Ag/Al_2O_3 nanopatterned substrates

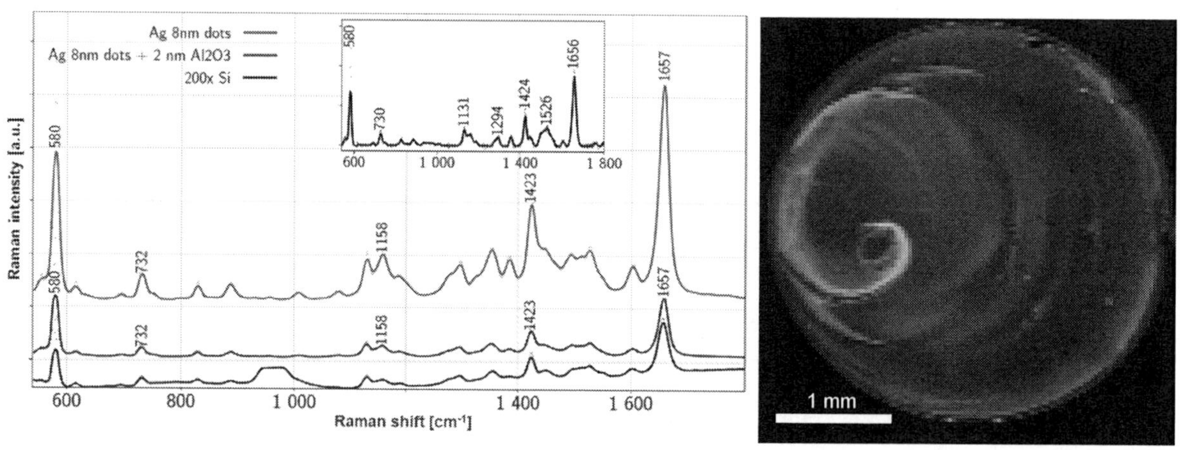

Figure 3 Background-corrected Raman spectra cumulated over the whole area of dried 3 µL droplets of 1 µM cresyl blue over Si based substrates (left panel); the inset shows spectrum of 100 mM aqueous cresyl blue solution for comparison. Distribution of the analyte across the dried droplet is shown in the right panel using 1657 cm^{-1} peak intensity mapping.

The Raman maps for dried droplets on Si-based substrates were taken with both Ag nanodots and Ag/Al_2O_3 nanopatterned film; the resulting cumulated Raman spectra are shown in **Figure 3.** Performing the measurement over the whole dried droplets, the overall SERS enhancement can be safely evaluated despite nonhomogeneous droplet coverage (cf. **Figure 3**). In particular, the SERS enhancement of the nanopatterned Si/Ag surface was found close to 10^3 (prior to its calculation, all spectra were corrected for a pure Si background). Although SERS enhancement for Ag/Al_2O_3 nanopatterned film is smaller than for Ag nanodots, alumina serves as both Ag capping agent to prevent aging of the pure silver surface due to its oxidation and scavenger of phosphate from the tested solution. The decrease of SERS enhancement to about 200x for Ag/Al_2O_3 nanopatterned film thus seems a reasonable compromise for efficient phosphate analysis.

3.3. Phosphate detection using Ag/Al_2O_3 nanopatterned substrates

To test the response of the kits towards phosphates, sapphire-based substrates were used together with 10 mM aqueous solution of K_3PO_4 as a model testing analyte; the concentration is comparable with expected range of phosphate abundance in the real pond water samples. The SERS results after soaking the substrates into the K_3PO_4 solution for 30 min are shown in **Figure 4**. The SERS spectra have been accumulated from an area about 100x100 μm under the collection conditions like measurement with the cresyl blue described above.

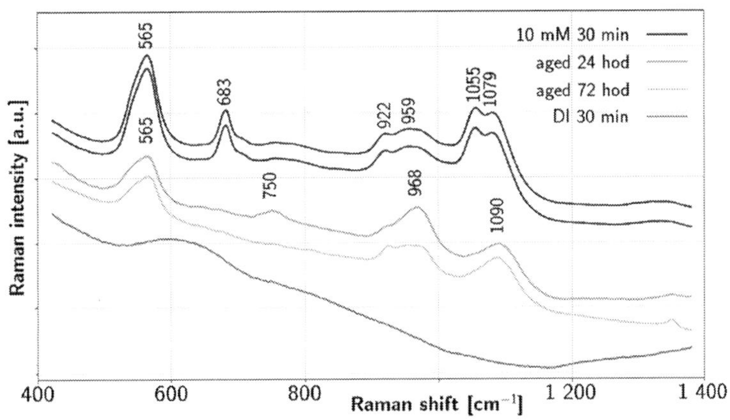

Figure 4 Cumulated Raman spectra (5 mW at 532 nm, 368 s) after soaking the sapphire/Ag/Al_2O_3 substrates into 10 mM K_3PO_4 solution for 30 min. To track the stability, the sample was measured also one and three days after deposition of the phosphate. In addition, the spectrum resulting from soaking the sapphire/Ag/Al_2O_3 substrates in water (DI) is shown.

The spectra show a combination of characteristic phosphate bands [11]: K_3PO_4 peak at 922 cm^{-1} (solid), H_2PO^{4-} peak at 1074 cm^{-1} (liquid), and a broad band for other phosphate ions in aqueous solution (989 cm^{-1} for HPO_4^{2}, 934 cm^{-1} for PO_4^{3-}). For comparison, **Figure 4** also shows the spectrum after soaking the substrate to pure water, proving that the observed peaks are justly assigned to the tested phosphate analyte. The SERS spectra re-measured 1 and 3 days after the phosphate deposition show that alumina shell deposited over silver ensured a sufficient stability of prepared SERS surface. This proves that our silver/alumina samples have a high potential in direct and reproducible analysis of phosphates.

4. CONCLUSION

The silver/alumina core/shell on substrates for SERS detection of phosphate in aqueous solution were successfully prepared and tested in detection of phosphates in the range of interest. The SERS activity was increased after the silver deposition about 1000 times. Although the intensity of Raman signal enhancement decreased after alumina deposition, the sample still allows the detection of phosphate in aqueous solution with sufficiently strong signal. In future research, we suggest optimizing alumina layer for better enhancement of

Raman signal and for selective scavenging of different phosphates from the real pond solutions. In addition, the presented testing kits should be optimized not only to qualitative but also quantitative detection of phosphates and organophosphates in water and water sediments.

ACKNOWLEDGEMENTS

This study was supported by a research project funded by the National Agency for Agricultural Research (QK1810161). The work was also carried out with the support of CzechNanoLab Research Infrastructure supported by MEYS CR (LM2018110).

REFERENCES

[1] WEZEL, A., ROBIN, J., GUERIN, M., ARTHAUD, F.VALLOD, D. Management effects on water quality, sediments and fish production in extensive fish ponds in the Dombes region, France. *Limnologica*, 2013, vol. 43, pp. 210-218.

[2] BOROVEC, J., HEJZLAR, J.VYHNALEK, V. Phosphorus cycling in a dimictic reservoir - the Sec Reservoir (Czech Republic). *International Review of Hydrobiology*. 1998, vol. 83, pp. 295-302.

[3] HEJZLAR, J., ŠÁMALOVÁ, K., BOERS, P.KRONVANG, B. Modelling Phosphorus Retention in Lakes and Reservoirs. *Water, Air, & Soil Pollution: Focus*. 2006, vol. 6, pp. 487-494.

[4] KRUSE, J., ABRAHAM, M., AMELUNG, W., BAUM, C., BOL, R., KUHN, O., LEWANDOWSKI, H., NIEDERBERGER, J., OELMANN, Y., RUGER, C., SANTNER, J., SIEBERS, M., SIEBERS, N., SPOHN, M., VESTERGREN, J., VOGTS, A.LEINWEBER, P. Innovative methods in soil phosphorus research: A review. *Journal of Plant Nutrition and Soil Science*. 2015, vol. 178, pp. 43-88.

[5] JI, W., SONG, W., TANABE, I., WANG, Y., ZHAO, B.OZAKI, Y.H. Semiconductor-enhanced Raman scattering for highly robust SERS sensing: the case of phosphate analysis. *Chemical Communications*. 2015, vol. 51, pp. 7641-7644.

[6] YANG, P.S., YIN, Y.T., LIN, P.C., CHEN, L.Y.CHEN, M.J. High chemical resistance and Raman enhancement in Ag/Al2O3 core-shell plasmonic nanostructures tailored by atomic layer deposition. *Materials Chemistry and Physics*. 2019, vol. 223, pp. 441-446.

[7] LING, X.Y., YAN, R.X., LO, S., HOANG, D.T., LIU, C., FARDY, M.A., KHAN, S.B., ASIRI, A.M., BAWAKED, S.M.YANG, P.D. Alumina-coated Ag nanocrystal monolayers as surfaceenhanced Raman spectroscopy platforms for the direct spectroscopic detection of water splitting reaction intermediates. *Nano Research*. 2014, vol. 7, pp. 132-143.

[8] LU, Z., RUAN, W., YANG, J., XU, W., ZHAO, C.ZHAO, B. Deposition of Ag nanoparticles on porous anodic alumina for surface enhanced Raman scattering substrate. *Journal of Raman Spectroscopy*. 2009, vol. 40, pp. 112-116.

[9] OLIVEIRA, M., RIBEIRO, D., NOBREGA, J.M., MACHADO, A.V., BRITO, A.G.NOGUEIRA, R. Removal of phosphorus from water using active barriers: Al2O3 immobilized on to polyolefins. *Environmental Technology*. 2011, vol. 32, pp. 989-995.

[10] HUANG, C.P.STUMM, W. Specific adsorption of cations on hydrous γ-Al2O3. *Journal of Colloid and Interface Science*. 1973, vol. 43, pp. 409-420.

[11] RUDOLPH, W.W. Raman- and infrared-spectroscopic investigations of dilute aqueous phosphoric acid solutions. *Dalton Transactions*. 2010, vol. 39, pp. 9642-9653.

COUPLED SEMICONDUCTORS AS PHOTO- ELECTROCHEMICAL CATALYST FOR WATER SPLITTING

[1]Denis ZABELIN, [1]Anna ZABELINA, [1]Anastasia TULUPOVA, [1]Roman ELASHNIKOV, [1]Vaclav SVORCIK, [1]Oleksiy LYUTAKOV

[1]Department of Solid State Engineering, University of Chemistry and Technology, Prague, Czech Republic, EU, lyutakoo@vscht.cz

https://doi.org/10.37904/nanocon.2021.4342

Abstract

Hydrogen is a perspective «green» fuel that produces only water when consumed in a fuel cell. The main method for hydrogen production is electrochemical water splitting. However, the water electrolysis is a rather energy-consuming process. Alternative method of the overall water splitting is photo-enhanced electrolysis under light illumination. The founding or new, photo-and electro-catalytically active materials, able to increase the efficiency of water splitting is a hot topic of actual research. In recent years, 2D materials have been considered as very promising materials for the water splitting. In this work, a hybrid material consisting of 2D WO_3-CdS coupled structure is created and introduced in water splitting process. Using the proposed design the high efficiency of hydrogen production can be potentially reached.

Keywords: Coupled semiconductors, 2D materials, water splitting, hydrogen production

1. INTRODUCTION

Hydrogen is a pure "green" fuel that produces only water when consumed in a fuel cell [1-4]. Today, governments and organizations are looking for alternative fuel sources for all types of energy consumers. While, for example, electric vehicles are generally seen as a response to the phasing out of fossil fuel vehicles, the application of hydrogen energy is increasingly proving to be a viable way of reducing carbon emissions [5,6].

It is known that there is a large amount of hydrogen in water. Today, there are many ways to split water molecules into hydrogen and oxygen [7-10]. However, most of them take place under the application of electric current, which makes overall process less energy effective. Thus, due to the increased attention to the environment and the excessive consumption of fossil energy, the founding of alternative ways for total decomposition of water into hydrogen and oxygen is a very important field of research [11-14].

In recent years, 2D materials have been among the most promising materials for the water splitting process, since the sophisticated design of 2D materials makes it possible to increase the efficiency of the water splitting process [15,16]. However, for the creation and separation of the electron-hole pair in 2D materials, it is necessary to apply relatively high energy, by, for example, materials illumination with $\lambda < 500$ nm. In order to be closer to full sunlight spectrum the NIR-light should be used.

In this work, a hybrid material consisting of a periodical metal electrode on the surface of which WO_3-CdS hybrid flakes was created. The combination of WO_3 and CdS semiconductors is one of the best for the photocatalytic formation of hydrogen and oxygen by water splitting based on the II-type semiconductors coupling. In turn, potential utilization of specific optical absorption from periodical metal electrode, which occurs due to excitation of collective electronic vibrations allows to introduced the apparent photo-contribution to water splitting efficiency [17,18].

2. EXPERIMENTAL

2.1. Materials

Sodium tungstate dihydrate (≥ 99%), tetrafluoroboric acid (48 wt. % in the water), cadmium chloride monohydrate (99.995%), sodium sulphide nonahydrate (≥ 99.99%) and deionized water were purchased from Sigma-Aldrich. Ethanol (p.a. 96%) and methanol (p.a. 99.96%) were purchased from Lachner.

2.2. Sample preparation

WO_3 nanoflakes were prepared by mixing of sodium tungstate dehydrate with tetrafluoroboric acid solution with subsequent transfer to a Teflon autoclave and kept at 180 °C for 8 hours. The resulting product was centrifuged and purified twice with deionized water and twice with ethanol and was dried in an oven at 65 °C for 8 hours. WO_3-CdS composite were prepared by adding consistently 0.1M $CdCl_2$ and 0.1M Na_2S solutions. The deposition of WO_3-CdS suspension onto the surface of periodic electrode was carried out by spin-coating method.

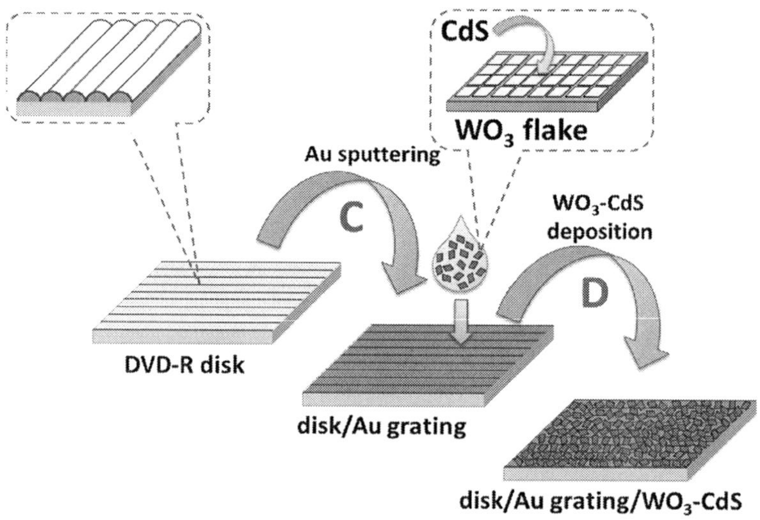

Figure 1 Schematic representation of sample preparation

2.3. Measurement techniques

X-ray diffraction microscopy spectra were recorded on a microXRD D8 Discover diffractometer for 30 min using Cu Kα radiation (1.5405 Å) 3 at 30 mA and 40 kV. HRTEM measurements were performed using an EFTEM Jeol 2200 FS microscope (Jeol, Japan). X-ray photoelectron spectroscopy (XPS) was performed using an Omicron Nanotechnology ESCAProbeP spectrometer with a monochromated Al K Alpha X-ray source operating at 1486.6 eV. The energy resolution was 0.4 eV for the survey study and 0.1 eV for the high-resolution XPS spectra measurements. The concentrations of elements were calculated in at.% using the sensitivity factors provided by the manufacturer.

3. RESULTS AND DISCUSSION

Figure 2 shows XRD measurements results of WO_3 nanoflakes and WO_3-CdS composites. The XRD pattern of pure WO_3 exhibits mostly a monoclinic phase. The main characteristic peaks acre located In the region of 23-24°. Looking at the XRD pattern of WO_3-CdS, the only independent peak of CdS at 33° can be seen. Thus, XRD analysis confirmed the successful combination (coupling) of two semiconductors.

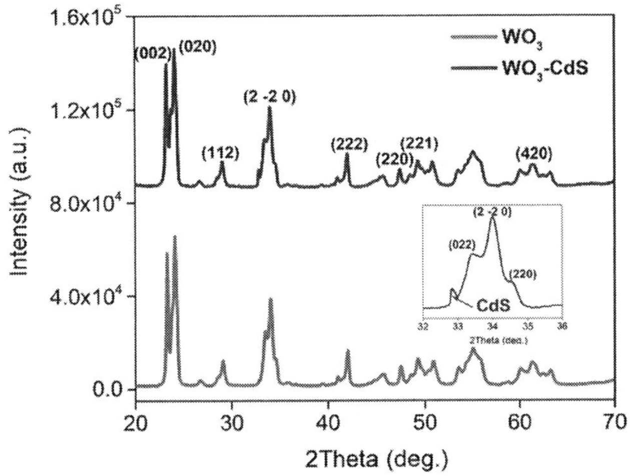

Figure 2 XRD patterns of WO_3 nanoflakes and WO_3-CdS composite

On the **Figure 3** the images from HRTEM measurements are presented. **Figure 3A** confirms that WO_3 flakes have thin 2D structure. After modification of WO_3 nanoflakes by CdS (**Figure 3B**) significant changes can be seen. Moreover, it is important to note that CdS particles are not occupy the entire surface area of the flakes, which is great for the overall water splitting process because both WO_3 and CdS should be in contact with water for HER and OER processes to proceed.

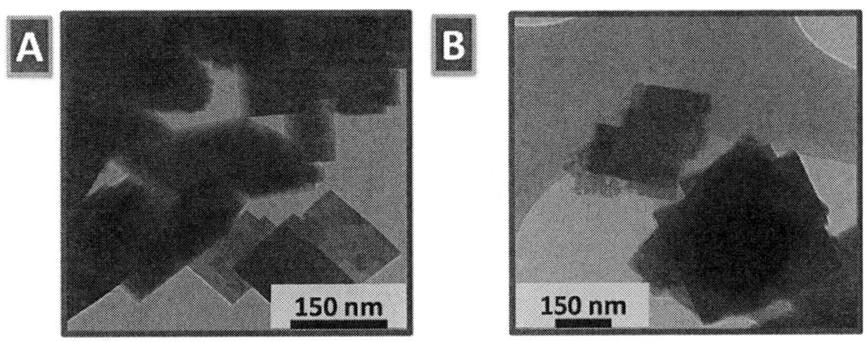

Figure 3 TEM images of pristine WO_3 (A), WO_3-CdS composite (B) and HRTEM image of WO_3 - CdS composite (C)

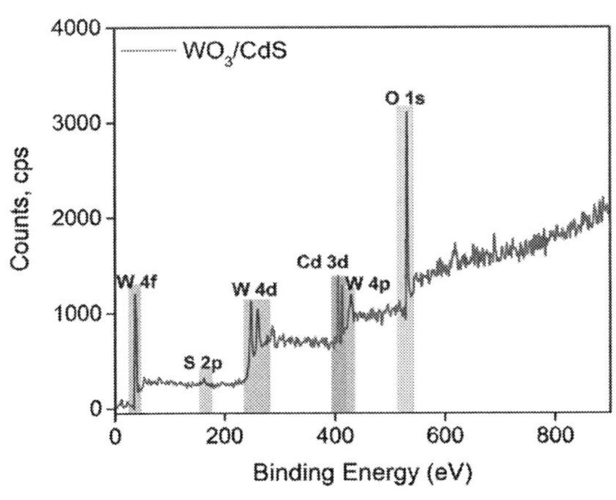

	W, at%	O, at%	Cd, at%	S, at%
WO_3	29,39	70,61	-	-
WO_3-CdS	26,17	66,62	3,68	3,53

Figure 4 Results of XPS analysis

Figure 4 shows the results of XPS analysis. The main thing that should be note is that the elemental quantity relationship between Cd and S on the surface of WO_3 is approximately 1:1. It means that the structure of cadmium sulfide doesn't have any defects and thus it has a possibility to product hydrogen from water.

4. CONCLUSION

In this work, a hybrid material consisting of a plasmon active gold lattice, on the surface of which are WO_3-CdS hybrid flakes, is created. The created materials represents an II-type coupled semiconductors. Moreover, both used materials are known to have the high efficiency in water splitting half-processes - OER and HER. The coupling of materials was confirmed using XRD, TEM and XPS techniques. Proposed structures can be considered as an interesting candidate for photo-electrochemically induced water splitting.

ACKNOWLEDGEMENTS

This work was supported by the project Internal Grant Agency UCT № A2_FCHT_2021_086

REFERENCES

[1] GHASEMZADEH, Kamran, ABBAS Aghaeinejad-Meybodi, and BASILE Angelo. Hydrogen production as a green fuel in silica membrane reactor: Experimental analysis and artificial neural network modeling. *Fuel.* 2018, vol. 222, pp. 114-124.

[2] HOSSEINI, S.E. and WAHID, M.A. Hydrogen production from renewable and sustainable energy resources: Promising green energy carrier for clean development. *Renewable and Sustainable Energy Reviews.* 2016, vol. *57*, pp. 850-866.

[3] AHMED, A., AL-AMIN, A.Q., AMBROSE, A.F. and SAIDUR, R. Hydrogen fuel and transport system: A sustainable and environmental future. *International journal of hydrogen energy.* 2016, vol. 41, no. 3, pp. 1369-1380.

[4] ALI., I. Water photo splitting for green hydrogen energy by green nanoparticles. *International Journal of Hydrogen Energy.* 2019, vol. 44, no. 23, pp. 11564-11573.

[5] LASHGARI, M. and GHANIMATI, M. Supported polyoxometalates as emerging nanohybrid materials for photochemical and photoelectrochemical water splitting. *Chemical Engineering Research and Design.* 2020, vol. 162, pp. 85-93.

[6] LASHGARI, M. and GHANIMATI, M. An excellent heterojunction nanocomposite solar-energy material for photocatalytic transformation of hydrogen sulfide pollutant to hydrogen fuel and elemental sulfur: A mechanistic insight. *Journal of colloid and interface science.* 2019, vol. 555, pp.187-194.

[7] CHI, J. and YU, H. Hierarchical coral-like $FeNi(OH)_x/Ni$ via mild corrosion of nickel as an integrated electrode for efficient overall water splitting. *Chinese Journal of Catalysis.* 2018, vol. 39, no.3, pp. 390-394.

[8] VINCENT, I. and BESSARABOV, D. Low cost hydrogen production by anion exchange membrane electrolysis: A review. *Renewable and Sustainable Energy Reviews.* 2018, vol. 81, pp. 1690-1704.

[9] NGUYEN, T., ABDIN, Z., HOLM, T. and MÉRIDA, W.. Hydrogen costs from water electrolysis at high temperature and pressure. *Energy Conversion and Management*, 2019, 200, p. 112108.

[10] KIM, J., JUN, A., GWON, O., YOO, S., LIU, M. et al. Hybrid-solid oxide electrolysis cell: A new strategy for efficient hydrogen production. *Nano Energy.* 2018, vol. 44, pp. 121-126.

[11] NALDONI, A., ALTOMARE, M., ZOPPELLARO, G., LIU, N., KMENT, S., ZBOŘIL, R. and SCHMUKI, P. Photocatalysis with Reduced TiO_2: From Black TiO_2 to Cocatalyst-Free Hydrogen Production. *ACS catalysis.* 2018, vol. 9, no. 1, pp. 345-364.

[12] WANG, M., ZUO, Y., WANG, J., WANG, Y., SHEN, X., QIU, B., CAI, L., ZHOU, F., LAU, S.P. and CHAI, Y. Remarkably Enhanced Hydrogen Generation of Organolead Halide Perovskites via Piezocatalysis and Photocatalysis. *Advanced Energy Materials*, 2019, vol. 9, no. 37, p. 1901801.

[13] BHARATVAJ, J., PREETHI, V. AND KANMANI, S. Hydrogen production from sulphide wastewater using Ce^{3+}-TiO_2 photocatalysis. *International Journal of Hydrogen Energy.* 2018, vol. 43, no. 8, pp. 3935-3945.

[14] YE, L. and WEN, Z. $ZnIn_2S_4$ nanosheets decorating WO_3 nanorods core-shell hybrids for boosting visible-light photocatalysis hydrogen generation. *International Journal of Hydrogen Energy*. 2019, vol. 44, no. 7, pp. 3751-3759.

[15] LOW, J., CAO, S., YU, J and WAGEH, S. Two-dimensional layered composite photocatalysts. *Chem. Commun.* 2014, vol. 50, pp. 10768 -10777

[16] LIN, L., OU, H., ZHANG, Y. and WANG, X. Tri-s-triazine-Based Crystalline Graphitic Carbon Nitrides for Highly Efficient Hydrogen Evolution Photocatalysis. *ACS Catalysis*. 2016, vol. 6, no. 6, pp. 3921-3931.

[17] ZABELINA, A., ZABELIN, D., MILIUTINA, E., LANCOK, J., SVORCIK, V., CHERTOPALOV, S. and LYUTAKOV, O. Surface plasmon-polariton triggering of $Ti_3C_2T_x$ MXene catalytic activity for hydrogen evolution reaction enhancement. *Journal of Materials Chemistry A*. 2021, vol. 9, no. 33, pp. 17770-17779.

[18] GUSELNIKOVA, O., TRELIN, A., MILIUTINA, E., ELASHNIKOV, R., SAJDL, P., POSTNIKOV, P., KOLSKA, Z., SVORCIK, V. and LYUTAKOV, O. Plasmon-induced water splitting-through flexible hybrid 2D architecture up to hydrogen from seawater under NIR light. *ACS Applied Materials & Interfaces*. 2020, vol. 12, no. 25, pp. 28110-28119.

SESSION C
BIONANOTECHNOLOGY, NANOMATERIALS IN MEDICINE

Chairmen

Prof. Dr. Anton FOJTIK	Czech Technical University in Prague, Czech Republic, EU
Prof. Dr. Jiri HOMOLA	Institute of Photonics and Electronics of the CAS, Prague, Czech Republic, EU

INDUSTRIAL MAGNETRON SPUTTERING OF ZrN/Cu NANOSTRUCTURED COATINGS FOR ANTI-BACTERIAL PURPOSES

[1]Sahand BEHRANGI, [2]Gabriela SUKOVÁ, [1]Vilma BURŠÍKOVÁ, [1]Pavel SOUČEK, [2]Ivo SEDLÁČEK, [3]Vjačeslav SOCHORA, [1]Petr VAŠINA

[1]*Masaryk University, Faculty of Science, Department of Physical Electronics, Brno, Czech Republic, EU,* behrangi@mail.muni.cz

[2]*Masaryk University, Faculty of Science, Department of Experimental Biology, Czech Collection of Microorganisms, Brno, Czech Republic, EU,* gsukova@gmail.com

[3]*SHM, s.r.o, Šumperk, Czech Republic, EU,* sochora@shm-cz.cz

https://doi.org/10.37904/nanocon.2021.4314

Abstract

In the present study, the anti-bacterial and mechanical properties of the ZrN/Cu nanostructured coatings were studied. Copper was introduced as the anti-bacterial agent owing to its substantial activity against bacteria. Industrial reactive magnetron sputtering was utilized to synthesize the films with varying copper contents. Mechanical properties were measured using nanoindentation and discussed in relation to the chemical composition. The inhibitory efficiency of the films was tested against the two most common bacteria, including gram-negative (E. coli) and gram-positive (S. aureus) bacteria. We observed that the anti-bacterial efficiency of the samples significantly improved with increasing the copper amount. After 80 min exposure, the coatings with copper amount >12 at% showed 100 % activity against both types of bacteria. The highest hardness measured was 28.5 GPa for the low copper coating followed by a gradual decrease with increasing copper.

Keywords: Magnetron sputtering, bacteria, anti-bacterial efficiency, hardness, chemical composition

1. INTRODUCTION

Bacterial infection could cause tremendous problems in human societies. The bacteria-related issues have been recognized in many areas such as medical tools, food industry, health facilities and so forth. The bacterial adhesion to the surface, colonization and biofilm formation are ascribed for the consequent issues. Furthermore, most of the hazardous bacteria could survive on the surfaces for a long time. If such surfaces are touched, these adherent bacteria could transform to the human body and cause serious illnesses. Therefore, many efforts have been put into solving this problem. Amongst, production of some anti-bacterial films has been attracted much interest in the last years. The development of such a layer not only leads to the killing of bacteria on the surface but also improves the mechanical properties of the substrate as to slow down the degradation rate [1,2].

Two well-known elements with impressive anti-bacterial activity are silver (Ag) and copper (Cu). Both mentioned metals are able to kill the bacteria in contact with them. Thus, production and characterization of the films containing either Cu [2-5] or Ag [6-10] have been studied in the recent years. However, Cu-containing films are preferable to Ag as Cu is an essential trace element and the human body could release it [11]. Copper can kill bacteria by releasing Cu ions which destroy their cell walls and cell membranes [4].

Zirconium nitride (ZrN) is a well-known hard and wear-resistant material which have been studied by numerous researchers to synthesize coatings with improved mechanical properties. Therefore, by adding a third element one can obtain the tailored properties e.g. mechanical and anti-bacterial properties in the case of Cu addition

[11]. Thanks to its high production rate and ability to produce films on the substrates with different geometries, reactive magnetron sputtering has become a promising method to produce such coatings like ZrN/Cu which will be discussed in this paper.

Cu is immiscible in ZrN structure and crystallizes separately as the second phase to form the so-called nanocomposite. Nanocomposites are composed of at least two discrete phases which are mutually immiscible. One phase form fine dispersion within the other phase. It has been observed that nanocomposites can exhibit better mechanical, magnetic and electronical properties than the ordinary coatings [12-14].

The current work focuses on the ZrN/Cu nanocomposite films synthesized by industrial magnetron sputtering to be used as anti-bacterial protective films. The films were prepared with different copper contents to observe the effect of copper concentration on the resultant properties. Chemical composition and mechanical properties of the films were measured. Furthermore, the killing efficiency of the films were measured against a gram-negative (*Escherichia coli*) bacterium and a gram-positive (*Staphylococcus aureus*) bacterium. The main privilege of our work is the much shorter exposure time compared to the literature. This ensures the killing of high number of bacteria colonies in a shorter time and having a cleaner surface.

2. MATERIALS AND METHODS

All the coatings were produced by SHM company utilizing the industrial sputtering machine. A hybrid method was applied as the combination of arc evaporation (for Zr target) and magnetron sputtering (for Cu target). The amount of copper altered through the varying magnetron power. The arc curent was 150 A and magnetron power varied from 0.5 kW to 3 kW. All the coating parameters are presented in the **Table 1**.

Table 1 Process parameters for the synthesis of ZrN/Cu films

Sample	Arc current (A)	Magnetron power (kW)	Deposition time (min)	Deposition temperature (°C)	N_2 pressure (Pa)	Bias (V)
Cu0	150	0.5	60	400	1.5	-50
Cu6	150	1	60	400	1.5	-50
Cu12	150	1.3	60	400	1.5	-50
Cu25	150	2.2	60	400	1.5	-50
Cu29	150	3	60	400	1.5	-50

Polished Si (100) was used as substarte for the coatings. Prior to process, degreasing of samples was performed using ultrasonication of the Si substrates in the ethanol medium. Furthermore, before the depsotion, the samples underwent a sputter clening process to roughen their surface and to eliminate any surface scale or contamination. After the deposition process completed, the samples were cooled inside the chamber down to the ambient temperature.

Tescan Mira3 Scanning Electron Microscope (SEM) equipped with X-ray Energy Dispersive Spectroscopy (EDS) detector was employed to measure the elemental composition of the films. X-ray Diffractometry (XRD) was employed to characterize the phases of the microstructure. This was done utilizing Rigaku Smartlab X-ray diffractometer using Cu Kα radiation with a wavelength of 0.15418 nm.

Hysitron TI950 Triboindenter equipped with a Berkovich tip was employed to measure the hardness and elastic modulus of the coatings.

For anti-bacterial testing, the common bacteria present normally on human skin either as contamination (*Escherichia coli* CCM3988) or as a commensal (*Staphylococcus aureus* CCM4516) were picked. The sampled bacteria were obtained from the Czech Collection of Microorganisms (CCM) in Brno, Czech Republic. From the original 24 h culture on trypton soya agar (TSA) at 37 °C, a 1° McFarland suspension in saline solution was prepared. 1 ml of this suspension was pipetted into 9 ml of saline solution for creating a final inoculum with the concentration of 3×10^7 CFU/ml for *S. aureus*. For *E. coli*, the inoculum was diluted 1:10 twice into final inoculum (1:10 into primary inoculum and then 1:10 again into final inoculum) with the concentration of 3×10^6 CFU/ml. These concentrations were chosen after optimization, where it was found that without two-step dilution of the *E. coli* suspension, the final plate counting was made impossible due to overcrowding of the plate with a high number of growing colonies. 10 ml of final inoculum was then pipetted onto sterilised coated samples, covered with cover slip and incubated at room temperature and humidity for 80 min. After incubation, the plates were put into 5 ml of saline solution and shaken at 37 °C for 10 min. Following shaking, 100 ml of the suspension was pipetted onto TSA plates in triplicates, spreaded with a spreader and incubated at 37 °C for 24 h. After 24 h the plates were taken out and colonies were counted.

Anti-bacterial efficiency (E) of the films was calculated by dividing the number of killed colonies on the coatings over the number of survived colonies on the control samples according to the equation (1).

$$E(\%) = \frac{\text{Number of killed bacteria after the exposure time}}{\text{Number of bacteria on the control sample after the exposure time}} = \frac{N_C - N_S}{N_C} \times 100 \quad (1)$$

where:

N_c - number of the colonies survived on the control sample

N_s - number of the colonies survived on the test sample

3. RESULTS AND DISCUSSION

The results of the EDS measurement are shown in **Table 2**. As can be seen in this table, the copper content is changing from ~0 at% to ~30 at% by varying magnetron power. Also, the amount of Zr and N are changing accordingly because nitrogen is only consumed by Zr while, Cu does not form any stable bond with nitrogen.

Table 2 Chemical composition of the ZrN/Cu coatings measured by EDS

Sample	Cu (at%)	Zr (at%)	N (at%)
Cu0	0.7	50.79	48.51
Cu6	6.74	47.04	46.22
Cu12	12.82	43.20	43.98
Cu25	25.86	36.76	37.38
Cu29	29.84	34.85	35.32

XRD was employed to have a better view of the microstructure of the coatings as well as to characterize the phases formed during sputtering process. The XRD spectra are demonstrated in **Figure 1**. The estimation of the grain size of the films are presented in **Table 3.** The XRD results are in consistence with the EDS results as only peak corresponding to Cu and ZrN are present. The two-phase nanocomposite films composed of Cu and ZrN characterized by clearly separated reflections of Cu (1 1 1), ZrN (1 1 1), ZrN (2 0 0) and ZrN (2 2 0). The formation of a separate copper grains is more distinguished with increasing copper content. At lower Cu amounts, the absence of Cu reflections implies that the Cu is either in an amorphous state or the size of Cu grains and their quantity is so small that their diffraction intensity is below the detection limit [13]. It can be seen that the ZrN/Cu films have polycrystalline structures. The crystallinity of films slightly decreases with the incorporation of Cu as can be implied in **Table 3.**

Hardness and elastic modulus of the films are plotted in **Figure 2**. As can be seen in this graph, mechanical properties of ZrN/Cu films are strongly affected by Cu concentration. The hardness and elastic modulus both decrease with the copper content of the coatings. This is a consequent of the growing portion of the soft phase (Cu) and its segregation as a separate phase in the microstructure. The hardness drops from 28.5 GPa for Cu0 to 15.1 GPa for Cu29. Likewise, the elastic modulus decreases from 336 GPa for Cu0 to 242 GPa for Cu29.

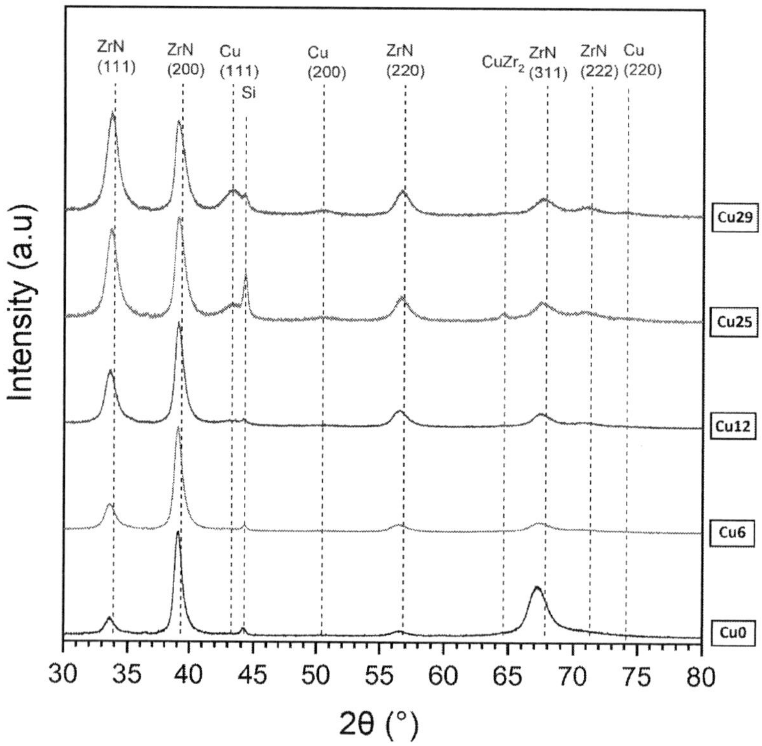

Figure 1 XRD spectra of the ZrN/Cu coatings

Table 3 Grain size of the ZrN phase in the microstructure of the ZrN/Cu coatings

Sample	Cu0	Cu6	Cu12	Cu25	Cu29
Grain size (nm)	8.40	7.76	7.82	7.69	7.49

Anti-bacterial efficiency of the ZrN/Cu coatings with different Cu concentrations against E. coli and S. aureus are plotted in **Figure 3**. The efficiency values are calculated using equation (1) after 80 min contact time of the surface with bacterial solution. From this figure, it is clear that the *E* value of the films strongly depends on the concentration of Cu. The killing efficiency of the ZrN films without any Cu is insignificantly low. However, addition of Cu even in low concentrations leads to the substantial improving of anti-bacterial efficiency. Consequently, the samples with >12 at% Cu show almost 100% killing efficiency against both bacteria. This indicates that for the excellent anti-bacterial efficiency, the ZrN/Cu films must contain >12 at% Cu. In this research, the contact time of bacteria with the surface of the samples was significantly lower compared to the literature [3,11,15]. This was intentionally done to make the results closer to the real working conditions. For example, if these films are to be applied on some parts like door handles, the killing of the bacteria in shorter times results in having an almost bacteria-free surface after being touched by people.

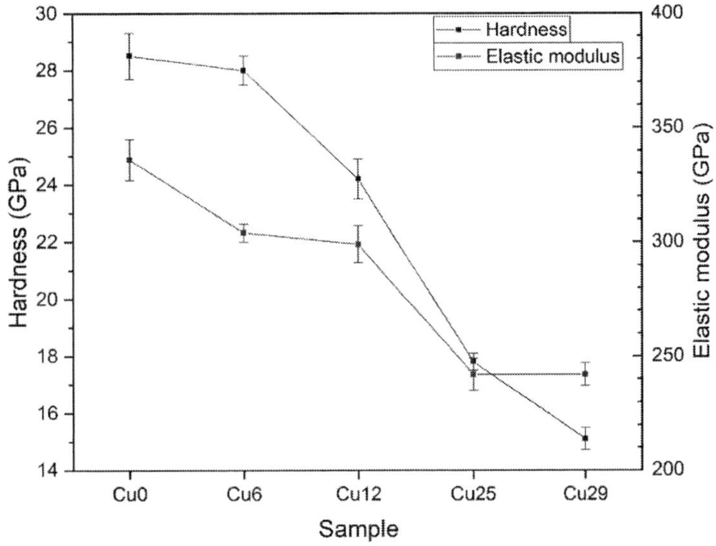

Figure 2 Hardness and elastic modulus of the ZrN/Cu coatings

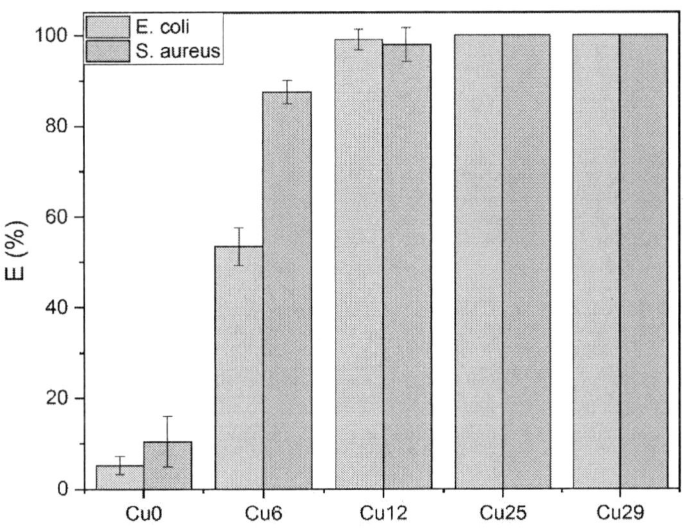

Figure 3 Anti-bacterial efficiency of the ZrN/Cu coatings against two kinds of bacteria after 80 min exposure time

4. CONCLUSIONS

The Cu-containing ZrN films were synthesized utilizing reactive magnetron sputtering method. According to the XRD measurements, ZrN and Cu phases were formed separately in the microstructure leading to the formation of nanocomposite thin films. Mechanical properties were observed to fall with increasing copper content. The bacteria-killing efficiency of the films were also evaluated. The presence of Cu in the films led to the significant improvement of the anti-bacterial efficiency so that 100 % of the bacteria killed after 80 min exposure time for the medium to high Cu contents.

ACKNOWLEDGEMENTS

This research has been supported by project FV30262, which is funded through the TRIO program of the Ministry of Industry and Trade, and project LM2018097 funded by the Ministry of Education, Youth and Sports of the Czech Republic.

REFERENCES

[1] CRAWFORD, R.J., WEBB, H.K., TRUONG, V.K., HASAN, J., IVANOVA, E.P. Surface topographical factors influencing bacterial attachment. *Adv. Colloid Interface Sci.* 2012, vol. 179-182, pp. 142-149.

[2] SHARIFAHMADIAN, O., SALIMIJAZI, H.R., FATHI, M.H., MOSTAGHIMI, J., PERSHIN, L. Relationship between surface properties and antibacterial behavior of wire arc spray copper coatings. *Surf. Coatings Technol.* 2013, vol. 233, pp. 74-79.

[3] WANG, J., ZHAO, B., LI, G., YANG, K., et al. Study of TiCu/TiCuN multilayer films with antibacterial activity. *Mater. Technol.* 2020, vol. 35, pp. 475-482.

[4] LUO, J., HEIN, C., MÜCKLICH, F., SOLIOZ, M. Killing of bacteria by copper, cadmium, and silver surfaces reveals relevant physicochemical parameters. *Biointerphases.* 2017, vol. 12, pp. 020301.

[5] VINCENT, M., DUVAL, R.E., HARTEMANN, P., ENGELS-DEUTSCH, M. Contact killing and antimicrobial properties of copper. *J. Appl. Microbiol.* 2018, vol. 124, pp. 1032-1046.

[6] SHIMAZAKI, T., MIYAMOTO, H., ANDO, Y., NODA, I., et al. In vivo antibacterial and silver-releasing properties of novel thermal sprayed silver-containing hydroxyapatite coating. *J. Biomed. Mater. Res. - Part B Appl. Biomater.* 2010, vol. 92, pp. 386-389.

[7] JU, H., YU, D., YU, L., DING, N., et al. The influence of Ag contents on the microstructure, mechanical and tribological properties of ZrN-Ag films. *Vacuum.* 2018, vol. 148, pp. 54-61.

[8] FIELDING, G.A., ROY, M., BANDYOPADHYAY, A., BOSE, S. Antibacterial and biological characteristics of silver containing and strontium doped plasma sprayed hydroxyapatite coatings. *Acta Biomater.* 2012, vol. 8, pp. 3144-3152.

[9] CHEN, W., LIU, Y., COURTNEY, H.S., BETTENGA, M., et al. In vitro anti-bacterial and biological properties of magnetron co-sputtered silver-containing hydroxyapatite coating. *Biomaterials.* 2006, vol. 27, pp. 5512-5517.

[10] KELLY, P.J., LI, H., WHITEHEAD, K.A., VERRAN, J., et al. A study of the antimicrobial and tribological properties of TiN/Ag nanocomposite coatings. *Surf. Coatings Technol.* 2009, vol. 204, pp. 1137-1140.

[11] MUSIL, J., ZÍTEK, M., FAJFRLÍK, K., ČERSTVÝ, R. Flexible antibacterial Zr-Cu-N thin films resistant to cracking. *J. Vac. Sci. Technol. A Vacuum, Surfaces, Film.* 2016, vol. 34, pp. 021508.

[12] ZHANG, S., SUN, D., FU, Y., DU, H. Recent advances of superhard nanocomposite coatings: A review. *Surf. Coatings Technol.* 2003, vol. 167, pp. 113-119.

[13] MUSIL, J., ZEMAN, P., HRUBÝ, H., MAYRHOFER, P.H. ZrN/Cu nanocomposite film - A novel superhard material. *Surf. Coatings Technol.* 1999, vol. 120-121, pp. 179-183.

[14] PATSALAS, P., ABADIAS, G., MATENOGLOU, G.M., KOUTSOKERAS, L.E., LEKKA, C.E. Electronic and crystal structure and bonding in Ti-based ternary solid solution nitrides and Ti-Cu-N nanocomposite films. *Surf. Coatings Technol.* 2010, vol. 205, pp. 1324-1330.

[15] PENG, C., ZHAO, Y., JIN, S., WANG, J., et al. Antibacterial TiCu/TiCuN Multilayer Films with Good Corrosion Resistance Deposited by Axial Magnetic Field-Enhanced Arc Ion Plating. *ACS Appl. Mater. Interfaces.* 2019, vol. 11, pp. 125-136.

SENSITIVE AMPEROMETRIC NANOBIOSENSOR FOR DETECTION OF SARCOSINE - POTENTIAL PROSTATE CANCER MARKER - IN URINE SAMPLES

Štefánia HRONČEKOVÁ, Tomáš BERTÓK, Ján TKÁČ

Institute of chemistry, Slovak Academy of Sciences, Bratislava, Slovakia, EU,
chemshrn@savba.sk

https://doi.org/10.37904/nanocon.2021.4313

Abstract

In this work, two-dimensional layered material $Ti_3C_2T_X$ MXene has been used to immobilize enzyme sarcosine oxidase to fabricate a nanostructured biosensor. The device was applied for detection of sarcosine, a potential prostate cancer biomarker, in urine samples for the first time. The morphology and structures of MXene have been characterised by atomic force microscopy (AFM) and scanning electron microscopy (SEM). Electrochemical, SEM and AFM results revealed that MXene along with chitosan is a good candidate for enzyme immobilization to fabricate a sensitive nanobiosensor. The prepared device exhibited a good analytical performance with a low detection limit of 18 nM and linear range from 0.09 to 7.8 µM. The proposed biosensing method also demonstrated high recovery index for sarcosine detection in urine samples in clinically relevant range (102.6 %).

Keywords: MXene, sarcosine, biosensor, prostate cancer

1. INTRODUCTION

Cancer is now widely recognized as a global problem that unfortunately lacks a global solution. Prostate cancer (PCa) is the most diagnosed cancer type among men and one of the most prevalent cancer types in general. It is a leading cause of deaths among men with an incidence of 1.1 million cases a year resulting in 366,000 deaths annually [1]. The gold standard in early-stage PCa diagnostics/screening is analysis of the level of prostate-specific antigen (PSA) in blood serum. PSA is a glycoprotein produced by the prostate gland which is responsible for seminal fluid liquefaction. Although widely used in connection with cancer, elevated levels of PSA will identify prostate disease, but not necessarily PCa. Hence, PSA is rather prostate-specific than cancer-specific and should not be used as reliable PCa diagnostic biomarker [2]. The key to effective prostate cancer management is early and accurate diagnostics, which includes the ability to differentiate between men with aggressive disease that require immediate treatment and those with slow-growing cancer that may not cause any significant difficulties within their lifetime. Utilizing a panel of diverse biomarkers is the most appropriate method. Potential PCa biomarkers include sarcosine, zinc-alpha-2-glycoprotein, prostate cancer antigen 3, Cofilin-1, annexin A3 as well as transmembrane protease serine 2-ETS-related gene [3,4]. This work is focused on sensitive electrochemical detection of sarcosine, which can provide significant discrimination between PCa patients and non-cancer individuals, with an area under curve characteristic (AUC) of 0.833 [5]. Sarcosine, also known as N-methylglycine, is a non-proteinogenic amino acid naturally occurring in muscles and various other body tissues [6]. Although the measurements in blood have been reported as unremarkable, significantly elevated levels in urine have been recorded in some studies, for patients with PCa. Normally, the level of sarcosine in the urine of the PCa patients is very low, approximately 20 nM to 5 µM [7]. This work describes construction of a novel sensitive sarcosine biosensor based on immobilization of sarcosine oxidase (SOx, EC 1.5.3.1) on the surface of pre-treated glassy carbon electrodes (GCEs). Two-dimensional layered nanomaterial $Ti_3C_2T_X$ (a member of the MXene family) was used to immobilise enzyme SOx to fabricate a nanostructured biosensor. MXenes have attracted considerable amount of interest due to their unique properties, morphology and important applications. With the excellent physical and chemical properties, such as graphene like structure, metallic conductivity, large electrochemically active surface, high stability, excellent mechanical properties and water solubility, MXene has been used in metal ion batteries, supercapacitors, fuel cells, electronic devices and absorbents [8-12]. This work indicates that the immobilization of proteins onto the MXene-based surfaces is a novel efficient method for the development of a new class of promising and sensitive nanostructured electrochemical biosensors with use in cancer diagnostics.

2. MATERIALS AND METHODS

2.1. Materials

Sarcosine oxidase from Bacillus sp. (25-50 $U \cdot mg^{-1}$), sarcosine, glutaraldehyde solution, chitosan, acetic acid, phosphate buffer (PB) components (KH_2PO_4 and K_2HPO_4) and Surine™ (a negative urine control for toxicology) were purchased from Sigma-Aldrich (USA). Micropolish alumina powder and polishing pads for disc electrodes were purchased from Buehler (USA). MXene was prepared using a previously described protocol [13].

2.2. MXene-based sarcosine biosensor construction

Prior to modification, all glassy carbon electrodes (GCEs) were mechanically polished by alumina powder, washed with distilled water and sonicated in ethanol and water successively. The biosensor was prepared by a simple drop-casting method. Firstly, chitosan solution (0.1 %) and MXene dispersion (3 $mg \cdot ml^{-1}$) were mixed and shaken overnight at a temperature of 20 °C and at a rotation speed of 1500 rpm. After preparation, the above-prepared mixture was cast onto the surface of polished GCEs. After drying, 5 %,1 % or no glutaraldehyde solution was applied to the surface for 10 minutes. Finally, SOx solution was applied onto the surface and allowed to dry at room temperature in a laminar box. Before electrochemical measurements, all the as-prepared electrodes were immersed in PB pH 7.4 to remove residual components. To optimise biosensor preparation and its performance, different configurations and nanocomposite preparation protocols were tested. Among all these configurations, from the stability point of view, only the configuration with mixed chitosan + MXene (overnight, 20 °C, 1500 rpm) prior to drop-casting, subsequently modified by layer-by-layer using desalted and diluted SOx with no GA activation, was chosen.

2.3. Electrochemical measurements

All electrochemical procedures were carried out on a laboratory potentiostat/galvanostat Autolab PGSTAT 302 N with an impedimetric module (Metrohm, Netherlands) with a GCE (Bioanalytical systems, USA) used as a working electrode. An Ag/AgCl/3 M KCl reference electrode and a counter platinum electrode (Bioanalytical systems, USA) were applied in a three-electrode cell system. Chronoamperometry measurements were performed on a rotating disc electrode employed as a working electrode (Metrohm, Netherlands). Cyclic voltammetry (CV) and chronoamperometry were performed in 0.1 M PB pH 7.4 purged with nitrogen for 15 minutes prior to use. CV measurements were run in the potential window from 0.1 V (volt) to −1.0 V at a scan rate of 0.1 $V \cdot s^{-1}$. Chronoamperometry measurements were run at a fixed potential value of −0.7 V. All electrochemical measurements were run under Nova Software 1.10 (Metrohm, Netherlands) and data acquired were evaluated using Origin Pro 9.1 (Origin Lab, USA).

2.4. Contact Angle Measurements

Contact angle measurements for MXene-based interfaces were run on a portable instrument System E (Advex Instruments, Czech Republic) to obtain wetting angle and free surface energy. As the testing liquids, distilled water and diiodomethane were used. Free surface energy was determined using the two-liquid Owens-Wendt method. For each sample, the water and diiodomethane wetting angle was obtained as an average value of assays performed using 3 parallel measurements.

2.5. Atomic force microscopy (AFM) and scanning electron microscopy (SEM)

The morphology and structure of MXene and MXene-chitosan nanocomposite were characterized by scanning electron microscopy (SEM) and atomic force microscopy (AFM). SEM images of MXene and MXene-chitosan composite were recorded using Nova NanoSEM 450 microscope (FEI, USA) applying accelerating voltage. A peak force tapping mode AFM in air was carried out on a Bioscope Catalyst instrument and Olympus IX71 microscope (Bruker, USA) in conjunction with NanoScope 8.15 software.

3. RESULTS AND DISCUSSION

3.1. Electrochemical measurements

To evaluate the bioelectrochemical activity of the biosensor and to construct the corresponding calibration curves, the tested sarcosine concentration ranged from 2.5 to 50 μM (**Figure 1**). Electrochemical reduction of H_2O_2 produced by catalytic decomposition of sarcosine by SOx present onto the surface of electrode started at -0.4 V with the peak maximum present at -0.7 V. The cathodic peak for H_2O_2 reduction increased with the increasing biomarker concentration. Calibration curves were constructed by plotting sarcosine concentration against current response in the 1st CV scan after subtraction of a blank. Aiming to choose the best enzyme immobilization procedure an optimization of SOx immobilization was carried out using glutaraldehyde solution (**Figure 1, inset**).

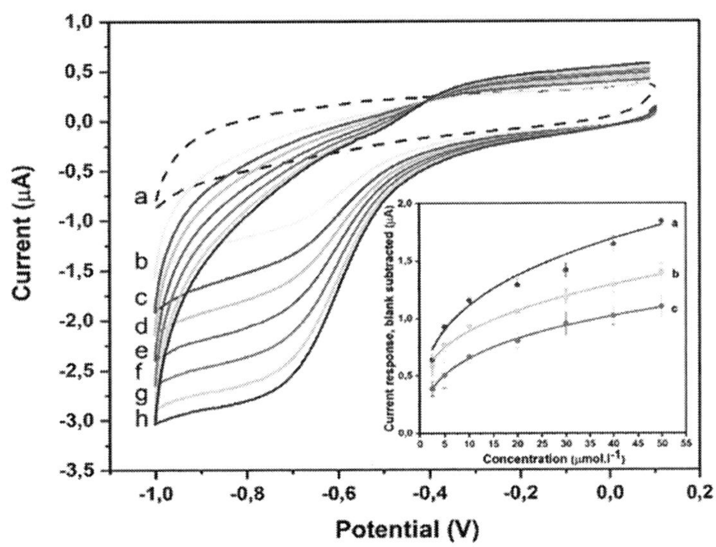

Figure 1 Cyclic voltammograms (CVs) of the prepared MXene-based biosensor in N_2-purged 0.1 M PB pH 7.4 containing 0 (a), 2.5 (b), 5 (c), 10 (d), 20 (e), 30 (f), 40 (g), 50 (h) μM sarcosine. **Inset**: Calibration plots obtained when using no glutaraldehyde (a), 1% glutaraldehyde (b) and 5% glutaraldehyde (c) during fabrication of the biosensor

However, the results obtained when using glutaraldehyde solution weren't satisfactory since glutaraldehyde application resulted in a decrease of a current responses. Calibration curves using a Hill model and Michaelis-Menten kinetics renders the following maximal current outputs: (I_{max}): (2.61 ± 0.15) μA (0% GA), (1.96 ± 0.26) μA (1% GA), and (1.42 ± 0.10) μA (5% GA). Glutaraldehyde solution has a negative effect on biosensor performance which can be explained by the formation a multi-point attachment of SOx to the surface, leading to enzyme inactivation. When no glutaraldehyde was present, the LOD of prepared SOx/MXene-chi/GCE biosensor (based on signal to noise ratio of 3) was calculated to be 91.4 nM using CV. Compared to the CV, chronoamperometry is more sensitive electrochemical method for indirect sarcosine detection. To obtain low LOD values, chronoamperometry on a rotating disc GCE at fixed potential of -0.7 V and 900 rpm was performed (**Figure 2**). The detection limit was calculated to be 18.0 nM (based on a signal-to-noise ratio of 3).

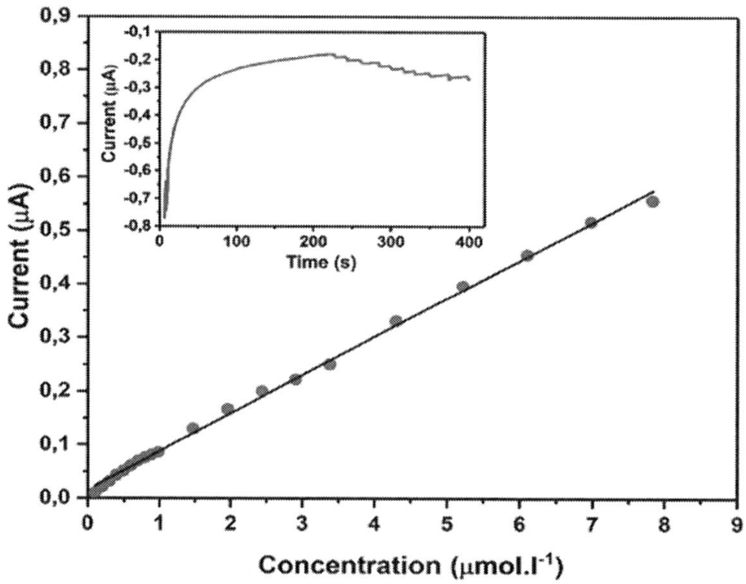

Figure 2 Plot of current responses after subtraction of blank of the prepared biosensor measured on a rotating disc GCE at −0.7 V and 900 rpm versus sarcosine concentration. **Inset**: a current versus time response of the prepared biosensor at −0.7 V to successive addition of stock sarcosine solution in stirred N_2-purged 0.1 M PB pH 7.4

3.2. Contact Angle Measurements

Wetting angle measurements were applied to investigate the surface modifications during individual steps of biosensor preparation. The hydrophilicity increased from bare GCE (89.99 ± 2.269°), MXene-modified GCE (71.87 ± 4.968°) and MXene/chitosan-modified GCE (69.51 ± 3.604°). After SOx adsorption, the biosensor surface becomes more hydrophobic again (87.44 ± 4.898°) (**Figure 3**).

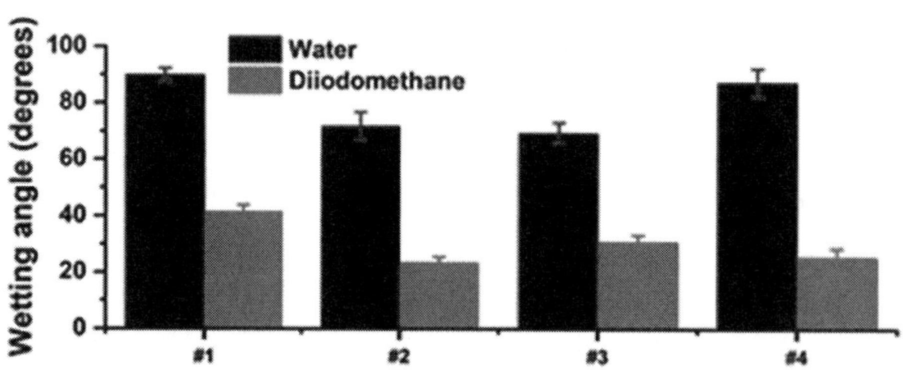

Figure 3 Graphical presentation of wetting angle measurements for: bare GCE (#1), MXene-modified GCE (#2), MXene/chitosan-modified GCE (#3) and SOx-MXene/chitosan-modified GCE (#4)

3.3. Attomic force microscopy (AFM) and scanning electron microscopy (SEM)

SEM and AFM analysis was used to observe the microstructures of the samples and to investigate and see the surface morphology of prepared MXene and MXene-chitosan composite layers before enzyme immobilization took place. Studied MXene dispersion contains microscale-sized sheets with a multilayer nanostructure similar to exfoliated graphene oxide. Various sized MXene-chitosan composite aggregates with porous structure were also detected (**Figure 4**). Height profile analysis results revealed that MXene-chitosan flakes were about 10 times thicker than those consisting of MXene. Average thickness of one MXene nanosheet was calculated to be (20.2 ± 3.1) nm (**Figure 5**).

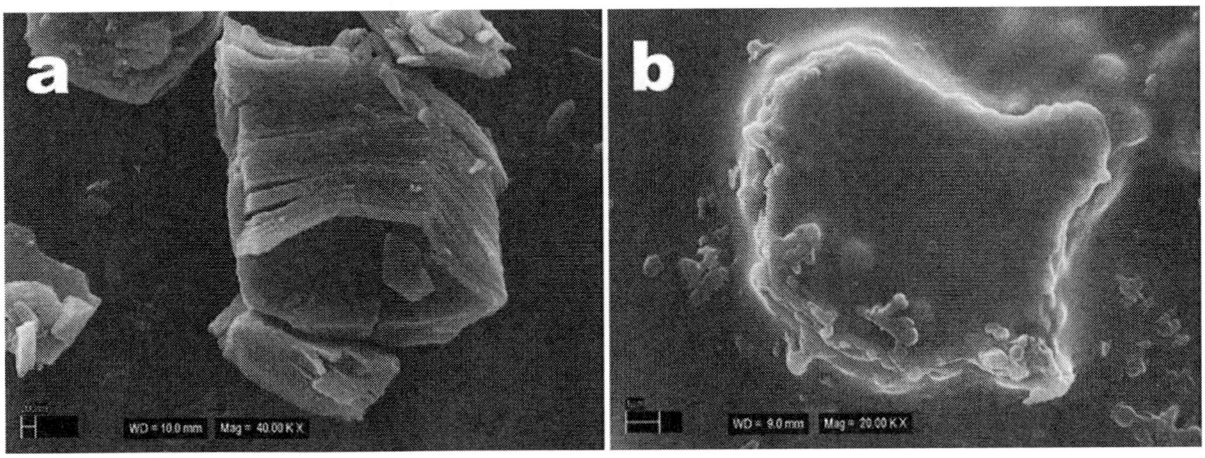

Figure 4 SEM images showing an unmodified $Ti_3C_2T_x$ MXene flakes, 40k magnification (a) and $Ti_3C_2T_x$ MXene/chitosan nanocomposite using 20k magnification (b)

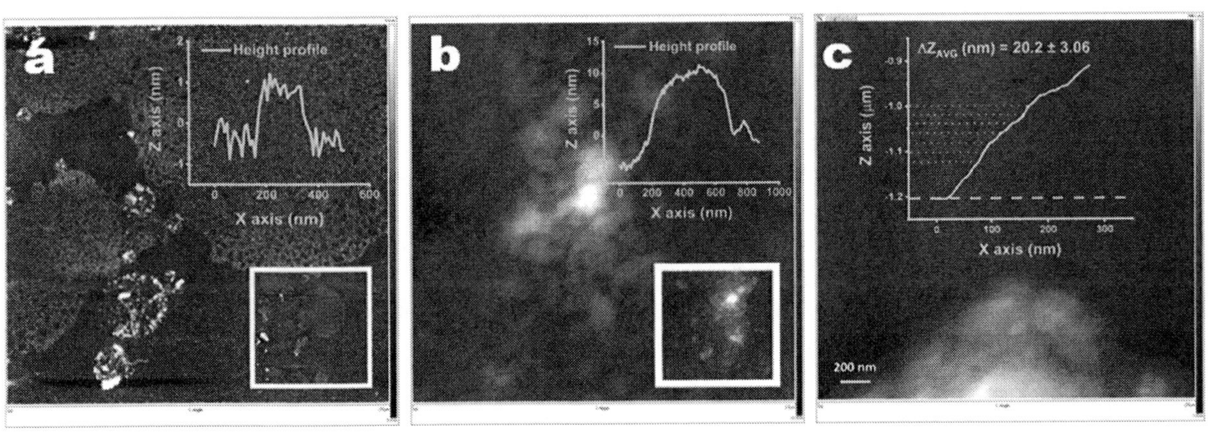

Figure 5 AFM images showing individual MXene sheets observed in MXene solution after a week of storage in an aqueous solution (a), individual MXene flake enwrapped in chitosan (b) and the edge of an unmodified MXene (c)

4. CONCLUSION

In this paper, a MXene-based biosensing platform was successfully fabricated for indirect sensitive detection of sarcosine. The sarcosine biosensor displayed low detection limit of 18 nM and a linear range up to 7.8 µM for sarcosine. The device was further used for sarcosine determination in artificial urine with a recovery index of 102.6 %. This work indicates that MXene is an excellent platform for protein-based biosensors applicable in medical diagnostics.

ACKNOWLEDGEMENTS

We would like to acknowledge the support received from the Slovak Research and Development Agency APVV 17-0300 and the support from the ERC Proof of Concept grant (No. 825586).

REFERENCES

[1] Global burden of disease cancer collaboration. FITZMAURICE, C., ABATE, D., ABBASI, N. et. al. Global, Regional, and National Cancer Incidence, Mortality, Years of Life Lost, Years Lived with Disability, and Disability-Adjusted Life-Years for 29 Cancer Groups, 1990 to 2017: A Systematic Analysis for the Global Burden of Disease Study. *JAMA Oncology*. [online]. 2019, vol. 5, no. 12, pp. 1749 - 1768. Available from: https://doi.org/10.1001/jamaoncol.2019.2996

[2] ABLIN, R. J., PIANA, R.: *The Great Prostate Hoax: How Big Medicine Hijacked the PSA Test and Caused a Public Health Disaster*. New York: St. Martin's Press, 2014.

[3] KATAFIGIOTI, A., KATAFIGIOTIS, I., SFOUNGARISTOS, S., et. al.: In the search of novel urine biomarkers for the early diagnosis of prostate cancer. Intracellular or secreted proteins as the target group? Where and how to search for possible biomarkers useful in the everyday clinical practice. *Archivio Italiano di Urologia e Andrologia*. [online]. 2016, vol. 88, no. 3, pp. 195 - 200. Available from: https://doi.org/10.4081/aiua.2016.3.195.

[4] JAKOBSEN, N. A., HAMDY, F. C., BRYANT, R. J.: Novel biomarkers for the detection of prostate cancer. *Journal of Clinical Urology*. [online]. 2016. Vol.9, pp. 3 - 10. Available from: https://doi.org/10.1177/2051415816656121.

[5] MARKIN, P. A., BRITO, A., MOSKALEVA, A., et. al.: Plasma Sarcosine Measured by Gas Chromatography-Mass Spectrometry Distinguishes Prostatic Intraepithelial Neoplasia and Prostate Cancer from Benign Prostate Hyperplasia. *Laboratory Medicine*. [online]. 2020, vol. 51, no. 6, pp. 566-573. Available from: https://doi.org/10.1093/labmed/lmaa008.

[6] SREEKUMAR, A., POISSON, L. M., RAJENDIRAN, T. M., et. al.: Metabolomic profiles delineate potential role for sarcosine in prostate cancer progression. *Nature*. [online]. 2009, vol. 457, no. 7231, pp. 910 -914. Available from: https://doi.org/10.1038/nature07762.

[7] HU, J., WEI, W., KE, S., et. al.: A novel and sensitive sarcosine biosensor based on organic electrochemical transistor. *Electrochimica Acta*. [online]. 2019, vol. 307, pp. 100 - 106. Available from: https://doi.org/10.1016/j.electacta.2019.03.180.

[8] DONG, Y., WU, Z. S., ZHENG, S., et. al.: Ti_3C_2 MXene-Derived Sodium/Potassium Titanate Nanoribbons for High-Performance Sodium/Potassium Ion Batteries with Enhanced Capacities. *ACS Nano*. [online]. 2017, vol. 11, no. 5, pp. 4792-4800. Available from: https://doi.org/10.1021/acsnano.7b01165.

[9] LUKATSKAYA, M. R., MASHTALIR, O., REN, C. E., et. al.: Cation intercalation and high volumetric capacitance of two-dimensional titanium carbide. *Science*. [online]. 2013, vol. 341, no. 6153, pp. 1502-1505. Available from: https://doi.org/10.1126/science.1241488.

[10] XIE, X., CHEN, S., DING, W., et. al.: An extraordinarily stable catalyst: Pt NPs supported on two-dimensional $Ti_3C_2X_2$ (X = OH, F) nanosheets for oxygen reduction reaction. *Chemical Communications*. [online]. 2013, vol. 49, no. 86, pp. 10112-10114. Available from: https://doi.org/10.1039/c3cc44428g.

[11] MASHTALIR, O., NAGUIB, M., MOCHALIN, V. N., et. al.: Intercalation and delamination of layered carbides and carbonitrides. *Nature Communications*. [online]. 2013, vol. 4, art. no. 1716. Available from: https://doi.org/10.1038/ncomms2664.

[12] WANG, F., YANG, C., DUAN, M., et. al.: TiO_2 nanoparticle modified organ-like Ti_3C_2 MXene nanocomposite encapsulating hemoglobin for a mediator-free biosensor with excellent performances. *Biosensors and Bioelectronics*. [online]. 2015, vol. 74, pp. 1022-1028. Available from: https://doi.org/10.1016/j.bios.2015.08.004.

[13] ALHABEB, M., MALESKI, K., ANASORI, B., et. al.: Guidelines for Synthesis and Processing of Two-Dimensional Titanium Carbide ($Ti_3C_2T_x$ MXene). *Chemistry of Materials*. [online]. 2017, vol. 29, no. 18, pp. 7633 - 7644. Available from: https://doi.org/10.1021/acs.chemmater.7b02847.

THE THEORETICAL AND EXPERIMENTAL WETTING OF POLYCAPROLACTONE NANOFIBRE LAYERS BY GELATIN HYDROGEL

Radek JIRKOVEC, Pavel HOLEC, Tomas KALOUS, Jiri CHVOJKA

Department of Nonwovens and Nanofibrous Materials, Faculty of Textile Engineering, Technical University of Liberec, Czech Republic, EU, radek.jirkovec@tul.cz

https://doi.org/10.37904/nanocon.2021.4351

Abstract

The study is focused on the wetting of nanofibrous layers of polycaprolactone. In the study, we proceeded to compare layers from direct current spinning and alternating current spinning. During the study, it was found that direct current spinning produces hydrophilic layers, and alternating current spinning produces hydrophobic layers. The study also found that the contact angle is affected by the solvent system. Due to the widespread use of polycaprolactone in tissue engineering, wetting of the fibrous layers with a gelatin hydrogel was also performed in the experiment. During the study, the Kwok-Neumann and Li-Neumann models were used to calculate the surface energy of the fibrous layers, and these models were also used to calculate the theoretical contact angles that the gelatin hydrogel will make on the fibrous layers. The study showed that the used models could predict how the fibre layers will be wetted with the prepared hydrogel.

Keywords: Nanofibres, polycaprolactone, gelatin, electrospinning, hydrogel, wetting

1. INTRODUCTION

Cell adhesion to a material is affected by many different influences, one of them is the wettability of the material. [1] Wetting is the ability of a liquid to spread over a solid surface. [2] One of the basic parameters describing the wetting of the material is the value of the contact angle that the selected liquid makes on the given material. The value of the contact angle describes the shape of the applied drop, its value as well as the shape of the drop, however, depends on the surface tension of the given liquid, on the free surface energy of the solid and on the interfacial energy between the liquid and the solid. The relationship between these quantities is described using Young's equation (1). [3]

$$\gamma_{sv} - \gamma_{sl} = \gamma_{lv} * \cos\theta \qquad (1)$$

Where:

θ - contact angle

γ_{lv} - surface tension of the liquid

γ_{sv} - surface energy of the solid

γ_{sl} - interfacial energy between the liquid and the solid

To ensure optimal cell adhesion to the material, the test material needs to be slightly hydrophilic. It is thus a material on which the applied liquid, e.g. medium, will form a contact angle of about 65°. [1] In order to obtain a slightly hydrophilic surface, it is therefore necessary to find a match between the surface tension of the liquid and the surface energy of the solid.

At present, porous scaffolds are used in tissue engineering, which can be prepared from both fibre systems and hydrogels, or from a combination of a fibre system and a hydrogel. [4-6]

Both natural polymers and synthetic polymers are used to prepare fibrous scaffolds. The most widely used natural polymers in tissue engineering include collagen, gelatin, chitosan, alginate, or hyaluronic acid. Synthetic polymers used to prepare scaffolds include polyvinyl alcohol, polylactic acid, polyglycolic acid, or polycaprolactone. [13-16] The last mentioned material, polycaprolactone (PCL), was used in this study to produce fibrous layers. It is a semicrystalline hydrophobic polyester. [7,8] PCL, as a biocompatible and biodegradable polymer, is used in tissue engineering to form skin covers [9], to form bone tissue [10], or, for example, for the production of small-diameter vessels [11].

Hydrogels can be defined as hydrophilic, three-dimensional, physically or chemically crosslinked polymer networks that have the ability to absorb large amounts of water or biological fluids. [12,13] Hydrogels are similar in composition and mechanical properties to extracellular matrix (intercellular mass), so they can serve as a support material for cells during tissue regeneration and allow the diffusion of nutrients, metabolites and growth factors. [14,15] Hydrogels are used in tissue engineering primarily to transport drugs or growth factors, such as grafts of irregularly shaped defects, or to fill pores in a three-dimensional network that can sufficiently stabilize the repair site. [14]

2. MATERIALS AND METHODS

2.1. Materials

PCL with a molecular weight of Mn 80,000 (Sigma-Aldrich, Germany) was chosen for the preparation of nanofibre layers. Three solvent systems were chosen to dissolve the PCL. The first was composed of chloroform and ethanol (both Sigma-Aldrich, Germany), the second solvent system was composed of chloroform, ethanol and acetic acid (Sigma-Aldrich, Germany) and the third solvent system was composed of acetic acid, formic acid and acetone (all Sigma-Aldrich, Germany). Pork skin gelatin (Sigma-Aldrich, Germany) was chosen for the preparation of hydrogel. Distilled water was chosen as the solvent system.

2.2. Preparation

PCL was prepared at a concentration of 8 wt% in three solvent systems. Chloroform and ethanol were in ratio 8:2, chloroform, ethanol and acetic acid were in ratio 8:1:1, and the solvent system of acetic acid, formic acid and acetone was in ratio 1:1:1. Gelatin was prepared at a concentration of 10 wt% in distilled water. The prepared PCL solutions were spun by direct current (DC) spinning and alternating current (AC) spinning. Nanospider NS 1WS500U was used for DC spinning, and a device consisting of ABB KGUG 36 high-voltage transformer and a Thalheimer-Trafowerke ESS 104 variable autotransformer was used for AC spinning.

2.3. Surface tension

The surface tension of the gelatin hydrogel was measured on a PocketDyne (Krüss) by measuring the bubble pressure. The measurement was performed at a temperature of 37 °C.

2.4. Scanning electron microscopy and analysis

The PCL nanofibre layers were gilded using Quorum Q150R ES, and images of the layers were obtained using a scanning electron microscope (SEM, Tescan Vega3, Czech Republic) at an accelerated voltage of 20kV.

2.5. Contact angle and surface energy of the fibre layers

The surface energy was determined via the measurement of the contact angle using a See System E instrument (Advex Instruments, Czech Republic). The liquid used for the measurement of the contact angle comprised glycerol with a surface tension of 64 mN/m. The Kwok-Neumann and Li-Neumann model was used to calculate the surface energy.

3. RESULTS AND DISCUSSION

3.1. Preparation and spinning of the solutions

The prepared polymer solutions were spun via the DC spinning at an electrode voltage of 50 kV and a collector voltage of -10 kV. The distance between the collector and the electrode was 170 mm. The polymer solution was also spun via the AC spinning method at an effective voltage of 37 kV and a frequency of 50 Hz. Diagrams of the used devices are shown in **Figure 1**.

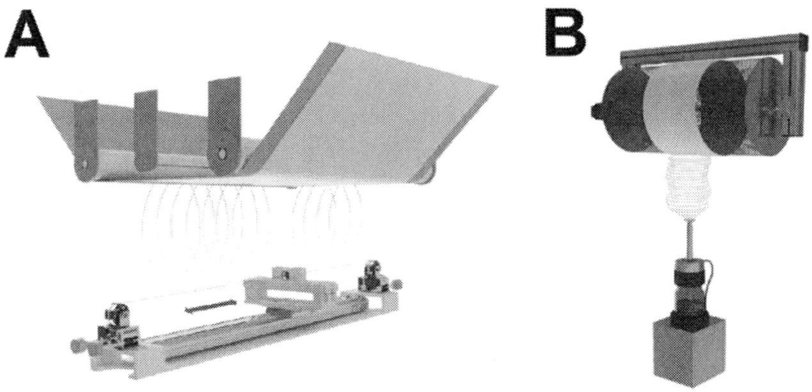

Figure 1 Diagram of the spinning devices. A) the Nanospider spinning device, B) the AC spinning device

3.2. Fibre morphology and diameter analysis

During spinning, homogeneous fibrous layers with occasional bead defects were produced. **Figure 2** shows SEM images and fibre diameter of the various fibre layers produced via DC spinning and AC spinning.

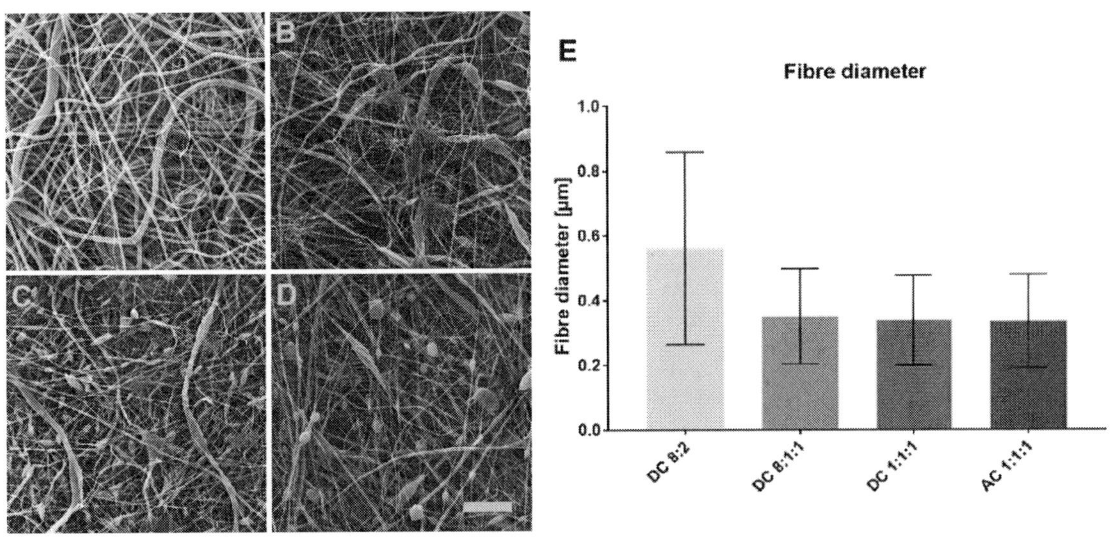

Figure 2 SEM images of the fibre layers. A) DC 8:2, B) DC 8:1:1. C) DC 1:1:1, D) AC 1:1:1, and E) diameter of the produced fibre layers. 95% confidence interval. **** $p < 0.0001$

3.3. Surface tension

The surface tension of the prepared gelatin hydrogel was measured at 37 °C due to the possible combination with the cell suspension. The measured surface tension was 65.62 mN/m.

3.4. Contact angle and surface energy of the fibre layers

The surface energy was determined by measuring the contact angle with glycerol. The models for calculating surface energy were Kwok-Neumann (2) and Li-Neumann (3). The contact angle and surface energy values are shown in **Figure 3**.

$$\gamma_{sl} = \gamma_{lv} + \gamma_{sv} - 2\sqrt{\gamma_{lv}\gamma_{sv}}(1 - 0.0001057(\gamma_{lv}-\gamma_{sv})^2) \quad (2)$$

$$\gamma_{sl} = \gamma_{lv} + \gamma_{sv} - 2\sqrt{\gamma_{lv}\gamma_{sv}}e^{-0.0001247(\gamma_{lv}-\gamma_{sv})^2} \quad (3)$$

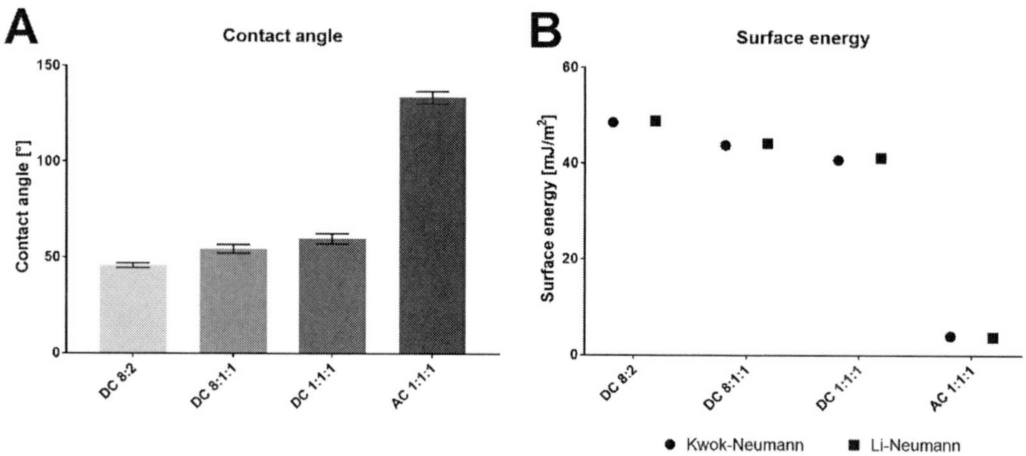

Figure 3 Values of the A) contact angle, and B) surface energy of the fibre layers

3.5. Theoretical contact angle

The theoretical contact angle that the gelatin hydrogel will form on the fibrous layers was calculated from the values of the surface energies of the individual fibrous layers and the surface tension of the prepared hydrogel.

By combining the Kwok-Neumann (2) and Li-Neumann (3) relationship with Young's equation (1), we obtain the relations for the calculation of contact angles by the Kwok-Neumann (4) and Li-Neumann model (5).

$$\cos\alpha = -1 + 2\sqrt{\frac{\gamma_{sv}}{\gamma_{lv}}}(1 - 0.0001057(\gamma_{lv}-\gamma_{sv})^2) \quad (4)$$

$$\cos\alpha = -1 + 2\sqrt{\frac{\gamma_{sv}}{\gamma_{lv}}}e^{-0.0001247(\gamma_{lv}-\gamma_{sv})^2} \quad (5)$$

Table 1 shows the values of the theoretical contact angles obtained by the Kwok-Neumann and Li-Neumann methods.

Table 1 Calculated values of theoretical contact angles using Kwok-Neumann and Li-Neumann methods

Spinning technology	*Solvent system*	*Theoretical contact angle [°]*	
		Kwok-Neumann	*Li-Neumann*
DC Spinning	8:2	48.2	48.2
	8:1:1	56.6	56.6
	1:1:1	61.9	61.9
AC Spinning	1:1:1	134.8	134.5

3.6. Verification of the theoretical contact angles

To compare the theoretical contact angles, the real contact angle was measured using a prepared gelatin hydrogel. The measured values of contact angles are given in **Table 2**. **Figure 4** is a graphical comparison of theoretical and real contact angles.

Table 2 Measured values of contact angles

Spinning technology	*Solvent system*	*Contact angle [°]*
DC Spinning	8:2	42.1 ± 2.2
	8:1:1	49.1 ± 2.1
	1:1:1	55.6 ± 2.4
AC Spinning	1:1:1	136.9 ± 2.8

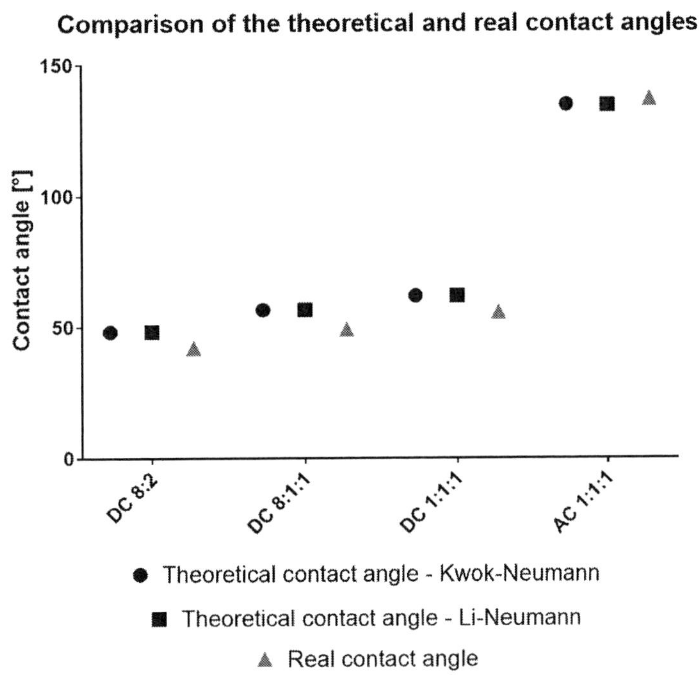

Figure 4 Comparison of theoretical and real contact angles

From the obtained values it is clear that the theoretical contact angles are close to the actually measured ones. Both models used are therefore suitable for predicting whether a given material will be wetted and what contact angle the selected liquid will make. **Figure 5** then captures images of wetting the fibrous layers with a gelatin hydrogel.

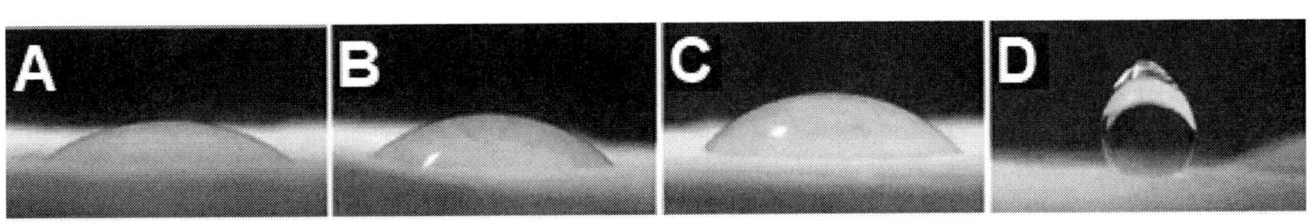

Figure 5 Wetting of the fibrous layers with a gelatin hydrogel. DC spinning layers A) 8:2, B) 8:1:1, C) 1:1:1. D) Fibre layer from AC spinning 1:1:1

4. CONCLUSION

The study described the wetting of nanofibre layers depending on their surface energy and surface tension of the prepared gelatin hydrogel. In the study, fibre layers were prepared by DC spinning and AC spinning. During the experiments, significant differences in wetting were found between the individual fibre layers. The DC spinning layers produced by the string electrode have high surface energy and are therefore hydrophilic. The fibre layers from the AC spinning have a low surface energy value and are therefore hydrophobic. As part of the study of wetting of the fibrous layers, calculations of the theoretical contact angles that the gelatin hydrogel will form on the fibrous layers were performed. The calculation was performed using the values of surface energies of fibre layers and surface tension of the prepared hydrogel. The resulting real values of the contact angle differ slightly from the theoretical ones, but both models are sufficient for the basic idea of wetting the produced materials.

ACKNOWLEDGEMENTS

This study was supported by the research of nanofibrous materials for the treatment of glaucoma disease (reg. no. TUL: PURE-2021-6005).

REFERENCES

[1] DOWLING, D.P.; MILLER, I.S.; ARDHAOUI, M.; GALLAGHER, W.M. Effect of Surface Wettability and Topography on the Adhesion of Osteosarcoma Cells on Plasma-Modified Polystyrene. *J Biomater Appl.* [online]. 2011, vol. 26, pp. 327-347. Available from: https://doi.org/10.1177/0885328210372148.

[2] MAO, N. 6 - Methods for characterisation of nonwoven structure, property, and performance. In *Advances in Technical Nonwovens*. KELLIE, G., Ed. Woodhead Publishing Series in Textiles; Woodhead Publishing, 2016, pp. 155-211. ISBN 978-0-08-100575-0.

[3] HEBBAR, R.S.; ISLOOR, A.M.; ISMAIL, A.F. Chapter 12 - Contact Angle Measurements. In *Membrane Characterization*; HILAL, N., ISMAIL, A.F., MATSUURA, T., OATLEY-RADCLIFFE, D., Eds. Elsevier, 2017; pp. 219-255 ISBN 978-0-444-63776-5.

[4] DHANDAYUTHAPANI, B.; YOSHIDA, Y.; MAEKAWA, T.; KUMAR, D.S. Polymeric Scaffolds in Tissue Engineering Application: A Review. [online]. Available from: https://www.hindawi.com/journals/ijps/2011/290602/ [accessed on 2 September 2020].

[5] PINA, S.; RIBEIRO, V.P.; MARQUES, C.F.; MAIA, F.R.; SILVA, T.H.; REIS, R.L.; OLIVEIRA, J.M. Scaffolding Strategies for Tissue Engineering and Regenerative Medicine Applications. *Materials (Basel).* [online]. 2019, vol. 12. Available from: https://doi.org/10.3390/ma12111824.

[6] MOHABATPOUR, F.; KARKHANEH, A.; SHARIFI, A.M. A Hydrogel/Fiber Composite Scaffold for Chondrocyte Encapsulation in Cartilage Tissue Regeneration. *RSC Adv.* [online]. 2016, vol. 6, pp. 83135-83145. Available from: https://doi.org/10.1039/C6RA15592H.

[7] MCKEEN, L. 12 - Renewable Resource and Biodegradable Polymers. In *The Effect of Sterilization on Plastics and Elastomers (Third Edition)*; MCKEEN, L., Ed. Plastics Design Library; William Andrew Publishing: Boston, 2012, pp. 305-317. ISBN 978-1-4557-2598-4.

[8] SAAD, B.; SUTER, U.W. Biodegradable Polymeric Materials. In *Encyclopedia of Materials: Science and Technology*. BUSCHOW, K.H.J., CAHN, R.W., FLEMINGS, M.C., ILSCHNER, B., KRAMER, E.J., MAHAJAN, S., VEYSSIÈRE, P., Eds. Elsevier: Oxford, 2001; pp. 551-555. ISBN 978-0-08-043152-9.

[9] KLICOVA, M.; KLAPSTOVA, A.; CHVOJKA, J.; KOPRIVOVA, B.; JENCOVA, V.; HORAKOVA, J. Novel Double-Layered Planar Scaffold Combining Electrospun PCL Fibers and PVA Hydrogels with High Shape Integrity and Water Stability. *Materials Letters.* [online]. 2020, vol. 263, p. 127281. Available from: https://doi.org/10.1016/j.matlet.2019.127281.

[10] RAMPICHOVÁ, M.; CHVOJKA, J.; JENČOVÁ, V.; KUBÍKOVÁ, T.; TONAR, Z.; ERBEN, J.; BUZGO, M.; DAŇKOVÁ, J.; LITVINEC, A.; VOCETKOVÁ, K.; et al. The Combination of Nanofibrous and Microfibrous Materials for Enhancement of Cell Infiltration and in Vivo Bone Tissue Formation. *Biomed Mater.* [online]. 2018, vol. 13, p. 025004. Available from: https://doi.org/10.1088/1748-605X/aa9717.

[11] YALCIN, I.; HORAKOVA, J.; MIKES, P.; SADIKOGLU, T.G.; DOMIN, R.; LUKAS, D. Design of Polycaprolactone Vascular Grafts. *Journal of Industrial Textiles.* [online]. 2016, vol. 45, pp. 813-833, Available from: https://doi.org/10.1177/1528083714540701.

[12] BATISTA, R.A.; OTONI, C.G.; ESPITIA, P.J.P. Chapter 3 - Fundamentals of chitosan-based hydrogels: elaboration and characterization techniques. In *Materials for Biomedical Engineering*; HOLBAN, A.-M., GRUMEZESCU, A.M., Eds. Elsevier, 2019, pp. 61-81. ISBN 978-0-12-816901-8.

[13] AHMED, E.M. Hydrogel: Preparation, Characterization, and Applications: A Review. *Journal of Advanced Research.* [online]. 2015, vol. 6, pp. 105-121. Available from: https://doi.org/10.1016/j.jare.2013.07.006.

[14] HESSE, E., HEFFERAN, T.E., TARARA, J.E., HAASPER, C., MELLER, R., KRETTEK, C., LU, L., YASZEMSKI, M.J. Collagen Type I Hydrogel Allows Migration, Proliferation and Osteogenic Differentiation of Rat Bone Marrow Stromal Cells. *J Biomed Mater Res A.* [online]. 2010, vol. 94, pp. 442-449. Available from: https://doi.org/10.1002/jbm.a.32696.

[15] KOETTING, M.C., PETERS, J.T., STEICHEN, S.D., PEPPAS, N.A. Stimulus-Responsive Hydrogels: Theory, Modern Advances, and Applications. *Materials Science and Engineering: R: Reports.* [online]. 2015, vol. 93, pp. 1-49. Available from: https://doi.org/10.1016/j.mser.2015.04.001.

TOWARD SERS-BASED DYNAMIC DETECTION OF INSULIN DIFFUSING THROUGH HYDROGEL MATRICES

[1]Farnoush SALEHTASH, [1,2]Adriana ANNUŠOVÁ, [1]Vojtech NÁDAŽDY, [1]Yuriy HALAHOVETS, [3]Paweł ALBRYCHT, [1,2]Peter ŠIFFALOVIČ, [1,2]Matej JERGEL, [1,2]Eva MAJKOVÁ, [4]Zuzana KRONEKOVÁ, [4]Michal PELACH, [2,4]Igor LACÍK

[1]Institute of Physics, Slovak Academy of Sciences, Bratislava, Slovakia, EU, farnoush.salehtash@savba.sk

[2]Centre for Advanced Materials Application, Slovak Academy of Sciences, Bratislava, Slovakia, EU, eva.majkova@savba.sk

[3]Institute of Physical Chemistry, Polish Academy of Sciences, Warsaw, Poland, EU, palbrycht@ichf.edu.pl

[4]Polymer Institute, Slovak Academy of Sciences, Bratislava, Slovakia, EU, igor.lacik@savba.sk

https://doi.org/10.37904/nanocon.2021.4363

Abstract

Insulin is a peptide hormone produced by beta cells of pancreatic islets. In type 1 diabetes mellitus, these islets are destroyed by the body's own immune system, no insulin is produced and the blood glucose level is increased. Nowadays, efforts in combatting type 1 diabetes focus on the transplantation of islets immunoprotected in microspheres made of non-covalently crosslinked hydrogels. The functionality of the encapsulated islets is retained, while the hydrogel matrix allows permeation of the produced insulin into the bloodstream. The applicability of these microspheres has been extensively studied in vivo. However, prior to biological models, a dynamic detection method to monitor the production of insulin and its diffusion through the microspheres is still missing. Herein, we apply the Surface Enhanced Raman Scattering (SERS) technique to detect physiologically relevant concentrations of insulin using planar Ag SERS substrates, while considering their implementation for monitoring insulin diffusion through alginate matrices. Insulin was detected after drying SERS planar substrates in a concentration range of 10^{-3}- 10^{-12} M. Additionally, we demonstrated the decrease in the deposition time using an alternating electric field. Moreover, the in situ monitoring of the SERS signal from insulin molecules has certain limitations when conducting experiments for SERS substrates submerged in water. As the secretion of insulin and its diffusion across the immunoprotective microspheres is a dynamic process, the development of an adequate detection method is expected to lead to a better understanding of these processes as a function of time, matrix composition, and glucose intake.

Keywords: Insulin, alginate-based hydrogel matrices, Surface Enhanced Raman Scattering

1. INTRODUCTION

One of the most sensitive and powerful bioanalytical techniques used in a wide range of applications is Surface Enhanced Raman Scattering (SERS). It is particularly useful for the sensing of bioactive molecules [1,2], i.e. compounds that have biological effects on the human body. The development of quick and reliable techniques for in situ detection of bioactive molecules is important in biomedical research and drug discovery.

In SERS, the otherwise relatively weak Raman signals of a target molecule are amplified by several orders of magnitude when trapped between the nanogaps of a noble metal nanostructure surface (e.g. Au or Ag) [3]. The signal enhancement is attributed to different effects. The largest amplification comes from the electromagnetic near fields. The charge transfer processes or chemical enhancement happens due to the chemical adsorption of molecules to SERS surfaces [4,5]. SERS is one of the most valuable methods for ultrahigh-resolution detection and analysis of bioactive compounds [5,6].

Insulin is necessary to maintain the physiological carbohydrate, protein, and fat metabolisms and plays a crucial role in controlling many diseases that can affect human health. Insulin is a peptide hormone synthetized by the beta cells of the pancreatic islets [7]. It consists of two peptide chains, A with 21 amino acid residues and B with 30 residues, linked by two disulfide connecting cysteine A7 and B7 and residues A20 and B19 [8]. The disruption of insulin secretion by the beta cells leads to diabetes mellitus [9]. Type 1 diabetes mellitus is an autoimmune deficiency that arises when the beta cells are destroyed by the body's own immune system. Low levels of insulin increase blood glucose and cause hyperglycemia. Patients then depend on exogenous insulin for survival, and there is currently no cure. One method currently under research and development involves the encapsulation of beta cells in a protective coating that is biocompatible and semipermeable, but does not induce an immune response upon their transplantation. Thus, insulin secreting cells could function properly and no exogenous insulin administration would be required. The most commonly employed are noncovalently crosslinked alginate hydrogels as they provide favorable physiological conditions for islets during the encapsulation processes [10].

Several studies described the detection of low insulin concentrations at quantities that can satisfy the requirements for monitoring of insulin levels in the human body. Hassanian et al. [1] realized SERS detection of reduced human insulin, where the disulfide bonds are broken. This allows the insulin molecules to chemically adsorb onto the SERS substrates. Thus, Au-S bonds are formed, allowing both electromagnetic and chemical enhancements to arise. They measured nine different concentrations of the protein, from 10 fM to 300 pM. Another study by Cho et al. [11] presented a columnar 3D Au SERS substrate with a dynamic range from 100 pM to 50 nM and detection limit 35 pM. Insulin concentrations in pancreatic islet secretions were characterized under low and high glucose conditions with high spatial uniformity of the Raman signal. The peak at 1002 cm^{-1}, corresponding to the ring-breathing mode of aromatic phenylalanine, was used to monitor the insulin signal. Gholami et al. reported a novel SERS quenching method for the detection of proteins and biomolecules [12,13]. They used Au nanoparticles functionalized by benzothiazole azo dye. The latter is displaced from the SERS substrates surface after its interaction with biomolecules with reduced disulfide bonds. This Raman probe was used to quantify insulin in the concentration range of 10^{-8}-10^{-14} M.

Although a lot of research has focused on low level insulin detection using SERS, there is still a need for the development of dynamic detection methods. Precise quantification of insulin secreted by beta cells is of primary importance in determining the degree of functionality of polymeric matrices employed in encapsulation procedures. The above-mentioned works studied the signal of insulin in a dried, presumably crystalline state. However, the hydrogel microcapsules are prone to instantaneous dehydration in non-aqueous environment, which makes the in situ insulin permeation monitoring more difficult. Our previous study successfully reported the application of label-free Confocal Raman Microscopy (CRM) to monitor structural changes in the hydrogel matrices of alginate-based microspheres [14]. The present work investigates the possibility of monitoring the diffusion process of insulin molecules through alginate-based hydrogel matrices using planar SERS substrates and under water immersion. First, the limits of detection for insulin molecules in our diagnostic system are presented. Second, we demonstrate the successful acceleration of the deposition time of insulin under the influence of an alternating electric field. Finally, the insulin SERS signal is monitored from the SERS substrate in the aqueous conditions.

2. EXPERIMENTAL PART

2.1. Instrumentation

All SERS experiments were carried out at a controlled room temperature of 20 °C, using a WITec alpha 300 R+ confocal Raman microscope equipped with a WITec UHTS300 spectrometer (1200 lines/mm grating) coupled to an EMCCD camera. The 785 nm laser line with the power of 5 and 30 mW was employed to excite the Raman signal. A water immersion objective (Carl Zeiss 20×/1NA) and a 100 μm diameter optical fiber were used to collect the spectra. We performed single-spot acquisitions with integration times (IT) of 10 or 20 s and accumulation number of 1 or 5, respectively (longer IT were used for measurements performed on SERS substrates immersed in water). Additional raster scanning images were done to monitor the spatial homogeneity of the insulin signal (50 × 50 μm, 100 × 100 lines, IT 1s).

2.2. SERS substrates

We used planar SERS substrates provided by SERSitive (Institute of Physical Chemistry, Polish Academy of Sciences, Warsaw), manufactured using electrodeposition of silver nanoparticles on an Indium-tin oxide (ITO) glass surface. The Ag hydrophilic substrates had an active area of 5×4 mm (Premium-Silver SERS) and 3.5×3.5 mm (S-Silver SERS). **Figure 1** presents photos of the substrates, and also the SEM and AFM images of the deposited Ag on the surface. These SERS substrates are capable of amplifying the normal Raman signals of about 100M times.

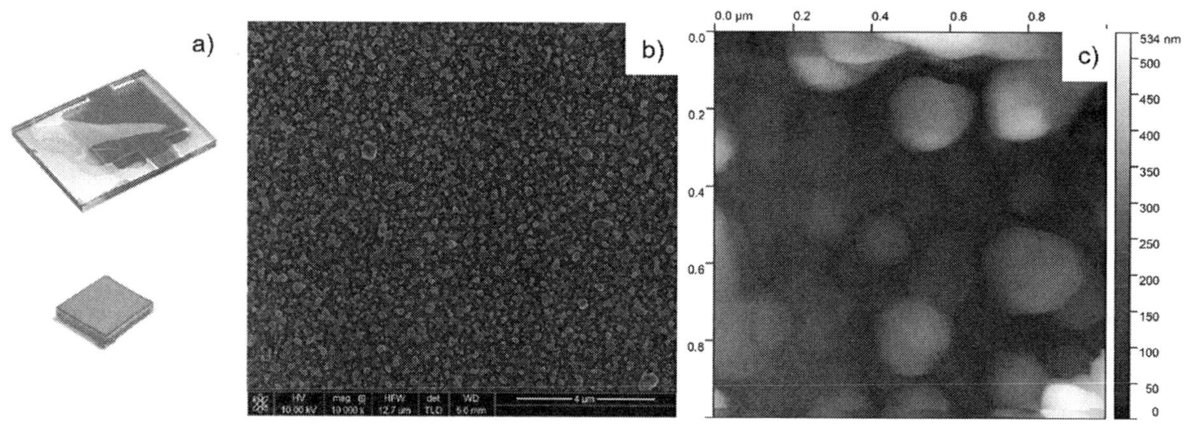

Figure 1 Optical image of the SERSitive substrates: Premium-Silver and S-Silver (a) [15,16], SEM (b) [17] and, AFM (c) image of Ag nanoparticles electrodeposited on the surface of the ITO glass

2.3. Deposition of insulin on planar substrates

A human insulin (Sigma-Aldrich) solution (1.7×10^{-3} M) was used to prepare a series of different concentrations of insulin solutions in the concentration range of 10^{-3}- 10^{-12} M by dilution in filtered deionized water (pH of 7.5). In this step, the smaller substrates (S-Silver SERS) were immersed in 200 μl of the indicated concentrations in an Eppendorf tube for 24 h, followed by drying the substrates in air, without a rinsing step.

To investigate the influence of electrical field on the time of insulin deposition on the substrates surface, we followed the work of Richter et al. [18]. The larger Premium-Silver SERS substrate served as one of the electrodes in a cell, realized according to [18] using a 3D printer. On the counter electrode (ITO substrate without Ag nanoparticles), a glass-PTFE foil of 250 μm thickness was placed to avoid Faraday currents flow. A volume of 50 μl of 1 mM insulin solution was pipetted into the cell. An alternating electrical field of +10 V (square waveform with oscillations from 0 to +10 V) and 1 kHz was applied on the electrodes for 5 min (Keithley 3390 waveform generator, Tektronix 2236 A oscilloscope). The chosen polarity and frequency were determined based on Zeta potential measurements (Malvern Zetasizer Nano ZS90) of 1 mM insulin solution [18]. The Zeta potential and the electrophoretic mobility are equal to -26.6 mV and 2.084 μm.cm/Vs, respectively. After deposition, no rinsing step was applied, and the substrates were dried in air. In some experiments, the substrates were then immersed under water during Raman measurements. Afterwards a second drying step followed, without excessive rinsing of the substrates.

3. RESULTS AND DISCUSSION

3.1. Low concentration insulin detection using planar SERS substrates

Figure 2a shows spectra obtained from different spots of four different S-Silver substrates previously immersed in mM, μM, nM, and pM insulin solutions. We were able to detect insulin in the nM-pM range, which corresponds to the insulin levels in human blood. The most pronounced peaks were identified as vibrational modes of phenylalanine (Phe) and tyrosine (Tyr) molecules [19]. It should be noted that no Raman peak of insulin was observed from substrates not containing Ag nanoparticles after immersion in 1 mM insulin solution.

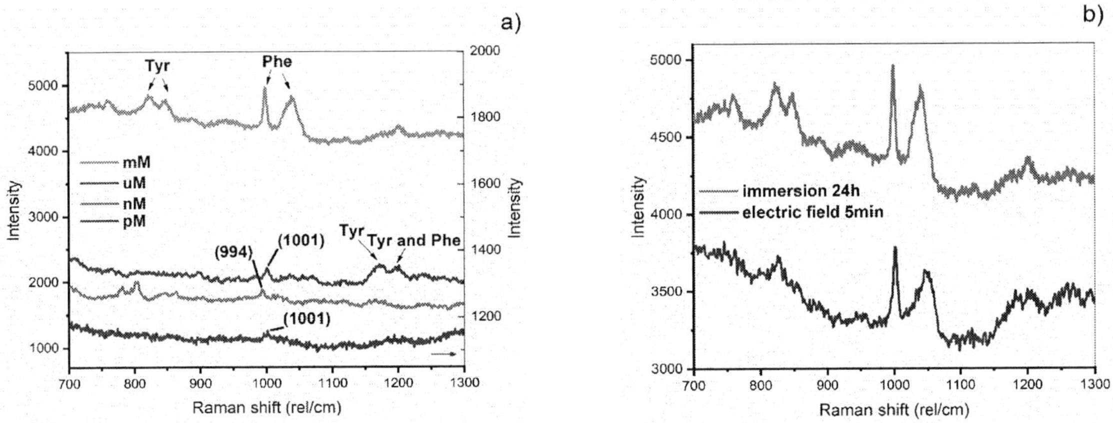

Figure 2 A comparison between SERS spectra of different concentrations of insulin solution (mM, μM, nM, and pM) from substrates prepared by the immersion deposition method (a) and a comparison of a SERS spectrum of 1 mM insulin solution deposited on the substrate using an alternating electric field to the 24 h immersion deposition method (b).

3.2. Insulin deposition on SERS substrates using an alternating electric field

In the work of Richter et al. [18], a shortening of deposition time was reported for different proteins and small molecules after applying an alternating electric field with a polarity depending on the charge of the analyte. Our results confirm that this method is also applicable in the case of 1 mM human insulin solution. The recorded spectra after only 5 min of electric field application resulted in homogenous and comparable signals to those obtained from substrates immersed in the same insulin concentration solution for 24 h (**Figure 2**b). The spatial uniformity of the absorbed molecules was also confirmed by the substrates raster images (not included here).

3.3. SERS signal of insulin from substrate immersed in water

Figure 3a depicts the spectrum measured from a dried substrate prepared under the influence of an electric field. As mentioned in the previous section, highly uniform and pronounced insulin peaks were detected on the surface. After these measurements, the substrate was immersed in 1.5 ml of deionized water (pH=7.5) in a plastic Petri dish. The substrate was scanned again using the same acquisition parameters as for dried substrate. However, no insulin signal was detected using these conditions. Insulin peaks, uniformly distributed on the surface, appeared only after increasing the power from 5 to 30 mW. However, their intensity decreased significantly compared to the signal from the dried substrate. This behavior was reproducible. As insulin is present in water in a dissolved form, the functional groups are charged depending on the pH. This in turn affects their ability to remain bonded with the substrate and influences the signal enhancement. The change in signal intensity compared to the signal from the crystalline insulin is rather significant. The intensity is comparable to a signal from a dried substrate previously immersed in pM concentration solution (see **Figure 2**a). This puts limitations on in situ dynamic detection of insulin in water. **Figure 3**b shows the comparison of spectra from substrates dried before and after the immersion test (i.e. dried substrate after deposition of insulin using electric field and substrate after the second drying step, respectively). Insulin did not dissolve from the substrate completely into water but remained on the SERS substrate surface, as pronounced insulin peaks are again visible. By raster imaging, we confirmed that insulin is still homogeneously distributed on the surface. However, there is a difference in the excitation of the phenylalanine vibrational modes, manifested by a change in the intensity ratios between 1001 and 1035 rel. cm^{-1}. Further study will involve lower concentration insulin solutions and pH dependent measurements, as well as the adaptation of this method for in situ detection of insulin diffusing through hydrogel matrices.

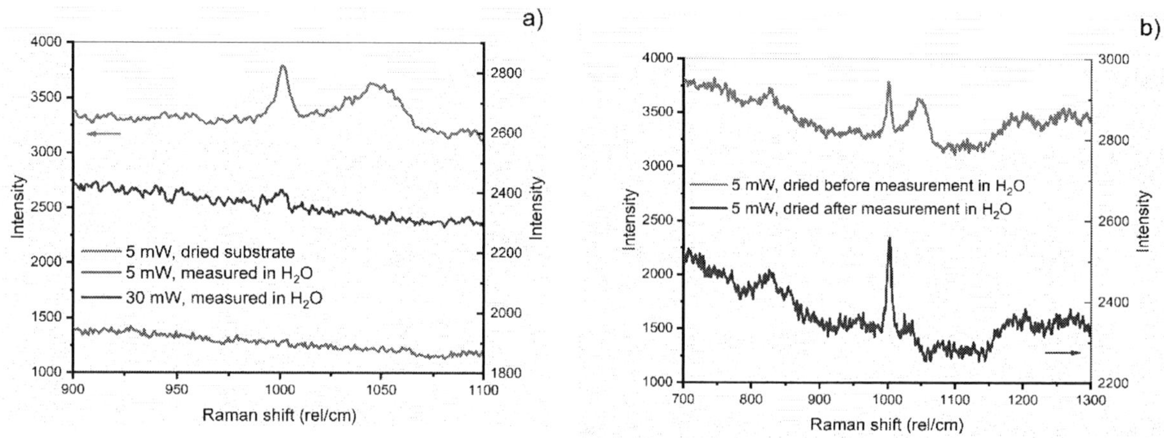

Figure 3 Comparison of spectra from substrates prepared using an alternating electric field, measured under different conditions (a) and comparison of spectra from substrates dried before and after the immersion test (b).

4. CONCLUSION

We studied the possibilities of detecting SERS signal for human insulin from planar Ag SERS substrates, dried as well as immersed in water. In the next step, this method will be applied for in situ dynamic monitoring of insulin diffusing through alginate-based hydrogel matrices. A detection limit of around nM to pM concentrations was achieved for a dried substrate. However, certain limitations were encountered in the detection of insulin SERS signal for a substrate immersed in water.

ACKNOWLEDGEMENTS

Financial support of the projects APVV-14-0120 and APVV-18-0480 is acknowledged. This work was performed during the implementation of the project Building-up Centre for advanced materials application of the Slovak Academy of Sciences, ITMS project code 313021T081 supported by the Integrated Infrastructure Operational Programme funded by the ERDF. We would like to express our appreciation to professor Marek Procházka (Institute of Physics of Charles University in Prague) for helpful discussions.

REFERENCES

[1] W. A. HASSANAIN and E. L. IZAKE, "Toward Label-Free SERS Detection of Proteins through Their Disulfide Bond Structure." *SLAS Discov.* [online]. 2020, vol. 25, no. 1, pp. 87-94, 2020. Available from: https://doi.org/10.1177/2472555219875102.

[2] D. CIALLA *et al.* "Surface-enhanced Raman spectroscopy (SERS): Progress and trends." *Anal. Bioanal. Chem.* [online]. 2012, vol. 403, no. 1, pp. 27-54. Available from: https://doi.org/10.1007/s00216-011-5631-x.

[3] S. Y. DING, E. M. YOU, Z. Q. TIAN, and M. MOSKOVITS. "Electromagnetic theories of surface-enhanced Raman spectroscopy." *Chemical Society Reviews.* [online]. 2017, vol. 46, no. 13. Royal Society of Chemistry, pp. 4042-4076. Available from: https://doi.org/10.1039/c7cs00238f.

[4] M. ŠUBR, A. KUZMINOVA, O. KYLIÁN, and M. PROCHÁZKA, "Surface-enhanced Raman scattering (SERS) of riboflavin on nanostructured Ag surfaces: The role of excitation wavelength, plasmon resonance and molecular resonance." *Spectrochim. Acta - Part A Mol. Biomol. Spectrosc.* [online]. 2018, vol. 197, pp. 202-207. Available from: https://doi.org/10.1016/j.saa.2018.01.055.

[5] M. F. MROZEK, D. ZHANG, and D. BEN-AMOTZ, "Oligosaccharide identification and mixture quantification using Raman spectroscopy and chemometric analysis." *Carbohydr. Res.* [online]. 2004, vol. 339, no. 1, pp. 141-145. Available from: https://doi.org/10.1016/j.carres.2003.09.019.

[6] B. SHARMA, R. R. FRONTIERA, A. I. HENRY, E. RINGE, and R. P. VAN DUYNE, "SERS: Materials, applications, and the future." *Mater. Today.* [online]. 2012, vol. 15, no. 1-2, pp. 16-25. Available from: https://doi.org/10.1016/S1369-7021(12)70017-2.

[7] C. ORTIZ, D. ZHANG, Y. XIE, V. J. DAVISSON, and D. BEN-AMOTZ. "Identification of insulin variants using Raman spectroscopy." *Anal. Biochem.* [online]. 2004, vol. 332, no. 2, pp. 245-252. Available from: https://doi.org/10.1016/j.ab.2004.06.013.

[8] J. BRANGE and L. LANGKJOER. "Insulin structure and stability." *Pharm. Biotechnol.* [online]. 1993, vol. 5, no. 1922, pp. 315-350. Available from: https://doi.org/10.1007/978-1-4899-1236-7_11.

[9] P. RORSMAN and M. BRAUN. "Regulation of insulin secretion in human pancreatic islets." *Annu. Rev. Physiol.* [online]. 2013, vol. 75, pp. 155-179. Available from: https://doi.org/10.1146/annurev-physiol-030212-183754.

[10] T. ANDERSEN, P. AUK-EMBLEM, and M. DORNISH, "3D Cell Culture in Alginate Hydrogels." *Microarrays.* [online]. 2015, vol. 4, no. 2, pp. 133-161. Available from: https://doi.org/10.3390/microarrays4020133.

[11] H. CHO *et al.*, "Surface-Enhanced Raman Spectroscopy-Based Label-Free Insulin Detection at Physiological Concentrations for Analysis of Islet Performance." *ACS Sensors.* [online]. 2018, vol. 3, no. 1, pp. 65-71. Available from: https://doi.org/10.1021/acssensors.7b00864.

[12] M. D. GHOLAMI, P. SONAR, G. A. AYOKO, and E. L. IZAKE. "A highly sensitive SERS quenching nanosensor for the determination of tumor necrosis factor alpha in blood." *Sensors Actuators, B Chem.* [online]. 2020, vol. 310, p. 127867. Available from: https://doi.org/10.1016/j.snb.2020.127867.

[13] M. D. GHOLAMI, P. SONAR, G. A. AYOKO, and E. L. IZAKE. "A SERS quenching method for the sensitive determination of insulin." *Drug Test. Anal.* [online]. 2021, vol. 13, no. 5, pp. 1048-1053. Available from: https://doi.org/10.1002/dta.2808.

[14] Z. KRONEKOVÁ *et al.* "Structural changes in alginate-based microspheres exposed to in vivo environment as revealed by confocal Raman microscopy." *Sci. Rep.* [online]. 2018, vol. 8, no. 1, pp. 1-12. Available from: https://doi.org/10.1038/s41598-018-20022-y.

[15] "SERSitive hydrophilic Premium-Silver SERS substrate. [online]." Available from: https://sersitive.eu/shop/silver-m/ (accessed Oct. 18, 2021).

[16] "SERSitive hydrophilic S-Silver SERS substrate. [online]. " https://sersitive.eu/shop/silver-s/ (accessed Oct. 18, 2021).

[17] "SEM image of S-Silver SERS substrate. [online]. " Available from: https://sersitive.eu/shop/silver-s/ (accessed Oct. 18, 2021).

[18] Ł. RICHTER *et al.*, "Fast and efficient deposition of broad range of analytes on substrates for surface enhanced Raman spectroscopy." *Biosens. Bioelectron.* [online]. 2020, vol. 156. Available from: https://doi.org/10.1016/j.bios.2020.112124.

[19] S. MANGIALARDO, F. PICCIRILLI, A. PERUCCHI, P. DORE, and P. POSTORINO. "Raman analysis of insulin denaturation induced by high-pressure and thermal treatments." *J. Raman Spectrosc.* [online]. 2012, vol. 43, no. 6, pp. 692-700. Available from: https://doi.org/10.1002/jrs.3097.

THE CO-CULTURE OF KERATINOCYTES AND FIBROBLASTS ON A MULTI-LAYERED POLYESTER NANOFIBROUS MEMBRANE ENRICHED WITH PLATELET LYSATE

[1]Andreu BLANQUER, [1]Elena FILOVA, [2]Vera JENCOVA, [1]Antonin BROZ,
[2]Eva KUZELOVA KOSTAKOVA, [2]Maxim LISNENKO, [3]Renata PROCHAZKOVA,
[1]Lucie BACAKOVA

[1]Institute of Physiology of the Czech Academy of Sciences, Prague, Czech Republic, EU, andreu.blanquerjerez@fgu.cas.cz

[2]Technical University of Liberec, Faculty of Sciences, Liberec, Czech Republic, EU, vera.jencova@tul.cz

[3]Regional Hospital in Liberec, Liberec, Czech Republic, EU, renata.prochazkova@nemlib.cz

https://doi.org/10.37904/nanocon.2021.4364

Abstract

The prevalence of chronic wounds is increasing due to the population ageing and specific illnesses like diabetes mellitus and vascular diseases. Nanofibrous membranes fabricated using synthetic polymers are promising materials to enhance skin wound healing. PCL and PVA membranes are being studied to be used as scaffolds for skin tissue engineering and hydrogels for controlled drug delivery, respectively. The present study considers the development of a multi-layered membrane made of PCL and PVA loaded with platelet lysate (PL). PCL nanofibers allowed cell adhesion and growth, whereas PVA acted as a hydrogel that releases the bioactive compounds of platelet lysate. The cytocompatibility of the membranes containing PL and without it was demonstrated on two cell types involved in wound healing, i.e. keratinocytes and fibroblasts. Both cell types were able to adhere and proliferate on the membranes. In addition, the membrane containing PL enhanced the proliferation of fibroblasts. A co-culture study was also performed by seeding each cell type on one side of the membrane. The cells were co-cultured for 7 days and the results showed that PL increased the proliferation of cells achieving a monolayer of keratinocytes or fibroblasts on each side of the membrane. Thus, the beneficial effect of PCL-PVA+PL membranes on monocultures and co-cultures of skin cells was demonstrated, and these membranes can be considered potential scaffolds for treatment of chronic wounds.

Keywords: Skin wound healing, electrospun nanofibers, platelet lysate, keratinocytes, fibroblasts

1. INTRODUCTION

Wound healing is a physiological process that involves multiple cell types, growth factors, cytokines and other metabolites and ions [1]. Among the cell types present in skin wound healing, keratinocytes and fibroblasts are two of the most important ones. Keratinocytes are the main cell type in epidermis, whereas fibroblasts are the main cell type in dermis. Both cell types proliferate and migrate to the wound site during the proliferation phase of wound healing. However, in chronic wounds, the proliferation of skin-resident cells is significantly reduced [2,3].

The increasing number of chronic wounds is becoming burdensome in terms of both human health and socio-economic consequences [4]. The prevalence of chronic wounds increases along with vascular diseases, diabetes mellitus and other systemic factors like advanced age. Indeed, many adult patients suffering from type 2 diabetes mellitus develop ulcers, including the diabetic foot ulcer [5]. It is therefore necessary to develop new wound dressings or skin scaffolds to improve wound healing.

Synthetic biocompatible polymers are widely used in tissue engineering. Specifically, nanofibrous membranes fabricated by electrospinning are considered promising materials for the regeneration of damaged skin. One of the advantages of nanofibrous membranes is that they mimic the fibrous component of the natural extracellular matrix, as well as they protect the wound against microbes. Moreover, these nanofibrous membranes are able to support cell adhesion, migration, growth and differentiation [6]. Among synthetic polymers, polycaprolactone (PCL) is widely used due to its high degree of solubility in a variety of solvents and its tunable mechanical properties. Poly(vinyl alcohol) (PVA) is another type of synthetic polymer that can be used to develop nanofibrous scaffolds. PVA is a highly hydrophilic and chemically resistant polymer, and allows the incorporation of bioactive substances which are soluble in water.

In the present study, we prepared a multi-layered nanofibrous membrane of PCL and PVA, and we incorporated platelet lysate (PL) into PVA. PCL nanofibers allowed cell adhesion and growth, whereas PVA acted as a hydrogel releasing bioactive compounds. The multilayer membrane was fabricated as a scaffold and a drug-releasing membrane with an interesting potential for skin wound healing. The cytocompatibility of the membrane and the effect of PL was analyzed using keratinocytes and fibroblasts on day 1 and 7 after cell seeding. Moreover, keratinocytes and fibroblasts were seeded on both sides of the membrane and co-cultured for 7 days.

2. METHODS

2.1. Platelet lysate preparation

A platelet concentrate was prepared via the centrifuge separation of platelets from human buffy coats obtained from 4 patients. The experimental procedure was previously described by Blanquer *et al.* [7]. Briefly, the buffy coats from the 4 patients were mixed and centrifuged to separate the platelet concentrate from the red blood cells and leukocytes. The final platelet concentration was determined to be in the range 600 x 10^9 to 850 x 10^9 platelets/L. The PL were then prepared using the freeze-thaw method. The supernatant, i.e. PL, was stored at -80 °C.

2.2. Nanofibrous membranes fabrication and characterization

A multi-layer membrane of poly(ε-caprolactone) (PCL) and poly(vinyl alcohol) (PVA) was fabricated by the direct current needleless electrospinning technique (Nanospider 1WS500U, Elmarco, Liberec, Czech Republic). The electrospinning solution for PCL contained 16% (w/w) PCL (Mn 45,000 g/mol) in chloroform/ethanol solution (8:2). For the PVA electrospinning solution, PVA with a molecular weight of 125,000 and degree of hydrolysis of 98% was electrospun from a 10% (w/w) polymer solution in a water/ethanol solution (9:1). PVA containing PL and without PL were prepared. The PVA+PL was prepared from PVA by adding 10% w/w of PL in the spinning solution immediately before electrospinning. The morphology of nanofibers was analysed by scanning electron microscopy and the diameter of the fibres was evaluated from at least 130 measurements using NIS Elements software (Nikon).

2.3. Cell culture

Human HaCaT keratinocytes (CLS Cell Lines Services) were cultured in Dulbecco's Modified Eagle's Medium (DMEM; ThermoFisher Scientific) with 10% fetal bovine serum (FBS; ThermoFisher Scientific) under standard conditions (37 °C and 5 % CO_2). Normal human dermal fibroblasts from neonatal foreskin (NHDF; Lonza) were cultured until passage 4 in DMEM with 10% FBS under standard conditions.

For monocultures of HaCaT cells and NHDF, membranes were cut into pieces of one cm^2 area and placed on a 24-well plate. Both cell types were seeded at a density of 20,000 cells per well and cultured over time. For co-culture experiments, a density of 20,000 NHDF cells per well was seeded on one side of the membranes, and after 1 hour required for cell adhesion, the membranes were placed upside down in a fresh cell culture plate. Then, HaCaT cells were seeded at a density of 50,000 cells per well on the opposite side of the membranes and cultured for 7 days in DMEM with 10%FBS. Cell culture medium was replaced after 4 days.

2.4. Metabolic activity/Cell proliferation

The metabolic activity, which is considered to be related to the cell number, was determined by CellTiter 96® AQueous One Solution Cell Proliferation Assay (MTS, Promega Corporation). The membranes with cells were transferred to fresh 24-well plates for the MTS analysis. The samples were incubated with 500 µl of phenol red-free DMEM supplemented with 10% FBS and 100 µl of MTS for 1 h. The absorbance was measured using a VersaMax ELISA Microplate Reader spectrophotometer (Molecular Devices Corporation) at a wavelength of 490 nm, and normalized per well.

2.5. Cytoskeleton visualization

Following the MTS assay, the cells were rinsed with phosphate-buffered saline (PBS) and fixed with 4% paraformaldehyde in PBS for 30 min. The initial cell adhesion and morphology were analyzed via the staining of the cells with phalloidin after 1 day in culture. In addition, cells grown for 7 days were stained with phalloidin. The samples were incubated with Atto 488-conjugated phalloidin (1:500; Sigma-Aldrich), i.e. a dye for actin cytoskeleton, and Hoechst 33258 (5 µg/ml in PBS, Sigma-Aldrich), i.e. a dye for the cell nuclei, for 20 min at room temperature (RT). Images of the adherent cells were captured under an IX-50 microscope equipped with a DP 70 digital camera (both from Olympus) and a spinning disk confocal microscope equipped with a Zyla 4.2 PLUS sCMOS camera (Andor).

3. RESULTS AND DISCUSSION

3.1. Characterization of the nanofibrous membranes

The measurements of the fiber diameters revealed that electrospun PCL-PVA membranes contained fibers with a wide range of diameters (approx. from 140 nm to 1.2 µm). The mean diameter was 340 ± 170 nm. Similar fiber diameters were observed for PCL-PVA+PL membranes, the mean diameter being 320 ± 210 nm.

3.2. Cell adhesion and cytoskeleton organization

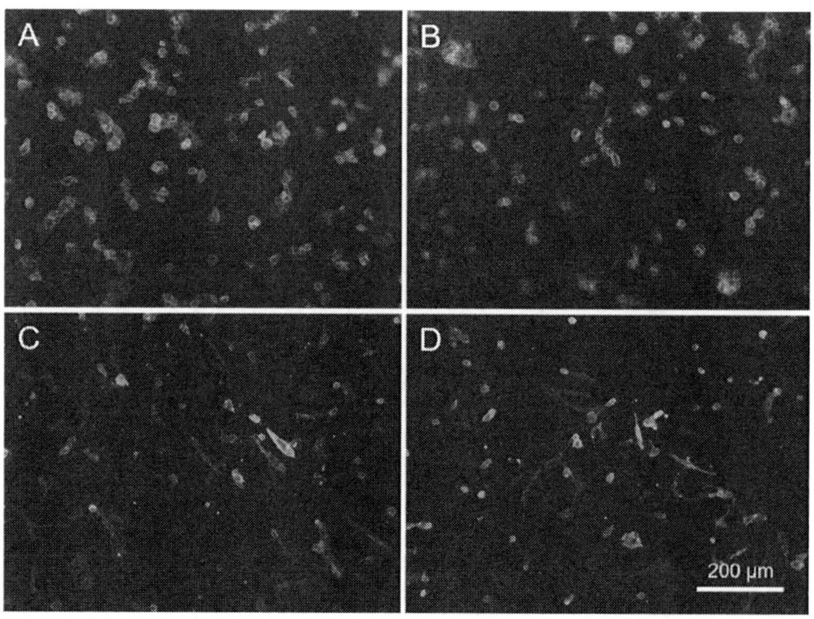

Figure 1 HaCaT cells (A, B) and NHDF (C, D) that adhered to PCL-PVA+PL (A, C) and PCL-PVA (B, D) nanofibrous membranes on day 1 after seeding. Images of the cells stained with phalloidin-Atto488 (green); Olympus IX50 microscope equipped with DP70 digital camera.

The cell adhesion and cytoskeleton organization of the cells grew on the membranes were evaluated after 24 hours in culture. The results indicated that HaCaT cells were able to adhere on both samples, with and without PL. The cells showed a rounded shape and created small island-like clusters in certain regions, which is the normal morphology and behavior of keratinocytes after 24 h of culturing (**Figure 1**). The fibroblasts were also able to adhere and spread on nanofibrous membranes. The cell morphology of fibroblasts was polygonal or spindle-shape, although few cells presented a rounded shape (**Figure 1**). No differences between samples were observed for fibroblasts adhesion and morphology after day 1 in culture. The initial adhesion of cells on the nanofibrous membranes is a necessary step for anchorage-dependent cells, including keratinocytes and fibroblasts. PCL is more hydrophobic than other synthetic polymers, and therefore it has been considered less supportive for cell adhesion [8]. This disadvantage has been mitigated by electrospinning of PCL with other polymers or bioactive additives. The results concerning the initial adhesion and morphology of cells indicated that the developed multi-layered nanofibrous membranes provided a suitable growth support for the basic skin cells types, i.e. keratinocytes and fibroblasts .

3.3. Proliferation of monocultured HaCaT cells and NHDF

The metabolic activity of the HaCaT cells and NHDF cultured on the membranes for 1 and 7 was quantified. The cells were found to proliferate on both membranes tested without any signs of cytotoxic or cytostatic effects (**Figure 2**). On day 1 following seeding, the metabolic activity of HaCaT cells and NHDF did not show significant differences between PCL-PVA and PCL-PVA+PL nanofibrous membranes. On day 7, similar metabolic activity was observed for HaCaT cells. However, the metabolic activity of NHDF cultured for 7 days was observed to be higher on membranes containing PL than on membranes without PL. Previously, we demonstrated that PVA nanomats containing PL enhanced the metabolic activity of 3T3 fibroblasts [9]. PL contains a mixture of cytokines, growth factors and other bioactive molecules, which can influence the behavior of keratinocytes and fibroblasts. Indeed, PL has been demonstrated to accelerate the proliferation of fibroblasts, keratinocytes and endothelial cells [10,11].

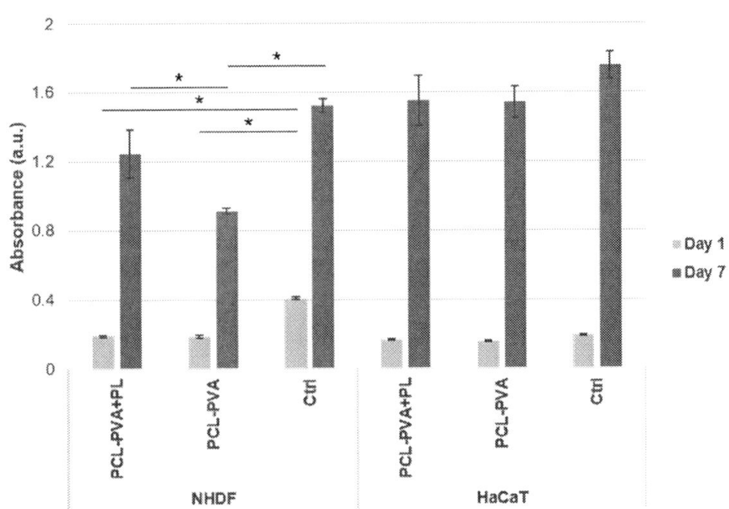

Figure 2 Metabolic activity of the NHDF and HaCaT cells that grew on the PCL-PVA+PL and PCL-PVA nanofibrous membranes in monoculture for 1 and 7 days. The asterisks indicate significant differences between samples for each cell type and time point ($p<0.05$).

3.4. HaCaT cells and NHDF co-cultured on membranes

The nanofibrous membranes allowed a co-culture of keratinocytes and fibroblasts on both sides of the membrane. The membrane was used a barrier to segregate each cell type. In addition, it was used as an artificial basement membrane, which constitutes the anchorage of the epithelium and the connective tissue in skin tissue [12]. Under *in vitro* conditions, the membranes created a physical barrier to avoid the contact between the two cell types and the potential overgrowth of one of them. However, the nanofibrous membranes allowed the molecular exchange by sharing the cell culture medium and all the bioactive compounds secreted by cells.

The morphology and cytoskeleton organization of the cells co-cultured on the membranes were analyzed by actin stress fiber staining on day 7 after cell seeding (**Figure 3**). Images showed that HaCaT cells were able to proliferate on both PCL-PVA and PCL-PVA+PL membranes and the morphology was also similar on both membranes. The fibroblasts proliferated on membranes and oriented in parallel. Both cell types created a monolayer on membranes containing PL, but several void spaces were observed on samples without PL. The results were in agreement with the previous results obtained on monocultures. The PL enhanced the proliferation of fibroblasts in monocultures, and it also enhanced the proliferation of keratinocytes in co-cultures. We hypothesize that the presence of fibroblasts could influence positively the proliferation of keratinocytes.

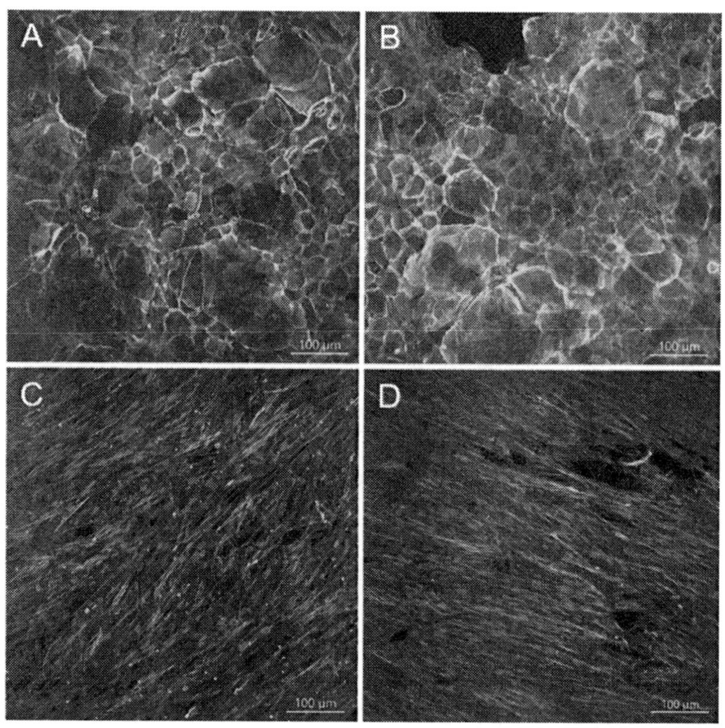

Figure 3 HaCaT cells (A, B) and NHDF (C, D) co-cultured on PCL-PVA+PL (A, C) and PCL-PVA (B, D) nanofibrous membranes for 7 after cell seeding. Images of the cells stained with phalloidin-Atto488 (green) and Hoechst (blue); spinning disk confocal microscope equipped with a Zyla 4.2 PLUS sCMOS camera.

4. CONCLUSION

The PCL-PVA and PCL-PVA+PL membranes allowed the adhesion and proliferation of keratinocytes and fibroblasts without exerting any cytotoxic effect. The presence of PL in PVA enhanced the proliferation of fibroblasts when cultured in monoculture. The growth of keratinocytes and fibroblasts in co-culture was also enhanced on membranes containing PL. The multi-layered membrane demonstrate to be safe for co-culture of the two skin cell types. Thus, nanofibrous membrane loaded with PL have the potential to be used as bioactive scaffolds in chronic wound treatment.

ACKNOWLEDGEMENTS

This project has received funding from the European Union's Horizon 2020 research and innovation programme (grant agreement No. 101003407-ELECTROSKIN) and the Czech Health Research Council, Ministry of Health of the Czech Republic (project No. NV18-01-00332).

REFERENCES

[1] SANON, S.; HART, D.A.; TREDGET, E.E. *Molecular and Cellular Biology of Wound Healing and Skin Regeneration.* Elsevier Inc., 2016. ISBN 9780128016541.

[2] YAN, J.; TIE, G.; WANG, S.; TUTTO, A.; DEMARCO, N.; KHAIR, L.; FAZZIO, T.G.; MESSINA, L.M. Diabetes impairs wound healing by Dnmt1-dependent dysregulation of hematopoietic stem cells differentiation towards macrophages. *Nat. Commun.* [online]. 2018, vol. 9, Available from: https://doi.org/10.1038/s41467-017-02425-z.

[3] HOLL, J.; KOWALEWSKI, C.; ZIMEK, Z.; FIEDOR, P.; KAMINSKI, A.; OLDAK, T.; MONIUSZKO, M.; ELJASZEWICZ, A. Chronic Diabetic Wounds and Their Treatment with Skin Substitutes. *Cells.* [online]. 2021, vol. 10, pp. 1-21, Available from: https://doi.org/10.3390/cells10030655.

[4] SEN, C.K.; GORDILLO, G.M.; ROY, S.; KIRSNER, R.; LAMBERT, L.; HUNT, T.K.; GOTTRUP, F.; GURTNER, G.C.; LONGAKER, M.T. Human skin wounds: A major and snowballing threat to public health and the economy: PERSPECTIVE ARTICLE. *Wound Repair Regen.* [online]. 2009, vol. 17, pp. 763-771. Available from: https://doi.org/10.1111/j.1524-475X.2009.00543.x.

[5] FRYKBERG, R.G.; BANKS, J. Challenges in the Treatment of Chronic Wounds. *Adv. Wound Care* [online]. 2015, vol. 4, pp. 560-582. Available from: https://doi.org/10.1089/wound.2015.0635.

[6] BACAKOVA, L.; ZIKMUNDOVA, M.; PAJOROVA, J.; BROZ, A.; FILOVA, E.; BLANQUER, A.; MATEJKA, R.; STEPANOVSKA, J.; MIKES, P.; JENCOVA, V.; et al. Nanofibrous Scaffolds for Skin Tissue Engineering and Wound Healing Based on Synthetic Polymers. In: *Applications of Nanobiotechnology*; IntechOpen, 2020.

[7] BLANQUER, A.; MUSILKOVA, J.; FILOVA, E.; TABORSKA, J.; BRYNDA, E.; RIEDEL, T.; KLAPSTOVA, A.; JENCOVA, V.; MULLEROVA, J.; KOSTAKOVA, E.K.; et al. The Effect of a Polyester Nanofibrous Membrane with a Fibrin-Platelet Lysate Coating on Keratinocytes and Endothelial Cells in a Co-Culture System. *Nanomaterials* [online]. 2021, vol. 11, p. 457. Available from: https://doi.org/10.3390/nano11020457.

[8] SCHUHLADEN, K.; RAGHU, S.N.V.; LIVERANI, L.; NEŠČÁKOVÁ, Z.; BOCCACCINI, A.R. Production of a novel poly(ε-caprolactone)-methylcellulose electrospun wound dressing by incorporating bioactive glass and Manuka honey. *J. Biomed. Mater. Res. - Part B Appl. Biomater.* [online]. 2020, vol. 1-13. Available from: https://doi.org/10.1002/jbm.b.34690.

[9] FILOVA, E.; BLANQUER, A.; KNITLOVA, J.; PLENCNER, M.; JENCOVA, V.; KOPRIVOVA, B.; LISNENKO, M.; KOSTAKOVA, E.K.; PROCHAZKOVA, R.; BACAKOVA, L. The Effect of the Controlled Release of Platelet Lysate from PVA Nanomats on Keratinocytes. *Endothelial Cells and Fibroblasts.* [online]. 2021. Available from: https://doi.org/10.3390/nano 11040995.

[10] TENCI, M.; ROSSI, S.; BONFERONI, M.C.; SANDRI, G.; BOSELLI, C.; DI LORENZO, A.; DAGLIA, M.; ICARO CORNAGLIA, A.; GIOGLIO, L.; PEROTTI, C.; et al. Particulate systems based on pectin/chitosan association for the delivery of manuka honey components and platelet lysate in chronic skin ulcers. *Int. J. Pharm.* [online]. 2016, vol. 509, pp. 59-70. Available from: https://doi.org/10.1016/j.ijpharm.2016.05.035.

[11] BAIK, S.Y.; LIM, Y.A.; KANG, S.J.; AHN, S.H.; LEE, W.G.; KIM, C.H. Effects of platelet lysate preparations on the proliferation of hacat cells. *Ann. Lab. Med.* [online]. 2014, vol. 34, pp. 43-50. Available from: https://doi.org/10.3343/alm.2014.34.1.43.

[12] LIN, H.Y.; CHEN, S.H.; CHANG, S.H.; HUANG, S.T. Tri-layered chitosan scaffold as a potential skin substitute. *J. Biomater. Sci. Polym. Ed.* [online]. 2015, vol. 26, pp. 855-867. Available from: https://doi.org/10.1080/09205063.2015.1061350.

EVALUATION OF FIBER ORIENTATION OF AC AND DC ELECTROSPUN PLCL NANOFIBROUS LAYERS

[1]Nikifor ASATIANI, [1]Jan VALTERA, [1]Maxim LISNENKO, [1]Kristyna HAVLICKOVA, [1]Jan KOVACICIN, [1]David LUKAS, [1]Eva KUZELOVA KOSTAKOVA, [2]Jiří MOLÁČEK

[1]*Technical University of Liberec, Liberec, Czech Republic, EU, nikifor.asatiani@tul.cz, eva.kostakova@tul.cz*

[2]*Faculty of Medicine in Pilsen, Charles University, Prague, Czech Republic, EU*

https://doi.org/10.37904/nanocon.2021.4353

Abstract

Electrospun PLCL nanofibrous materials produced by direct current (DC) or alternating current (AC) electrospinning are good candidates for absorbable reinforcement for cardiovascular surgery composite bandages. For an appropriate description of the fiber layers and prediction of its relevant biomechanical characteristics, the knowledge of fiber orientation in the electrospun layers is essential. In this study, the different orientation of fibres in AC and DC electrospun materials was achieved only by changing the rewinding speed of the supporting material, as the easy way for the production of relatively large samples with controlled fiber orientation. The fiber orientation was measured on scanning electron microscope images at different magnification using an open-source image processing software FIJI/Image J with the implementation of the "OrientationJ" plugin. Subsequently, the fiber layers were compared in terms of fiber orientation. Optimal image parameters for the measurement and the best unidirectional oriented-aligned samples were determined

Keywords: Electrospinning, biodegradable nanofibers, aligned nanofibers, image analysis, scanning electron microscopy

1. INTRODUCTION

Electrospun fibrous materials are widely used in medical applications nowadays [1,2]. One of the important applications of nanofibrous materials are scaffolds for cardiovascular systems. The different electrospun nanofibrous structures are used. Thus, outer circumferentially aligned electrospun nanofibers are also used as a parts of small-diameter vascular grafts [3] or bandages, which application is based on the mechanical interaction of the arterial wall and the external support. The electrospun nano or microfibrous material morphological testing is necessary. However, not only fiber diameter distribution is essential to observe. It is necessary to check also fiber orientation. The fiber orientation or alignment of fibers in electrospun layer can dramatically influence mainly mechanical properties but also biological properties such as cell orientation, adhesion, migration etc. In order to evaluate the suitability of electrospun materials, it is necessary not only morphological analyses and spinnability analyses, etc., but it is also necessary to look at the orientation of the fibres in the materials produced, and this paper deals with this.

Today, both a high-voltage direct current (DC) source and an alternating current (AC) source are used to form polymer nanofibers. The use of a DC source requires an electrically active collector, but the AC source does not. The nature of both processes also indicates differences in the resulting morphologies of nanofiber layers, even in terms of fiber orientation controlled only by the rate of withdrawal of the base material. The orientation of fibres in DC electrospun materials with the special collectors using was studied several times for medical applications mainly [4-8]. The difference between the two technologies with changes in towing speed is described in this paper. The selection of a suitable magnification of images from a scanning electron microscope is also taken into account and the optimal procedure for evaluating such images for their comparison is set. This study shows how these electrospinning technologies used for the industrial production of nanofibrous materials find it difficult or almost unintentional to change the orientation of the fibers during the electrospinning process.

2. MATERIALS AND METHODS

2.1. Fabrication of nano-/microfibrous layers

A copolymer of L-lactide and ε-Caprolactone in a 70/30 molar ratio and with an inherent viscosity midpoint of 1.5 dl/g (Purasobr PLCL 7015, Corbion) was used for fabrication of the materials. The PLCL copolymer is biocompatible, biodegradable, and has very good mechanical properties for the potential application in cardiovascular surgery as a vascular biodegradable bandage. For the DC and AC electrospinning, the polymer solution was prepared with solvent system acetic acid/formic acid/acetone in weight ratio 1:1:1. The final polymer concentration was 10wt%. All the materials were electrospun from the freshly prepared solution after 4 hours at the magnetic stirrer. The electrospinning was done by DC needleless electrospinning by Nanospider™ machine, where static wire spinning electrode is used. The AC device with a cylindrical spinning electrode and a rotating cylinder as an electrically inactive collector was applied to create a second set of electrospun materials. The schematic difference in between these two electrospinning technologies is introduced in **Figure 1**.

DC electrospinning *AC electrospinning*

Figure 1 Scheme and photos of used DC and AC electrospinning devices and examples of electrospun nanofibrous layers on the nonwoven supporting materials

All the nanofibrous materials were DC electrospun on a needless electrospinning device type NS 1WS500U (Elmarco, CZ) based on a stationary wire spinning electrode. The spinning electrode was charged with direct current high voltage source positively (40 kV), and the collector was charged negatively (-10 kV). The distance between the spinning electrode and the collector was 180 mm. The polymer solution was fed onto stationary wire spinning electrode by moving carriage module filled with a polymer solution. Metal insert orifice 0.7 mm in diameter was used for control of polymer solution layer on the spinning electrode. The DC electrospinning device is complemented by an air conditioning unit and thanks to that the exact values of temperature and air humidity in the spinning space were used. The temperature was 22 °C and the relative humidity was 40 %. The used supporting material was spunbond Pegatex S (PFNonwovens, CZ) with 20 g/m^2 and the withdrawal speeds were changed according to **Table 1**.

The second set of samples was prepared on a device using a high voltage AC source. This device was designed and constructed at the Technical University of Liberec [9,10]. The distance between the spinning electrode and the collector was 180 mm. The perforated rotating cylinder with 30 cm in diameter coated with a supporting spunbond nonwoven Pegatex S (PFNonwovens, CZ) with 20 g/m^2 was used as an electrically inactive collector capturing the formed nanofibers. The withdrawal speed introduced in **Table 1** represents the circumferential speed of rotation of the cylinder collector. The value of the effective electric voltage was 35 kV. The device is not equipped with an air conditioning unit, but the ambient conditions were measured. The temperature during the AC electrospinning was 21 °C, and the relative humidity was 43 %. The minimal peripheral speed of rotating perforated cylinder as electrically inactive collector was 30 mm/min, slower speed does not allow sufficient catching of fibres on the roller surface. The maximal peripheral speed was 100 mm/min. When the higher speed was used, the fibres broke. The nanofibrous materials immediately removed from the cylindrical collector after spinning showed large precipitation along the length. For this reason, the sample was also tested and the sample was left on the cylinder for 24 hours to dry the residual solvents completely and to relax the stress in the sample.

Table 1 Withdrawal speeds set up for DC and AC electrospinning to produce nanofibrous layer

	Withdrawal speed of supporting nonwoven material [mm/min]				
DC electrospinning	0	6	20	100	215
AC electrospinning	30	80	80 (left on roller for 24 hours)	100	

2.2. Characterization of produced fibres

Scanning electron microscopy (SEM) was used to analyse the alignments of obtained fibrous layers using Orientation J Distribution plugin in the FIJI/ImageJ software (NIH). Each submitted SEM image was first de-framed so that the entire field of view could be considered the area of interest. Some of the images have been rotated 90° to make the histograms easier to read. In particular, the methodology of measuring several selected regions of interest on one image was also tested. If necessary, contrast and sharpness of an image were adjusted. Before each measurement, it was important to set three parameters - 1) sigma, σ, or a rough estimate of the size of the structure of interest (i.e., the fiber diameter in pixels); 2) the desired coherence (i.e., uniformity or continuity) and energy (this parameter allow to "distinguish" fiber from the background), %. The resulting fiber orientation is then judged on the basis of a color-coded map, where hue corresponds to the angle of local fiber orientation and saturations reflect the coherence. The angle of local orientation can range from -90° to -90° relative to the horizontal. Supplementary video accompanies this paper at https://youtu.be/u7b1_uKloOM.

3. RESULTS AND DISCUSSION

Using the OrientationJ plugin, images of nanofibrous layers fabricated with different withdrawal speeds were analysed. The main focus was on the feasibility and suitability of using the plugin to obtain qualitative and quantitative information to assess fiber orientations based simply on SEM images. Images at different magnifications and their influence on the resulting measurements were tested. The effect of image threshold settings, contrast and sharpness of individual images was also investigated. Finally, different input parameters such as coherence, energy and sigma were tested. It was experimentally discovered that 10 % energy and 10 % coherence are optimal for most of the SEM images as input values. The most sensitive parameter of the measurement, however, is σ, or a rough estimate of the size of the structure of interest. As this parameter corresponds to the size of individual fibres, the correct estimation of σ before the measurement was crucial. The essential problem with an estimation of σ can occur when the fibrous layer has a multimodal distribution of fiber diameters.

The analyzed SEM images of DC electrospun fibrous layers revealed that it was nearly impossible to control the fiber orientation with rewinding speed of the substrate. It can be seen from **Figure 2** that even at high speed (100 or 215 mm/min) it was difficult to collect fibers in a controlled manner. The main difference between the materials fabricated at low (6 or 20 mm/min) or even zero speed was in the thickness and density of the obtained layers. As **Figure 2** indicates, the high density can negatively affect the orientation measurement. With increasing density of fibers "packing" within the fibrous layer, dramatically decreases distinguishability of individual fibers. Thus, images with higher magnifications are more appropriate for the dense layers. As shown from **Figure 2** the densest fibrous layers obtained had the narrowest distribution peaks of orientations at 500x magnifications. Conversely, when analysing the images with a magnification of 5000x, a significant change towards polydispersion of orientation distribution was observed. Furthermore, one should also consider the fact, that Orientation J cannot "separate" two overlapping fibers. This drawback has a significant effect on estimated orientation distribution when the magnification of the image is too high and low number of fibers is presented on the view field.

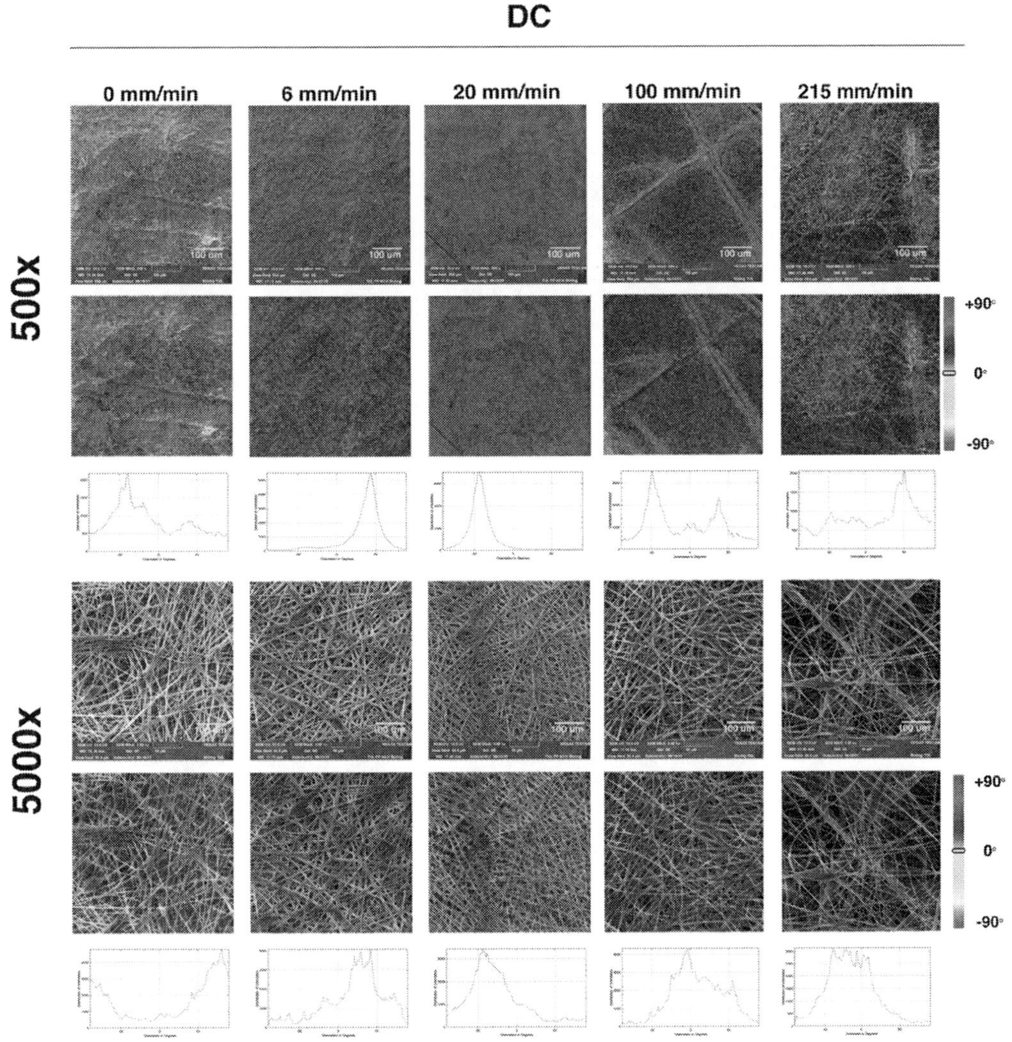

Figure 2 SEM images of fabricated DC electrospun nanofibrous layers and corresponding color-coded analyzed image with histogram of distribution of fiber orientations within the layers

Analysis of the SEM images of AC electrospun fibrous layers revealed that in contrast to DC electrospun layers the orientation was mainly anisotropic at both high and low rewinding speeds. The lowest rewinding speed (30 mm\min) was comparable to the higher speeds (80 and 100 mm\min). This can be partially explained by the nature of AC-spinning process, where the first collected fibers on a collector generally control the orientation of the following fibers. The sample rewound at 80 mm/min and left on the roller afterwards to completely solidify had the highest heterogeneity among the AC-spun materials. This is very interesting and shows changes in the fibrous structure when it is dried in the off state. Apparently, certain types of stress transfer occur, and this changes the arrangement of the fibers in the sample with respect to their orientation. This effect should be further studied and explained in detail. Comparing to the SEM images of DC-spun materials, it can be seen from **Figure 3**, that the apparent isotropy of the layers was increasing with increasing the magnification. This is mainly due to the drawback of Orientation J described above when dealing with a relatively small number of fibers having a large number of intersections, Orientation J cannot consider the "bottom" fibers as a whole and analysis the fiber as a set of separated segments, thus greatly distorting the resulting distribution of orientations.

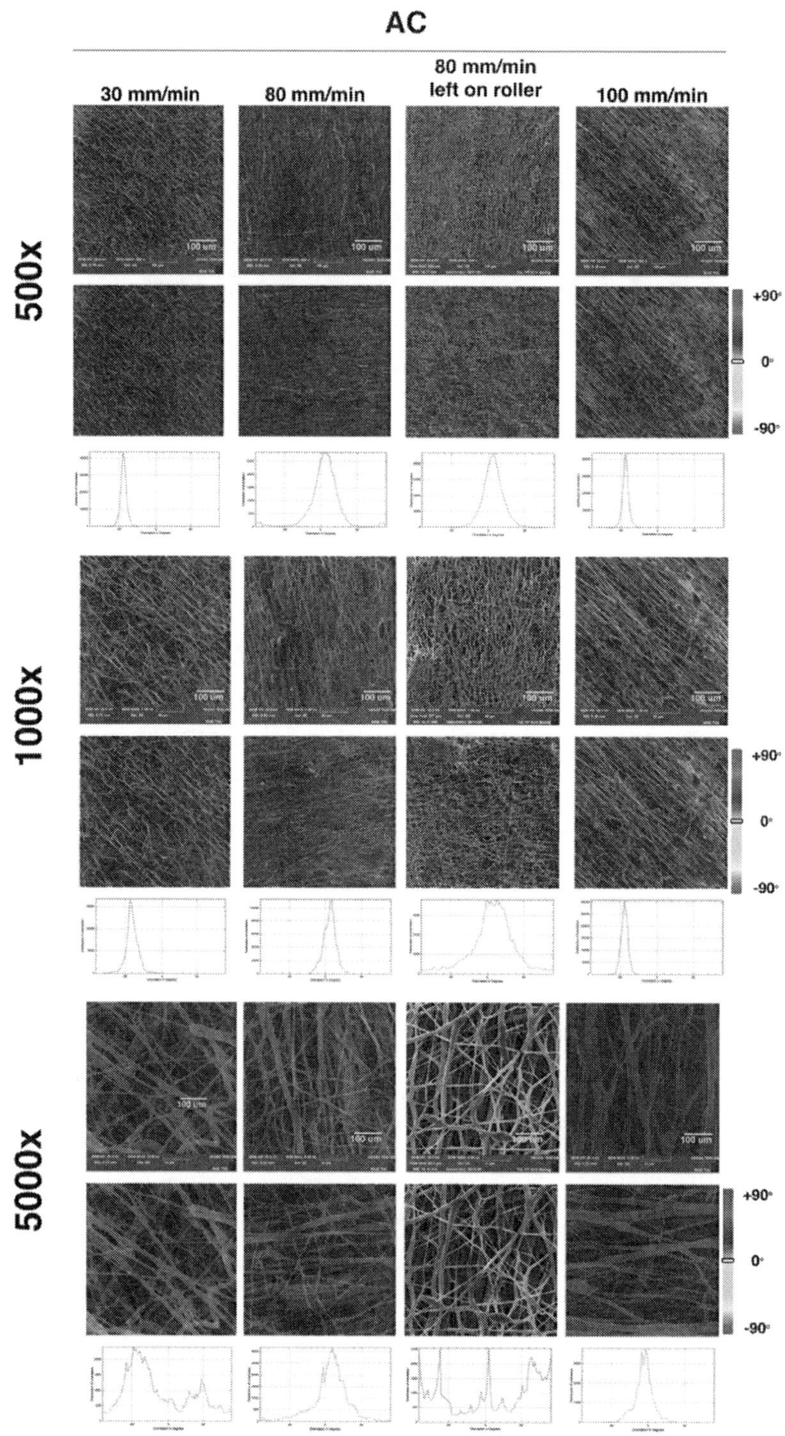

Figure 3 SEM images (different magnifications) of fabricated DC electrospun nanofibrous layers and corresponding color-coded analysed image with histogram of distribution of fiber orientations within the layers

4. CONCLUSION

The orientation distributions of the materials obtained by AC and DC electrospinning differed significantly from each other. The AC fibers had a well-defined prevailing fiber orientation in their majority, while the DC fibers were inherently isotropic at all rewinding speeds. For materials fabricated by DC-spinning, which were characterized by more dense fibrous layers, images at a magnification of 1000x or higher were more suitable for analysis. On the other hand, in the case of AC-spun fibers, high magnifications distorted the analysis because poorly distinguishable fibers with significantly different σ - parameter (structure of interest size) came into play. However, it was shown that at high speeds of material rewinding, a higher level of fiber orientation is achieved, preserving monodisperse decomposition histograms even at high magnifications of the images. Conversely, keeping the fibrous layers on the collector may lead to fiber relaxation, which in turn results in higher isotropy. The established parameters of the measurement could be helpful for the further design of alignment-dependent electrospun materials.

ACKNOWLEDGEMENTS

This work was supported by the Czech Health Research Council, project No NU20-02-00368.

REFERENCES

[1] JAHANGIRI, A., ADIBKIA K. Applications of electrospinning/electrospraying in drug delivery. *Bioimpacts*. [online]. 2016, vol. 6, no. 1, pp. 1-2. Available from: https://doi.org/10.15171/bi.2016.08.

[2] RAHMATI, M. et al. Electrospinning for tissue engineering applications. *Progress in Materials Science*. [online]. 2021, vol. 117, p. 100721. Available from: https://doi.org/10.1016/j.pmatsci.2020.100721.

[3] BAI, S., ZHANG, X., ZANG, L., YANG, S., CHEN, X., YUAN, A X. Electrospinning of Biomaterials for Vascular Regeneration. *Chem. Res. Chin. Univ.* [online]. 2021, vol. 37, no. 3, pp. 394-403. Available from: https://doi.org/10.1007/s40242-021-1125-3.

[4] Oriented nanofibrous membranes for tissue engineering applications_ Electrospinning with secondary field control. *Elsevier Enhanced Reader*. [online]. https://reader.elsevier.com/reader/sd/pii/S1751616115002325?token=1DB51A87DF92D115AAAEBF43E6DCACD56EB773CFEFA662D4CCF34C9CAE43C865EC1DE84864C796ABAC5229083080541A&originRegion=eu-west-1&originCreation=20210920134909 (Accessed 20 September 2021).

[5] ABUDULA, T., SAEED, U., N. SALAH, N., A. MEMIC, A., H. AL-TURAIF, H. Study of Electrospinning Parameters and Collection Methods on Size Distribution and Orientation of PLA/PBS Hybrid Fiber Using Digital Image Processing. *Journal of Nanoscience and Nanotechnology*. [online]. 2018, vol. 18, no. 12, pp. 8240-8251. Available from: https://doi.org/10.1166/jnn.2018.15885.

[6] CHIVERTON, J. P., KAO, A., ROLDO, M., TOZZI, A G. Automatic diameter and orientation distribution determination of fibrous materials in micro X-ray CT imaging data. *Journal of Microscopy*. [online]. 2018, vol. 272, no. 3, pp. 180-195. Available from: https://doi.org/10.1111/jmi.12719.

[7] WALSER, J., FERGUSON, A S. J. Oriented nanofibrous membranes for tissue engineering applications: Electrospinning with secondary field control. *Journal of the Mechanical Behavior of Biomedical Materials*. [online]. 2016, vol. 58, pp. 188-198. Available from: https://doi.org/10.1016/j.jmbbm.2015.06.027.

[8] ZHANG, L., YU, A W. Orientation image analysis of electrospun submicro-fibers based on Hough transform and Regionprops function. *Textile Research Journal*. [online]. 2017, vol. 87, no. 18, pp. 2263-2274. Available from: https://doi.org/10.1177/0040517516669070.

[9] POKORNY P., *et al.* Effective AC needleless and collectorless electrospinning for yarn production. *Phys. Chem. Chem. Phys.* [online]. 2014, vol. 16, no. 48, pp. 26816-26822. Available from: https://doi.org/10.1039/C4CP04346D.

[10] VALTERA J., *et al.* Fabrication of dual-functional composite yarns with a nanofibrous envelope using high throughput AC needleless and collectorless electrospinning. *Sci Rep*. [online]. 2019, vol. 9, no. 1, p. 1801. Available from: https://doi.org/10.1038/s41598-019-38557-z.

DEGRADATION OF POLYCAPROLACTONE ELECTROSPUN MATERIALS - METHODS OF ANALYSIS

[1]Kristýna HAVLÍČKOVÁ, [2]Michal ŘEZANKA, [3]Libor KOBERA, [1]Eva KUŽELOVÁ KOŠŤÁKOVÁ, [1]David LUKÁŠ, [1]Věra JENČOVÁ

[1]Technical University of Liberec, Faculty of Science, Humanities and Education, Departmnt of Chemistry, Liberec, Czech Republic, EU, kristyna.havlickova@tul.cz

[2]Technical University of Liberec, Institute for Nanomaterials, Advanced Technology and Innovation, Department of Nanochemistry, Liberec, Czech Republic, EU, michal.rezanka@tul.cz

[3]Institute of Macromolecular Chemistry CAS, Czech Academy of Sciences, Prague, Czech Republic, EU, kobera@imc.cas.cz

https://doi.org/10.37904/nanocon.2021.4354

Abstract

Biodegradable electrospun materials are widely used for medical application. Polycaprolatone is polymer suitable for electrospinning technology and is very often used to create nanofibrous scaffolds for tissue engineering. The time to disintegration or biodegradation of such materials is very important here. However, testing is not entirely easy. It is not possible to create exactly body-like conditions in vitro. Moreover, it is not easy to find suitable analytical methods that would show exactly what happens in the nanofibrous polycaprolactone electrospun samples at certain stages of degradation, ie how the internal structure of decaying nanofibers changes. This paper describes the traditional use of methods for testing polycaprolatone nanofibers by enzymatically catalysed degradation. Morphological changes are studied using scanning electron microscope images. However, it also offers a non-traditional analysis of polycaprolactone electrospun materials using the ssNMR method.

Keywords: Electrospinning, nanofibers, enzymatic degradation, solid state NMR

1. INTRODUCTION

Polymeric materials are the most widely used building units for the production of biocompatible materials for medical applications. In the last three decades, the attention in the area of polymer synthesis focused on the theme of biodegradable and biocompatible polymers. Much attention is paid in literature to the preparation of nanofibers from polycaprolactone (PCL) [1-4]. The polymer is semicrystalline linear aliphatic polyester that is biocompatible and biodegradable. PCL has outstanding physical-chemical properties (glass transition temperature -60 °C; melting temperature 60 °C, temperature of most rapid crystallization 30 °C), solubility and sensitivity to biodegradation.

The production of polymer nanofibers can be ensured by several different technologies such as melt blown, centrifugal spinning, flash spinning, air-jet spinning, drawing, etc. [5,6]. One of the most used technologies for the production of polymer nanofibers is electrospinning. Electrospinning is used mainly with a DC power supply, but it is also possible to use AC power supplies [7,8]. The DC needle-less continuous electrospinning produces homogeneous nanofibrous layers from polymeric solutions.

Biodegradation of electrospun polymeric especially PCL nanofibers are nowadays widely studied topics [9], which can be observed by different methods focused on the study of changes in the internal structure of nanofibers and also on morphological changes. This paper presents the observation of morphological changes in enzymatically catalysed degradation using a scanning electron microscope and changes in the internal arrangement unconventionally using the solid state NMR method.

2. MATERIALS AND METHODS

2.1. Materials

PCL 45 (Mw 45 000 g·mol^{-1}; Merck, Germany) was use for preparation of micro/nanofiber layers. A mixed solvent was used consisting of chloroform and ethanol (Penta, Czech Republic) in a ratio of 8:2 by weight. For enzymatic degradation of nanofibers, was used Lipase form *Pseudomonas cepacia* (specific activity 30 ≥ U·mg^{-1}; Merck, Germany). Sodium azide (NaN_3) was obtained from Merck (Germany).

2.2. Nanofibrous materials preparation

Mikro/nanofibrous layers were electrospun from 16 wt% solutions of PCL 45. An appropriate amount of polymer was added to the mixed solvent and stirred at room temperature until the polymer was completely dissolved. Materials were prepared using electrospinning technology (Nanospider™ 1WS500U, Elmarco, Czech Republic). The prepared polymer solutions were electrospun at humidity 50 % and temperature 22 °C. The applied voltage was -10 kV (collector) and +40 kV (spinning string). The electrode distance was 180 mm, the EMW speed was 450 mm/sec and the rewinding speed was 30 mm/min. The diameter of the string was 0.4 mm and the size of the hole was 0.7 mm. After that were materials stored at 4 °C at fridge.

Prepared materials were cut to testing samples and sterilized by means of Anprolene AN-74i (Anprolene; Great Britain) according standard (ČSN EN ISO 11135) applying a low-temperature cycle (at room temperature; 12 hours), then were samples ventilated for 1 week at room temperature.

2.3. Enzymatic degradation

For the degradation tests, the electrospun layers (30 g·$^{-2}$) were cut into small samples (n=10 + 2 negative control (NC) for testing day; weight of each sample 50 ± 5 mg) and placed into 5 ml vials and sterilized. (it was mentioned above). Enzymatic degradation was carried out at 37 °C in phosphate buffer solution (PBS with 0.02 % NaN_3, pH 7.4) to simulate biological fluids. These solutions contained 5 ml PBS+0.02 % NaN_3 along with enzyme Lipase (5 U·ml^{-1}); for NC was prepared only pure buffer solution. PBS solution with enzymes was changed every day, at the same time 10 samples replicates and 2 NCs were removed, washed twice with distilled water and dried at room temperature a week before further investigation.

2.4. Morphological analysis

Morphological analysis was performed by scanning electron microscopy (Vega Tescan 3, Tescan, Czech Republic) after 10 nm of gold sputter coating (Quorum Q50ES, Quorum Technologies, Great Britain). Samples were analysed at an accelerating voltage of 20 kV. The average diameter and the diameter distribution were evaluated using Image J (300 randomly measured value from 4-5 images per each material).

2.5. Weight loss analysis

After properly drying samples, were immediately weighed using a digital balance (PA224C four-range analytical balance, Ohaus, Switzerland). To determine the percentage weight loss, weights of samples before and after degradation were determined and then subjected to calculation by Eq. (1). Four replicate samples were prepared and the values were averaged.

$$w_{loss} = \frac{w_b - w_a}{w_b} \cdot 100 \qquad (1)$$

where:

w_{loss} - weight loss of samples during the degradation (%)

w_b - dry weight before degradation (mg)

w_a- dry weight after degradation (mg)

2.6. ssNMR

Solid-state NMR (ssNMR) spectra were collected using a 500 MHz Bruker Avance III HD NMR spectrometer (B0 = 11.7 T) at Larmor frequency v(13C) = 125.833 MHz using a double-resonance 3.2-mm magic angle spinning (MAS) probe. All MAS NMR experiments were recorded at a 20 kHz spinning speed. The number of scans collected for each 13C MAS NMR spectrum was 1024 at recycle delays 10 s. In case of 13C MAS NMR experiment, SPINAL64 decoupling sequence was applied for suppression of strong dipolar interactions. The 13C chemical shift was calibrated using glycine (13C: 176.03 ppm -high field signal). All dried samples were packed and kept in ZrO2 rotors.

3. RESULTS AND DISCUSSION

Crystallinity of PCL nanofibers during enzyme-catalysed degradation was studied. Due to the enzymatically catalysed degradation of polyester nanofibrous materials, changes in the fibber structure occur (see **Figure 1**). These changes are evident from electron microscopy images; according to available sources, there is, among other things, a change in the crystal structure of the polymer. Nanostructuralization of the surface is evident, which is probably due to the recrystallization of macromolecules during degradation (the macromolecules are broken down, the release of soluble degradation products is created and this creates free space and changes the character of interacting molecules). The experiment was focused on the analysis of the crystallinity of nanofibrous materials from PCL during enzymatically catalysed degradation. Degradation was catalysed by lipase from *Pseudomonas cepacia*, degradation was monitored by evaluating weight-loss (see **Figure 2**) and morphology.

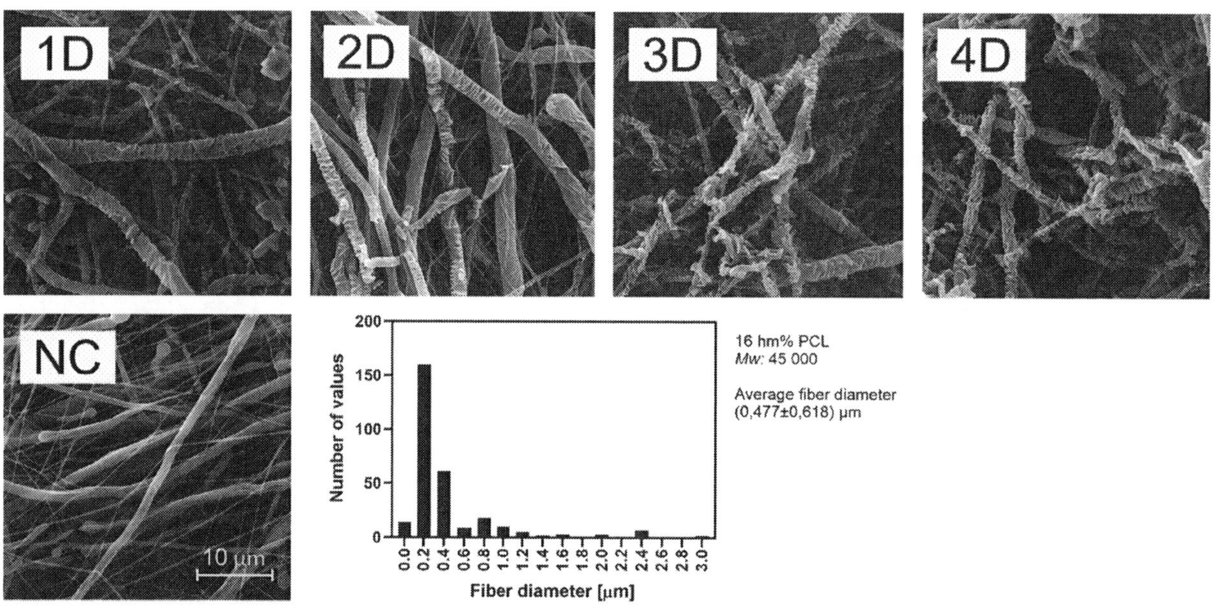

Figure 1 SEM images of PCL electrospun nanofibers capturing test PCL 45 samples after 1, 2, 3 and 4 days of degradation experiment in comparison with negative controls (NC). Scale 10 micrometres

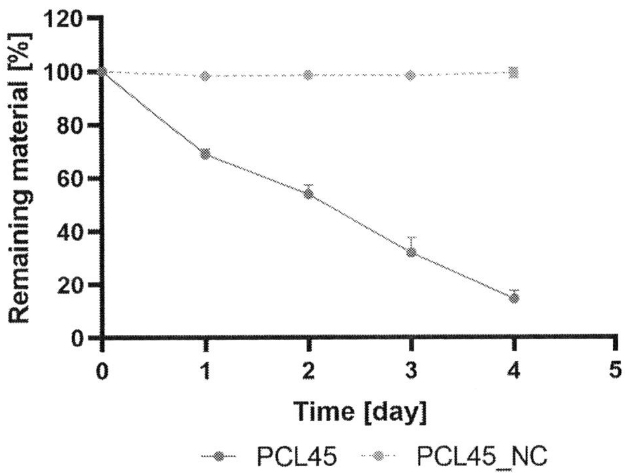

Figure 2 Graph of enzymatic degradation process of PCL 45 samples by Lipase from *Pseudomonas cepacia.* Weight loss or remaining material during enzymatic degradation process (1 - 4 days) is represented by red line in comparison of negative control (material without enzyme solution) represented by green line. After 2 days was remaining material (53.75 ± 3.59) %, at the end of experiment it was (14.40 ± 3.11)%.

The ^{13}C MAS NMR spectroscopy was used for investigation of enzymatic degradation processes of polycaprolactone in range of four days, see **Figure 3**.

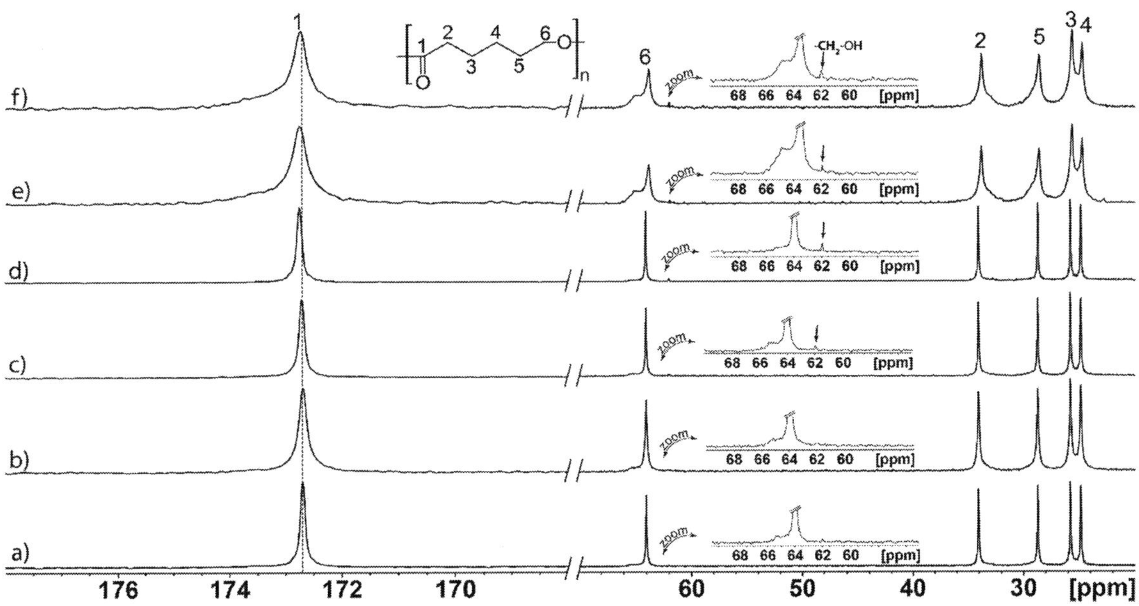

Figure 3 Experimental 13C MAS NMR spectra of investigated polycaprolactones (PCL 45) at distinct degradation times - a) fresh, b) control sample, c) one day degraded, d) two days degraded, e) three days degraded and f) four days degraded.

From comparison of ^{13}C MAS NMR spectra of fresh and control sample (**Figure 2**) and b)) it is evident that all detected signals of control sample are slightly broadened. However, no changes in signal shift or new spectral line(s) were detected. This signal broadening can be result of moderate structural differences. Control sample can be defined as a material with a higher degree of disorder. (In other words, the crystallinity of control sample is lower as compare to fresh sample; signal broadening is maybe caused by swelling of sample).

In case of enzymatic degradation, the changes in ^{13}C MAS NMR spectra are more significant (compare spectra in **Figures 1 a-f)**). Firstly, one new small peak, at ca. 62 ppm, was observed in all treated systems (**Figures 1 c-f)**). This peak was attributed to methylene groups neighbouring with hydroxyl groups ($-CH_2-OH$) at the end of macromolecular chains. Their slightly increasing intensity suggests some degree of depolymerization process. Moreover, the structural changes induced by enzymatic degradation are clearly visible in case of three and four days treated samples in ^{13}C MAS NMR spectra (**Figure 1 e)** and **f)**. The significant broadening of all detected signals and appearance of new relatively broad signal at 65 ppm is evident in **Figure 2**. These changes in ^{13}C MAS NMR spectra suggest a formation of disorder/amorphous phase after three/four days of enzymatic treatment. Furthermore, small change was observed in case of carbonyl carbons (C=O), see **Figure 2**. This slight downfield shift is caused by formation of stronger inter- or intra- molecular bond(s) [10,11].

4. CONCLUSION

Study of biodegradation process is very complex phenomenon not only for nanofibrous materials. PCL electrospun nanofibers showed interesting morphological and internal structure changes as it is obvious from SEM images and ssNMR spectra. Nanostructulization of the surface is observed, probably due to recrystallization of macromolecules during degradation. Penetration of water (PBS) into the polymer caused hydrolytic degradation, when amorphous domains are attacked primary and it probably cause a higher chain flexibility in comparison of crystalline regions. [12,13] Than is possible observed formation of stronger molecular bonds and possibly increasing of crystallinity. The result interpretation is not simple and there is necessary to study by these method different biodegradable materials and to compare them with other analytical methods.

ACKNOWLEDGEMENTS

Authors thank for the financial support of the grant "Study of physical networks in polymer materials" of the Technical University of Liberec Grant Program „PURE", No. PURE-2020-4007.

REFERENCES

[1] BAJI, A., MAI, Y.-W., WONG, S.-C., ABTAHI, M., CHEN, A P. Electrospinning of polymer nanofibers: Effects on oriented morphology, structures and tensile properties. *Composites Science and Technology*. [online]. 2010, vol. 70, no. 5, pp. 703-718. Available from: https://doi.org/10.1016/j.compscitech.2010.01.010.

[2] LOWERY, J. L., DATTA, N., RUTLEDGE, A G. C. Effect of fiber diameter, pore size and seeding method on growth of human dermal fibroblasts in electrospun poly(ε-caprolactone) fibrous mats. *Biomaterials*. [online]. 2021, vol. 31, no. 3, pp. 491-504. Available from: https://doi.org/10.1016/j.biomaterials.2009.09.072.

[3] VAN DER SCHUEREN, L., DE SCHOENMAKER, B., KALAOGLU, Ö. I., DE CLERCK, A K. An alternative solvent system for the steady state electrospinning of polycaprolactone. *European Polymer Journal*. [online]. 2011, vol. 47, no. 6, pp. 1256-1263. Available from: https://doi.org/10.1016/j.eurpolymj.2011.02.025.

[4] WANG, X., ZHAO, H., TURNG, L.-S., LI, A Q. Crystalline Morphology of Electrospun Poly(ε-caprolactone) (PCL) Nanofibers. *Ind. Eng. Chem. Res.* [online]. 2013, vol. 52, no. 13, pp. 4939-4949. Available from: https://doi.org/10.1021/ie302185e.

[5] NAYAK, R., PADHYE, R., KYRATZIS, I. L., TRUONG, Y. B., ARNOLD, A L. Recent advances in nanofibre fabrication techniques. *Textile Research Journal*. [online]. 2012, vol. 82, no. 2, pp. 129-147. Available from: https://doi.org/10.1177/0040517511424524.

[6] A. BARHOUM, K. PAL, H. RAHIER, H. ULUDAG, I. S. KIM, A M. BECHELANY, „Nanofibers as new-generation materials: From spinning and nano-spinning fabrication techniques to emerging applications", *Applied Materials Today*. 2019, vol. 17, pp. 1-35. Available from: https://doi.org/10.1016/j.apmt.2019.06.015.

[7] POKORNY, P. *et al.*, Effective AC needleless and collectorless electrospinning for yarn production. *Phys. Chem. Chem. Phys.* [online]. 2014, vol. 16, no. 48, pp. 26816-26822. Available from: https://doi.org/10.1039/C4CP04346D.

[8] VALTERA, J. *et al.* Fabrication of dual-functional composite yarns with a nanofibrous envelope using high throughput AC needleless and collectorless electrospinning. *Sci Rep.* [online]. 2019, vol. 9, no. 1, p. 1801. Available from: https://doi.org/10.1038/s41598-019-38557-z.

[9] DONG, Y., LIAO, S., NGIAM, M., CHAN, C. K., RAMAKRISHNA, A S. Degradation Behaviors of Electrospun Resorbable Polyester Nanofibers. *Tissue Engineering Part B: Reviews*. [online]. 2009, vol. 15, vol. 3, pp. 333-351. Available from: https://doi.org/10.1089/ten.teb.2008.0619.

[10] MACIEL, G. E., RUBEN, A G. C. Solvent Effects on the C ^{13} Chemical Shift of the Carbonyl Group of Acetone. *J. Am. Chem. Soc.* [online]. 1963, vol. 85, no. 23, pp. 3903-3904. Available from: https://doi.org/10.1021/ja00906a049.

[11] TAKEGOSHI, K., NAITO, A., MCDOWELL, A C. A. Intermolecular hydrogen-bonding effects on the 13C NMR shielding tensor of the carbonyl carbon nucleus in a single crystal of dimedone. *Journal of Magnetic Resonance (1969)*. [online]. 1985, vol. 65, no. 1, pp. 34-42. Available from: https://doi.org/10.1016/0022-2364(85)90373-7.

[12] NATU, M. V., DE SOUSA, H. C., GIL, A M. H. Influence of polymer processing technique on long term degradation of poly(ε-caprolactone) constructs. *Polymer Degradation and Stability*. [online]. 2013, vol. 98, no. 1, pp. 44-51. Available from: https://doi.org/10.1016/j.polymdegradstab.2012.10.030.

[13] ANTHEUNIS, H., VAN DER MEER, J.-C., DE GEUS, M., HEISE, A., KONING, A C. E. Autocatalytic Equation Describing the Change in Molecular Weight during Hydrolytic Degradation of Aliphatic Polyesters. *Biomacromolecules*. [online]. 2010, vol. 11, no. 4, pp. 1118-1124. Available from: https://doi.org/10.1021/bm100125b.

BRAIDED THREADS WITH AC ELECTROSPUN NANOFIBERS FOR HYGIENIC AND MEDICAL APPLICATIONS - PRODUCTION AND PROPERTIES

[1]Divyabharathi MADHESWARAN, [1]Šárka HAUZEROVÁ, [1]Jan VALTERA, [1]Maxim LISNENKO, [1]Manikandan SIVAN, [1]Ondřej BAŤKA, [1]Věra JENČOVÁ, [1]David LUKÁŠ, [2]František JANDA, [1]Eva KUŽELOVÁ KOŠŤÁKOVÁ

[1]Technical University of Liberec, Liberec, Czech Republic, EU, divyabharathi.madheswaran@tul.cz

[2]Wikinomist, s.r.o., Prague, Czech Republic, EU

https://doi.org/10.37904/nanocon.2021.4355

Abstract

Threads are the basic textile linear structure used in many areas. The variability of the thread structure made possible by special constructions brings great possibilities of variability. It is also possible to introduce nanofiber material into the yarn construction as a cover for central microfiber yarns, and these can be further intertwined into more complex yarn constructions. Continuous production of so-called composite yarns containing central microfiber yarns and nanofiber wrappers is made possible by alternating current electrospinning. Subsequent braiding of these composite yarns allows the protection of nanofiber packages and the variability of the overall construction of the resulting yarns. This paper introduces the creation of yarns and various types of basic as well as braided yarns with nanofibers. Selected sutures were further subjected to in-vitro testing to verify cytocompatibility for application in hygiene such as dental sutures or medicine as surgical sutures.

Keywords: AC electrospinning, composite nanofibrous yarns, braiding, cytocompatibility, dental floss

1. INTRODUCTION

Electrospinning is one of the widely used techniques for the fabrication of polymer-based nanofibers [1,2]. Usually, this technology uses a DC high voltage source to create nanofibrous planner materials [3]. However, for the formation of linear structures, *i.e.*, nanofibrous yarns or composite yarns with a nanofibrous sheath wounded around the classic yarn, it is facile to use an alternating current voltage (AC) source over DC source. During the AC electrospinning process [4], the emerging fibrous plume from the AC electrode can be deposited on an electrically inactive substrate due to the AC nature of the power source. Hence it is much easier to collect the nanofibers on the classic yarns to construct the core-nanofibrous sheath structure using AC electrospinning. In this work, polycaprolactone (PCL) and polyvinyl alcohol (PVA) have been selected due to their excellent biocompatibility and electrospinnability. In addition, low toxic solvents were used to prepare the ideal spinning solutions [5] of PCL or PVA. Composite yarns consisting of a micro-fibrous core and nanofibrous shell (PCL or PVA) can be used in medical or hygiene applications. Despite the inherent biocompatibility of the polymer, the cytotoxicity of the end product needs to be analysed. Hence, the objective of the following study is to prepare PCL or PVA nanofibrous sheath-core (PA6) composite yarn using the combinations of AC electrospinning and braiding technology. Subsequently, we examined the cytotoxicity of the composite yarns and braided composite yarns.

2. MATERIALS AND METHODS

For the composite yarn production, core yarn in multifilament form from polyamide 6 (PA6) twelve trilobal fibers with linear density 44 dtex (Odetka a.s. CZ) was used. The nanofibrous shell was produced from polycaprolactone (PCL; M_n 80.000) or polyvinyl alcohol (PVA; M_n 125.000), both purchased from Merck CZ.

Acetic acid, formic acid, acetone, and HPLC grade water were obtained from Penta CZ. Polymeric solutions were prepared by dissolving the PCL pellets in acetic acid/formic acid/acetone (1/1/1 v/v/v) and PVA in HPLC grade water to prepare 10 wt% solutions of PCL and PVA, respectively.

The fabrication of composite nanofibrous yarn using industrial-scale AC electrospinning is shown in **Figure1**. This device was designed and constructed at the Technical University of Liberec using the following components: 1) high voltage AC power source, 2) solution reservoir which was embedded with screw pump and electrode, and 3) continuously running core yarn (as a collector), 4) twirling devices and 5) heating zones (for removing the residual solvents) [6,7]. The distance between the spinning electrode and the core yarn was set to 250 mm. As mentioned earlier, the moving core yarn was used as an electrically inactive collector for capturing the emerging nanofibrous plume from the electrode. The feeding rate of the core yarn or production rate of the composite yarn was maintained at 10 m/min. From each material, 200 m of composite nanofibrous yarn was produced. All the resultant composite yarns were fabricated using the applied effective AC high voltage of 34.5 kV. Furthermore, these composite yarns were subsequently braided using 8 or 9 individual composite yarns from each material. It is worth mentioning that the braiding machine was slightly modified in order to render the damage to the nanofibrous envelope.

The Linear density of the materials (pristine and braided) was calculated in Decitex using the following formula $(mass\ of\ the\ yarn/length\ of\ the\ yarn) \times 10000$ and corresponding results have been listed in **Table 1**.

Figure 1 Photo of composite yarn production by AC electrospinning (left): 1 - connection to an AC high voltage source; 2 - spinning electrode with a polymer solution and feeding system; 3 - emerging nanofibers; 4 - feeding the core yarn; 5 - withdrawing the resulting composite yarn. Photo of detail to braiding zone, in braiding machine during braiding of 9 yarns (the middle yarn indicated by the arrow).

Surface morphology of the composite (**Figure 2**) and braided yarn (**Figure 3**) were evaluated using TESCAN Vega 3 (Czech Republic) scanning electron microscopy (SEM) and UHR-FE-SEM Zeiss Ultra Plus Microscope (Germany). Prior to SEM analysis, all the samples were gold coated.

Evaluation of the cytotoxicity of the materials was tested (*in-vitro*) with the cell line 3T3-A31 mouse fibroblasts (ATCC). The standard determines the material as cytotoxic if the viability of cells incubated with the extracts of the monitored material in the culture medium reaches values lower than 70% viability of negative control cells (cells incubated in the complete medium=DMEM) [8].

The cells were incubated in complete culturing medium DMEM High Glucose (Dulbecco's Modified Engles Medium, Merck) with 10% Fetal bovine serum (Biosera) and 1% antibiotic - Pen / Strep Amphotericin B (Lonza). For testing, the sterilized samples (using ethylene oxide) with three different lengths were prepared from each material which were 2, 4, and 6 cm. The samples were placed in the 96 well plates, and subsequently, a glass ring was placed on the sample to fix at the bottom of the well.

On the first day, cells were seeded in 96 wells plates (passage 19). The concentration of cells suspension was 10 000 cells per 100 µl (per one well). For preparing the extracts of the sample,1 ml of complete culture medium was added to the sample-loaded well plates (10 wells/sample). The well plates with cells and samples were incubated at 37 °C in a CO_2 incubator for 24 hours. On the second day, the culture medium from the cell-seeded well plate was removed, and subsequently, 100 µl of an extract from the samples was added and then incubated for 24 hrs in a CO_2 incubator. On the third day, the extract from the wells with cells was removed, and subsequently, 100 µl of MTT solution was added and incubated at 37 °C in the CO_2 incubator for 2 hours. After incubation, the MTT solution was removed, and add 100 µl of isopropyl alcohol and mixed by pipette. The UV absorbance was measured at the wavelength of 470 nm (reference was 650 nm) by spectrophotometer (Tescan).

3. RESULTS AND DISCUSSION

The polymeric solutions of PCL and PVA were spinnable using the AC electrospinning. In addition, linear density analysis of the samples reveals that PCL-based composite and braided yarns result in better spinnability than PVA (**Table 1**). From the SEM images (**Figure 2**), it is thus clear that PCL and PVA-based electrospun nanofiber were completely wounded around the PA6 core yarn. Moreover, the utilized braiding process as well did not damage the nanofibrous envelope of the braided composite yarns, which was also seen from the SEM images (**Figure 3**). Hence, the SEM and linear density analysis of the composite and braided composite yarn shows that the prime objective of the following study was successfully achieved in terms of fabricating the composite and braided composite yarn without damaging the nanofibrous envelope.

Table 1 Overview of the tested yarns with their final linear densities

Sample	Linear density [dtex]
PA6 core yarn	46±0.7
PA6 8 braided yarns	362±0.7
PA6/PCL composite yarn	112.6±0.5
PA6/PCL 8 braided composite yarns	1040±1.1
PA6/PCL 9 braided composite yarns	1121.8±1.6
PA6/PVA composite yarn	70.6±1.5
PA6/PVA 8 braided composite yarns	563±2

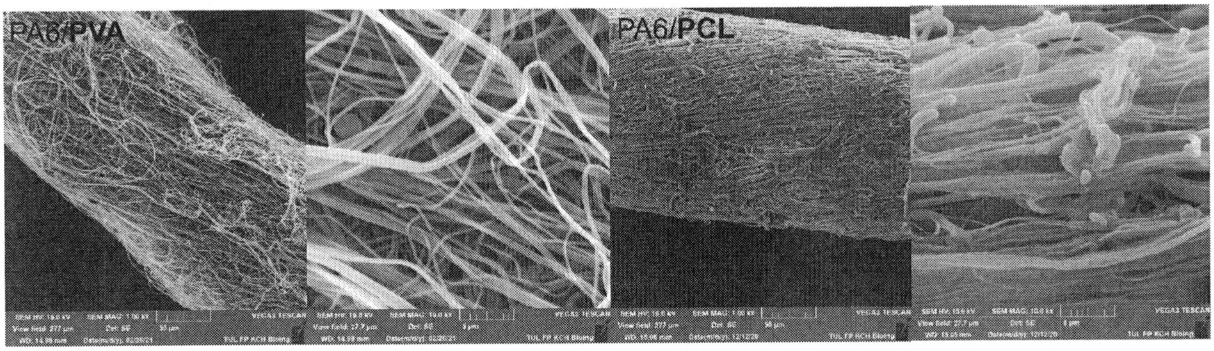

Figure 2 SEM images of the surface of composite yarns with AC electrospun nanofibrous sheath from PVA (left) and PCL (right)

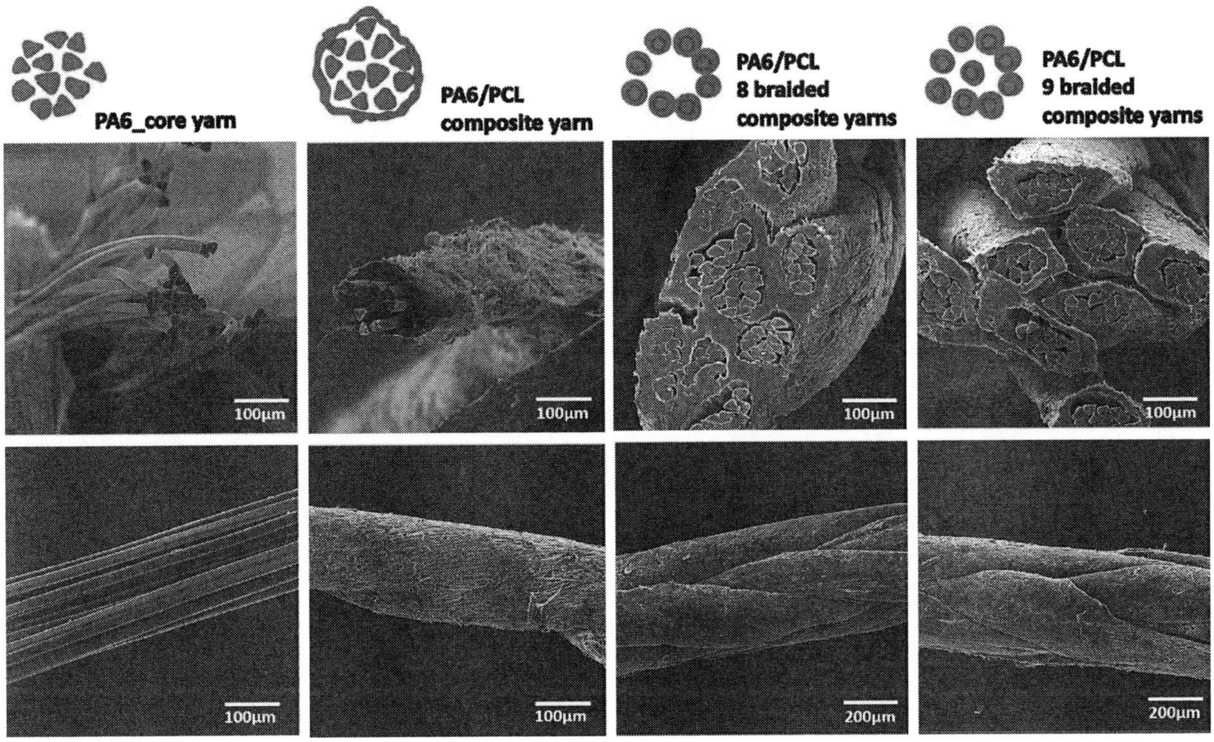

Figure 3 Scheme (above) and SEM images (cross-section view in middle and longitudinal view bellow) of microfibrous multifilament core yarn (PA6) and composite yarn (PA6/PCL) and braided composite yarns from eight and nine composite yarns

It is well known that the PCL and PVA were biocompatible. However, the MTT metabolic assay was performed whether the used polymer processing technologies (solution, AC electrospinning, and braiding) is induced any cytotoxicity on the fabricated materials. The outcomes of the MTT assay were evaluated according to the standard ČSN EN ISO 10993-5 (85 5220), and the results are shown in **Figure 4.** All the fabricated samples were not shown cytotoxic effects on the 3T3-SA mouse fibroblasts cell with more than 80% cell viability. In that, 2 cm length of all the samples were shown higher cell viability than that of 4 and 6 cm. It could be attributed to the leaching of the small amount of residual solvent from the length of the 2 cm sample. In contrast, a higher sample length may leach out a probably higher amount of residual solvents, which may slightly reduce the viability of the cells. Thus, the MTT assay shows that all the fabricated and braided composite yarn was biocompatible with the 3T3-SA mouse fibroblasts cell. Adhesion and proliferation of the cells on these materials for a more extended time will be evaluated in the future. Due to its biocompatibility, these samples could be used as dental threads or surgical sutures in medical fields.

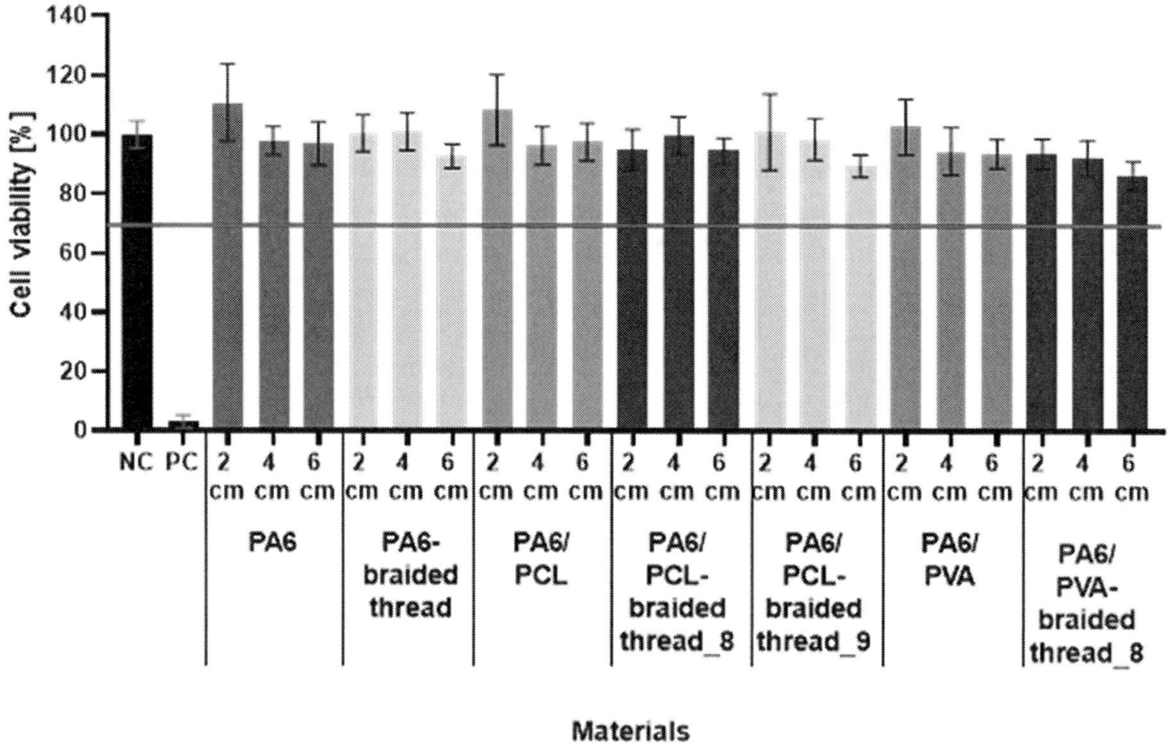

Figure 4 Bar graph of cell viability measured by metabolic assay (MTT) after culturing 3T3-SA mouse fibroblasts with the extract of pristine core yarn, composite nanofibrous yarns, and braided composite yarns. From each material were tested three different lengths (2cm, 4cm, and 6cm). NC is negative control and PC positive control.

4. CONCLUSION

The article presented the manufacturing and testing of composite yarns and braided composite yarns formed by a microfiber core and a nanofiber sheath in several material combinations. The resultant yarns were subsequently tested for possible cytotoxicity or cytocompatibility through cell viability testing using metabolic MTT assay. The results of the tests significantly show that the technological steps of the production of composite yarns and the subsequent braiding did not bring any cytotoxic effect to the resulting yarns; hence all the materials were biocompatible. Therefore, the fabricated materials can be considered as suitable carriers for specific active substances according to specific applications, such as surgical sutures or dental flosses in the medical field.

ACKNOWLEDGEMENTS

This work was supported by the project Technology Agency of the Czech Republic TACR Trend No. FW02020042.

REFERENCES

[1] BARHOUM, A., PAL, K., RAHIER, H., ULUDAG, H., KIM, I. S., BECHELANY, A M. Nanofibers as new-generation materials: From spinning and nano-spinning fabrication techniques to emerging applications. *Applied Materials Today*. 2019, vol. 17, pp. 1-35. Available from: https://doi.org/10.1016/j.apmt.2019.06.015.

[2] KENRY, LIM, C. T. Nanofiber technology: current status and emerging developments. *Progress in Polymer Science*. 2017, vol. 70, pp. 1-17. Available from: https://doi.org/10.1016/j.progpolymsci.2017.03.002.

[3] LIN, T., WANG, A X. *Needleless electrospinning of nanofibers: technology and applications*. Singapore: Pan Stanford Publications, 2014.

[4] POKORNY, P. *et al.* Effective AC needleless and collectorless electrospinning for yarn production. *Phys. Chem. Chem. Phys.* 2014, vol. 16, no. 48, pp. 26816-26822. Available from: https://doi.org/10.1039/C4CP04346D.

[5] SIVAN, M. *et al.* Plasma treatment effects on bulk properties of polycaprolactone nanofibrous mats fabricated by uncommon AC electrospinning: A comparative study. *Surface and Coatings Technology*. 2020, vol. 399, p. 126203. Available from: https://doi.org/10.1016/j.surfcoat.2020.126203.

[6] VALTERA, J. *et al.* Fabrication of dual-functional composite yarns with a nanofibrous envelope using high throughput AC needleless and collectorless electrospinning. *Sci Rep*.2019, vol. 9, no. 1, p. 1801. Available from: https://doi.org/10.1038/s41598-019-38557-z.

[7] AMLER, E. *et al.* Method for Production of Polymeric Nanofibers by Spinning of Solution or Melt of Polymer in Electric Field, and a Linear Formation from Polymeric Nanofibers Prepared by This Method. 2014. WO2014094694A1.

[8] HRAZDIL, J., RYCHNOVSKY, A L. Biological evaluation of medical devices - Part 1: Evaluation and testing; ČSN EN ISO 10993-1. Český normalizační institut. [Online]. 2004. [Accessed 13 July 2021]. Available from: https://doi.org/http://csnonlinefirmy.agentura-cas.cz/html_nahledy/85/70508/70508_nahled.htm

FACILE FABRICATION OF Ti_3C_2 MXENE NANOSHEETS AND THEIR PHOTOTHERMAL PROPERTIES

[1]Israt ALI, [1]M. FARAZ UD DIN, [1,2]Anna KÁLOSI, [1,2]Adriana ANNUŠOVÁ, [3]Martina LABUDOVÁ, [4]Yaryna SOYKA, [4]Mária OMASTOVÁ, [1,2]Matej JERGEL, [1,2]Peter ŠIFFALOVIČ, [1,2]Eva MAJKOVÁ

[1]Institute of Physics, Slovak Academy of Sciences, Bratislava, Slovakia, EU, fyziisra@savba.sk

[2]Centre for Advanced Material Application, Slovak Academy of Sciences, Bratislava, Slovakia, EU, eva.majkova@savba.sk

[3]Institute of Virology, Biomedical Research Center, Slovak Academy of Sciences, Bratislava, Slovakia, EU, martina.labudova@savba.sk

[4]Polymer Institute, Slovak Academy of Sciences, Bratislava, Slovakia, EU, maria.omastova@savba.sk

https://doi.org/10.37904/nanocon.2021.4359

Abstract

Recently discovered two-dimensional (2D) transition metal carbides and nitrides (MXenes) have received tremendous attention because of their unique electrical, optical and chemical properties. These exceptional properties make them a suitable candidate for a variety of applications including multimodal tumor therapy by photothermal effect. In this work, we demonstrate how to reduce the size of 1-5 μm large Ti_3C_2 monolayer MXene sheets into ultrasmall 100-160 nm nanosheets by applying consecutive ultrasonication processes. Different microscopic techniques have been used to visualize the formation of ultrasmall single-layer Ti_3C_2 nanosheets. The as-prepared MXene nanosheets have shown good solubility in water and ethanol. Further, (3-aminopropyl) triethoxysilane (APTES) and poly(3,4-ethylene dioxythiophene) polystyrene sulfonate (PEDOT:PSS) were utilized for surface modification of the MXene nanosheets to open the possibility of subsequent antibody bio-conjugation. Moreover, PEDOT:PSS improved the photothermal conversion performance of the nanosheets as documented by increasing their temperature from 48.6 °C to 58.1 °C on irradiation by 808 nm wavelength laser. Further in vivo and in vitro studies will be necessary to optimize the photothermal properties of Ti_3C_2 nanosheets.

Keywords: Ti_3C_2 nanosheets, ultrasonication, surface modification, photothermal effect, cell viability

1. INTRODUCTION

Photothermal therapy (PTT) emerged as a non-invasive therapeutic strategy that can kill cancer cells through hyperthermia by converting photon energy into heat energy [1]. Different noble metal nanoparticles, nanorods, nanostars and 2D nanosheets of graphene, reduced graphene and black phosphorous have been utilized as photoabsrobers for PTT application [2-4]. MXenes first described in 2011 are a new class of 2D transitional metal carbides, nitrides, and carbonitrides obtained by etching and delamination of MAX phases [5,6]. MXenes have the formula $M_{n+1}X_nT_x$ where M is the transition metal, X is C and/or N and T is a functional group e.g., -O, -F, -OH [5]. During this short time, different materials of MXenes have been discovered. Owing to ease in surface modification and excellent near-infrared light-absorbing capability, Ti_3C_2 nanosheets distinguish themselves from other MXene nanomaterials [7].

Several synthetic methods have been developed for the production of the biocompatible size range (50-160 nm) of Ti_3C_2 nanosheets. Xuan et al, utilized an organic base tetramethylammonium hydroxide (TMAOH) to etch and delaminate multilayered Ti_3C_2 MXene sheets [8], while Yang et al obtained Ti_3C_2 nanosheets with the hydrodynamic size range of 91.7 nm by using Tetrapropylammonium hydroxide (TPAOH) organic base [9].

The usage of organic bases can cause complications in biomedical applications. It is still a great challenge to reduce the size of Ti_3C_2 MXene nanosheets by only physical means within a short time while preserving the effect of PTT.

Further, compared to traditionally used gold nanomaterials for PTT, Ti_3C_2 MXene nanosheets have also shown the Localized Surface Plasmon Resonance (LSPR) effect. Ti_3C_2 (MXene) nanosheets, having transition metal element (titanium), exhibit a strong NIR absorption as well as subsequent light-to-heat conversion property enabling effective application in PTT. In this regard, Liu et al. have proved that the LSPR effect of Ti_3C_2 MXene nanosheets can be enhanced by introducing Al-containing moieties in them [10]. Lin et al, in their pioneering work, presented preparation of few-layer T_3C_2 MXene nanosheets and their successful application as PTT agent. This work points at the great potential of Ti_3C_2 nanosheets (MXenes) as a novel photothermal agent used for cancer therapy [11]. Complexation in synthetic methods and increase in toxicity of Ti_3C_2 nanosheets required complex surface modifications that cause limitation of such a novel material in biomedical application. Taking these effects into account, cytotoxicity and stability are the two major concerns for the application of Ti_3C_2 nanosheets in PTT. Hence, a suitable Ti_3C_2 surface modification is required to attain improved LSPR effect, lower toxicity, and high stability in physiological systems [12].

Herein, we have successfully achieved a 100-160 nm size range of Ti_3C_2 nanosheets without the addition of any organic bases. Ultrasonication methods were utilized for the cutting of 1-5 µm single layer Ti_3C_2 MXene sheets into 100-160 nm Ti_3C_2 nanosheets. Further, (3-aminopropyl) triethoxysilane (APTES) were functionalized on the surface of Ti_3C_2 MXene nanosheets that further attached with NHS-PEG-Biotin through a chemical bond. This surface modification not only reduces the toxicity of Ti_3C_2 nanosheets but also causes great stability in different solvents including water, ethanol, and phosphate buffer solution (PBS). Interestingly, the LSPR effect of Ti_3C_2 MXene nanosheets is enhanced by utilizing poly(3,4-ethylenedioxythiophene) polystyrene sulfonate (PEDOT:PSS), which can further be attached with poly(ethylene glycol) moieties and can be utilized for the NIR-II PTT effect.

2. EXPERIMENTAL

2.1. Preparation of Tl_3C_2 nanosheets

Single layer micro size Ti_3C_2 MXene sheets (1-5 µm) were synthesized by using mixed acids (HCl+LiF) to etch Aluminum (Al) from Ti_3AlC_2 [13]. The cutting of as prepared larger Ti_3C_2 sheets into smaller ones was performed by utilizing subsequent bath and probe sonication. First, bath sonication (Sonorex RK 510 H, 35 kHz, Bandelin) was applied for 24 hfollowed by 4 h probe sonication (Sonopuls HD 2070, 20 kHz, Bandelin) placed in the ice bath. The nanosheets were centrifuged (Model 3-30K, Sigma centrifuge) with 8000 RPM for 40 min at 20 °C. The residue was collected and vacuum dried overnight. Next, the fully dried nanosheets were dissolved in 5 ml ethanol. The nanosheets were characterized by Atomic Force Microscopy (Bruker, Multimode 8), Scanning electron microscopy (manufacturer), Zetasizer Nano ZS 90 (Malvern Panalytical).

2.2. Surface modification of Ti_3C_2 nanosheets

First, we have prepared amino-functionalized MXene sheets (MXene-NH_2). 5 ml MXene solution (0.4 mg/mL) and 10 µl APTES (Sigma Aldrich) dissolved in 1 ml ethanol were mixed on a laboratory shaker for 24h. Next, 1 mg of NHS-PEG-Biotin (Sigma Aldrich, 1469.72 MW) was dissolved in 1 ml ethanol and poured into the MXene-NH_2 solution. The mixture was sonicated for 24 h, followed by vacuum drying overnight. The obtained solvent-free MXene-PEG-biotin powder was dissolved in PBS solution and stored for further use at 4 °C.

Secondly, we have prepared PEDOT:PSS modified nanosheets (MXene-PEDOT:PSS). Similarly as in the previous functionalization procedure, the PEDOT:PSS (50 µL, OSSILA) was mixed with Ti_3C_2 nanosheets in deionized water and left on a laboratory shaker for 24 h. Next, the solution was centrifugated at 10.000 RPM for 20 min, then the residue was collected and dissolved in DI water for further use.

2.3. Photothermal effect

The absorbance measurements were realized by a SolidSpec-3700 UV-VIS-NIR spectrophotometer (Shimadzu, Japan), using an integrating sphere. Deionized water was used for the baseline correction before the measurements and also as a reference sample during the measurements. 100 µl of pure and surface modified Ti_3C_2 nanosheets were used in a transparent quartz cell and irradiated with an 808 nm laser (0.430 W). The temperature changes were recorded with a K-type thermocouple coupled to an Extech SDL200 data-logging thermometer.

2.4. Cell viability test

The HCT116 (CCL-247, human colorectal carcinoma) cells were seeded into 96 well plates in a concentration of 10 thousand cells per well. For each sample, we have used 6 parallel wells. Cells were cultivated in DMEM medium with stable 2 mM L-glutamine (Biochrom) supplemented with 10% fetal calf serum (FCS, Biochrom) and gentamicin 80 µg/ml (Lek). Cells were incubated in a humidified atmosphere at 37 °C in the presence of 5 % CO2 for 24 h. The cells were moved to hypoxic workstation Ruskinn Invivo2 300 (Ruskin) with 1 % O_2 and 5 % CO_2 for 24 h to induce the expression of CAIX protein and other hypoxia-inducible proteins. After 24 h in hypoxia, 40 µl of the samples were added to the media and incubated further for 24 h or 48 h. Cell viability assay was done by Cell Titer blue viability assay (Promega) according to manufacturer's instructions. 20 µl of Cell titer blue reagent were added and incubated for 1 hour at 37 °C in the presence of 5 % CO_2. The fluorescence was measured at 590 nm (Synergy H4 microplate reader, BiTek) and the fluorescence and percentage of viable cells were determined. Each sample analysis was done in hexaplets and the standard deviation and T-test were determined.

3. RESULTS AND DISCUSSION

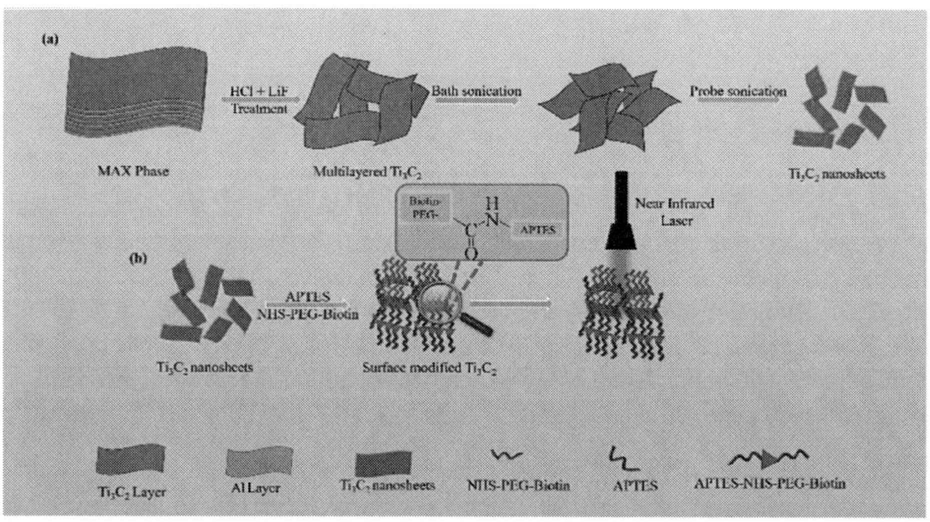

Figure 1 (a) Schematic illustration of a synthetic process for Ti_3C_2 MXene nanosheets including mixed acids-based etching, ultrasonication exfoliation, and cutting. (b) Subsequent APTES-PEG-Biotin surface modification of Ti_3C_2 nanosheets and their photothermal response to NIR light.

The scheme of the preparation of the MXene nanosheets is illustrated in **Figure 1(a)**. To achieve smaller size Ti_3C_2 MXene nanosheets compatible with the biomedical application, two subsequent ultrasonication processes were applied on the MXene sheets. Traditionally the liquid phase exfoliation method uses sonication to create the force necessary to exfoliate 2D materials. Since, in our experiments, we apply sonication on already single- and few-layer MXene sheets. Hence, further mechanical agitation by sonication is leading to size reduction. Firstly, the bath sonication of the larger Ti_3C_2 MXene sheets (1-5 μm) causes the cutting of the sheets into 500-700 nm diameter sheets **Figure 2(a)**. Then probe sonication further reduces the nanosheets into considerably smaller sizes (100-160 nm). The size of the nanosheets both after bath and probe sonication was confirmed by AFM (see **Figure 2 (b)**). The size distribution was estimated from SEM micrographs taken from drop-casted samples on the silica substrate of the MXene solution (see **Figures 2(c-d)**). The SEM images were evaluated by ImageJ software to acquire the lateral size of the sheets. The mean value of the lateral size is 100-160 nm **Figure 2(d)**. From the size distribution study of the sonicated MXene solution, we can conclude that it was required to combine both bath and probe sonication for proper size reduction.

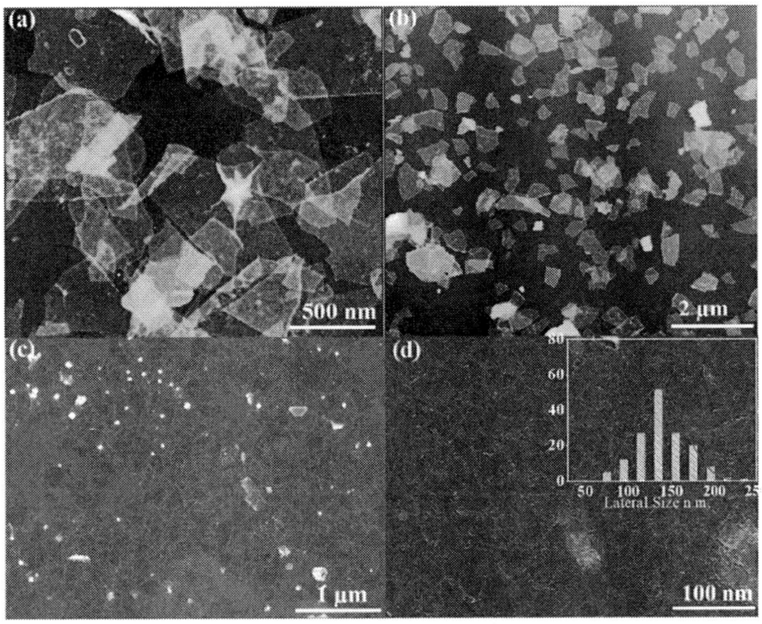

Figure 2 AFM images of (a) after 24 h bath sonicated Ti_3C_2 MXene nanosheets, and (b) bath (24 h) and probe (4 h) sonicated Ti_3C_2 nanosheets. SEM images of (c) 24 h bath sonicated Ti_3C_2 MXene nanosheets, and (d) 24 h bath and 4 h probe sonicated Ti_3C_2 MXene nanosheets: Inset is showing the size distribution of 150 nanosheets done by using ImageJ software.

The MXene nanosheets were functionalized to further explore their potential in biomedical application (see **Figure 1(b)**). We performed amino-functionalization via APTES, which resulted in $-NH_2$ terminals on the surface. Further, the MXene-NH_2 was PEGylated since polyethylene glycol is well known for its protein resistance and biocompatibility. Moreover, the biotin on the PEG terminal offers the opportunity for subsequent antibody bio-conjugation. Furthermore, we also prepared PEDOT:PSS functionalized nanosheets. The prepared MXene nanosheets before and after functionalization have shown high stability. The zeta-potential of as-prepared and surface modified MXene nanosheets was -50.8 mV and -45.3 mV respectively.

To verify that our treatment did not diminish the optical and photothermal properties of the MXene sheets, we studied them with UV-VIS spectrophotometry and checked the photothermal performance at 808 nm. From the UV-vis absorption spectra, it can be seen clearly that Ti_3C_2 nanosheets show absorption in the NIR region **Figure 3(a)**. Further, the optical absorption scaled linearly with a concentration upon the dilution of pure Ti_3C_2 MXene nanosheets. The absorption properties of optically active materials are dependent on size and surface chemistry. The modification of the surface of Ti_3C_2 MXene nanosheets with APTES/NHS-PEG-Biotin and PEDOT:PSS resulted in the broadening of the absorbance peak in the NIR region **Figure 3(b)**. These absorbance properties of pure nanosheets and surface-modified Ti_3C_2 nanosheets suggest that the presented materials could be potential photothermal agents.

The strong absorption of Ti_3C_2 nanosheets in the NIR range made possible their application as a photothermal agent in cancer therapy [8]. It can be seen clearly in **Figures 3(c,d)** that both pure MXene and MXene-PEDOT:PSS nanosheets have very good photothermal cycling stability under 808 nm laser irradiation. The temperature of pure Ti_3C_2 nanosheets dispersion was increased to about 20°C within the first 4 min of irradiation **Figure 3(e)**, and the highest temperature achieved was 48.6°C. However, it was surprising to observe that by mixing PEDOT:PSS with Ti_3C_2 nanosheets the temperature increase was faster and higher to about 10°C as compared to pure Ti_3C_2 nanosheets **Figure 3(f)**. Hence, the manipulation of photothermal properties of MXene nanosheets can be done by modifying their surfaces with different surfactants.

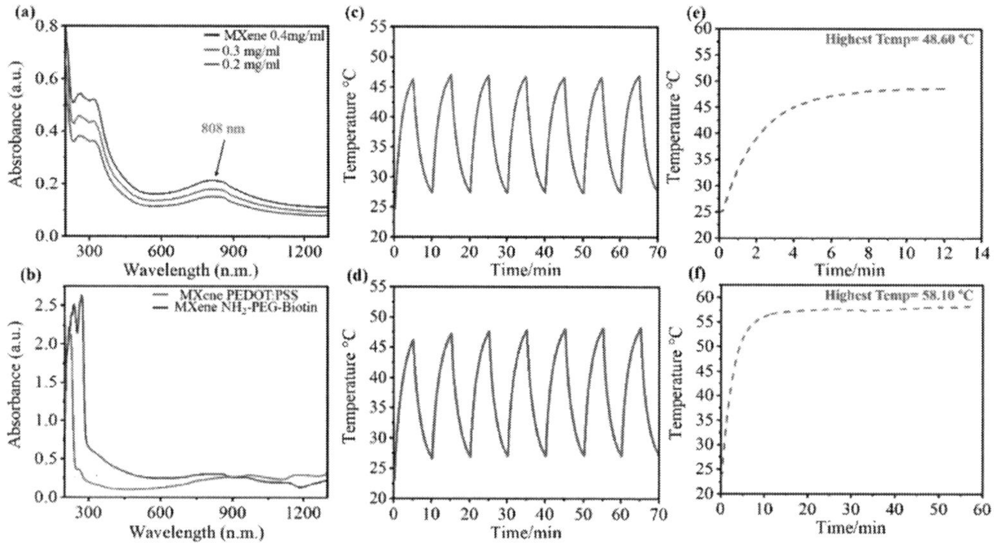

Figure 3 UV-visible absorption spectra of (a) MXene nanosheets and with two dilutions, (b) surface modified MXene with PEDOT:PSS and APTES/NHS-PEG-Biotin. Photostability test of (c) pure, and (d) MXene-PEDOT:PSS nanosheets in DI water under 808 nm laser irradiation. Photothermal performance of (e) pure, and (f) MXene-PEDOT:PSS nanosheets.

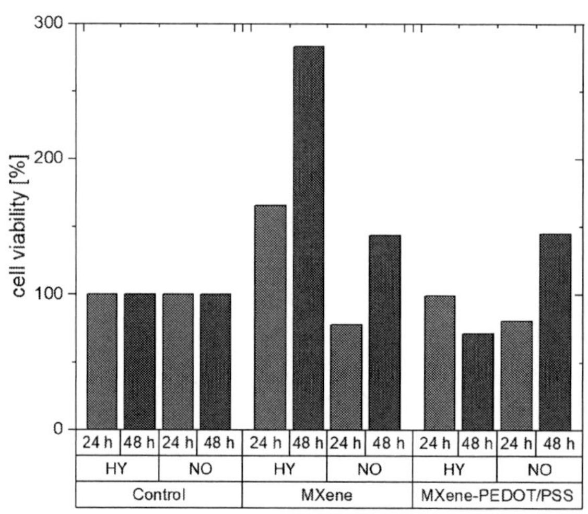

Figure 4 Cytotoxicity of MXene and MXene-PEDOS:PSS nanosheets regarding HCT-116 cell line

Besides having outstanding photothermal properties, the other key feature for biomedical nanomaterial design is the cytotoxicity of the material. Our results in **Figure 4** revealed that pure Ti_3C_2 and MXene-PEDOS:PSS nanosheets do not cause any damage to the cells and therefore are nontoxic. In normoxic (21 % O_2) conditions after 24 h, the viability for pure Ti_3C_2 nanosheets is 78 % while for PEDOT:PSS modified Ti_3C_2 nanosheets results out 80 %, after 48 h it is even more than the control. In hypoxic (1 % O_2) conditions after 24 h and 48 h, pure Ti_3C_2 nanosheets have even better viability than the control, and the values of MXene-PEDOT:PSS are similar to the control.

4. CONCLUSION

In summary, an organic solvent-free ultrasonication-based approach was designed to get a highly uniform and biocompatible size range of Ti_3C_2 nanosheets (100-160 nm). The cutting of larger size (1-5 µm) Ti_3C_2 MXene sheets into the anticipated smaller size was achieved by consecutive bath and probe sonication. The attained Ti_3C_2 MXene nanosheets were stable in water for more than three weeks. The surface modification of Ti_3C_2 nanosheets by NHS-PEG-Biotin and PEDOT:PSS were studied. While the usage of NHS-PEG-Biotin can later easily facilitate a specific antibody's binding, in the case of PEDOT:PSS surface functionalization, we observed enhanced photothermal conversion performance under 808 nm irradiation. Cell viability tests have shown that Ti_3C_2 nanosheets and PEDOT:PSS modified Ti_3C_2 nanosheets are nontoxic to the tested cell lines. We believe that this work will provide a feasible approach for the fabrication of biomedical application-worthy Ti_3C_2 nanosheets.

ACKNOWLEDGEMENTS

Financial support of the project APVV-19-0465 is acknowledged. This work was performed during the implementation of the project Building-up Centre for advanced materials application of the Slovak Academy of Sciences, ITMS project code 313021T081 supported by the Integrated Infrastructure Operational Programme funded by the ERDF.

REFERENCES

[1] HUANG, X., JIAN, P.K., El-SAYED, I.H., El-SAYED, M.A. Plasmonic photothermal therapy (PPTT) using gold nanoparticles. *Lasers in medical science*. 2008, vol. 23, no. 3, pp. 217-228.

[2] LIU, S., PAN X., LIU, H. Two-dimensional nanomaterials for photothermal therapy. *Angewandte Chemie*. 2020, vol. 132, no. 15, pp. 5943-5953.

[3] VERGER, L., NATU, V., CAREY, M., BARSOUM, M.W. MXenes: an introduction of their synthesis, select properties, and applications. *Trends in Chemistry*. 2019, vol. 1, no. 7, pp. 656-669.

[4] ANASORI, B., GOGOTSI, Y. Introduction to 2D transition metal carbides and nitrides (MXenes). In: 2D Metal Carbides and Nitrides (MXenes). *Springer*. 2019, pp. 3-12.

[5] SALIM, O., MAHMOUD K.A., PANT, K.K., JOSHI, R.K. Introduction to MXenes: synthesis and characteristics. *Materials Today Chemistry*. [online]. 2019, vol. 14. Available from: https://doi.org/10.1016/j.mtchem.2019.08.010.

[6] ZHAN, X., SI, C., ZHOU, J., SUN, Z. MXene and MXene-based composites: synthesis, properties and environment-related applications. *Nanoscale Horizons*. 2020, vol. 5, no. 2, pp. 235-258.

[7] CHEN, J., HUANG, Q., HUANG, H. Recent progress and advances in the environmental applications of MXene related materials. *Nanoscale*. 2020, vol. 12, no. 6, pp. 3574-3592.

[8] XUAN, J., WANG, Z., CHEN, Y. DUJUAN, L.L.C., XIAOJING, C., ZHUANG, L., RENZHI, M., TAKAYOSHI, S., FENGXIA, G. Organic-Base-Driven Intercalation and Delamination for the Production of Functionalized Titanium Carbide Nanosheets with Superior Photothermal Therapeutic Performance. *Angewandte Chemie*. [online]. 2016, vol. 128, no. 47, pp. 14789-14794. Available from: https://doi.org/10.1002/ange.201606643.

[9] LI, Z., ZHANG, H., HAN, J., CHEN, Y., LIN, H., YANG, T. Surface Nanopore Engineering of 2D MXenes for Targeted and Synergistic Multitherapies of Hepatocellular Carcinoma. *Advanced Materials*. [online]. 2018, vol. 30, no. 25. Available from: https://doi.org/10.1002/adma.201706981.

[10] LIU, G., ZOU, J., TANG, Q., XIAOYAN, Y., YEWEI, Z., QI, Z., WEI, Huan., PENG, C., JINJUN, S., XIAOCHEN, D. Surface Modified Ti_3C_2 MXene Nanosheets for Tumor Targeting Photothermal/Photodynamic/Chemo Synergistic Therapy. *ACS Applied Materials and Interfaces*. [online]. 2017, vol. 9, no. 46, pp. 40077-40086. Available from: https://doi.org/10.1021/acsami.7b13421.

[11] LIN, H., WANG, X., YU, L., CHEN, Y., SHI, J. Two-dimensional ultrathin MXene ceramic nanosheets for photothermal conversion. *Nano letter*. 2017, vol. 17, no. 1, pp. 384-391.

[12] MALAKI, M., MALEKI, A., VARMA, R.S. MXenes and ultrasonication. *Journal of Materials Chemistry A*. [online]. 2019, vol. 7, no. 18, pp. 10843-10857. Available from: https://doi.org/10.1039/c9ta01850f.

[13] LIU, F., ZHOU, A., CHEN, J., JIA, J., ZHOU, W., WANG, L., HU, Q. Preparation of Ti_3C_2 and Ti_2C MXenes by fluoride salts etching and methane adsorptive properties. *Applied Surface Science*. 2017, vol. 416, pp. 781-789.

EFFECTS OF CHITOSAN/CHONDROITIN SULFATE POLYSACCHARIDE LIDS ON AMOXICILLIN DRUG RELEASE FROM MICROCONTAINERS

[1,2]Ilkay KARAKURT, [3]Khorshid KAMGUYAN, [3]Carmen Milian GUIMERA, [1]Kadir OZALTIN, [3]Anja BOISEN, [1,4]Marian LEHOCKY

[1]*Centre of Polymer Systems, University Institute, Tomas Bata University in Zlin, Zlin, Czech Republic, EU, ykarakurt@utb.cz, ozaltin@utb.cz, lehocky@utb.cz*

[2]*Department of Health Technology, DTU Health Tech, Technical University of Denmark, Lyngby, Denmark, EU, ilkara@dtu.dk*

[3]*The Danish National Research Foundation and Villum Foundation's Center for Intelligent Drug Delivery and Sensing Using Microcontainers and Nanomechanics (IDUN), Technical University of Denmark, Lyngby, Denmark, EU, khokam@dtu.dk, aboi@dtu.dk*

[4]*Faculty of Technology, Tomas Bata University in Zlín, Zlín, Czech Republic, EU, lehocky@utb.cz*

https://doi.org/10.37904/nanocon.2021.4352

Abstract

Microcontainers are micrometer-sized cylindrical reservoir-based devices serving as drug carriers for oral administration systems with ensuring a high drug concentration at the desired site of absorption and providing a sustained resident time. These polymeric micro-cylinders consist of drug-filled cavities and can be functionalized by sealing with polymeric lids to protect the active pharmaceutical ingredient against degradation before reaching the absorption site. The present study aimed to develop biopolymer-based lids for microcontainers using polysaccharides for the prolonged release of an antibiotic. Amoxicillin was chosen as a model drug and loaded into microcontainers using the polydimethylsiloxane masking technique, followed by spray coating with chitosan (CS) and chondroitin sulfate (ChS) as polymeric lids onto the cavity of the devices. The structural and physicochemical properties of the microcontainers were investigated. Moreover, the effect of polysaccharide-based lids on the drug release behavior was further analysed and discussed. The thickness and morphology of the lids were characterized by profilometry and SEM before and after spray coating. The *in vitro* release profiles revealed a fast and an immediate release of amoxicillin from only ChS-coated microcontainers as in the uncoated ones, and slow, prolonged release from only CS-coated ones. However, the microcontainers with CS-ChS lids showed a biphasic release pattern that consists of a lesser burst release in the early stage of treatment, followed by a more sustained release. Therefore, the formation of polyelectrolytes between CS and ChS plays an important role in controlling the release process. The results suggest that the CS-ChS based lids for microcontainers can be useful in biodegradable, tailor-made controlled drug delivery systems.

Keywords: Microdevices, polysaccharide lids, spray coating, drug delivery, sustained release

1. INTRODUCTION

Bacterial infections have been one of the most prevalent and serious threats to human health for decades [1]. Faced with such a severe situation, controlled use of antibiotics still seems to be the most effective method in infection control [2]. Amoxicillin (α-amino-hydroxybenzylpenicillin), an aminopenicillin antibiotic with broad-spectrum antibacterial activity, is used to treat several infections resulted from most gram-positive and gram-negative bacterial pathogens [3]. However, the main limitations associated with this antibiotic are possessing a short half-life and the need for higher doses to effectively show a therapeutic activity which causes severe side effects in turn [4,5]. In order to avoid overuse of antibiotics and to improve their pharmacokinetic properties, encapsulation into liposomes, emulsions, polymeric micro-and nanoparticles are being actively developed [6].

Recently, the advances in reservoir-based microdevices such as microcontainers held great promises as oral dosage forms toward delivering the pharmaceuticals to the absorption site, providing a unidirectional release, and improving oral bioavailability [7]. In addition, with the functionalization by polymeric lids, controlled drug release profiles and additional mucoadhesion can be achieved [8].

Among natural polysaccharides, chitosan and chondroitin sulfate have been widely used as release modifiers in drug delivery applications owing to their unique properties, including biodegradability, antibacterial activity, and excellent biocompatibility [9]. Chitosan acts as a polycation due to the free NH_2 groups in its structure, which in turn reacts with negatively charged substances such as chondroitin sulfate, that consists of SO_4^{2-} and COO^- groups, making it an optimal candidate to form a polyelectrolyte complex (PEC) [10,11].

In this study, a BCS class I antibiotic was loaded into microcontainers through the polydimethylsiloxane (PDMS) masking technique and were coated with the polysaccharides by using the Layer-by-layer (LBL) spraying technique. The microcontainers were functionalized by applying two layers as lids on top of the microcontainers, the first layer was either chitosan or chondroitin sulfate. The amoxicillin loaded and polysaccharide-coated microcontainers were evaluated in vitro for morphology and drug release properties.

2. MATERIALS AND METHODS

Silicon (Si) wafers (4-in, b100N n-type) were obtained from Okmetic (Vantaa, Finland), and SU-8 2075 and SU-8 developers were purchased from Microresist Technology GmbH (Berlin, Germany). Low molecular weight chitosan (50-190 kDa) with 75-85% degree of deacetylation, chondroitin sulfate A sodium salt (60% balance is chondroitin sulfate C), acetic acid (100.5%) and PBS tablets were purchased from Sigma Aldrich (Prague, Czech Republic). Sylgard® 184 two-part Silicone Elastomer (viscosity (mixed) = 3500 cP, specific gravity (cured) = 1.03) was obtained from Dow Corning (MI, USA). Amoxicillin trihydrate was provided from Tokyo Chemical Industry (Tokyo, Japan) and Milli-Q deionized water was supplied by a MilliQ Integral Water Purification System for Ultrapure Water, produced by Merck Millipore (Burlington, MA, USA).

2.1. Microcontainers Fabrication

Microcontainers were produced on silicon wafers using the negative epoxy-based photoresist SU-8 with two steps of photolithography technique [12]. Briefly, SU-8 was dispensed onto Si wafers and spin coated in two steps. To remove the solvent from the resulting SU-8 film, a soft-baked step was involved and followed by selective exposure to UV light through a chromium mask. The polymerization of exposed walls was induced by post-exposure baking. Defining the sidewalls of microcontainers were achieved with a second spin coating of SU-8, UV lithography, and baking steps. After immersion in propylene glycol methyl ether acetate (PGMEA), and rinsing in isopropyl alcohol, microcontainers containing wafers were cut into 1.2 × 1.2 cm chips using an automatic dicing saw (DAD 321, DISCO, Tokyo, Japan).

2.2. Drug Loading

The microcontainers were manually filled with amoxicillin using the PDMS-masking technique [13] (**Figure 1**). Briefly, resin and curing agent of Sylgard® 184 were mixed (10:1 w/w ratio) and slowly poured in microcontainer-containing wells in the chip tray. After the gaps between microcontainers fully covered by PDMS, the chip tray was put into the oven and kept 30 min at 70 °C for curing. Amoxicillin was filled into the microcontainers using a spatula to press onto the powder and excess drug was swept using a soft brush. The PDMS mask was gently peeled off by pulling through the edges and detached from the Si wafer bearing the drug-loaded microcontainers. The amount of the drug was calculated by weighing chips before and after loading.

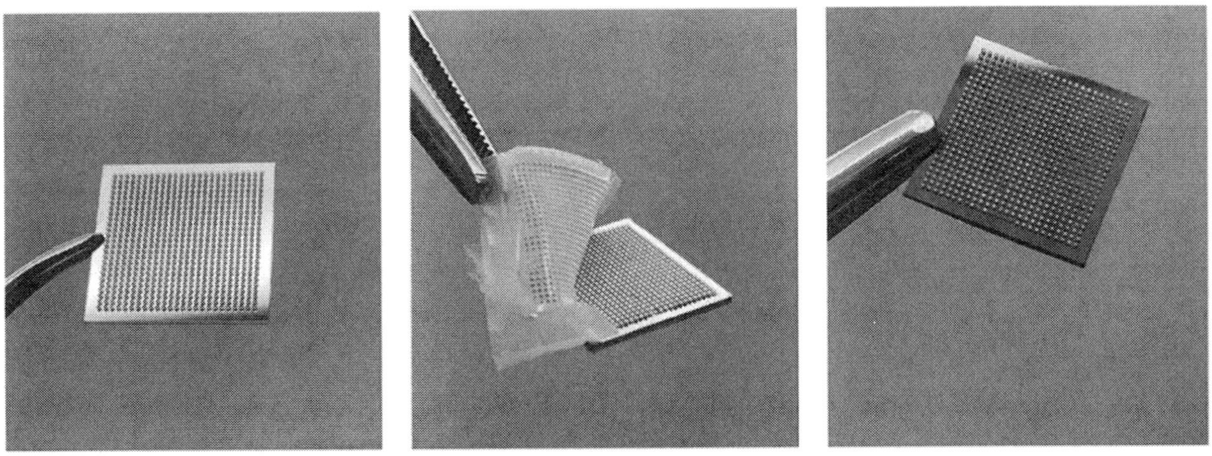

Figure 1 Empty microcontainers on Si chip (a), peeling off PDMS masking after drug loading (b), drug-loaded microcontainers (c)

2.3. Spray Coating

Amoxicillin-loaded microcontainers were coated with either CS, ChS or their combination by using an ultrasonic spray coater (ExactaCoat, Sono Tek, USA) with an Accumist nozzle. For spray coating, 0.5% (w/v) concentration of chondroitin sulfate in Milli Q water and 0.5% (w/v) chitosan solution in 0.1M acetic acid were prepared separately. The solutions were sprayed onto the microcontainers with a 1.3 W generator power and coated in a total of 100 loops.

2.4. Morphological Characterization

The microcontainers were characterized before and after loading and coating processes using a TM3030Plus tabletop scanning electron microscope (SEM, Hitachi High Technologies Europe GmbH, Germany) with mounting on a metallic holder with a tilt of 30° and analyzed at a voltage of 15 keV.

The dimensions of the empty microcontainers and the thickness of the coating were measured using optical and contact profilometry using a 3 mg tip force with a scan speed of 100 µm/s.

2.5. Drug Release

Amoxicillin release from the coated and uncoated microcontainers was measured using a µ-Diss Profiler (Pion Inc. Woburn, MA, USA) with a temperature-controlled water bath at 37 °C. Each chip was attached to a cylindrical magnetic stirrer and placed into a 10mL of PBS-filled glass vial. The absorbance was measured at 270 nm. The release profiles were monitored in real-time through in situ UV probes with a path length of 5mm and each channel was calibrated with drug concentrations of 0.05 to 0.5 mg/mL. The cumulative release profiles were calculated at each time point and presented as mean ± SD.

3. RESULTS

Two different types of profilometry were used to acquire topographical properties of empty microcontainers and step height measurements of coating. As can be seen from the three-dimensional profiles of the empty microcontainers in **Figure 2**, they have an inner diameter of 232 ± 1 µm and an inner height of 272 ± 3 µm (mean ± SD, n = 3).

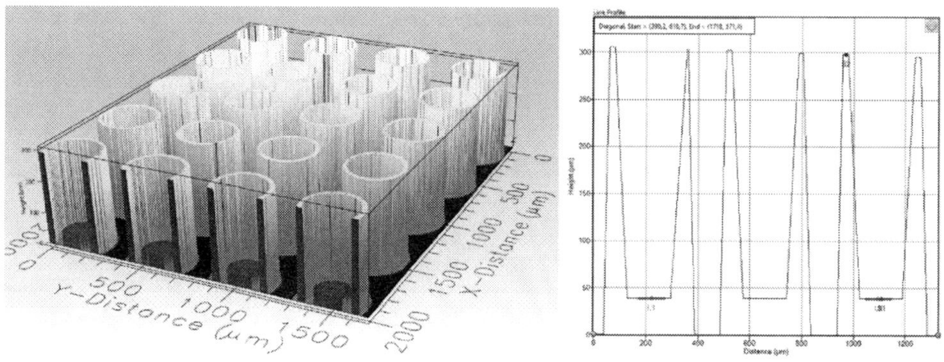

Figure 2 2D and 3D profiles of empty microcontainers

The microcontainers were subjected to spray coating of CS or ChS solutions separately and in combination with the parameters described in **Table 1**. For the layer-by-layer coating, the first layer was either CS or ChS, and the next layer consisted of 50 loops of the counterpart polysaccharide that resulted in a total of 100 loops coating. The thicknesses of the different coatings were measured using a stylus profilometer. The highest lid thickness was obtained with CS due to having a higher density, whereas ChS coating was resulted in a lower value.

Table 1 The parameters of the spray coating and the thicknesses of the different coatings

Parameters	Coatings		
	CS	ChS	CS/ChS
Path speed [mm s−1]	20	20	20
Infusion rate [mL min−1]	0.05	0.04	0.05/0.04
Generator power [W]	1.3	1.3	1.3
Shaping air pressure [kPa]	0.012	0.01	0.012/0.01
Heat plate temperature [°C]	50	35	50/35
No. of loops	100	100	50/50
Thickness (µm)	6.37 ± 1.33	4.49 ± 1.15	5.50 ± 0.74

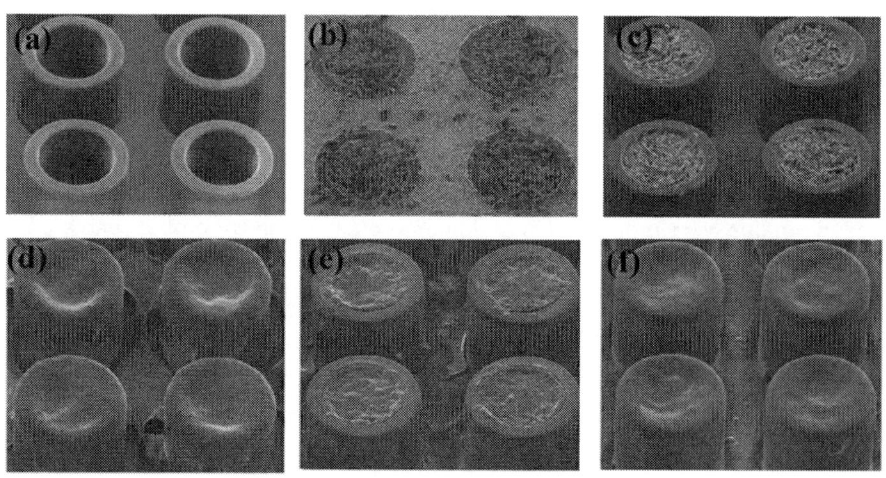

Figure 3 SEM images of empty (a), PDMS-masked drug loaded (c), after peeling of the masking (d), CS (e), ChS (f) and CS/ChS coated microcontainers

Figure 3 shows SEM images of uncoated, PDMS-masked drug loading and spray coated microcontainers. As can be seen, the drug loadings were efficient, the coatings were homogenous and successfully sealed the drug-loaded microcontainers.

In vitro release profile of uncoated and polysaccharides-coated amoxicillin loaded microcontainers is shown in **Figure 4**. A cumulative release of 97.3 ± 10.3% for uncoated, 89.1 ± 3.1% for CS coated and 91.5 ± 9.2% for ChS coated microcontainers were obtained in 15h. Within the first hour, an initial burst release of over 60 % of amoxicillin occurred for both uncoated and ChS containing microcontainers. However, CS ones showed slower release as can be clearly seen in the inset of **Figure 4**. Nevertheless, CS microcontainers continued their fast release profile due to the hydration and swelling, reached and passed uncoated and ChS coated ones within 2h. Microcontainers with CS-ChS and ChS-CS as lids, followed significantly slower and sustained release patterns and reached 74.7± 4.8% and 73.8 ± 2.9% after 14h. This behavior can be attributed to the fact that PEC formation between CS and ChS would cause the lids to be less susceptible to swelling and restrain the release of amoxicillin [14].

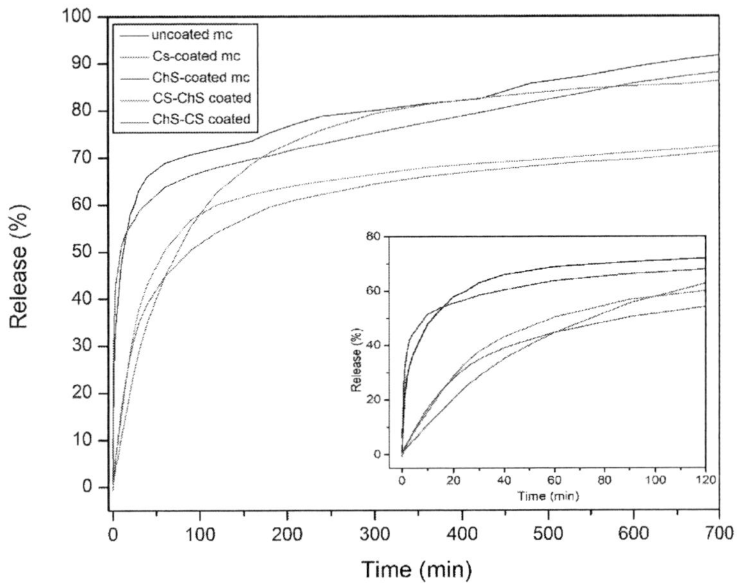

Figure 4 Cumulative release profile of amoxicillin from uncoated and coated microcontainers (mc)

4. CONCLUSION

In this study, the influence of the combination of polysaccharides as polymeric lids for microcontainers was evaluated for an oral drug delivery system. Following the successful drug loading through the PDMS masking technique, the microcontainers were coated with either CS, ChS or their combination using spray coating. SEM images and thickness measurements with profilometer showed homogenous coatings of microcontainers. The results of in vitro release studies confirm that the microcontainers with CS-ChS lids showed a slower and prolonged release (regardless of the layer order) compared to the individual counterparts, which proves the PEC formation between the two polysaccharides. It is expected that using CS-ChS as release controlling polymer lids could be a promising therapeutic alternative to enhance the oral delivery of antibiotics by increasing the residence time.

ACKNOWLEDGEMENTS

This research funded by Internal Grant Agency of Tomas Bata University in Zlín, Czech Republic (IGA/CPS/2021/001).

REFERENCES

[1] WANG, X., DENG, A., CAO, W., LI, Q., WANG, L., ZHOU, J., HU, B., XING, X. Synthesis of chitosan/poly (ethylene glycol)-modified magnetic nanoparticles for antibiotic delivery and their enhanced anti-biofilm activity in the presence of magnetic field. *Journal of Materials Science*. 2018, vol. 53, no. 9, pp. 6433-6449.

[2] SMITH, K. J., MOSHREF, A. R., JENNINGS, J. A., COURTNEY, H. S., HAGGARD, W. O. Chitosan sponges for local Synergistic infection Therapy: A pilot study. *Clinical Orthopaedics & Related Research*. 2013, vol. 471, no. 10, pp. 3158-3164.

[3] TRUCILLO, P., FERRARI, P., CAMPARDELLI, R., REVERCHON, E., PEREGO, P. A supercritical assisted process for the production of amoxicillin-loaded liposomes for antimicrobial applications. *The Journal of Supercritical Fluids*. 2020, vol. 163, p. 104842.

[4] LOPES-DE-CAMPOS, D., M. PINTO, R., COSTA LIMA, S. A., SANTOS, T., SARMENTO, B., NUNES, C., REIS, S. Delivering amoxicillin at the infection site- a rational design through lipid nanoparticles. *International Journal of Nanomedicine*, 2019, vol. 14, pp. 2781-2795.

[5] XU, J., XU, B., SHOU, D., XIA, X., HU, Y. Preparation and evaluation of vancomycin-loaded n-trimethyl chitosan nanoparticles. *Polymers*. 2015, vol. 7, no. 9, pp. 1850-1870.

[6] NIJLAND, R., HALL, M., BURGESS, J. Dispersal of BIOFILMS BY Secreted, Matrix degrading, bacterial dnase. *Biofilm Control and Antimicrobial Agents*. 2014, pp. 27-43.

[7] BIRK, S. E., HAAGENSEN, J. A., JOHANSEN, H. K., MOLIN, S., NIELSEN, L. H., BOISEN, A. Microdevices: Microcontainer delivery of Antibiotic improves treatment of Pseudomonas aeruginosa biofilms (adv. Healthcare Mater. 10/2020). *Advanced Healthcare Materials*. 2020, vol. 9, no. 10, 2070027.

[8] ABRUZZO, A., GIORDANI, B., MITI, A., VITALI, B., ZUCCHERI, G., CERCHIARA, T., BIGUCCI, F. Mucoadhesive and mucopenetrating chitosan nanoparticles for glycopeptide antibiotic administration. *International Journal of Pharmaceutics*. 2021, vol. 606, 120874.

[9] AMRUTKAR, J. R., GATTANI, S. G. Chitosan-Chondroitin sulfate based Matrix tablets for Colon specific delivery of Indomethacin. *AAPS PharmSciTech*. 2009, vol. 10, no. 2, pp. 670-677.

[10] SHARMA, S., SWETHA, K. L., ROY, A. Chitosan-Chondroitin sulfate based polyelectrolyte complex for effective management of Chronic wounds. *International Journal of Biological Macromolecules*. 2019, vol. 132, pp. 97-108.

[11] LAYEK, B., MANDAL, S. Natural polysaccharides for controlled delivery of oral therapeutics: A recent update. *Carbohydrate Polymers*. 2020, vol. 230, 115617.

[12] NIELSEN, L. H., KELLER, S. S., GORDON, K. C., BOISEN, A., RADES, T., MÜLLERTZ, A. Spatial confinement can lead to increased stability of amorphous indomethacin. *European Journal of Pharmaceutics and Biopharmaceutics*. 2012, vol. 81, no. 2, pp. 418-425.

[13] KAMGUYAN, K., THAMDRUP, L. H., VAUT, L., NIELSEN, L. H., ZOR, K., BOISEN, A. Development and characterization of a pdms-based masking method for microfabricated oral drug delivery devices. *Biomedical Microdevices*. 2020, vol. 22, no. 2.

[14] KAUR, G., RANA, V., JAIN, S., TIWARY, A. K. Colon delivery of Budesonide: Evaluation of Chitosan-Chondroitin Sulfate interpolymer complex. *AAPS PharmSciTech*. 2009, vol. 11, no. 1, pp. 36-45.

SURFACE FUNCTIONALIZATION OF THE BIOLOGICAL GOLD NANOPARTICLES FOR MICRO-RNA TARGETING

[1]Parastoo POURALI, [1]Oldrich BENADA, [1]Veronika BENSON

[1]Institute of Microbiology, Czech Academy of Sciences, Czech Republic, EU,
parastoo.pourali@biomed.cas.cz, benada@biomed.cas.cz, benson@biomed.cas.cz

https://doi.org/10.37904/nanocon.2021.4358

Abstract

Among non-viral gene carriers with low toxicity and high transfection efficiency, the use of gold nanoparticles (AuNPs) is of particular interest due to their biocompatibility and special properties. This is the first time we attempted to functionalize the surface of the biological AuNPs in order to conjugate them with antimiR-135b through electrostatic interactions and knockdown the microRNA-135b gene expression inside the cells. A fungal strain, *Fusarium oxysporum*, was cultured in Sabouraud Dextrose Broth (SDB), centrifuged, and the mycelium-free supernatant was challenged with 1 mmol final concentration of $HAuCl_4.3H_2O$ and incubated for 24 h at 37 °C in a shake flask. AuNPs were characterized by visible spectrophotometry, Transmission Electron Microscopy (TEM), Scanning Electron Microscopy (SEM), Energy-Dispersive X-ray spectroscopy (EDS), and a zetasizer. The washed and sterilized AuNPs were used for cytotoxicity and conjugation assays. First transferrin (Tf) and then polyethylenimine (PEI) were used to functionalize and change the surface charge of the AuNPs and then antimiR-135b was conjugated to the AuNPs trough electrostatic interactions. Their association was confirmed by visible spectrophotometry and electrophoresis. Confocal microscopy was used to investigate the internalization of the AuNPs-antimiR-135b complex. The results proved the formation of AuNPs with a maximum absorption peak at 528 nm, round and oval shapes (15-20 nm), and average zeta potential of -21.02 mV. The AuNPs-antimiR-135b showed delayed electrophoresis unlike antimiR-135b or AuNPs alone. Functionalized AuNPs did not cause any toxicity in cell culture and confocal microscopy showed successful transfection of AuNPs-antimiR-135b into the vast majority of 4T1 cells. We concluded that the biological AuNPs were non-toxic and they could carry antimiR-135b to enable gene silencing.

Keywords: Gold nanoparticles (AuNPs), Biological method, Conjugation, AntimiR-135b

1. INTRODUCTION

Various viral and non-viral carriers have been introduced for gene delivery applications. The use of viral vectors has some disadvantages, such as their immunogenicity, toxicity, and problems in scaling up for use in the medical field [1]. In addition, viral vectors cannot infect non-dividing cells. The second group, non-viral carriers such as nanoparticles (liposomes, magnetic nanoparticles and carbon nanotubes) [2-5] have been studied more because of their safety, ease of modification, long circulation time, and good utility [6,7]. Their disadvantages are high cytotoxicity and that they are hindered by several mechanisms leading to their low transfection efficiency. Therefore, researchers have attempted to develop cationic molecules that can form complexes with the negatively charged nucleic acids [8]. Among non-viral carriers with low toxicity and high transfection efficiency, the use of gold nanoparticles (AuNPs) is of particular interest due to their biocompatibility and special properties. It has been shown that the surface of AuNPs can be easily modified to achieve better gene delivery. Although, for example, the cationic polymer polyethylenimine (PEI) is usually considered the gold standard for gene delivery [9], it was shown that the use of PEI coated with AuNPs showed a six fold higher transfection property in the monkey kidney cell line (COS -7) than PEI-25 KDa as control [10]. There are some other studies that showed similar results [11-13]. AuNPs modified with cationic quaternary ammonium groups or 2-aminoethanethiol were able to bind to plasmid DNA by electrostatic binding [5,14,15].

AuNPs modified with cetyltrimethylammonium bromide (CTAB) to bind to siRNA have also been reported. AuNPs have been modified for targeted gene transfer so that they can bind to the specific target on the surface of the cells and carry the genes into the target cells. This type of modification increases the efficiency of gene delivery compared to normal gene delivery [16]. For example, AuNPs conjugated to transferrin (Tf, for cell targeting) and a DNA plasmid have been shown to have a high transfection rate in vitro and in vivo [17]. In addition to the numerous advantages of AuNPs used in this field, the main disadvantage is their toxicity. This type of nanoparticles is mainly produced by chemical and then by physical methods. In the production of AuNPs for gene delivery, the other method called biological technique has not been explored yet. In this approach, the nanoparticles are produced by different types of microorganisms as well as plants. The biologically produced nanoparticles can be formed inside or outside the cells, depending on the method used. It has been shown that some active groups of enzymes as well as polymers or polysaccharides are responsible for the reduction of toxic ions, introduced into the microbial cells, while turning them into less toxic nanoparticles. The biological technique for the production of nanoparticles is safe, fast, simple, environmentally friendly and inexpensive. The main advantage of biologically produced nanoparticles is their non-toxicity [18]. In this study, we used a fungal strain, *Fusarium oxysporum*, for the extracellular production of AuNPs. The zeta potential of the produced AuNPs was analyzed and to promote the interaction of AuNPs with antimiR-135b (synthetic ssRNA complementary to endogenous target microRNA-135b), PEI was used to change the surface charge of AuNPs. We conjugated Tf to the AuNPs surface to accelerate targeted gene delivery. This was the first time that attempts were made to alter the surface charge of biological AuNPs to conjugate them with antimiR-135b through electrostatic interactions. The accomplishment of this technique was demonstrated by the detection of antimiR-135b in the cells, indicating successful gene transfer. The antimiR-135b alone could not transfect the cells. Following the AuNPs-antimiR-135b internalization, target microRNA-135b is found and silenced by antimiR-135b.

2. MATERIALS AND METHODS

2.1. AuNPs Production and Characterization

Fusarium oxysporum (CCF 3732, Czech Republic) was cultured in Sabouraud Dextrose Broth (SDB, Sigma Aldrich, USA) at 30 °C in a shake flask for 1 week. The culture was centrifuged (2000g for 5 min) and the mycelia-free supernatant was used to prepare AuNPs [19]. In order to AuNPs production, the pH of the mycelium-free supernatant was adjusted to 10. 100 ml of the mycelium-free supernatant was mixed with 1 mmol final concertation of $HAuCl_4.3H_2O$ (Sigma Aldrich, USA) and incubated for 24 h at 37 °C in a shake flask. The negative control flask (sterile SDB containing the same concentration of $HAuCl_4.3H_2O$ was also incubated [19].

The color change of the reaction mixture is the first sign of AuNPs production due to the localized surface plasmon resonance (LSPR) of AuNPs [20]. The produced AuNPs were washed three times with RNase-free ddH_2O (pH 7.4, Qiagen, Hilden, Germany). The tube was centrifuged at 20000 rpm for 30 min and the pellet was dissolved in 1000 µl RNase-free ddH_2O [20].

2.1.1. Visible Spectrophotometry

The maximum absorption peak of AuNPs was determined using the Nanodrop spectrophotometer (Thermo Fisher Scientific, Massachusetts, USA). The sample was tested between 400-750 nm (blank sample was ddH_2O). The production would be proved if the maximum absorption peak was seen between 500-550 nm due to the LSPR of AuNPs [20].

2.1.2. Transmission Electron Microscopy (TEM)

Five microliters of AuNPs suspension were applied onto a glow discharge activated (30 s, 1 kV, 10 mA, [21]) carbon/formvar coated 300 mesh copper grids and let adsorb for 30 s. Next followed the blotting the excess of the sample with filter paper and air-drying the grids at room temperature. Dried grids were examined at 80 kV in the Philips CM100 TEM (Philips EO, Eindhoven, The Netherlands) equipped with a Veleta slow-scan CCD camera (EMSIS, Muenster, Germany).

2.1.3. Scanning Electron Microscopy (SEM) and Energy-Dispersive X-ray Spectroscopy (*EDS)*

Ten microliters of AuNPs suspension were dried out overnight at 56 °C on a glow discharge activated silicon wafer. SEM and EDS analyses were performed using FEI Nova Nano SEM 450 scanning electron microscope (FEI, Brno, Czech Republic) equipped with an Ametek® EDAX Octane, SDD detector, and TEAM™ EDS Analysis Systems (AMETEK B. V.; Tilburg, The Netherlands) essentially as described in [22].

2.1.4. Zetasizer Analysis

The average size and zeta potential of AuNPs were determined using Zetasizer Nano ZS90 (Malvern Panalytical, Malvern, UK) and DLS (Dynamic Light Scattering, by noninvasive backscattering technique) instruments in the presence of ddH_2O as dispersant and 25 °C [23].

2.2. Cytotoxicity Test

Before the cytotoxicity test the nanoparticles were sterilized using the Tyndallization technique according to Pourali et al [19]. MTT assay was used to determine the toxicity of AuNPs before any other experiments. Two different cell cultures, NIH/3T3 (ATCC CRL-1658) and 4T1 (ATCC CRL-2539), were used as control (normal mouse fibroblasts) and test (BALB/c mouse breast cancer cell line), respectively. NIH/3T3 and 4T1 cells were cultured in Dulbecco's Modified Eagle Medium (DMEM, Sigma Aldrich, USA) and Roswell Park Memorial Institute Medium 1640 (RPMI-1640, Sigma Aldrich, USA), respectively. Both in the presence of 10% Fetal Bovine Serum (FBS, Gibco, Massachusetts, USA) and 44 µg/ml gentamicin (Sandoz, Novartis Company, Prague, Czech Republic). In each well of 96 wells bottom tissue culture plate (JETBiofil, Guangzhou, China), 2×10^5 cells were seeded and after reaching 80% confluence monolayer cells, cells were washed with phosphate buffered saline (PBS) and all wells examined were filled with 200µl of the fresh medium. AuNPs were ½ diluted by titration in the B2 line in fresh medium and then the diluted samples were used. In well A1, 50 µL of AuNPs were added and the same amount of AuNPs, but diluted, was added in A2 to A11. Well A12 was the positive control without addition of AuNPs. The plate was incubated overnight in a humid atmosphere, 5% CO_2 incubator at 37 °C and then 20 µL of 5mg/mL (3-4,5-dimethylthiozol-2-yl) 2,5-diphenyltetrazolium bromide (MTT, EMD Millipore, CA, USA) in PBS (pH 7.4) was added to all wells. After an incubation period of 4 hours, the mixture was removed and 100 µL of dimethyl sulfoxide (DMSO, Sigma Aldrich, USA) was added. The absorbance of the wells was measured using a Tecan spectrophotometer (Thermo Fisher Scientific, Massachusetts, USA) at a wavelength of 570 nm. The percentage of half-maximal inhibitory concentration (IC_{50}) was calculated for each cell line using the following formula: IC_{50} = OD of maximum inhibition – 50 % (OD of maximum inhibition - OD of minimum inhibition) [20].

2.3. Functionalization and Conjugation

To functionalize AuNPs with Tf and PEI, 200 µl AuNPs (at a non-toxic concentration according to MTT assay result) were incubated with 200 µl Tf (0.9 µg/ml; labeled with Texas Red, Life Technologies, Prague, Czech Republic) for 1 h with shaking at room temperature (RT). 400 µl of low weight PEI 800 (Sigma-Aldrich, Prague, Czech Republic) was added to the sample and the mixture was incubated overnight at RT. Unbound molecules (Tf and PEI) were washed with RNase-free ddH_2O and removed by centrifugation twice at 9000g for 60 min. The pellet was dissolved in 400 µl of RNase-free ddH_2O. To conjugate the antimiR-135b to the surface-functionalized AuNPs, 36 µl of antimiR-135b: 5'UptCptAptCAUAGGAAUGAAAAGCCptAptUptA 3' (labeled with Alexa Fluor 488 at 5') was added and incubated at RT for 1 h under shaking conditions. Unbound antimiR-135b molecules were removed by centrifugation at 9000g for 60 min. The pellet was suspended in 400 µl RNase-free ddH_2O and used for in vitro assay [24]. To detect the conjugation of antimiR-135b to AuNPs, different methods were used. Firstly, the maximum absorption peak of antimiR-135b-AuNPs was checked by spectrophotometry. The change in the curve compared to AuNPs as control was evidence of successful functionalization and conjugation. The blank value used was ddH_2O [20]. Second, agarose gel electrophoresis was performed for antimiR-135b-AuNPs, antimiR-135b (as control) and MassRuler low range DNA Ladder (Thermo Fisher Scientific, Massachusetts, USA). Tris-acetate-EDTA buffer (TAE) was used as buffer and the gel of 2% w/v agarose (Sigma Aldrich, USA) was prepared. 5 µL of each sample was mixed with 1 µL of 6X Mass Ruler DNA loading dye (Thermo Fisher Scientific, Massachusetts, USA) and added to the wells of the gel. The gel was run at a voltage of 100 V for 45 min and visualized under gel documentation. Successful conjugation results in a delay in the migration of antimiR-135b-AuNPs compared to antimiR-135b as a control [25-27]. Third, the zeta potential and size distribution of antimiR-135b-AuNPs were determined using the same technique as described previously.

2.4. Confocal Microscopy

4T1 cells were cultured in 35-mm dishes with 20-mm glass bottoms (Cellvis, Ontario, Canada) for 12 h under culture conditions described previously. After the cells were 80% confluent, the medium from each plate was discarded and 1500 µL of fresh medium was added, and 60 µL of antimiR-135b-AuNPs sample was added for the test plate. After 24 hours of incubation, cells were washed three times with PBS and immediately analyzed live under an Olympus FluoView FV1000 confocal microscope (objectives 20X/ NA 0.75 and 40X/ NA 0.95) without fixation or staining. Excitation/emission parameters: Alexa Fluor 488 473/520 nm; TexasRed 595/615 nm. Results were analyzed using Olympus FluoView 2.0 software [24].

3. RESULTS AND DISCUSSION

3.1. AuNPs Production and Characterization

After 24 hours of incubation, the mixture turned from yellow to red due to the LSPR of AuNPs. **Figure 1 B** shows the color change compared to the negative control (A). As mentioned earlier, the pH of the mycelium-free supernatant was adjusted to 10 because an alkaline environment is reportedly preferred, as the repulsion between Au^+ and H^+ ions at acidic pH values reduces the probability of reduction and nucleation of Au^+ ions to AuNPs [28].

3.1.1. Visible Spectrophotometry

The maximum absorption peak of AuNPs was reached at 528 nm (the blank sample was ddH_2O), which was the evidence of the production of AuNPs in the sample. **Figure 1 C** shows the results obtained.

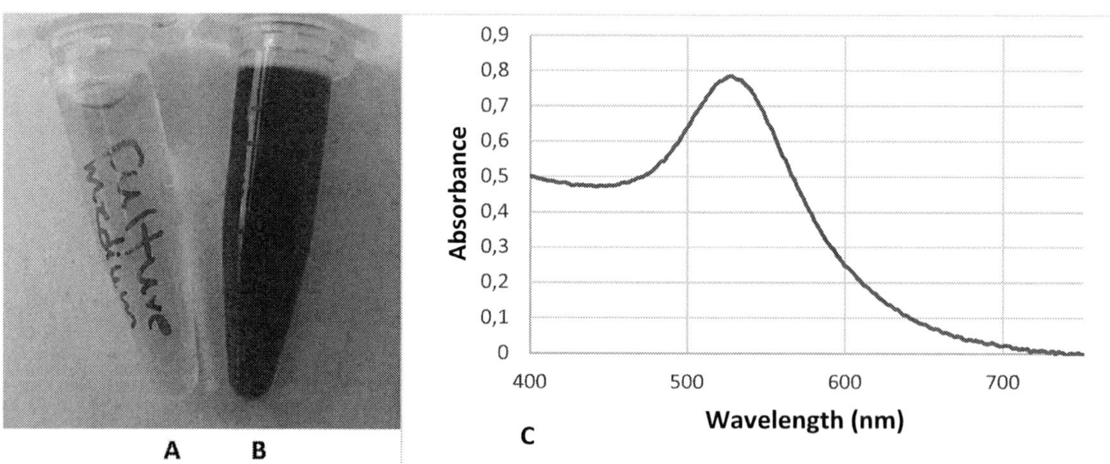

Figure 1 Production and characterization of biologically produced AuNPs. A) Control B) The color-changed mycelia-free supernatant. C) The obtained AuNPs spectrum with maximum absorption peak at 528nm

3.1.2. Electron Microscopy of AuNPs

Figure 2 shows the representative frame of TEM AuNPs imaging that proved the spherical to oval shape and almost the same size (15-20 nm) of analyzed particles. TEM analysis also proved the purity of the prepared AuNPs sample.

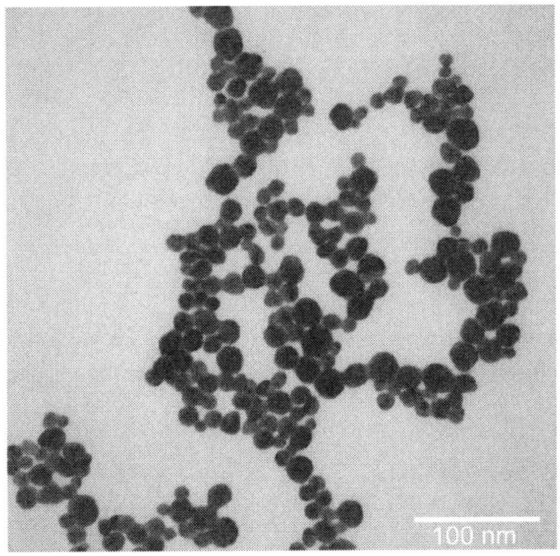

Figure 2 TEM image of AuNPs on glow discharge activated carbon/formvar support film. The particles possess a spherical shape with an approximate size of 20 nm. Scale bar = 100 nm.

Figure 3 A shows a backscattered electron SEM image of well-spread AuNPs with not many aggregated clusters on the wafer surface. The glow-discharge activation of the silicon wafer surface worked quite well and minimized the artificial clustering of AuNPs.

The EDS spectrum (**Figure 3 B**) confirmed the Au content in the imaged nanoparticles showing characteristic Au Mα and Mβ peaks at the energies of 2.123 keV and 2.203 keV. The insert in **Figure 3 B**, an SEM secondary electron image, shows the approximate location of the electron beam during the EDS spectrum acquisition, marked with a white letter X.

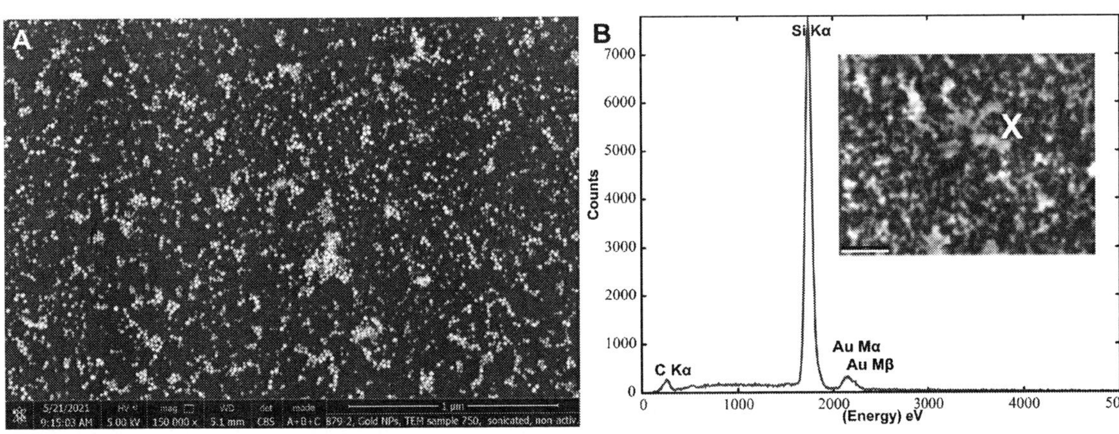

Figure 3 SEM and EDS analyses of Au NPS. A) SEM back-scattered electron image and B) EDS spectrum and spot location in secondary electron image. Spectrum acquisition conditions - Spot size: 5; Accelerating voltage: 6 kV; Magnification: 65000 ×; Take-off angle: 35; Live Time: 30 s; Amp Time: 7.68 μs; Resolution: 128.1 eV. Scale bar = 1 μm.

The other peaks in the spectrum represent the Si Kα signal at the energy of 1.740 keV, which originates from silicon wafer support, and the C Kα signal at the energy of 0.277 keV. The carbon signal probably reflects the presence of organic capping on the AuNPs' surface. The AuNPs sample was prepared from mycelium-free supernatant and extensively washed before sample preparation for SEM. It minimized the contamination of AuNPS with culture medium or the remnants of fungal mycelium.

3.1.3. Zetasizer analysis

The results show that the hydrodynamic diameter (Z average) of AuNPs is 49.81±4 nm and the zeta potential is -21.02. The obtained Z-average was higher than the results of TEM, 49.81±4 nm compared to 20 nm. Thus, the Z-average result from the digital images of TEM is different from DLS because DLS measures the hydrodynamic diameter and TEM measures only the diameter of AuNPs [27].

3.2. Cytotoxicity assay

MTT assay was used to determine the toxicity of AuNPs before any other experiments. The results showed that the AuNPs did not induce toxic effects in NIH/3T3 and 4T1 cell cultures. The result is consistent with previous studies that reported that this type of nanoparticles produced by microorganisms has dose-dependent and low toxic effects [18].

3.3. Functionalization and Conjugation

During the conjugation process, no significant color change of the samples was observed after the addition of the single molecules to the AuNPs, indicating that the nanoparticles were stable during the functionalization process. After functionalization of AuNPs with Tf and PEI and conjugation with antimiR-135b, the maximum absorption peak was determined. **Figure 4** shows the differences between the maximum absorption peaks of AuNPs (528 nm) versus antimiR-135b-AuNPs (533 nm). As can be seen (**Figure 4 A**), the peak of the control was sharper in contrast to the antimiR-135b-AuNPs. Moreover, the absorption maximum of the AuNPs after conjugation with these molecules was at a higher wavelength, but in the range of 500-550 nm. This means that the AuNPs were stable during the conjugation process and remained in their nanoform. The broader peak and the change in peak position (**Figure 4 A**) indicate that the AuNPs conjugated with these molecules.

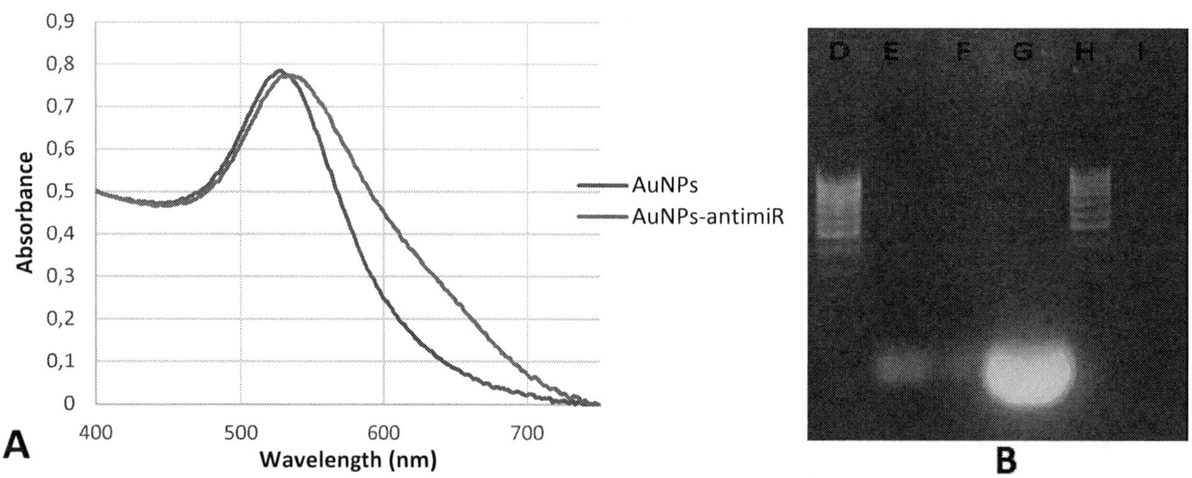

A B

Figure 4 Detection of conjugation by visible spectrophotometry and agarose gel electrophoresis. Blue peak is AuNPs as control and red peak belongs to antimiR-135b AuNPs. (B) Agarose gel electrophoresis results. Lanes D and H are DNA ladder, lanes E and F are antimiR-135b-AuNPs with two different loads, lane G is antimiR-135b and lane I is AuNPs.

Agarose gel electrophoresis was performed for antimiR-135b-AuNPs and antimiR-135b. The results show the delayed migration of antimiR-135b-AuNPs compared to antimiR-135b as control. **Figure 4 B** shows the results obtained. A short delay can be seen in lanes E and F (10 and 5 µl were loaded respectively) in contrast to the control (lane G). This means that after conjugation of antimiR-135b with AuNPs, since this complex is heavier than antimiR-135b as control, it is seen in the higher position in the gel. The zetasizer results showed that the Z average of antimiR-135b-AuNPs was 53±7 nm with a zeta potential of +10.53 mV. The obtained Z average was higher than that of AuNPs, indicating the changes on the surface of AuNPs after conjugation with the corresponding molecules.

All in all, the obtained data confirmed that the surface charge of the biologically prepared aunps changed from negative to positive after pei treatment and the nanoparticles were ready for conjugation with antimir-135b. Thus, the functionalization of the aunps was successful. We have previously reported that direct conjugation of the biological aunps with various drugs such as antibiotics is possible [19,20]. In the current study, we have shown that the surface charge of the nanoparticles can be easily changed so that they are ready to bind to the nucleic acids through electrostatic interactions and without additional linkers.

3.4. Confocal Microscopy

Confocal microscopy results showed successful transfection of antimiR-135b-AuNPs into the vast majority of 4T1 cells (**Figure 5**).

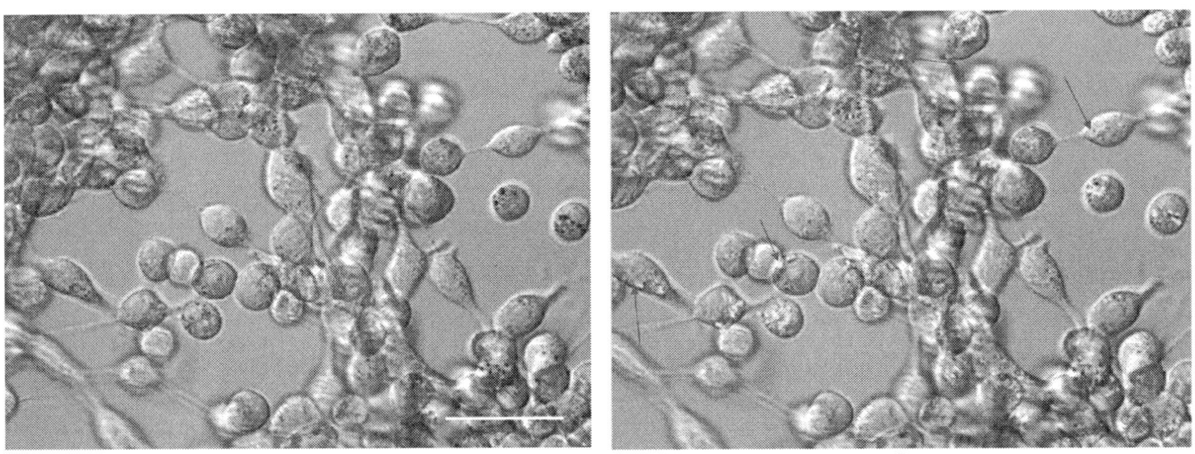

A B

Figure 5 Results of confocal microscopy of live 4T1 cells after transfection with antimiR-135b-AuNPs. A) the cells as background and B) the internalized complexes in the cells. The red signal is antimiR-135b and the green signal is Tf. The arrows show some of the complexes. Scale bar= 100 µm.

4. CONCLUSION

This was the first study of possible functionalization of AuNPs produced by the biological method, which was completely unknown before. The biological AuNPs were non-toxic and can be functionalized to bind antimiR-135b to enable gene silencing. Our results showed that antimiR-135b could attach to the positively charged nanoparticles after the AuNPs were functionalized with Tf and PEI. Confocal microscopy results showed that the nanoparticles were taken up into the target cells. Thus, the functionalization of the biological AuNPs made them a good candidate for targeted gene silencing in the cells. Further studies are needed to investigate the other capabilities of the biological AuNPs.

ACKNOWLEDGEMENTS

This work was supported by ESF " International Mobility of Researchers - MSCA-IF IV (Institute of Microbiology of CAS, v. v. i.) No. CZ.02.2.69/0.0/0.0/20_079/0017812. We thank CMS-Biocev ("Biophysical techniques, crystallization, diffraction, structural mass spectrometry") of CIISB, center Instruct-CZ Centre, supported by MEYS CR (LM2018127).

REFERENCES

[1] VERMA, I.M., et al. Gene therapy: promises, problems and prospects. *Genes and Resistance to Disease*. 2000, pp. 147-157.

[2] BOYER, C., et al. Anti-fouling magnetic nanoparticles for siRNA delivery. *Journal of Materials Chemistry*. 2010, vol. 20, no. 2, pp. 255-265.

[3] GAO, L., et al. Carbon nanotube delivery of the GFP gene into mammalian cells. *ChemBioChem*. 2006, vol. 7, no. 2, pp. 239-242.

[4] SUZUKI, R., et al. Effective gene delivery with novel liposomal bubbles and ultrasonic destruction technology. *International journal of pharmaceutics*. 2008, vol. 354, no. 1-2, pp. 49-55.

[5] MCINTOSH, C.M., et al. Inhibition of DNA transcription using cationic mixed monolayer protected gold clusters. *Journal of the American Chemical Society*. 2001, vol. 123, no. 31, pp. 7626-7629.

[6] BRANCO, M.C. SCHNEIDER, J.P. Self-assembling materials for therapeutic delivery. *Acta biomaterialia*. 2009, vol. 5, no. 3, pp. 817-831.

[7] LIU, F. HUANG, L. Development of non-viral vectors for systemic gene delivery. *Journal of controlled release*. 2002, vol. 78, no. 1-3, pp. 259-266.

[8] MINTZER, M.A. SIMANEK, E.E. Nonviral vectors for gene delivery. *Chemical reviews*. 2009, vol. 109, no. 2, pp. 259-302.

[9] BOUSSIF, O., et al. A versatile vector for gene and oligonucleotide transfer into cells in culture and in vivo: polyethylenimine. *Proceedings of the National Academy of Sciences*. 1995, vol. 92, no. 16, pp. 7297-7301.

[10] THOMAS, M. KLIBANOV,A.M. Conjugation to gold nanoparticles enhances polyethylenimine's transfer of plasmid DNA into mammalian cells. *Proceedings of the National Academy of Sciences*. 2003, vol. 100, no. 16, pp. 9138-9143.

[11] NOH, S.M., et al. Enhanced cellular delivery and transfection efficiency of plasmid DNA using positively charged biocompatible colloidal gold nanoparticles. *Biochimica et Biophysica Acta (BBA)-General Subjects*. 2007, vol. 1770, no. 5, pp. 747-752.

[12] HU, C., et al. Low molecular weight polyethylenimine conjugated gold nanoparticles as efficient gene vectors. *Bioconjugate chemistry*. 2010, vol. 21, no. 5, pp. 836-843.

[13] RAHME, K., et al. Bioconjugated gold nanoparticles enhance siRNA delivery in prostate cancer cells. *RNA interference and cancer therapy*. 2019, pp. 291-301.

[14] HAN, G., et al. Stability of gold nanoparticle-bound DNA toward biological, physical, and chemical agents. *Chemical biology & drug design*. 2006, vol. 67, no. 1, pp. 78-82.

[15] NIIDOME, T., et al. Preparation of primary amine-modified gold nanoparticles and their transfection ability into cultivated cells. *Chemical communications*. 2004, vol. 17, pp. 1978-1979.

[16] CHEN, P.C., et al. Gold nanoparticles: from nanomedicine to nanosensing. *Nanotechnology, science and applications*. 2008, vol. 1, pp. 45.

[17] SALEM, A.K., et al. Multifunctional nanorods for gene delivery. *Nature materials*. 2003, vol.. 2, no. 10, pp. 668-671.

[18] POURALI, P., et al. Biosynthesis of gold nanoparticles by two bacterial and fungal strains, *Bacillus cereus* and *Fusarium oxysporum,* and assessment and comparison of their nanotoxicity in vitro by direct and indirect assays. *Electronic Journal of Biotechnology*. 2017, vol. 29, pp. 86-93.

[19] POURALI, P., et al. Bio-synthesis of gold nanoparticles by *Fusarium oxysporum* and assessment of their conjugation possibility with two types of β-lactam antibiotics without any additional linkers. *Microbiology*. 2018, vol. 87, no. 2, pp. 229-237.

[20] YAHYAEI, B., POURALI, P. One step conjugation of some chemotherapeutic drugs to the biologically produced gold nanoparticles and assessment of their anticancer effects. *Scientific reports*. 2019, vol. 9, no. 1, pp. 1-15.

[21] BENADA, O. POKORNY, V. Modification of the Polaron sputter-coater unit for glow-discharge activation of carbon support films. *Journal of electron microscopy technique*. 1990, vol. 16, no. 3, pp. 235-239.

[22] POURALI, P., et al. Response of Biological Gold Nanoparticles to Different pH Values: Is It Possible to Prepare Both Negatively and Positively Charged Nanoparticles? *Applied Sciences*. 2021, vol. 11, no. 23, pp. 11559-11579.

[23] EATON, P., et al. A direct comparison of experimental methods to measure dimensions of synthetic nanoparticles. *Ultramicroscopy*. 2017, vol. no. 182, pp. 179-190.

[24] KRIVOHLAVA, R., et al. Knockdown of microRNA-135b in mammary carcinoma by targeted nanodiamonds: potentials and pitfalls of in vivo applications. *Nanomaterials*. 2019, vol. 9, no. 6, pp. 866.

[25] CUI, J., et al. Application of PEI-modified magnetic nanoparticles as gene transfer vector for the genetic modification of animals. *Advances in Materials Science and Engineering*. 2012, vol. 2012, pp. 1-6.

[26] GUO, C. YARGER, J.L. Characterizing gold nanoparticles by NMR spectroscopy. *Magnetic Resonance in Chemistry*. 2018, vol. 56, no. 11, pp. 1074-1082.

[27] SHAWKY, S.M., et al. Direct detection of unamplified hepatitis C virus RNA using unmodified gold nanoparticles. *Clinical biochemistry*. 2010, 43, no.13-14, pp. 1163-1168.

[28] GURUNATHAN, S., et al. Biosynthesis, purification and characterization of silver nanoparticles using *Escherichia coli*. *Colloids and Surfaces B: Biointerfaces*. 2009, vol. 74, no. 1, pp. 328-335.

TAILORING THE INTERNAL MICROSTRUCTURE OF THE HYDROGELS BASED ON POLY-HEMA TARGETED FOR DRUG DELIVERY SYSTEMS

[1]Monika TRUDICOVA, [1]Hana PAPEZIKOVA, [1]Petr SEDLACEK, [1]Miloslav PEKAR

[1]Brno University of Technology, Faculty of Chemistry, Brno, Czech Republic, EU,
xctrudicova@fch.vut.cz

https://doi.org/10.37904/nanocon.2021.4350

Abstract

Transport and barrier properties are important material characteristics in terms of the development of the targeted drug delivery systems. Therefore, this work focused on the study of the possibility of influencing these properties and internal architecture in a model hydrogel system formed by a semi-interpenetrating polymer network with an incorporated polyelectrolyte component. Due to its application potential and its extensive use in medicine, poly(2-hydroxyethyl methacrylate) was chosen as the model hydrogel system for this work. Sodium polystyrene sulfonate was used as the interpenetrating linear polymer. Scanning electron microscopy was used to examine the structure of the three-dimensional hydrogel network. Transport properties were monitored by diffusion tests. Furthermore, diffusion through the hydrogel (barrier properties) was monitored using diffusion cells (Franz cells, horizontal cells). The usability of the semi-interpenetrating polymer network concept for the preparation of materials with tunable relevant properties was confirmed on the proposed materials.

Keywords: Hydrogel, poly(2-hydroxyethyl methacrylate), transport properties, drug delivery systems

1. INTRODUCTION

Hydrogels; also defined as systems consisting of a three-dimensional network of polymer chains and water filling the space between these macromolecules; have acquired a key position in diverse fields of biomedicine, such as in wound dressings, controlled release, or as scaffolds in tissue engineering. The properties of hydrogels must vary according to the specific needs of individual applications, e.g., for drug delivery systems, suitable transport properties are crucial to their application. Recent studies [1-4] have focused on various strategies to improve these properties in hydrogels using a process such as the use of interpenetrating networks, composite structures, and dual crosslinking. Overall, hydrogel preparation methods that simultaneously mimic the hydration, strength, and stiffness of soft and supporting tissues have the potential to be used in a much wider range of biomedical applications. Using semi-interpenetrating polymer networks (semi-IPN) is a suitable way to modify the material properties by manipulation of the internal chemical and morphological architecture [2-4].

Hydrogels based on poly(2-hydroxyethyl methacrylate) (poly-HEMA) were the first synthetic hydrogels for pharmaceutical applications, they are biologically compatible hydrogels with tunable mechanical properties [5]. They are usually synthesized by the polymerization of methacrylate with free radicals. Due to their properties, poly-HEMA hydrogels are suitable to produce hydrophilic contact lenses and artificial corneas [6]. Poly-HEMA is also an interesting material for drug delivery systems, e.g. chitosan blends with this polymer are commonly used in drug delivery systems [7,8]. For these reasons, poly-HEMA was chosen as the model hydrogel system in this work.

2. MATERIALS AND METHODS

In this work, the synthesis took place by radical polymerization by photoinitiation. All chemicals used in this work were purchased from Sigma Aldrich (97% 2-hydroxyethyl methacrylate (HEMA) CAS: 868-77-9; 98%; ethylene glycol dimethacrylate (EGDMA) CAS: 97-90-5; 98% Irgacure 2959 CAS: 106797-53-9). All components of the solution were mixed in a vial in a ratio of 2:3 by volume deionized water and 2-hydroxyethyl methacrylate monomer, with the addition of EGDMA as a crosslinker and Irgacure 2959 as an initiator. A thin layer of the liquid mixture was poured into a petri dish, then placed under the UV lamp so that the resulting radiation intensity was about 2.4 mW /cm^2. At this radiation intensity, the synthesis time was 30 minutes. The samples containing an incorporated component in the form of a polyelectrolyte were also prepared. Sodium polystyrene sulfonate (PSS, Sigma-Aldrich, CAS: 25704-18-1) was added to the above polymer solution in such an amount that its final concentration in the reaction mixture was 0.1 g/l. After the solidification of hydrogels, it was necessary to remove a residual monomer. The hydrogels were immersed in ethanol for 24 hours and then in deionized water or only in deionized water. During the immersion in ethanol, changes in structural properties of hydrogels were evident, which were further investigated in the subsequent experimental set.

Determining the structural character of a hydrogel is a crucial part of understanding its transport properties. The literature [5] and previous experiments have shown that upon contact with water (after removing residual monomer), the optical properties of the hydrogel (turbidity) can change significantly, which is related to the change in the internal structure. To determine the change in the internal structure, hydrogels were subjected to direct visualization by scanning electron microscopy (SEM). For SEM observations small parts of the specimens were taken. These specimens were subsequently gold-coated in a sputtering device and investigated using a scanning electron microscope ZEISS EVO LS 10 in the mode of secondary electrons (SE). The accelerating voltage was set to 5 kV.

To investigate transport properties, Rhodamine 6G (R6G) was used as a diffusion probe since its structure (aromatic cation) models some real active substances (e.g. analgesics) and at the same time allows an easy spectroscopic detection (photometry, fluorimetry). Transport properties were monitored by diffusion tests (absorption of the dye into the internal structure of the gel, subsequent release from the gel). Throughout the experiments, the absorbance of dye solutions was measured at certain times on a UV-VIS spectrometer in the wavelength range from 800 to 300 nm. Furthermore, diffusion through the gel (barrier properties) was monitored using diffusion cells (vertical Franz cells, horizontal side-by-side cells).

3. RESULTS AND DISCUSSION

3.1. Structural analysis

Freeze-drying using a regular freezer is usually associated with significant structural artifacts or collapse of the internal structure of hydrogels caused by the formation and growth of large ice crystals. On contrary, rapid freezing (e.g. with the use of liquid nitrogen) that supports vitrification can be a way to preserve the internal structure of the hydrogels without artifacts [9]. In this case, the results showed that for a hydrogel containing only about 30 wt% water, it does not matter whether liquid nitrogen or a freezer is used.

The analysis was performed on a total of six samples. Three with the addition of polyelectrolyte and three without. The first pair of hydrogels was frozen immediately after hydrogel synthesis and is shown in **Figure 1**. The hydrogel structure is smooth, and no pores are visible. Occasional clusters may indicate some polymers not incorporated into the structure (oligomers) which remained on the surface of the hydrogel. Water content and surface suggest that it is a microporous or even non-porous type of HEMA hydrogel.

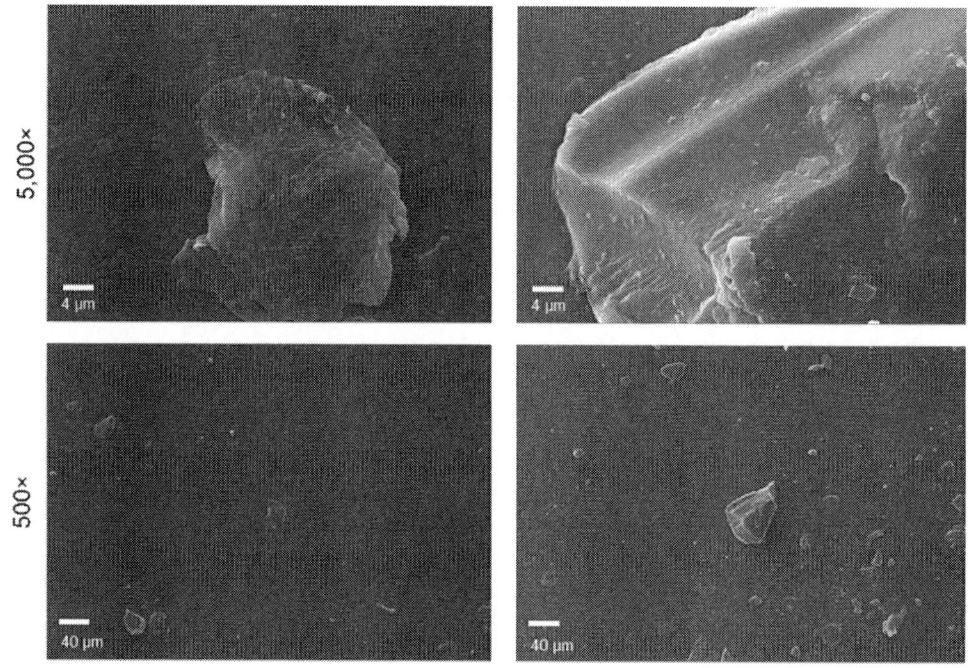

Figure 1 Poly-HEMA hydrogels frozen immediately after synthesis: without (left) and with (right) the addition of PSS

During immersion of the fresh hydrogels in ethanol (while washing the samples to remove the uncrosslinked components), anisotropic swelling of the sample occurred preferentially in the horizontal direction and changes in structural properties (indicated by reduced turbidity) were observed (shown in **Figure 2**). Therefore, the effect of ethanol on the structure was investigated further. For this purpose, another pair of samples was placed in ethanol for 24 hours before freezing and is shown in **Figure 3A**. The swelling of the samples can be demonstrated by the wrinkling of the surface of the samples. At a magnification of 5,000×, it could be said that there are small cracks between the individual parts.

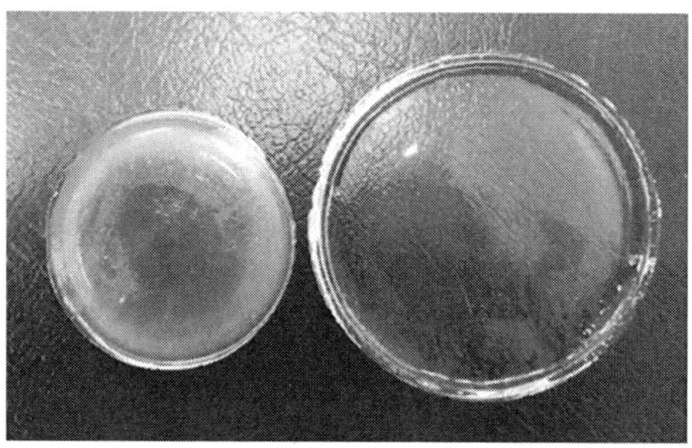

Figure 2 Demonstration of swelling and changes in the structure of the hydrogel (sample with the addition of PSS) after 24 hours in water (left) and after 24 hours in ethanol (right)

The third pair of samples was immersed in ethanol for 24 hours and then in water for 24 hours before freezing and its structure is shown in **Figure 3B**. Occasional pores to holes are visible in these samples. One of the possible causes is the reduction of the gel volume, where these pores formed from previous cracks. Another possibility may be that during the elution, ethanol has ripped some part of the network out of the structure.

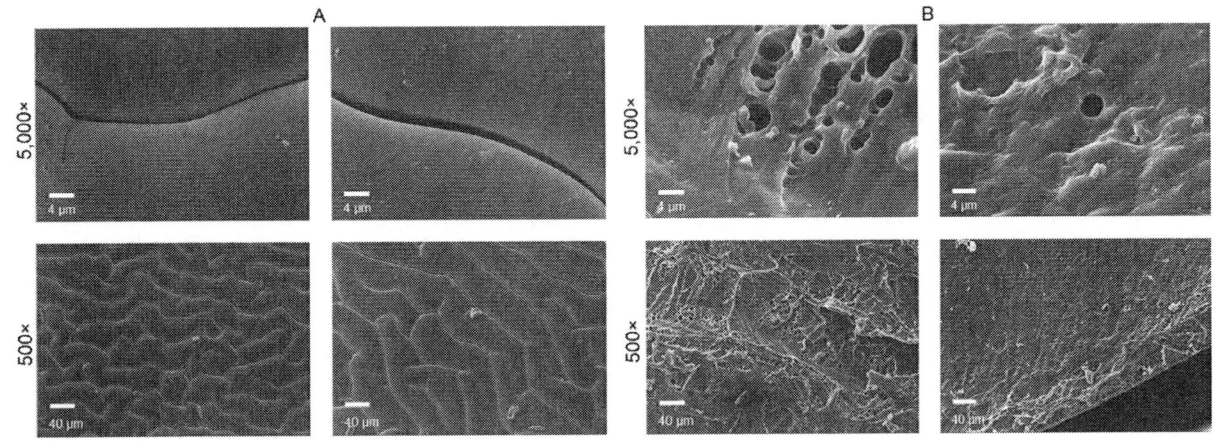

Figure 3 Poly-HEMA hydrogels frozen after 24 hours in ethanol (part A): without (left) and with (right) the addition of PSS and poly-HEMA hydrogels after 24 hours in ethanol and the following 24 hours in the water (part B): without (left) and with (right) the addition of PSS

3.2. Study of transport properties

During the first experiment, uptake of the dye from solutions with different concentrations into the internal structure of the hydrogel was monitored. From the attached graph (**Figure 4**), which shows the average absorbance values at 526 nm in the dye solution over time, the samples with PSS have a greater ability to absorb the dye solution into the hydrogel structure, most likely due to an electrostatic attraction between the cationic dye and the opposite charge on PSS. Nevertheless, as can also be seen in **Figure 4**, the initial uptake tendency of the two gels has the opposite trend, i.e. more dye is absorbed by the gel without PSS soon after the start of the experiment. This may be explained by an effect of PSS on the effective porosity and tortuosity of the gel since the addition of polyelectrolyte may reduce the number of diffusion trajectories available for the dye. The final absorbance value is similar in both cases. The result is due to the total rhodamine concentration that was initially chosen. After 24 hours during the experiment, almost all the dye was diffused into the hydrogel. This trend was observed for samples immersed in all prepared rhodamine concentrations.

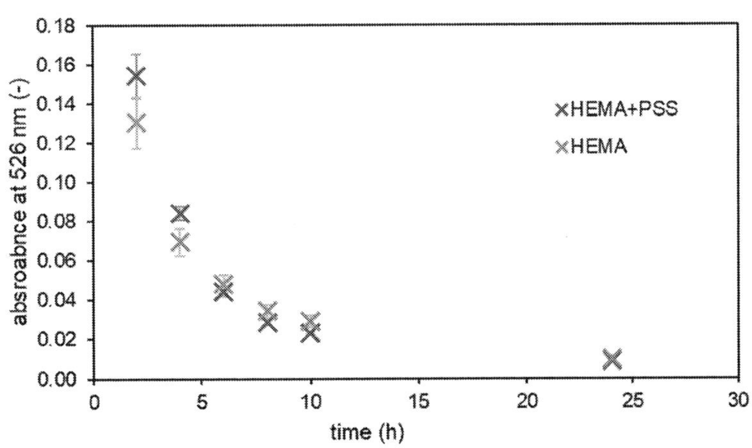

Figure 4 Example of graph for Experiment 1 of absorbance versus experiment time (for immersing in solution with highest rhodamine concentration)

In the second experiment, the release of the dye (after its absorption) from the hydrogel into the environment was monitored. This time, higher concentrations of rhodamine were used than in the previous experiment.

First, rhodamine is released into the external environment of the hydrogel, but at some point, the hydrogel begins to resorb the dye back into its structure. Another interesting phenomenon observed in this experiment is the fact that PSS not only absorbs more dye but also, interacts with it and the dye remains in the structure, which can be seen in **Table 1** (the rhodamine concentration is lower in the hydrogel samples with PSS throughout this part of the experiment).

Table 1 Experiment 2 (release of dye from hydrogels after absorption): rhodamine concentration in water over 72 hours (X- poly-HEMA hydrogel, PSS - with the addition of PSS, 1 - absorption from 0.025 g/l R6G, 2 - absorption from 0.05 g/l R6G and 3 - absorption from 0.1 g/l R6G)

	X-1	**PSS-1**	**X-2**	**PSS-2**	**X-3**	**PSS-3**
t (h)	**c (µg/l)**		**c (µg/l)**		**c (µg/l)**	
12	9.434	6.261	19.26	12.56	44.22	30.24
24	4.937	3.705	11.62	6.970	33.33	15.82
36	2.566	2.904	5.430	4.213	18.72	8.802
48	1.811	2.088	3.459	3.305	8.356	5.445
60	1.826	1.996	2.612	2.859	5.784	4.737
72	1.765	1.749	2.134	2.412	4.306	3.782

Next, a series of experiments to monitor the resorption of dye after immersing hydrogel with absorbed dye into the water was performed. The results showed that after the water change every 12 hours, the diffusion of rhodamine happens continuously from the hydrogel into the external environment and no resorption occurs. The results suggest that changes in the arrangement and binding of the polymer chains may occur due to the diffusion of rhodamine.

To prevent the resorption of the transported substance, in the next experiments, hydrogels were prepared directly with rhodamine in their structure (during synthesis). The gels had to be synthesized for 45 minutes instead of 30. A possible explanation is that rhodamine enters the polymerization reaction and forms bonds with free radicals or that it removes UV radiation due to its fluorescence. These experiments again showed the effect of a polyelectrolyte, which retained more dye in the hydrogel than the hydrogel without it.

Finally, another set of experiments was performed to monitor the effect of immersion of the samples in ethanol on the subsequent diffusion. The samples were first immersed for 24 hours in ethanol, then in water to wash thoroughly, and then in the dye solution. For comparison, the experiment was performed with the same concentration of rhodamine in samples without immersion in ethanol. The results showed that the hydrogels immersed in ethanol absorbed far more Rhodamine 6G during the diffusion of the dye into the internal structure than the samples immersed in water. One possible explanation is that the higher absorption capacity is caused by the pores, which were evident in **chapter 2.1**., where it was shown that their formation occurs with the use of ethanol. These hydrogels again showed the effect of a polyelectrolyte, which absorbed more dye.

The barrier properties were investigated using diffusion cells, where a solution of the rhodamine dye was present in the donor chamber and pure deionized water was present in the receiving chamber. First, experiments were performed with vertically oriented Franz cells. The hydrogel has a water content of about 30 wt% and with decreasing water content in the hydrogel, the free volume of hydrogel available for diffusion transport decreases, and the curvature of the diffusion trajectory increases. The first non-zero absorbance in the characteristic spectral region was thus recorded on the 12th day. Next, horizontal diffusion cells were subsequently used for this experiment. Triplicates, three samples of poly-HEMA hydrogel, and three with the addition of PSS were prepared. During the experiment, the average values of the concentrations again showed a higher barrier ability of hydrogels containing polyelectrolyte.

4. CONCLUSION

This work focused on monitoring the influence of an incorporated polyelectrolyte component on the resulting transport properties and internal architecture of a model hydrogel system in terms of the development of the targeted drug delivery systems. The dried samples were subjected to morphological analysis by scanning electron microscopy. The results showed that the freezing process did not affect these hydrogels. The results show that during freezing, the pores are reduced, and the surface of the gels is smooth. During the immersion in ethanol, changes in the structural properties of hydrogels occurred.

Experiments to determine the transport properties have demonstrated the influence of the incorporation of a semi-interpenetrating polyelectrolyte component (PSS). The hydrogels with the addition of PSS were able to absorb a larger amount of diffusion probe (Rhodamine 6G) into their structure. Certainly, the electrostatic interaction of the polyelectrolyte and the dye, which are oppositely charged and therefore attract each other, had a key role. The barrier properties of the hydrogels were investigated using diffusion cells. The influence of the incorporated polyelectrolyte on diffusion through the hydrogel was confirmed.

From the obtained results it can be stated that the use of semi-IPN can modify the transport and barrier properties of the model poly-HEMA hydrogels, which are the key properties for the use of these materials in controlled drug release systems.

REFERENCES

[1] MEANS, A.K., GRUNLAN, M.A. Modern strategies to achieve tissue-mimetic, mechanically robust hydrogels. *ACS Macro Letters.* [online]. 2019, vol. 8, no. 6, pp. 705-713. Available from: https://doi.org/10.1021/acsmacrolett.9b00276.

[2] MYUNG, D., WATERS, D., WISEMAN, M., DUHAMEL, P.E., NOOLANDI, J. TA, C.N., FRANK, C.W. Progress in the development of interpenetrating polymer network hydrogels. *Polymers for Advanced Technologies.* [online]. 2008, vol. 19, no. 6, pp. 647-657. Available from: https://doi.org/10.1002/pat.1134.

[3] TRUDICOVA, M., SMILEK, J., KALINA, M., SMILKOVA, M., ADAMKOVA, K., HRUBANOVA, K., KRZYZANEK, V., SEDLACEK, P. Multiscale experimental evaluation of agarose-based semi-interpenetrating polymer network hydrogels as materials with tunable rheological and transport performance. *Polymers.* [online]. 2020, vol. 12, no. 11, 2561. Available from: https://doi.org/10.3390/polym12112561.

[4] LIN, C.C., METTERS, A.T. Hydrogels in controlled release formulations: network design and mathematical modeling. *Advanced Drug Delivery Reviews.* [online]. 2006, vol. 58, no. 12-13, pp. 1379-1408. Available from: https://doi.org/10.1016/j.addr.2006.09.004.

[5] BAKER, M.V., BROWN, D.H., CASADIO, Y.S., CHIRILA, T.V. The preparation of poly(2-hydroxyethyl methacrylate) and poly{(2-hydroxyethyl methacrylate)-co-[poly(ethylene glycol) methyl ether methacrylate]} by photoinitiated polymerisation-induced phase separation in water. *Polymer.* [online]. 2009, vol. 50, no. 25, pp. 5918-5927. Available from: https://doi.org/10.1016/j.polymer.2009.10.047.

[6] ROTARU, I., OLARU, D. Mechanical behaviour of p(HEMA) hydrogel for disc prosthesis on lumbar spine. *Journal of Optoelectronics and Advanced Materials.* 2014, vol. 16, no. 7-8, pp. 881-886.

[7] GARCIA, J., RUIZ-DURANTEZ, E., VALDERRUTEN, N.E., Interpenetrating polymer networks hydrogels of chitosan and poly(2-hydroxyethyl methacrylate) for controlled release of quetiapine. *Reactive and Functional Polymers.* [online]. 2017, vol. 117, pp. 52-59. Available from: https://doi.org/10.1016/j.reactfunctpolym.2017.06.002.

[8] PEPPAS, N. Hydrogels in pharmaceutical formulations. *European Journal of Pharmaceutics and Biopharmaceutics.* [online]. 2000, vol. 50, no. 1, pp. 27-46. Available from: https://doi.org/10.1016/S0939-6411(00)00090-4.

[9] MARMORAT, C., ARINSTEIN, A., KOIFMAN, N., TALMON, Y., ZUSSMAN, E., RAFAILOVICH, M. Cryo-imaging of hydrogels supermolecular structure. *Scientific Reports.* [online]. 2016, vol. 6, 25495. Available from: https://doi.org/10.1038/srep25495.

HIGH ASPECT RATIO PARYLENE C MICROPILLARS FORMED BY MOLDING AND ION-BEAM ETCHING METHOD

[1,2]Zdenka FOHLEROVA, [2]Peter FECKO, [1,2]Imrich GABLECH, [2]Zuzana KOSELOVA

[1]*Central European Institute of Technology, Brno, Czech Republic, EU, zdenka.fohlerova@ceitec.vutbr.cz*

[2]*University of Technology, Faculty of Electrical Engineering and Communication, Brno, Czech Republic, EU*

https://doi.org/10.37904/nanocon.2021.4349

Abstract

This paper represents the molding and plasma etching method for developing a high aspect ratio parylene C pillar array. The silicon mold was fabricated using the modified Bosch process. The parylene was deposited by chemical vapor deposition into the silicone holes, followed by the etching of silicon to obtain the transparent membrane of hollow parylene C pillars. The etching method was initialized by transferring the pattern on the hard titanium mask via standard photolithography, followed by titanium etching. The oxygen plasma ion-milling of parylene C produced the flexible pillars standing on the silicon substrate. Both pillar arrays were characterized by scanning electron microscopy. Arrays of micropillars could be applicable for measuring cellular forces or as the bioinspired platform with modulated surface chemistry.

Keywords: Parylene C, micropillars, molding, etching

1. INTRODUCTION

A high aspect ratio polymeric micro- and nanopillar arrays have attracted great attention in the fields of optics due to the ability to influence the propagation of light, in surface chemistry by modifying the bulk substrate wettability, controlling the cell attachment or estimation of cellular force in the field of biology [1]. The polymeric pillars have been prepared from various materials, including PDMS, photocurable, or thermoplastic polymers [2]. The properties of the individual polymer and the final high aspect ratio of pillars predetermine the fabrication technique and route of pillar array development. The common techniques for the fabrication of micro- and nanopillars are lithography techniques (UV, X-ray, or electron/ion-beam), molding and imprinting methods, plasma etching or laser ablation [3,4]. They differ in many aspects such as resolution, high aspect ratio, directionality, cost-effectiveness, low/high throughput or processed polymeric material, etc. For example, the molding technique of high aspect ratio polymeric pillars is challenging as the demolding process can damage pillars. Moreover, a high pillar density may lead to the structure collapse. Imprinting techniques require UV or thermally processable polymers, the etching process is limited to the availability of selective polymer etchant, lithographic techniques are expensive, etc.

Recently, parylene C got attention in biomedicine as the biocompatible polymer with high Young´s modulus (≈4 GPa) [5]. Compared to the other polymers, parylene C must be deposited by a technique such as chemical vapor deposition, which has the disadvantage of higher requirement on the instrumentation. The fabrication of a high aspect ratio SiO_2 decorated micropillar array by parylene C plasma etching has been introduced previously by our group [6]. Such pillar development requires a high degree of etching control and anisotropy as they are more prone to loss of features due to undercutting. Moreover, the durability of the etch mask in the plasma process is a key factor for achieving high aspect ratios. Thus, the metallic hard mask is utilized in most cases.

On the other hand, molding methods have been frequently used to develop lower high aspect ratio PDMS pillars. The development of parylene C pillars by molding technique appears challenging, but we previously showed the fabrication of bioinspired high aspect ratio SiO_2 coated parylene pillars by molding technique [7].

More complex etch processes have to be utilized to prepare silicon mold to increase the aspect ratio. For such a purpose, the Bosch process introduces the solution of many repeated cycles between an isotropic SF_6 etch and the deposition of a passivation polymer layer from C_4F_8 gas to provide anisotropy.

This paper introduces the high aspect ratio parylene C micropillars developed by the ion-beam etching method and replica molding technique combined with the modified Bosch process. Micropillars geometry was characterized by scanning electron microscopy (SEM), and both techniques were briefly discussed.

2. METHODS

2.1. Dry etching method

A ≈10 µm thick layer of parylene C was deposited on the Si (100) wafer using the chemical vapor deposition (CVD) method. A 300 nm thin titanium layer was deposited using the electron beam evaporation technique. UV photolithography was performed using the ≈1.4 µm thick layer of photoresist (PR) AZ 5214E via mask aligner in vacuum contact mode (**Figures 1-1A**). The pattern of hexagonally arranged features with a diameter of 2 µm and center-to-center distance of 4 µm was developed. Then, the titanium was etched using reactive ion etching (RIE) in Cl_2 plasma (**Figures 1-1B**). Finally, the wafer was placed into the ion beam etching (IBE) instrument employing Kaufman ion-beam source with collimated grids using pure O_2 plasma for etching the parylene C from the areas uncovered by Ti. The PR was completely removed during the IBE process, and the Ti residues were additionally removed using chlorine-based RIE (**Figures 1-1C**). The wafer was cut into the samples with dimensions of (2×2) cm^2 using a femtosecond laser to eliminate the generation of impurities.

2.2. Molding method

The parylene micropillars have been fabricated by molding method using Si mold (**Figures 1-2A**). The pattern was transferred to the Si wafer with ≈300 nm thermal SiO_2 film and ≈1.5 µm positive photoresist (PR) using contact lithography. The PR was baked at 120 °C for 1 min and developed in a tetramethylammonium hydroxide solution. The SiO_2 was then etched by the deep reactive ion etching (DRIE) method in inductively coupled plasma (C_4F_8/O_2) for 1 minute (**Figures 1-2B**). The circular holes with a diameter and pitch of 1.6 µm and 4 µm, respectively, were created by etching of ≈0.535 mm thick Si (100) substrate with the Bosch process (**Figures 1-2C**). This process provided a rigid mold of hexagonally arranged holes of a depth of ≈11 µm (**Figures 1-2D**). PR residues were then removed by O_2 plasma, and silicon oxide was removed with 49% HF solution. The Si platform served as the mold for parylene C deposition under low pressure ≈1.6 Pa to improve coating uniformity (**Figures 1-2E**). The Si mold was etched in XeF_2 vapors for 10 s to obtain the silicon chamber with exposed micropillar membrane (**Figures 1- 2F**).

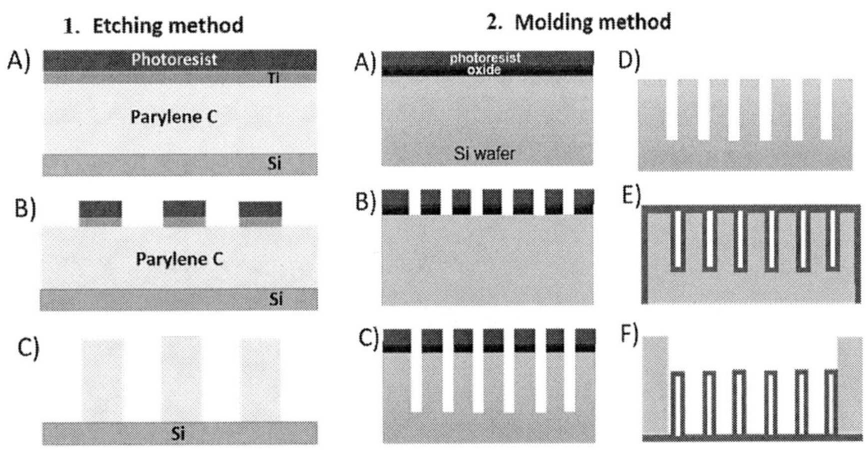

Figure 1 Scheme of fabrication of parylene C micropillars via molding (1A-C) and etching (2A-F) techniques

3. RESULTS AND DISCUSSION

3.1. Etching technique

Utilization of chemical dissolvent is possible for etching organic polymers in anisotropic way, but the parylene C, which is chemically inert in practice, the only option that remains is plasma-based dry etching. Preliminary experiments on ion-beam etching systems (IBE) using O2 plasma were carried out. The parylene C deposited on Si substrate was covered with evaporated titanium film and micropatterned using standard photolithography. The hard titanium mask was subsequently etched with chlorine-based plasma followed by oxygen plasma etching of parylene (**Figure 2A**) [6]. In **Figure 2B**, the nano-forest formation can be visible on the bottom of the substrate, which is supposed to arise from the micro-/nano-masking effect of the sputtered titanium metal mask. To overcome this problem, introducing SF_6 gas into the plasma source can counteract this masking phenomenon. However, the SF6 ion-beam suffered from poor directionality even at maximum allowed accelerator voltages, which yielded poor results [8]. However, inorganic hard masks lead to significant problems such as debris, sidewall roughening and grass formation, interaction with a polymer layer, scumming, and pattern distortion due to stress. Next, the slightly conical shape of pillars has been observed (**Figure 2A**) due to the ion-beam directionality of IBE. Nevertheless, the ideal cylindrical shape is beyond the equipment's capabilities we used. The top-down process of micropillar development provided the array of the hexagonally ordered parylene pillars with a ≈1.6 µm diameter, ≈4 µm center-to-center distance, ≈9 µm length, and rod-like morphology, as confirmed by SEM.

3.2. Molding method

The hexagonally arranged hollow parylene C micropillars have been developed by molding method as shown in **Figures 2C, D**. The pillar geometry was obtained of the diameter of ≈2 µm, pitch ≈4 µm, and the length of ≈11 µm. Using the Bosch process, a significant problem when applying this process can be the characteristic scalloping on the sidewall due to the isotropic etch step [9]. From SEM images is evident that we managed to minimalize this effect to a minimum. However, a slight scallop is visible (**Figure 2D**). Further, during the DRIE process of SiO_2 and Si, controlling the taper angle plays an important role. We optimized the angle during the Bosch process, and we got etched holes with a positive taper angle of ≈89° instead of ≈90°, which resulted in the narrowing of holes to ≈1,6 µm. Tapering in the HAR Bosch process can be controlled by balancing the passivation step and the breakthrough step, i.e., decreasing passivation time or increasing breakthrough time to reduce positive tapering and vice versa. Thus, further optimization of the Bosch process is necessary to obtain a perfectly cylindrical hole of a high depth-to-diameter ratio. The optimization process with the development of high aspect ratio pillars complicates the effectiveness of high-energy ions. During the breakthrough step, ion energy decreases with depth due to particle collisions in plasma and an opposite electrostatic charge of vertical walls. Therefore, the ion energy can be more precisely controlled by advanced modulating of the plasma parameters of the Bosch process.

Figure 2 SEM image of parylene C micropillars fabricated by etching method (A). Nanoforest from hard mask (B). SEM image of micropillars fabricated by molding method (C). Scallop effect on pillar surface (D), insert shows hollowness of broken pillars.

4. CONCLUSION

This paper introduced two approaches for developing a parylene C micropillar array for various surface chemistry or biology applications. The etching technique produced high aspect ratio micropillars (HR≈6) with slightly conical pillars due to the limitation in the capabilities of the equipment we used. The ion-beam etching is dependent on the beam energy and space charge, which is locally changing due to high polymer thickness. The non-transparency of the silicon substrate also limits this platform's use for transmission microscopy. Additionally, the residues from the hard mask are a disadvantage that is hard to overcome without losing original shape. On the other hand, the development of parylene C micropillars by molding technique can create a transparent micropillar membrane. The proper optimization of the Bosch process can develop high aspect ratio parylene pillars (HR≈10) with minimal scalloping effect. However, micropillars become more fragile and collapse. To conclude, the proper optimization of both techniques must be performed to know the limitations of methods used in this work to develop cylindrical HAR micropillars.

ACKNOWLEDGEMENTS

The authors wish to acknowledge the financial support of the GACR Project Number GA19-04270Y from the Czech Republic and the internal grant of BUT (FKT-S-20-6215).

REFERENCES

[1] ALAPAN, Y., ICOZ, K., GURKAN, U. A. Micro- and nanodevices integrated with biomolecular probes. *Biotechnology Advances.* 2015, vol. 33, no. 8, pp.1727-1743.

[2] SCHOEN, I., HU, W., KLOTZSCH, E., VOGEL, V. Probing Cellular Traction Forces by Micropillar Arrays: Contribution of Substrate Warping to Pillar Deflection. *Nano Letters.* 2010, vol. 10, no. 5, pp. 1823-1830.

[3] SHINOHARA, H., GOTO, H., KASAHARA, T., MIYUNO, J. Fabrication of a Polymer High-Aspect-Ratio Pillar Array Using UV Imprinting. *Micromachines.* 2013, vol.4, no. 2, pp. 157-167.

[4] PAPADOPOULOS, P., PINCHASIK, B.-E., TRESS, M., DORIS, V., KAPPL, M., BUTT, H.-J. Wetting of soft superhydrophobic micropillar arrays. *Soft Matter.* 2018, vol. 14, pp. 7429-7434.

[5] DE LA OLIVA, N., MUELLER, M., STIEGLITZ, T., NAVARRO, X., DEL VALLE, J. On the use of Parylene C polymer as substrate for peripheral nerve electrodes. *Scientific Reports.* 2018, vol. 8, no. 1, pp. 5965-5977.

[6] FOHLEROVA, Z., GABLECH, I., OTAHAL, A., FECKO, P. SiO2-Decorated Parylene C Micropillars Designed to Probe Cellular Force. *Advanced Materials Interfaces.* 2021, vol. 8, no. 6, pp. 2001897-2001907.

[7] LIU, X., FECKO, P., FOHLEROVA, Z., KARASEK, T., PEKAREK, J., NEUZIL, P. Parylene micropillars coated with thermally grown SiO2. *Journal of Vacuum Science & Technology B.* 2020, vol. 38, no. 6, pp. 063001.

[8] ZHANG, L., LIU, Y., LI, Z., WANG, W. SF6 Optimized O2 Plasma Etching of Parylene C. Micro*machines.* 2018, vol. 9, no. 4, pp. 162-170.

[9] PARK, J. S., KANG, D.-H., KWAK, S. M., KIM, T. S., PARK, J. H., KIM, T. G., BAEK, S.-H., LEE, B. C. Low-temperature smoothing method of scalloped DRIE trench by post-dry etching process based on SF6 plasma. *Micro and Nano Systems Letters.* 2020, vol. 8, no. 1, pp. 14-22.

CONSTANT CURRENT CHRONOPOTENTIOMETRY STUDY OF DNA FOR THE DETECTION OF AFRICAN SWINE FEVER VIRUS

[1]Ondrej RYCHLY, [1,2,5]Bozena HOSNEDLOVA, [3]Arli Aditya PARIKESIT, [2]Milan JAKUBEK, [4]Marta KEPINSKA, [5]Petra VASICKOVA, [1,2,4,5]Rene KIZEK

[1]CONEM Metallomics Nanomedicine Research Group (CMNRG), Vinohrady, Prague, Czech Republic, EU, rychly@seznam.cz

[2]BIOCEV, First Faculty of Medicine, Charles University, Vestec, Czech Republic, EU, bozena.hosnedlova@post.cz; milan.jakubek@lf1.cuni.cz

[3]Department of Bioinformatics, School of Life Sciences, Indonesia International Institute for Life Sciences, Jakarta Timur, Indonesia, pakresit@el.ind

[4]Department of Pharmaceutical Biochemistry, Division of Biomedical and Environmental Analyses, Faculty of Pharmacy, Wroclaw Medical University, Wroclaw, Poland, EU, marta.kepinska@umw.edu.pl

[5]Veterinary Research Institute, Brno, Czech Republic, EU, kizek@sci.muni.cz

**Correspondence:* kizek@sci.muni.cz; *Tel./Fax: +42-05-4156-2820*

https://doi.org/10.37904/nanocon.2021.4357

Abstract

African Swine Fever Virus (ASFV) is a DNA virus of the Asfivirus genus of the Asfarviridae family that is found in blood, body fluids, and internal organs. ASFV was described more than 40 years ago. This virus spreads pandemically and the mortality rate of the virus-related disease ranges from 90 to 100 %. The aim of this study was to propose the detection of the specific nucleic acid of ASFV using electrochemical detection. Determination of DNA was conducted by chronopotentiometric stripping analysis (time of accumulation 120 s, stripping current 20, 10, 5, 1 μA, measurement time 240 s). The volume of the analyzed sample was 10 μL. Sufficient analysis duration (200–600 s) is required to fully record the H peak signal of DNA at a given applied current. Extending the DNA analysis time at a given current significantly improved the dt/dE records of the H peak and the reproducibility of the analysis (RSD up to 10%). It was found that the suitable time at which the H peak is plotted reliably is around 240 s. The detected dsDNA concentration dependence (0 to 60 μg/mL) on the catalytic signal of the H peak was described according to the equation $y = 260.4 + 366.7x$, $r = 0.961$, RSD 23.1%. Subsequently, the SPION particle was modified with dsDNA. After the thermal release of DNA from SPION, DNA was immediately determined by chronopotentiometric stripping analysis method.

Keywords: Emergency biosensor, nucleic acid, oligonucleotide, superparamagnetic nanoparticle

1. INTRODUCTION

African swine fever (ASF) is a highly dangerous swine disease occurring on various continents. The African swine fever virus (ASFV) is known mainly for its infectivity and high mortality of the Suidae family. ASF first appeared in the sub-Saharan region of Africa [1] and from there the disease spread through the Transcaucasian region to Russia, Europe and Asia. ASF is already spreading in the Pacific. There is a growing concern that the disease will spread to other areas as a result of travel and imports of pork products from countries where ASF is present. In wild boars, the situation is out of control in many countries. The causative agent of ASFV is an enveloped cytoplasmic double-stranded DNA arbovirus (*Asfavirus*) (genome size 170-

193 kbp), which is the only member of the Asfarviridae family (**Figure 1**). Anti-epidemiological measures include disinfection regimes, partial or complete quarantine in the areas concerned and the consistent eradication of susceptible individuals at the outbreak. Early diagnosis is key to performing these measures. Commonly used virus identification techniques are based on the detection of viral nucleic acid by PCR or are based on the detection of viral antigens by immunological techniques. The specificity and sensitivity of these methods is very good, but they require laboratory facilities and qualified personnel. Electrochemical detection of nucleic acids is based on the redox properties of the nucleic bases present in DNA or RNA [2,3]. Until almost the late 1950s, nucleic acids were considered electrochemically inert and too large to be analyzed by electrochemistry. In 1958 Professor Emil Paleček found that the nuclear bases adenine and cytosine are electrochemically reduced [4,5]. Typical reduction signals of adenine and cytosine [6] are observable as a signal with a maximum in the region of about -1.30 V [7]. Later, Professor Trnkova described the detailed electrochemical behavior of guanine [6]. Both these discoveries helped to develop a completely new field of bioanalytical chemistry: sensors and biosensors. In contrast, electrochemical biosensors (do not require qualified BSL-2 laboratories) and can also speed up the detection of dangerous viruses [8,9]. The aim of this work was to propose the use of chronopotentiometric stripping analysis (CPSA) for sensitive detection of dsDNA.

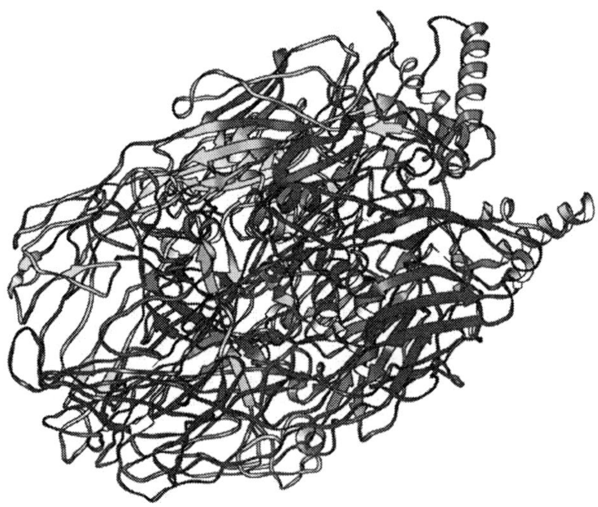

Figure 1 The most important capsid virion protein of the African Swine Fever Virus (ASFV) is depicted from the Protein Data Bank (PDB repository (https://www.rcsb.org/). P72 is the most abundant capsid virion antigenic in swine and considered the most important biomarker for the disease diagnosis (PDB ID: 6I2t). The 3-D visualization of the ASFV capsid virion protein p72 in the resolution of 4.1 Å, where the hydrogen bond could be examined accordingly [10,11].

2. MATERIALS AND METHODS

2.1. Chemicals and chemical analysis

All chemicals used in this study such as Trizma base, trisodium citrate, hydrogen tetrachloroaurate(III) ($HAuCl_4$), acetic acid (HAc), sodium acetate (NaOAc), tris(2-carboxyethyl)phosphine hydrochloride (TCEP), ethylenediaminetetraacetic acid (EDTA), disodium hydrogen phosphate (Na_2HPO_4), potassium dihydrogen phosphate (KH_2PO_4), sodium chloride (NaCl), potassium chloride (KCl), and HCl were purchased from Sigma-Aldrich (St. Louis, MO, USA), in ACS purity. Deionised water was prepared by using the reverse osmosis equipment Aqual 25 (Brno, Czech Republic), and was further purified by using an ELGA apparatus equipped with a UV lamp (Lane End, United Kingdom). The resistance was 18 MΩ and the pH was measured using the pH meter VWR (Radnor, USA).

2.2. Spectrometry measurement

The absorption spectra, in the wavelength range of 200 to 800 nm was measured. For fluorescence spectra, there was an excitation wavelength of 250 nm and an emission wavelength in the range from 350 to 800 nm (Varioskan, Thermo Fisher Scientific, USA). The optical properties of nanoparticles were studied by spectral analysis.

2.3. Electrochemical measurement

Electrochemical measurements were performed with an AUTOLAB Analyser connected to VA-Stand 663, using a standard cell with three electrodes. The working electrode was a hanging mercury drop electrode (HMDE) with a drop area of 0.4 mm^2. The reference electrode was an Ag/AgCl/3M KCl electrode, and the auxiliary electrode was a graphite electrode. The supporting electrolyte containing 0.2 M McIlvaine buffer (pH 7.0) was used. CPSA: time of accumulation 120 s, stripping current 20, 10, 5 and 1 µA, measurement time 240 s.

2.4. Zetasizer analysis

The size distribution (i.e., the hydrodynamic diameter, DH) was determined by dynamic light scattering (DLS) using the Zetasizer Nano ZS ZEN3600 (Malvern Instruments, Malvern, UK) with the detection angle of 173° in optically homogeneous square polystyrene cells. The samples have diluted a hundredfold with deionized water. All measurements were performed at 25 °C. Each value was obtained as an average of 5 runs with at least 10 measurements. Version 7.10 of the Zetasizer Software was applied for data evaluation. The particle charge (ζ-potential) was measured by the microelectrophoretic method using a Malvern Zetasizer Nano ZS ZEN3600 (Malvern Instruments, Malvern, UK). All the measurements were performed at 25 °C in polycarbonate cuvettes. Each value was obtained as an average of 5 subsequent runs of the instrument with at least 20 measurements.

2.5. Synthesis of SPIONs and SPION/DNA

1.3 g of $Fe(NO_3)_3 \cdot 9H_2O$ was dissolved in 80 mL of water (18 MΩ). A quantity of 1.4 mL 25% NH_3 was diluted in 8.6 mL water and 0.2 g $NaBH_4$ was dissolved in this mixture and, subsequently, the mixture was stirred (300 rpm, 10 min at 22 °C). The solution colour turned dark brown. The mixture was heated to 100 °C for 2 h and stirred (300 rpm, stirring overnight at room temperature). The magnetic particles were separated from the solution via a magnet and washed several times in water. The solution of $HAuCl_4$ (20 mL, 1 mM) was added to the magnetic particles and stirred (300 rpm 3 h at 22 °C). Subsequently, the solution of $C_6H_5Na_3O_7 \cdot 2H_2O$ (0.5 mL, 0.265 g/10 mL) was added to the mixture and stirred overnight. Finally, the gold SPIONs were separated via a magnet and dried (40 °C).

2.6. Statistical data analysis

Available experimental data were processed and evaluated mathematically and statistically directly in the Qinslab database. The exclusion of extreme values for data sets was performed by calculation in the Grubbs test. Experimental work was performed in at least three independent experiments (n_1). Each sample in the experiments was analyzed at least five times (n_2). The data presented in this work are presented as average values. LOD (limit of detection) values were determined according to the work of Hubaux and Vos [12] at a level of significance of 95%. Data visualization was performed using the Qinslab database.

3. RESULTS AND DISCUSSION

Electrochemical biosensing of viral pathogens is summarized in a study by Campuzano et al. [13]. A viral pathogen can be detected by the presence of its nucleic acid. The designed biosensor allows recognition very easily [3]. The concept proposed by us is summarized in **Figure 2**. The SPION particle is modified with a specific ASFV antibody (most often to the p78 protein). The virus is captured on SPION and after the necessary washing steps is mechanically and physically inactivated and disrupted (thus guaranteeing specificity). The nucleic acid is released into solution and analyzed by CPSA. In view of the direct detection of the presence of a pathogen in a sample, it is necessary for these purposes to obtain a significant sensitivity of DNA detection, which is guaranteed by CPSA which is highly sensitive in the order of up to fg in the analyzed sample.In several previous experimental works, we used mainly redox signals of nucleic acid bases (adenine, cytosine) [14]. The sensitivity of these approaches varies in submicrograms per mL. Catalytic signals are known to increase the sensitivity of analyte detection [15]. Catalytic signals of hydrogen excretion have been described for proteins but also for nucleic acids [16,17]. We focused on the description of the basic behavior of the H peak induced by the presence of dsDNA (in McIlvaine buffer, pH 7.0).

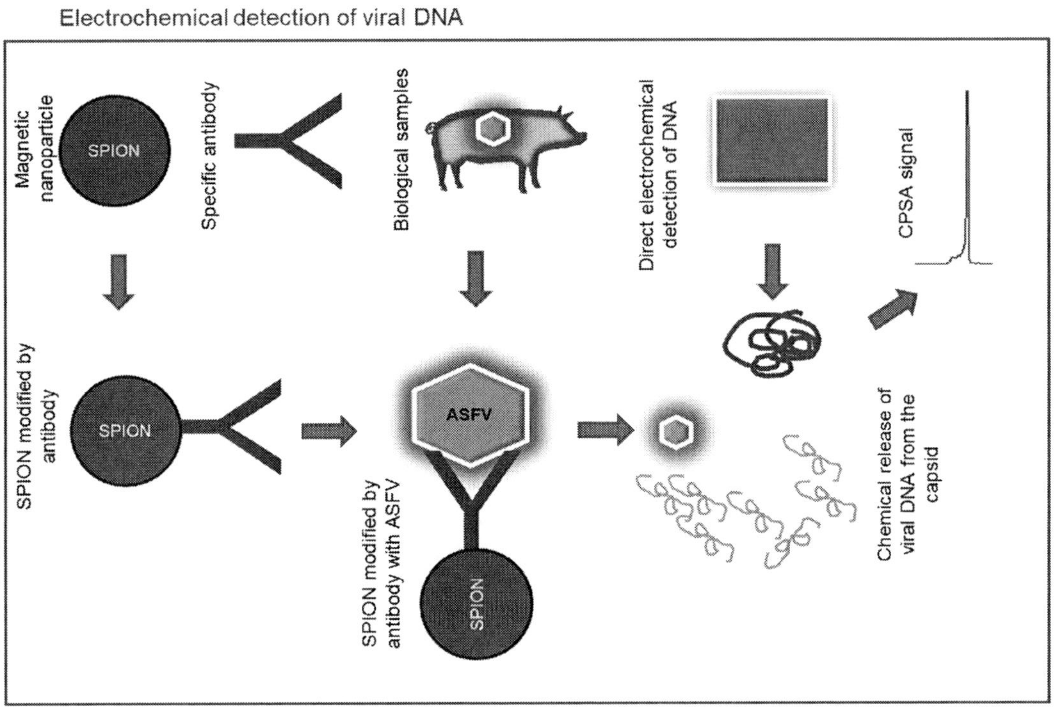

Figure 2 Scheme of presumed direct detection of viral DNA using electrochemical detection of CPSA. The procedure uses gold magnetic particles (SPIONs) modified with a specific antibody to the ASFV. A biological sample (homogenate) containing ASFV is incubated with SPION. Bound ASFV to SPION is subsequently disintegrated to release genomic DNA. The obtained sample is analyzed electrochemically by CPSA.

Gold is known to adsorb to DNA through affinity interactions [18]. Some forms of interactions between ds-DNA and citrate coated gold nanoparticles have been reported [19-25]. Different electrointeractions between the surface of gold nanoparticles (having a positive charge) and DNA that has a negative charge, were demonstrated. Gold nanoparticles are known for their excellent biocompatibility with biomolecules and possess remarkable structural, electronic, magnetic, optical and catalytic properties [26]. Caballero et al. [27] found that α-lipoic acid/aminoanthraquinone functionalized gold nanoparticles caused notable coiling/wrapping of DNA despite their anionic surface charge. The functionalization of magnetic nanoparticles with gold was demonstrated through the conjugation of thiolated DNA to the gold shell [28]. Synthesis Deghan et al. reported that SPIONs doped with 2H-chromene via dopamine as cross linker bound to calf thymus DNA via groove binding mode [29]. Regarding ssDNA and dsDNA interactions with AuNPs, dsDNA has (due to its double helix structure, which exposes the negative charges of the phosphate groups to the outside) greater rigidity than ssDNA, which makes it difficult to interact with the surface of anionic AuNPs. In addition to affecting the electrostatic component of this interaction, Van der Waals forces are also affected by the presence of non-polar regions in DNA. In the case of cationic AuNPs, electrostatic forces prefer the molecular approach [30-32]. The presence of cationic NPs causes partial denaturation of the DNA helix, resulting in a number of intermediate structures between ss and dsDNA [31]. Electrostatic interactions lead to DNA bending, whereas hydrophobic interactions with bases in the grooves deform the DNA structure [30,31].

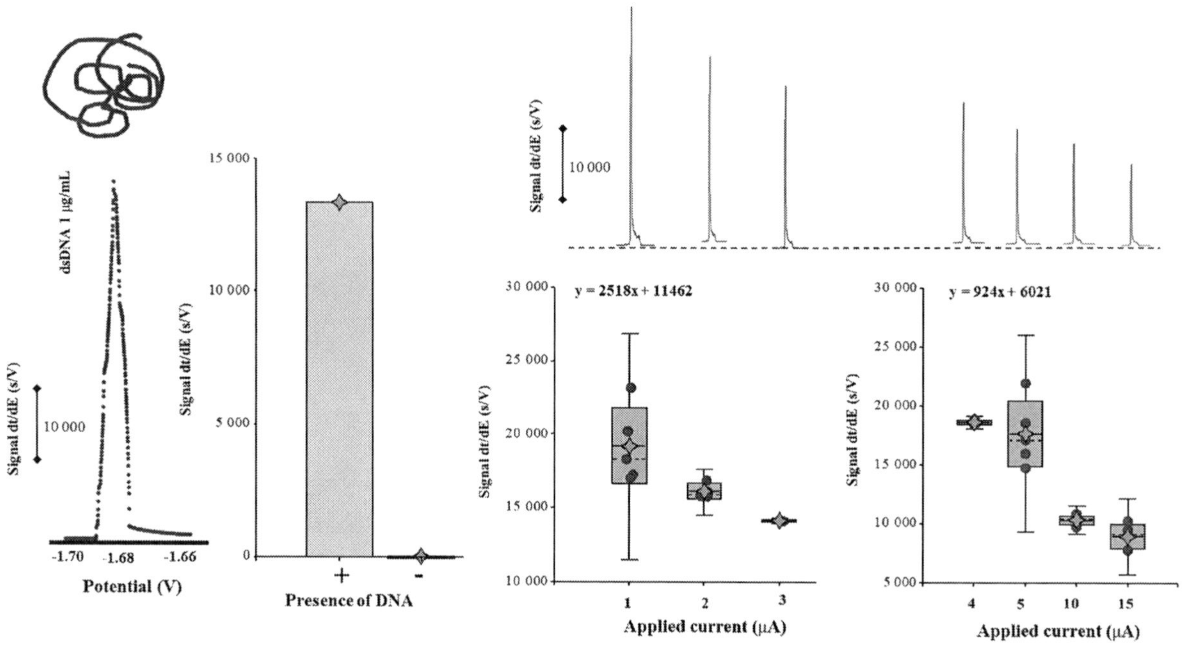

Figure 3 Typical CPSA signal (peak H) of genomic dsDNA (1 µg/mL), basic electrolyte (black line) at a measurement time of 120 s. CPSA peak H was tested in the presence and absence of dsDNA (n = 10). Changes of the H peak signal on the applied current (at measurement times 180 and 360 s). At the top of the figure, there are typical peaks of H at a given current.

Several different signals appear on the obtained chronopotentiogram of DNA (10 µg/mL) (**Figure 3**). We verified that the presence of DNA in the sample leads to the catalytic signal of the H peak (n = 10). We further studied under which conditions (concentration of DNA in the sample, applied current used for analysis and maximum duration of analysis) different catalytic signals are generated and how the H catalytic peak of DNA changes. The applied current is essential for the subsequently detected dt/dE signal (**Figure 3**). The applied current suitable for analysis varies according to the amount of dsDNA present in the sample. We tested different applied currents (20, 10, 5 and 1 µA). At a DNA concentration of less than 1 µg/mL, we achieved the best-plotted H peaks (dt/dE) at a current of 5 µA. Different current values were applied in two independent experiments: 10 µA for DNA concentrations of 1, 5 and 10 µg/mL, and 15 µA for concentrations of 10, 15 and 20 µg/mL. For DNA concentrations above 20 µg/mL, the best responses were obtained when applying a current of 20 µA. Sufficient analysis duration (dependence of a potential E on a time t) is required to fully record the signal of the H peak of dsDNA at a given applied current. Extending the DNA analysis time at a given current significantly improved the dt/dE records of the H peak and the reproducibility of the analysis (RSD up to 10%). It was found that the suitable time for which the H peak is reliably plotted is around 240 s. The observed concentration dependence of DNA (0 to 60 µg/mL) on the catalytic signal of the H peak was described according to the equation $y = 260.4 + 366.7x$, $r = 0.961$, RSD 23.1 % (**Figure 4**). To simulate the detection of dsDNA from the sample, SPIONs/DNA modified particles were prepared: 500 µg of SPIONs was added to 100 µg/mL of dsDNA, the interaction was performed for 30 min, 25 °C with gentle stirring; phosphate buffer (pH 5.0, 0.2 M, 200 µL) was used for the washing step. The tube was placed on a shaker (25 °C, 90 s, 300 rpm). The individual steps were repeated 3 times. The SPIONs/DNA was placed on a magnet and the liquid was aspirated, and then the elution phosphate buffer (pH 5.0, 0.2 M) was added. Due to the direct detection of DNA in the sample, DNA fragmentation may occur, which we do not consider to be a defect. The tube was placed in a thermoblock (100 °C, 15 min, 350 rpm). Subsequently, the sample was centrifuged at 16,000 *g*, 5 min, 4 °C, the tube with the sample was placed on a magnet, the liquid was pipetted and transferred to a clean tube.) The prepared final sample was analyzed by CPSA. A signal was recorded in the presence of DNA in the analyzed sample, while no signal appeared in its absence (**Figure 4**).

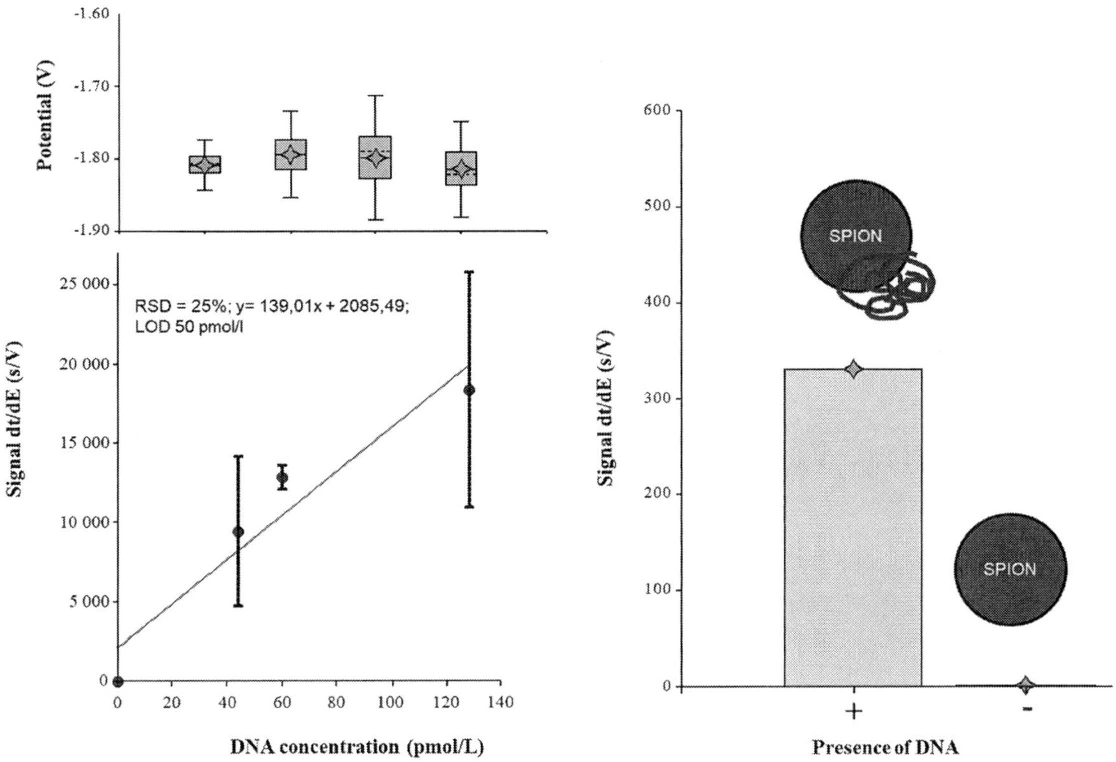

Figure 4 Mean CPSA signals dt/dE dsDNA as a function of concentration and the potential of peak H (upper part of the figure). Confirmation of the presence of dsDNA captured on SPION particles after their interaction for 240 s at 300 rpm and 25 °C in hybridization buffer.

4. CONCLUSION

Research in the field of sensors and biosensors for the rapid diagnosis of dangerous pathogens is a highly topical issue. We have found that CPSA can be used to detect ASFV DNA using the catalytic signal of the H peak. The sensitivity of the catalytic signals is significantly higher than the redox signals of nucleic acids. This fact could bring a significant increase in the sensitivity of direct detection of the detected viral pathogen.

ACKNOWLEDGEMENTS

The study was supported by the project AMOR QK1920113, and by the project of Charles University in Prague (Progress Q26/LF1).

REFERENCES

[1] KRZYZANKOVA, M., KRASNA, M., BENA, M.,VASICKOVA, P. Virus afrického moru prasat a moznosti jeho sireni masem a masnymi vyrobky. *Maso.* 2020, vol., no. 1, pp. 32-38.

[2] PALECEK, E.,JELEN, F. Electrochemistry of nucleic acids and development of DNA sensors. *Crit. Rev. Anal. Chem.* 2002, vol. 32, no. 3, pp. 261-270.

[3] WANG, J., CAI, X. H., RIVAS, G., SHIRAISHI, H.,DONTHA, N. Nucleic-acid immobilization, recognition and detection at chronopotentiometric DNA chips. *Biosens. Bioelectron.* 1997, vol. 12, no. 7, pp. 587-599.

[4] PALECEK, E. Oscillographic polarography of highly polymerized deoxyribonucleic acid. *Nature.* 1960, vol. 188, no. 4751, pp. 656-657.

[5] PALECEK, E., OSTATNA, V.,PECHAN, Z. Fifty Five Years of Nucleic Acid Electrochemistry. *Chem. Listy.* 2014, vol. 108, no. 5, pp. 490-499.

[6] TRNKOVA, L., STUDNICKOVA, M.,PALECEK, E. Electrochemical-behavior of guanine and its derivatives .1. Fast cyclic voltammetry of guanosine and calf thymus DNA. *Bioelectrochem. Bioenerg.* 1980, vol. 7, no. 4, pp. 643-658.

[7] BOUSSICAULT, F.,ROBERT, M. Electron transfer in DNA and in DNA-related biological processes. Electrochemical insights. *Chem. Rev.* 2008, vol. 108, no. 7, pp. 2622-2645.

[8] GARGULAK, M., SEHNAL, K., HOSNDLOVA, B., HOAI, N., DOCEKALOVA, M., OFOMAJA, A., STANKOVA, M., UHLIROVA, D., KEPINSKA, M., MILNEROWICZ, H., FERNANDEZ, C., VACIKOVA, P.,KIZEK, R. 3D-printed CdTe QDs-based sensor for sensitive electrochemical detection of viral particles. In: *IEEE International Conference on Sensors and Nanotechnology*, 2019, pp. 1-4.

[9] BANAS, D., RYCHLY, O., SALMISTRARO, S., AKSU, A. D., KRZYZANKOVA, M., KIZEK, R.,LTD, T. 2021. Optimization of nucleic acid binding to magnetic particles with the aim of detection of dangerous viruses. In: *12th International Conference on Nanomaterials - Research and Application (NANOCON).* Brno, Czech Republic: Tanger, 2020, pp. 324-329.

[10] LIU, Q., MA, B. T., QIAN, N. C., ZHANG, F., TAN, X., LEI, J. L.,XIANG, Y. Structure of the African swine fever virus major capsid protein p72. *Cell Res.* 2019, vol. 29, no. 11, pp. 953-955.

[11] WANG, N., ZHAO, D. M., WANG, J. L., ZHANG, Y. L., WANG, M., GAO, Y., LI, F., WANG, J. F., BU, Z. G., RAO, Z. H.,WANG, X. X. Architecture of African swine fever virus and implications for viral assembly. *Science.* 2019, vol. 366, no. 6465, pp. 640-+.

[12] HUBAUX, A.,VOS, G. Decision and detection limits for calibration curves. *J. Anal. Chem.* 1970, vol. 42, no. 8, pp. 849-855.

[13] CAMPUZANO, S., YANEZ-SEDENO, P.,PINGARRON, J. M. Electrochemical Biosensing for the Diagnosis of Viral Infections and Tropical Diseases. *Chemelectrochem.* 2017, vol. 4, no. 4, pp. 753-777.

[14] BANAS, D., AKSU, D. A., NOGUERA, V. M., MERT, P., HOSNEDLOVA, B.,KIZEK, R. Electrochemical Study of Quantum Dots-Labeled Oligonucleotide Probes for Detecting Nucleic Acid of African Swine Fever Virus. *Chem. Listy.* 2020, vol. 114, no. 11, pp. 778-783.

[15] MASARIK, M., KIZEK, R., KRAMER, K. J., BILLOVA, S., BRAZDOVA, M., VACEK, J., BAILEY, M., JELEN, F.,HOWARD, J. A. Application of avidin-biotin technology and adsorptive transfer stripping square-wave voltammetry for detection of DNA hybridization and avidin in transgenic avidin maize. *Anal. Chem.* 2003, vol. 75, no. 11, pp. 2663-2669.

[16] KIZEK, R., TRNKOVA, L.,PALECEK, E. Determination of metallothionein at the femtomole level by constant current stripping chronopotentiometry. *Analytical Chemistry.* 2001, vol. 73, no. 20, pp. 4801-4807.

[17] PALECEK, E.,BARTOSIK, M. Intrinsic Electrocatalysis in DNA. *Chemelectrochem,* 2018, vol. 5, no. 6, pp. 936-942.

[18] KOO, K. M., SINA, A. A., CARRASCOSA, L. G., SHIDDIKY, M. J.,TRAU, M. DNA-bare gold affinity interactions: mechanism and applications in biosensing. *Analytical Methods.* 2015, vol. 7, no. 17, pp. 7042-7054.

[19] CASTILLO, P. M., JIMENEZ-RUIZ, A., CARNERERO, J. M.,PRADO-GOTOR, R. Exploring Factors for the Design of Nanoparticles as Drug Delivery Vectors. *ChemPhysChem.* 2018, vol. 19, no. 21, pp. 2810-2828.

[20] GEARHEART, L. A., PLOEHN, H. J.,MURPHY, C. J. Oligonucleotide adsorption to gold nanoparticles: a surface-enhanced Raman spectroscopy study of intrinsically bent DNA. *The Journal of Physical Chemistry B.* 2001, vol. 105, no. 50, pp. 12609-12615.

[21] SANDSTRÖM, P., BONCHEVA, M.,ÅKERMAN, B. Nonspecific and thiol-specific binding of DNA to gold nanoparticles. *Langmuir.* 2003, vol. 19, no. 18, pp. 7537-7543.

[22] ZHANG, X., SERVOS, M. R.,LIU, J. Surface science of DNA adsorption onto citrate-capped gold nanoparticles. *Langmuir.* 2012, vol. 28, no. 8, pp. 3896-3902.

[23] CARNERERO, J. M., JIMENEZ-RUIZ, A., GRUESO, E. M.,PRADO-GOTOR, R. Understanding and improving aggregated gold nanoparticle/dsDNA interactions by molecular spectroscopy and deconvolution methods. *Physical Chemistry Chemical Physics.* 2017, vol. 19, no. 24, pp. 16113-16123.

[24] CARNERERO, J. M., MASUOKA, S., BABA, H., YOSHIKAWA, Y., PRADO-GOTOR, R.,YOSHIKAWA, K. Decorating a single giant DNA with gold nanoparticles. *RSC advances.* 2018, vol. 8, no. 47, pp. 26571-26579.

[25] ZINCHENKO, A.,YOSHIKAWA, K. Compaction of double-stranded DNA by negatively charged proteins and colloids. *Current Opinion in Colloid & Interface Science.* 2015, vol. 20, no. 1, pp. 60-65.

[26] CARNERERO, J. M., JIMENEZ-RUIZ, A., CASTILLO, P. M., PRADO-GOTOR, R. Covalent and Non-Covalent DNA-Gold-Nanoparticle Interactions: New Avenues of Research. *Chemphyschem.* 2017, vol. 18, no. 1, pp. 17-33.

[27] CABALLERO, A. B., CARDO, L., CLAIRE, S., CRAIG, J. S., HODGES, N. J., VLADYKA, A., ALBRECHT, T., ROCHFORD, L. A., PIKRAMENOU, Z.,HANNON, M. J. Assisted delivery of anti-tumour platinum drugs using DNA-coiling gold nanoparticles bearing lumophores and intercalators: towards a new generation of multimodal nanocarriers with enhanced action. *Chemical science.* 2019, vol. 10, no. 40, pp. 9244-9256.

[28] ROBINSON, I., TUNG, L. D., MAENOSONO, S., WÄLTI, C.,THANH, N. T. Synthesis of core-shell gold coated magnetic nanoparticles and their interaction with thiolated DNA. *Nanoscale.* 2010, vol. 2, no. 12, pp. 2624-2630.

[29] DEHGHAN, G., MEHDIPOUR, M., HOSSEINPOUR FEIZI, M. A., TARASI, R., KHOOBI, M.,DASTMALCHI, S. Spectroscopic investigation on the interaction of DNA with superparamagnetic iron oxide nanoparticles doped with chromene via dopamine as cross linker. *Nanomedicine Journal.* 2018, vol. 5, no. 1, pp. 36-45.

[30] HAN, G., MARTIN, C. T.,ROTELLO, V. M. Stability of gold nanoparticle-bound DNA toward biological, physical, and chemical agents. *Chemical Biology & Drug Design.* 2006, vol. 67, no. 1, pp. 78-82.

[31] RAILSBACK, J. G., SINGH, A., PEARCE, R. C., MCKNIGHT, T. E., COLLAZO, R., SITAR, Z., YINGLING, Y. G., MELECHKO, A. V. Weakly Charged Cationic Nanoparticles Induce DNA Bending and Strand Separation. *Advanced Materials.* 2012, vol. 24, no. 31, pp. 4261-+.

[32] GOODMAN, C. M., CHARI, N. S., HAN, G., HONG, R., GHOSH, P.,ROTELLO, V. M. DNA-binding by functionalized gold nanoparticles: Mechanism and structural requirements. *Chemical Biology & Drug Design.* 2006, vol. 67, no. 4, pp. 297-304.

MINERALISED POLYLACTIDE AND POLYCAPROLACTONE SOFT FOAMS WITH HIERARCHICAL MICRO-MACRO POROUS STRUCTURE FOR TISSUE ENGINEERING

[1]Miloš BERAN, [1]Eliška BERKOVÁ, [2]Jana MUSÍLKOVÁ, [2]Antonín SEDLÁŘ, [3]Petr SLEPIČKA, [3]Dominik FAJSTAVR

[1]Food Research Institute Prague, Prague, Czech Republic, EU, m.beran@vupp.cz

[2]Institute of Physiology CAS, Prague, Czech Republic, EU, jana.musilkova@fgu.cas.cz

[3]University of Chemistry and Technology, Prague, Czech Republic, EU, petr.slepicka@vscht.cz

https://doi.org/10.37904/nanocon.2021.4346

Abstract

The purpose of the study was development of soft foams from resorbable polymers with unique micro-macro porous interconnected hierarchical structure specially designed as scaffold for engineering of soft tissues.
The foams have been prepared by freeze-drying of solutions of polylactide (PLA) and polycaprolactone (PCL) in 1,4-dioxane. The foams prepared by freeze-drying had interconnected porous aerogel characteristics. The hierarchical structure with bimodal micro-macro pore size distribution were obtained after addition of sucrose or NaCl porogens with defined crystal size distributions to the solutions before freeze-drying and leaching the porogen crystals from the freeze-dried foams with demineralised water. Polyethyleneimine was chemically conjugated to the alkali-treated foams followed by conjugation of citric acid using carbodiimide chemistry. Finally, they were mineralised by immersing and incubating in a simulated body fluid with ionic concentration similar to that of human blood plasma, to obtain tissue engineering scaffolds. To verify their biocompatibility, the scaffolds were seeded with adipose-derived stem cells (ASC) and sarcoma osteogenic-2 (SaOs-2) human osteoblast-like cells. Morphology of the cells attached to the scaffolds was evaluated and their viability was verified by a metabolic test.
Biocompatibility and usability of the scaffolds was successfully verified by incubation with adipose-derived stem cells and SaOs-2 human osteosarcoma cell line. Mineralised scaffolds are more suitable growth supports for both the cell types than unmineralized collagen scaffolds. The scaffolds have been specially designed for engineering of soft tissues, but they can be used in other categories of tissue engineering, too.

Keywords: Tissue engineering, polylactide, polycarolactone, soft foams, hierarchical structure

1. INTRODUCTION

Tissue engineering combines isolated functional cells and a biodegradable biomaterial to promote theregeneration of repair diseased or injured tissues. A highly porous scaffold plays a critical role inaccommodating cells and guiding their growth and tissue regeneration in three dimensions. Experiencesand research data have suggested that a pore size between 200 and 300 μm might be necessary for the cellseeding efficiency and the homogeneity of the tissue engineered. In the last two decades, a number of foaming techniques have been developed to fabricate highly interconnective, porous scaffolds for tissue engineering applications. Artificial tissue substitute prepared either from nanostructured biocompatible materials or by decellularized non-autologous pericardium combined with autologous ASCs seems to be a promising candidate for cardiovascular grafts capable of accelerating in situ endothelialization. ASCs differentiation could be facilitated by sustained release of growth factors [1]. An advantage of this approach is that ASCs can easily be collected from the patient by liposuction. In this work, we have developed a method of preparation of polymeric foams with hierarchical micro-macro porous structure by freeze-drying of polymeric solutions in 1,4-dioxane with addition of water soluble porogens.

2. MATERIALS AND METHODS

2.1. Preparation of the soft foams with hierarchical micro-macroporous structure from PLA, PCL and their mixtures

Polylactide (PLA) Ingeo 4043D, Mn 160.000 g.mol^{-1} (NatureWorks, USA) and polycaprolectone (PCL), average Mr 80 000 (Sigma-Aldrich, Czech Republic) and their different mixtures were used to prepare the soft foams. Single polymers were dissolved in 1,4-dioxane (DO). Concentrations of the polymers, or their mixtures, were in the range 1-5% (w/w). Porogens in the form of commercial caster sugar (crystal size in the range 0.16-0.63 mm) or finely ground NaCl (0.25-0.5 mm) were added to the solutions aligned to the solution surface level. After freezing at - 75 °C the suspensions were freeze dried in Lyovac GT 2 freeze dryer (LH Leybold, Germany). All chemicals, except PLA, were purchased from Sigma-Aldrich (Czech Republic).

2.2. Mineralisation of the prepared foams

The prepared foams were functionalized by mineralization with simulated body fluid (SBF) solution, using carbodiimide chemistry, as described by Jaidev and Kaushik Chatterjee [2]. Composition and preparation of the SBF solution were described by Kobuko and Takadama [3]. All chemicals were purchased from Sigma-Aldrich (Czech Republic).

2.3. Microstructure visualization in SEM

SEM scans were acquired by a scanning electron microscope (FIB-SEM, LYRA3 GMU, Tescan, Brno,Czech Republic). The applied acceleration voltage was 10 kV. The examined samples were covered with a Pt conductive layer of 20 nm thickness.

2.4. Sterilization of the prepared foams

Before the experiments, the samples of foams were sterilized by gamma-irradiation, using the basicsterilization dose (Microtron MT25, Nuclear Physics Institute, Academy of Sciences of the Czech Republic, Řež).

2.5. Metabolic aktivity assay

The metabolic activity of adipose derived stem cells (ASCs) and sarcoma osteogenic-2 (SaOs-2) human osteoblast-like cells (European Collection of Cell Cultures, Cat. No. 89050205) seeded in soft foams scaffolds, incubated in 24 or 96-well plates was measured with a resazurin assay according to Sedlář, et al. [4]. ASCs were isolated from lipoaspirates and characterized according earlier study [5]. ASCs were incubated in DMEM medium with 10% (w/w) of fetal serum (FS) and 10 ng/ml of growth factor FGF 2. The medium was inoculated with 150 thousand ACS cells in 1,5ml of the medium per sample. SaOs-2 cells were incubated in McCoys medium with 15% (w/w) of FS. The medium was inoculated with 200 thousand SaOs-2 cells in 1,5ml of the medium per sample. The cell metabolic activity was regarded as an indirect marker of the cell proliferation and the cell number. The results of the metabolic activity assay were compared with results obtained with polystyrene (PS) substrate (reference control).

2.6. Determination of morphology and cell growth

Morphology and cell growth were determined by a standard method of staining of cytoskeleton with fluorescent and biotinylated derivative of phalloidin. Nuclear DNA was stained with fluorescent4',6-diamidin-2-fenylindol (DAPI). Images obtained by fluorescent microscopy with Olympus IX71 microscope equipped with DP71 digital camera allowed to distinguish cytoskeleton with red colour and cell nuclei with blue colour. Dragonfly confocal microscope (Andor Technology, Great Britain) have been used to visualize ingrowth of the cells into the scaffolds. The results were compared with results obtained with polystyrene (PS) substrate (reference control).

3. RESULTS AND DISCUSSION

The foams prepared by freeze-drying had interconnected microporous aerogel characteristics. This primary pore sizes can be regulated by changing the concentration of the polymers in solution in the range from several µm to several tens of µm. The hierarchical structure with bimodal micro-macro pore size distribution were obtained after addition of water soluble porogens with defined crystal size distributions to the solutions before freeze-drying and leaching the crystals from the freeze-dried foams with demineralised water. Photography of the soft foams prepared from PCL/PLA mixture (5:3), with total concentration of polymers in DO solution 4% w/w, after leaching the caster sugar porogen is shown in **Figure 1A**. SEM image of inner structure of the foam is shown in **Figure 1B**.

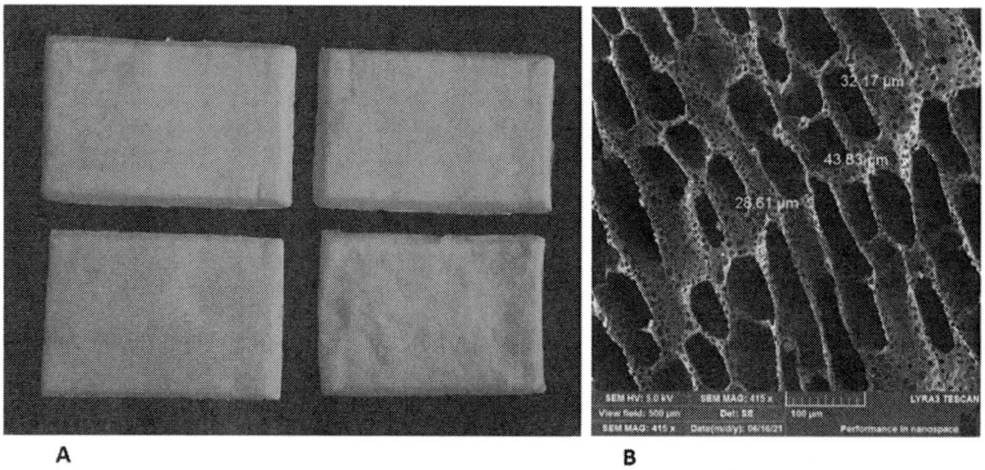

Figure 1 A - photography of preprared PLA/PCL (3:5, total concentration of the polymers in DO 4% w/w) foams; **B** - SEM image of microstructure of the mineralised scaffold

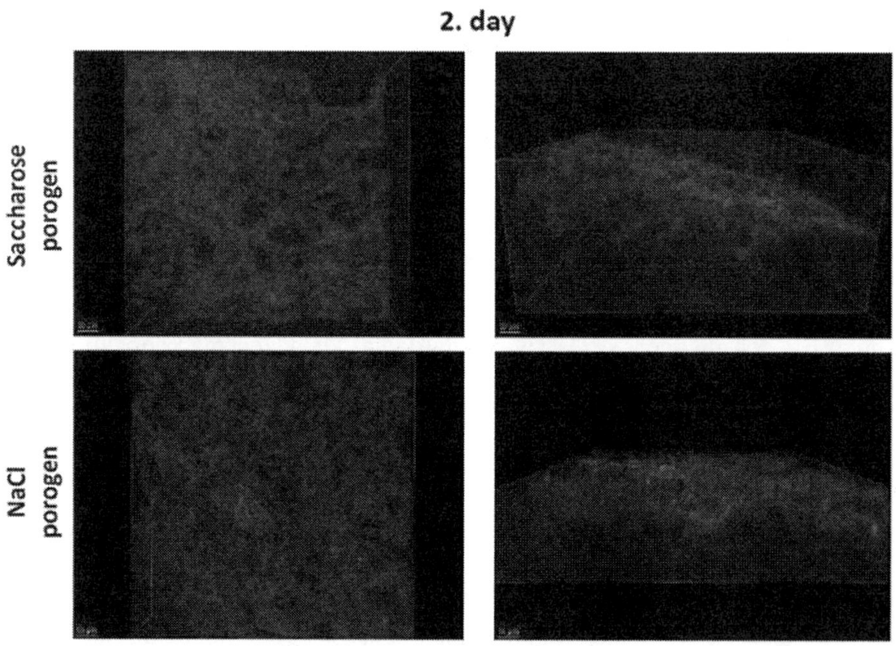

Figure 2 3D photography from the Drangonfly confocal microscope - ACS cells ingrowth and proliferation through the unmineralised PLA/PCL foams

The inner structures of the foams prepared from PLA and PCL, or their different mixtures, were almost identical at defined concentrations of the polymers. The primary micropores, visible at the image, was formed by dioxane sublimation from the freezed structure. The macropores, formed after the porogen leaching, are slightly smaller than porogen crystal sizes because of partial polymer foam shrinkage after the leaching and drying. The aeras of cross-section of the primary and secondary micropores of the PLA/PCL (3:5) foams prepared with the total concentration of the polymers in DO 4% w/w after leaching the caster sugar porogen were in the range 3.1 - 314 μm^2 and 2400 - 7800 μm^2, respectively. **Figure 2** shows visualisation of ingrowth and proliferation of ASCs into the unmineralised scaffolds. The cells are mainly localised at the surface of the scaffold with a limited ingrowth into the foam interior. We suppose, that the main reason of the limited ingrowth of the cells is hydrophobicity of the surface resulting in restricted suction of aqueous media.

The metabolic activity of ASC and SaOs-2 cells cultivated on different PLA/PCL foam types was tested. For the unmineralised foams the ASCs as a tested cells were chosen. For the testing of the mineralisation effectiveness of both the stem cells (ASCs) and the osteoblast-like cells (SaOs-2) were used. The metabolic activity of ACS cells cultivated on unmineralised PLA/PCL foams in comparison with polystyrene (PS) as a positive control taken on 10. day after seeding are shown at **Figure 3A**. We found out, that the metabolic activity of the cells cultivated on tested scaffolds was considerably high, around 40 % of the value of the positive control (PS). **Figure 3B** shows the comparison of metabolic activity the both ASCs and SaOs-2 cells cultivated on unmineralised (C) and mineralized (M) PLA/PCL foams on 2. day (blue) and 6. day (orange) in comparison with polystyrene (PS) as a positive control. The ASCs and SaOS-2 cells metabolic activities after 6 days of cultivation on the mineralised scaffolds were comparable despite the fact, that SaOS-2 grow usually slowly than ASCs cells, as was shown on PS.

The cells growing on the PLA/PCL foams were visualized by fluorescence staining. **Figure 4** shows results of fluorescent staining of SaOs-2 cells (blue colour) incubated in unmineralised and mineralised scaffolds prepared with the saccharose porogen and **Figure 5** results of fluorescent staining of ASCs (blue colour) incubated on mineralised scaffolds prepared with the saccharose porogen in comparison with PS substrate control. Both the cell types formed rather single layer on the scaffold surface on day 2. In the following days the cells formed more layers and started to penetrate into the scaffold interior, as noticeable from higher density of the cell nuclei and morphology of the cells at higher magnification. More distinct fibrillar structure (red colour) testify in favour of ASCs differentiation towards smooth muscle. The metabolic activities of both the cell types cultivated on the both mineralized and unmineralized PLA/PCL foams are rising during the time interval between the day 2 and day 6. The metabolic activities of both the cell types were getting closer to the metabolic activities on the reference PS substrate surface, despite the fact, that a part of the cells is hidden inside 3D structure of the foams. That means, that PLA/PCL foams of both types are biocompatible and appropriate for the cell cultivation.

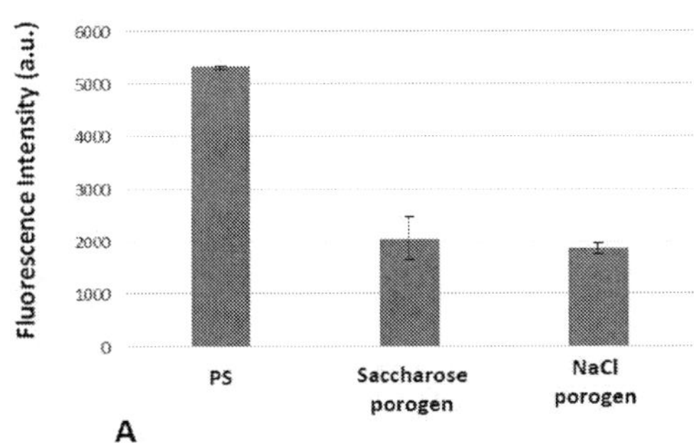

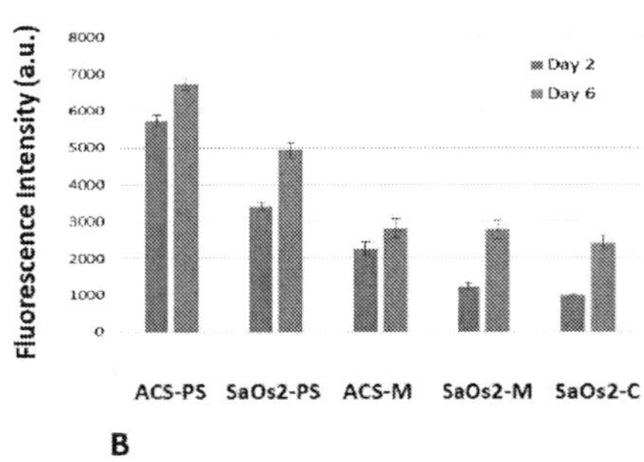

Figure 3 A - metabolic activity of ASCs cultivated on unmineralised PLA/PCL foams prepared with saccharose NaCl porogens in comparison with ASCs cultivated on polystyrene (PS) support on day 10. after cell seeding; **B** - metabolic activity of ASCs or SaOS-2 cells cultivated on mineralised (M) and unmineralised (C) PLA/PCL foams prepared with saccharose porogen with in comparison with polystyrene (PS) support. Samples tested on 2. day (blue) and 6. day (orange) after cell seeding.

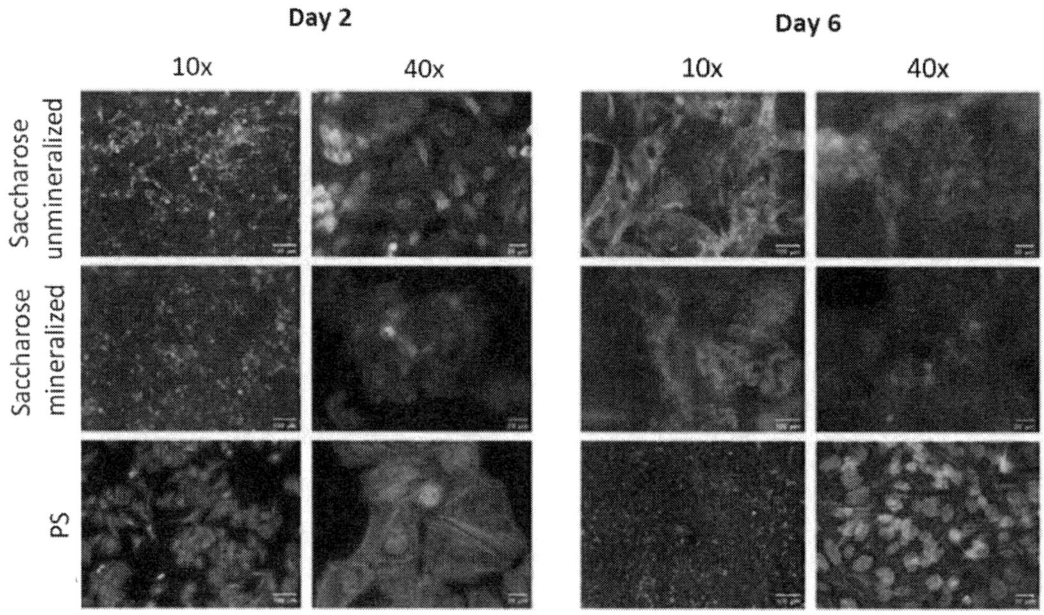

Figure 4 Fluorescent staining of SaOs-2 cells (blue colour) incubated in native and mineralised scaffolds prepared with saccharose porogen and PS control substrate

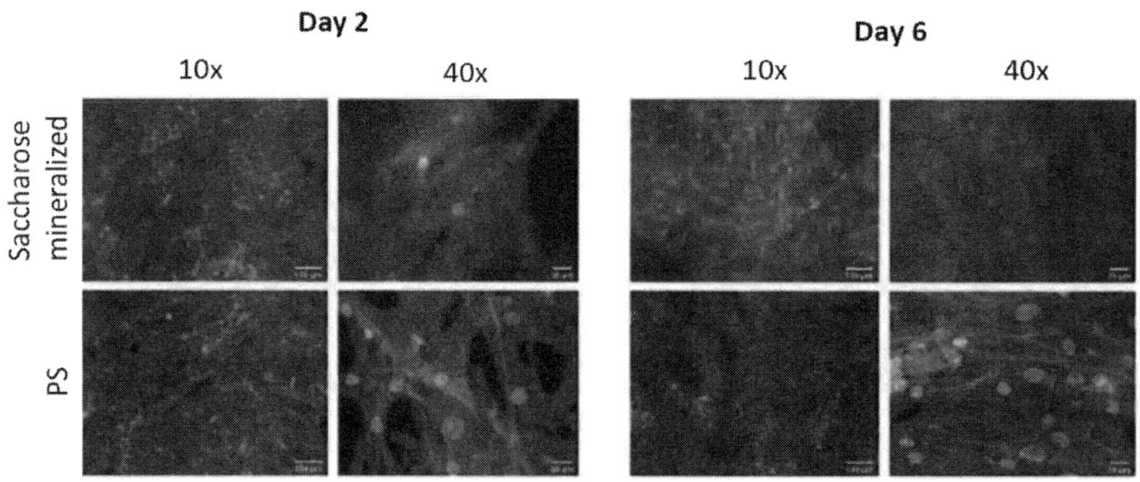

Figure 5 Fluorescent staining of ACS cells (blue colour) incubated in mineralised scaffolds prepared with saccharose porogen and PS control substrate

Polymeric foams find uses in variety of applications including packaging, thermal and acoustic insulation, energy dissipation, shock protection, filtration and separation. Hierarchically structured porous materials have attracted much attention because of their diversity and good performance, and are largely used as an important family of functional materials in catalysis, photocatalysis, adsorption, separation, energy conversion and storage, sensing, and biomedicine in recent years [6]. We suppose, that low density and well controlled and interconnected hierarchical porosity at different length scales will facilitate in-vivo resorption of the foam scaffolds for tissue engineering and improve nutrients diffusion into and metabolites diffusion outside the foam scaffolds. DU, Y., et al. [7] already described preparation of PCL nanofibrous scaffolds with hierarchical pore structure for tissue engineering. Macro/microporous poly(lactic-co-glycolic acid)/polyethylene glycol composite scaffold architectures have been developed using a hybrid solid freeform fabrication/thermally induced phase separation technique [8]. PCL foams with a well-defined micrometric and bimodal open-pore dimension distribution, suitable as scaffolds for tissue engineering, were prepared by SALERNO, A., et al. [9]. They prepared porous network pathway by combining gas foaming and selective polymer extraction techniques with thermoplastic gelatin. Dual porosity protein-based scaffolds with enhanced cell infiltration and proliferation were prepared by other authors [10] using the combination of foaming and freeze-drying. In our study, we have described simple and fast alternative method to prepare polymeric foams with hierarchical micro-macro porous structure suitable for tissue engineering.

4. CONCLUSIONS

Soft foams with micro-macro porous hierarchical structure have been prepared by freeze-drying of solutions of PLA and PCL mixtures in 1,4-dioxane with or without addition of sucrose or sodium chloride porogens. The foams prepared by freeze drying had interconnected porous aerogel characteristics. The hierarchical structure with bimodal micro-macro pore size distribution were obtained after addition of the porogens with defined crystal size distributions to the solutions before freeze drying and leaching the porogen crystals from the freeze-dried foams with demineralised water. The freeze dried foams were mineralised using carbodiimide chemistry. Biocompatibility and usability of the scaffolds was successfully verified by incubation with adipose-derived stem cells and SaOs-2 human osteosarcoma cell line. Mineralised scaffolds are more suitable growth supports for both the cell types than unmineralized collagen scaffolds. The scaffolds have been specially designed for engineering of soft tissues, but they can be used in other categories of tissue engineering, too.

ACKNOWLEDGEMENTS

This work was supported by the Czech health research council project NU20-08-00208 and by the Institutional Development Fund of FRIP provided by the Ministry of Agriculture of the Czech Republic.

REFERENCES

[1] FILOVÁ, E., et al. Accelerated in vitro recellularization of decellularized porcine pericardium for cardiovascular grafts. *Biomedical Materials*. 2021, vol. 16, 025024.

[2] JAIDEV, L.R., KAUSHIK CHATTERJEE. Surface functionalization of 3D printed polymer scaffolds to augment stem cell response. *Materials & Design*. 2019, vol. 61, pp. 44-54.

[3] KOKUBO, T., TAKADAMA, H. How useful is SBD in predicting in vivo bone aktivity? *Biomaterials*. 2006, vol. 27, pp. 2907-2915.

[4] SEDLÁŘ, A., et al. Growth Factors VEGF-A165 and FGF-2 as Multifunctional Biomolecules Governing Cell Adhesion and Proliferation. *Int. J. Mol. Sci.* 2021, vol. 22, p. 1843.

[5] TRÁVNÍČKOVÁ, M., et al. Differentiation of adipose tissue-derived stem cells towards vascular smooth cells on modified poly(L-lactide) foils. *Biomedical Materials*. 2021, vol. 16, 025016.

[6] LIANG, W., ZHENGYI, F., BAO-LIAN, S. Hierarchically structured porous materials: synthesis strategies and applacations in energy storage. *National Science Review*. 2020, vol. 7, pp. 1661-1701.

[7] DU, Y., CHEN, X., LEI, B. Facilely Fabricating PCL nanofinrous scaffolds with hierarchical pore structure fot tissue engineering. *Materials Letters*. 2014, vol. 14, pp. 62-65.

[8] AKBARZADEH, R., et al. Hierarchical polymeric scaffolds support the growth of MC3T3-E1 cells. *J. Mater. Sci: Mater. Med.* 2015, vol. 26, pp. 116.

[9] SALERNO, A., et al. Design and Preparation of μ-Bimodal Porous Scaffold for Tissue Engineering. *Journal of Applied Polymeric Science*. 2007, vol. 106, pp. 3335-3342.

[10] RASOULIANBOROUJENI, M., et al. Dual Porosity Protein-based Scaffolds with Enhanced Cell Infiltration and Proliferation. *Sci Rep*. 2018, vol. 8, 14889.

3D-PRINTED EXPERIMENTAL DEVICE FOR THE PRODUCTION OF BIODEGRADABLE FIBROUS MATERIALS FOR BIOMEDICAL SCAFFOLD PURPOSES VIA THE WET ELECTROSPINNING METHOD

Jan KOVACICIN, Sarka HAUZEROVA, Lubos BEHALEK, David LUKAS, Eva KUZELOVA KOSTAKOVA

Technical University of Liberec, Liberec, Czech Republic, EU, Jan.Kovacicin@tul.cz

https://doi.org/10.37904/nanocon.2021.4356

Abstract

The wet electrospinning method involves the use of a liquid instead of a solid or other type of collector for the collection of fabricated materials. This paper provides a description of an experimental laboratory wet electrospinning device, most of the various components of which were 3D printed by means of Fused Filament Fabrication (FFF) technology. The various morphologies of the biodegradable electrospun materials produced via the wet electrospinning technique are compared with those obtained via the classic needle electrospinning method employing SEM image analysis. The paper also provides a comparison with the structures and in-vitro viability of samples produced using the classic electrospinning approach and the wet electrospinning method with drying under both laboratory and freeze-drying conditions. The wet electrospinning freeze-drying method provides for the production of both high porosity and bulky materials. The in-vitro cell viability tests demonstrated that the wet electrospinning samples dried via the freeze-drying method evinced the highest rate of fibroblast proliferation. The novelty of the approach lies in its simplicity, the relatively low costs involved and the ease of the replication of the device for laboratory and low-quantity production purposes. Moreover, the various experiments conducted served both to prove the usefulness of the device and to determine that the resulting structure is suitable for cell growth, thus enabling the production of candidate materials for tissue engineering scaffolds.

Keywords: Wet electrospinning, scaffold, tissue engineering, 3D printer, cell viability

1. INTRODUCTION

The aim of the study was to compare 2D- and 3D-electrospun materials as candidate materials for use in tissue engineering applications and to evaluate their cell proliferation potential. Wet electrospinning (WET ES) technology represents a promising approach in terms of the production of materials suitable for tissue engineering purposes in the 3D form with larger pores and more voluminous structures than those produced via the classic method [6,7]. While the use of a classic solid electrospinning collector produces flat (2D) nanofiber layers that do not allow enough space for cells, and that are unable to replace the volume of larger tissues, WET ES produces a 3D structure with a higher degree of porosity that enables the proliferation and migration of cells, the exchange of nutrients and the waste disposal.

The wide-ranging experimental research led to the design and fabrication of a device comprising mainly components that were printed using a low cost 3D FFF printer [1], which allowed for the accurate replication, and the potential for the rapid and universal adaptation [2], of the device. Moreover, it was also designed to be user friendly and ready for use in a short time, including a simple wire connection system. The research of the base parameters for the experiments was inspired by a literature search [3-7] followed by the conducting of experiments aimed at determining the most suitable candidate materials. Once the final four materials had been selected, samples were produced for biological testing purposes and prepared for the cell viability testing phase. CCK8 cell viability tests were performed supplemented by scanning electron microscopy (SEM) and fluorescent microcopy analyses that enhanced the investigation of cell behavior within the material.

2. MATERIALS AND METHOD

The project consisted of four phases, the first of which comprised the design and construction of the WET ES experimental device, the second the determination of the candidate materials, followed by the fabrication of samples and, finally, the compilation of the biological testing plan. Following the testing of several materials, polymeric biocompatible and biodegradable PCL (polycaprolactone) was selected for experimentation due to its proven suitability for use in medical applications and its broad spinnability range [3-5,10,11]. The PCL was purchased from Merck (Germany) with two molecular weights (average Mn of 45,000 and Mn 80,000). Chloroform, ethanol, acetic acid, formic acid and acetone purchased from Penta (CZ) were used as the solvent systems (see **Table 1**); ethanol was also used as the liquid collector.

Table 1 The PCL solutions (designated G1 to G4) selected for biological testing. The first three differed in terms of the solvent, concentration and molecular mass Mn; G4 was a modified version of G3 with an increased polymer concentration. The average fiber diameters and standard deviations were calculated for each electrospun sample.

Sample designation	G1	G2	G3	G4
Material	PCL	PCL	PCL	PCL
Concentration (w/w %)	20	10	12	15
Solvent	Chloroform	Chloroform/ethanol	formic acid/acetic acid/acetone	formic acid/acetic acid/acetone
Solvent ratio (w/w)	-	8:2	1:1:1	1:1:1
Average molecular mass [Mn $*10^3$]	45	80	45/80 (w/w 1:3)	45/80 (w/w 1:3)
Average fiber diameter [nm]	6 864	9 722 625 *(bimodal values)	481	662
Standard deviation of the fiber diameter [nm]	479	5 185 129 *(bimodal values)	241	142

2.1. Device design and fabrication

In order to simplify the construction of the device, most of the components were fabricated, replicated and modified via 3D printing [2] using a Prusa MK3S 3D printer [1]. The use of conductive materials was minimized so as to ensure the maximum level of electrical safety in the vicinity of the High Voltage (HV) supply and to limit any potential electric field distortion. The device was constructed so as to facilitate simple one-piece transportation, rapid electric connection and the broadest possible range of setup parameters. The universal grip [8] allowed for the use of a wide range of spinning needles and other spinning electrodes (tips). The device included the relevant cable and tube arrangements, a non-conductive 3D printed lab platform jack [9] and a liquid bath dish, which was printed in black with rounded sides; the black color allowed for the enhanced observation of the lighter colored fiber structures in the dish and the rounded sides helped to minimize charge cumulation. The stainless electrode positioned in the middle enabled the connection of the ground potential from below the dish to the dish volume. Images of the device are shown in **Figures 1** and **2**.

Figure 1 Images of the experimental wet electrospinning device. The image on the left shows the whole of the device, the middle image shows a side view of the platform with the polymeric 3D-printed jack [9] and the bath dish, and the image on the right shows the universal grip [8] and the hydraulic pressure syringe transmitter with a needle spinning tip.

2.2. Biological testing

Batch samples G1-G4 were subjected to cell viability testing using a 10 % solution of CCK-8 counting test assay with 3T3 mouse embryonic fibroblasts (MEFs). Four samples were tested on each testing day (days 1, 3 and 7) with one negative control (NC) without seeded cells. The samples were transferred to fresh sterile microtitration plates prior to the CCK-8 testing procedure so as to ensure that only the cells in the samples were counted. Since the cell viability test compared the metabolic activity of the cells in the samples, it served for the evaluation of the suitability of the samples for cell proliferation purposes.

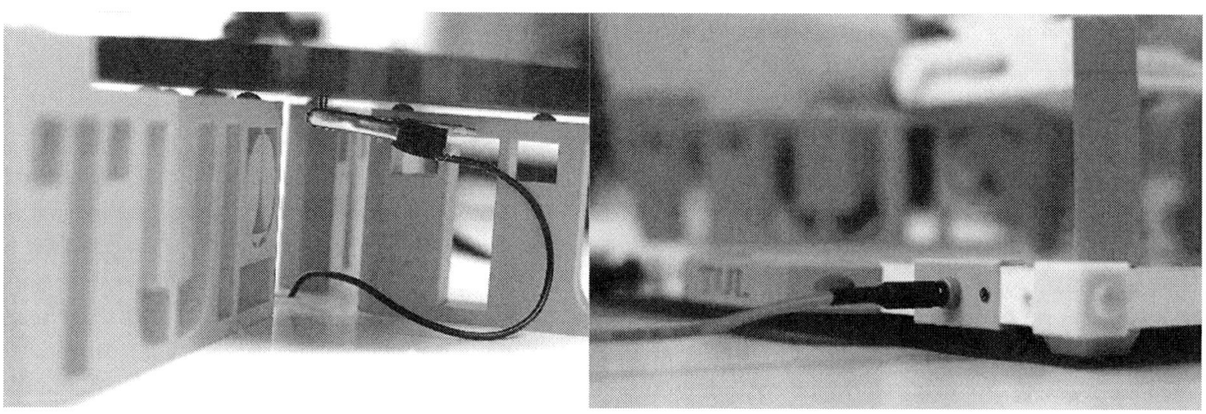

Figure 2 Further images of the experimental wet electrospinning device. The image on the left shows the platform, the grounding access point and the connection of the ground potential to the bath dish. The image on the right shows the ground input connector of the platform.

3. RESULTS

The next stage involved the testing of the PCL with the selected solvents, supplemented by the evaluation of the spinnability and the submersion quality and the SEM image analysis. Using a scoring system and the SEM image evaluation, the best-performing candidate materials were selected for the production of the final samples (see **Table 1** and **Figure 3**). The criteria for the selection of the samples for the biological testing phase comprised optimum spinnability performance, wettability by the collector liquid during WET ES and the variability of the structure (varying fiber diameters, the bimodal distribution of the fiber diameters in the sample, the smoothness or curliness of the fibers in the sample, etc.).

Batches of 15 samples were produced for each material (G1-G4). Batches were prepared of the WET samples along with batches of samples produced on a hard-flat metal collector. The WET samples were produced in an ethanol bath; the ethanol was replaced with water for lyophilization (freeze drying) purposes. With concern to the G4 material, a 3rd batch of 15 samples was produced in addition to the initial two batches; these samples were fabricated via the WET ES method but the ethanol base was allowed to evaporate naturally at room temperature in a fume hood. The batches of the WET lyophilized samples were observed to be both bulky and 3D symmetrical; the ethanol-evaporated WET ES samples were also seen to be 3D symmetrical but more compact than their lyophilized counterparts. The classic electrospinning samples were observed to be very flat. A comparison of the shapes and sizes of the samples is shown in **Figure 4**. All the samples were sterilized using ethylene oxide, which was allowed to evaporate for fourteen days.

Table 2 List of the mean absorbances and calculated growth rates. The cell metabolic activity as represented by the absorbance indicates that the planar samples exhibited the best initial seeding performance, whereas the WET 3D samples evinced the best growth rates.

	Absorbance	Absorbance	**Growth rate**	Absorbance	**Growth rate**
	Day 1	**Day 3**	**Days 1 to 3**	**Day 7**	**Days 3 to 7**
PC	0.193263	1.186075	**514 %**	1.743300	**47 %**
G1 ES	0.007838	0.024300	**210 %**	0.154538	**536 %**
G2 ES	0.010000	0.032375	**224 %**	0.304763	**841 %**
G3 ES	0.037438	0.101538	**171 %**	0.393975	**288 %**
G4 ES	0.032250	0.103388	**221 %**	0.539800	**422 %**
G1 WET ES	0.005050	0.004100	**-19 %**	0.080850	**1 872 %**
G2 WET ES	0.000010	0.000375	**3 650 %**	0.081200	**21 553 %**
G3 WET ES	0.000325	0.007425	**2 185 %**	0.182825	**2 362 %**
G4 WET ES	0.006125	0.008450	**38 %**	0.096625	**1 043 %**
G4 WET ES MOD	0.003950	0.013700	**247 %**	0.309950	**2 162 %**

The results of the biological testing revealed lower initial cell counts in the bulky samples fabricated using the WET ES method, **Figure 5** and **Table 2**. The absorbance-cell activity values were observed to be higher for the planar samples produced using the classic ES method; this was most likely due to the initial seeding of a higher number of cells in structures with smaller pores and the fact that the samples lay flat upon the bottom of the cultivation well plates, thus allowing cells to migrate into the sample from the bottom. However, over time, the cells in the planar samples did not proliferate well due to the small size of the pores and/or a lack of space. Conversely, the WET samples initially seeded to a significantly lower extent; the initial cell count based on the absorbance was lower due to the presence of large pores which failed to physically "hook" the cells during the initial cell seeding period. However, over time, the WET ES samples evinced rapid cell proliferation with a very rapid growth rate, i.e. 21 553% compared to a maximum growth rate of 422% recorded for the flat samples produced via the classic collector electrospinning method (**Figure 6**). The growth rate was calculated as an increase in the cell number represented by the absorbance, and was calculated between the 1st and 3rd days and the 3rd and 7th days. The explanation for the rapid growth rate of the WET bulky samples lies in their higher pore sizes and large 3D structures, which allowed the cells to proliferate and migrate within the bulky structures. The 3D structures contained more space for the cells and allowed for the exchange of nutrients and the disposal of waste. It is believed that the cell count was lower for the WET samples on the first day since the larger pores did not retain the cells in the samples, which resulted in their remaining on the bottom of the microtitration plates, i.e. the main reason comprised the large pores of the WET samples. On the other hand, the larger pores subsequently allowed the cells to proliferate and migrate freely into the volume of the samples; hence, the enormous growth rates recorded for the WET samples. It is recommended that further experiments should be conducted on 3D WET ES materials in the future that allow for cell viability testing over longer time periods of up to 14 days; moreover, other cell types should be considered and an efficient seeding technique developed specifically for bulky porous materials.

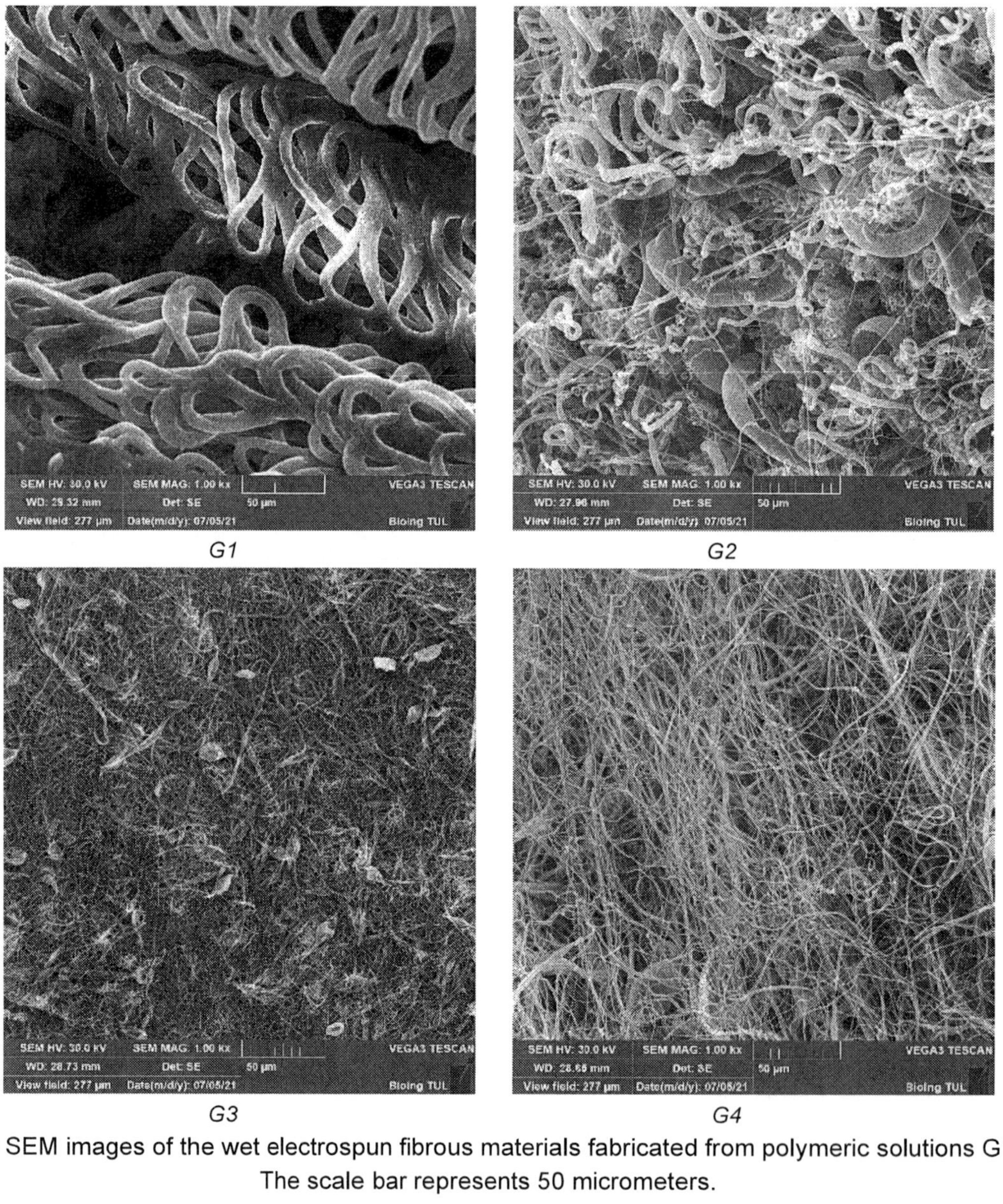

Figure 3 SEM images of the wet electrospun fibrous materials fabricated from polymeric solutions G1 - G4. The scale bar represents 50 micrometers.

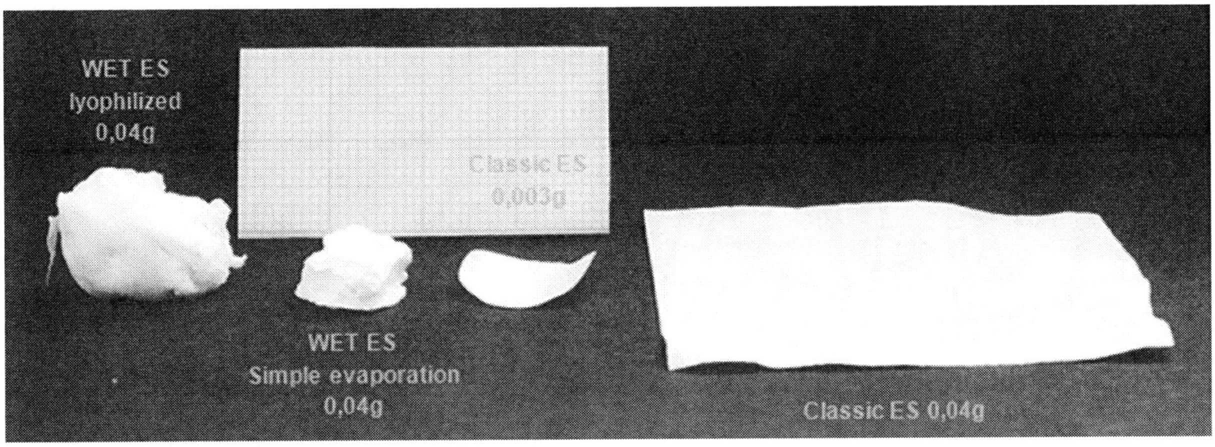

Figure 4 Images of the G4 material samples; from left to right: a WET electrospinning lyophilized bulky material, a WET ES material dried via ethanol evaporation, and a classic flat electrospinning sample trimmed to the testing size. The image on the right is included so as to provide a comparison of a classic flat electrospinning sample with the same weight as the bulky WET lyophilized sample shown on the far left.

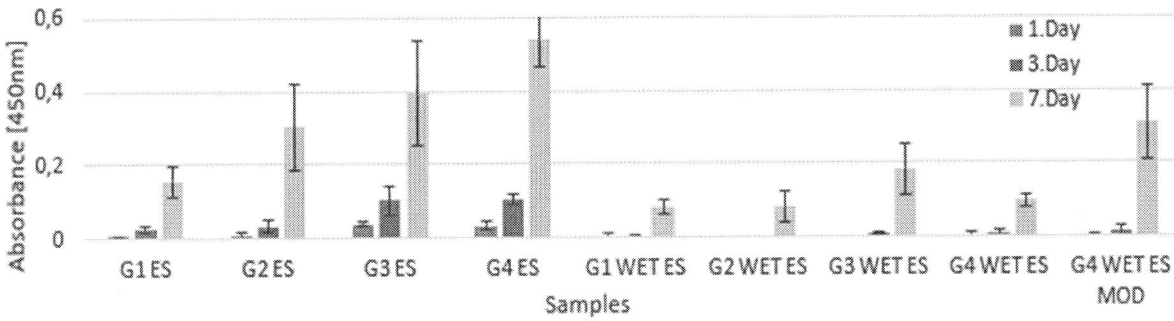

Figure 5 The cell count as represented by the spectrophotometry absorbance; the first four G1-4 ES samples were planar (flat) fabricated via the classic hard electrospinning technique; the next four G1-4 WET ES samples were fabricated via the WET electrospinning technique and dried by means of lyophilization; the final G4 WET ES MOD sample was dried via the natural evaporation of the ethanol.

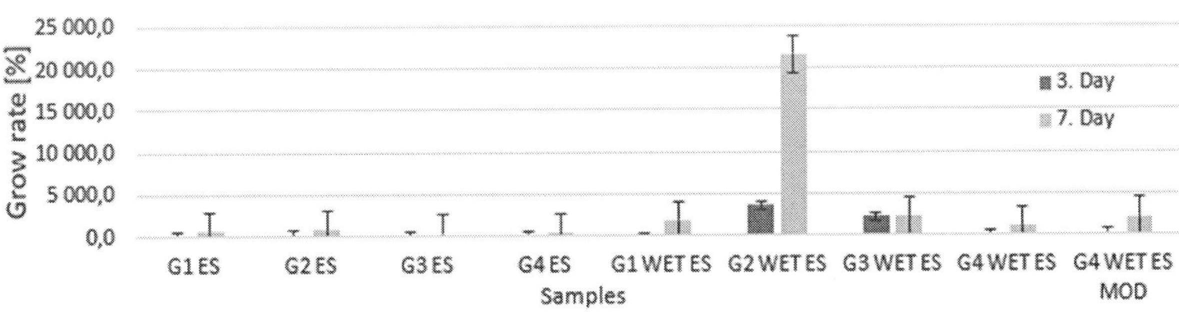

Figure 6 Growth rates representing the cell gain between the 1st and 3rd days (designated 3. Day) and the 3rd and 7th days (designated 7. Day); the first four G1-4 ES samples were planar (flat) fabricated via the classic hard collector electrospinning technique; the next four G1-4 WET ES samples were fabricated via the WET electrospinning technique and dried by means of lyophilization; the final G4 WET ES MOD sample was dried via the natural evaporation of the ethanol.

4. CONCLUSIONS

The electrospinning of fibrous materials in a liquid collector leads to significant differences in terms of the final structure of the material compared to materials electrospun using conventional flat collectors. Moreover, the three-dimensional character of wet electrospun materials revealed interesting differences in terms of the initial in-vitro testing results; in particular, a significant difference in the cell proliferation rate. While initial seeding was observed to be more effective on the planar structures, the cell proliferation rate was much more pronounced for the 3D structures fabricated via the WET ES method (a 21 553% increase from day 3 to day 7 for sample G2 WET ES) than for the best-performing planar sample (an 841% increase from day 3 to day 7). It is suggested that the vastly improved proliferation rate of the bulkier tissues indicates the true potential of the developed 3D materials for use in tissue engineering applications.

ACKNOWLEDGEMENTS

The authors express their gratitude for the financial support provided by the "Study of physical networks in polymer materials" grant project of the "PURE Technical University of Liberec Grant Program", Ref. PURE-2020-4007.

REFERENCES

[1] Original Prusa i3 MK3S+ - Prusa Research a.s.. - 3D print a 3D printers by Josef Průša [Online].

[2] SELLS, E., BAILARD, S., SMITH, Z., BOWYER, A., OLLIVER, V. RepRap: The Replicating Rapid Prototyper: Maximizing Customizability by Breeding the Means of Production. In: Handbook of Research in Mass Customization and Personalization [Online]. *World Scientific Publishing Company*. 2009, pp. 568-80.

[3] WOODRUFF, M.A., HUTMACHER, D.W. The return of a forgotten polymer-Polycaprolactone in the 21st century. *Prog Polym Sci*. Vol. 35, no. 10, pp.1217-56.

[4] KIM, M.S., KIM, G.H. Highly porous electrospun 3D polycaprolactone/β-TCP biocomposites for tissue regeneration. *Mater Lett*. Vol. 120, pp. 246-50.

[5] KOSTAKOVA, E., SEPS, M., POKORNY, P., LUKAS, D. Study of polycaprolactone wet electrospinning process. *Express Polym Lett*. 2014; vol. 8, no. 8, pp. 554-64.

[6] POKORNY, P., KOSTAKOVA, E., SANETRNIK, F., MIKES, P., CHVOJKA, J., KALOUS, T., et al. Effective AC needleless and collectorless electrospinning for yarn production. *Phys Chem Chem Phys*. 2014, vol. 16, no. 48, pp. 26816-22.

[7] YOKOYAMA, Y., HATTORI, S., YOSHIKAWA, C., YASUDA, Y., KOYAMA, H., TAKATO, T., et al. Novel wet electrospinning system for fabrication of spongiform nanofiber 3-dimensional fabric. *Mater Lett*. 2009; vol. 63, no. 9-10, pp. 754-6.

[8] MaH - Modular Arm Helper - No support - Easy print by b03tz - Thingiverse [Online].

[9] Platform Jack [Fully Assembled, No Supports] by Intentional3D - Thingiverse [Online].

[10] DOMINGOS, M., INTRANUOVO, F., GLORIA, A., GRISTINA, R., AMBROSIO, L., BÁRTOLO, PJ., et al. Improved osteoblast cell affinity on plasma-modified 3-D extruded PCL scaffolds. *Acta Biomater*. 2013; vol. 9, no. 4, pp. 5997-6005.

[11] ZHANG, M., LIN, H., WANG, Y., YANG, G., ZHAO, H., SUN, D. Fabrication and durable antibacterial properties of 3D porous wet electrospun RCSC/PCL nanofibrous scaffold with silver nanoparticles. *Appl Surf Sci*. 2017, vol. 414, pp. 52-62.

BIOSYNTHESIS OF SILVER NANOPARTICLES USING EXTRACT FROM VINE CANES AND EVALUATION OF THEIR ANTIBACTERIAL ACTIVITY AGAINST *PSEUDOMONAS AERUGINOSA*

Anna MIŠKOVSKÁ, Alena ČEJKOVÁ

UCT - University of Chemistry and Technology, Prague, Czech Republic, EU, anna.miskovska@vscht.cz

https://doi.org/10.37904/nanocon.2021.4348

Abstract

One of the many interesting properties of silver nanoparticles is their antimicrobial activity. Since these properties are related to the morphology of the particles, it is interesting to pay attention to the antimicrobial effects of various metal nanostructures, which differ in the method of synthesis, size, or shape. One of the very interesting methods of synthesis of metal nanoparticles, which seems to be very advantageous because of its low cost, efficiency, and ecology, is the synthesis using plant extracts. For this work, woody parts of the *Vitis vinifera* were selected as material for the preparation of the extract. The canes of this plant are an agricultural waste containing many bioactive substances, which are not used for other significant purposes, making it a promising material for the preparation of metal nanoparticles with antimicrobial potential. The biosynthesized nanoparticles were detected and characterized using UV-Vis spectrophotometry, transmission electron microscopy (TEM) and dynamic light scattering (DLS). TEM showed that the obtained nanoparticles have a heterogeneous shape with a size 4-51 nm. The DLS analysis supported that the Z-average was 113.8 nm and 0.165 PDI value. The use of biosynthesized nanoparticles showed an excellent growth inhibition property against two strains of the opportunistic pathogenic bacteria *Pseudomonas aeruginosa* (ATCC 10145 and ATCC 15442). Furthermore, a bactericidal concentration (BC) 40% v/v of AgNPs was found for *P. aeruginosa* ATCC 15442, while no BC was found in the range of tested concentrations for *P. aeruginosa* ATCC 10145.

Keywords: Biosynthesis, plant extract, nanoparticles, antimicrobial activity

1. INTRODUCTION

Metal nanoparticles are very popular nanomaterials in both scientific and commercial spheres. Nanoparticles have unique chemical, physical, and biological properties, which are closely related to specific characteristics of the nanoparticle, such as shape, size, and distribution [1]. Among the most widely used NPs are silver nanoparticles, which have many interesting applications, for example, in biomedicine. For many applications of silver nanoparticles, their key property is antimicrobial activity. Even historically, silver has been known for its disinfecting effects. In recent times, many studies have reported the antimicrobial activity of AgNPs against a variety of clinically important microorganisms (e.g., Gram-positive methicillin-resistant *Staphylococcus aureus* [2], Gram-negative *Escherichia coli*, funghi *Fusarium oxysporum* [3] or yeast *Candida albicans* [4]).

As interest in metal nanoparticles has grown significantly in recent decades, it is important rational choice of methods for their production. Therefore, in recent years, biological methods are becoming increasingly popular, using the principle of so-called green chemistry, which uses natural raw materials for the synthesis of metal nanoparticles (plants, plant extracts, microorganisms, biomolecules...). The use of plant extracts is an extremely practical solution, which is often the case preferred for the simplicity of the process, rapid reaction time, and high efficiency. The ability of plant extracts to reduce metal ions and stabilize the resulting nanoparticles is caused by the presence of a large number of different active metabolites (e.g., phenols, ketones, carboxylic acids, aldehydes, amides, proteins and others) [5].

In this study, we biosynthesized silver nanoparticles using an extract from agricultural waste. *Vitis vinifera* is well-known plant which is nontoxic and has many applications. Specifically, "wooden parts" - canes from *Vitis vinifera* plant were used. This material was chosen because it is easily available (waste parts of vine) and additionally it is full of many active metabolites, which could mediate the formation and stabilization of the nanoparticles. Prepared nanoparticles were detected using UV-Vis, TEM, and DLS. Subsequently, the antimicrobial potential against the opportunistic pathogenic bacteria *Pseudomonas aeruginosa* was also studied.

2. EXPERIMENTAL

2.1. Chemicals and materials

Vitis vinifera canes (wine-growing region of Čechy), silver nitrate ($AgNO_3$, Sigma-Aldrich, USA), ethanol (96%, Penta, CZ), distilled water, Phosphate-buffered saline (pH 7.4, PBS), Lysogeny broth medium (LB) - 10 g/L Trypton (Oxoid, UK), 5 g/L Yeast extract (Carl Roth, DE), 10 g/L sodium chloride (NaCl, Penta, CZ), LB agar (addition of 20 g/L of agar to liquid LB medium).

2.2. *Vitis vinifera* extract production

The dried *Vitis vinifera* canes were cut into small pieces and then crushed to powder. After that, 75 g of the prepared material was mixed with 150 ml of 40 % ethanol and extracted in the dark for 24 h. The mixture was then filtered through a laboratory filter and additionally through a filter membrane (0.22 µm size) to give the final extract for the synthesis of nanoparticles. The extract was stored in the dark at 4 °C until further use.

2.3. Biosynthesis and detection of silver nanoparticles

For the nanoparticle synthesis, the ethanolic extract of *V. vinifera* prepared in the previous step was mixed with silver nitrate to give a 1 mM solution. The extract made up 10 % of the final volume (dilution with distilled water). Reaction occurred in a 50 ml Erlenmeyer flask. Subsequently, the mixture was heated in a water bath at 60 °C for 30 min. After heating, the flask was incubated for the next 48 h at room temperature in the dark. The final solution was stored at 4 °C for future use.

The formation of silver nanoparticles was detected preliminarily by the colour change of the solution. Furthermore, the formation of AgNPs was confirmed by UV-Vis spectrophotometry at 260-700 nm with 10 nm resolution. The morphology of the prepared nanoparticles was determined by transmission electron microscopy. The size of the particles was measured using ImageJ software. The hydrodynamic diameter and the index of polydisperzity (PDI) were observed using dynamic light scattering (DLS).

2.4. Evaluation of antimicrobial activity of the nanoparticles

The activity of biosynthesized nanoparticles was evaluated against two bacterial strains of *Pseudomonas aeruginosa* (ATCC 10145 and ATCC 15442). For this analysis, the cryopreserved bacterium was inoculated in sterile LB medium and incubated for 24 h at 150 rpm and 37 °C. The culture was then centrifuged (9,000 rpm, 10 min, 10 °C) and resuspended in fresh LB medium to prepare the final inoculum with the optical density (OD600 nm) 0.100 ± 0.020.

The antibacterial assay investigating the ability of nanoparticles to inhibit cell growth was performed using sterile 100-well microtiter plates for a BioscreenC microculture device. The wells were filled with antimicrobials at a concentration range of 1.25 to 40.0 % v/v (two-fold dilution). Subsequently, 160 µl of sterile LB medium and 30 µl of inoculum from the previous step were added to each well. A control was performed for each experiment without the addition of antimicrobials. The microtiter plates were incubated using the BioscreenC device (37 °C) for 24 h and the optical density of each well was measured in 30-min intervals. Each experiment was done in 10 parallels.

Moreover, the ability of nanoparticles to kill bacteria (bactericidal) was investigated. To determine the bactericidal concentration of the nanoparticles, 5 µl of the grown cell suspension from each well was transferred to sterile Petri dishes with LB agar, which were incubated for another 24 hours at 37 °C.

3. RESULTS AND DISCUSSION

3.1. Nanoparticles biosynthesis and detection

As shown in **Figure 1**, the formation of silver nanoparticles could first be detected by the colour change of the solution from light brown to dark orange.

Figure 1 Formation of silver nanoparticles accompanied by colour change: from the left - 1 mM silver nitrate, 10 % v/v V. vinifera extract, AgNPs (60 °C, 30 min)

The presence of silver nanoparticles was also detected by measuring the UV-Vis spectra of the colloid (**Figure 2**). The obtained absorbance band with maximum at 450 nm together with the mentioned colour change are characteristic for the formation of silver nanoparticles [6,7]. Moreover, λ_{max} at 450 nm corresponds usually to nanoparticles of sizes 70-80 nm (when spherical) [8].

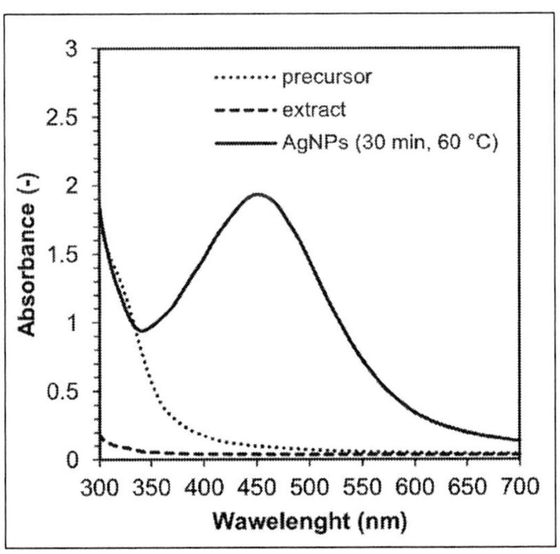

Figure 2 UV-Vis spectra of biosynthesized nanoparticles and controls

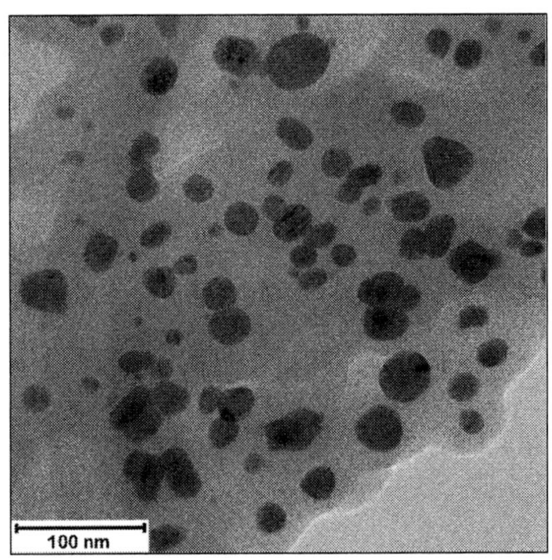

Figure 3 TEM image of silver nanoparticles biosynthesized using *V. vinifera* cane extract

Additionally, TEM was done for further characterization of the prepared nanoparticles (**Figure 3**). TEM analysis showed that the biosynthesized nanoparticles have a heterogeneous shape and are 4-51 nm in size. We supported our results by DLS measurement to observe the hydrodynamic diameter and PDI of the synthesized AgNPs. We found the Z-average 113.8 nm and 0.165 PDI value. The difference in size determined by TEM and DLS could be explained by the fact that the DLS method includes also an organic coating of the nanoparticle surface when determining the size. An electrostatic attachment of biomolecules from plant extract to the nanoparticle surface is desirable, because it improves the stability of nanoparticles. High stability of plant extract mediated metal nanoparticles is one of the advantages of this approach. However, for the possibility to use nanoparticles prepared in this way (e.g., in biomedicine), it is necessary to analyse the surface of nanoparticles and determine which biomolecules surround it. Techniques such as XPS or FTIR could be used for it [9], and it is also planned in our research.

3.2. Antimicrobial activity against planktonic bacteria

The results confirmed that the biosynthesized nanoparticles have excellent bacterial growth inhibition property against both strains of *P. aeruginosa*. The growth of planktonic cells of *P. aeruginosa* (ATCC 10145, ATCC 15442) was completely inhibited using 2.5 % v/v of AgNPs (**Figure 4** and **Figure 5**). However, even the lowest tested concertation of NPs (1.25 v/v) affected the growth dynamics in both strains of *P. aeruginosa* by prolongation the lag phase. Similar extend of the lag phase was observed in bacteria when using low concentrations of antimicrobials (e.g., antibiotics, metal nanoparticles) [10]. Similar results were investigated by Otari et al. who tested the antimicrobial activity of biologically synthesized silver nanoparticles (5-50 nm) against *P. aeruginosa*. The lowest concentration of NPs (10 mg/L) used by authors extended the lag phase from 5 to 10 hours, while by using higher concentrations of AgNPs (50 mg/L) the growth of bacteria was completely inhibited [11].

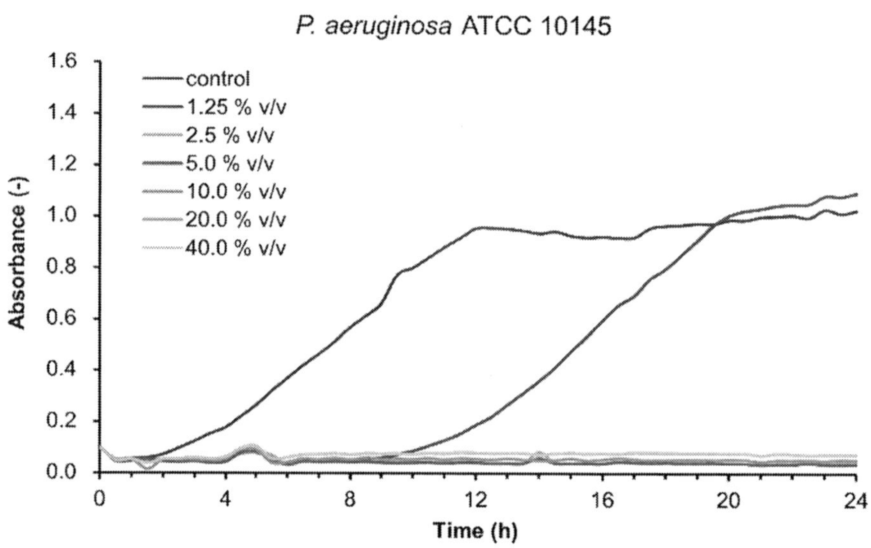

Figure 4 Growth curves of *Pseudomonas aeruginosa* ATCC 10145 in the presence of silver nanoparticles (% v/v) prepared using *V. vinifera* cane extract

We also studied the ability of biosynthesized AgNPs to kill bacteria by transferring an aliquot of grown bacterial culture from the microtiter plate to sterile LB agar, which was incubated for another 24 h. The results showed that the bactericidal concentration value of AgNPs against selected bacteria varied, which may be due to their different phenotypic characteristics. While no bactericidal concentration was found in *P. aeruginosa* ATCC 15442, the bactericidal concertation of the second microorganism tested (*P. aeruginosa* ATCC 10145) was 40 % v/v of the nanoparticles. It shows that AgNPs synthesized using viticultural waste are promising antibacterial agents.

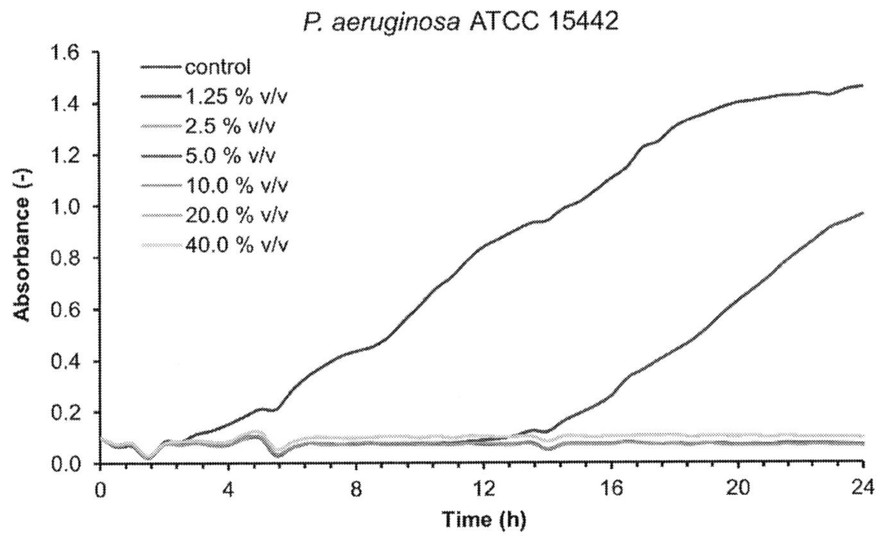

Figure 5 Growth curves of *Pseudomonas aeruginosa* ATCC 15442 in the presence of silver nanoparticles (% v/v) prepared using *V. vinifera* cane extract

Antibacterial activity of silver nanoparticles produced using plant extracts has been already described in literature. Using an extract from *Origanum vulgare* L. plant were prepared polydisperse silver nanoparticles with antibacterial activity against *P. aeruginosa, S. aureus, Staphylococcus epidermidis and E. coli* [12]. In another study, silver nanoparticles with heterogeneous shape and size from 5-20 nm were synthesized using *Murraya koenigii* extract and their antibacterial activity was proved against *S. aureus* and *E. coli* [13].

4. CONCLUSION

The present work describes the biosynthesis of silver nanoparticles using an extract from the waste parts of the agriculturally important plant *Vitis vinifera*. The extract from this material proved to be able to reduce silver ions and stabilize the resulting nanoparticles. TEM revealed that the prepared nanoparticles have a heterogeneous shape and size in the range of 4-51 nm. This data were supported by DLS analysis, which showed that the Z-average of nanoparticles was 113.8 nm and PDI was 0.165. Furthermore, it was found that these nanoparticles have the ability to inhibit the growth of the clinically important opportunistic pathogenic bacteria *P. aeruginosa*. This study shows that AgNPs synthesized using viticultural waste could find interesting applications as antimicrobial agents.

ACKNOWLEDGEMENTS

This project was supported from the grant of Specific university research - grant No. A2_FPBT_2021_011

REFERENCES

[1] SAVITHRAMMA, N., LINGA RAO, M., RUKMINI, K., SUVARNALATHA devi, P. Antimicrobial activity of silver nanoparticles synthesized by using medicinal plants. *International Journal of ChemTech Research*. 2011, vol. 3, pp. 1394-1402.

[2] KUNG, J.-C., WANG, W.-H., LEE, C.-L., HSIEH, H.-C., SHIH, C.-J. Antibacterial activity of silver nanoparticles (AgNP) confined to mesostructured, silica-based calcium phosphate against methicillin-resistant *Staphylococcus aureus* (MRSA). *Nanomaterials*. 2020, vol. 10, pp. 1264.

[3] ARITONANG, H.F., KOLEANGAN, H., WUNTU, A.D. Synthesis of silver nanoparticles using aqueous extract of medicinal plants' (*Impatiens balsamina* and *Lantana camara*) fresh leaves and analysis of antimicrobial activity. *International journal of microbiology*. 2019, vol. 2019, p. 8642303.

[4] SZWEDA P, GUCWA K, KURZYK E, ROMANOWSKA E, DZIERŻANOWSKA-FANGRAT K, ZIELIŃSKA JUREK A, KUŚ PM, MILEWSKI S. Essential oils, silver nanoparticles and propolis as alternative agents against fluconazole resistant *Candida albicans, Candida glabrata* and *Candida krusei* clinical isolates. *Indian journal of microbiology*. 2015, vol. 55, pp. 175-183.

[5] MORADI, F., SEDAGHAT, S., MORANDI, O., SALMANABADI, S. A. Review on green nano-biosynthesis of silver nanoparticles and their biological activities: with an emphasis on medicinal plants. *Inorganic and Nano-Metal Chemistry*. 2021, vol. 51, pp. 133-142.

[6] ZUAS, O., HAMIM, N., SAMPORA, Y. Bio-synthesis of silver nanoparticles using water extract of *Myrmecodia pendan* (Sarang Semut plant). *Materials Letters*. 2014, vol. 123, pp. 156-159.

[7] MARSLIN, G., SELVAKESAVAN, R. K., FRANKLIN, G., SARMENTO, B., DIAS, A. CP Antimicrobial activity of cream incorporated with silver nanoparticles biosynthesized from *Withania somnifera. International Journal of Nanomedicine*. 2015, vol. 10, p. 5955.

[8] NanoComposix. *Silver Nanoparticles: Optical Properties*. [online]. 2014. [viewed: 2021-09-11]. Available from: https://nanocomposix.com/pages/silver-nanoparticles-optical-properties#size.

[9] ZHANG, X.-F., LIU, Z.-G., SHEN, W., GURUNATHAN S. Silver nanoparticles: synthesis, characterization, properties, applications, and therapeutic approaches. *International journal of molecular sciences*. 2016, vol. 17, p. 1534.

[10] THEOPHEL, K., SCHACHT V. J., SCHLÜTER, M., SCHNELL, S., STINGU, C.-S., SCHAUMANN, R., BUNGE M. The importance of growth kinetic analysis in determining bacterial susceptibility against antibiotics and silver nanoparticles. *Frontiers in microbiology*. 2014, vol. 5, p. 544.

[11] OTARI, S., PATIL, R., GHOSH, S., THORAT, N., PAWAR S. Intracellular synthesis of silver nanoparticle by actinobacteria and its antimicrobial activity. *Spectrochimica Acta Part A: Molecular and Biomolecular Spectroscopy*. 2015, vol. 136, pp. 1175-1180.

[12] SHAIK, M. R., KHAN, M., KUNIYIL, M., AL-WARTHAN, A., ALKHATHLAN, H. Z., SIDDIQUI, M. R. H., SHAIK, J. P., AHAMED, A., MAHMOOD, A., KHAN, M., ADIL, S. F. Plant-extract-assisted green synthesis of silver nanoparticles using *Origanum vulgare* L. extract and their microbicidal activities. *Sustainability*. 2018, vol. 10, p. 913.

[13] QAIS, F.A., SHAFIQ, A., KHAN, H. M., HUSAIN, F. M., KHAN, R. A., ALENAZI, B., ALSALME, A., AHMAD, I. Antibacterial effect of silver nanoparticles synthesized using *Murraya koenigii* (L.) against multidrug-resistant pathogens. *Bioinorganic chemistry and applications*. 2019, vol. 2019, p. 4649506.

ANTIMICROBIAL PROPERTIES OF POLYMERIC NANOFIBROUS MEMBRANES CONTAINING FERROUS SULPHATE

Zuzana KRCHNACKOVA, Dusan KIMMER, Lenka LOVECKA, Miroslava KOVAROVA, Hana PISTEKOVA, Daniela VESELA, Ivo VINCENT, Muhammad YASIR, Vladimir SEDLARIK

Centre of Polymer Systems, Tomas Bata University in Zlin, Czech Republic, EU, krchnackova@utb.cz

https://doi.org/10.37904/nanocon.2021.4347

Abstract

This paper reposts on nanofibers of polymers doped with ferrous sulphate ($FeSO_4$) possibly in combination with quaternary ammonium salt (QAS) prepared by electrospinning. Three types of polymers (polyvinylidene fluoride/PVDF, polylactic acid/PLA and polyurethane/PU) with good electrospinning processability and good mechanical properties of nanofibers were chosen. The prepared nanofibrous membranes were characterised in terms of morphology of the fibres assessed by SEM, and the pore sizes were determined by porometry. The leaching test showed a firm anchoring of the additive in the nanofiber structure. Antimicrobial activity was monitored after 0 h, 4 h and 24 h using *Staphylococcus aureus (CCM 4516)* and *Klebsiella pneumoniae (CCM 4415)* strains. Furthermore, aerosol filtration efficycy was dermined and also quality factor of filtration.

The membrane prepared from PU doped with $FeSO_4$ showed the best antibacterial efficiency. The porosity and morphology of the nanofibrous membrane effectively contributed to the trapping of microorganisms. This system was also evaluated as the most suitable for the electrospinning process from the effectivity point of view. Spinning ang dopping process of preparation is easily applicatbole to the the industrial conditions of production which is very importatnt aspect of this submitted scientifique work.

Keywords: Antimicrobial properties, doped nanofibers, $FeSO_4$, filtration

1. INTRODUCTION

In recent years, polymeric nanofibers have been demonstrated as effective tool for aerosol filtration applications. They have a great ability to aerosol filtration and trapping microorganisms that have became extremely relevant due to the COVID-19 occurrence [1]. The size of SARS-COV-2 virus is 60-140 nm with nanospikes coated on its spherical viral capsid/envelop with heights of 9-12 nm [2]. The most beneficial properties in this field of study are: high surface area to weight ratio, low density, high pore volume and besides other also small pore size [3]. As increase requirement for antimicrobial effect across nanofiber applications it needs to be mention also another valuable advantage; possibility of their relatively easy surface coating or doping with various species, which can provide nanofibers with specific properties, including antimicrobial effects [4].

As demands for functionality of nanofibers membrane are more and more complex, many research groups have prepared already wide range of surface modified nanofibers, including antimicrobial [5]. However, it needs to be said that some of methods are very complicated, expensive or not stable during usage time of membrane. Therefore, in our study we developed easy procedure for antimicrobial nanofibers doping with $FeSO_4$ possibly in combination with QAS as antimicrobial agents. Our method of spinning and doping of nanofibers is easily applicable in conditons of industrial production.

In out work, after detailed characterization of prepared marials the antimicrobial effect was evaluated on two species of bacteria, *Staphylococcus aureus* and *Klebsiella pneumoniae*. Apart from the pathogenic effects of

these two bacterial strains, another reason for testing them is that they are fundamentally different in the arrangement of outer membranes. So obtained results provide general information on the interaction between a biocide compound and other Gram-positive and Gram-negative bacteria [6].
Furthermore, aerosol filtration efficacy at different particle sizes was proved via automated filter tester Model 3160. An ideal air filter should have high particle-removal efficiency, low-pressure drop, and a long lifetime as well as high quality factor number. [7]

Nanofibrous membranes doped with an active antibacterial component directly in the mass of nanofibers are the next step towards products that actively eliminate microorganisms and thus protect human health. This work focused on developing nanofibrous membrane doped with antimicrobial compounds for aerosol filtration, which is the next step towards products that actively eliminate microorganisms and thus protect human health.

2. MATERIALS AND METHODS

We prepared three polymeric nanofibrous membranes based on PVDF, PU, and PLA deposited on polypropylene (PP) non-woven as collecting support. All types of successfully electrospun nanofibers were doped via 1 % of $FeSO_4$. Then PU nanofibers were doped with both 1 % of $FeSO_4$ and also 7.5 % of QAS incorporated into PU chains. Basic weights of all prepared nanofibers were at a maximum of 1 g/cm^2, as shown in **Table 1**.

Table 1 Basic weights of all prepared nanofibrous materials

Sample	Basic weight (g/cm^2)
PVDF/$FeSO_4$	0.77
PLA/$FeSO_4$	0.56
PU/$FeSO_4$	0.45
PU/$FeSO_4$/QAS	1.00

Polymeric nanofibers were prepared by electrospinning device SpinLine 40 (SPUR, Czech Republic) equipped with two sets of moving nanofibers forming jets. The applied voltage during the spinning process was 75 kV, and the distance between electrodes was 19 cm for all samples. Other parameters for polymeric solution were different for each sample. Summarisation of these parameters is completed in **Table 2**.

Table 2 Summarization of parameters of polymeric solutions

Sample	Solvent	Basic weight (g/cm^2)	Viscosity (Pa.s)	Conductivity (μS/cm)
PVDF/$FeSO_4$	DMF	0.77	1.60	25.10
PLA/$FeSO_4$	DMF/aceton	0.56	0.75	110.3
PU/$FeSO_4$	DMF/H_2O	0.45	3.25	120.20
PU/$FeSO_4$/QAS	DMF/H_2O	1.00	0.80	310.00

The morphology of the electrospun nanofibers was observed using scanning electron microscopy (SEM; Vega 3, Tescan, Czech Republic). The average diameter of the electrospun nanofibers was determined by analysing the SEM images using a custom code image analysis program. As well as the pore size of nanofibrous materials was performed by porometry device (SPUR, Czech Republic) according to ASTM F316-03. The antimicrobial activity of the prepared nanofiber's mats was tested according to the testing method for the antibacterial activity of textiles using *Staphylococcus aureus (CCM 4516) and Klebsiella pneumoniae (CCM 4415)* strains. Then antimicrobial activity was calculated via ISO 22196:2011.

Antimirobial activity (A) calculation:

A = log N0 - log Nx

Where:

x - incubation time with bacterial suspension (0, ½, 4 a 24h)

A - antibacterial activity of sample

N - number of viable bacteria (CFU/cm^2)

The samples were cut into 20 x 20 mm squares and the polypropylene cover foil was also cut into 20 x 20 mm. Prior to testing, the samples were disinfected with UV radiation for 30 minutes on both sides (sterilization in an autoclave or disinfection with ethanol wasn't possible due to potential damage of sample). The cover foils were disinfected with 70 % ethanol. The steriled samples were prepared in triplicated and put in to the sterile glass petri dishes and covered byf 0.1 ml of bacterial suspension, and this was covered with PP foil. Prepared samples were incubated for 4h and 24h at 35 °C with a relative humidity of 95 %. Samples in time 0h was immediately processed. Triplicates of one sample including foil were transferred in to the sterile PP container. Neutralization medium (SCDLP) was added in a total volume 15 ml (5 ml per sample). Incubation time was 30 s. Then all saples were decimally diluted nad mixed with agar media (PCA) and incubated for 24h at 35 °C.

3. RESULTS AND DISCUSSION

The samples were tested after optimising the spinning process of all the polymer solutions leading to repetable structures. The SEM images in **Figure 1** show the porous structure of the prepared nanofibrous materials with randomly oriented nanofibers having a smooth surface without beads-like or any other morphological defects. Although the electrospinning process was adequately optimised, in the case of sample PU/$FeSO_4$ the produced nanofibers were not straight and were tangled. The fibres also had a larger diameter compared to the other materials. These deficiencies were corrected by adding 7.5 % QAS to the polymeric solution containing 1 % $FeSO_4$. On the contrary, PVDF and PLA polymeric nanofibers were straight even without adding QAS into the solution.

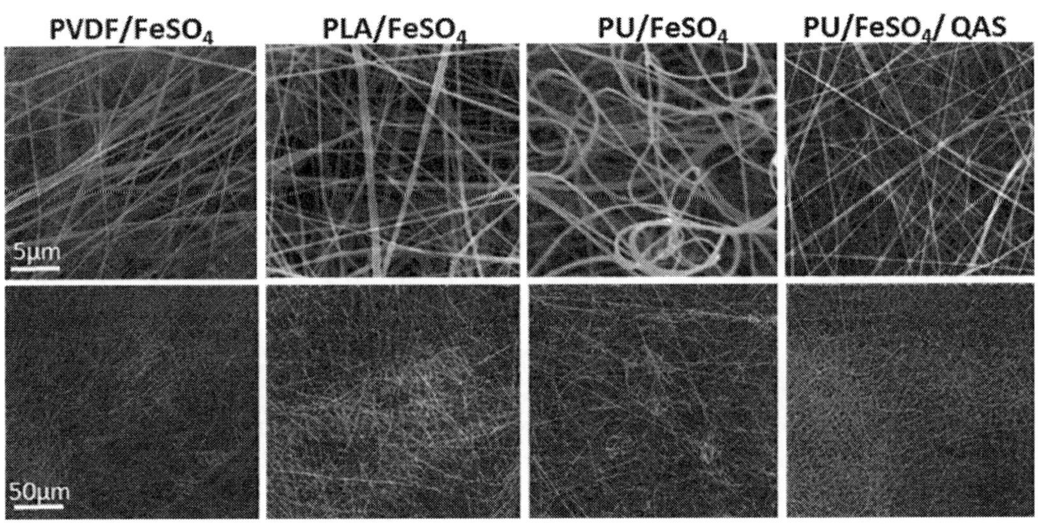

Figure 1 SEM images of four different types of electrospun nanofibers: PVDF doped with 1 % of $FeSO_4$, PLA doped with 1 % of $FeSO_4$, PU doped with 1 % of $FeSO_4$ and PU doped with 1 % $FeSO_4$ in combination with 7.5 % of QAS

Among other characterisation methods, the measurement of fibre diameter is crucial for predicting the properties of nanofiber structures. The diameter of fibres depends on the surface tension, flow rate, viscosity and electric conductivity of polymer solution [8]. **Figure 2** represents the fibre diameter distribution of the electrospun fibres onto the non-woven PP substrate, indicating a range of diameter between 0.16 µm and 0.33 µm. The PVDF/$FeSO_4$ sample shows the largest average nanofiber thickness, which is due to the low conductivity of the polymer solution. However, this sample also contains a significant proportion of fibres with a diameter of around 0.1 µm, which could improve the filtration properties of the material. For polyurethane samples, the addition of QAS shifts the distribution of fibre diameters towards lower values and the average fibre diameter value drops from 0.33 to 0.16 µm. Usage of nanometric fibres has proven advantageous in air filtration since their small diameter, and high surface to volume ratios can enhance the capture of particles through interception.

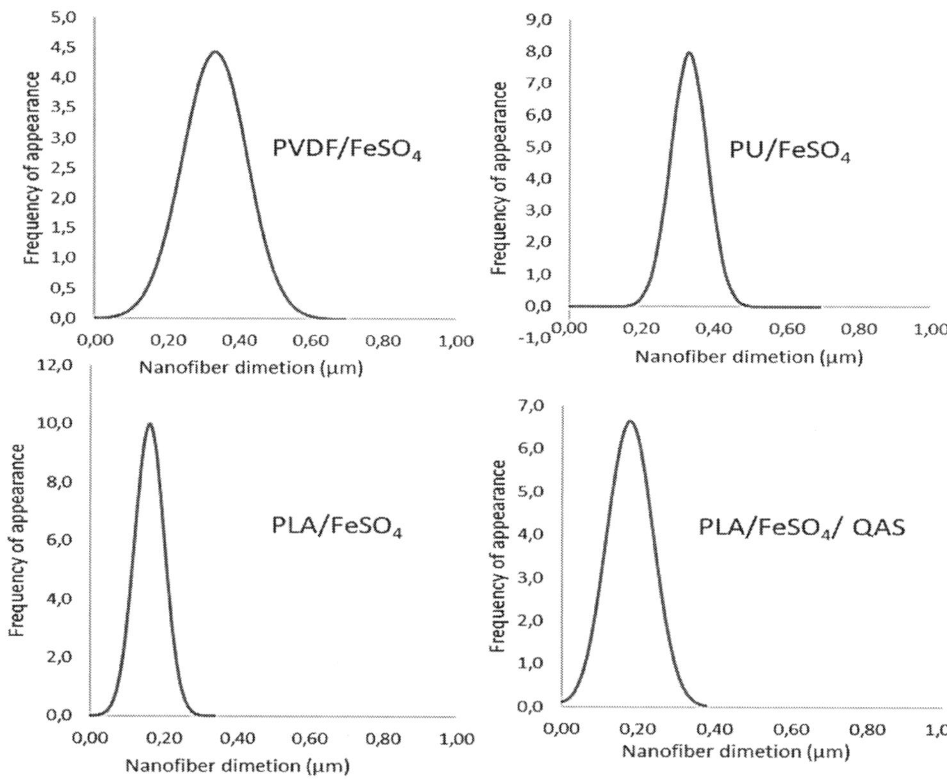

Figure 2 Nanofibers diameter distributions of PVDF/ FeSO4 (average dismeter 0.33 µm), PLA/ FeSO4 (0.16 µm), PU/ FeSO4 (0.33 µm) and PU/ FeSO4/QAS (0.18 µm)

Another important measurement for revealing the properties of prepared samples was porometry. The flow porometry technique is a useful tool for testing nanofiber structures. **Figure 3** presents the pore size distributions of all tested samples. While the distributions of PVDF and PLA have a width of 0.75 µm and 0.85 µm, respectively, and a mean pore size of 0.92 µm and 0.84 µm, the width of the distribution curve for PU/$FeSO_4$ is up to 1.6 µm and the mean pore size reaches 2 µm. However, using a combination of $FeSO_4$ and QAS results in a significant narrowing of the distribution (width of 0.64 µm) and a decrease in the mean pore size to 0.91 µm. The narrower pore size distribution suggests that the material will be more homogeneous. The more homogenous nanofiber layer means lower pressure drop, which is a characteristic playing a significant role in the application of electrospun nanofibers in air filtration [9].

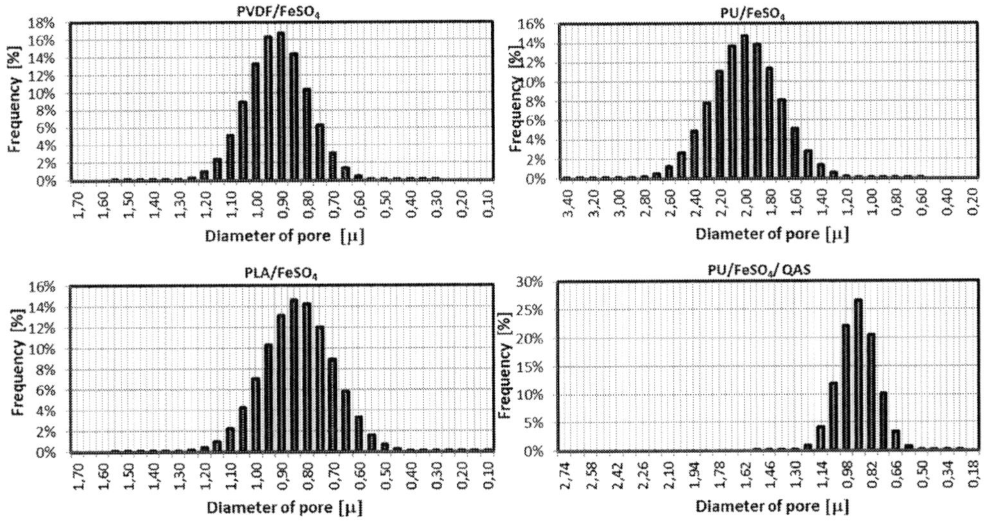

Figure 3 Flow porometry of spun nanofibers doped with $FeSO_4$

After detailed characterisation of prepared nanofibrous materials, we proceeded to antimicrobial activity tests according to -SO 22196:2011. Antimicrobial properties were investigated using *S. aureus* and *K. pneumoniae* as model microorganisms. After treatment time with bacterial solution (4h, 24h) final CFU of bacteria was counted and converted to antimicrobial activity (A) described above. Then it was possible to evaluate precisely the efficacy of antimicrobial activity of the tested materials. **Table 3** describes the exact rate of effectivity of antimicrobial activity according to the A values given in **Figure 4**. It seems that hydrophilicity/hydrophobicity has a significant influence on the antimicrobial activity of the polymer because electrospun nanofibers from PVDF and PLA polymers have low antimicrobial activity compared with more hydrophilic PU nanofibers. Furthermore, adding QAS to polymer solution shows an essential effect on bacteria elimination. The antimicrobial effect is strongest in the case of both bacterial strains ($A \geq 3$).

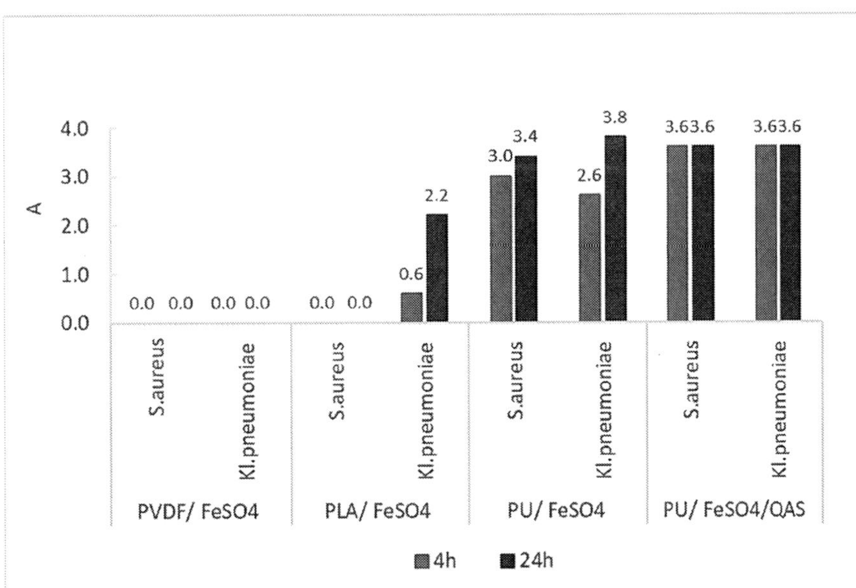

Figure 4 Antimicrobial activity of prepared nanofibrous materials using *S. aureus* and *K. pneumoniae* as model microorganisms, and evaluation of antimicrobial activity (A) after 4h and 24h

Table 3 Explanation of effectivity of antimicrobial activity (A) values

Efficacy of antibacterial properties	Value of antimicrobial activity A
Weak	$1 < A < 2$
Significant	$2 \leq A < 3$
Strong	$A \geq 3$

4. CONCLUSION

Electrospun nanofibers offer a number of advantages when used in air filters. There are extensive studies that demonstrate improvement in filteration efficiency without increasing pressure drop. In this research, we successfully doped various nanofibrous mats (PVDF, PLA, PU) designed for air filtration purposes with potential antimicrobial species $FeSO_4$ and QAS.

Characterisation analyses such as SEM, porometry or measurement of the diameter of nanofibers were performed. These characteristics may help to elucidate correlations between antimicrobial activity and the nature and structure of the nanofibrous material. Changes in conductivity of electrospinning polymer solutions significantly impact the diameter and the distribution of nanofiber's diameters as well as porosity of nanofibrous material and its distribution. Furthermore, we proved that adding QAS to PU/$FeSO_4$ polymer solution increased the antimicrobial effect, which was observed for a total period of 24h. Prepared nanofibrous materials with the best effect against bacteria, i.e. PU/$FeSO_4$ and PU/$FeSO_4$/QAS, was selected for a further study dealing with optimising of nanostructure properties for air filtration.

ACKNOWLEDGEMENTS

The authors gratefully acknowledge financial support from the contract research Plastr 2020 - VaV by companies Plastikářský klastr s. z. and SPUR a.s. and by the Ministry of Education, Youth, and Sports of the Czech Republic (grant no. RP/CPS/2020/002).

REFERENCES

[1] RAMAKRISHNA, Seeram. An introduction to electrospinning and nanofibers. *World scientific*, 2005.

[2] LEUNG, Wallace Woon-Fong; SUN, Qiangqiang. Charged PVDF multilayer nanofiber filter in filtering simulated airborne novel coronavirus (COVID-19) using ambient nano-aerosols. *Separation and purification technology*. 2020, vol. 245, p. 116887.

[3] SUBBIAH, Thandavamoorthy, et al. Electrospinning of nanofibers. *Journal of applied polymer science*. 2005, vol. 96, no.2, pp. 557-569.

[4] XUE, Jiajia, et al. Electrospun nanofibers: new concepts, materials, and applications. *Accounts of chemical research*. 2017, vol. 50, no. 8, pp. 1976-1987.

[5] RODRÍGUEZ-TOBÍAS, Heriberto; MORALES, Graciela; GRANDE, Daniel. Comprehensive review on electrospinning techniques as versatile approaches toward antimicrobial biopolymeric composite fibers. *Materials Science and Engineering: C*. 2019, vol. 101, pp. 306-322.

[6] CLERMONT, Olivier; GORDON, David; DENAMUR, Erick. Guide to the various phylogenetic classification schemes for Escherichia coli and the correspondence among schemes. *Microbiology*. 2015, vol. 161, vol. 5, pp. 980-988.

[7] RAJAK, Abdul, et al. Controlled morphology of electrospun nanofibers from waste expanded polystyrene for aerosol filtration. *Nanotechnology*. 2019, vol. 30, no.42, 425602.

[8] JENA, Akshaya; GUPTA, Krishna. Pore volume of nanofiber non-wovens. *International Nonwovens Journal*. 2005, vol. 2, 1558925005os-.

[9] LEUNG, Wallace Woon-Fong; SUN, Qiangqiang. Charged PVDF multilayer nanofiber filter in filtering simulated airborne novel coronavirus (COVID-19) using ambient nano-aerosols. *Separation and purification technology*, 2020, vol. 245, 116887.

DIFFERENTIATION OF MESENCHYMAL STEM CELLS ON FIBRIN ASSEMBLIES SUPPORTED BY IMMOBILIZED GROWTH FACTORS FGF2 AND VEGF

[1]Jana MUSILKOVA, [1]Elena FILOVA, [1]Ondrej KAPLAN, [1]Lucie BACAKOVA

[1]*Department of Biomaterials and Tissue Engineering, Institute of Physiology of the Czech Academy of Sciences, Czech Republic, EU,*
jana.musilkova@fgu.cas.cz, elena.filova@fgu.cas.cz, ondrej.kaplan@gmail.com, lucie.bacakova@fgu.cas.cz

https://doi.org/10.37904/nanocon.2021.4360

Abstract

Bioartificial heart valves and vascular grafts prepared from decellularized tissues could be recellularized with bone marrow-derived mesenchymal stem cells (MSCs) that are able to differentiate into both smooth muscle cells and endothelial cells. MSCs differentiation is facilitated by sustained release of growth factors. In our study assemblies based on fibrin, fibrin with heparin, fibrin with adsorbed or covalently-immobilized vascular endothelial growth factor A165 (VEGF) or basic fibroblast growth factor (FGF-2) via binding to heparin attached to fibrin have been prepared and were evaluated for their stimulation of MSCs differentiation. We estimated the mRNA expression of endothelial marker CD31 (PECAM1), smooth muscle marker α-actin (ACTA2), osteoblast markers osteocalcin (BGLAP) and alkaline phosphatase (ALP). The gene expression was estimated using RT-PCR on days 1, 7 and 21 after seeding. The cell morphology and viability was evaluated by LIVE/DEAD staining. VEGF, both adsorbed and covalently bound, increased significantly the expression of smooth muscle marker α-actin. The mRNA expression of ACTA2 on day 7 and 21 raised more than 200 times in comparison to control samples (undifferentiated cells before seeding). The ACTA2 gene expression significantly exceeded the expression of all other evaluated genes at all time intervals. Moreover, on day 21, the late smooth muscle marker desmin (DES) was steeply rising in cells cultivated on assemblies containing heparin and covalently bound VEGF. The expression of osteocalcin was minimal. We conclude that fibrin assembly containing covalently bound VEGF is the most convenient for MSCs differentiation towards smooth muscle cells.

Keywords: Stem cells, differentiation, growth factors, smooth muscle cells, fibrin assemblies

1. INTRODUCTION

Biological xenogeneic heart valve prostheses and patches are made of decellularized porcine heart valve or from bovine pericardium. These replacements undergo degeneration after 15 years. There is an effort to prepare prostheses with autologous cellular component with regeneration potential and to reduce immune response. Bioartificial heart valves and vessels prepared from decellularized tissues could be recellularized with bone marrow-derived mesenchymal stem cells (MSCs). The ability to differentiate into smooth muscle cells makes the MSCs to be a suitable candidate for recellullarization of pericardium scaffolds [1]. MSCs differentiation toward smooth muscle cells, endothelial cells or valve interstitial cells could be facilitated by sustained release of growth factors (GF) [2]. The choice of GF, the way of GF attachment onto the graft surface, their concentration and even their combination or sustained release are important factors for recellularization process of biological scaffolds. Sustained delivery of GF prolongs their life time, and allows to use lower concentrations of GF necessary for cell differentiation [3]. In the present study, various fibrin layers with adsorbed/attached FGF or VEGF have been developed; heparin was used for the long-term GF binding. We have estimated the presence of smooth muscle differentiation markers after a long-term cultivation in the presence of FGF2 and VEGF A165 immobilized on fibrin assemblies. Moreover we have tested the presence of endothelial or osteogenic markers to compare the prevailing type of differentiation and to make sure that the MSCs isolated from bone marrow do not differentiate towards osteoblast.

2. MATERIALS AND METHODS

2.1. Material preparation

The bottom of a 24- Well Glass Bottom Plate (Cellvis) was coated with a fibrin assemblies as described earlier [4] by a subsequent incubation of fibrinogen (5 µg/ml in Tris-HCl buffer, pH 7.4, TB, 4 °C, overnight) (Sigma-Aldrich), thrombin (2 U/ml in TB, 30 min, room temperature) (Sigma-Aldrich), and a mixture of antithrombin III (0.5 U/ml in TB) (Chromogenix) and fibrinogen (200 µg/ml in TB) for 2 hours. Heparin sodium salt from porcine intestinal mucosa was activated using a procedure described previously [5], i.e. heparin (14 mg/ml in PBS) was oxidized by NaIO4 (2.4 mg/ml) (Sigma-Aldrich) in the darkness for 90 min and then periodate was removed by ultrafiltration. Activated heparin solution was diluted in the citrate buffer, and was incubated in the wells overnight. 100 ng/ml of human FGF2 (FGF-basic 154aa) and human VEGF A 165 (GenScript) in PBS were added to the wells coated with only Fb or Fb with attached heparin) for 2 hours. The prepared layers were washed three times with PBS and UV irradiated. Glass cover slips, pure Fb and Fb+heparin were used as control samples.

2.2. Cell cultivation conditions

The MSCs (Lonza, passages 3) were seeded at the density of 5,000 cells/cm^2 onto six different types of fibrin layers: fibrin (Fb), fibrin+FGF2 adsorbed (FGF2), fibrin+VEGF adsorbed (VEGF), fibrin+heparin (H), fibrin+heparin+VEGF (H+VEGF), fibrin+heparin+FGF2 (H+FGF2) and glass control (glass). The cells were cultivated in α-MEM medium supplemented with 2.5% FBS and with aprotinin (3.73 mg/100 ml) for 21 days.

2.3. Evaluation of cell viability by LIVE/DEAD test

On day 1, MSCs viability on fibrin assemblies was assessed by LIVE/DEAD test. The fluorescent images were taken using Olympus IX 51 microscope, equipped with DP70 digital camera from 30 microscopic fields (from 3 individual samples), living cells are stained in green, the dead cells, in red.

2.4. Evaluation of cell differentiation by Real-time PCR (RT-PCR)

The mRNA expression of genes specific for smooth muscle cells ACTA2 (α-actin) and DES (desmin), endothelial marker PECAM1 (CD31), and osteogenic markers ALP (alkaline phosphatase) and BGLAP (osteocalcin) was tested to estimate the MSCs differentiation on fibrin assemblies. The total mRNA isolation was performed using Total RNA Purification Micro Kit (Norgen Biotek). The reverse transcription was performed using the Omniscript Reverse Transcription Kit (205113; Qiagen, Hilden, Germany) and random hexamers (New England Biolabs, Inc, Ipswich, MA, USA), and was carried out according to the manufacturer's protocol. The samples were stored at -20 °C. The mRNA level was quantified using 5×HOT FIREPol Probe qPCR Mix Plus (ROX) (Cat. No. 08-36-00001; Solis BioDyne, Tartu, Estonia) and by TaqMan Gene Expression Assays (Cat. No. 4331182; Thermo Fisher Scientific for genes ACTA2 (Hs 05005341_m1), DES (Hs 00157258_m1), PECAM1 (Hs 00169777_m1), ALP (Hs 01129144_m1) and BGLAP (Hs 01587814_g1). The experiments were performed with GAPDH (Hs 02786624_g1) as a reference gene.

For RT-PCR analysis, One way ANOVA, the Student-Newman Keuls method was used. Data are calculated as $2^{-\Delta\Delta Ct}$ values and expressed as a multiple of the control and presented as mean ± standard deviation; p value ≤ 0.05 was considered significant. As the control sample, the mRNA isolated from cells before seeding was taken.

3. RESULTS AND DISCUSSION

3.1. LIVE/DEAD test

Cell viability on fibrin assemblies was assessed by LIVE/DEAD test (**Figure 1**) on day 1. MSCs adhered well and spread on all tested fibrin assemblies. The cells were alive (stained green); only some individual cells were dead (i.e. stained red). On the pure heparin assembly the cell settlement was sparse on day 1. We can summarize that all assemblies are convenient for MSCs growth. In our previous study [7] we tested the metabolic activity of the HUVECs, seeded on crosslinked decellularized pericardium coated with fibrin mesh or with a mesh containing attached heparin and/or fibronectin. The lowest degree of viability was attained for samples that were coated by pure heparin. In another study, heparin immobilized on chitosan films decreased smooth muscle cells proliferation, while heparin immobilized on chitosan-collagen films improved their proliferation [7].

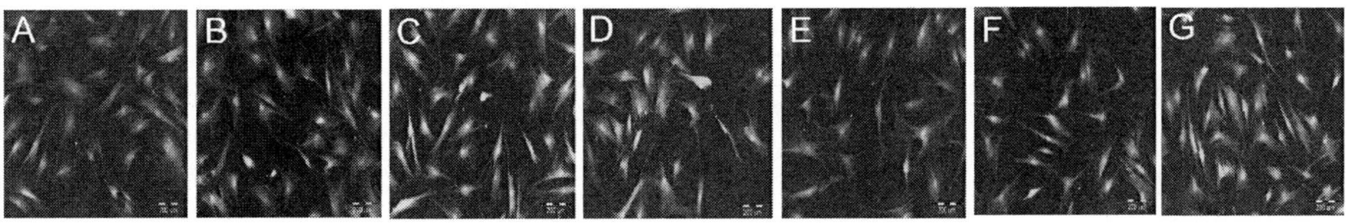

Figure 1 Viability of MSCs on day 1 after seeding on fibrin assemblies: fibrin (A), fibrin+FGF2 adsorbed (B), fibrin+VEGF adsorbed (C), fibrin+heparin (D), fibrin+heparin+VEGF(E), fibrin+heparin+FGF2 (F) and glass control (G). The cells were stained with a LIVE/DEAD Cell Viability Cytotoxicity kit (Life Technologies). Live cells are stained in green, and dead or dying cells are stained in red. Olympus IX 51 microscope, DP70 digital camera, obj. × 20, scale bar = 200 μm.

3.2. Real-time PCR

We evaluated the MSCs differentiation using Real-time PCR method. Differentiation towards smooth muscle cells and other cell types can be ascertained by the presence of early smooth muscle markers, i.e. smooth muscle alpha-actin, medium-term markers, such as calponin and H-caldesmon, and late markers, such as myosin heavy chain 11 and desmin [8,9].

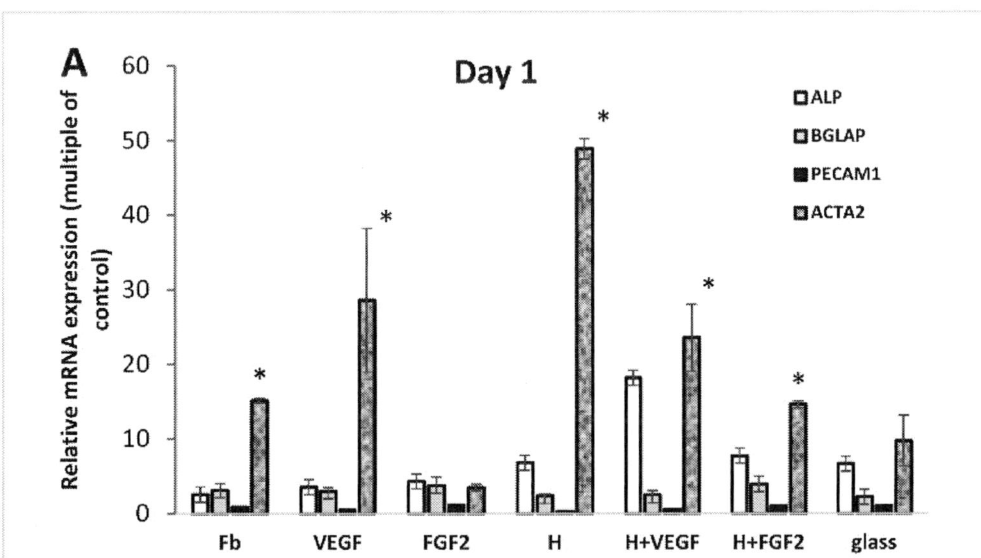

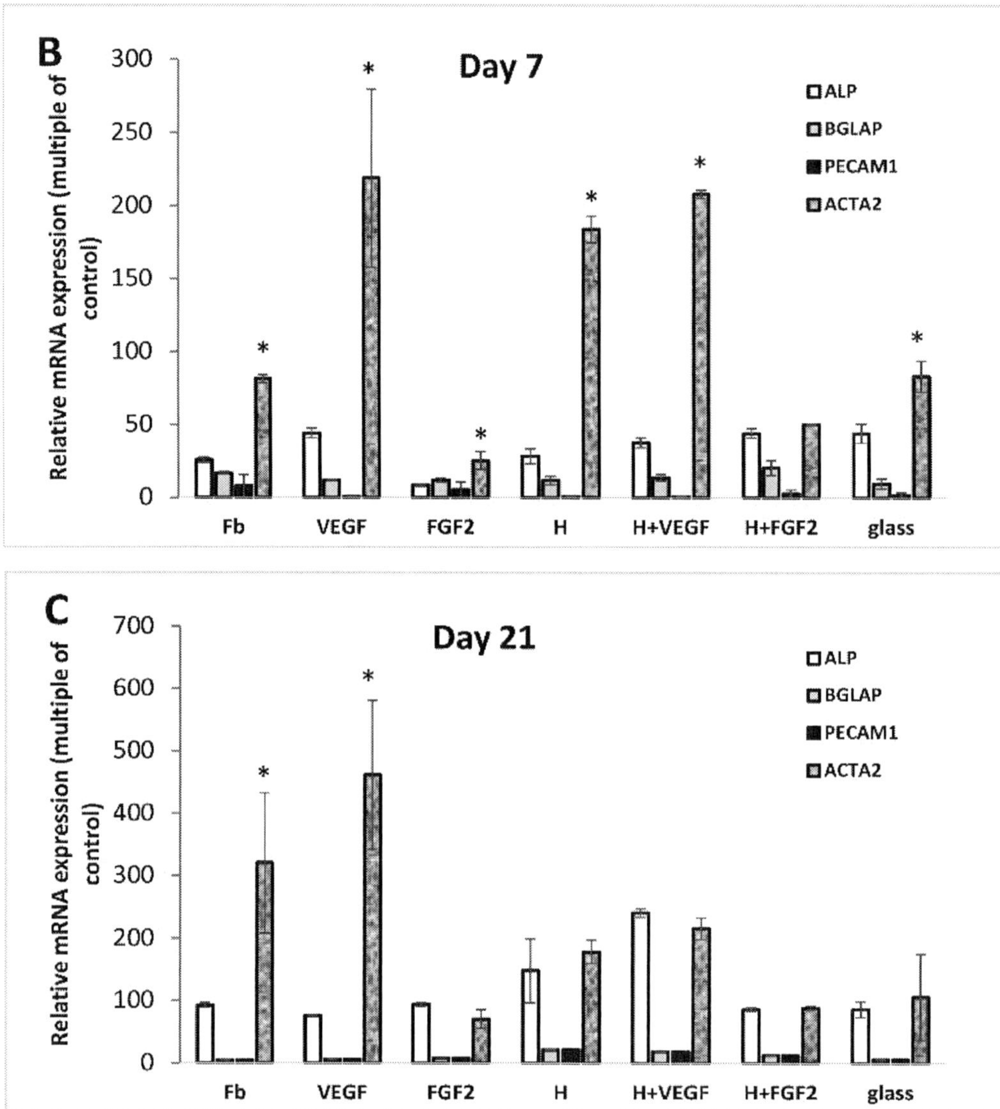

Figure 2 Relative mRNA expression of ALP (alkaline phosphatase), BGLAP (osteocalcin), PECAM1 (CD 31) and ACTA2 (smooth muscle α-actin) on fibrin assemblies in MSCs on day 1 (A), on day 7 (B), and on day 21 (C) after seeding. Fibrin assemblies: fibrin (Fb), fibrin+FGF2 adsorbed (FGF2), fibrin+VEGF adsorbed (VEGF), fibrin+heparin (H), fibrin+heparin+VEGF(H+VEGF), fibrin+heparin+FGF2 (H+FGF2) and glass control (glass). Statistical significance ($p \leq 0.05$) among genes on the same assembly type on the same day is marked by (*) above the columns.

The growth factor VEGF A 165 supported the differentiation of MSCs preferentially into smooth muscle cells (**Figures 2A-C**) on all fibrin assemblies. The presence of FGF2 reduced the expression of gens specific for smooth muscle cells (**Figure 2, Figure 3**). The expression of endothelial marker PECAM 1 is moderately rising on assemblies with VEGF (**Figure 2**). The differentiation into osteoblasts was very poor (**Figures 2A-C**). After a long-time cultivation on fibrin assemblies containing heparin or covalently bound VEGF A 165 the MSCs differentiation towards smooth muscle subtype was supported. There was extremely high expression of ACTA2 compared to all other genes at all time intervals. On day 21 the late smooth muscle marker DES expression increased markedly in MSCs cultivated on assemblies containing heparin and covalently bound VEGF. These results indicates that the MSCs differentiation was developing towards smooth muscle cells [9].

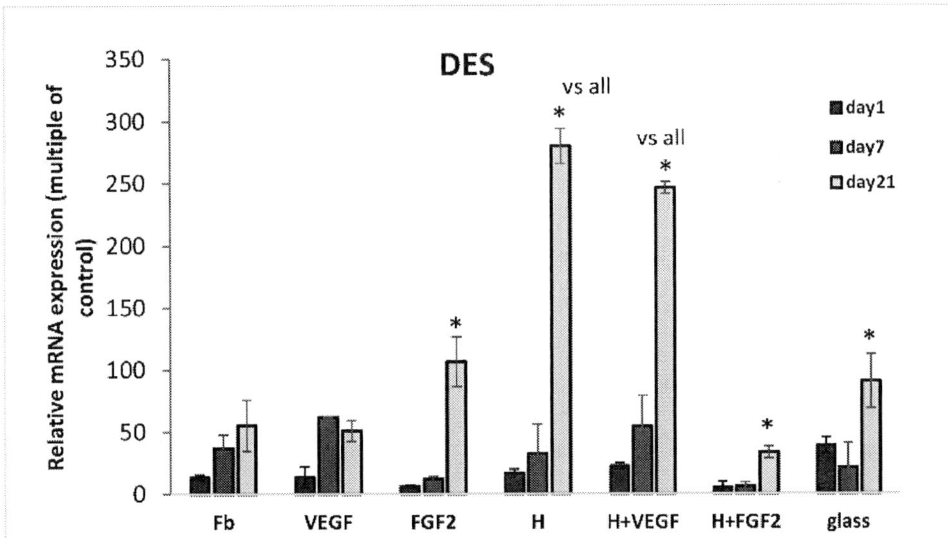

Figure 3 Relative mRNA expression of DES (desmin) on fibrin assemblies on day 1, 7 and 21 after seeding. Fibrin assemblies: fibrin (Fb), fibrin+FGF2 adsorbed (FGF2), fibrin+VEGF adsorbed (VEGF), fibrin+heparin (H), fibrin+heparin+VEGF(H+VEGF), fibrin+heparin+FGF2 (H+FGF2) and glass control (glass). Statistical significance ($p \leq 0.05$) of DES expression among different time intervals of the same assembly is marked by (*). The highest DES expression among all experimental groups (vs all) is depicted above the columns.

4. CONCLUSIONS

We have prepared various fibrin assemblies with adsorbed or covalently bound growth factors FGF2 and VEGF and we have evaluated differentiation of human MSCs on these surfaces in regards to their potential use in cardiovascular tissue engineering. MSCs differentiated mainly into the smooth muscle cells on fibrin assemblies containing VEGF. Moreover, the late smooth muscle marker DES was steeply rising in the cells cultivated on the assembly containing heparin and covalently bound VEGF. The MSCs differentiation into osteoblasts or into endothelial cells was negligible. We conclude that fibrin assembly containing covalently bound VEGF is the most convenient for MSCs differentiation towards smooth muscle cells and for the preparation of bioartificial vascular or heart valve replacements.

ACKNOWLEDGEMENTS

Supported by the Czech Health Research Council, the Ministry of Health of the Czech Republic (grants No. NV19-02-00068 and No. NV18-02-00422).

REFERENCES

[1] GU, W., HONG, X., LE BRAS, A., NOWAK, W.N., ISSA BHALOO, S., DENG, J., XIE, Y., HU, Y., RUAN, X.Z., XU, Q.J. Smooth muscle cells differentiated from mesenchymal stem cells are regulated by microRNAs and suitable for vascular tissue grafts. *Biological Chemistry*. 2018, vol. 293, pp. 8089-8102.

[2] FILOVA, E., STEINEROVA, M., TRAVNICKOVA, M., KNITLOVA, J., MUSILKOVA, J., ECKHARDT, A., HADRABA, D., MATEJKA, R., PRAZAK, S., STEPANOVSKA, J., KUCEROVA, J., RIEDEL, T., BRYNDA, E., LODEREROVA, A., HONSOVA, E., PIRK, J., KONARIK, M., BACAKOVA, L. Accelerated in vitro recellularization of decellularized porcine pericardium for cardiovascular grafts. *Biomedical Materials*. 2021, vol. 16, p. 025024.

[3] KUMOREK, M., KUBIES, D., FILOVÁ E., HOUSKA M., KASOJU, N., MÁZL CHÁNOVÁ, E., MATĚJKA R., KRÝSLOVÁ M., BAČÁKOVÁ L., RYPÁČEK F. Cellular Responses Modulated by FGF-2 Adsorbed on Albumin/Heparin Layer-by-Layer Assemblies. *PLoS One*. 2015, vol. 10, no. 5, e0125484.

[4] RIEDEL, T., BRYNDA, E., DYR, J.E., HOUSKA M. Controlled preparation of thin fibrin films immobilized at solid surfaces. *Journal of Biomedical Materials Research Part A.* 2009, vol. 88, no. 2, pp. 437-447

[5] Kaplan, O., HIERLEMANN, T., KRAJEWSKI, S., KURZ, J., NEVORALOVÁ, M., HOUSKA, M., RIEDEL, T., RIEDELOVÁ, Z., ZÁRUBOVÁ, J., WENDEL, H.P., BRYNDA, E. Low-thrombogenic fibrin-heparin coating promotes in vitro endothelialization. ***Journal** of **Biomedical** Materials **Research Part** A.* 2017, vol. 105, no. 11, pp. 2995-3005.

[6] MUSILKOVA, J., FILOVA, E., PALA, J., MATEJKA, R., HADRABA, D., VONDRASEK, D., KAPLAN, O., RIEDEL, T., BRYNDA, E., KUCEROVA J, KONARIK M, LOPOT F, JAN PIRK, BACAKOVA L. Human Decellularized and Crosslinked Pericardium Coated with Bioactive Molecular Assemblies. *Biomedical Material.* [online]. 2019, vol.15, no. 1, pp. 015008. Available from: https://doi.org/10.1088/1748-605X/ab52db.

[7] ROBU, I.S., WALTERS III, H.L., MATTHEW, H.W.T. Morphological and growth responses of vascular smooth muscle and endothelial cells cultured on immobilized heparin and dextran sulfate surfaces. *Journal of Biomedical Materials Research Part A.* 2017, vol. 105, no. 6, pp. 1725-1735.

[8] PRZEKORA, A., VANDROVCOVA M., TRAVNICKOVA, M., PAJOROVA, J., MOLITOR, M., GINALSKA, G., BACAKOVA, L. Evaluation of the potential of chitosan/β-1,3-glucan/hydroxyapatite material as a scaffold for living bone graft production in vitro by comparison of ADSC and BMDSC behaviour on its surface. *Biomedecal Materials.* [online]. 2017. vol 12, no.1, 015030, [viewed 2018-09-28]. Available from: https://doi.org/10.1088/1748-605X/aa56f9.

[9] BACAKOVA, L., ZARUBOVA, J., TRAVNICKOVA, M., MUSILKOVA, J., PAJOROVA, J., SLEPICKA, P., SLEPICKOVA KASALKOVA, N., SVORCIK, V., KOLSKA, Z., MOTARJEMI, H., MOLITOR, M. Stem cells: their source, potency and use in regenerative therapies with focus on adipose-derived stem cells - a review. *Biotechnology Advances.* [online]. 2018, vol. 36, no. 4, pp. 1111-1126 [viewed 2018-09-28]. Available from: https://doi.org/10.1016/j.biotechadv.2018.03.011.

BIOGENIC SYNTHESIS OF SILVER NANOPARTICLES USING *TRACHYSTEMON ORIENTALIS* L. AND THEIR ANTIBACTERIAL ACTIVITIES

[1]Hasan Ufuk CELEBIOGLU, [2]Rizvan IMAMOGLU, [1]Recep TAS

[1]*Department of Biotechnology, Faculty of Science, Bartin University, Bartin, Turkey*

[2]*Department of Molecular Biology and Genetics, Faculty of Science, Bartin University, Bartin, Turkey*

https://doi.org/10.37904/nanocon.2021.4361

Abstract

Nanotechnology is a scientific field for the characterization, production, and application of nanoscale particles of 1-100 nm size. Due to their large surface area and small size, nanoparticles have different properties compared to bulk materials such as mechanical, electrical, magnetic, and chemical properties. They can be synthesized by physical, chemical and biological methods. The biosynthesis of nanoparticles is the synthesis of nanoparticles using microorganisms or plants with biomedical applications. This synthesis approach stands out as a cost-effective, environmentally friendly, bio-compatible, safe, and green approach.

The present study aims to synthesize silver nanoparticles by green chemistry using *Trachystemon orientalis* L. extract and to investigate anti-bacterial and anti-fungal effects properties.

UV-Visible absorption, fourier-transform infrared spectroscopy, and scanning electron microscopy analyses confirmed the synthesis. The results showed that silver nanoparticles have inhibitory effect on both gram-negative (*E. coli, Klebsiella pneumonia* and *Pseudomonas aeruginosa*) and gram-positive (*Staphylococcus aureus* and *Enterococcus faecalis*) bacterial strains. Furthermore, they showed anti-fungal activities against *Candida utilis* and *Candida albicans.*

Silver nanoparticles can affect the cell walls of the microorganism, changing the structure of cell membranes and even causing cell death. They can also increase the permeability of cell membranes, generate reactive oxygen species, and interrupt the replication of deoxyribonucleic acid.

In conclusion, green synthesis of silver nanoparticles using *Trachystemon orientalis* L. extract can be an effective method for biomedical purposes.

Keywords: Antibacterial, antifungal, nanoparticle, silver, *Trachystemon orientalis* L.

1. INTRODUCTION

Nanotechnology is a scientific field for the characterization, production and application of nanoscale particles of 1-100 nm size. Technologically, it is a new and rapidly developing technology that aims to bring new physical, chemical and biological properties to matter at the atomic and molecular level [1-2]. Due to their large surface area and small size, nanoparticles have different properties compared to bulk materials such as mechanical, electrical, magnetic and chemical properties [1,3]. In recent years, studies on nano-sized materials have become important and raise awareness. Nano-sized structures are of interest for all applications of nanotechnology, where the shape and size of nanoparticles determine their characteristic properties [4]. The main reason why these nanoscale structures attract attention and attract attention is that the substances show functionality with their own characteristics different from their different volumetric dimensions and structures [5]. Due to the unique physical, chemical, optical and mechanical properties of nanoparticles, they have many applications in different fields, including coating and paint, food, textile, energy, automotive and agriculture, cosmetics, space field, MRI contrast, separation and purification of cell parts, cells and biological molecules, tissue engineering heterogeneous catalysis, antimicrobial additives, UV protection, LCD technology, medicine and genetics, biodetection of pathogens, detection of proteins, investigation of DNA structure [6-27]. Nanoparticles (Nps) can be synthesized by physical, chemical and biological methods. Although many traditional methods are used physically and chemically for the synthesis of Nps, these methods create expensive and toxic by-products at the same time. Moreover, it is difficult to control the surface chemistry, size and structure of nanoparticles synthesized by these methods [28]. The biosynthesis of nanoparticles is the synthesis of nanoparticles using microorganisms and plants with biomedical applications. This synthesis approach; stands out as a cost-effective, environmentally friendly, bio-compatible, safe and green approach [29]. These natural strains and plant extracts used in the synthesis of nanoparticles secrete some phytochemicals that act as both a reducing agent and a capping or stabilizing agent. Due to the special phytochemicals they produce, plant parts such as roots, stems, leaves, fruits and seeds were used to synthesize Ag nanoparticles. Using natural extracts of plant parts in nanoparticle synthesis is a very environmentally friendly and inexpensive process. Plants have become the most preferred natural resources for nanoparticle synthesis, as they lead the large-scale and stable production of nanoparticles of different shapes and sizes [30]. Thus, the aim of the present study is to synthesize silver nanoparticles from metal oxides by green chemistry using *Trachystemon orientalis* L. and to investigate their anti-bacterial and anti-fungal effects.

2. MATERIALS AND METHODS

2.1. Preparation of the plant extracts

Trachystemon orientalis L. Bartin was obtained from the local public market. *Trachystemon orientalis* L. was extracted according to the method suggested by Rao et al [31]. The plant was washed 3 times with deionized water. The washed leaves were dried and passed through a food processor and divided into small pieces. 10 g of the obtained herbal powder was mixed into 100 mL of distilled water and extracted by standing in a water bath for one hour at room temperature. The resulting mixture was extracted by filtering through Whatman filter paper and was freshly prepared and used in the synthesis stage. Stored in a dark glass bottle at +4 °C to be used as a stock.

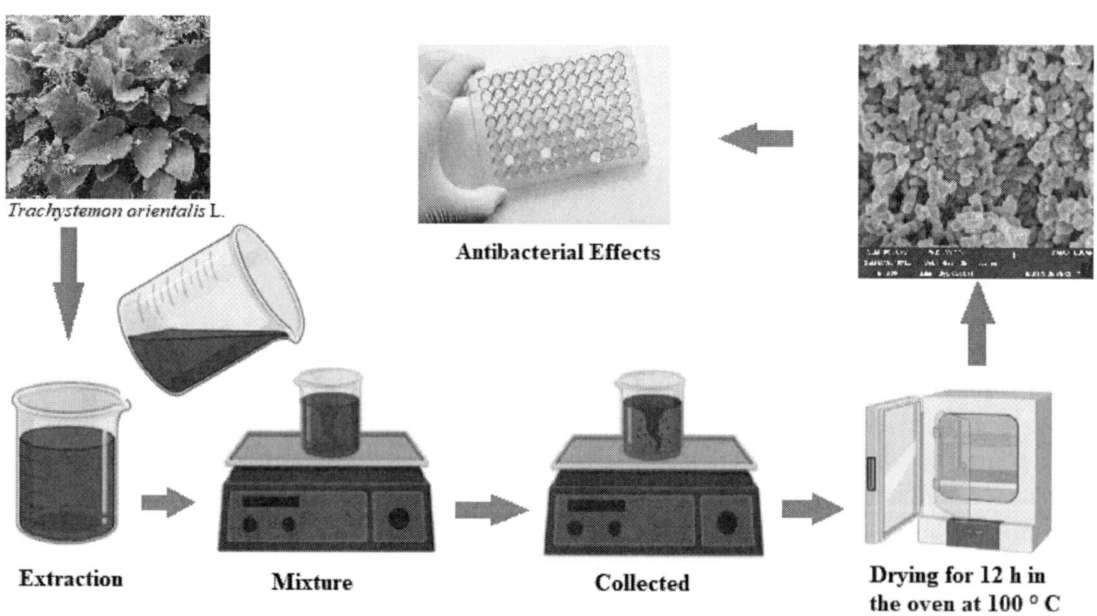

Figure 1 Schematic representation of AgNP synthesis

2.2. Synthesis of silver nanoparticles

To 10 mL of $AgNO_3$ solution at a concentration of 0.1 M, 50 mL of *Trachystemon orientalis* L. extract was added dropwise and stirred on magnetic stirrer at room temperature for 24 h at a constant stirring speed (500 rpm). The formation of AgNPs 15 min after the addition of the extract was observed by observing the color of the solution from green to blackish-gray (**Figure 1**). The resulting AgNPs were purified by centrifugation at 8,000 rpm for 15 minutes. The nanoparticles separated from the suspension by centrifugation were washed twice with pure water and dried in a vacuum oven at 100 °C for 12 hours. The nanoparticles obtained were stored in centrifuge tubes for use in antimicrobial studies.

2.3. Characterization of c Ag nanoparticles

The optical properties of AgNPs were characterized by taking UV-Vis region spectra (THERMO, Model Multiscaner spectrophotometer) between 200-800 nm. FTIR (Shimadzu Iraffinity-1) was used for functional group analysis. Particle morphology and size was measured by SEM (TESCAN, MAIA3 XM). The crystal structure and particle size of AgNPs was calculated using XRD analysis. XRD analysis was performed using a powder X-ray diffractometer (Rigaku, Smartlab).

2.4. Anti-bacterial and Antifungal Effects of Silver Nanoparticle

Broth Micro-dilution Assay was used in order to investigate antibacterial activities of the compounds (Brandt et al., 2010), with some modifications. Briefly, The antimicrobial activity of the synthesized TO-AgNPs was determined against gram-negative bacterial strains (*K. pneumonia, P. aeruginosa,* and *E.coli*), gram-positive bacterial strains (*E. faecalis* and *S. aureus*), and fungal strains (*C. albicans* and *C. utilis*) from frozen stocks for 24 h at 37°C. Then, new cultures were prepared until 0.5 McFarland Unit at 37°C. In total 200 µL of microtiter plate wells, 20 µL of bacterial cultures were inoculated with NB containing different concentrations of the compounds suspended in DMSO (0-30 mM). Negative controls were prepared using LB without bacteria. Positive controls did not contain any compounds, but respective amount of DMSO. Absorbances of microtiter plates were read at 600 nm using a micro-plate reader before (0th h) and after (24th h) the incubation at 37°C. Bacterial viability was measured as percentage of compound-treated bacterial groups to the positive control (bacterial viability of positive control was taken as 100%). Minimum inhibitory concentrations (MIC) were calculated using the plot of nanoparticle concentration versus relative bacterial growth.

Statistical analysis and comparable data sets were evaluated using GraphPad Prism 8.0.2 software with a two-way ANOVA test. Probability values of $p < 0.05$ were considered statistically significant.

3. RESULTS AND DISCUSSION

3.1. UV-Visible absorption analysis

The UV-Vis spectrum of TO-AgNPs obtained by using the extract of leaf parts of the plant *Trachystemon orientalis* L. with the green synthesis method is shown in **Figure 2**. The wavelength of maximum absorption in the UV spectrum of TO-AgNPs was determined as 419 nm.

3.2. Fourier-transform infrared spectroscopy (FTIR)

FTIR was performed to determine the nature of possible natural products responsible for the surface coating and efficient stabilization of AgNPs synthesized using the water extract of *Trachystemon orientalis*. FTIR spectrum of the sample is shown in **Figure 3**. The broad peak in the range of 3100-3400 cm^{-1} is caused by the O-H stresses of the alcoholic and phenolic groups found in the polymeric structure of the plant and the N-H stress of the amine groups. Peaks at 2920 and 2844 cm^{-1} belong to aliphatic C-H stresses.

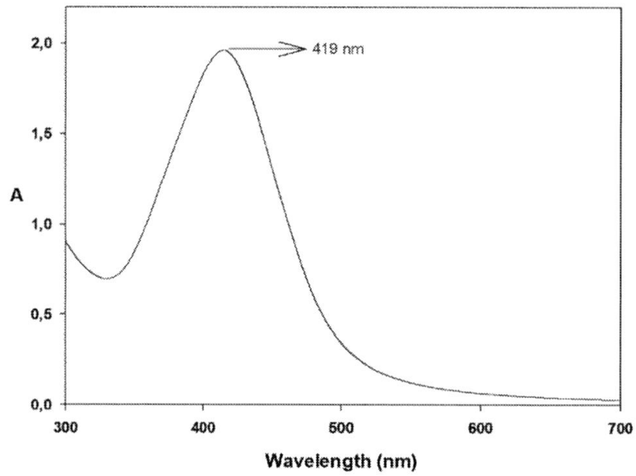

Figure 2 The UV-Vis spectrum of TO-AgNPs

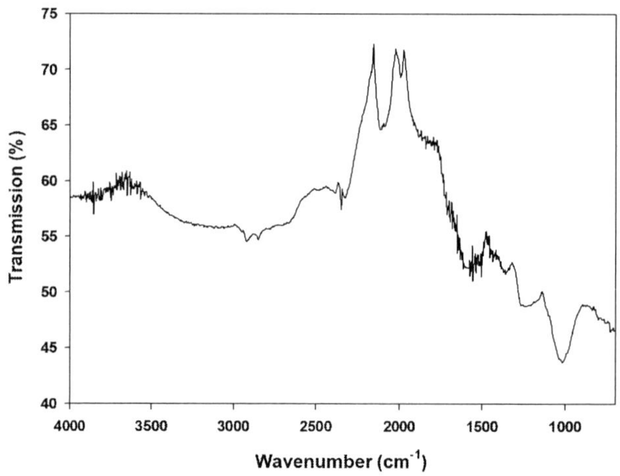

Figure 3 FTIR spectrum of TO-AgNPs

3.3. XRD analysis

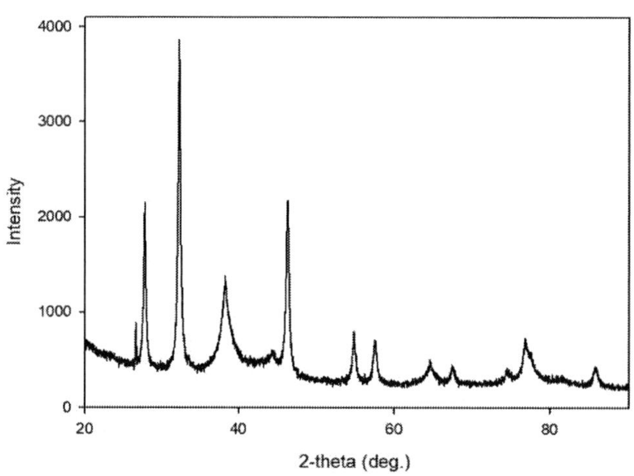

Figure 4 XRD of TO-AgNPs

3.4. Scanning electron microscopy (SEM) study of AgNPs

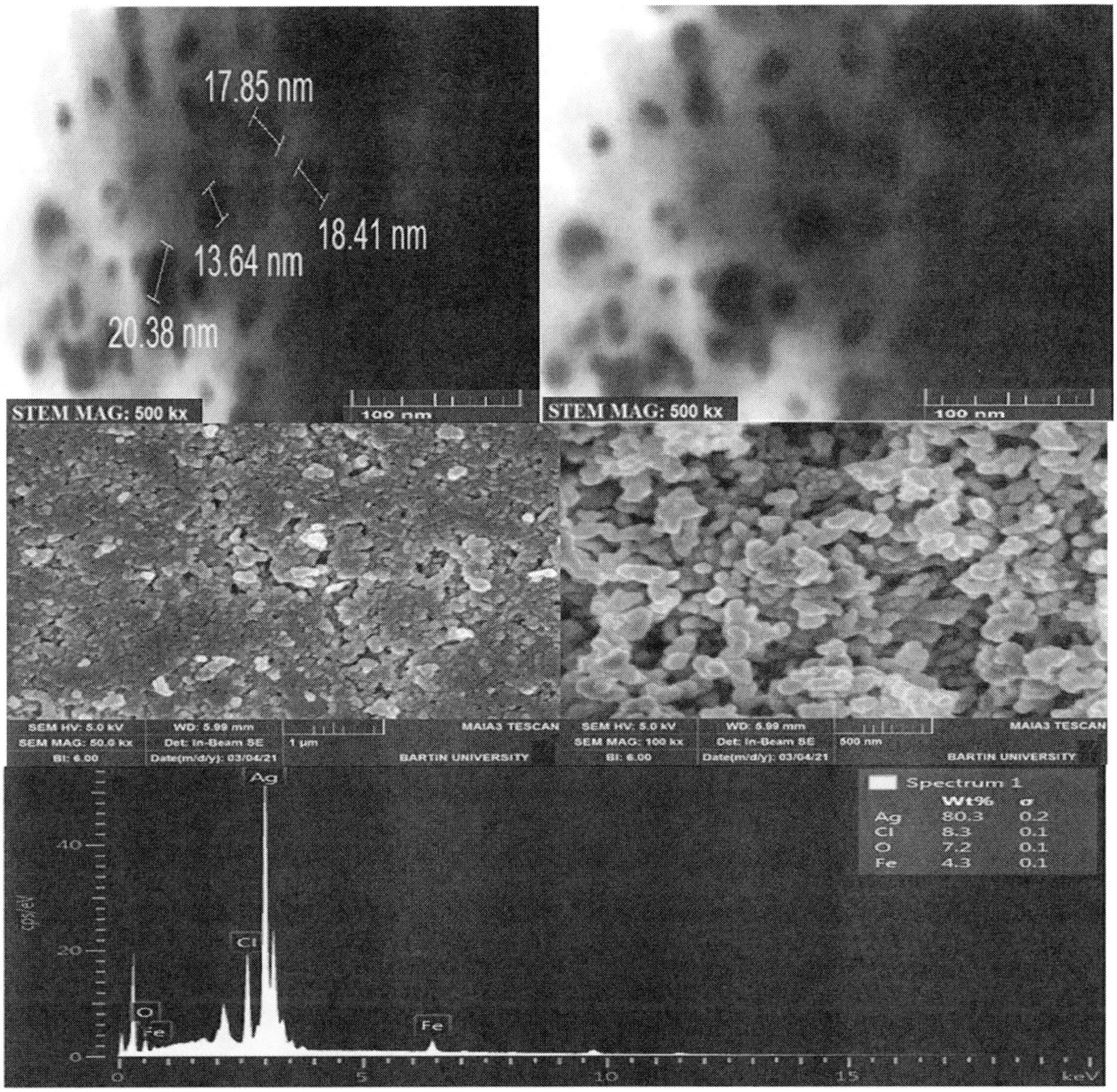

Figure 5 SEM results of TO-AgNPs

3.5. Antibacterial activity

In the present study, anti-bacterial effects of the nanoparticle were examined the. According to the experimental MIC data, the TO-AgNPs have shown a greater inhibitory effect on both gram-negative (*E. coli, Klebsiella pneumonia* and *Pseudomonas aeruginosa*) and gram-positive (*Staphylococcus aureus* and *Enterococcus faecalis*) bacterial strains TO-AgNPs *C. utilis*, and *C. albicans* fungi showed an inhibitory effect on it depending on the dose (**Figure 6**).

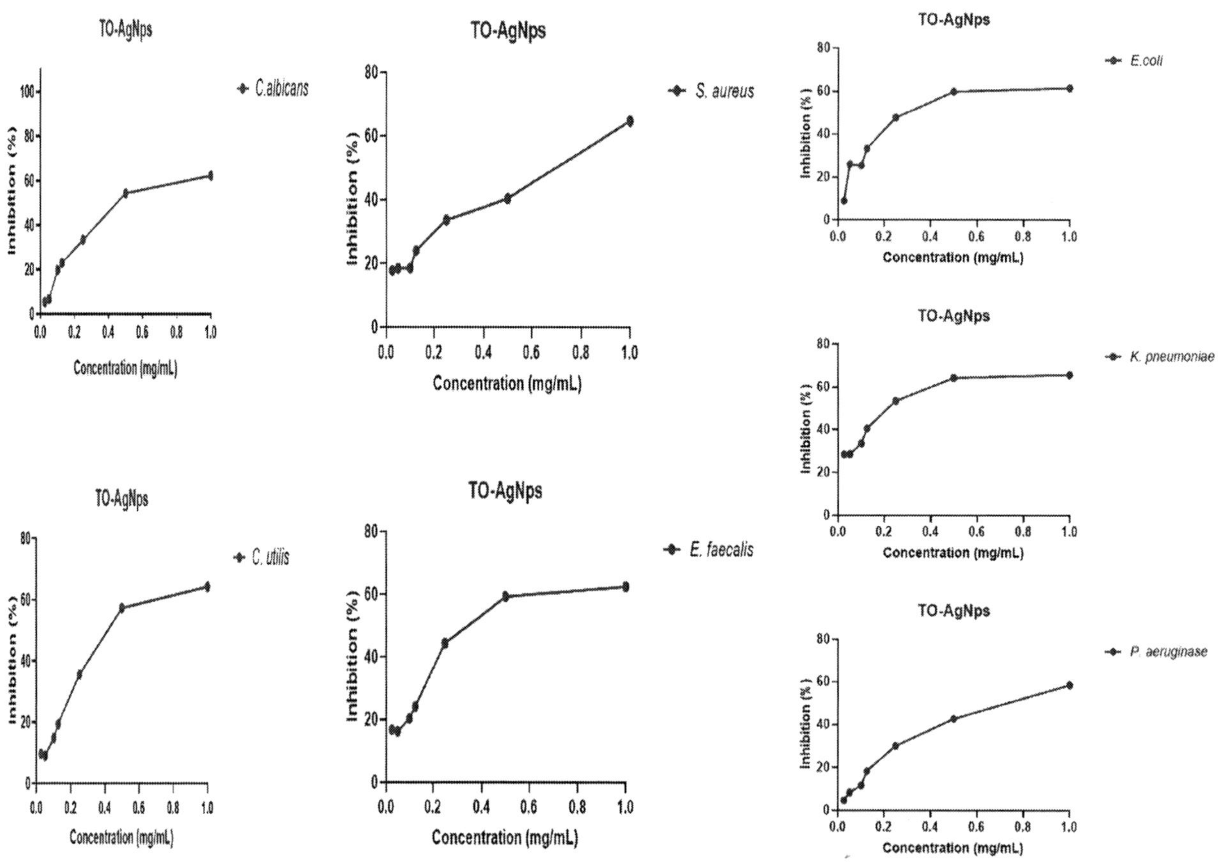

Figure 6 Growth inhibition graph of pathogenic microorganisms exposed to TO-AgNPs

Table 1 Minimum inhibitory concentrations of TO-AgNPs

Microorganisms	MIC Values of TO-AgNPs (mg/mL)
S. aureus	0.619±0.07
P. aeruginosa	0.676±0.04
K. pneumoniae	0.232±0.04
E.coli	0.352±0.06
E. faecalis	0.408±0.08
C. albicans	0.495±0.04
C. utilis	0.463±0.06

4. CONCLUSION

In this study, we reported the synthesis of silver nanoparticles from *Trachystemon orientalis* plant. We carried out antibacterial and antifungal studies of the synthesized nanoparticles. Synthesized TO-AgNPs showed significant antibacterial and antifungal properties depending on the dose.

ACKNOWLEDGEMENTS

The authors acknowledges Bartin University for supporting this study.

REFERENCES

[1] SINGH, A., DAR, M. Y., JOSHI, B., SHARMA, B., SHRIVASTAVA, S., and SHUKLA, S. Phytofabrication of silver nanoparticles: novel drug to overcome hepatocellular ailments. *Toxicology reports*. 2018, vol. 5, pp. 333-342.

[2] RAMSDEN, J. Applied nanotechnology: the conversion of research results to products. *William Andrew.* 2018. ISBN: 9780128133439.

[3] LLOYD, J. R., BYRNE, J. M., and COKER, V. S. Biotechnological synthesis of functional nanomaterials. *Current Opinion in Biotechnology*. 2011, vol. 22, no. 4, pp. 509-515.

[4] MAO, B. H., TSAI, J. C., CHEN, C. W., YAN, S. J., and WANG, Y. J. Mechanisms of silver nanoparticle-induced toxicity and important role of autophagy. *Nanotoxicology*. 2016, vol. 10, no. 8, pp. 1021-1040.

[5] RAO, SAPRA, S., SARMA, D. D., C. N. R., MULLER, A., and CHEETHAM, A. K. The Chemistry of Nanomaterials: Synthesis, Properties and Applications. *Chemistry of Nanomaterials.* [online]. 2004, 9783527602476. Available from: https://doi.org/10.1002/352760247X

[6] ANYAOGU, K. C., FEDOROV, A. V., AND NECKERS, D. C., Synthesis, characterization, and antifouling potential of functionalized copper nanoparticles, *Langmuir*. 2008, vol. 24, no. 8, pp. 4340-4346.

[7] KUMAR, A., VEMULA, P. K., AJAYAN, P. M., AND JOHN, G. Silver-nanoparticle embedded antimicrobial paints based on vegetable oil. *Nature Materials.* 2008, vol. 7, no. 3, p. 236.

[8] ESPITIA, P. J. P., SOARES, N. D. F. F., DOS REIS COIMBRA, J. S., DE ANDRADE, N. J., CRUZ, R. S., and MEDEIROS, E. A. A. Zinc oxide nanoparticles: synthesis, antimicrobial activity and food packaging applications. *Food and Bioprocess Technology.* 2012, vol. 5, no. 5, pp. 1447-1464.

[9] WEIR, A., WESTERHOFF, P., FABRICIUS, L., HRISTOVSKI, K., and VON GOETZ, N., Titanium dioxide nanoparticles in food and personal care products. *Environmental Science and Technology.* 2012, vol. 46, no. 4, pp. 2242-2250.

[10] KATHIRVELU, S., D'SOUZA, L., AND DHURAI, B., UV protection finishing of textiles using ZnO nanoparticles. *Indian Journal of Fibre and Textile Research.* 2009, vol. 34, pp. 267-273.

[11] LI, Y., WU, Y., AND ONG, B. S. Facile synthesis of silver nanoparticles useful for fabrication of high-conductivity elements for printed electronics. *Journal of the American Chemical Society.* 2005, vol. 127, no. 10, pp. 3266-326.

[12] FREY, N. A., PENG, S., CHENG, K., AND SUN, S. Magnetic nanoparticles: synthesis, functionalization, and applications in bioimaging and magnetic energy storage. *Chemical Society Reviews.* 2009, vol. 38, no. 9, pp. 2532-2542.

[13] IM, K., CHO, K., KIM, J., and KIM, S. Transparent heaters based on solution processed indium tin oxide nanoparticles. *Thin Solid Films.* 2010, vol. 518, no. 14, pp. 3960-3963,

[14] MÜLLER, R. H., RADTKE, M., and WISSING, S. A. Solid lipid nanoparticles (SLN) and nanostructured lipid carriers (NLC) in cosmetic and dermatological preparations. *Advanced Drug Delivery Reviews.* 2002, vol. 54, pp. 131-155.

[15] VOEVODIN, N. N., KURDZIEL, J. W., and MANTZ, R., Corrosion protection for aerospace aluminum alloys by Modified Self-assembled Nanophase Particle (MSNAP) sol-gel. *Surface and Coatings Technology.* 2006, vol. 201, no. 3-4, pp. 1080-1084.

[16] MULDER, W. J., STRIJKERS, G. J., VAN TILBORG, G. A., GRIFFIOEN, A. W., and NICOLAY, K.. Lipid-based nanoparticles for contrast-enhanced MRI and molecular imaging. NMR in Biomedicine. *An International Journal Devoted to the Development and Application of Magnetic Resonance In vivo.* 2006, vol. 19, no. 1, pp. 142-164.

[17] CHIANG, C. L., SUNG, C. S., WU, T. F., CHEN, C. Y., and HSU, C. Y. Application of superparamagnetic nanoparticles in purification of plasmid DNA from bacterial cells. *Journal of Chromatography B.* 2005, vol. 822, no. 1-2, pp. 54-60.

[18] LEE, I. S., LEE, N., PARK, J., KIM, B. H., YI, Y. W., KIM, T., and HYEON, T. Ni/NiO core/shell nanoparticles for selective binding and magnetic separation of histidine-tagged proteins. *Journal of the American Chemical Society.* 2006, vol. 128, no. 33, pp. 10658-10659.

[19] GOLDBERG, M., LANGER, R., and JIA, X. Nanostructured materials for applications in drug delivery and tissue engineering. *Journal of Biomaterials Science Polymer Edition.* 2007, vol. 18, no. 3, pp. 241-268.

[20] YU, C. H., TAM, K., and TSANG, E. S. Chemical methods for preparation of nanoparticles in solution. *Handbook of Metal Physics.* 2008, vol. 5, pp. 113-141.

[21] FONDEVILA, M., HERRER, R., CASALLAS, M. C., ABECIA, L., and DUCHA, J. J. Silver nanoparticles as a potential antimicrobial additive for weaned pigs. *Animal Feed Science and Technology.* 2009, vol. 150, no. 3-4, pp. 259-269.

[22] SHI, L., SHAN, J., JU, Y., AIKENS, P., and PRUD'HOMME, R. K. Nanoparticles as delivery vehicles for sunscreen agents. *Colloids and Surfaces A: Physicochemical and Engineering Aspects.* 2012, vol. 396, pp. 122-129.

[23] SHIRAISHI, Y., TOSHIMA, N., MAEDA, K., YOSHIKAWA, H., XU, J., and KOBAYASHI, S. Frequency modulation response of a liquid-crystal electro-optic device doped with nanoparticles. *Applied Physics Letters.* 2002, vol. 81, no. 15, pp. 2845-2847.

[24] CHO, K., WANG, X. U., NIE, S., and SHIN, D. M., Therapeutic nanoparticles for drug delivery in cancer. *Clinical Cancer Research.* 2008, vol. 14, no. 5, pp. 1310-1316.

[25] SANVICENS, N., PASTELLS, C., PASCUAL, N., and MARCO, M. P. Nanoparticle-based biosensors for detection of pathogenic bacteria. *TrAC Trends in Analytical Chemistry.* 2009, vol. 28, no. 11, pp. 1243-1252.

[26] YOU, C. C., MIRANDA, O. R., GIDER, B., GHOSH, P. S., KIM, I. B., ERDOGAN, B., SAI A. K., BUNZ U. H. F. and ROTELLO, V. M. Detection and identification of proteins using nanoparticle-fluorescent polymer 'chemical nose'sensors. *Nature Nanotechnology.* 2007, vol. 2, no. 5, p. 318.

[27] WANG, L., LIU, X., HU, X., SONG, S., and FAN, C. Unmodified gold nanoparticles as a colorimetric probe for potassium DNA aptamers. *Chemical Communications.* 2006, vol. 36, pp. 3780-3782.

[28] K. GUDIKANDULA, S.C. MARINGANTI. Synthesis of silver nanoparticles by chemical and biological methods and their antimicrobial properties. *J Exp Nanosci.* 2016, vol. 11, pp. 714-721.

[29] ABDUL, H., SIVARAJ, R. and VENCKATESH, R. Green Synthesis and Characterization of Zinc Oxide Nanoparticles from *Ocimum Basilicum* L. Var. Purpurascens Benth.-Lamiaceae Leaf Extract. *Mater Lett.* 2014, vol. 131, pp. 16-18.

[30] QU, J., YUAN, X., WANG, X., and SHAO, P. Zinc Accumulation and Synthesis of Zno Nanoparticles Using *Physalis alkekengi* L. *Environ Pollut*, 2011, no. 159, pp. 1783-1788.

[31] RAO, KG, ASHOK, CH, RAO, KV, CHAKRA, CHS, AKSHAYKRANTH, A. Eco-friendly synthesis of MgO nanoparticles from orange fruit waste. *Int J Appl Phys Sci.* 2015, vol. 2, no. 3, pp. 1-6.

PREPARATION OF ALUMINIUM OXIDE NANOPARTICLES USING GREEN SYNTHESIS

[1]Sivaranjani RAVICHANDRAN, [1]Ramya Devi DURAI, [1]Priya Dharshini KRISHNAN, [2]P. RAMESHKUMAR, [3]Tiju THOMAS, [4]Sathiya N. NARAYANAN, [5,6]Bozena HOSNEDLOVA, [7]Barbora HAVELKOVA, [5,8]Milan JAKUBEK, [7]Miroslava BEKLOVA, [5,6,7,9]Rene KIZEK, [1]Vedha Hari B NARAYANAN*

[1]Pharmaceutical Technology Lab, School of Chemical & Biotechnology, SASTRA Deemed University, Thanjavur, India, vedhahari@scbt.sastra.edu

[2]Mandapam Regional Centre of Central Marine Fisheries Research Institute, Marine Fisheries P. O. Mandapam Camp, Tamil Nadu, India

[3]Applied Nanostructures Engineering and Nanochemistry, ANEN, Department of Metallurgical and Materials Engineering, Indian Institute of Technology Madras, Chennai, India

[4]School of Mechanical Engineering, SASTRA Deemed University, Thanjavur, India

[5]BIOCEV, First Faculty of Medicine, Charles University, Vestec, Czech Republic, EU, bozena.hosnedlova@post.cz, Milan.Jakubek@lf1.cuni.cz

[6]CONEM Metallomics Nanomedicine Research Group (CMNRG), Vinohrady, Prague, Czech Republic, EU, kizek@sci.muni.cz

[7]Department of Ecology and Diseases of Zooanimals, Game, Fish and Bees, Faculty of Veterinary Hygiene and Ecology, University of Veterinary Brno, Czechia, EU, havelkovab@vfu.cz, beklovam@vfu.cz

[8]Department of Pediatrics and Inherited Metabolic Disorders, First Faculty of Medicine, Charles University, Prague, Czech Republic, EU

[9]Department of Biomedical and Environmental Analyses, Faculty of Pharmacy, Wroclaw Medical University, Wroclaw, Poland, EU

**Correspondence: kizek@sci.muni.cz; vedhahari@scbt.sastra.edu; Tel./Fax: +42-05-4156-2820*

https://doi.org/10.37904/nanocon.2021.4362

Abstract

Nanoparticles are widely used in industry as well as biomedicine. An interesting group are aluminium oxide nanaoparticles (alumina nanoparticles) belonging to the family of metal oxide nanoparticles, which can be prepared by various methods (the sol-gel, precipitation, micro-emulsion and hydrothermal methods). In this work, alumina nanoparticles were prepared by biological synthesis using plant extract from fresh neem (*Azadirachta indica*) leaves. The particles were confirmed by the SEM analysis. The distribution diagram showed a size from 60 to 1,500, with a maximum around 200-400 nm. Alumina nanoparticles possess many potential applications in biomedicine such as drug delivery, therapy of cancer and other diseases, and immunotherapy.

Keywords: Aluminium nanoparticles, green synthesis; biotechnology; nanomedicine

1. INTRODUCTION

The sol-gel technique is one of the optimistic and explored methods due to its ability to produce ultrafine sized particles with controlled size, high surface area, purity, and crystallinity. **Figure 1** shows a schematic representation of the synthesis of alumina nanoparticles by using an ethanol solution of iron nitrate as precursor through the simple sol-gel method [1].

Rogojan et al. prepared aluminium oxide nanopowders using 2 different precursors, organic and inorganic precursors to be used as a biomaterial in medical applications. The organic precursor used was aluminium chloride ($AlCl_3$) and the inorganic precursor was aluminium triisopropylate $Al[OCH(CH_3)_2]_3$. [2]. The precursors were chemically treated by using the sol-gel protocol, and the resulting gel was calcined for 2 hours at two different temperatures, 1,000 °C and 1,200 °C. The X-ray diffraction analysis revealed that thermal treatment at 1,000 °C leads to the formation of α + γ - Al_2O_3 mixture having relatively low degrees of crystallinity. An increase in the heat treatment temperature to 1200°C led to the formation of α-Al_2O_3 with higher crystallinity. The SEM and TEM analysis confirmed the production of alumina powder in nanometer-scale by the sol-gel method. Application of heat treatment at temperatures higher than 1,000 °C had led to the formation of crystalline α alumina as a single phase.

Li et al. synthesized nano-sized α-Al_2O_3 powder using citrate precursor that was derived from aluminium nitrate and citric acid precursor solution. The ratio of citric acid (CA) to metal nitrate (N) had a greater influence on the size of the nanoparticle prepared by the sol-gel method. Initially, the ratio was maintained as 1:1 and the precursors were heated to 400 °C. The precursors decomposed and resulted in the formation of amorphous nano-Al_2O_3, which further transformed into to γ-Al_2O_3 (size = 15 nm). Further heating to 800°C resulted in the transformation of γ -Al_2O_3 to α-Al_2O_3 with a crystalline size of 75 nm. The calcination process was completed at 1,000 °C with the formation of 200 nm-sized alumina nanoparticles. The increasing molar ratio of CA/N favored the rapid phase transition of γ-Al_2O_3 to α-Al_2O_3, whereas 1:1 CA/N molar ratio yielded a monodispersed ultrafine α-Al_2O_3 [3].

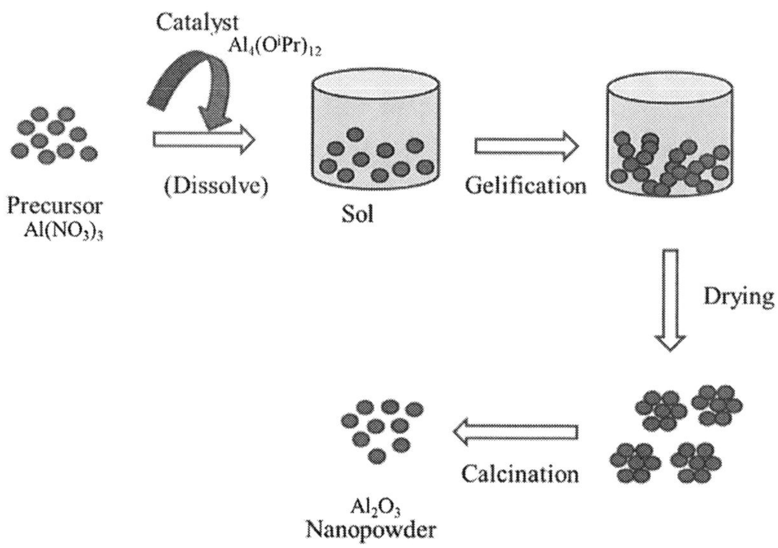

Figure 1 Scheme for the synthesis of alumina nanoparticles by the sol-gel method

Khazaei et al. synthesized porous γ-aluminium oxide nanoparticles using sodium aluminate liquor as a precursor. The effect of two surfactants namely, polyethylene glycol (PEG) and polyvinyl alcohol (PVA) were investigated for their stabilizing action on alumina nanoparticles (**Figure 1**). The calcination temperature was maintained at 80 °C. PVA as stabilizing agent offered better structural properties (average crystallite size of 2.063 nm, the average particle size of 46 nm, the specific surface area of 201.1 m^2/g and pore volume of ~ 0.246 cm^3/g) when compared with PEG (average crystallite size of 2.313 nm, the average particle size of 20 nm, the specific surface area (SSA) of 138.8 m^2/g, and pore volume of ~ 0.166 cm^3/g). PEG stabilized synthesis resulted in the formation of ultrafine nanoparticles (20 nm), whereas PVA stabilized nanoparticles are comparatively larger, thus the adhesion between the nanoparticles was less and remained in a more dispersed state [4].

Mirjalili et al. synthesized ultrafine α-alumina nanoparticles using aluminium isopropoxide, aluminium nitrate hydrate (0.5 mol/L) as precursors and sodium dodecylbenzene sulfonate (SDBS), sodium bis-2-ethylhexyl sulfosuccinate (Na-AOT) as stabilizing agents to prevent the aggregation of nanoparticles. The surfactants played a major role in controlling the size and shape of the nanoparticle and in addition the degree of aggregation has also been influenced by the two stabilizing agents. Using sodium dodecylbenzene sulfonate as a stabilizing agent resulted in the fine and well-dispersed nanoparticles (20−30 nm). The better stabilization action of SDBS can be attributed to the higher ionization of SDBS in comparison Na-AOT that resulted in negative charge adsorption in the positively charged precursor. This charge-charge interaction decreased particle aggregation [5].

The aim of this work was to use the sol-gel method for the green synthesis of alumina nanoparticles.

2. METHODS

2.1. Plant material and preparation of the extract

Generally, biosynthesis of nanoparticles occurs when plant aqueous extract undergoes bioreduction process with an aqueous solution of suitable metal salt. Fresh neem (*Azadirachta indica*) leaves were collected from the institute and washed twice with distilled water to remove all the debris present on it. About 20−30 g of leaves was crushed with mortar and pestle and the obtained extract was filtered with muslin cloth. The extract was then stored at room temperature for further experiments [6]. The leaves extract contains flavonoids and phenolic acids with various functional groups, which are capable of forming nanoparticles by acting as reducing and stabilizing agents.

2.2. Biosynthesis of Al_2O_3NPs

Aluminum chloride (Merck, India) was used as the primary chemical for the synthesis of aluminum oxide nanoparticles. The prepared extract was mixed with saturated solution of aluminum chloride and kept for two hours in magnetic stirrer (REMI, Laboratory Instruments, India) for appropriate mixing of the solutions. The pH change was observed after stirring process to check the reaction occurring between the phytochemicals present in the extract and the metal ions to produce metal oxide nanoparticle. The reaction mixture was then centrifuged at 2500 rpm for 20 minutes using cooling centrifuge (REMI C-24 Plus) to settle down the unused extract and the obtained supernatant was again ultra-centrifuged at 14,000 rpm for 30 minutes. The resulted supernatant was discarded and the pellets were repeatedly washed with water and ethanol and dried in hot air oven at 60 °C to obtain Al_2O_3NPs. The alumina nanoparticles were characterized by the SEM method.

3. RESULTS AND DISCUSSION

Aluminium oxide nanoparticles (Al_2O_3NPs) are class of metal oxide nanoparticles that have various biomedical applications due to their excellent physicochemical and structural properties (chemical and mechanical resistence, very good optical characteristics, a porous vast surface area etc.) as well as their low cost of preparation [7]. The synthesis of metal oxide nanoparticles is the major technological challenge. The behavior of aluminium oxide nanoparticles for different applications is determined by the physical, chemical, optical and magnetic properties, which, in turn, are influenced by the structural characteristics of the nanoparticles. Numerous production methods are developed such as sol-gel, precipitation, micro-emulsion, and hydrothermal method. Al_2O_3NPs have various potential biomedical applications − drug delivery, biosensing, cancer therapy, anti-microbial agens, treatment other diseases such as asthma, biomolecular preservation and stabilisation, and immunotherapy [7]. Due to their biomedical applications, it is needed test their environmental biotoxicity.

Table 1 summarizes the data on the synthesis of alumina nanoparticles using various methods and techniques. Thus, it was possible to prepare nanoparticles with sizes from 4 to 100 nm.

Table 1 Brief description of the different methods applied for the synthesis of the aluminium and aluminium oxide (alumina) nanoparticles used in various research studies

Method of preparation	Precursor	Nanoparticle	Particle size (nm)	Reference
Sol-gel method	aluminium chloride and aluminium triisopropylate	α-Al_2O_3	-	[2]
	aluminium nitrate and citric acid	α-Al_2O_3	75 nm	[3]
	sodium aluminate liquor		20 nm	[4]
	aluminium isopropyloxide, aluminium nitrate hydrate	α-Al_2O_3	20−30 nm	[5]
Precipitation method	SDS, PEG 6000	$Al(OH)_3$	80−100 nm	[8]
	polysulfone and polystyrene	$Al(OH)_3$		[9]
	ammonium bicarbonate, ammonium carbonate, sodium bicarbonate and sodium carbonate	γ-alumina	4.7-5.7 nm	[10]
	-	γ-Al_2O_3	4.5 nm	[11]
	polyethylene glycol	Al_2O_3	-	[12]
	-	*γ* and *α* phases of alumina nanoparticles	20−50 nm	[13]
Micro-emulsion method	Triton X-100 and n-butyl alcohol	-	-	[14]
	silane agent	Al_2O_3	-	[15]
Hydrothermal method	polyethyleneglycol (PEG)	γ-aluminium oxide	10−15 nm	[16]
	cationic (cetyltrimethylammonium tosylate, CTAT) and anionic (sodium dodecylbenzene, sulfonate, SDBS)	δ-Al_2O_3	-	[17]

A study has also been conducted to explain the complete process for producing aluminium hydroxide with improved whiteness. In the initial step, the colorant containing aluminium hydroxide was dissolved as the solution. The solution was treated with colorant-collecting reagent and separated as a solid phase. The solid phase was removed from the solution and the aluminium hydroxide was precipitated as a pure white powder.

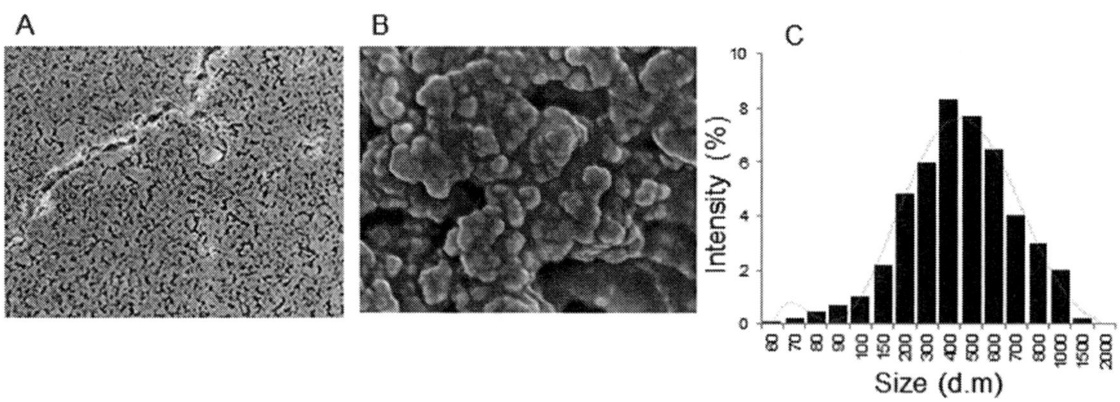

Figure 2 Characterization of alumina nanoparticles - A) SEM images 1000x; B) 30 000x; C) Size distribution

In this study, alumina nanoparticles were prepared biologically – by green synthesis from neem (*Azadirachta indica*) leaves. This plant belongs to the so-called ethnomedical herbs, which has antibacterial, antiviral and anticancer properties. Green-synthesized alumina nanoparticles were characterized by the SEM technique. The method showed the irregular arrangement, wherein the spherical shaped nanoparticles were aggregated (**Figures 2A,B**). Distribution diagram shown in **Figure 2C** displayed showed a size of nanoparticles from 60 to 1,500, with a maximum around 200-400 nm. The zeta potential of prepared alumina nanoparticles was +19.5 mV. The size of nanoparticles formed by the sol-gel method ranged from 20−200 nm. From the above results, it is also clear that maintaining the calcination temperature within 800−1,000 °C results in the formation of ultrafine, monodispersed alumina nanoparticles, which is important for its various applications.

Sutradhar et al. used microwave-assisted green chemistry and synthesized alumina nanoparticles using tea and coffee extracts with a size of 50-200 nm which were spherical in shape, and also oval shaped nanoparticles using triphala extract with an average size of 200-400 nm [18]. Hasanpoor et al. found that average size of clusters of alumina nanoparticles varied with different routes from of 60-300 nm. They used extracts from five different plants (*Syzygium aromaticum*, *Origanum vulgare*, *Origanum majorana*, *Theobroma cacao* and *Cichorium intybus*) for their synthesis [19]. Narayanan and Rakesh prepared alumina nanoparticles using lemon extract [20]. Recently, alumina nanoparticles were also synthesized using leaf extract of *Ocimum sanctum* [21]. Neem extraxt has been used in previous studies for the green synthesis of gold [22], silver [23,24], iron [25], zinc [26,27] and copper [28] nanoparticles. However, to the best of the authors' knowledge, alumina nanoparticles have not yet been synthesized using this plant.

4. CONCLUSION

Alumina nanoparticles were prepared by green synthesis from *Azadirachta indica*, containing a number of phenolic compounds (nimbin, azadirachin, nimbidin, salannin). The nanoparticles slightly aggregated with the most frequent size of 400 nm and zeta potential +19.5 mV. We expect their further potential in use in biomedical applications, especially those antibacterial and antiviral.

ACKNOWLEDGEMENTS

This work was supported by the grant FVHE/Pikula/ITA2021. We also thank Operational Programme Research, Development and Education, within the project Center for Tumor Ecology – Research of the Cancer Microenvironment Supporting Cancer Growth and Spread (reg. No. CZ.02.1.01/0.0/0.0/16_019/0000785). The authors would like to acknowledge the SASTRA Deemed University, Thanjavur-613401, India for infrastructural support in the manuscript.

REFERENCES

[1] FARAHMANDJOU, M.,GOLABIYAN, N. Synthesis and characterization of Alumina (Al2O3) nanoparticles prepared by simple sol-gel method. *Int. J. Bio-Inorg. Hybr. Nanomater.* 2016, vol. 5, no. 1, pp. 73-77.

[2] ROGOJAN, R., ANDRONESCU, E., GHITULICA, C.,VASILE, B. S. Synthesis and characterization of alumina nano-powder obtained by sol-gel method. *UPB Buletin Stiintific, Series B: Chemistry and Materials Science.* 2011, vol. 73, no. 2, pp. 67-76.

[3] LI, J., PAN, Y. B., XIANG, C. S., GE, Q. M.,GUO, J. K. Low temperature synthesis of ultrafine alpha-Al2O3 powder by a simple aqueous sol-gel process. *Ceramics International.* 2006, vol. 32, no. 5, pp. 587-591.

[4] NAZARI, S., KARIMI, G., GHADERI, E., MANSOURI MORADIAN, K.,BAGHERPOR, Z. Synthesis and characterization of γ-alumina porous nanoparticles from sodium aluminate liquor with two different surfactants. *International Journal of Nanoscience and Nanotechnology,* 2016, vol. 12, no. 4, pp. 207-214.

[5] MIRJALILI, F., MOHAMAD, H.,CHUAH, L. Preparation of nano-scale alpha-Al2O3 powder by the sol-gel method. *Ceramics-Silikaty.* 2011, vol. 55, no. 4, pp. 378-383.

[6] AHMED, S., SAIFULLAH, AHMAD, M., SWAMI, B. L.,IKRAM, S. Green synthesis of silver nanoparticles using Azadirachta indica aqueous leaf extract. *Journal of radiation research and applied sciences.* 2016, vol. 9, no. 1, pp. 1-7.

[7] HASSANPOUR, P., PANAHI, Y., EBRAHIMI-KALAN, A., AKBARZADEH, A., DAVARAN, S., NASIBOVA, A. N., KHALILOV, R.,KAVETSKYY, T. Biomedical applications of aluminium oxide nanoparticles. *Micro & Nano Letters.* 2018, vol. 13, no. 9, pp. 1227-1231.

[8] YU, B., TIAN, Z. H., XIONG, J.,XIANG, L. Synthesis of Al(OH)(3) Nanostructures from Al(OH)(3) Microagglomerates via Dissolution-Precipitation Route. *Journal of Nanomaterials.* 2013, vol. 2013.

[9] GOUDARZI, M., GHANBARI, D., SALAVATI-NIASARI, M.,AHMADI, A. Synthesis and Characterization of Al(OH)(3), Al2O3 Nanoparticles and Polymeric Nanocomposites. *Journal of Cluster Science.* 2016, vol. 27, no. 1, pp. 25-38.

[10] PARIDA, K. M., PRADHAN, A. C., DAS, J.,SAHU, N. Synthesis and characterization of nano-sized porous gamma-alumina by control precipitation method. *Materials Chemistry and Physics.* 2009, vol. 113, no. 1, pp. 244-248.

[11] POTDAR, H. S., JUN, K. W., BAE, J. W., KIM, S. M.,LEE, Y. J. Synthesis of nano-sized porous gamma-alumina powder via a precipitation/digestion route. *Applied Catalysis a-General.* 2007, vol. 321, no. 2, pp. 109-116.

[12] LASHANIZADEGAN, M., FARZI, G.,NIA, N. E. Synthesis and surface modification of aluminum oxide nanoparticles. *Journal of Ceramic Processing Research.* 2014, vol. 15, no. 5, pp. 316-319.

[13] RAJAEIYAN, A.,BAGHERI-MOHAGHEGHI, M. Comparison of sol-gel and co-precipitation methods on the structural properties and phase transformation of γ and α-Al 2 O 3 nanoparticles. *Advances in Manufacturing.* 2013, vol. 1, no. 2, pp. 176-182.

[14] HUANG, K., YIN, L., LIU, S.,LI, C. Preparation and formation mechanism of Al2O3 nanoparticles by reverse microemulsion. *Transactions of Nonferrous Metals Society of China.* 2007, vol. 17, no. 3, pp. 633-637.

[15] GHEZELBASH, Z., ASHOURI, D., MOUSAVIAN, S., GHANDI, A. H.,RAHNAMA, Y. Surface modified Al2O3 in fluorinated polyimide/Al2O3 nanocomposites: Synthesis and characterization. *Bulletin of Materials Science.* 2012, vol. 35, no. 6, pp. 925-931.

[16] TABATABAEE, M.,SABERI, N. A Simple Method for Synthesis of γ-Aluminum Oxide Nanostructure Using PEG as a Neutral Surfactant. *Int. J. Bio-Inorg. Hybd. Nanomat.* 2012, vol. 1, no. 4, pp. 253-256.

[17] YAACOB, II, BHANDARKAR, S.,BOSE, A. SYNTHESIS OF ALUMINUM HYDROXIDE NANOPARTICLES IN SPONTANEOUSLY GENERATED VESICLES. *Journal of Materials Research.* 1993, vol. 8, no. 3, pp. 573-577.

[18] SUTRADHAR, P., DEBNATH, N.,SAHA, M. Microwave-assisted rapid synthesis of alumina nanoparticles using tea, coffee and triphala extracts. *Advances in Manufacturing.* 2013, vol. 1, no. 4, pp. 357-361.

[19] HASANPOOR, M., FAKHR NABAVI, H.,ALIOFKHAZRAEI, M. Microwave-assisted synthesis of alumina nanoparticles using some plants extracts. *Journal of Nanostructures.* 2017, vol. 7, no. 1, pp. 40-46.

[20] NARAYANAN, M. V.,RAKESH, S. Synthesis of colloidal alumina nanoparticles using green method. *IOP Conference Series: Materials Science and Engineering.* IOP Publishing, 2018, p. 012150.

[21] SHARMA, P.,SHARMA, N. Antimicrobial activity of alumina nanoparticles synthesized by leaf extract of Ocimum sanctum. *International Journal of Engineering Applied Sciences and Technology.* 2020, vol. 5, no. 4, pp. 251-253.

[22] ANURADHA, J., ABBASI, T.,ABBASI, S. Green'synthesis of gold nanoparticles with aqueous extracts of neem (Azadirachta indica). *Research Journal of Biotechnology.* 2010, vol. 5, no. 1, pp. 75-9.

[23] CHAND, K., ABRO, M. I., AFTAB, U., SHAH, A. H., LAKHAN, M. N., CAO, D., MEHDI, G.,MOHAMED, A. M. A. Green synthesis characterization and antimicrobial activity against Staphylococcus aureus of silver nanoparticles using extracts of neem, onion and tomato. *RSC advances.* 2019, vol. 9, no. 30, pp. 17002-17015.

[24] KHAN, Z., HUSSAIN, J. I.,HASHMI, A. A. Shape-directing role of cetyltrimethylammonium bromide in the green synthesis of Ag-nanoparticles using Neem (Azadirachta indica) leaf extract. *Colloids and Surfaces B: Biointerfaces.* 2012, vol. 95, pp. 229-234.

[25] PATTANAYAK, M.,NAYAK, P. Green synthesis and characterization of zero valent iron nanoparticles from the leaf extract of Azadirachta indica (Neem). *World Journal of Nano Science & Technology.* 2013, vol. 2, no. 1, pp. 06-09.

[26] NOORJAHAN, C., SHAHINA, S. J., DEEPIKA, T.,RAFIQ, S. Green synthesis and characterization of zinc oxide nanoparticles from neem (Azadirachta indicia). *International Journal of Scientific Engineering and Technology Research.* 2015, vol. 4, no. 30, pp. 5751-5753.

[27] SOHAIL, M. F., REHMAN, M., HUSSAIN, S. Z., HUMA, Z.-E., SHAHNAZ, G., QURESHI, O. S., KHALID, Q., MIRZA, S., HUSSAIN, I.,WEBSTER, T. J. Green synthesis of zinc oxide nanoparticles by Neem extract as multi-facet therapeutic agents. *Journal of Drug Delivery Science and Technology.* 2020, vol. 59, p. 101911.

[28] ANSILIN, S., NAIR, J. K., ASWATHY, C., RAMA, V., PETER, J.,PERSIS, J. J. Green synthesis and characterisation of copper oxide nanoparticles using Azadirachta indica (Neem) leaf aqueous extract. *Journal of Nanoscience and Technology.* 2016, pp. 221-223.

AB INITIO STUDY OF SILVER NANOPARTICLES, GRAIN BOUNDARIES AND THEIR QUADRUPLE JUNCTIONS

[1,2]Svatava POLSTEROVÁ, [1,2]Monika VŠIANSKÁ, [1]Martin FRIÁK, [1]Naděžda PIZÚROVÁ, [3]Sergey SOKOVNIN, [2,1]Mojmír ŠOB

[1]*Institute of Physics of Materials, v.v.i., Czech Academy of Sciences, Brno, Czech Republic, EU,* polsterova@mail.muni.cz, 230038@mail.muni.cz, friak@ipm.cz, pizurova@ipm.cz

[2]*Department of Chemistry, Faculty of Science, Masaryk University, Brno, Czech Republic, EU,* mojmir@ipm.cz

[3]*Institute of Electrophysics Ural Branch RAS, Yekaterinburg, Russia*

[4]*Ural Federal University, Yekaterinburg, Russia,* sokovnin@iep.uran.ru

https://doi.org/10.37904/nanocon.2021.4365

Abstract

Motivated by our experimental research related to silver nanoparticles with various morphologies, we have employed quantum-mechanical calculations to provide our experiments with theoretical insight. We have computed properties of a 181-atom decahedral silver nanoparticle and two types of internal extended defects, $\Sigma5(210)$ grain boundaries (GBs) and quadruple junctions (QJs) of these GBs. We have employed a supercell approach with periodic boundary conditions. Regarding the thermodynamic stability of the decahedral nanoparticle, its energy is higher than that of a defect-free face-centered cubic (fcc) Ag by 0.34 eV/atom. As far as the $\Sigma5(210)$ GB is concerned, its energy amounts to 0.7 J/m^2 and we predict that the studied GBs would locally expand the volume of the lattice. Importantly, the system with GBs is found rather close to the limit of mechanical stability. In particular, the computed value of the shear-related elastic constant C_{66} is as low as 9.4 GPa with the zero/negative value representing a mechanically unstable system. We thus predict that the $\Sigma5(210)$ GBs may be prone to failure due to specific shearing deformation modes. The studied GBs have also the value of Poisson's ratio for some loading directions close to zero. Next, we compare our results related solely to $\Sigma5(210)$ GBs with those of a system where multiple intersecting $\Sigma5(210)$ GBs form a network of quadruple junctions. The value of the critical elastic constant C_{66} is higher in this case, 13 GPa, and the mechanical stability is, therefore, better in the system with QJs.

Keywords: Nanoparticles, silver, grain boundaries, junctions, quantum-mechanical calculations, elasticity

1. INTRODUCTION

The dependence of the biological activity of nanoparticles on their morphology is well known, see, e.g., Ref. [1]. The key aspect is the number of {111} surface facets that exhibit the highest atomic density as well as the highest electronic charge density [2]. The nanoparticles containing the highest number of {111} facets are formed with decahedral or icosahedral shapes, described as 5 or 20 tetrahedra joined by twin boundaries. Because the tetrahedra do not fill the space completely, grain boundaries or elastic strains are required to complete the nanoparticle shape [3]. The nanoparticle morphologies described above are known as multiply-twinned nanoparticles (MTPs) with fivefold symmetry axes. Our experimental research focused on the method of synthesis and morphological analysis. We used the spark erosion method described in Berkowitz et al. [4] to prepare silver nanoparticles. Morphological analyzes were performed using a JEM2100F transmission electron microscope. We received various nanoparticle morphologies (**Figure 1**), including typical fivefold MTPs. Inspired by these results, we performed quantum mechanical calculations to evaluate the stability of nanoparticle morphology containing quadruple junctions (QJs).

First, we have determined the internal atomic structure and the energy of a 181-atom decahedral nanoparticle of silver. Second, we have used the calculations to characterize the properties of other internal extended defects, in particular, the Σ5(210) GBs as well as their quadruple junctions (QJs), building upon the expertise related to our previous work [5].

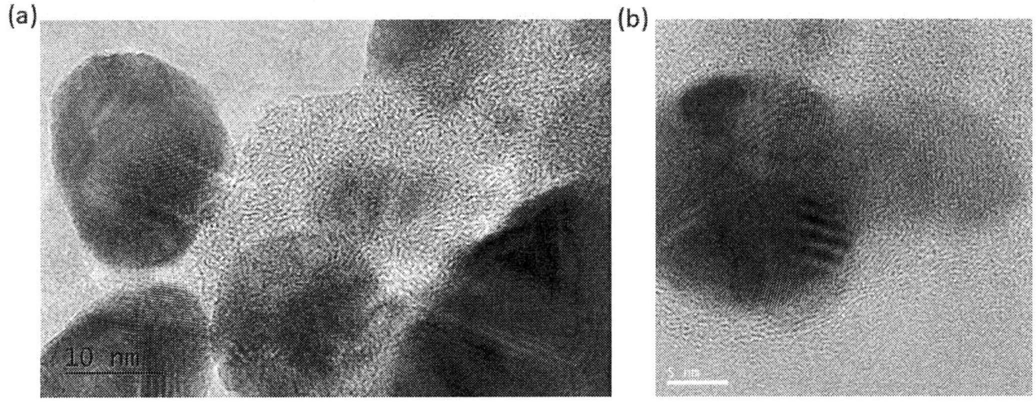

Figure 1 Transmission electron microscopy (TEM) images of Ag nanoparticles illustrating their morphologies (a), especially fivefold MTPs (b)

2. COMPUTATIONAL METHODOLOGY

When modeling decahedral silver nanoparticles as well as GBs and their junctions, we have employed a supercell approach. The decahedral nanoparticle is shown in **Figure 2**. It contains 181 Ag atoms and it was surrounded by a vacuum within a bigger supercell so as to eliminate spurious interaction between an individual nanoparticle and its periodic images in the neighboring cells generated by the periodic boundary conditions. The nanoparticle contains both internal grain boundaries and their fivefold-symmetry junction. Further, supercells modeling Σ5(210) grain boundaries as well as their quadruple junctions (QJs) are shown in **Figure 3**. The application of periodic boundary conditions in the case of these supercells results in two GBs per supercell visualized in **Figure 3a** and eight GBs together with four QJs per 180-atom supercell in **Figure 3b**.

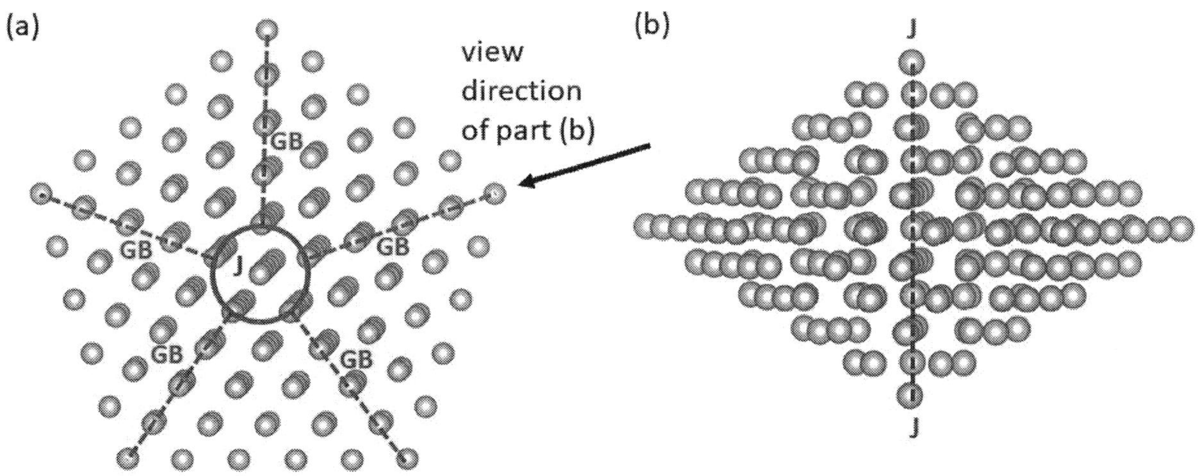

Figure 2 Schematic visualizations of our 181-atom computational model of a decahedral Ag nanoparticle (a) together with its side-view (b). The nanoparticle contains grain boundaries (GBs), see red dashed lines, and junctions (see J letters) of GBs indicated by blue circles in part (a) and a blue dashed line in part (b).

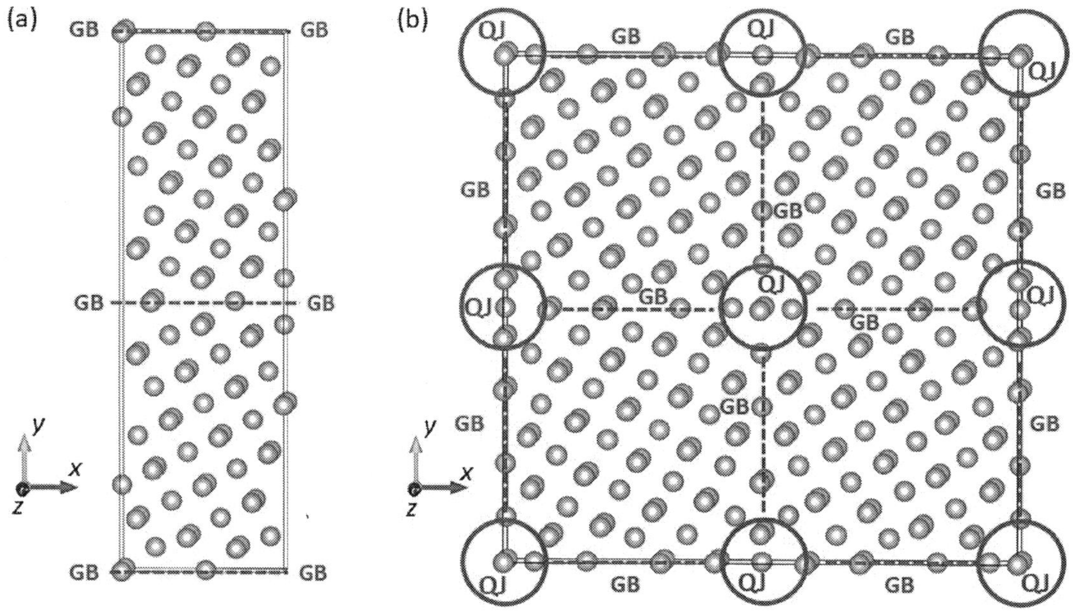

Figure 3 Schematic figures of 64-atom computational supercell modeling Σ5(210) grain boundary (GB) (a) and a 180-atom supercell utilized in our calculations of quadruple junctions (QJ) of Σ5(210) GBs (b). The GB planes are shown as red dashed lines while the regions of quadruple junctions are indicated by blue circles.

Our quantum-mechanical calculations were performed employing the Vienna Ab initio Simulation Package (VASP) [6,7] within the framework of density functional theory [8,9] using projector-augmented-wave (PAW) pseudopotentials [10,11]. The exchange and correlation energy were treated in the generalized gradient approximation (GGA) as parametrized by Perdew, Burke and Erhzerhof (PBE'96) [12]. The plane-wave energy cut-off was equal to 450 eV and different Monkhorst-Pack k-point meshes were used for sampling the Brillouin zone in the reciprocal space. In particular, the 3x3x4 k-point mesh was applied in the case of the supercell with the lattice parameters 33.8 Å x 32.6 Å x 24.5 Å containing the 181-atom decahedral nanoparticle. When computing the Σ5(210) GB, a 6x2x14 k-point mesh was utilized and 2x2x12 k-points were used for 180-atom supercell modeling QJs of Σ5(210) GBs. Lastly, we applied a 14x14x14 k-point mesh for calculations of a defect-free face-centered cubic (fcc) elemental silver modeled by a cubic-shape 4-atom cell.

3. RESULTS

The 181-atom decahedral nanoparticle has an internally distorted atomic structure as seen in **Figure 2**. Its energy is higher than that of a defect-free fcc Ag by 0.34 eV/atom. This increase in the energy is caused by the occurrence of a number of extended defects that define the nanoparticle, in particular different surfaces, internal grain boundaries and their central junction, as well as elastic strain field, and further mutual interactions of these defects. It is rather difficult to decompose the energy increase into these individual components. Instead, we use our calculations to assess properties of selected extended defects, Σ5(210) grain boundaries and their junctions, in order to test their thermodynamic and mechanical stability. We start with Σ5(210) GBs that we model with the help of the 64-atom supercell shown in **Figure 3a**. In order to compute the energy increase due to the studied GBs, we evaluate the GB interface energy. In order to do so, we determine the difference between the energy of 64-atom supercell and that of 64 atoms of the defect-free fcc Ag. The GB interface energy is then equal to this difference divided by twice the area of the GB interface (because there are two GBs per supercell in **Figure 3a).** The Σ5(210) GB interface energy amounts to 0.7 J/m^2 (or 0.04 eV/Å^2). Regarding the structure, the Σ5(210) GB expands locally the volume of the lattice by 2.4 %.

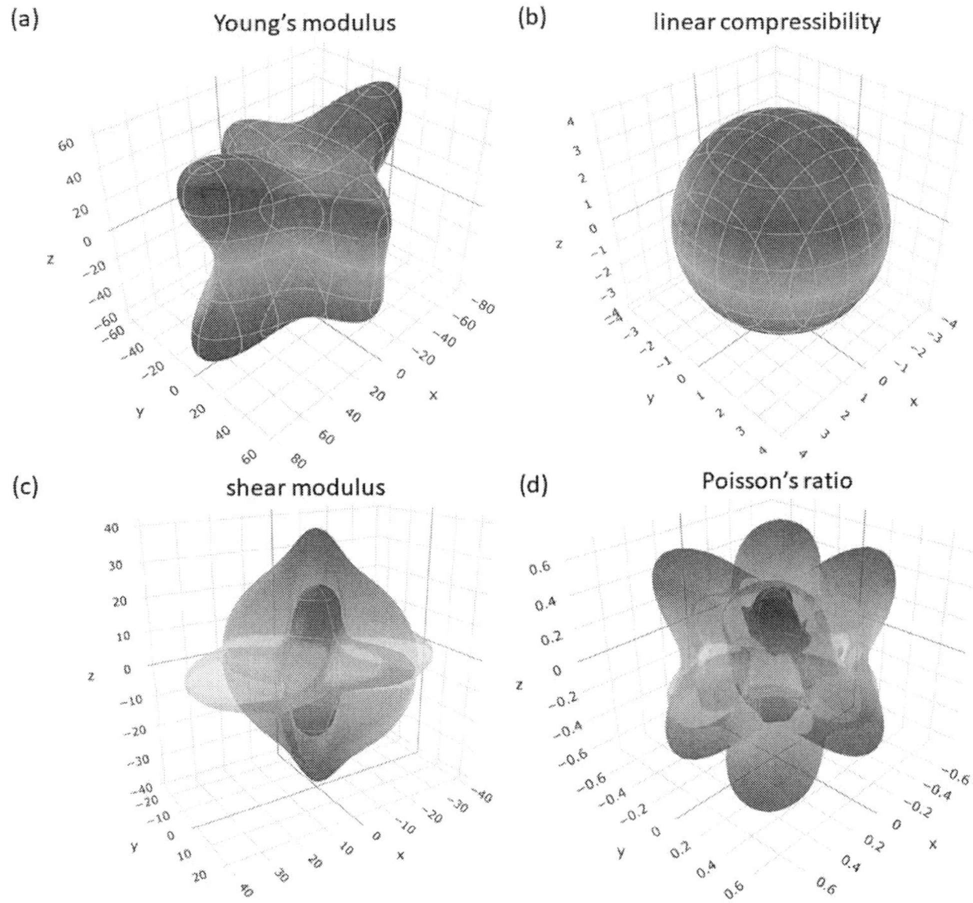

Figure 4 Calculated anisotropic elastic properties of 64-atom computational supercell modeling the Σ5(210) grain boundary visualized in the form of directional dependences of Young's modulus (a), linear compressibility (b), maximum and minimum values of shear modulus (c) and maximum and minimum values of Poisson's ratio (d). For details see Ref. [13] related to the ELATE software (http://progs.coudert.name/elate).

Next we determine elastic properties of crystalline systems containing the Σ5(210) GBs using the stress-strain method [14]. The elastic properties are characterized by the following set of elastic constants C_{11} = 119 GPa, C_{12} = 61 GPa, C_{13} = 79 GPa, C_{22} = 119 GPa, C_{23} = 77 GPa, C_{33} = 102 GPa, C_{44} = 23 GPa, C_{55} = 39 GPa and C_{66} = 9.4 GPa. The elasticity is visualized in **Figure 4**. The Young's modulus is very strongly anisotropic, see **Figure 4a**, with the minimum and maximum values equal to 29 and 93 GPa, respectively. In contrast, the directional dependence of the linear compressibility, see **Figure 4b**, is very close to a spherical shape with the minimum and maximum values equal to 3.81 and 3.94 TPa^{-1}, respectively, i.e. it is nearly isotropic. It is worth noting that systems with a cubic symmetry exhibit perfectly isotropic linear elasticity due to their high symmetry. It is quite interesting that the linear compressibility of the supercell with GBs is so close to the isotropic directional dependence despite of a very low symmetry of its internal configuration of atoms (see **Figure 3a).** Regarding shear deformations, the maximum and minimum values of shear modulus for different shear directions (for details see Ref. [13]) are shown in **Figure 4c**. The highest and lowest value out of those for all directions, 39 and 9.4 GPa, respectively, are equal to the values of C_{55} and C_{66}, respectively. If the lowest one would reach zero, the system would become mechanically unstable [15], and the value of C_{66} is close to this limit. Regarding the maximum and minimum values of Poisson's ratio for different directions shown in **Figure 4d**, the absolute lowest and highest values are equal to 0.001 and 0.855, respectively, with the former remarkably close to zero.

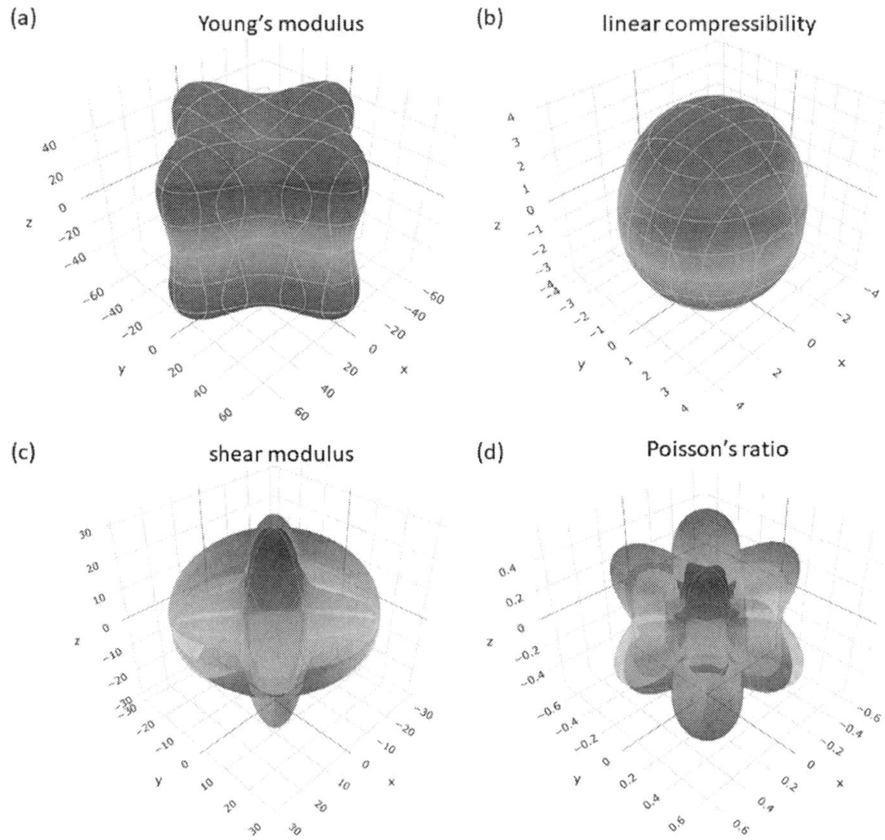

Figure 5 Computed anisotropic elastic properties of 180-atom computational supercell modeling a quadruple junction of four $\Sigma 5(210)$ grain boundaries visualized in the form of directional dependences of Young's modulus (a), linear compressibility (b), maximum and minimum values of shear modulus (c) and maximum and minimum values of Poisson's ratio (d). The figures were obtained using the ELATE software [13].

Regarding the system with quadruple junctions of $\Sigma 5(210)$ GBs, its energy is higher than that of a defect-free fcc Ag by 0.06 eV/atom. It is, unfortunately, difficult to decompose this energy increase into those due to QJs and GBs because it is unclear where exactly the regions associated with GBs and QJs start and finish, see **Figure 3b**. Therefore, it is not possible to evaluate the corresponding energy contributions. The elastic properties of the system with QJs are characterized by elastic constants C_{11} = 118 GPa, C_{12} = 56 GPa, C_{13} = 72 GPa, C_{22} = 118 GPa, C_{23} = 74 GPa, C_{33} = 103 GPa, C_{44} = 30 GPa, C_{55} = 30 GPa and C_{66} = 13 GPa. They are visualized in **Figure 5**. When comparing the elasticity of $\Sigma 5(210)$ GB in **Figure 4** with that of QJ in **Figure 5** it is obvious that the Young's modulus, see **Figure 5a**, is less anisotropic. Its minimum and maximum value equals to 38 and 81 GPa, respectively. In contrast, the linear compressibility of the studied QJs is more anisotropic, see **Figure 5b**, than that of $\Sigma 5(210)$ GB, see **Figure 4b**, and the minimum and maximum values amount to 3.86 and 4.19 TPa^{-1}, respectively. Most of the values of the above listed elastic constants for a system with QJs are very similar to those that we obtained for the system containing solely GBs but the value of the shear-related parameters, see also **Figures 4c and 5c**, are different. In particular, the value of C_{66} is higher (equal to 13 GPa) in the case of the supercell with QJs when compared with the supercell with GBs (C_{66} = 9.4 GPa). It means that the elasticity of the supercell with QJs is further from the limit of the mechanical instability (when C_{66} would be zero or negative). Regarding the minimum and maximum value of Poisson's ratio (see **Figure 5d**), the lowest and highest values are equal to 0.050 and 0.699, respectively. As far as the structure is concerned, the studied system with QJs expands the lattice volume by 3.8 %.

4. CONCLUSION

We have performed a series of quantum-mechanical calculations complementing our experimental research focused on silver nanoparticles with various morphologies. In particular, we have computed properties of a 181-atom decahedral silver nanoparticle and two types of internal extended defects, $\Sigma 5(210)$ grain boundaries (GBs) and quadruple junctions (QJs) of these GBs. As far as the thermodynamic stability of the decahedral nanoparticle is concerned, its energy is higher than the energy of a defect-free elemental Ag by 0.34 eV/atom. Regarding the properties of the $\Sigma 5(210)$ GB, its computed energy is equal to 0.7 J/m^2 and we also predict that the volume of the lattice would be locally expanded due these GBs. Importantly, the system with GBs is predicted to be rather close to the limit of mechanical instability. Specifically, the computed value of the elastic constant C_{66}, that is related to shear deformations, is as low as 9.4 GPa (with the zero or negative values representing a mechanically unstable system). We thus predict that the $\Sigma 5(210)$ GBs may be prone to a failure due to specific shearing deformation modes. The $\Sigma 5(210)$ GBs have also the lowest values of Poisson's ratio close to zero for some loading directions. Interestingly, when comparing our results related to solely $\Sigma 5(210)$ GBs with those of a system where multiple intersecting $\Sigma 5(210)$ GBs form a network of quadruple junctions, the properties are significantly different. The value of the critical elastic constant C_{66} is higher, 13 GPa, in the system with QJs and its mechanical stability is therefore better. Also the lowest value of Poisson's ratio is not close to zero in the system containing QJs.

ACKNOWLEDGEMENTS

The authors acknowledge the Czech Science Foundation for the financial support received under the Project No. 21-31852J. Computational resources were made available by the Ministry of Education, Youth and Sports of the Czech Republic under the Project of the IT4Innovations National Supercomputer Center (Project "e-Infrastructure CZ-LM2018140") within the program Projects of Large Research, Development and Innovations Infrastructures and via the CESNET (Project No. LM2015042) and CERIT-Scientific Cloud (Project No. LM20150850). M.F., N.P. and M.Š. acknowledge the institutional support provided by the Czech Academy of Sciences (Project No. UFM-A-RVO:68081723). The computational supercells in Figs. 2 and 3 were visualized using the VESTA package [16].

REFERENCES

[1] WANG, Z. L. (ed.), *Characterization of Nanophase Materials*, Weinheim, WILEY-VCH Verlag GmbH, 2000.

[2] SONG, M. *et al.* Influence of morphology and surface exchange reaction on magnetic properties of monodisperse magnetite nanoparticles. *Colloids and Surfaces A: Physicochemical and Engineering Aspects*. 2012, vol. 408, pp. 114-121.

[3] MARKS, L. D., HOWIE, A. Multiply-twinned particles in silver catalysts. *Nature*. 1979, vol. 282, pp. 196-198.

[4] BERKOWITZ A. E., WALTER J. L. Spark erosion: A method for producing rapidly quenched fine powders. *Journal of Materials Research*. 1987, vol. 2, pp. 277-287.

[5] POLSTEROVÁ, S., FRIÁK, M., VŠIANSKÁ, M., ŠOB, M. Quantum-Mechanical Assessment of the Energetics of Silver Decahedron Nanoparticles, *Nanomaterials*. 2020, vol. 10, Art. No. 767.

[6] KRESSE, G., HAFNER, J. Ab initio molecular dynamics for liquid metals. *Physical Review B*. 1996, vol. 47, pp. 558-561.

[7] KRESSE, G., FURTHMŰLLER, J. Efficient iterative schemes for ab initio total-energy calculations using a plane-wave basis set. *Physical Review B*. 1996, vol. 54, pp. 11169-11186.

[8] HOHENBERG, P., KOHN, W. Inhomogeneous electron gas. *Physical Review B*. 1964, vol. 136, pp. B864-B871.

[9] KOHN, W., SHAM, L. J. Self-consistent equations including exchange and correlation effects. *Physical Review A*. 1965, vol. 140, pp. A1133-A1138.

[10] BLŐCHL, P. E. Projector augmented-wave method. *Physical Review B*. 1994, vol. 50, pp. 17953-17979.

[11] KRESSE, G., JOUBERT, D. From ultrasoft pseudopotentials to the projector augmented-wave method. *Physical Review B*. 1999, vol. 59, pp. 1758-1775.

[12] PERDEW, J. P., BURKE, K., ERNZERHOF, M. Generalized gradient approximation made simple. *Physical Review Letters*. 1996, vol. 77, pp. 3865-3868.

[13] GAILLAC, R.; PULLUMBI, P.; COUDERT, F.X. ELATE: an open-source online application for analysis and visualization of elastic tensors. *Journal of Physics: Condensed Matter*. 2016, vol. 28, Art. No. 275201.

[14] ZHOU, L.; HOLEC, D.; MAYRHOFER, P.H. First-principles study of elastic properties of cubic Cr1−xAlxN alloys. *Journal of Applied Physics*. 2013, vol. 113, Art. No. 043511.

[15] MOUHAT, F.; COUDERT, F.X. Necessary and sufficient elastic stability conditions in various crystal systems. *Physical Review B*. 2014, vol. 90, Art. No. 224104.

[16] MOMMA, K., IZUMI, F. VESTA 3 for three-dimensional visualization of crystal, volumetric and morphology data. *Journal of Applied Crystalography*. 2011, vol. 44, pp. 1272-1276.

ANTIMICROBIAL PROPERTIES OF TiO_2 NANOCOMPOSITE COATING

[1]Tomislav HUDIKA, [1]Tomislav CIGULA, [1]Marina VUKOJE

[1]*Univesity of Zagreb, Faculty of Graphic arts, Zagreb, Croatia, EU,* thudika@grf.hr

https://doi.org/10.37904/nanocon.2021.4345

Abstract

Packaging plays important part of the visual communication and in consumer's choice of purchasing goods. To enhance visual appearance, packaging material is often coated. Beside enhancement of visual appearance, additional coating often improves other packaging properties. The COVID-19 pandemic stressed the importance of the antimicrobial properties of goods that encounter consumers. During purchasing, consumer first meets the packaging making it significant in the consumer's protection. The aim of this research is to determine antimicrobial properties of nanocomposite coating which includes nanosized TiO_2. For the purpose of the research a set of offset cardboard prints was coated with nanocomposite coating composed of water-based varnish (WD) and nanoscale TiO_2 particles. The prepared samples were characterized by determining *CIE L*a*b** coordinates of primary colours (CMYK), detecting colour fading after the accelerated ageing process by density measurements and by determining inhibition of microorganisms' growth by using smear test.

The change in chroma affected by UV radiation (accelerated ageing) is most visible on yellow samples while both, cyan and magenta proved to be more resistant to UV radiation. UV radiation did not cause significant change on the *L** coordinate of black, although its values were affected with initial varnishing as TiO_2 is also used as a white pigment. Although increase of the TiO_2 concentration in nanocomposite causes increase of the colour change, only the one with the highest concentration (2%) proved to be unacceptable. On the other hand, as the beneficial effects of nanocomposites increase with increase of the TiO_2 concentration, the nanocomposite with 1% of TiO_2 should be the choice.

Keywords: Cardboard packaging, nanoparticles, titan dioxide, functional coating, antimicrobial properties

1. INTRODUCTION

The packaging sector is the part of printing industry that connects protection of the goods and packaging as well as the visual communications [1]. To achieve this role some packaging materials are coated. To enhance inclusion of the coating process in the packaging production it is beneficial if the coating can provide more then one advantage i.e., to protect printed surface from colour degradation, to help with the barrier properties, to have antimicrobial potential, to be sustainable, etc. To upgrade existing varnish benefits, it is possible to introduce certain compounds into the composition, which are known to have protective potential. Those kinds of compounds are mostly nanosized (<100 nm) and could be added in the commercial varnishes [2]. Recent studies show that metal oxides such as titanium dioxide (TiO_2) and zink oxide (ZnO) demonstrate strong antifungal and antibacterial potential against wide range of known bacteria [3][4][5][6]. Moreover, the TiO_2 has the ability when UV exposed, to develop photocatalyst state in which oxygen free radicals (OFRs) develop. Those OFRs can affect the bacteria/fungus cell wall and help stop its growth [7]. Due to overuse of antibiotics, humans are more vulnerable to bacterial and fungus [8]. Even more, it is not uncommon for the customers to wear the gloves during shopping as a means of protection against COVID19 and other microbes which also leads to human's lower immunity against microbes [9][10]. In the light of the recent pandemic COVID19, consumers have become more aware of microbial threat in general [10]. Although, microbes are all around us and help maintain the natural balance, not all of them are desirable [8]. The printing industry and its packaging sector has a new challenge; how to provide antimicrobial properties on the packaging surface while maintaining the quality of visual communications. On the other hand, this is also significant market opportunity as it is projected that packaging alone is to have growth of 2.85% and reach 1.05 trillion US dollars (2019-2024) [11]. The antimicrobial packaging alone is projected 5.89% (2021-2026), mostly attributed with previously mentioned pandemic [12]. The aim of this research is to determine antimicrobial properties and diminished colour fading of the prepared cardboard prints due to the applied with TiO_2 nanocomposites.

2. MATERIALS AND METHODS

For this research, the gloss art print paper (coated) with production name UPM Finesse gloss paper (300g/m^2) was used as a printing substrate. The samples were printed via sheetfed offset printing technique using quickset process inks (Novavit Supreme Bio by Flintgroup) in compliance with FOGRA PSO 2016, i.e., ISO 12647-2:2013 [13].

Nanocomposite coatings were prepared by dispersing nano sized TiO_2 (Sigma Aldrich TiO_2, rutile) in commercial water-based varnish (TerraWet High Gloss Coating G9/285, ACTEGA, NY) further denominated as WD. Nanoparticles were added in designated weight ratios 0.25%, 0.5%, 1% and 2%.

The homogenisation process of nanoparticles into the WD varnish was done using ultrasound dispenser Hielscher UP100H for 30 minutes at 100% amplitude and 100% power [14]. The samples were cooled during the homogenisation by immersing containers containing the mixtures into a cooled water at 7 °C. Following the process, nanocomposites were applied onto printed samples using K202 Control Coater in compliance with ISO 187:1990, using coating bar 1 which leads to approx. 6 µm of wet film thickness [15].

After drying and characterization the samples were exposed to an artificial ageing process in the Cofomegra Solarbox 1500e Xenon Test chamber for 30 hours period at irradiation energy of 550 W/m^2 and 50 °C. The indoor filter was used for the simulation of internal environment i.e. sunrays filtered through a window pane [16].

To analyse and characterize prepared samples, optical density and colour coordinates were determined at three stages: original (uncoated) prints, coated prints and aged coated prints. The colorimetric measurements were conducted using Techckon SpectroDens spectrophotometer with colorimetric measurements' settings: D50 illuminant, 2° standardized observer, M1 filter, no polarisation filter and calibrated before measurements on absolute white. Instrument settings for density measurements were density status E, illuminant D50, no polarization filter and calibrated on plain paper (original print).

The influence of TiO_2 nanoparticles in varnish on growth of microorganisms was evaluated by examination of the total number of microorganisms of on coated samples (S2) and artificially aged samples (S3). For the microbiological testing of samples by smear method, the procedure described by Guzińska et. al was used [17]. The microorganisms were isolated (washed) from the samples surface by horizontal sampling method by means of a swab wetted in 9% saline solution. The swab was then transferred to a 9% saline solution and shaken, after which that solution was used for the preparation of series of dilutions from 10^{-1} - 10^{-5}. Petri dishes were inoculated with 1 mL of the prepared dilutions and flooded with nutrient agar (NA) medium. Incubation of the Petri dishes was made at 37 °C for 48 to 72 hours and at 30 ± 1 °C. The number of microorganisms per 1 mL of the tested cardboard surface was calculated.

3. RESULTS AND DISCUSSION

3.1. Results and discussion

To check the printing process, the ΔE_{ab} colour difference of the printed sample in relation to ISO 12674-2:2013 standard (Fogra PSO 2016) by was calculated. The measured values of Cyan (ΔE_{ab} = 1.5) show that is the closes to the standard of all presented samples, while Black (ΔE_{ab} = 3.1) and Magenta (ΔE_{ab} = 3.7) and Yellow (ΔE_{ab} = 4.2) are within the tolerance. Yelow is closest to allowed tolerance (ΔE_{ab}<5). To determine the influence of the coating process on the colour's perception between two sets of prints *(printed samples (S1), coated samples (S2))* colour difference ΔE_{00} was calculated from measured *L*a*b** values (**Table 1**). The black ink is sensitive to change as the ink is dark, so contrast is easy to spot. The biggest change occurred on the black coated with nanocomposite including 2% TiO_2 (*ΔE_{00}=4.32*). This result could be attributed to the fact that the TiO_2 is used as white pigment [18].

Table1 ΔE_{00} calculated difference

	C	M	Y	K
S1-S2 wd	0.54	0.84	0.40	2.47
S1-S2 0.25%	0.68	1.36	0.26	1.58
S1-S2 0.5%	0.83	1.18	0.32	1.55
S1-S2 1%	0.86	1.52	0.21	1.60
S1-S2 2%	1.18	3.13	1.60	4.32

3.2. Influence of accelerated ageing on prints' colour

Chroma in colours is mostly connected with intensity of colour, meaning that the low chroma colour can appear dull to the human eye. In order to detect behaviour of the colour intensity C_{ab}, was calculated using formula (1):

$$C_{ab} = \sqrt{a^2 * b^2} \quad (1)$$

where: *a - is a* coordinate from CIE L*a*b** and *b - is b* coordinate from CIE L*a*b**

In the **Figure 1**, Croma (C_{ab}) is presented for cyan (right) and magenta (left). The cyan samples (**Figure 1 left**) show insignificant change after accelerated ageing. The increment in the WD (varnished samples without TiO_2) can be attributed to the spectrophotometer measurements error since the change is so small. The magenta (**Figure 1 right**) shows very similar behaviour on all samples after except 2% TiO_2 where the values are lower, but the trend is similar. It could be noted that on both colours the accelerated ageing is causing diminishing of the colour chroma, i.e. colour intensity.

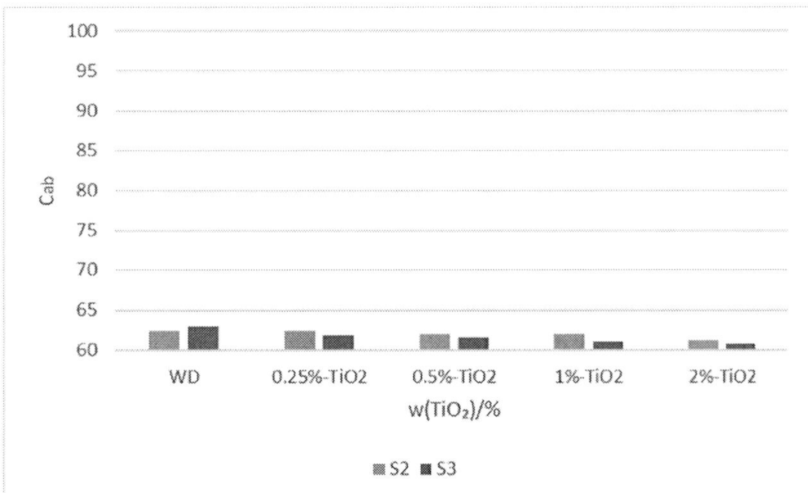

Figure 1/1 Cyan (up) and magenta (down) C_{ab} values

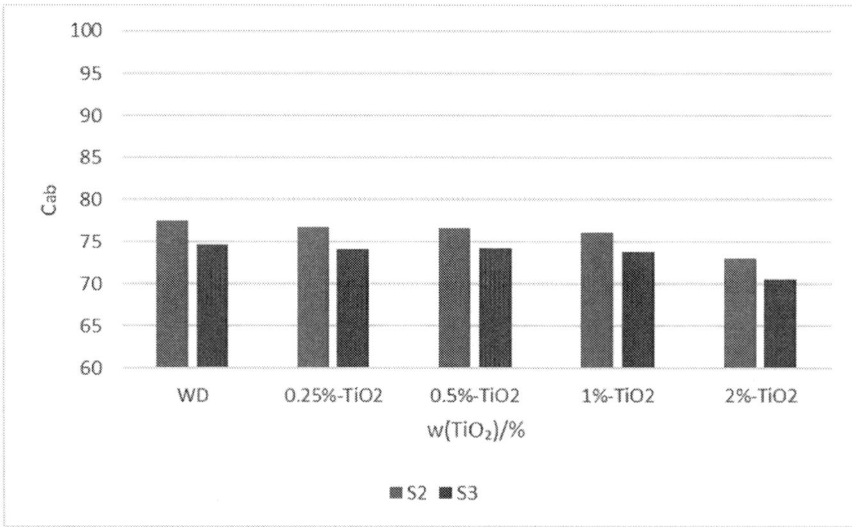

Figure 1/2 Cyan (up) and magenta (down) C_{ab} values

Chroma of yellow printed samples (**Figure 2**) is most affected with the UV radiation exposure (*ageing process*). Although it appears that nanocomposite coatings are keeping chroma at the same level, please note that increasing the weight ratio of the TiO_2 is decreasing chroma of coated samples (before accelerated ageing process). The chroma decrease with increase of the TiO_2 could be attributed to the addition of achromatic (TiO_2 is used as a white pigment) while one the other hand, it is preserving pigment from degradation by absorbing part of the UV light.

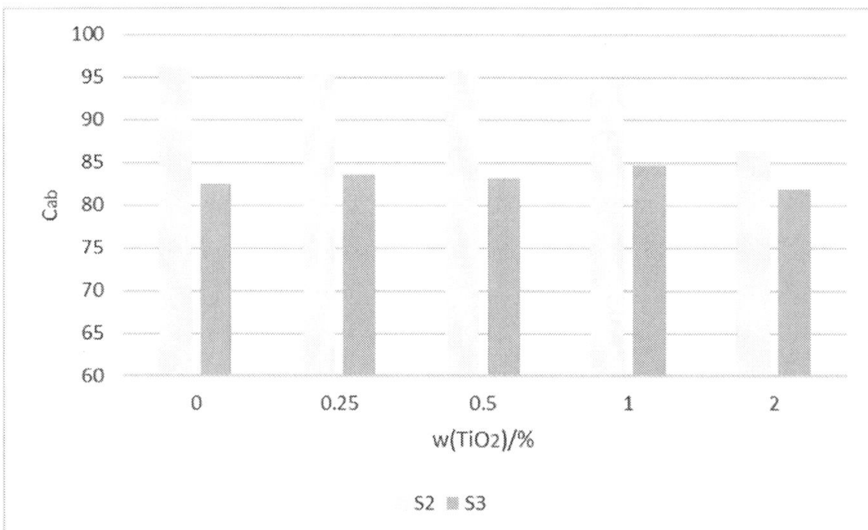

Figure 2 Yellow (left) C_{ab} values

The change that occurs with the black samples can be best observed via *L** coordinate due to fact that black is achromatic and therefore change in *a** and *b** is less visible. In the **Figure 3**, measured lightness (L* coordinate) values can be seen. It can be noted that with the increase of TiO_2 weight ratio, the lightness increases since the TiO_2 is also used as a white pigment. The change from S2 to S3 is below 0,5 meaning it is insignificant. This shows that although TiO_2 rises initial values after coating it does protect the ink from additional fading.

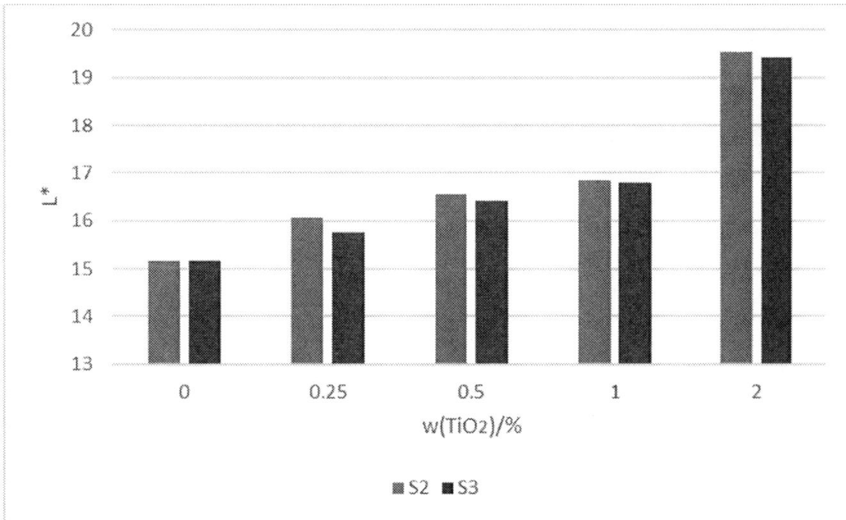

Figure 3 Black *L'* values

To further assess the influence of UV exposure during accelerated ageing, ink' density was measured. To enable better image of the coating and aging influence, In **Figure 4**, one can see diagrams for density change (*D'*). The cyan (*left*) and magenta (*right*) samples are presented. As the colours are printed to match colorimetric expectation, printing is performed in different densities. Therefore, to enable comparison between inks, density change (*D'*) was intorduced. The density change is calculated using Equation (2):

$$D' = \frac{D_0}{D_i} \quad (2)$$

where, D_o - density value of coated samples(*S2*), D_i - density value of coated samples after aging(*S3*).

Observing **Figure 4**, it can be noted that cyan's and black's density is constant and nearly 1 which means it was unchanged after accelerated aging. Magenta is also almost unchanged in the 0-1% weight ratio, where a growing trend can be spotted in the higher ratios. This can be attributed to the TiO_2 photoluminescence which occurs both in polymorphs (*rutile and anatase*) [19]. Yellow, as the most sensitive one show that increase of the TiO_2 weight ratio in nanocomposite lead to the increased UV resistance (the lowest D' is at 2% of TiO_2).

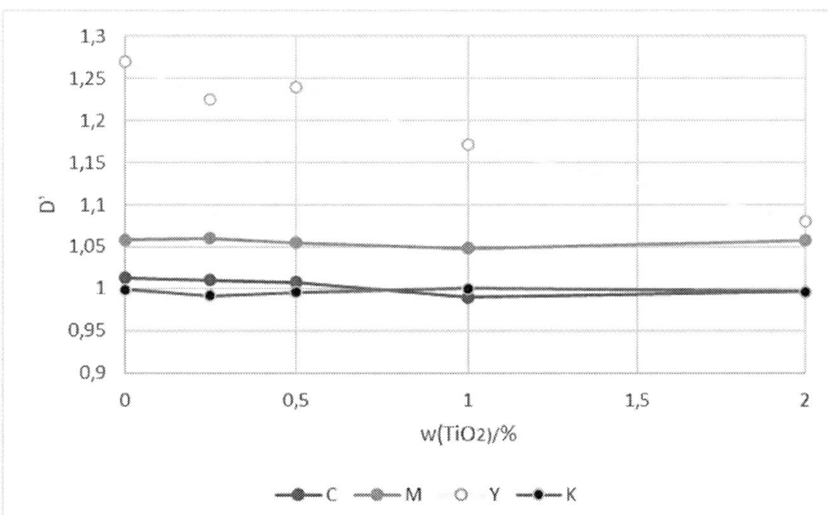

Figure 4 Ink density change (D') - magenta (left) and yellow (right)

3.3. Antimicrobial properties of prepared samples

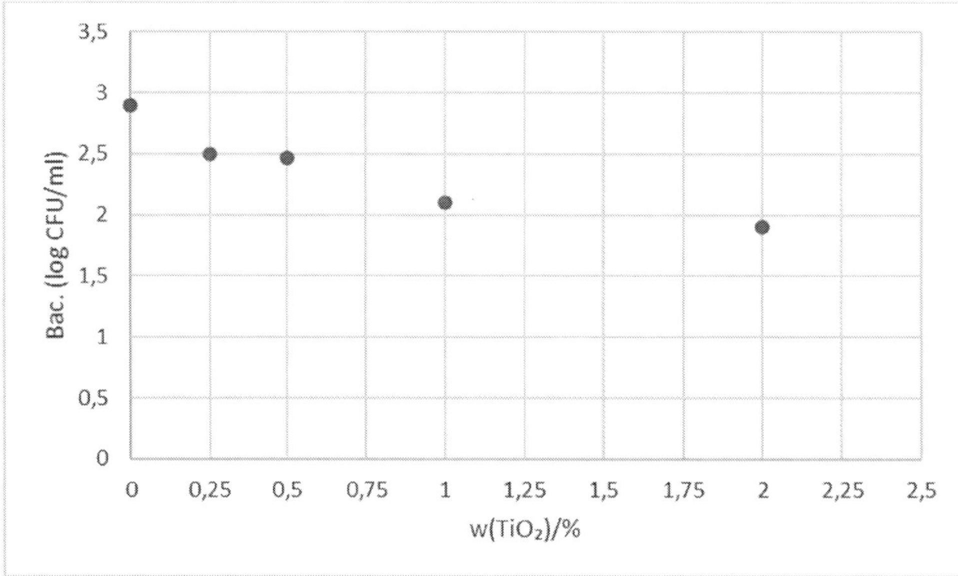

Figure 5 Total number of bacteria in the tested aged samples

The applied smear method did not reveal the bacterial contamination of the unaged uncoated and coated samples surface probably due to the presence of antibacterial additives in varnish or by the presence of the bacteria in the deeper layer [17]. The problem related to the growth of various microorganisms occurs in almost all water-based inks. Since water-based varnishes used in printing have similar composition as printing inks, only without pigment, this problem can be related to them as well. In order to prevent or inhibit microbial growth, different biocides and fungicides can be added in their formulation [20]. When it comes to printing inks, additives in a function of biocides/fungicides are added in proportion of <0.5 %, based on the total quantity of the ink formulation. For this reason, on the unaged samples no bacteria occurred. On the contrary, bacterial growth was obtained on aged samples. This leads to a conclusion that UV irradiation most probably caused the breakdown of biocide composition. On the aged samples, it can be seen that by increasing the content of TiO_2 in varnish, the total number of bacteria decreases which may lead to a conclusion that nanoparticles have antibacterial effect (**Figure 5**), i.e. higher share of TiO_2 nanoparticles has higher antimicrobial effect.

4. CONCLUSION

The growing packaging market demands new solutions that can cope with numerous challenges at the same time. The change in the colour during product storage and sales and, due to the recent pandemic of Covid19, antimicrobial behaviour are, among other, two factors that can be a powerful tool in demanding market. The aim of this research was to investigate how does application of nanocomposites composed of commercial water based varnish and nanoscale TiO_2 is replying to those challenges.

To summarize, although increase of the TiO_2 concentration in nanocomposite causes increase of the colour change, only the one with the highest concentration (2%) proved to be unacceptable. On the other hand, as the beneficial effects of nanocomposites increase with increase of the TiO_2 concentration, the nanocomposite with 1 % of TiO_2 should be the choice. Nevertheless, to further evaluate application of the proposed nanocomposites, future research will be focused on the coating process, barrier properties, adhesion on the substrate etc.

ACKNOWLEDGEMENTS

This work was supported by the Croatian science foundation [grant number uip-2017-05-4081,2017].

REFERENCES

[1] KIPPHAN H. *Handbook Of Print media.* [online]. Berlin-Heidelberg: Springer-Verlag Berlin Heidelberg, 2001. Available from: https://doi.org/10.1007/978-3-540-29900-4.

[2] CIGULA T., HUDIKA T., VUKOJE M. Modulation of water based commercial varnish by adding ZnO and SO2 nanoparticles to enhance protective function on printed packaging. In: 2. *Int. Circ. Packag. Conf., ICP.* [online]. FTPO: Slovenia, Ljubljana, Slovenia, 2021, pp. 249-260.

[3] SALLA J., PANDEY K.K., SRINIVAS K. Improvement of UV resistance of wood surfaces by using ZnO nanoparticles. *Polym. Degrad. Stab. 97.* [online]. 2012, pp. 592-596. Available from: https://doi.org/10.1016/j.polymdegradstab.2012.01.013.

[4] ZVEKIC D., SRDIC V., KARAMAN M., MATAVULJ M. Antimicrobial properties of ZnO nanoparticles incorporated in polyurethane varnish. *Process. Appl. Ceram. 5.* [online]. 2011, pp. 41-45. Available from: https://doi.org/10.2298/pac1101041z.

[5] DICASTILLO C.L., CORREA M.G., MARTÍNEZ F.B., STREITT C., GALOTTO M.J. Antimicrobial Effect of Titanium Dioxide Nanoparticles. *InTechOpen.* [online]. 2020. Available from: https://doi.org/10.5772/INTECHOPEN.90891.

[6] KUBACKA A., DIEZ M.S., ROJO D., BARGIELA R., CIORDIA S., ZAPICO I., Understanding the antimicrobial mechanism of TiO 2 -based nanocomposite films in a pathogenic bacterium. *Sci. Rep. 4.* [online]. 2014, pp. 1-9. Available from: https://doi.org/10.1038/srep04134.

[7] KRUMDIECK S.P., BOICHOT R., GORTHY R., LAND J.G. Nanostructured TiO 2 anatase-rutile-carbon solid coating with visible light antimicrobial activity. *Sci. Rep. 9.* [online]. 2019, pp. 1-11. Available from: https://doi.org/10.1038/s41598-018-38291-y.

[8] Science News for Students. *Bacteria are all around us - and that's okay.* [online]. [viewed: 2021-09-16]. Avaiable from: https://www.sciencenewsforstudents.org/article/bacteria-are-all-around-us-and-thats-okay.

[9] RUSSELL C.D. Co-infections, secondary infections, and antimicrobial use in patients hospitalised with COVID-19 during the first pandemic wave from the ISARIC WHO CCP-UK study: a multicentre, prospective cohort study. *The Lancet Microbe.* [online]. 2021, vol. 2, pp. 354-365. Available from: https://doi.org/10.1016/S2666-5247(21)00090-2.

[10] *Gloves Won't Reduce Your Risk of COVID-19 at the Grocery Store.* [online]. [viewed: 2021-09-14]. Avaiable from: https://www.healthline.com/health-news/wearing-gloves-while-grocery-shopping-doesnt-prevent-covid-19#Why-gloves-arent-medically-necessary.

[11] Smithers Pira Forecasts Global Packaging Market To $269 Billion By 2024. *Identifies 7 Key Drivers And Trends For Flexible Packaging - Reusable Packaging News.* [online]. [viewed: March 8, 2021-09-08]. Avaiable from: https://packagingrevolution.net/smithers-pira-forecasts-global-packaging-market-to-269-billion-by-2024-identifies-7-key-drivers-and-trends-for-flexible-packaging/

[12] Antimicrobial Packaging Market | 2021 - 26 | Industry Share, Size, Growth - Mordor Intelligence, (n.d.). [online]. [viewed: 2021-09-14]. Avaiable from: https://www.mordorintelligence.com/industry-reports/antimicrobial-packaging-market

[13] ISO 12647-2:2013. *Graphic technology - Process control for the production of half-tone colour separations, proof and production prints - Part 2: Offset lithographic processes*, ISO Europe, 2013.

[14] Hielscher. *UP100H - Compact Ultrasonic Laboratory Device - Hielscher Ultrasound Technology.* [online]. 2020, [viewed: 2021-10-19]. Avaiable from: https://www.hielscher.com/100h_p.htm

[15] ISO 187:1990. *Paper, board and pulps - Standard atmosphere for conditioning and testing and procedure for monitoring the atmosphere and conditioning of samples*, ISO International, 1990.

[16] ASTM D3424-11. *Standard Practice for Evaluating the Relative Lightfastness and Weatherability of Printed Matter.* W.C.PA USA, ASTM, 2011.

[17] GUZIŃSKA K., OWCZAREK M., DYMEL M. Investigation in the microbiological purity of paper and board packaging intended for contact with food, Fibres Text. *East. Eur.* 2012, vol. 96, pp. 186-190.

[18] WANG Y., LI J., WANG L., XUE T., T. QI. Preparation of rutile titanium dioxide white pigment via doping and calcination of metatitanic acid obtained by the NaOH molten salt method. *Ind. Eng. Chem. Res.* [online]. 2010, vol. 49, pp. 7693-7696. Available from: https://doi.org/10.1021/ie1007147.

[19] PALLOTTI D.K., PASSONI L., MADDALENA P., DI FONZO F., LETTIERI S. Photoluminescence Mechanisms in Anatase and Rutile TiO2. *J. Phys. Chem. C.* [online]. 2017, vol. 121, pp. 9011-9021. Available from: https://doi.org/10.1021/acs.jpcc.7b00321.

[20] ZOŁEK-TRYZNOWSKA Z. Additives for Ink Manufacture. *Print. Polym. Fundam. Appl.* [online]. 2015, pp. 57-66. Available from: https://doi.org/10.1016/B978-0-323-37468-2.00004-X.

SESSION D
MONITORING AND TOXICITY OF NANOMATERIALS

Chairman

Dr. Jan TOPINKA Institute of Experimental Medicine of the CAS, Prague, Czech Republic, EU

METHODOLOGY FOR DECREASING THE LEVEL OF RISK IN WORKPLACES WITH THE OCCURRENCE OF FINE AND ULTRAFINE PARTICLES

[1]Klaudia KÖBÖLOVÁ, [1]Vladimír ADAMEC, [1]Michal URBÁNEK, [2]Tomáš ZEMAN, [1]Jana DRBOHLAVOVÁ

[1]Brno University of Technology, Institute of Forensic Engineering, Brno, Czech Republic, EU, Klaudia.Kobolova@vutbr.cz, Vladimir.Adamec@vut.cz, Michal.Urbanek2@vutbr.cz, xcchovancova@vut.cz

[2]University of Defence, Faculty of Military Leadership, Brno, Czech Republic, EU, tomas.zeman2@unob.cz

https://doi.org/10.37904/nanocon.2021.4315

Abstract

With the development of new technologies, the requirements for safety and health at workplaces are also increasing. One of these sectors are also industrial plants which process or use fine and ultrafine particles (<2.5 µm) during their activities or where these particles are created unintentionally as a by-product of the production processes. In the case of risk assessment of these particles, it is not only a matter of health risks associated with direct exposure to these particles but can have negative effects on mental health of employees. This could lead to stress, increase mental discomfort, reduce attention, performance, and motivation to work. The aim of the methodology is to evaluate the health and safety management system, increase its level and efficiency in operations loaded with fine and ultrafine particles. The methodology includes both the assessment of risks associated with the exposure of workers and the rate of perception of risks by employees in the field of information about the risks associated with fine and ultrafine particles. It also includes perception of the adequacy of existing safety measures to reduce worker exposure to these particles. The proposal of basic measures to reduce exposure of workers is also an integral part of methodology. The proposed methodology is in line with OECD guidelines focused on nanosafety.

Keywords: OSH, PM1, UFP, health, workplace

1. INTRODUCTION

The aim of occupational safety and health (OSH) is to ensure that the working environment is such that the safety risks to employees are minimised. The general principles of OSH include the overall company strategy (legislation, tasks and objectives, programmes, systems of evaluated objects), identification, analysis and evaluation of hazards, control, and corrective measures (monitoring, inspections, prevention, records, audits, etc.), continuous identification of achieved results (external and internal) and their comparison with expected results [1,2]. A functional safe work environment policy is also based on ensuring the socio-technical system, which includes people (employees) with all their characteristics (education, ambition, reliability, etc.) [3-5].

With the development of technology, the demand for the level of OSH is also increasing and new areas are emerging, especially in industry, where sufficient safety and preventive measures are not yet provided. One of these sectors are industrial plants that process or use fine and ultrafine particles (size up to 2.5 µm) in their operations, or where these particles are unintentionally generated as a by-product of the manufacturing processes used [6]. Outside of industrial plants, there are work processes where nanomaterials are applied, such as spraying or coating titanium dioxide and nanosilver. Direct exposure to these particles can have a severe impact on human health [7,8]; thanks to their small size they can carry toxic substances and reach deeper into the human respiratory system [9]. But in the case of risk assessment of fine and ultrafine particles (FP and UFP), it is not only the health risks associated with direct exposure to these particles. Working in an environment where FP and UFP are present can also have negative effects on the mental health of the employee [10]. These psychological effects are mainly related to the employee's subjective perception of the level of safety in a work environment with FP and UFP. If this level is inadequate from the employee's point of view, it can lead to stress, increased psychological discomfort, reduced attention, performance, and motivation to work. However, secondary risks should also be considered in the context of OSH, such as the impact on the operation of the company and its economy associated with reduced employee efficiency and productivity due to safety measures, increased sickness, and absenteeism [11].

One major and important challenge is to reliably assess the risks and safety of FP and UFP. The main problem is the lack of systematic knowledge about the exposure of these particles, the lack of limit values, and the lack of knowledge about the impact on human health and the environment. Currently, there is no specific guideline that addresses the OSH management system in workplaces burdened by FP and UFP. The Occupational Safety Research Institute in the Czech Republic has developed a certified methodology for the provision of personal protective equipment in nanoparticle hazardous environments [12], which, as the name implies, addresses only the allocation of personal protective equipment for respiratory protection. Due to these facts, a methodology for better OSH management was developed mainly for the Czech Republic.

2. BACKGROUND

Development a suitable methodology for FP and UFP is very difficult because the health effect of these particles is determined by their several properties, not just by their concentration. In contrast to particles with larger dimensions, the sedimentation level of UF and UFP in the air due to gravity is very low to negligible [13]. For this reason, these particles may persist in the workplace air for several days. Thus, the Organisation for Economic Co-operation and Development (OECD) [14] recommends for the monitoring of FP and UFP in the work environment their long-term measurement. The interaction of these particles with other airborne substances may not be stable over the long term and may become re-release from these bonds.

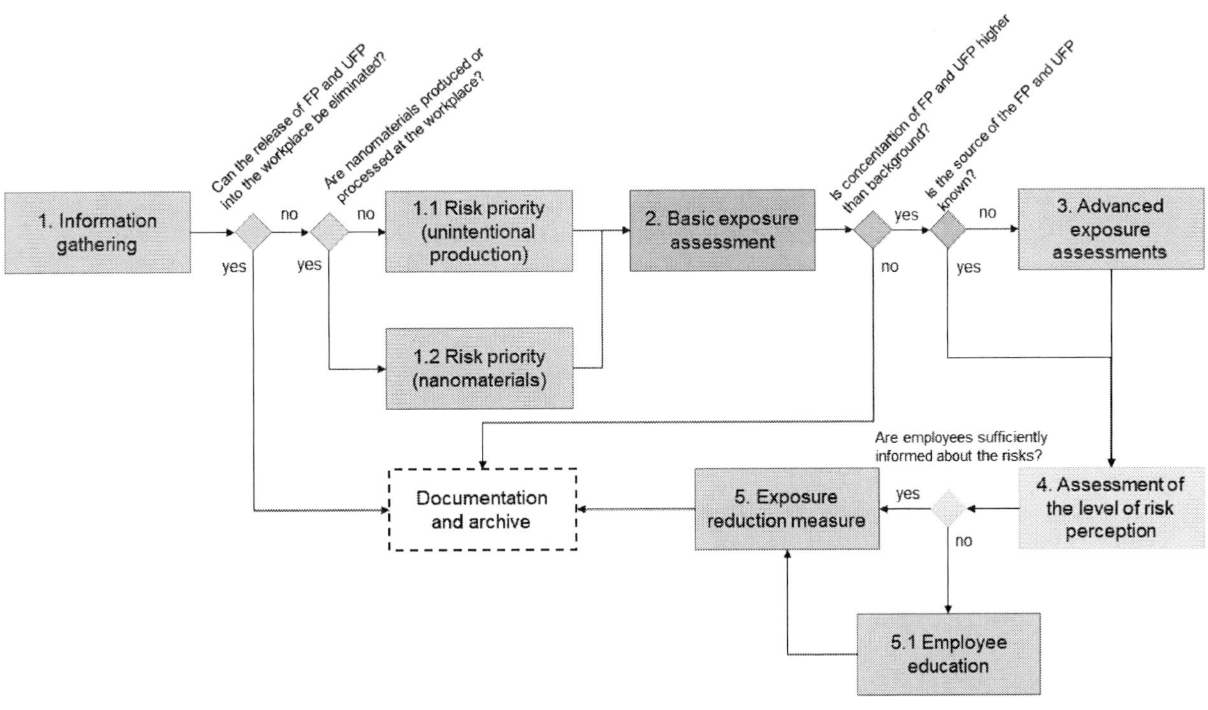

Figure 1 Scheme of methodology to manage risks of FP and UFP

Our methodology is based on the OECD nanosafety procedure [14]. The methodology aim is to evaluate the OSH management system, to increase its level and efficiency in companies dealing with particles with aerodynamic diameter <2.5 μm. It consists of several chapters, which first focus on workplace characterisation and risk prioritisation based on the knowledge obtained from the workplace documentation. The next two sections deal with basic and advanced exposure assessment according to the OECD methodology. According to the assessment, the level of risk perception of the workers is assessed, and a summary risk assessment is produced. The last section describes proposals for measures to reduce employee exposure.

3. METHODOLOGY FOR OSH MANAGEMENT

The methodology based on a five-step procedure (see in **Figure 1**) consists of the following steps: information gathering, basic exposure assessment, advanced exposure assessments, assessment of the level of risk perception by employees, and measures.

3.1. Information gathering

The first step is the characterisation of the workplace. It should be based primarily on a study of the documentation, or a personal inspection of the workplace, to identify possible sources of FP and UFP at the workplace. These main indicators should be investigated depending on the type of production processes as information on the workplace, the size of the workplace and production areas, the characteristics and properties of the produced or processed FP and UFP, an overview of workplace processes that may generate FP and UFP as a by-product, an overview of workplace processes that may affect the measurement of FP and UFP concentrations, the current status of exposure reduction measures (personal protective equipment (PPE) or the ventilation technology) and risk characterisation (synthesis of the knowledge gained in this step).

The first step aim to answer the question if there is a possibility to eliminate the release of FP and UFP into the work environment by prioritising the risk. Further, it is necessary to determine if nanomaterials are produced or processed at the workplace or FP and UFP are expected to occur during their unintentional production. If nanomaterials are produced or processed in the workplace, it is recommended to use the Risk and Control Banding tool for risk prioritising, such as the Stoffenmanager Nano tool [15,16]. The tool assesses the exposure level and the risk level, and risk priority according to a combination of these levels (see **Figure 2**). If determined priority is 1 or 2, that must be continuing step 2. On the other hand, nanomaterials with priority 3 proceed to the assessment of the level of risk perception of employees. In the case of the possible generation of FP and UFP by unintentional processes proceeds to step 2.

Hazard band / Exposure band	A	B	C	D	E
1	3	3	3	2	1
2	3	3	2	2	1
3	3	2	2	1	1
4	2	1	1	1	1

Hazard band: A= lowest hazard, E= highest hazard. Exposure band: 1=lowest exposure, 4=highest exposure.
Overall result: 1= highest priority, 3= lowest priority

Figure 2 Example for risk priority determination in Stoffenmanager Nano tool [17]

3.2. Basic exposure assessment

This part contains an initial measurement of FP and UFP in the workplace. The measurements shall be carried out continuously for at least one working shift, starting at least 30 minutes before the start of the processes. At the same time, all processes and other activities carried out during the measurement, including the use of PPE by workers, must be recorded. For each employee, the time of entry to and exit from the workplace shall be recorded. For measurements could use any instrument with a valid certified calibration capable of continuously recording the number concentrations of FP and UFP within the selected particle size range. It is recommended that the measured particle size range be selected based on an evaluation of the information on the nanomaterial or the manufacturing activity obtained in the previous. If the size range cannot be determined, it is recommended to use a size range between 10 nm and 500.

If measured concentrations of FP and UFP are higher compared to background concentrations, then proceeds step 3, which is advanced exposure assessment. Step 3 should have proceeded if the difference between the mean concentration of FP and UFP (MC_1) in the work environment and the mean background concentration (MC_0) is greater than three times the standard deviation of the background concentrations (SD_0) of FP and UFP (see equation **1**).

$$\mathrm{MC_1} - \mathrm{MC_0} > 3\mathrm{SD_0}, \tag{1}$$

The output of basic exposure assessment is a record that contains the time series of the measured data, the means and standard deviations of the concentrations of FP and UFP at the measurement site and in the background, information activities that could affect the distribution, photo documentation and basic descriptions of microclimatic parameters, information on the times of entry and exit of workers from workplaces and statistical tests of the difference in concentrations in the working environment compared to background.

3.3. Advanced exposure assessment

The objective of advanced exposure measurement is to identify the source of FP and UFP in the workplace and to characterise these particles (composition, size, and shape of the identified particles). For this purpose, it is necessary to use advanced instruments for measuring FP and UFP, in particular [14]:

- instruments using particle mobility measurements in an electric field, e.g. Scanning Mobility Particle Sizer (SMPS);
- optical or aerodynamic particle counters;
- Electrical Low-Pressure Impactor (ELPI);
- Condensation Particle Counter (CPC).

This step always needs to be customised to the situation at the workplace, thus no universal procedure can be established in advance. To determine the correct procedure, it is necessary to rely on the previous baseline exposure assessment and the characteristics of the workplace. Alternatively, it is possible to follow the standard ISO 14644:2015 [18].

If the advanced exposure assessment leads to the successful identification of the source of FP and UFP in the workplace and the characterisation of these particles is carried out, the results obtained can be used to design exposure reduction measures. Regardless of whether the identification and characterisation of the source of FP and UFP in the workplace is successful in this step, an assessment of the level of risk perception by workers is then undertaken.

3.4. Assessment of the level of risk perception of employees

This section has two parts, the identification of the level of the risk via questionnaire and the overall risk assessment. The questionnaire is used to assess the level of risk perception associated with inhaled FP and UFP by employees working in operations where these particles are present. The questionnaire asks for the following topics: level of awareness of employees of the risks of exposure to FP and UFP during work, adequacy of existing safety measures to reduce worker exposure FP and UFP in the workplace. Further step depends on the employees' answers if they have knowledge about FP and UFP. If all the employees answer 'YES' to this question, the risk assessment then proceeded to an overall risk assessment. If any of the employees' answer 'NO' to this question, the implementation of the measure proceeds directly.

The overall risk assessment aggregate information about the level of employee awareness of the risks of exposure and level of perception of the adequacy of existing safety measures. The priority risk level ranges from 1 to 3. If the identified risk level is 2 and 3 it is recommended to introduce regular training/education of workers about risks associated with exposure to FP and UFP. If the priority is 1 no action is required in concerning the training of employees. Regardless of the outcome of the overall risk assessment, it should be an implementation of exposure reduction measures.

3.5. Measures and controlling

If measured concentrations of FP and UFP are higher (according to equation **1**), the associated risks can be reduced by taking appropriate measures. The following measures are the most effective in terms of inhaled FP and UFP:

- process isolation or local ventilation;
- training of workers;
- use of personal protective equipment;
- operational measures.

After implementation of the selected exposure reduction measures, it is recommended to return to basic exposure measurements (see **Figure 1**). If, as a result of the measures taken, the number has been reduced to an acceptable level according to equation **1**, it is recommended that the information obtained in each step of this methodology document and archive. Otherwise, the process should be repeated until the number of concentrations is reduced below the required limit according to equation **1**. Once this objective is achieved, it is recommended, by the OECD recommendation [14], to repeat the whole procedure every two years or in case of a change in the processes at the workplace.

4. CONCLUSION

The methodology is focused on OSH management to address the need to increase the level and efficiency of OSH in operations with particulate matter size up to < 2.5 μm. The methodology can also be used by employers in workplaces exposed to FP and UFP to develop measures to reduce employee exposure and to assess the level of employee risk perception.

ACKNOWLEDGEMENTS

This paper is supported by TA ČR within the TL02000240 project of Increasing the Level of OSH Management in Workspace with Occurrence of Fine and Ultrafine particles.

REFERENCES

[1] AVEN, T. "Risk assessment and risk management: Review of recent advances on their foundation." *European Journal of Operational Research*. 2016, vol. 253, no. 1, pp. 1-13.

[2] ISO 45001:2018. *Occupational health and safety management systems - Requirements with guidance for use*, 2018.

[3] AVEN, T. and YLÖNEN, M. "A risk interpretation of sociotechnical safety perspectives." *Reliab. Eng. Syst. Saf.* 2018, vol. 175, pp. 13-18.

[4] AVEN, T. "How some types of risk assessments can support resilience analysis and management." *Reliab. Eng. Syst. Saf.* 2017, vol. 167, pp. 536-543.

[5] CARAYON, P., HANCOCK, P., LEVESON, N., NOY, I., SZNELWAR, L. and VAN HOOTEGEM, G. "Advancing a sociotechnical systems approach to workplace safety - developing the conceptual framework." *Ergonomics.* 2015, vol. 58, no. 4, pp. 548-564.

[6] VAN BROEKHUIZEN, Pieter, VAN BROEKHUIZEN, Fleur, CORNELISSEN Ralf a Lucas REIJNDERS. Workplace exposure to nanoparticles and the application of provisional nanoreference values in times of uncertain risks. *Journal of Nanoparticle Research* [online]. 2012, 14, vol. 4. Available from: https://doi.org/10.1007/s11051-012-0770-3. ISSN 1388-0764.

[7] SCHRAUFNAGEL, Dean E. The health effects of ultrafine particles. *Experimental & Molecular Medicine.* [online]. 2020, vol. 52, no. 3, pp. 311-317. ISSN 1226-3613. Available from: https://doi.org/10.1038/s12276-020-0403-3

[8] SHARMA, Shubham, CHANDRA, Mina a KOTA, Sri Harsha. Health Effects Associated with PM2.5: a Systematic Review. *Current Pollution Reports.* [online]. 2020, vol. 6, no. 4, pp. 345-367. ISSN 2198-6592. Available from: https://doi.org/10.1007/s40726-020-00155-3

[9] TIAN, Guoxiong, WANG, Juan, LU, Zhongbing, WANG, Hongyun, ZHANG, Wei, DING, Wenjun a ZHANG, Fang. Indirect effect of PM1 on endothelial cells via inducing the release of respiratory inflammatory cytokines. *Toxicology in Vitro.* [online]. 2019, vol. 57, pp. 203-210. ISSN 08872333. Available from: https://doi.org/10.1016/j.tiv.2019.03.013

[10] BRUGGE, Doug a FULLER, Christina H. *Ambient Combustion Ultrafine Particles and Health.* 1. Nova Science Publishers, 2021. ISBN 978-1-53618-831-8.

[11] BAYRAM, Metin a BURGAZOGLU, Huseyin. The Relationships Between Control Measures and Absenteeism in the Context of Internal Control. *Safety and Health at Work.* [online]. 2020, vol. 11, no. 4, pp. 443-449. ISSN 20937911. Available from: https://doi.org/10.1016/j.shaw.2020.07.007

[12] Occupational Safety Research Institute. *Methodology for providing personal protective equipment in an environment with a risk of nanoparticles.* [online]. Prague, 2016, p. 23. Certificated methodology. In: https://www.mpsv.cz/documents/20142/225499/Metodika_pro_poskytovani_OOP_v_prostredi_s_rizikem_vyskytu_nanocastic.pdf/260c679b-9a39-4734-12cd-79736892d095

[13] HINDS, William C. *Aerosol technology: properties, behavior, and measurement of airbome particles.* 2nd ed. New York: John Wiley, c1999. A Wiley-Interscience publication. ISBN 978-0-471-19410-1.

[14] OECD: *Harmonized tiered approach to measure and assess the potential exposure to airborne emissions of engineered nano-objects and their agglomerates and aggregates at workplaces*, 2015. [online]. In: https://www.oecd.org/officialdocuments/publicdisplaydocumentpdf/?cote=env/jm/mono(2015)19&doclanguage=en

[15] MARQUART, H. HEUSSEN, H. LE FEBER, M. NOY, D. TIELEMANS, E. SCHINKEL, J. WEST, J. and VAN DER SCHAAF, D. "'Stoffenmanager', a Web-Based Control Banding Tool Using an Exposure Process Model." *The Annals of Occupational Hygiene.* 2008, vol. 52, no. 6, pp. 429-441.

[16] VAN DUUREN-STUURMAN, B., VINK, S. R., VERBIST, K. J. M., HEUSSEN, H. G. A., BROUWER, D. H., KROESE, D. E. D., VAN NIFTRIK, M. F. J., TIELEMANS, E. and FRANSMAN, W. "Stoffenmanager Nano Version 1.0: A Web-Based Tool for Risk Prioritization of Airborne Manufactured Nano Objects." *The Annals of Occupational Hygiene.* 2012, vol. 56, no. 5, pp. 525-541.

[17] VÄÄNÄNEN V, RYDMAN E, ILVES M, HANNUKAINEN K, NORPPA H, VON WRIGHT A, et al. Evaluation of the suitability of the developed methodology for nanoparticle health and safety studies. Sunpap conference; [online]. 2012. http://sunpap.vtt.fi/pdf/SUNPAP_WP10_DEL10_5_20120827_FIOH.pdf.

[18] ISO 14644-1:2015 *Cleanrooms and associated controlled environments - Part 1: Classification of air cleanliness by particle concentration*, 2015.

ECOTOXICITY OF $CuCl_2$ AND CUO NANOPARTICLES ON SELECTED BIOINDICATORS

Barbora HAVELKOVA, Michal PRIBYL, Lucie TESAROVA, Miroslava BEKLOVA

Department of Ecology and Diseases of Zoo animals, Game, Fish and Bees, Faculty of Veterinary Hygiene and Ecology, University of Veterinary Sciences Brno, Brno, Czech Republic, EU, havelkovab@vfu.cz

https://doi.org/10.37904/nanocon.2021.4370

Abstract

Metal oxide nanoparticles are increasingly being produced and will inevitably and up in the aquatic environment. In this work, we focus on the metal oxide nanoparticles CuO. These nanoparticles are used in many commercial products (e.g. CuO nanoparticles in catalysts, semiconductors and chemical sensors).

We focused on acute and adverse effects of $CuCl_2$ and CuO NPs <50 nm, in particular. The aim of this study was to investigate the acute toxicity of the nanoparticles to *Daphnia magna* and *Tubifex tubifex*. Selected representatives are an important part of the fish food chain. The effect of the $CuCl_2$ salt and CuO NPs was tested on the growth *D. magna* in an acute test scenario according to OECD guidelines 202. Tests with *T. tubifex* were performed according to method ASTM E1706-04. The concentrations of $CuCl_2$ and CuO NPs tested were as follows: 0.0005, 0.01, 0.05, 0.1, 0.5, 1 $mg \cdot l^{-1}$ and 0.1, 0.25, 0.5, 0.75, 1 $mg \cdot l^{-1}$ to *D. magna* and *T. tubifex*, respectively. Graphpad Prism was used for data visualization and statistics. Dose-response curves were constructed and EC50/LC50 values were calculated. In particular, changes in the behaviour of test organisms were noted. The impairment of Daphnids movement and holding at the bottom and *Tubifex tubifex* burrowing behaviour may lead to ecologically detrimental effects, such as an increase in the susceptibility of invertebrate species to predation. This could lead to possibly biomagnifying NPs up to food chain, thereby affecting the entire ecosystem.

Keywords: Nano, metal, aquatic toxicity, *Daphnia magna, Tubifex tubifex*

1. INTRODUCTION

Engineered metal nanoparticles (NPs), such as copper oxide (CuO) NPs, have unique properties because of their small size (1-100 nm) and high surface-to-volume ratio compared to their larger counterparts [1]. CuO NPs and other Cu-containing NPs have vast applications, including antifouling paint, bioactive coatings, cosmetics, electronics, health products, inks, lubricants, plastics, solar cells, and batteries [2]. Therefore, there has been a dramatic increase in the use of Cu-containing particles over the past decades. These nanoparticles are used in many commercial products (e.g. CuO nanoparticles in catalysts, semiconductors and chemical sensors). As the production of CuO NPs increases the release of these particles via wastewater and agricultural runoff, release from weathered surfaces treated with NP-coatings and antifouling paints will likely cause increased exposure in the aquatic environment [3]. Aquatic sediments have been recognized as a major compartment for metal NP accumulation. CuO accumulation in the sediment compartment may lead to concentrations in the ng to $\mu g \cdot kg^{-1}$ range [4]. Once in the aquatic environment, NPs are highly affected by their surroundings and consequently undergo transformations (e.g., agglomeration, aggregation, dissolution, sulfidation). The fate and behaviour of NPs depends both on their physical-chemical properties and on the characteristics of the receiving environment, including pH, temperature, concentration of natural organic matter, ionic strength and salinity and water hardness [5]. As a result of these dynamics, aquatic organisms are not only exposed to NPs but also to their dissolution and/or aggregation products. In aquatic sediments, arrange of transformation processes will occur, but the complex sediment matrix makes it very difficult to predict NP transformations in this compartment [3].

The freshwater crustacean *Daphnia magna* is commonly used for standardised toxicity testing of 24-48 h acute immobility and 21-day chronic effects regarding reproduction, growth and mortality. Daphnids have high ecological relevance and a relatively short life-cycle that enables chronic studies within a practical and reasonable testing timeframe of 21 days, in addition to being easily managed in the laboratory [6]. The sediment-dwelling Oligochaete, *Tubifex tubifex* was selected as model organism, because they are widely distributed in the freshwater eco-system, and have a feeding behaviour that includes ingesting large amounts of fine particles (<60 µm) and extract organic matter associated with ingested sediment. Due to their bioturbation activities, high tolerance for polluted ecosystems, intermediate position in the trophic network and the easy of breeding in the laboratory, they are widely used as a standard model organism in ecotoxicological studies [7].

The aim of this study was to investigate the acute toxicity and adverse effects of the nanoparticles to important representatives of the aquatic environment *Daphnia magna* and *Tubifex tubifex*. Since benthic invertebrates represent a major food source for fish, it is hypothesized that CuO NPs may be further transferred along the food chain.

2. METHODS

The effect of the $CuCl_2$ salt and CuO NPs (<50 nm) was tested on the mortality and growth of *Daphnia magna* in an acute test scenario according to OECD guidelines 202 [8]. Tests with *Tubifex tubifex* were performed according to method ASTM E1706-04 [9].

The CuO NPs were added to pure water to obtain a stock suspension of 1 $g \cdot l^{-1}$, which was sonicated for 5-6 h in a sonication bath. Test solutions were prepared by diluting the initial dispersion of $CuCl_2$/CuNPs in culture medium to the desired concentration. The concentrations of $CuCl_2$ and CuO NPs tested were as follows: 0.0005, 0.01, 0.05, 0.1, 0.5, 1 $mg \cdot l^{-1}$ and 0.1, 0.25, 0.5, 0.75, 1 $mg \cdot l^{-1}$ to *D. magna* and *T. tubifex*, respectively. These concentrations were chosen on the basis of preliminary study. Both $CuCl_2$ and CuO were purchased from Sigma-Aldrich Co. and were used without any purification. The particle sizes were indicated by the manufacturer as <50 nm for CuO. For statistical analysis, the EC50/LC50 values (the median effect concentration/the median lethal concentration), as well as their associated 95% confidence intervals (95% CI), were determined by Graphpad Prism version 6.05 software (San Diego, CA, USA). Nonlinear regression analysis was performed on dose-response immobilization/mortality and a curve with variable slope was placed on each of the data sets.

Test organism (*Daphnia magna*) originated in cultures from the Ecotoxicological laboratory of the University of Veterinary Sciences Brno (Czech Republic). Acute immobilization tests were conducted with 48 h continuous exposure in accordance with OECD Guideline 202. Immobility was recorded after 24 and 48 h. The test include control and six exposure concentrations with four replicates, each comprised of ten neonates (<24 h old) in 50 ml test suspension contained in 100 ml beakers. After 24/48 hour exposure to the test solutions at 20 °C with a photoperiod of 16 h light and 8 h darkness, immobilized neonates were counted. Neonates that were not able to swim within 15 s were considered to be immobilized after gentle agitation of the vessel. The validity criteria were met for all acute tests, with no control mortalities and an oxygen content >8 $mg \cdot l^{-1}$ [8].

Test organism (*Tubifex tubifex*) originated in cultures from the Ecotoxicological laboratory of the University of Veterinary Sciences Brno (Czech Republic). The effect of the $CuCl_2$ salt and CuO NPs (<50 nm) was tested on the mortality and growth of *Tubifex tubifex.* Acute tests were conducted for 96 h into an incubator at 22±1 °C, in the dark, and without water renewal or feeding. The number of dead worms in each vessel was counted at least once every 24 h. A worm was considered to be dead when there was no response in the 10 s after receiving a slight disturbance with a bar [10]. After 5 days of exposure to CuO/CuCl2, surviving worms were transferred to beakers containing 2 cm uncontaminated natural sediment and 40 ml *T. tubifex* media. Burrowing behaviour has been subjectively recorded at 24 hour.

3. RESULTS AND DISCUSSION

Ecotoxicity of $CuCl_2$ and CuO was determined in two aquatic tests. The influence of selected matters on the observed effect in *Daphnia magna* and *Tubifex tubifex* is shown in **Figure 1**. In this study, Daphnids was more sensitive to copper than *T. tubifex*. This is confirmed in other work, where algae, molluscs, crustaceans, insects, other invertebrates and fish were included and crustaceans were the most sensitive to copper [11].

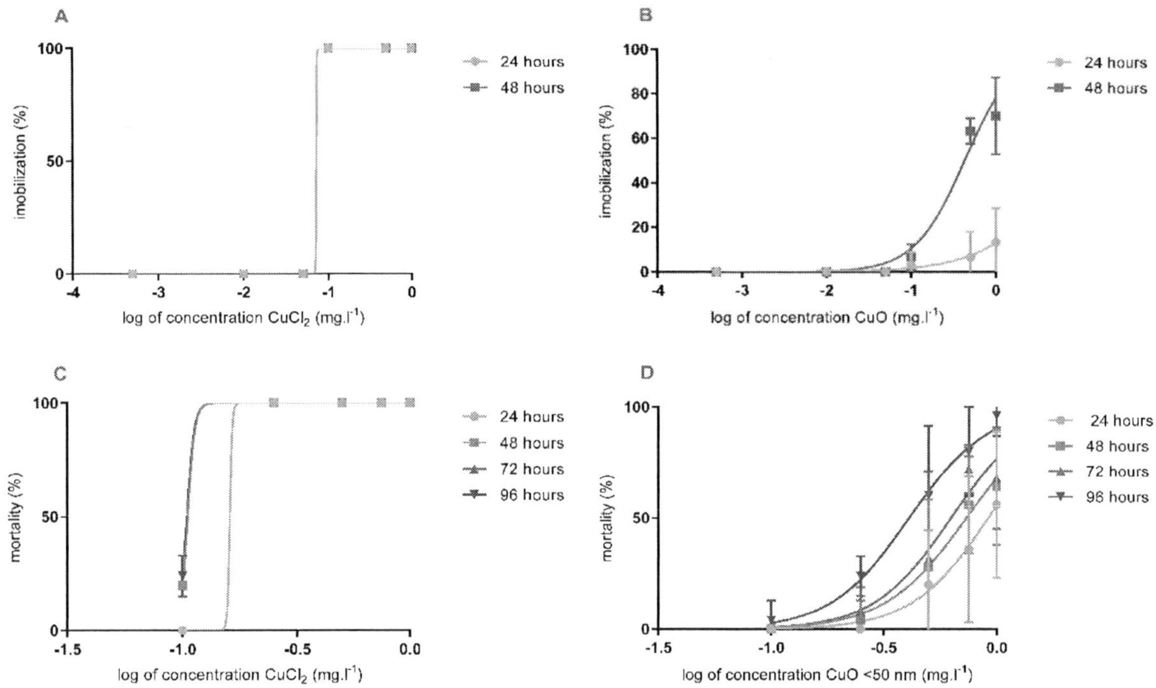

Figure 1 Dose-response curves of tested substances for *D. magna* and *T. Tubifex*. Data points are the average of four replicates. A the EC50 for $CuCl_2$ in *D. magna* was determined at 48 h 0.048 (CI 0.024 to 0.121) mg·l^{-1} B the EC50 for CuO in *D. magna* were determined at 24 and 48 h 7.89 (CI 0.21 to 297) and 0.43 (0.34 to 0.54) mg·l^{-1}, respectively C the LC50 for $CuCl_2$ in *T. tubifex* was determined at 96 h 0.125 (CI 0.84 to 1.60) mg·l^{-1} D the LC50 for CuO in *T. tubifex* were determined at 24, 48, 72 and 96 h 0.92 (CI 0.72 to 1.17), 0.73 (0.61 to 0.89), 0.63 (0.51 to 0.77) and 0.41 (0.34 to 0.50) mg·l^{-1}, respectively

We noticed nanoparticles in the intestine of *D. magna*. It is obvious that nanoparticles or nanoparticle aggregates are ingested by the Daphnids (**Figure 2 B**). This ingestion of nanoparticles aggregates is possible through the filter feeding mechanism of *D. magna*. Nanoparticle aggregates that are taken in may occur as dispersed nanoparticles or aggregates in the gut or dissolve in the gut or in the cells (e.g. after uptake by endocytosis) due to lower pH values [11]. Subsequently, the nanoparticles (or their derivates) can be either incorporated or eliminated from the body. In this study, we observed as a side effect changes in movement and holding at the bottom of the vessel was subjectively observed from the concentration of 0.1 mg·l^{-1}.

The length of the Daphnids (mm) was monitored for the (unexposed) blank Daphnids and the ones exposed to the CuO and $CuCl_2$ during the 21-day experiment. No significant differences were observed in length between the blank and exposed Daphnids (one-way ANOVA indicated no significant differences at most time points). Here as well, no effect of the exposure (nanoparticle or metal salt) could be seen on the Daphnia length.

In this study, after 48 h of exposure to immobilization EC50 values of the copper salt $CuCl_2$ 0.048 (CI 0.024 to 0.121) $mg \cdot l^{-1}$ and CuO nanoparticles 0.43 (0.34 to 0.54) $mg.l^{-1}$ were measured (**Figures 1 A, B**). The results indicate that the copper from the nanoparticles is taken less efficiently than the copper from the copper salt. Other authors [11,12,13,14] mention, that in the nanoparticle exposure, the Daphnids were mostly exposed to CuO aggregates and only to very low concentrations of dissolved copper, whereas in the metal salt exposure, the Daphnids were solely exposed to dissolved copper. Whilst some studies have reported that the toxicity of CuO on aquatic organisms were due to dissolved Cu^{2+} ions, others have reported that the observed toxicity is due to the metal oxide NPs [14].

In this study with *T. tubifex*, after 96 h of exposure to mortality LC50 values of the CuO nanoparticles 0.41 (0.34 to 0.50) $mg \cdot l^{-1}$ and copper salt $CuCl_2$ 0.125 (CI 0.84 to 1.60) $mg \cdot l^{-1}$ were measured (**Figures 1 C, D**). The mortality of *T. tubifex* was 0 % in the control. The burrowing behavior of surviving *T. tubifex* in clean sediment was significantly affected (decreased) following exposure in the present study. Zhang et al. (2017) mention that burrowing behavior of benthic invertebrates may be a more sensitive endpoint in behavior tests, and likely of particularly importance when considering NP effects [7]. Since *T. Tubifex* plays an important role in biogeochemical processes through its burrowing and irrigation activity, the impairment of burrowing behavior may lead to ecologically detrimental effects, such as an increase in the susceptibility of sediment-dwelling species to predation [15-17].

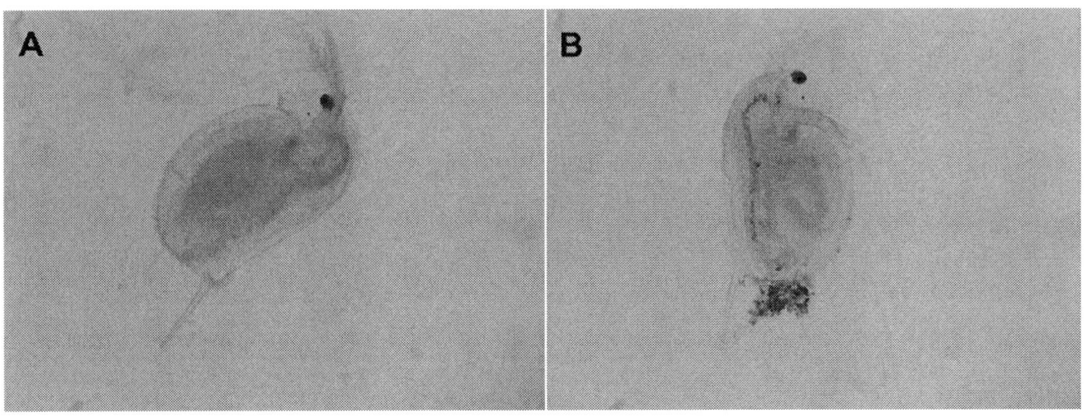

Figure 2 A *Daphnia magna* (control) **B** *Daphnia magna* (1 $mg \cdot l^{-1}$ CuO)

4. CONCLUSION

Metal nanoparticle bioaccumulation in benthic invertebrates in exposed aquatic ecosystem has been studied in a number of benthic invertebrates and these studies suggested that benthic invertebrates could accumulate nanoparticles like CuO NPs. The impairment of movement and holding at the bottom (*Daphnia magna*) and burrowing behavior (*Tubifex tubifex*) may lead to ecologically detrimental effects, such as an increase in the susceptibility of invertebrates species to predation. This could lead to an increased predation of contaminated Daphnids and worms by fish, possibly biomagnifying NPs up to food chain, thereby affecting the entire ecosystem.

ACKNOWLEDGEMENTS

This work was supported by the grant IGA 219/2021/FVHE.

REFERENCES

[1] NEL, A., XIA, T., MADLER, L., LI, N. Toxic potential of materials at the nanolevel. *Science*. 2006, vol. 311, no.5761, pp. 622-627.

[2] REN, G., HU, D., CHENG, E.W.C., VARGAS-REUS, M.A., REIP, P., ALLAKER, R.P. Characterisation of copper oxide nanoparticles for antimicrobial applications. *International Journal of Antimicrobial Agents*. 2009, vol. 33, no. 6, pp. 587-590.

[3] THIT, A., SANDGAARD, M.H., STURVE, J., MOUNEYRAC, C., BAUN, A., SELCK, H. Influence of aging on bioaccumulation and toxicity of copper oxide nanoparticles and dissolved copper in the sediment-dwelling oligochaete *Tubifex tubifex*: A long-term study using a stable copper isotope. *Frontiers in Toxicology*. 2021, vol. 3, article 737158.

[4] GARNER, K.L., SUH, S., KELLER, A.A. Assessing the risk of engineered nanomaterials in the environment: development and application of the nanofate model. *Environmental Science & Technology*. 2017, vol. 51, pp. 5541-5551.

[5] SELCK, H., HANDY, R.D., FERNANDES, T.F., KLAINE, S.J., PETERSEN, E.J. Nanomaterials in the aquatic environment: a European union-united states perspective on the status of ecotoxicity testing, research priorities, and challenges ahead. *Environmental Toxicology and Chemistry*. 2016, vol. 35, no. 5, pp. 1055-1067.

[6] SORENSEN, S.N., LUTZHOFT, H.CH.H., RASMUSSEN, R., BAUN, A. Acute and chronic effects from pulse exposure of *D. magna* to silver and copper oxide nanoparticles. *Aquatic Toxicology*. 2016, vol. 180, pp. 209-217.

[7] ZHANG, P., SELCK, H., TANGAA, S.R., PANG, CH., ZHAO, B. Bioaccumulation and effects of sediment-associated gold- and graphene oxide nanoparticles on *Tubifex tubifex*. *Journal of Environmental Sciences*. 2017, pp. 138-145.

[8] OECD 202. *OECD Guideline for testing of chemicals Daphnia Sp. Acute immobilization test.* Organisation for economic co-operation and development: Paris, France, 2004.

[9] ASTM E1706-05. *Standard Test Method for Measuring the Toxicity of Sediments-Associated Contaminants with Freshwater Invertebrates*. ASTM: Philadelphia, 2005.

[10] MAESTRE, Z., MARTINEZ-MADRID, M., RODRIGUEZ, P. Monitoring the sensitivity of the oligochaete *Tubifex tubifex* in laboratory cultures using three toxicants. *Ecotoxicology and Environmental Safety*. 2009, vol. 72, pp. 2083-2089.

[11] ADAM, N., LEROUX, F., KNAPEN, D., BLAS, S., BLUST, R. The uptake and elimination of ZnO and CuO nanoparticles in nanoparticles in Daphnia magna under chronic exposure scenarios. *Water research*. 2015, vol. 68, pp. 249-261.

[12] BLINOVA, I., IVASK, A., HEINLAAN, M., MORTIMER, M., KAHRU, A. Ecotoxicity of nanoparticles of CuO and ZnO in natural water. *Environmental pollution*. 2010, vol. 158, pp. 41-47.

[13] KARLSSON, H.L., CRONHOLM, P., GUSTAFSSON, J., MÖLLER, L. Cooper oxide nanoparticles are highly toxic: a comparison between metal oxide nanoparticles and carbon nanotubes. *Chemical Research in Toxicology*. 2008, vol. 21, no. 9, pp. 1726-1732.

[14] MWAANGA, P., CARRAWAY, E.R., HURK, P. The induction of biochemical changes in *Daphnia magna* by CuO and ZnO nanoparticles. *Aquatic Toxicology*. 2014, vol. 150, pp. 201-209.

[15] BUFFET, P.E., RICHARD, M., CAUPOS, F., VERGNOUX, A., PERREIN-ETTAJANI, H., LUNA-ACOSTA, A., AKCHA, F., AMIARD, J.C., AMIARD-TRIQUET, C., GUIBBOLINI, M., RISSO-DE FAVERNEY, CH., THOMAS-GUYON, H., REIP, P., DYBOWSKA, A., BERHANU, D., VALSAMI-JONES, E., MOUNEYRAC, C. A mesocosm study of fate and effects of CuO nanoparticles on endobenthic species (*Srobicularia plana*, *Hediste diversicolor*). *Environmental Science & Technology*. 2013, vol. 47, no. 3, pp. 1620-1628.

[16] BOLDINA-COSQUERIC, I., AMIARD, J.C., AMIARD-TRIQUET, C., DEDOURGE-GEFFARD, O., METAIS, I., MOUNEYRAC, C. Biochemical, physiological and behavioural markers in the endobenthic bivalve *Scrobicularia plana* as tools for the assessment of estuarine sediment quality. *Ecotoxicology and Environmental Safety*. 2010, vol. 73, no. 7, pp. 1733-1741.

[17] THIT, A., SELCK, H. Biodynamics and adverse effects of CuO nanoparticles and $CuCl_2$ in the oligochaete *T. tubifex*: Cu form influence biodynamics in water, but not sediment. *Nanotoxicology*. 2021, vol. 15, no. 5, pp. 673-689.

MEASURING THE INCIDENCE OF NANOPARTICLES IN THE WORKPLACE ATMOSPHERE WITH A FOCUS ON THE WOODWORKING AND ENGINEERING INDUSTRIES

[1]Petra ROUPCOVÁ, [1,2]Karel KLOUDA, [1]Ondřej CHROMEČKA, [1]Simona SLIVKOVÁ, [3]Hana KUBÁTOVÁ

[1]*VSB- Technical University of Ostrava, Ostrava, Czech Republic, EU,* *petra.roupcova@vsb.cz, simona.slivkova@vsb.cz*

[2]*Occupational Safety Research Institute, Prague, Czech Republic, EU, klouda@vubp-praha.cz*

[3]*State Office for Nuclear Safety, Prague, Czech Republic, EU, hana.kubatova@sujb.cz*

https://doi.org/10.37904/nanocon.2021.4366

Abstract

Nanoparticles are present in everyday life including industry, and this article focuses mainly on their occurrence in the work environment. We measured the concentration of nanoparticles in anthropogenic activity, where nanoparticles are generated as unwanted waste. Two plants were selected for such measurements. In the first plant we measured nanoparticles based on metals, while in the second we measured nanoparticles of organic origin, specifically wood. This article also deals with the possible toxic effect of nanoparticles on the environment. The procedures of ISO / TR 27 628 - workplace atmospheres - ultrafine, nanoparticle and nanostructured aerosols - inhalation exposure characterization and assessment were applied to identify the key sites in both plants that produced the largest number of nanoparticles. We suggest the provision of protective work funds in an environment with a risk of nanoparticles according to the certified methodology of the Ministry of Labour and Social Affairs.

Keywords: Nanoparticles, aerosols, wood, metal, toxicity, safety measures

1. INTRODUCTION

Nanoparticles are formed in natural processes (volcanoes, forest fires, etc.), in anthropogenic activities, and as "waste" in many industrial productions. However, they are being increasingly produced in artificial ways, for example in the production processing of nanofibers. [1,2]. This paper presents the measurements of nanoparticle concentrations in the size 10-700 nm produced at two plants that work respectively with steel and wood:

1) Metrostav a.c. Division 3 - steel (source of nanoparticles from steel),
2) Iktus Furniture-wood (source of nanoparticles from wood).

In assessing the measurement results, we followed the ISO/TR 27 628 standard - workplace atmospheres - ultrafine, nanoparticle and nano-structured aerosols - inhalation exposure characterization and assessment, with the aims of:

- determining the expected composition of the aerosol particulate matter and estimating/determining its toxic effect (wood vs steel). [3,4]
- registering the airflow with aerosol transport in the workplace (doors, windows, controlled ventilation),
- analysing work activities, technologies used, work habits/behaviour of employees.

Chemical composition of wood may have plenty of variations. Apart from its main components (cellulose, hemicellulose and lignin) it consists of e.g. phenols, tannins, alkaloids which may cause adverse biological activity. Exotic woods (pointed shape) resemble asbestos in their behavior towards the organism [5]. In metallurgical operations it is important to count with heavy metals which inhibit cell division (oxidative stress) [6].

Subsequently, we addressed the employer/operator with questions/suggestions on:

- modifications to the working atmosphere (natural ventilation, central, local, mobile exhaust, air curtains, air filtration through hepa filters, and ideally the extraction of emissions and ventilation,
- measures of a technological or technical nature (change of technology and production method, installation of CNC machines, mobile walls, air curtains, placing production or part of it in a hermetic box or fume cupboard, change of raw material,
- modifications of an organisational nature (continuous prevention policy, control zones, precautionary principle, cleanliness of the workplace - regular cleaning, maintenance, reduction of staff at the main source of nanoparticles),
- introduction of respiratory protective equipment (possibility of using the certified methodology of the Ministry of Labour and Social Affairs, having a good and sufficient quantity of PPE and OPDO [7]; respiratory protection with PPE should be applied as the last option of the measures to protect against nanoparticles in the working environment;
- introduction of specialised occupational medical examinations (e.g. biological monitoring - SZU, Department of Occupational Medicine, 1st Faculty of Medicine, Prof. Daniela Pelclová - Measurement of unsaturated aldehydes in breath.

2. MEASUREMENT OF THE HALL SPACE DURING THE PRODUCTION OF A DOOR COMPONENT

The measurement of nanoparticle concentrations was carried out in a unique production process, namely the production of chamber gates.

The Gabčíkovo waterworks chamber gate production site consists of three interconnected halls (**Figure 1**) and a paint shop. The basis of the production technology is the application of welding techniques* for welding steel plates into complex technological units whose size and shape are influenced by their transport to the site and their assembly in the chamber of the waterworks.

*Technologies used in hall welding:

- manual arc welding with fusion electrode
- arc welding with filled electrode in active gas
- submerged arc welding with multiple wire electrodes
- flame welding
- oxygen cutting
- arc fusion welding in active gas
- submerged arc welding with wire electrode
- tungsten arc welding in inert gas
- stroke welding of bolts
- plasma cutting
- electric arc grooving with compressed air

Figure 1 Illustrative photographs of the three halls of the machine area

The first type of measurement, which aimed to map the working environment in the halls, was divided into two days at 09:00 - 10:00 o'clock when the majority of welders and turners (about 20 employees) were carrying out their specific work activities.

The measuring instrument The Testo Disc mini 133 device was used for measuring a whole passage between the technologies in three halls and a height of 1 m from the floor. The obtained nanoparticle concentration results and mean particle diameters have been processed in graphs (**Figures 2, 3, 4, 5**).

The measured values were influenced by the location in the hall in relation to the technological operation that took place near the measuring instrument. In general, it can be concluded that the air in the halls as a whole fails to be optimal. Ventilation of the halls is provided through air vents - on the columns and naturally through windows located around the perimeter of the hall or through open doors.

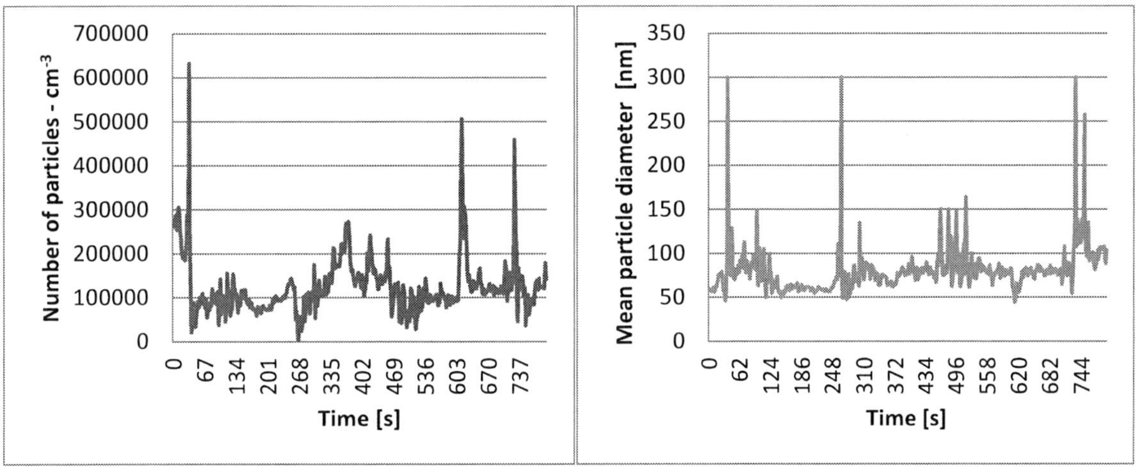

Figure 2 Concentration and mean diameter of nanoparticles as they pass through the halls

3. MEASUREMENTS IN SELECTED TECHNOLOGICAL (WELDING) OPERATIONS

After measurements at unspecified locations in the halls, we measured nanoparticle concentrations at selected welding operations, namely at the hole of the closed segment (max. 400,000 #-cm^{-3}), at the open segment (max. 700,000 #-cm^{-3}), at each welder (max. 120,000 #-cm^{-3}), and at the combination of grinding and welding (max. 160,000 #-cm^{-3}).

4. HALL VENTILATION

Hall ventilation is provided through the central air conditioning vents and naturally through windows located around the perimeter of the hall and through open doors. Ventilation in the halls is provided by supplying "filtered" air to the vents located next to the supporting columns.

The aim of this measurement was to investigate the effect of the air handling system and to compare the concentration of nanoparticles at the exhaust (measuring instrument located about 20 cm from the mesh) and the hall area (about 2 m from the mesh). The measurements were performed by moving the instrument in the space to a distance of 2 m from the exhaust after about 1 min. Measurements were taken at two separate vents located in the centre of the hall.

We assumed the same values of nanoparticle concentration at 20 cm from the mesh for both vents and higher values after displacement (after 1 min) compared to the values measured at the mesh. Our assumption was not confirmed since the same concentration pattern of nanoparticles was measured over time and distance (20 cm versus 2 m). The measured nanoparticle concentration was higher for the first vent (80,000-100,000

#-cm^{-3}) compared to 60,000 #-cm^{-3} for the second measurement, but here the mean particle concentration was lower (**Figure 3**).

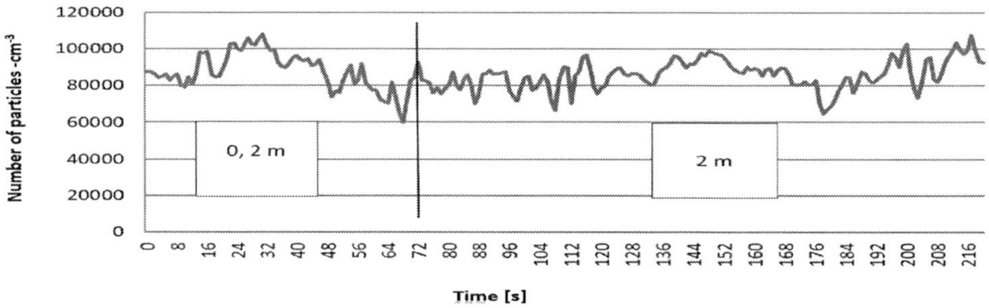

Figure 3 Concentration of nanoparticles at different distances from the exhaust ([1st] exhaust)

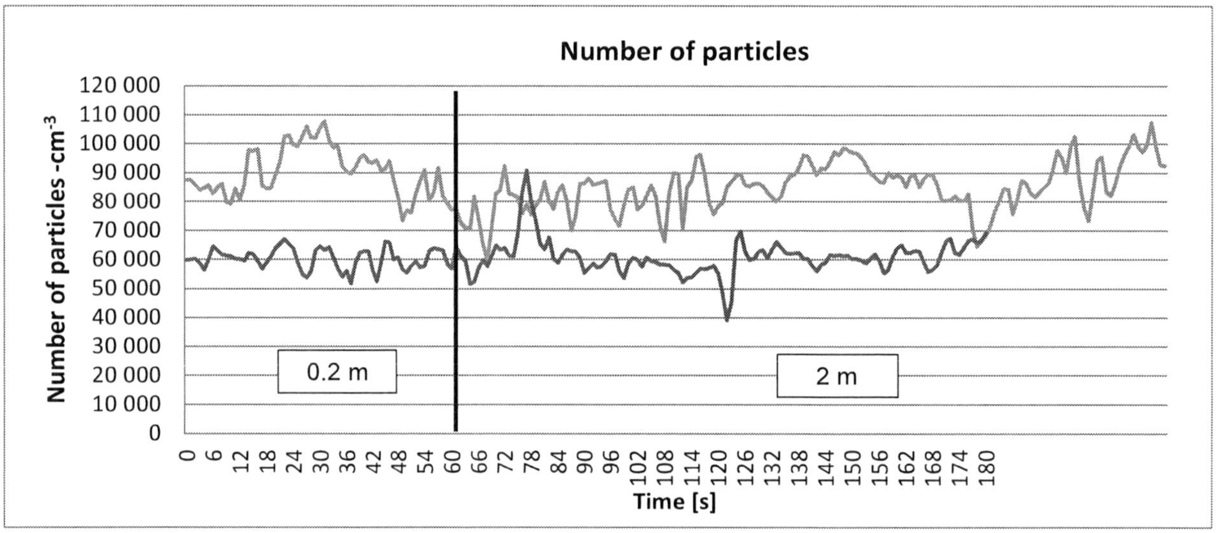

Figure 4 Comparison of the particle number concentration of two forced ventilation flows

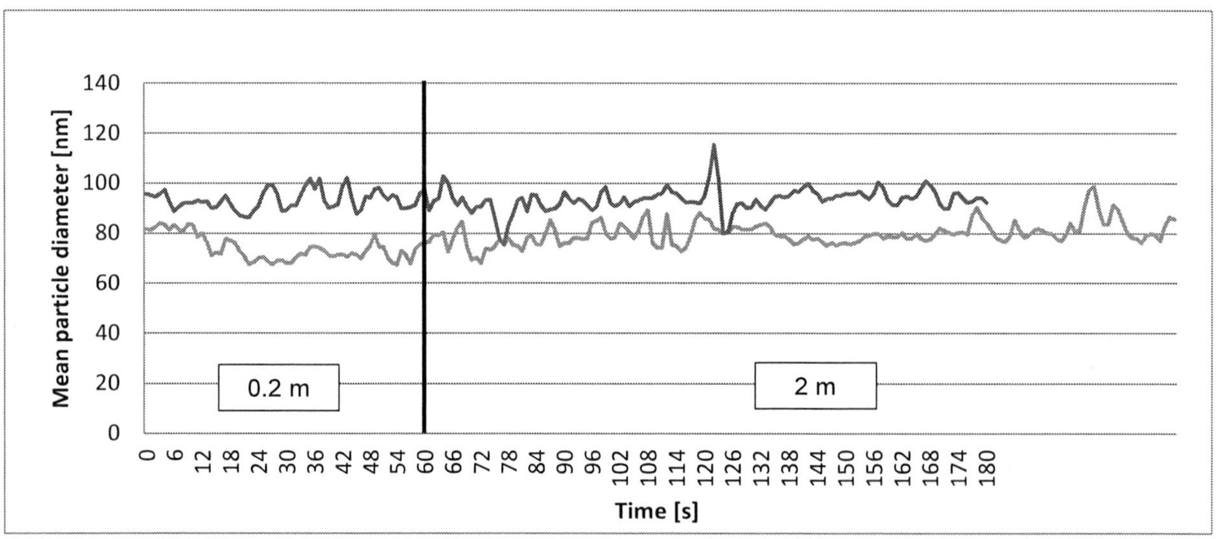

Figure 5 Comparison of mean particle diameters of two forced ventilation flows

5. MEASUREMENT OF NANOPARTICLE CONCENTRATION IN FURNITURE PRODUCTION

Furniture production covers many areas, yet the production of classic corner benches, dining sets, wardrobes, beds, chests of drawers predominates. Most technological operations are carried out with the use of CNC machines. For example, cutting saws, taping machines and doweling machines are used for drilling, milling, and cutting. Most of the CNC machines are located in the production Halls 1 and 3.

The measurements were performed while walking with the measuring device through the halls (1-4), with the measuring device placed at a height of about 1 m above the floor surface. The obtained partial results are presented in the form of graphs (**Figures 6** and **7**) as the change in the number of particles per unit volume of the atmosphere over time and the change in the size of the diameter over time in each production hall.

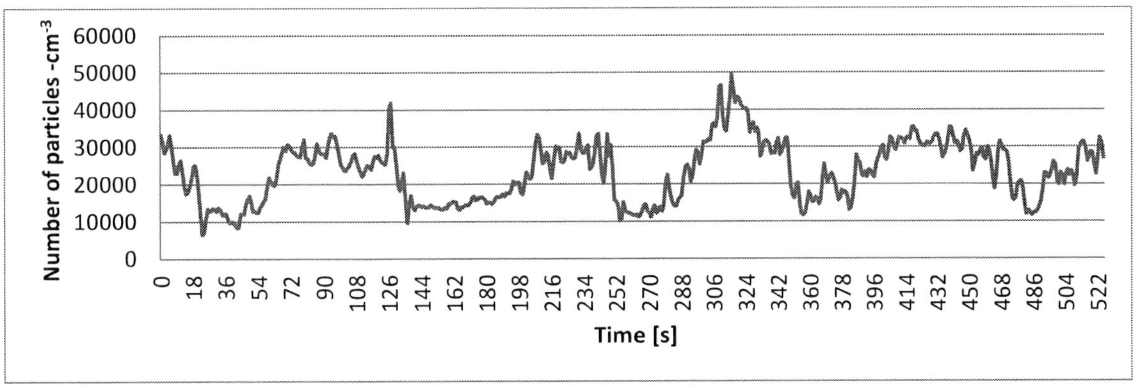

Figure 6 Nanoparticle concentration in Hall 3

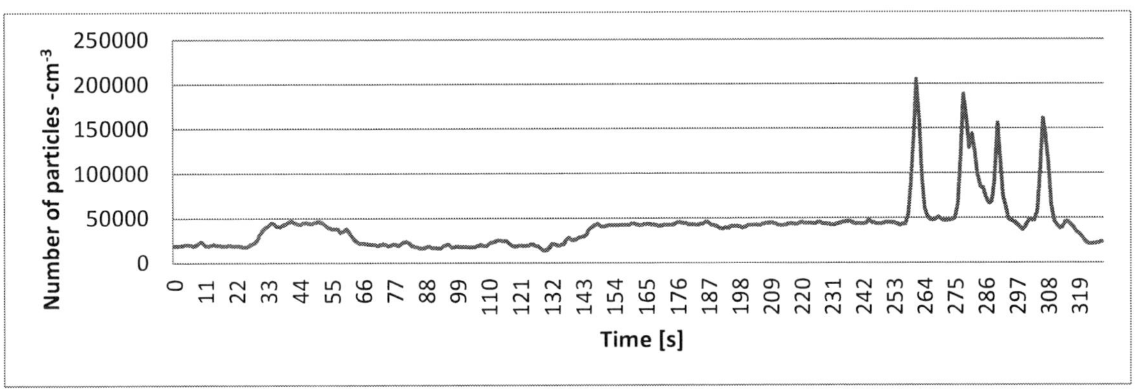

Figure 7 Nanoparticle concentration in Hall 4

HALL 1

The jump increase in nanoparticle concentration was measured at the CNC forming saw, where the values obtained reached a magnitude of about 150,000-200,000 #-cm^{-3} as the semi- finished product exited the machine.

HALL 2

In this hall, we identified from the measurement values the area where the nanoparticles are localized, namely at the lower cutter near the free suction necks. Here their concentration ranged from 200,000 to 300,000 #-cm^{-3}.

HALL 3

In this hall, where the CNC machines are located, there is a fairly regular pulsation of nanoparticle concentration values (**Figure 6**) from 10,000 #-cm^{-3} to 35,000-50,000 #-cm^{-3} with stable values of mean particle diameter - 40-50 nm.

HALL 4

Hall 4 houses press doweling machines and a circular saw, which is fitted with a plastic guard with an exhaust connection. However, concentrations with pulsating values of up to 200,000 #-cm^{-3} were measured in the vicinity of the operating worker (**Figure 7**) and with a decrease in the mean particle diameter from 28 nm to 17 nm.

6. CONCLUSION

The measurements in Hall 1 at the exit of the semi-finished product from the machine show that even an efficient CNC machine does not prevent the accumulation of nanoparticle concentrations. By failing to design the hall exhaust accurately, a significant increase in the concentration of nanoparticles in the work area can occur. In particular, the circular saw, even when connected to an exhaust, is a source of high nanoparticle concentrations. A surprising finding at MetroState Division's was the discovery that the air in the halls circulates without filtration. If we want to compare the possible toxic effect of wood against steel, we must consider not only the main role of concentration but also the composition, shape of nanoparticles and the ambient conditions with which they enter the working area [4].

ACKNOWLEDGEMENTS

The publication was written within the Student grant competition "Release of Nano and Microplastics from Nanotextiles and Determination of Their Effect on the Environment" - project number SP2021/69.

REFERENCES

[1] ALBA, J., ROUPCOVA, P., KLOUDA, K. Nanotechnology Safety and Security: Nanoparticles and Their Impact on the World. In *Safety and Security Issues in Technical Infrastructures*. Edited by REHAK, D., BERNATIK, A., DVORAK, Z., HROMADA, M. HERSHEY, PA: IGI Global, 2020, pp. 321-352. Available from: https://doi.org/10.4018/978-1-7998-3059-7.ch012.

[2] ROUPCOVA, P., KLOUDA, K., NECHVATAL, M., BATRLOVA, K., BERNATIKOVA, S. New risks in OHS with focus on selected nanotechnological workplaces. *Proceedings of the 29th European Safety and Reliability Conference, ESREL 2019*. Hannover: ESREL, 2020, pp. 3911-3916.

[3] KLOUDA, K., SINAY, J., CEJPEK, J., LACH, K. Identification of Nano- and Microparticle Aerosols Generated in Selected Operations. *Acta Mechanica Slov.* 2013, vol. 17, no. 2, pp. 90-102. Available from: https://doi.org/10.21496/ams.2013.025.

[4] NEL, A., XIA, T., MADLER, L., LI, N. Toxic potential of materials at the nanolevel. *Science*. 2006, vol. 311, pp. 622-627.

[5] JANKOVSKY, M., LACHMAN, J., STASZKOVA, L. *Wood chemistry*. Prague: Czech University of Agriculture, 1999. ISBN 80-213-0559-2.

[6] KAFKA, Z., PUNCOCHAROVA, J. Heavy metals in nature and their toxicity. *Chemicke listy*. 2002, vol. 96, pp. 611- 617.

[7] *Certified methodology for the provision of personal protective equipment in an environment with a risk of nanoparticles (aimed at employers in addressing the need to assign personal protective equipment for respiratory protection)*. Prague: Research Institute of Occupational Safety, v. V. I, 2016.

MEASUREMENT OF NANOPARTICLES IN 3D PRINTING USING FFF / FDM TECHNOLOGY

[1,3]Karel KLOUDA, [2]Hana KUBÁTOVÁ, [1]Marek NECHVÁTAL, [1]Kateřina BÁTRLOVÁ, [3]Petra ROUPCOVÁ

[1]*Occupational Safety Research Institute, Prague, Czech Republic, EU,*
klouda@vubp-praha.cz; nechvatal@vubp-praha.cz; batrlova@vubp-praha.cz

[2]*State Office for Nuclear Safety, Prague, Czech Republic, EU, hana.kubatova@sujb.cz*

[3]*VSB- Technical University of Ostrava, Ostrava, Czech Republic, EU, petra.roupcova@vsb.cz*

https://doi.org/10.37904/nanocon.2021.4367

Abstract

For additive production, where the material is prepared by applying thin layers, a simplified designation 3D printing has been used. During this production volatile organic compounds and ultrafine particles are emitted into the air. A number of measurements for a given type of printer with a specific type of thermoplastic are published on this topic. The procedure of our measurement of the concentration of nanoparticles (10-700 nm) and their mean diameters can be called "field", because it took place in a real environment with mass deployment of 3D printing on the principle of FFF / FDM, with the measuring device moving between printers in various premises and activities:

- in the production of parts intended for the assembly of printers,
- in the manufacture of components for shields during the lockdown period associated with Covid-19,
- in the creative workshops and laboratories of Prusa Research a.s.

The aim of these measurements in real situations was to find measures to reduce the concentration of nanoparticles. Evaluation of the results was proceeded according to the ISO / TR 27 628 standard and determined a proposal of measures of a technological and organizational nature to reduce concentrations. Based on the results of our measurements these measures include (a) using central extract ventilation of the entire workplace, (b) separating the printer area from the workplace and using extract ventilation, and (c) filtering the air in the entire workplace using an air purifier. When adopting these measures, the financial requirements and technical feasibility must always be considered.

Keywords: Nanoparticles, 3D printing, FFF/FDM technology, concentrations of nanoparticles, safety measures

1. INTRODUCTION

Over the past two decades, 3D printing from thermoplastics using Fused Filament Fabrication (FFF), or Fused Deposition Modeling (FDM) has become a very popular manufacturing method. The international RepRap project launched in 2005 [1] has contributed significantly to its popularity. However, subsequent studies on the emission of substances and materials during 3D printing have shown that volatile organic substances (VOS) and ultrafine particles (UFP) are released into the environment during the manufacturing process [2-4].

Previous studies have usually compared the measured concentrations of UFPs, their averages or their emission rates into the environment depending on the different types of filaments used in the printing process [5-16]. These measurements have shown that the composition of emissions from 3D printing is influenced by a number of factors, such as the chemical composition of the filament, its colour, the operating temperature of the print, the effect of the nozzle used, and the functionality of the printer [5-16]. The observed results are also influenced by the measurement scenario [11] or the type of instrumentation used [12].

Recent research has also compared the composition of emissions for a larger range of filaments, e.g. between ABS, PLA, PVA, HIPS, NYLON and PET [6], the number of particles in the emissions when printing between a pair of filaments, e.g. between ABS and PC, where PC generates more particles than ABS [13], or between ABS and PLA, where a strong concentration of styrene was found in the emissions [14].

Importantly, previous experiments were performed under reproducible conditions, where external influences were eliminated, for example, by placing the 3D printer in a sealed chamber of defined volume (e.g. 2.5 m3), supplying air to the chamber through a HEPA filter, and positioning a particle counter and spectrometer inside the chamber to measure the concentration of particles in the environment near the printer [15]. These exact measurements were performed for filaments made from materials most commonly used for 3D printing, such as ABS, PLA and HIPS. The average total concentration of UFP was 217,000 #.cm-3 for the HIPS material under the given conditions, 20,000 #.cm-3 for ABS and 14,000 #.cm-3 for PVA. [15].

In our research, the chosen method to measure nanoparticle concentrations and their mean diameter can be described as "field" because it was carried out in the real environment of selected workplaces. This fact is one of the reasons why our results differ from past experiments in which external influences were excluded.

2. MEASUREMENTS WHEN RUNNING MULTIPLE 3D PRINTERS USING FFF/FDM TECHNOLOGY

The Testo Disc mini 133 device was used for all measurements - measuring range 10-700 nm.

2.1. Large-scale production of 3D printers based on thermoplastic deposition

The measurements in the large-scale production of 3D printers took place at the workplace of one of the largest manufacturers of this type of printers, Prusa Research a.s. The measurements were carried out at three different workplaces of the company, namely at a workplace called "The Farm", in the area intended for the work of designers, and in the development laboratories.

2.2. The Farm

The Farm is a room in which rows of printers are placed on racks three levels above each other. The room in which the printers are located is partially separated from the surroundings by a plastic film (**Figure 1A**).

A B C

Figure 1 A) Production of parts for assembling 3D printers using 3D printing at Prusa Research, a.s.; B) Design workplace at Prusa Research, a.s.; C) Small laboratory

At the time of the measurements, the Farm had 480 printers printing parts for the assembly of 3D printers made of filaments of the chemical composition polyethylene terephthalate-glycol (PET-G; trade name Prusament PETG Prusa Orange). The actual measurements were performed while walking between the stands with the measuring equipment placed at a height of 1 m above the floor surface. The results obtained are presented as the change in the number of particles per unit volume of atmosphere over time (**Figure 2**) and the change in particle size over time. During the measurements, the concentration of nanoparticles varied in the range of 50,000-70,000 #-cm^3, with a mean particle diameter of 29-30.5 nm.

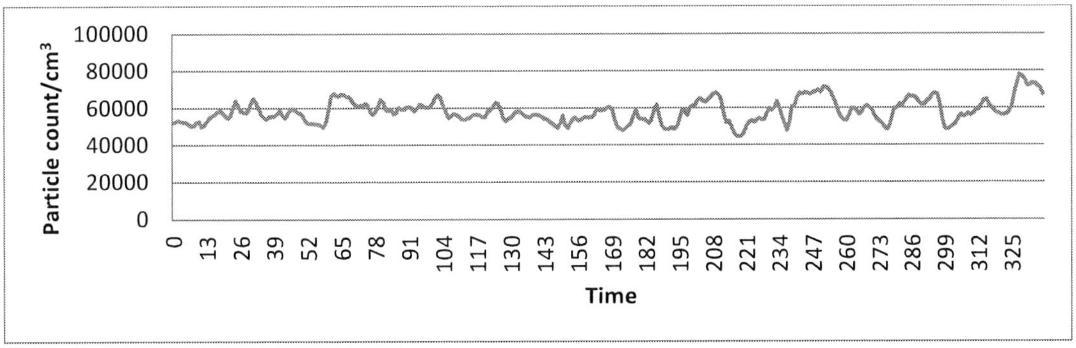

Figure 2 Concentration of nanoparticles measured at the Farma workplace during thermoplastic 3D printing with 480 printers running simultaneously

2.3. The Designers' Workplace

The designers' workplace contains the designers' workbenches (**Figure 1B**) and the 3D printers are freely placed on tables. The designers' workplace is not equipped with central extract ventilation, and air exchange is only provided through windows. During the measurements, the windows were closed, 9 designers were present at the workplace, and seven 3D printers were running simultaneously (chemical composition of PETG and PLA filaments). The measurements took place while walking between the tables, with the measuring equipment placed at a height of 1m above the floor surface. The results obtained are presented as the change in the number of particles per unit volume of the atmosphere over time (**Figure 3**). The concentration of nanoparticles in the engineers' room had a pulsed character and reached a value of up to 200,000 #-cm^{-3} (**Figure 3**). In the rooms with the maximum concentrations, the mean diameter of nanoparticles decreased from 30 nm to about 20 nm.

During simultaneous thermoplastic 3D printing on multiple printers, the concentration of nanoparticles in the working environment exceeded the values identified in the methodology of the Ministry of Labour and Social Affairs of the Czech Republic [17] as hazardous and thus requiring the application of protective equipment.

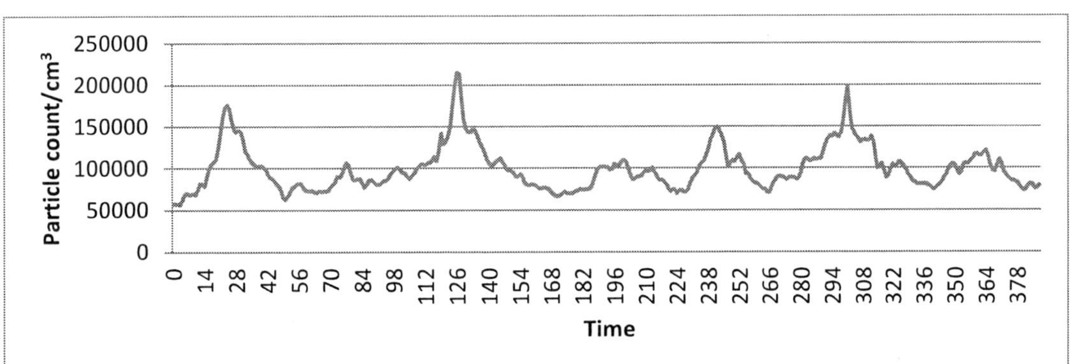

Figure 3 Concentration of nanoparticles in the designers' workplace

2.4. The Development Laboratories

At Prusa Research a.s., several measurements were also carried out in the laboratories. Their activities are focused on further development, but at the same time they search for optimal solutions for the protection of employees and the possibility of reducing the concentration of nanoparticles in the environment of the existing production facilities. For example, they test the effect of the use of different air purifiers (performance in the range of 100-400 m^3/h) in short- and long-term cleaning of the laboratory air. **Figures 4** and **5** show the changes in the concentration and mean diameter of nanoparticles measured during the normal operation of six 3D printers in a small laboratory environment (**Figure 1C**). The effect of air purification can be compared with the values recorded in **Figures 5** and **6** which were measured after the ETA Nubela 2569 air purifier was activated and filtered the laboratory air for 15 minutes. Filtration of the air resulted in a significant decrease in the concentration of nanoparticles even with the printers running.

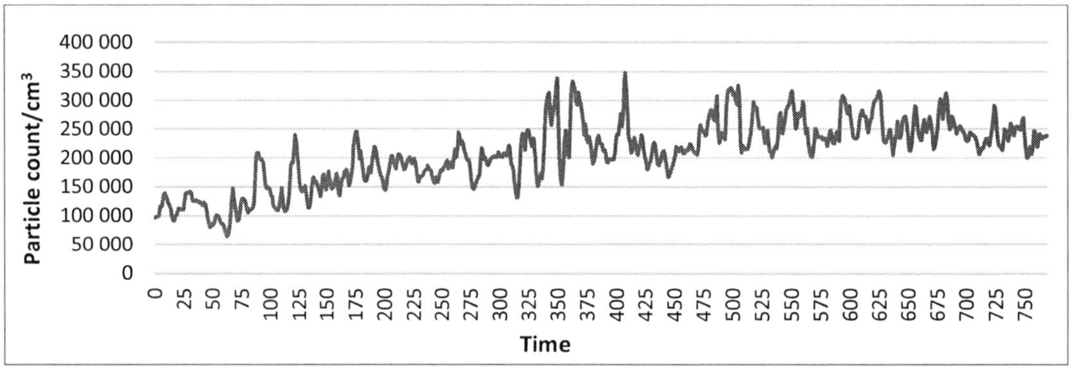

Figure 4 Nanoparticle concentration in a small laboratory with 6 3D printers running

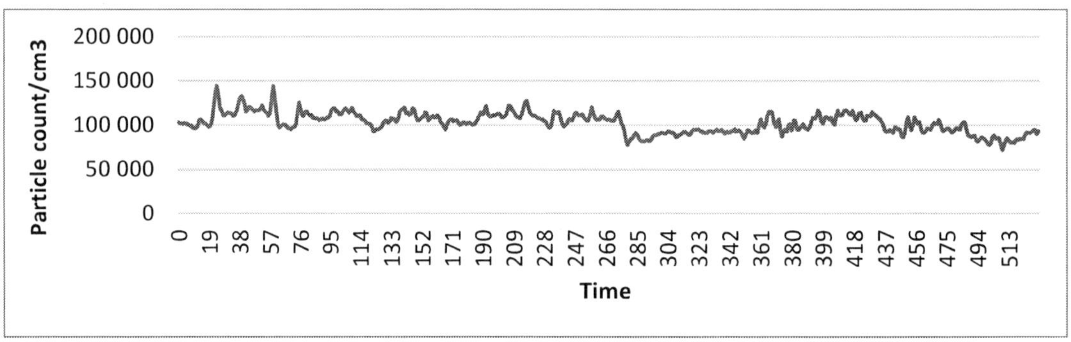

Figure 5 Concentration of nanoparticles in a small laboratory after 15 min of air filtration

2.5. Production of components for protective shields

The COVID-19 disease pandemic has activated efforts to use 3D printing to produce protective devices in both the private (Prusa Research a.s.) and academic spheres. At the Technical University of Ostrava (VŠB-TUO), the production of components for protective shields took place at the Protolab workplace (the business incubator of VŠB-TUO) where, in cooperation with the Ostrava City Council, all the available 3D printers based on FFF/FDM technology were concentrated. The temporary production facility was set up in a meeting room with about 60 3D printers placed on tables (**Figure 6**). The meeting room is equipped with ceiling ventilation, which was switched on during 3D printing. The measurements followed the same scenario as in Prusa Research Inc.

Figure 6 Production of components for protective shields at VŠB-TUO

The pulsed nature of the values on the graph in **Figure 7** is a consequence of the chosen measurement method, since the measuring instrument was placed at different distances from the particle sources when passing between the tables. During the measurements, the concentration of nanoparticles ranged from 22,000-55,000 #-cm^{-3}, with a mean particle diameter of 21-27 nm. Also in this measurement, a response (not a direct proportion) of the change in mean particle diameter to the particle concentration in space was observed (**Figure 7**).

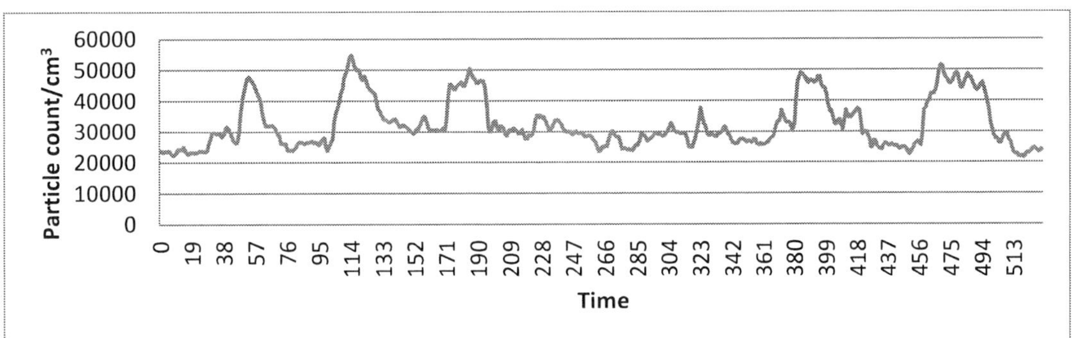

Figure 7 Concentrations of nanoparticles produced in the manufacturing of shielding components

3. CONCLUSION

During 3D printing with FFF/FDM technology, nanoparticles are emitted into the environment around the printers. On the basis of the measurements carried out, it was found that the fastest and most affordable solution is to filter the workplace air with air cleaners of various capacities with HEPA and C-filters. Their use does not require structural modifications, but additional operating costs, in particular the replacement and disposal of filters.

REFERENCES

[1] All3DP. *The Official History of the RepRap Project.* [online]. 2016 [viewed: 2021-06-16]. Available from: https://all3dp.com/history-of-the-reprap-project/.

[2] KIM, Y., YOON, C., HAM, S., PARK, J., KIM, S., KWON, O., TSAI, P. Emissions of Nanoparticles and Gaseous Material from 3D Printer Operation. *Environmental Science & Technology.* [online]. 2015, vol. 49, no. 20, pp. 12044-12053. Available from: https://doi.org/10.1021/acs.est.5b02805.

[3] GU, J., WENSING, M., UHDE, E., SALTHAMMER, T. Characterization of particulate and gaseous pollutants emitted during operation of a desktop 3D printer. *Environment International.* [online]. 2019, Volume 123, pp. 476-485. Available from: https://doi.org/10.1016/j.envint.2018.12.014.

[4] WOJTYŁA, S., KLAMA, P., BARAN, T. Is 3D printing safe? Analysis of the thermal treatment of thermoplastics: ABS, PLA, PET, and nylon. *Journal of Occupational and Environmental Hygiene.* [online]. 2017, Volume 14, Issue 6, pp. D80-D85. Available from: https://doi.org/10.1080/15459624.2017.1285489.

[5] AZIMI, P., ZHAO, D., POUZET, C., CRAIN, N. E., STEPHENS, B. Emissions of Ultrafine Particles and Volatile Organic Compounds from Commercially Available Desktop Three-Dimensional Printers with Multiple Filaments. *Environmental Science & Technology.* [online]. 2016, vol. 50, no. 3, pp. 1260-1268. Available from: https://doi.org/10.1021/acs.est.5b04983.

[6] FLOYD, E.L., WANG, J., REGENS, J. L. Fume emissions from a low-cost 3-D printer with various filaments. *Journal of Occupational and Environmental Hygiene.* [online]. 2017, vol. 14, no. 7, pp. 523-533. Available from: https://doi.org/10.1080/15459624.2017.1302587.

[7] SEEGER, S., BRÖDNER, D., JACOBI, T., RASCH, F., ROTHHARDT, M., WILKE, O. Emissions of fine and ultrafine particles and volatile organic compounds from different filament materials operated on a low-cost 3D printer. *Gefahrstoffe Reinhaltung der Luft.* 2018, vol. 78, no. 3, pp. 79-87.

[8] STABILE, L., SCUNGIO, M., BUONANNO, G., ARPINO, F., FICCO, G. Airborne particle emission of a commercial 3D printer: the effect of filament material and printing temperature. *Indoor Air.* [online]. 2016, vol. 27, no. 2, pp. 398-408. Available from: https://doi.org/10.1111/ina.12310.

[9] STEINLE, P. Characterization of emissions from a desktop 3D printer and indoor air measurements in office settings. *Journal of Occupational and Environmental Hygiene.* [online]. 2016, vol. 13, no. 2, pp. 121-132. Available from: https://doi.org/10.1080/15459624.2015.1091957.

[10] VANCE, M.E., PEGUES, V., VAN MONTFRANS, S., LENG, W., MARR, L.C. Aerosol Emissions from Fuse-Deposition Modeling 3D Printers in a Chamber and in Real Indoor Environments. *Environmental Science & Technology.* [online]. 2017, vol. 51, no. 17, pp. 9516-9523. Available from: https://doi.org/10.1021/acs.est.7b01546.

[11] ZONTEK, T.L., OGLE, B.R., JANKOVIC, J.T., HOLLENBECK, S.M. An exposure assessment of desktop 3D printing. *Journal of Chemical Health and Safety.* [online]. 2017, vol. 24, no. 2, pp. 15-25. Available from: https://doi.org/10.1016/j.jchas.2016.05.008.

[12] YI, J., LEBOUF, R.F., DULING, M.G., NURKIEWICZ, T., CHEN, B.T., SCHWEGLER-BERRY, D., VIRGI, M.A., STEFANIAK, A.B. Emission of particulate matter from a desktop three-dimensional (3D) printer. *Journal of Toxicology and Environmental Health,* [online]. Part A. 2016, vol. 79, no. 11, pp. 453-465. Available from: https://doi.org/10.1080/15287394.2016.1166467.

[13] FARCAS, M.T., STEFANIAK, A.B., KNEPP, A.K., BOWERS, L., MANDLER, W.K., KASHON, M., QIAN, Y. Acrylonitrile Butadiene Styrene (ABS) and Polycarbonate (PC) Filaments Three-Dimensional (3-D) Printer Emissions-Induced Cell Toxicity. *Toxicology Letters.* [online]. 2019, vol. 15, no. 317, pp. 1-12. Available from: https://doi.org/10.1016/j.toxlet.2019.09.013 .

[14] STEPHENS, B., AZIMI, P., EL ORCH, Z., et al. Ultrafine particle emissions from desktop 3D printers. *Atmospheric Environment.* [online]. 2013, Vol. 79, pp. 334-339. Available from: https://doi.org/10.1016/j.atmosenv.20 13.06.050

[15] KWON, O., YOON, C., HAM, S., PARK, J., LEE, J., YOO, D., KIM, Y. Characterization and Control of Nanoparticle Emission during 3D Printing. *Environmental Science & Technology.* [online]. 2017, vol. 51, no. 18, pp. 10357-10368. Available from: https://doi.org/10.1021/acs.est.7b01454 .

[16] BERNATIKOVA, S., DUDACEK, A., PRICHYSTALOVA, R., KLECKA, V., KOCURKOVA, L. Characterization of Ultrafine Particles and VOCs Emitted from a 3D Printer. *International Journal of Environmental Research and Public Health.* [online]. 2021, vol. 18, no. 3, p. 929. Available from: https://doi.org/10.3390/ijerph18030929.

[17] *Certified methodology for the provision of personal protective equipment in an environment with a risk of nanoparticles* (aimed at employers in addressing the need to assign personal protective equipment for respiratory protection). Prague: Research Institute of Occupational Safety, v. V. I, 2016.

CONTROLLED NANOPARTICLE GROWTH BY VAPOUR CONDENSATION

[1]Tomáš SITEK, [1]Klaudia KÖBÖLOVÁ, [1]Ján POLÁČIK, [1]Stela VALOVIČ

[1]*Faculty of Mechanical Engineering, Brno University of Technology, Brno, Czech Republic, EU,*
Tomas.Sitek@vutbr.cz, Klaudia.Kobolova@vutbr.cz, Jan.Polacik@vutbr.cz, Stela.Valovic@vutbr.cz

https://doi.org/10.37904/nanocon.2021.4369

Abstract

Domestic boilers are generally characterized by higher emissions of airborne dust. A commonly used secondary method of reducing emissions in the energy sector is a cyclone. However, its wider expansion in households is limited by, among other things, the low efficiency of particle capture below 1 micrometre in diameter, and it is these sizes that dominate in the flue gas of domestic heating devices. By sharply lowering the temperature of the flue gas below the dew point of the vapour, it condenses on all available surfaces. This effect could increase the diameter of the particles, which could be separated with higher efficiency. A change in the numerical distribution of the fine particles with a temperature and thus the supersaturation of the flue gas was sought. The flue gas passed through an impinger filled with water and isopropyl alcohol at three different temperature regimes. The impinger also served to capture the condensate, which was then subjected to morphology analysis using an electron microscope and determination of particle distribution in the condensate.

Keywords: Biomass, particulate matter, combustion, SEM, SMPS

1. INTRODUCTION

In the last few decades, in connection with the environmental impact of combustion processes, attention was primarily paid to gas emissions, but it is currently shifting towards fine particle emissions. Fine particles are a small liquid and solid fraction up to 2.5 μm in diameter. In the last few years, the researchers were devoted to submicron and ultrafine particles (up to 0.1 μm), which affect human health significantly. They are so small that the human respiratory tract does not intercept them and they penetrate directly into the alveoli [1,2]. Due to an active surface, ultrafine particles can be much more dangerous than the same mass of larger particles [3,4]. Many researchers mention the necessity of changing the current way of emission assessment [5]. The crucial aspects in relation to the impact on human health are the active surface of the particles [6], and their ability to bind toxic substances [7] and penetrate into the lower respiratory tract.

The technologies for fine particle reduction are different. By using a suitable combustion chamber design and selecting the operating parameters we ensure that all the volatile combustible matter remains at a high temperature for a sufficient time so that it gets combusted. However, we know from practice that in real combustion conditions, a certain amount of non-combusted substances get into the atmosphere. Therefore, we are introducing secondary separation of these particles using suitable methods. In the biomass combusting boilers in power or heating plants, secondary separation of dust particles is a matter of course. On the other hand, small biomass boilers in households do not usually use separation technology and their contribution to the total air pollution is considerable [8].

Thus, it is appropriate to look for ways to minimize the number of particles emitted into the air by small combustion devices as well. However, PM_1 particles (size up to 1 μm) are difficult to separate by the available methods. HEPA filters can separate particles of this size, but the operation of these devices is associated with high-pressure loss, high cost and frequent maintenance requirements due to intensive clogging [9]. From the group of dry separators, an electrostatic precipitator [10] is most often used to capture fine particles. For small combustion plants, however, a major technical complication is to supply the electrostatic precipitator with electricity or provide water for wet separators [11]. A bubble column filled with water is used for wet separation in experimental devices of laboratory scale, as in this study. In a saturated vapour environment, particles may serve as condensation nuclei [12]. In such an environment, these particles grow, and their separation becomes more efficient.

2. EXPERIMENTAL METHODS

An automatic wood pellet boiler EcoScroll ALFA, specially adapted for experimental purposes, was used as the experimental combustion device. The nominal thermal output of the boiler is 25 kW. As a fuel, spruce wood pellets ENplus A1 were burned. The moisture content of used fuel was around 6 %, ash content was 0.3 % of raw biomass. The proximate analysis was conducted by the authors according to the European standards EN ISO 18134-3 and 18122. During the tests, the boiler worked at its nominal output with continuous fuel feed. A sample of flue gas from the boiler at approximately 140 °C was taken using a sampling probe located in the flue gas stack, at a height of 1 m above the outlet of the combustion device (see **Figure 1**). This point was selected in compliance with the standard requirements (EN 13240) for measuring emissions in small combustion devices. Oxygen content in the flue gas fluctuated from 10.0 to 11.5 %, and thus the results were converted to reference oxygen content (10 %).

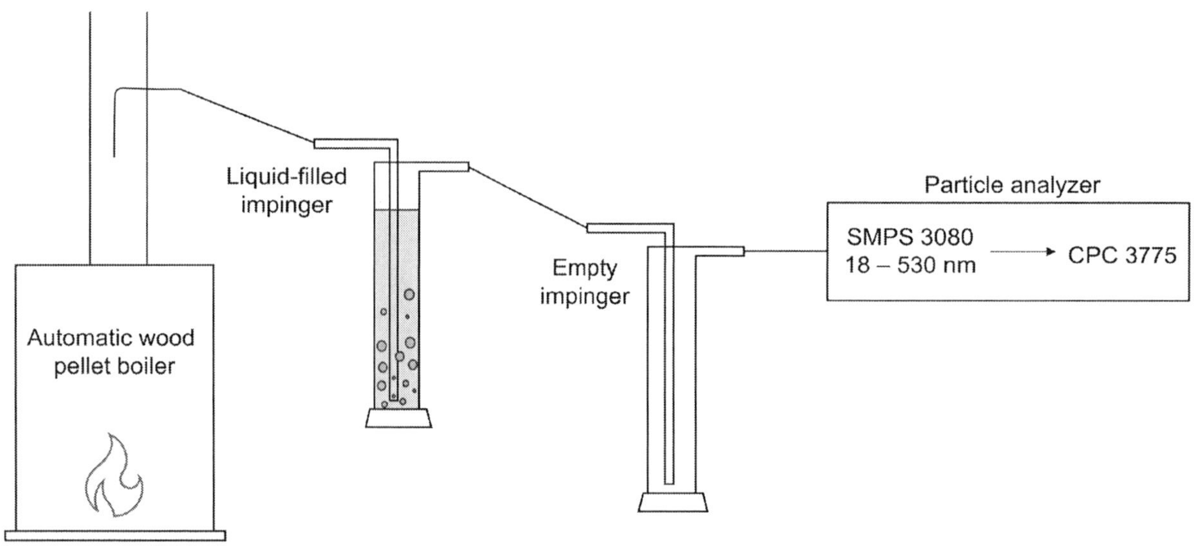

Figure 1 Scheme of experimental setup

The flue gas with dispersed fine particles was carried from the stack through the short tube directly to the first impinger. This 250 ml impinger was filled with 200 ml of water or isopropyl alcohol (so the gas had to pass through approximately 12 cm of liquid) and submerged into the water bath. To cool flue gas down, brine solution (−20 °C at the beginning, but gradually warming up to −8 °C) was used. To warm flue gas up, hot water was used, which progressively cooled down from 85 °C to 55 °C. Due to the very low sampling volume (0.3 litres per minute), it was arduous to directly measure the temperature of the flue gas past the impinger, as it cooled rapidly at the outlet. Nevertheless, we assume the minimal temperature difference between the water bath and flue gas. Subsequently, the gas was carried downwards to second (again 250 ml), empty impinger to collect potential condensate. After that, flue gas was carried directly to the measuring device. Empty impinger without any liquid was used for comparison as well.

For particle concentration and distribution measurements, Scanning mobility particle sizer (SMPS 3080) and Condensation particle counter (CPC 3775) devices by TSI Inc. were used. Unseparated particles with electrical mobility diameter from 18 to 530 nm were monitored for 60 minutes in each setting.

The morphology of the particles was observed using a scanning electron microscope (SEM), particularly the LYRA 3 (Tescan) electron microscope. For the SEM measurement, the sample was covered with a conductive layer of gold with a thickness of 20 nm using a planetary rotation of the samples. A detector of secondary electrons (SE) was used for the measurements. The elemental composition of the particles in the condensate was analysed by the ICP-OES spectrometer ARCOS by Spectro.

3. RESULTS AND DISCUSSION

Initially, the empty impinger setting at room temperature was tested to obtain the boiler's original particulate emissions. Due to the randomness of combustion processes, 60 one-minute measurements were averaged. Results of average particle number concentration are shown in **Figure 2,** the uncertainty is expressed as a single standard deviation.

At room temperature, both water and isopropanol cause a decrease of number concentration by 22 % compared with empty impinger. Different temperature leads to another decrease with ambiguous trends. In the case of water, cooling the gas down is more effective and results in supersaturation and particles with condensed water vapour are collected in the brine solution. On the other hand, hot water creates gas with higher absolute but lower relative humidity. In this case, supersaturation is achieved as soon as flue gas cools down, which is somewhere between the two impingers.

Cold isopropanol does not seem to do much of a difference while warming it is the most effective method regarding particle separation. Hot isopropanol separated 65 % of the original particle number concentration. The temperature of the hot water bath was very close to the boiling point of isopropanol, which caused a considerable amount of isopropanol to evaporate and condensate in the tube leading downwards to the second impinger. The wet inner surface of the tube and condensation of liquid after cooling are probably the main factors.

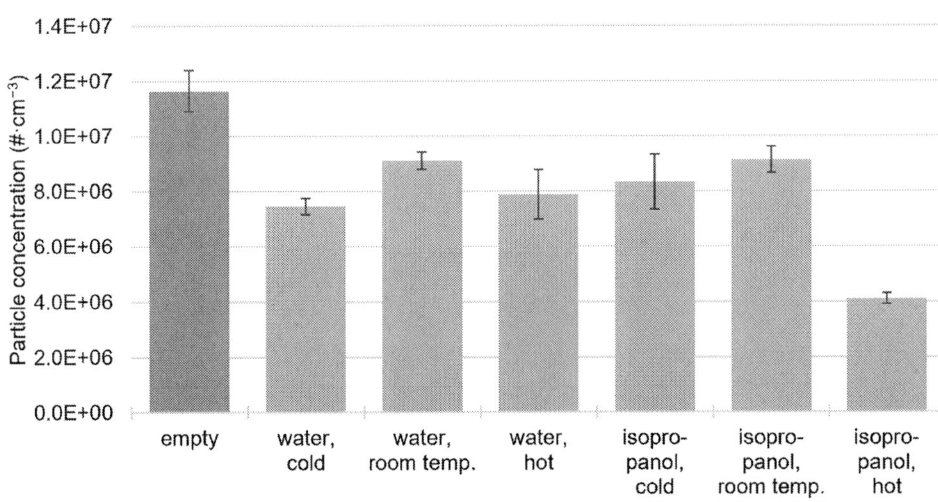

Figure 2 Average particle number concentrations

Particle diameters slightly increased after passing through the water compared to the empty impinger (**Figure 3**). In the cold bath, the particles grew more than in the hot bath or at room temperature conditions; the latter two setups were rather similar.

Isopropanol and water curves at room temperature are almost indistinguishable. Cold bath again caused the biggest increase in size, while the hot bath is the only setup, that affected the distribution noticeably and caused diameters to decrease. When the effect of lower particle concentration is disregarded, it is clearly visible, that much more of the larger particles were separated. There is even a decrease in the number of smallest particles, which were unaffected by other settings. Temperature does not appear to have a significant impact on particle sizes and the size changes are more likely due to the more probable separation of larger particles.

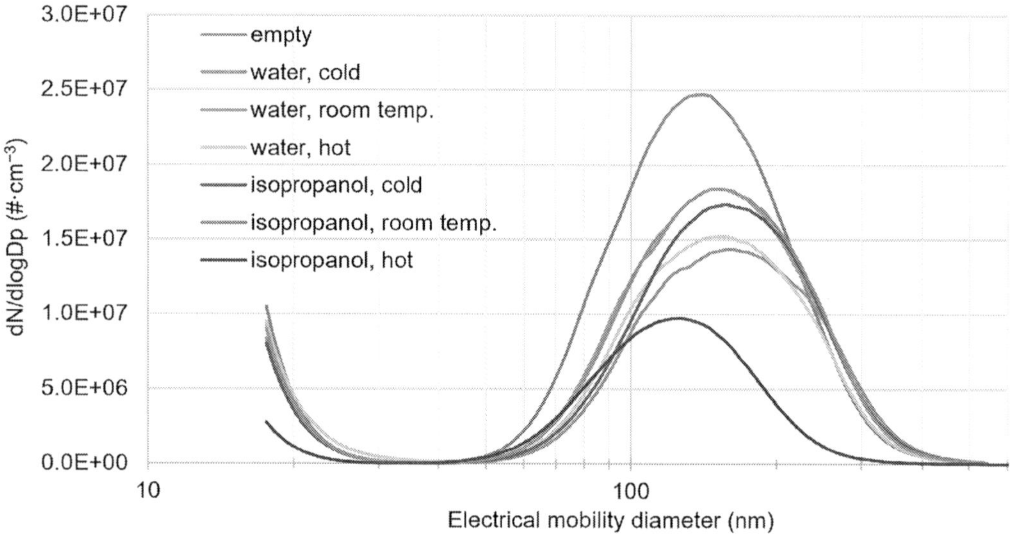

Figure 3 Average particle distributions

Figure 4 Morphology of particles in condensate with a magnification of 138 000×, measuring scale is in the pictures

A sample of the condensate with captured particles from the second impinger was slowly dried and analysed by SEM. There are two types of particles observed in the condensate - spherical to oval nanoparticles and regular spherical macroparticles (see **Figure 4**). The size of the nanoparticles ranges from 45 to 70 nm and that of the spherical macroparticles ranges from 300 to 350 nm. Both types of particles are agglomerated together into a three-dimensional structure, but nanoparticles are sometimes fused together almost beyond recognition. It is not known, how much was the morphology affected by the thickening of the condensate, i.e., how particles looked before being separated. The number of spherical macroscopic particles in the sample is considerably smaller. The exact composition of spherical particles is unknown, but their shape indicates phase change (solidification) in flight.

Concerning metals content, the most dominant was potassium and calcium, in smaller amounts were detected also manganese, magnesium and sodium. These results are no surprise, as they account for most of the inorganic content in used biomass. Note that the used measuring device is unable to detect carbon and gases (e.g., oxygen, nitrogen, and hydrogen), and carbon and oxygen in particular normally constitute a considerable part of particle mass [13].

4. CONCLUSION

The experiments proved that the number of fine particles of 18 to 530 nm in diameter in the flue gas can be reduced by passing the gas through a liquid-filled impinger with significant efficiency. There is almost no difference when brine solution or isopropanol is used as a scrubbing medium at room temperature. Both heating and cooling these liquids results in additional improvement of particle separation. The highest particle separation efficiency of 65 % was achieved with isopropanol heated to temperatures just below its boiling point. Condensation of isopropanol on particles and intense wetting of surfaces caused even the smallest particles to be captured. In a practical operation, however, usage of a flammable and irritant substance is not optimal and usage of regular water is much more promising. Heating water to its boiling point could have a similar effect.

Regarding morphology of particles collected in the condensate, the typical soot structures around 60 nm in diameter were found as well as larger spherical particles with diameter up to 350 nm. By analysing metal content, similar elements as in the original biomass were found, such as potassium and calcium.

The original idea of increasing particle diameters by condensing liquid in their surfaces turned out to be more complicated than expected. Condensation did occur, but too little for any practical application. It could be caused by preferred condensation on surfaces of tubes, which are colder than aerosol particles. If this is the case, usage of larger diameter tubes would be beneficial.

ACKNOWLEDGEMENTS

This paper was supported by TA ČR within the TJ04000415 project of System for Separation of Fine Particles from Flue Gas Stream During Biomass Combustion and Research Centre for Low-Carbon Energy Technologies funded as project No. CZ.02.1.01/0.0/0.0/16_019/0000753 by Czech Republic Operational Programme Research, Development and Education, Priority 1: Strengthening capacity for high-quality research and the collaboration. This research was further funded by projects of Brno University of Technology: FSI-S-20-6280.

REFERENCES

[1] ROM, William N., MARKOWITZ, Steven MD. Environmental and Occupational Medicine. 4. Philadelphia, PA: Lippincott Williams & Wilkins, 2007. ISBN 9780781762991.

[2] U.S. EPA. Air Quality Criteria for Particulate Matter (Final Report, April 1996). U.S. Environmental Protection Agency, Washington, D.C., EPA 600/P-95/001. Available from: http://cfpub.epa.gov/ncea/risk/recordisplay.cfm?deid=2832.

[3] WANG, Kun, WANG, Weiye, LI, Lili et al. Seasonal concentration distribution of PM1.0 and PM2.5 and a risk assessment of bound trace metals in Harbin, China: Effect of the species distribution of heavy metals and heat supply. *Scientific Reports.* [online]. 2020, vol. 10, no. 1. ISSN 2045-2322. Available from: https://doi.org/10.1038/s41598-020-65187-7.

[4] MAINKA, Anna and ZAJUSZ-ZUBEK, Elwira. PM1 in Ambient and Indoor Air-Urban and Rural Areas in the Upper Silesian Region, Poland. *Atmosphere.* [online]. 2019, vol. 10, no. 11. ISSN 2073-4433. Available from: https://doi.org/doi.org/10.3390/atmos10110662.

[5] MERKISZ, Jerzy, PIELECHA, Jacek. The Relationship between Particle Mass and Particle Number. In: *Nanoparticle Emissions From Combustion Engines.* [online]. 2015, pp. 93-108. ISBN 978-3-319-15927-0. Available from: https://doi.org/10.1007/978-3-319-15928-7_8.

[6] PARK, Minhan, JOO, Hung Soo, LEE, Kwangyul, et al. Differential toxicities of fine particulate matters from various sources. *Scientific Reports.* [online]. 2018, vol. 8, no. 1. ISSN 2045-2322. Available from: https://doi.org/10.1038/s41598-018-35398-0.

[7] SCHMID, Otmar, STOEGER Tobias. Surface area is the biologically most effective dose metric for acute nanoparticle toxicity in the lung. *Journal of Aerosol Science.* [online]. 2016, vol. 99, pp. 133-143. ISSN 00218502. Available from: https://doi.org/10.1016/j.jaerosci.2015.12.006.

[8] WEIMER, S., ALFARRA, M. R., SCHREIBER, D., MOHR, M., PRÉVÔT, A. S. H. and BALTENSPERGER, U. Organic aerosol mass spectral signatures from wood-burning emissions: Influence of burning conditions and wood type. *Journal of Geophysical Research.* [online]. 2008, vol. 113, no. D10. ISSN 0148-0227. Available from: https://doi.org/10.1029/2007JD009309.

[9] HWANG, Sungho, ROH, Jaehoon and PARK, Wha Me. Comparison of the relative performance efficiencies of melt-blown and glass fiber filter media for managing fine particles. *Aerosol Science and Technology.* [online]. 2018, vol. 52, no. 4, pp. 451-458. ISSN 0278-6826. Available from: https://doi.org/10.1080/02786826.2017.1423274.

[10] YUAN, Fangyang, YU, Jianfeng and ZHENG, Xiangyang. Experimental study on fine particle separation in a wet electrostatic classifier. *Chemical Engineering and Processing - Process Intensification.* [online]. 2020, vol. 156. ISSN 02552701. Available from: https://doi.org/10.1016/j.cep.2020.108108.

[11] BIANCHINI, Augusto, PELLEGRINI, Marco, ROSSI, Jessica and SACCANI, Cesare. Theoretical model and preliminary design of an innovative wet scrubber for the separation of fine particulate matter produced by biomass combustion in small size boilers. *Biomass and Bioenergy.* [online]. 2018, vol. 116, pp. 60-71. ISSN 09619534 Available from: https://doi.org/10.1016/j.biombioe.2018.05.011.

[12] CUI, Lin, SONG, Xiangda, LI, Yuzhong, WANG, Yang, FENG, Yupeng, YAN, Lifan a DONG, Yong. Synergistic capture of fine particles in wet flue gas through cooling and condensation. *Applied Energy.* [online]. 2018, vol. 225, pp. 656-667. ISSN 03062619. Available from: https://doi.org/10.1016/j.apenergy.2018.04.084.

[13] JONES, Jenny M, LEA-LANGTON, Amanda R., MA, Lin, POURKASHANIAN Mohamed a WILLIAMS, Alan. Pollutants Generated by the Combustion of Solid Biomass Fuels. [online]. 2014. Available from: https://doi.org/10.1007/978-1-4471-6437-1.

TRANSPORT AND INTERACTIONS OF SULFAPYRIDINE IN AGAROSE HYDROGELS ENRICHED BY HUMIC ACIDS

[1]Martina KLUČÁKOVÁ

[1]Brno University of Technology, Faculty of Chemistry, Brno, Czech Republic, EU, klucakova@fch.vutbr.cz

https://doi.org/10.37904/nanocon.2021.4368

Abstract

Knowledge of behaviour of pharmaceuticals as contaminants in nature is essential for their mobility and toxicity in the soils and surface- and ground-water environments. Their migration ability and toxicity can be affected by their interactions with organic matter. In this contribution, sulfapyridine (a sulfanilamide drug used for antibacterial medication) is studied from the point of view of its transport in hydrogels containing humic substances. Agarose hydrogel was used as a model medium for the transport experiments. Standard Elliot soil humic acid (4S102H) purchased from International Humic Substances Society was incorporated into the hydrogel as an active nanomaterial for interactions with drug and potential suppression of their migration and toxicity in soil systems. The method of diffusion couple was applied on the diffusion experiments. Donor part was saturated by sulfapyridine and acceptor part was free of drug. After connecting of both parts, the time development of concentration profiles in the couple was monitored. The effective diffusion coefficients for the transport of sulfapyridine in hydrogel enriched by humic acids and pure agarose hydrogel were determined and compared. It was found that the transport of drug is influenced by interactions with humic acids which can immobilize partially sulfapyridine in humic structure. Simultaneously, sorption/desorption experiments were realized. Their results confirmed high degree of drug immobilization (60 - 90 % in the dependence on drug concentration).

Keywords: Sulfapyridine, humic acids, agarose hydrogel, diffusion, interaction

1. INTRODUCTION

The use of many different pharmaceuticals in human and veterinary medicine and their persistence can result in potential dangerous behaviour in soil and aqueous environments [1-5]. Drugs are not completely eliminated, therefore their residues leave the body and pass into water, sediments and soils [6-12]. The degradation of drugs are very different in the dependence on their chemical character and physicochemical circumstances. An important role is attributed to soil properties and quality of organic matter. The migration ability and toxicity of pharmaceuticals in nature can be significantly affected by their interactions with soil components.

Humic substances as important constituent of soil organic matter can form complexes with drugs of different strengths therefore strongly bond, ion exchangeable and extractable fractions of complexes co-exist in nature [5,13,14]. The binding character varies between electrostatic interactions, hydrogen bonds and hydrophobic interactions [5,13-18]. Sulfonamides are one of the most commonly used antibiotic in human and veterinary medicine. Their overuse and misuse led to their widespread occurrence in nature [19-21], e.g. their representative - sulfapyridine - is one of the most frequently antibiotics detected in wastewater effluents and surface waters [22-24]. They reach agricultural soils via contaminated sewage sludge, wastewater, and manure used for fertilization [25-27]. The occurrence, fate, ecotoxicity and uptake of antibiotics by edible crops were described in detail in thorough reviews [28-30]. Studies on interactions of, soil, organic matter and humic substances with antibiotics are focused mainly on adsorption processes [31-35]. Chen et al. [31] studied the adsorption of sulfamethoxazole and sulfapyridine on peat soil and composted manure. They resulted that content of organic carbon greatly enhanced their adsorption on the soil and suggested the hydrophobic partition as the major mechanism. The pyridine group was mainly responsible for the adsorption at high pH values (in the case of sulfapyridine). Haham et al. [32] investigated sorption-desorption behavior of sulfapyridine with three different soils. The showed important roles of soil mineral matrices and dissolved organic matter. Similarly, the important effect of dissolved organic matter on the adsorption of sulfonamides on graphene oxides was observed by Liu et al. [33]. Thiele-Bruhn et al [34] studied sorption of sulfonamides on whole soils and their particle-size fractions. According to their results, the adsorption increased with aromaticity and electronegativity of functional groups attached to the sulfonyl-phenyl-amine core. Modeling and molecular mechanics calculations showed preferred site-specific sorption via hydrogen bonds and van der Waals interactions. Thiele [35] investigated the adsorption of sulfapyridine on Chernozem. They observed that the adsorption was substantially higher in moist soil in comparison with dried sample. They concluded also that the N-heterocycle significantly contributed to the adsorption of sulfapyridine.

Zhi et al. [36] stated that traditional adsorption models (Langmuir, Freundlich) cannot exhibit their behavior under different influencing conditions. Thus, they are not able to reflect the precise adsorption mechanism of antibiotics in soil. Schwarz et al. [25] concluded that the mechanism of immobilization of sulfapyridine varied from reversible sorption to the formation of non-extractable bound residues. The character of interactions, immobilization degree and mobility of antibiotics in nature thus are significantly influences by soil properties and character of organic matter [31,34,37].

Our approach is different. The contribution is focused on the diffusivity of sulfapyridine in model hydrogel matrix and the effect of humic substances incorporated in hydrogel on the migration of the antibiotics. The study is based on our previous experiences with diffusion experiments of different pollutants in hydrogel matrices containing humic substances [38,39] and the effect of the interactions on the mobility [40-43]. The method of diffusion couple [38] was chosen for the investigation of humic-drug interactions effect on the mobility of sulfapyridine in model hydrogel based on agarose enriched by humic acids [39,41].

2. MATERIALS AND METHODS

Sulfapyridine (as sulfonamide antibotics) and agarose (routine use class) were purchased from Sigma-Aldrich. Elliot soil humic acids (4S102H) were purchased from the International Humic Substances Society. The main characteristics of humic acids (e.g. elemental composition or the contents and properties of acidic functional groups) can be found on the website of the International Humic Substances Society.

The preparation of hydrogels was based on the thermo-reversible gelation of agarose solution. It was dissolved in deionized water (1 % wt.), heated at 80 °C and stirred, and finally sonicated to remove gasses. Afterwards, the solution was slowly poured into the PMMA spectrophotometric cuvette. Its orifice was immediately covered with pre-heated plate of glass to prevent drying and shrinking of gel. Flat surface of the boundary of resulting hydrogels was provided by wiping an excess solution away. Gentle cooling of cuvettes at the laboratory temperature led to the gradual gelation of the mixture. Hydrogels enriched by humic acids were prepared from 1 % wt. agarose solution containing 0.01 % wt. of humic acids. Donor hydrogels were prepared by means of agarose (1 % wt.) dissolved in 1 mM sulfapyridine solution.

The diffusion couple was assembled from two parts: donor and acceptor hydrogels. Donor hydrogel was formed from agarose containing sulfapyridine as diffusing drug particles. Acceptor hydrogels were based on agarose or agarose enriched by humic acids without antibiotics (their initial concentrations in acceptor hydrogels were equal to zero).

The distributions of sulfapyridine in donor and acceptor hydrogels were determined in selected time intervals. The cuvettes were disconnected and the UV-VIS spectra were measured at various distances from the orifice by means of Varian Cary 50 UV-VIS spectrophotometer equipped with the special accessory providing controlled fine vertical movement of the cuvette in the spectrophotometer. Using the collected UV-VIS spectra, the concentrations of drug were determined at different positions in hydrogels. The obtained data were used to compute the concentration profiles of drug in the donor and acceptor parts of diffusion couple. The diffusion fluxes in given times were determined as the total contents of sulfapyridine in acceptor hydrogel (normalized on the area of interface). The content of drug in acceptor hydrogel at given time should be the same as the decrease in drug content in the donor part of diffusion couple.

Sorption experiments were realized with sulfapyridine solutions with concentration range 1-10 mg.dm^{-3}. Solid humic hamic (in form of powder) were mixed with the drug solution (1 g / 50 cm^3) and stirred. After equilibration (48 h), the powder was centrifuged and mixed with deionized water in order to determine the amount of extractable (mobile) phase of sulfapyridine. The decrease in concentration of drug solutions as well as the concentration of leachates were determined by means of UV/VIS spectrometry (see above).

All experiments were triplicated and performed at laboratory temperature (25 ± 1 °C). Data are presented as average values with standard deviation bars.

3. RESULTS AND DISCUSSION

In this work, the effect of humic-drug interaction on the diffusivity of sulfapyridine in agarose hydrogels were studied. In **Figure 1** the example of experimental data obtained after 4 days are shown. We can see that the crossing of drug particles from donor hydrogel into acceptor hydrogels with and without of humic acids are different. As assumed, the presence of humic acids as active substance can cause interactions between their binding sites and sulfapyridine which can result in a partial immobilization of drug particles. The effect of humic-drug interactions should be more noticeable for longer times of diffusion. The mathematical description of realized experiments were based on second Fick's law valid for non-stationary diffusion [38,40-43]. In the case of diffusion couple, the boundary conditions of semi-infinite mediums and equality between the decrease in drug content in donor hydrogel and the increase in drug content in acceptor hydrogel in given time were applied. The concentration at interface between both hydrogels was equal to the half of initial donor concentration. The initial boundary condition was the homogeneous initial concentration of drug in donor hydrogel (c_0) and zero initial concentrations of sulfapyridine in acceptor hydrogels.

The mathematical solution of the second Fick's law is

$$c(x,t) = \frac{1}{2} c_0 erfc \ \frac{x}{\sqrt{4D_{\mathrm{ef}}t}} \quad (1)$$

and the equation for the total diffusion flux m_t which goes through the interface between donor and acceptor hydrogels (x = 0) in time t can be expressed as

$$m_t = c_0 \sqrt{\frac{D_{\mathrm{ef}}t}{\pi}} \quad (2)$$

The effective diffusion coefficient D_{ef} is the main parameter characterizing rate of the diffusion process. It allows for two main factors: the tortuous movement of the diffusing particles in the porous structure of hydrogel and interactions of diffusing particles with active components incorporated in hydrogels. If we assume a local equilibrium between free mobile drug and immobilized particles, its equilibrium constant K can be expressed as the radio between concentrations of immobilized and free sulfapyridine and involved into the value of D_{ef}:

$$D_{ef} = \frac{D_0}{1+K} \quad (3)$$

Diffusion coefficients for of drug for pure agarose hydrogel and hydrogel enriched by humic acids were determined by means of equation 2. In this contribution, D_0 thus represents the diffusivity of drug particles in non-reactive agarose hydrogel (without humic acids) and D_{ef} involves the influence of interactions. Their values were calculated as D_0 = 1.01×10^{-10} m^2.s^{-1} and D_{ef} = 9.85×10^{-11} m^2.s^{-1}. The equilibrium constant K is then (according to equation 3) equal to 0.37.

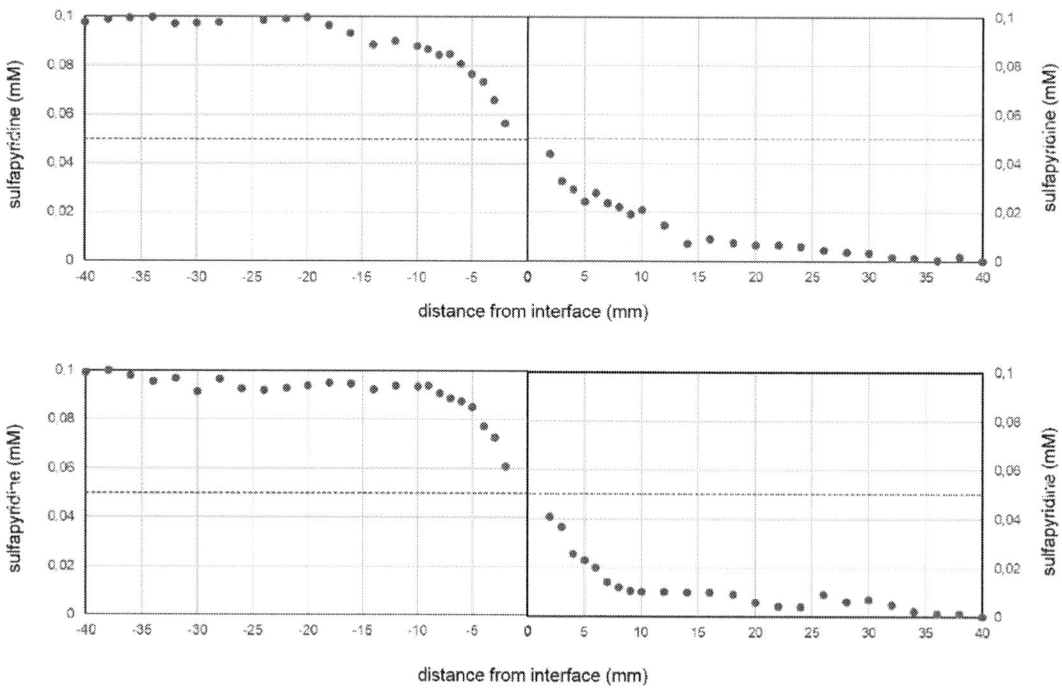

Figure 1 The concentration profiles of sulfapyridine in the diffusion couple formed from donor hydrogel and acceptor hydrogels based on pure agarose (blue) and agarose enriched by humic acids (green) after 4 days. Dashed lines (red) represent theoretical homogeneous concentration of sulfapyridine after attainment of equilibrium.

Adsorption/desorption experiments showed that humic acids are able to immobilize sulfapyridine with adsorption capacity 0.28 mg.g^{-1} (based on Langmuir adsorption isotherm [44]). The desorption experiments showed that the content of mobile fraction increased from 16.4 % (for initial concentration of 1 mg.dm^{-3}) up to 38.5 % (for 10 mg.dm^{-3}). It means that humic acids were able to strongly bind a majority of drug and its percentage decreased as the capacity of humic acids were gradually saturated.

4. CONCLUSION

In this work, the diffusivity of sulfapyridine in model hydrogels was studied. The diffusion couple was formed by donor agarose hydrogel with incorporated drug particles and acceptor hydrogel with initially zero drug concentration. Two types of acceptor hydrogel (with and without humic acids) were used in order to investigate the influence of humic-drug interactions on the mobility of sulfapyridine. It was found that drug particles were partially immobilized by the interactions which resulted in lower value of effective diffusion coefficient. The percentage of non-leachable drug fraction achieved 60-80 % in the dependence on the total drug content.

REFERENCES

[1] RICHARDSON, S.D., TERNES, T.A. Water analysis: emerging contaminants and current issues. *Analytical Chemistry.* [online]. 2005, vol. 77, pp. 3807-3838. Available from: https://doi.org/10.1021/acs.analchem.7b04577.

[2] FENT, K., WESTON, A.A., CAMINADA, D. Ecotoxicology of human pharmaceuticals. *Aquatic Toxicology.* [online]. 2006, vol. 76, pp. 122.159. Available from: https://doi.org/10.1016/j.aquatox.2005.09.009.

[3] SCHLÜSENER, M.P., BESTER, K. Persistence of antibiotics such as macrolides, tiamulin and salinomycin in soil. *Environmental Pollution.* [online]. 2006, vol. 143, no. 3, pp. 565-571. Available from: https://doi.org/10.1016/j.envpol.2005.10.049.

[4] CUNNINGHAM, V.L., BINKS, S.P., OLSON, M.J. Human health risk assessment from a presence of human pharmaceuticals in the aquatic environment. *Regulatory Toxicology and Pharmacology.* [online]. 2009, vol. 53, pp. 39-45. Available from: https://doi.org/10.1016/j.yrtph.2008.10.006.

[5] KLAVINS, M., ANSONE, L., PURMALIS, O., ZICMANIS, A. Characterization of interaction between tricyclic structures containing pharmaceuticals, their model and humic substances. *Water Science & Technology.* [online]. 2011, vol. 63, pp. 845-852. Available from: https://doi.org/10.2166/wst.2011.119.

[6] SACHER, F., LANGE, F., BRAUCH, H., BLANKENHORN, I. Pharmaceuticals in groundwaters: Analytical methods and results of a monitoring program in Baden-Württemberg, Germany. *Journal of Chromatography A.* [online]. 2001, vol. 938, pp. 199-210. Available from: https://doi.org/10.1016/S0021-9673(01)01266-3.

[7] JUX, U., BAGINSKI, R., ARNOLD, H.G., KRÖNKE, M., SENG P.N. Detection of pharmaceutical contaminations of river, pond, and tap water from Cologne (Germany) and surroundings. *International Journal of Hygiene and Environmental Health.* [online]. 2002, vol. 205, pp. 393-398. Available from: https://doi.org/10.1078/1438-4639-00166.

[8] KIM, S.C., CARLSON, K. LC-MS2 for quantifying trace amounts of pharmaceutical compounds in soil and sediment matrices. *Trends in Analytical Chemistry.* [online]. 2005, vol. 24, pp. 635-644. Available from: https://doi.org/10.1016/j.trac.2005.04.006.

[9] SIM, W., LEE, J., OH, J. Occurrence and fate of pharmaceuticals in wastewater treatment plants and rivers in Korea. *Environmental Pollution.* [online]. 2010, vol. 158, pp. 1938-1947. Available from: https://doi.org/10.1016/j.envpol.2009.10.036.

[10] ÁLVAREZ-RUIZ, R., ANDRÉS-COSTA, J.M., ANDREU, V., PICÓ, Y. Simultaneous determination of traditional and emerging illicit drugs in sediments, sludges and particulate matter. *Journal of Chromatography A.* [online]. 2015, vol. 1405, pp. 103-115. Available from: https://doi.org/10.1016/j.chroma.2015.05.062.

[11] KOBA, O., GOLOVKO, O., KODEŠOVÁ, R., FÉR, M., GRABIC, R. Antibiotics degradation in soil: A case of clindamycin, trimethoprim, sulfamethoxazole and their transformation products. *Environmental Pollution.* [online]. 2017, vol. 220, pp. 1251-1263. Available from: https://doi.org/10.1016/j.envpol.2016.11.007.

[12] MONTEIRO, S.C., BOXALL, A.B.A. Factors affecting the degradation of pharmaceuticals in agricultural soils. *Environmental Toxicology and Chemistry.* [online]. 2009, vol. 28, pp. 2546-2554. Available from: https://doi.org/10.1897/08-657.1.

[13] KHIL´KO, S.L., SEMENOVA, R.G. Interaction of humic acid salts with drug preparations. *Solid Fuel Chemistry.* [online]. 2016, vol. 50, pp. 60-64. Available from: https://doi.org/10.3103/S0361521916060057.

[14] LIN, K., GAN, J. Sorption and degradation of wastewater-associated non-steroidal anti-inflammatory drugs and antibiotics in soils. *Chemosphere.* [online]. 2011, vol. 83, pp. 240-246. Available from: https://doi.org/10.1016/j.chemosphere.2010.12.083.

[15] MARGON, A., PASTRELLO, A., MOSETTI, D., CANTONE, P., LEITA, L. Interaction between diclofenac and soil humic acids. *Soil and Sediment Contamination.* [online]. 2009, vol. 18, pp. 489-496. Available from: https://doi.org/10.1080/15320380902962353.

[16] MARTÍNEZ-MEJÍA, M. J., SATO, I., RATH, S. Sorption mechanism of enrofloxacin on humic acids extracted from Brazilian soils. *Environmental Science and Pollution Research.* [online]. 2017, vol. 24, pp. 15995-16006. Available from: https://doi.org/10.1007/s11356-017-9210-3.

[17] LIU, X., LU, S., LIU, Y., MENG, W., ZHENG, B. Adsorption of sulfamethoxazole (SMZ) and ciprofloxacin (CIP) by humic acid (HA): characteristics and mechanism. *RCS Advances.* [online]. 2017, vol. 7, pp. 50449-50458. Available from: https://doi.org/10.1039/C7RA06231A.

[18] PARK, J., CHO, K.H., LEE, E., LEE, S., CHO, J. Sorption of pharmaceuticals to soil organic matter in a constructed wetland by electrostatic interaction. *Science of the Total Environment.* [online]. 2018, vol. 435, pp. 1345-1350. Available from: https://doi.org/10.1016/j.scitotenv.2018.04.212.

[19] KÜMMERER, K. Antibiotics in the aquatic environment - A review – Part I. *Chemosphere.* [online]. 2009, vol. 75, pp. 417-434. Available from: https://doi.org/10.1016/j.chemosphere.2008.11.086.

[20] MANZETTI, S., GHISI, R. The environmental release and fate of antibiotics. *Mar. Pollut. Bull.* [online]. 2014, vol. 79, pp. 7-15. Available from: https://doi.org/10.1016/j.marpolbul.2014.01.005.

[21] MARTÍNEZ, J.L. Antibiotics and antibiotic resistance genes in natural environments. *Science* [online]. 2008, vol. 321, pp. 365-367. Available from: https://doi.org/10.1126/science.1159483.

[22] GARCÍA-GALÁN, M.J., VILLAGRASA, M., DIAZ-CRUZ, M.S., BARCELO, D. LC-QqLIT MS analysis of nine sulfonamides and one of their acetylated metabolites in the Llobregat River basin. Quantitative determination and qualitative evaluation by IDA experiments. *Anal. Bioanal. Chem.* [online]. 2010, vol. 397, pp. 1325-1334. Available from: https://doi.org/10.1007/s00216-010-3630-y.

[23] PENG, X., WANG, Z., KUANG, W., TAN, J., LI, K. A preliminary study on the occurrence and behavior of sulfonamides, ofloxacin and chloramphenicol antimicrobials in wastewaters of two sewage treatment plants in Guangzhou, China. *Sci. Total Environ.* [online]. 2006, vol. 371, pp. 314-322. Available from: https://doi.org/10.1016/j.scitotenv.2006.07.001.

[24] LIU, H., ZHANG, B., LI, Y., FANG, Q., HOU, Z., TIAN, S., GU, J. Effect of radical species and operating parameters on the degradation of sulfapyridine using a UV/chlorine system. *Ind. Eng. Chem. Res.* [online]. 2020, vol. 59, pp. 1505-1516. Available from: https://doi.org/10.1021/acs.iecr.9b06228.

[25] SCHWARZ, J., KNICKER, H., SCHAUMANN, G.E., THIELE-BRUHN, S. Enzymatic transformation and bonding of sulfonamide antibiotics to model humic substances. *J. Chem.* [online]. 2015, vol. 2015, ID 892708. Available from: https://doi.org/10.1155/2015/829708.

[26] BOXALL, A.B., FOGG, L.A., BLACKWELL, P.A., KAY, P., PEMBERTON, E.J., CROXFORD, A. Veterinary medicines in the environment. *Rev. Environ. Contam. Toxicol.* [online]. 2004, vol. 180, pp. 1-91. Available from: https://doi.org/10.1007/0-387-21729-0_1.

[27] KIM, K.R., OWENS, G., KWON, S.I., SO, K.H., LEE, D.B., OK, Y.S. Occurrence and environmental fate of veterinary antibiotics in the terrestrial environment. *Water Air Soil Pollut.* [online]. 2011, vol. 214, pp. 163-174. Available from: https://doi.org/10.1007/s11270-010-0412-2.

[28] PAN, M., CHU, L.M. Fate of antibiotics in soil and their uptake by edible crops. *Sci. Total Environ.* [online]. 2017, vol. 599-600, pp. 500-512. Available from: https://doi.org/10.1016/j.scitotenv.2017.04.214.

[29] DU, L. LIU, W. Occurrence, fate, and ecotoxicity of antibiotics in agro-ecosystems. A review. *Agron. Sustain. Dev.* [online]. 2012, vol. 32, pp. 309-327. Available from: https://doi.org/10.1007/s13593-011-0062-9.

[30] WANG, S., WANG, H. Adsorption behavior of antibiotic in soil environment: a critical review. *Front. Environ. Sci. Eng.* [online]. 2015, vol. 9, pp. 565-574. Available from: https://doi.org/10.1007/s11783-015-0801-2.

[31] CHEN, K.L., LIU, L.C., CHEN, W.R. Adsorption of sulfamethoxazole and sulfapyridine antibiotics in high organic content soils. *Environ. Pollut.* [online]. 2017, vol. 231, pp. 1163-1171. Available from: https://doi.org/10.1016/j.envpol.2017.08.011.

[32] HAHAM, H., OREN, A., CHEFETZ, B. Insight into the role of dissolved organic matter in sorption of sulfapyridine by remiarid Soils. *Environ. Sci. Technol.* [online]. 2012, vol. 46, pp. 11870-11877. Available from: https://doi.org/10.1021/es303189f.

[33] LIU, F.F., ZHAO, J., WANG, S., XING, B. Adsorption of sulfonamides on reduced graphene oxides as affected by pH and dissolved organic matter. *Environ. Pollut.* [online]. 2016, vol. 210, pp. 85-93. Available from: https://doi.org/10.1016/j.envpol.2015.11.053.

[34] THIELE-BRUHN, S., SEIBICK, T., SCHULTEN, H.R., LEINWEBER, P. Sorption of sulfonamide pharmaceutical antibiotics on whole soils and particle-size fractions. *J. Environ. Qual.* [online]. 2004, vol. 33, pp. 1331-1342. Available from: https://doi.org/10.2134/jeq2004.1331.

[35] THIELE, S. Adsorption of the antibiotic pharmaceutical compound sulfapyridine by a long-term differently fertilized loess Chernozem. *J. Plant Nutr. Soil Sci.* [online]. 2000, vol. 163, pp. 589-594. Available from: https://doi.org/10.1002/1522-2624(200012)163:6<589::AID-JPLN589>3.0.CO;2-5.

[36] ZHI, D., YANG, D., ZHENG, Y., YANG, Y., HE, Y., LUO, L., ZHOU, Y. Current progress in the adsorption, transport and biodegradation of antibiotics in soil. *J. Environ. Manage.* [online]. 2019, vol. 251, ID 109598. Available from: https://doi.org/10.1016/j.jenvman.2019.109598.

[37] BIALK, H.M., PEDERSEN, J.A. NMR investigation of enzymatic coupling of sulfonamide antimicrobials with humic substances. *Environ. Sci. Technol.* [online]. 2008, vol. 42, pp. 106-112. Available from: https://doi.org/10.1021/es070779d.

[38] KLUČÁKOVÁ, M., PEKAŘ, M. Transport of copper(II) ions in humic gel - New results from diffusion couple. *Colloid. Surface. A* [online]. 2009, vol. 349, pp. 96-101. Available from: https://doi.org/10.1016/J.COLSURFA.2009.08.001

[39] SEDLÁČEK, P., SMILEK, J., KLUČÁKOVÁ, M. How Interactions with polyelectrolytes affect the mobility of ionic dyes in hydrogels - Results from diffusion cells. *React. Funct. Polym.* [online]. 2013, vol. 73, pp. 1500-1509. Available from: https://doi.org/10.1016/j.reactfunctpolym.2013.07.008.

[40] KLUČÁKOVÁ, M., KALINA, M., SMILEK, J., LAŠTŮVKOVÁ, M. The transport of metal ions in hydrogels containing humic acids as active complexation agent. *Colloid. Surface. A* [online]. 2018, vol. 557, pp. 116-122. Available from: https://doi.org/10.1016/j.colsurfa.2018.02.042.

[41] KLUČÁKOVÁ, M. Agarose hydrogels enriched by humic acids as complexation agent. *Polymers* [online]. 2020, vol., 12, pp. 687-698. Available from: https://doi.org/10.3390/polym12030687.

[42] KLUČÁKOVÁ, M., KALINA, M. Diffusivity of Cu(II) ions in humic gels - influence of reactive functional groups of humic acids. *Colloid. Surface. A* [online]. 2015, vol. 483, pp. 162-170. Available from: https://doi.org/10.1016/j.colsurfa.2015.05.041.

[43] KLUČÁKOVÁ, M., KALINA, M., SEDLÁČEK, P., GRASSET, L. Reactivity and transport mapping of Cu(II) ions in humic hydrogels. *J. Soil. Sediment.* [online]. 2014, vol. 14, pp. 368-376. Available from: https://doi.org/10.1007/s11368-013-0730-2.

[44] KLUČÁKOVÁ, M., PEKAŘ, M. New model for equilibrium sorption of metal ions on solid humic acids. *Colloid. Surface. A* [online]. 2006, vol. 286, pp. 126-133. Available from: https://doi.org/10.1016/J.COLSURFA.2006.03.013.

SESSION E
ADVANCED METHODS OF PREPARATION AND CHARACTERIZATION OF NANOMATERIALS

Chairmen

Assoc. Prof. Dr. Pavel JELINEK	Institute of Physics of the CAS, Prague, Czech Republic, EU
Dr. Petr KLAPETEK	Czech Metrology Institute, Brno, Czech Republic, EU

PLASMA-SYNTHESISED ZINC OXIDE NANOPARTICLE BEHAVIOR IN LIQUIDS

[1*]David RUTHERFORD, [1]Jaroslav JÍRA, [2]Kateřina KOLÁŘOVÁ, [3]Iva MATOLÍNOVÁ, [2]Zdenek REMEŠ, [1]Jaroslav KULIČEK, [4]Dilli PADMANABAN, [4]Paul MAGUIRE, [4]Davide MARIOTTI, [1]Bohuslav REZEK

[1]*Department of Physics, Czech Technical University, Prague, Czech Republic, EU, *ruthedav@fel.cvut.cz*

[2]*Institute of Physics, Czech Academy of Sciences, Prague, Czech Republic, EU*

[3]*Department of Surface and Plasma Science, Charles University, Prague, Czech Republic, EU*

[4]*School of Engineering, Ulster University, Belfast, N. Ireland*

https://doi.org/10.37904/nanocon.2021.4318

Abstract

Zinc oxide nanoparticles have been synthesized using non-thermal atmospheric pressure plasma (ZnO-NTP). We investigated the behavior of these ligand-free as a colloid suspension using different solvents, from deionized water to physiological saline and microbial culture broth. We found that the zeta potential of ZnO-NTP became more negative after exposure to microbial culture broth relative to water, which suggests increased colloid stability. Photoluminescence spectra of ZnO-NTP were similar regardless of liquid type, yet optical and fluorescent images of samples showed different agglomeration behavior depending on liquid type. Scanning electron microscopy images revealed large agglomerates of ZnO-NTP interacting with the surface of bacteria cells, ranging in size from 200 nm up to 2 µm. We also studied effect of sub-lethal concentrations of ZnO-NTP on bacteria under illumination. There was no significant difference in viable bacteria concentration after 24h exposure to 10 µg/mL ZnO-NTP relative to untreated control irrespective of sample illumination.

Keywords: Zinc oxide, nanotechnology, colloid suspension, bacteria

1. INTRODUCTION

Non-thermal plasma (NTP) have emerged as a promising technology for the synthesis of highly crystalline nanoparticles [1,2]. Commercially available nanoparticles produced on an industrial scale can result in the introduction of unwanted impurities that can negatively impact their integration into downstream applications. For example, according to the material safety data sheet for spherical zinc oxide nanoparticles (ZnO NP), there may be traces of aluminum oxide present [3]. NTP can be described as a green synthesis technique, in that no toxic chemicals are required in the generation of ZnO NP. There are a number of examples where NTP have been used to synthesize ZnO NP [4-7]. However, their interaction with complex liquids of biological relevance remains yet to be fully explored.

Nanoparticles are greatly influenced by their surrounding environment due to the high surface area to volume ratio making them highly responsive to the water chemistry of the solvent when in a colloidal suspension. Thus, their behavior in the colloid as well as inherent properties can be suppressed or enhanced by surface interaction with other molecules that are also present in the liquid such as proteins or carbohydrates. Our prior computer simulations showed that the interaction between amino acids and zinc oxide surface involved covalent bonding and not merely due to passive adsorption [8].

Here, we synthesize ZnO nanoparticles by using non-thermal atmospheric-pressure plasma (ZnO-NTP) and characterize its material properties under specific conditions. The nanoparticles are then introduced in liquids of varying ionic and biomolecule content and we measure the resulting optical and colloidal and microbiological effects.

2. MATERIALS AND METHODS

2.1. Nanoparticle synthesis

ZnO nanoparticles were synthesized using a non-thermal plasma reactor (**Figure 1 A**) consisting of a hollow stainless steel housing with a quartz tube (1 mm outer diameter, 0.7 mm inner diameter) and a pure zinc wire (0.5 mm, 99.99%, Alfa Aesar) positioned inside the housing attached to ground. The wire along with the quartz tube was placed inside a copper plate which was connected to a radio-frequency (40W, 13.56 MHz) power source connected using a matching network. A mass flow controller delivered helium (He, 0.3 slm) into the housing and a gas-phase microplasma ignited within the quartz tube and interacted with the zinc wire. Generated nanoparticles (ZnO-NTP) were transported away from the microplasma by the gas flow and collected downstream on a clean Si (intrinsic) substrate at a rate of 0.3 mg/s. The dry powder was then transferred to a clean vial and used for characterization analyses.

2.2. Nanoparticle characterization

The crystallographic profile of the synthesized ZnO-NTP was explored by high-resolution transmission electron microscope (TEM, Jeol JEM02100 F) coupled to selected area electron diffraction (SAED) to determine particle morphology, phase and size distribution after the synthesis. The transmission microscopy images and SAED patterns were analyzed using ImageJ (1.48v) software for particle size distribution and d-spacing calculation. ZnO-NTP optical properties were probed using UV-visible spectroscopy (UV-vis, PerkinElmer LAMDA 365), and photoluminescence (PL, WITec, alpha300 RAS) after exposure to liquids: demineralized water (dH_2O, Resta), 0.9% sodium chloride (NaCl, Penta) and Mueller Hinton broth (MHB, Oxoid). The size distribution and zeta potential of ZnO-NTP after exposure to different liquids was analyzed by dynamic light scattering (ZetaSizer, Malvern).

A stock solution of ZnO-NTP and dilutions thereof were made using demineralized H_2O (<1 µS/cm) and subjected to sonication prior to analysis (160 W for 30 min, DT31, Sonorex Digitec). For PL and zeta potential measurements, ZnO-NTP were first removed from dH_2O by centrifugation (30min at 13k rpm, MPW-150R) followed by resuspending in the same volume of a different liquid type. ZnO-NTP were subsequently vortexed in either dH_2O, 0.9 % NaCl or MHB (8000 rpm, 1 min) before sonication under the same conditions previously used. The liquid was removed by centrifugation again and ZnO-NTP were re-suspended in dH_2O for a final round of sonication before analysis.

2.3. Nanoparticle interaction with bacteria

Escherichia coli (*E. coli*, CCM 3954, Brno) were resuscitated on Mueller Hinton agar (MHA, Carl Roth) plates and cultivated overnight (37 °C). The following day, 1 colony was removed and resuspended in 100 mL MHB and placed on an orbital shaker located inside an incubator (150 rpm for 24 h at 37 °C). Next, the bacteria were adjusted to McFarland's Density 1.0 in fresh MHB using a densitometer (Biosan), then subsequently diluted 1:1000 in fresh MHB to an approximate concentration of $1x10^5$ cfu/mL. 1 mL of ZnO-NTP in dH_2O (100 µg/mL) was added to 9 mL *E. coli* in a 50 mL Falcon tube and placed inside reverse spin bioreactors (2000 rpm/1s, RTS-1, BioSan) with and without illumination (Reflekta). The white light source was selected based on overlapping spectral output and UV-vis absorbance profile of ZnO-NTP (i.e. between 400-500 nm). The concentration of viable bacteria was measured at the start and again after 24 hours by performing a 1:10 dilution series using 0.9% NaCl and inoculating onto MHA before incubating (37° C for 24 h). 100 µL of the bacteria-ZnO-NTP mixture was removed after 24 h and diluted 1:100 in dH_2O then 10 µL was drop-casted onto silicon wafer for SEM analysis (MIRA III, Tescan). It was not possible to image ZnO-NTP in 0.9% NaCl or MHB using SEM due considerable crystallization of salts and other components on the substrate.

3. RESULTS

3.1. Morphology, UV-vis and phase

Morphology of the synthesized ZnO-NTP collected in ethanol was probed using TEM and the images showed uniform grain-like shaped particles (**Figure 1 (B)**). One can see that the particles are well dispersed with individual primary particles clearly visible. Analysis of TEM images revealed the mean particle size was 18.6 ± 6.5 nm and there was a log normal distribution with regards to particle diameter and number (**Figure 1 (C)**). **Figure 1 (D)** shows a Tauc plot derived from UV-vis absorbance data of ZnO-NTP in ethanol and the energy band gap (Eg) calculated to be approximately 3.22 eV. **Figure 1 (E)** shows the SAED profile of ZnO-NTP that displays a hexagonal wurzite phase (space group $P_{63}mc$, indexed based on JCPDF file # 65-3411).

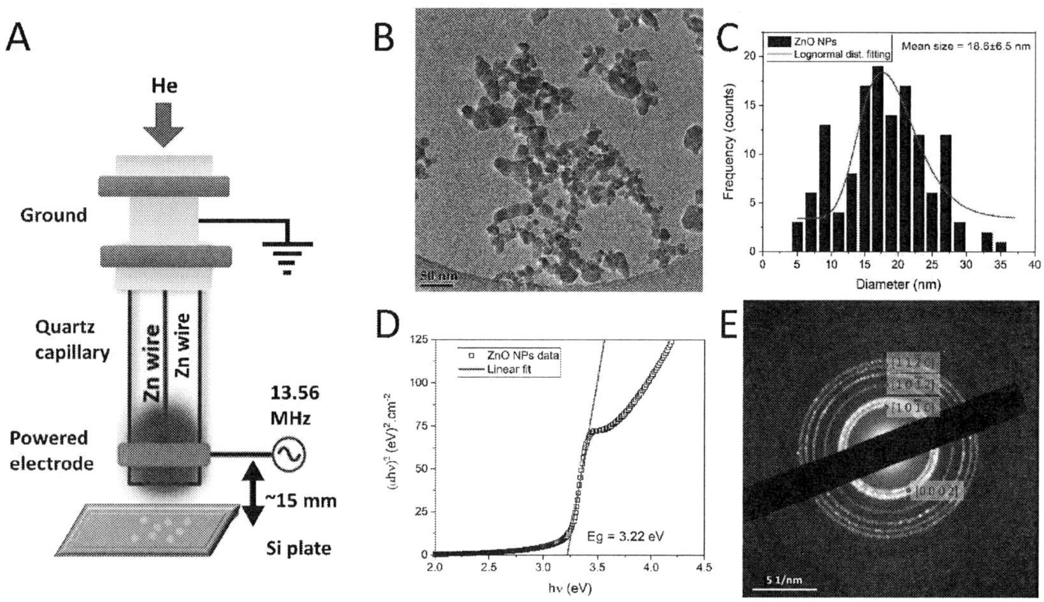

Figure 1 (A) Schematic diagram of the non-thermal atmospheric pressure plasma system with zinc wire used for ZnO nanoparticle synthesis (ZnO-NTP). (B) TEM image of ZnO-NTP. (C) Histogram showing size distribution analysis from TEM images. (D) Tauc plot derived from ZnO-NTP absorbance data obtained using UV-vis spectroscopy (E) SAED profile of ZnO-NTP.

3.2. Zeta potential and DLS

ZnO-NTP were first exposed to dH_2O and then suspended in 3 liquid types during sonication before being resuspended in dH_2O for size distribution and zeta potential analysis. From **Figure 2**, one can see three main size fractions from in MHB-exposed ZnO-NTP, specifically at 79 nm, 712 nm, and 5560 nm. This correlates well with the zeta potential measurements, where there were also three peaks detected, at -36.6 mV, -23.4 mV and -7 mV. There were also three distinct size fractions observed from dH_2O-exposed ZnO-NTP, at 396 nm, 1484 nm and 4801 nm. However, the different sized particles generated a single peak for zeta potential measurement at -12.4 ± 0.4 mV. ZnO-NTP exposed to 0.9% NaCl became more positive with an observed value of 5.3 ± 0.6 mV, however DLS analysis was unable to detect particles below 10 µm even after extensive sonication prior to measurement which makes the value unreliable.

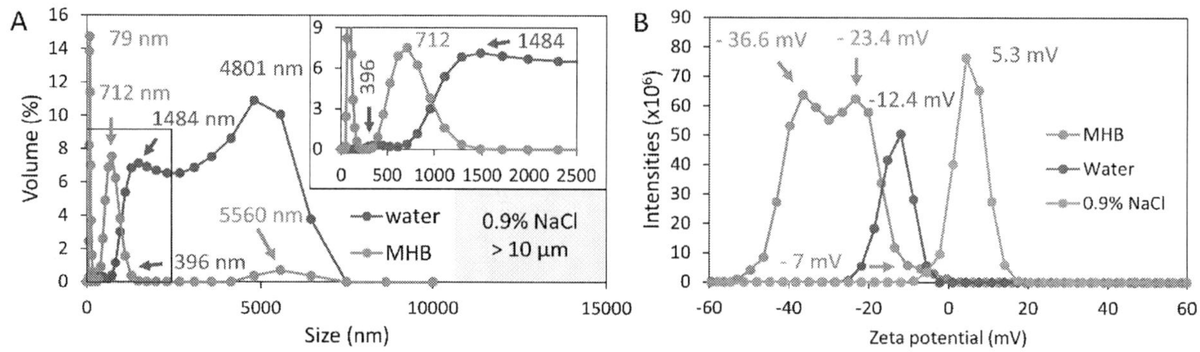

Figure 2 ZnO-NTP size distribution according to volume (A) and intensities with corresponding zeta potential (B). Numbers inset indicate measured values and arrows point to the specific data

3.3. Photoluminescence (PL)

Room temperature PL spectra from ZnO-NTP dried under ambient conditions on Si wafer and illuminated using blue laser (430-435 nm) and filter can be seen in **Figure 3 (A)**. The line profile of each PL spectra is similar regardless of liquid type, with a dominant peak between 600-610 nm and two troughs located either side at 590 nm and 615 nm. No new peaks were detected from any ZnO-NTP and the maximum PL intensity was recorded at the same wavelength in each sample (606 nm).

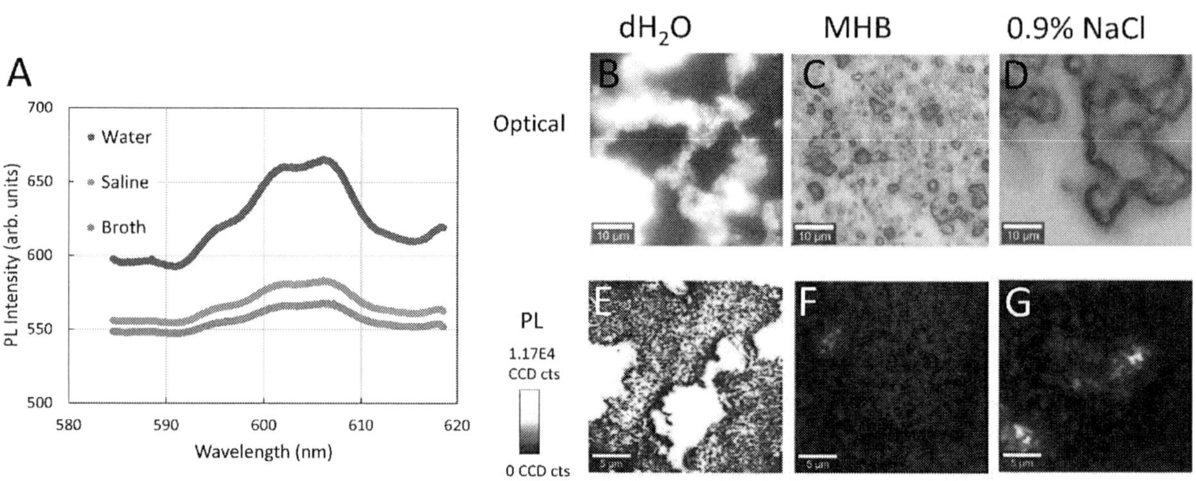

Figure 3 PL spectra of ZnO-NTP illuminated by blue light (430-435 nm) using 570 nm filter (A). Optical (B-D) and fluorescent (E-G) images of ZnO-NTP that generated the spectra in A

Optical image of ZnO-NTP exposed to dH_2O showed large agglomerates (**Figure 3 (B)**) with strong fluorescent emission (**Figure 3 (E)**). ZnO-NTP exposed to 0.9 % NaCl also showed agglomeration (**Figure 3 (D)**), but without the strong fluorescent emission seen previously for dH_2O-exposed ZnO-NTP (**Figure 3 (G)**). The optical image of MHB-exposed ZnO-NTP showed the most dispersed arrangement of particles (**Figure 3 (C)** with the least amount of fluorescence (**Figure 3 (F)**).

3.4. Interaction with bacteria (cfu/mL & SEM)

Reverse spin bioreactors enabled interaction between a low concentration of ZnO-NTP (10 µg/mL) and *E. coli* in MHB. From our previous work, 10 µg/mL ZnO did not inhibit growth of *E. coli* in MHB. Here, we employed yet another different type of ZnO nanoparticles (in shape and size) and studied also the effect of illumination with a non-bactericidal white light source having spectral output overlapping with UV-vis absorbance profile of ZnO-NTP (i.e. between 400-500 nm).

There was no significant difference in *E. coli* concentration after 24 h exposure to ZnO-NTP relative to unexposed (control) regardless of illumination (P>0.05, **Figure 4 (A)**. SEM image of *E. coli*-ZnO-NTP taken after 24 h exposure revealed small particles resting on the cell surface, as well porous superstructures that were large enough to interact with whole bacteria cells (**Figure 4 (B)**).

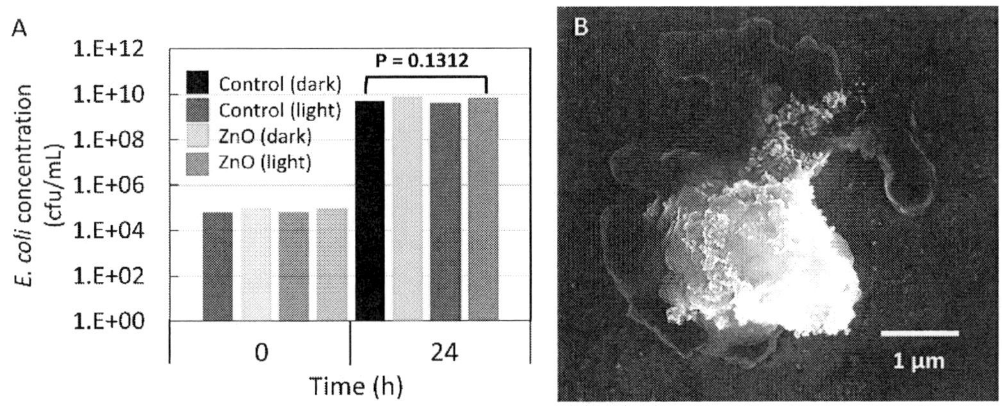

Figure 4 (A) Bar chart displaying the concentration of viable *E. coli* at the start of the experiment (0) and after 24 h exposure to 10 µg/mL ZnO-NTP in reverse spin bioreactors. (B) Scanning electron microscopy image showing a size range of ZnO-NTP aggregates interacting with multiple *E. coli* cells.

4. DISCUSSION

Novel non-thermal plasma technique has been demonstrated to synthesize ZnO nanoparticles of specifically small size, shape and quality at room temperature without any harsh chemicals or large vacuum systems. The non-equilibrium nature inside the discharge region causes reactions occur rapidly and continue in the afterglow where atmospheric gases and humidity gain influence that can lead to the formation of reactive oxygen and nitrogen species (RONS). Nanoparticle formation arises from plasma interaction with nanoparticle precursor atom/molecule in either the feed gas mixture, liquid solution, or solid substrate. Here, a solid zinc wire was placed inside an rf-driven non-thermal plasma using He as the feed gas which carried ZnO-NTP away from the core plasma region for collection. TEM image of ZnO-NTP showed uniform grain morphology with a log normal size distribution and mean particle diameter of 19 nm, however there were also many particles at around 10 nm and 30 nm. The same plasma system was also capable of synthesizing smaller ZnO quantum dots with a mean particle diameter of 1.9 nm when operating under the same power conditions but using argon instead of helium [7]. This suggests that it is possible to control the size of the nanoparticle synthesized using non-thermal plasma by varying the feed gas composition.

ZnO-NTP showed typical UV light absorbance behavior, with a sharp peak at the maximum absorbance wavelength of 370 nm and the calculated energy band gap from Tauc plot derived from absorbance data was 3.22 eV, which is in good agreement with the literature [9]. ZnO-NTP also exhibited photoluminescence when illuminated with light of smaller energy than the band gap (λ=430-435 nm). The PL spectra are similar, and the maximum PL intensity remained at 606 nm in all samples irrespective of liquid type. Thus, the PL spectra are robust in biological environment. Such PL emission arises due to ZnO surface defects [9,10].

The observed differences in PL intensities between ZnO-NTP exposed to different liquid types are most likely due to variation in sample thickness which is influenced in part by material agglomeration as shown in the optical images in **Figure 3**. The PL microscopic images showed high particle agglomeration in water, which could account for the higher PL signal compared to 0.9% NaCl and MHB. Nevertheless, prior reports showed that the PL emission can be quenched by amino acid residues from proteins bound to the surface of ZnO [11]. Quenching of the fluorescence of BSA due to interaction with ZnO NP through the formation of stable ground state complexes was also observed [12], highlighting the fact that both ZnO NP and BSA can influence each other's properties and it is not solely the action of one or the other. Such complex interaction could be occurring in our case, where biomolecules contained in MHB may coat ZnO-NTP, forming a protein/carbohydrate corona that results in PL emission quenching.

Protein coronas also tend to increase particle aggregation due to enhanced particle-particle interaction thereby creating a less well dispersed population. Yet the PL images showed that MHB-exposed ZnO-NTP to be the most dispersed material compared to dH_2O and 0.9% NaCl-exposed ZnO-NTP.

Zeta potential analysis of MHB-exposed ZnO-NTP also implied a more dispersed population compared to dH_2O and 0.9% NaCl-exposed ZnO-NTP. There were three distinct peaks in both zeta potential measurement and size distribution for MHB-exposed ZnO-NTP, two pronounced peaks and one much smaller. The values of the two pronounced peaks (-36.6 mV and -23.4 mV) were more negative than the single peak from dH_2O-exposed ZnO-NTP of -12.4 mV. A higher absolute value of zeta potential can result in a more disperse population of particles in the suspension due to better electrostatic repulsion of nanoparticles. Zeta potential of ZnO also became more negative after exposure to biomolecules such as BSA [13], in agreement with our data in MHB. The third peak from zeta potential measurement was less negative than the prior two (-7 mV) which may have been generated from the larger size fraction at 5560 nm which accounted for a small percentage of the total volume of particles. For MHB-exposed ZnO-NTP, one can assume that the three separate peaks observed in zeta potential analysis were generated from the three different size fractions. However, this was not the case for dH_2O-exposed ZnO-NTP where there were three size fractions observed yet only one zeta potential peak. This suggests that the effect of MHB on the size and zeta potential of ZnO-NTP is more pronounced than for dH_2O. The two smaller size fractions from MHB-exposed ZnO-NTP (79 nm and 712 nm) were smaller than the two smallest size fractions from dH_2O-exposed ZnO-NTP (396 nm and 1484 nm) which also suggests a more dispersed population of particles with smaller aggregates. However, even the smallest size fraction observed (79 nm) was larger than the primary particle size seen from TEM image in **Figure 1** (19 nm), therefore some aggregation of primary particles occurs on MHB-exposed ZnO-NTP. Upon exposure to 0.9% NaCl, the zeta potential of ZnO-NTP shifted to a less negative value of 5.3 mV. This trend was also observed for ZnO NP exposed to increasing NaCl concentrations, from -28 mV to -2 mV, however the highest concentration of NaCl used in that study and produced the least negative zeta potential was x30 less than what was used in our experiments [14].

It has been reported in the literature that the smaller the nanoparticle, the greater the antibacterial effect [15]. This is also true for aggregates of particles and one might expect greater antibacterial effect from more dispersed colloidal suspension. However, we have previously shown that particle morphology is also important in the antibacterial effect of ZnO in MHB, and particles with uniform rounded morphology were less antibacterial than particles composed of many needle-like clusters [16]. At the same concentration used here (10 µg/mL), there was no significant difference in viable bacteria relative to unexposed bacteria regardless of ZnO size or shape. At higher concentrations (≥100 µg/mL), the needles were still sufficiently sharp even after being coated with components of MHB to inactivate bacteria by puncturing the cell membrane, whereas the more rounded particles tended to aggregate on the cell surface. A non-bactericidal white light source with overlapping spectral output and UV-vis absorbance profile of ZnO-NTP was chosen to try and enhance the antibacterial effect (i.e. between 400-500 nm). Visible light photoinactivation of bacteria mediated by ZnO has been reported that was mainly attributed to the bactericidal effect of zinc ions [17]. We have previously shown that without illumination the concentration of zinc ion produced in MHB would not be great enough to become cytotoxic due to the components in MHB scavenging zinc ion before interacting with bacteria [16]. However, hydrogen peroxide was measured from illuminating a much higher concentration of commercially available ZnO (2 mg/mL) using the same experimental set up and illumination source used here (unpublished data).

5. CONCLUSION

Variation in the physicochemical properties of ZnO-NTP nanoparticles due to the effect of exposure to various liquids was observed. Interestingly, the best dispersion was observed in cell culture MHB medium, providing smallest aggregates (79 nm) and most negative zeta potential (-36.6 mV). The PL spectra remained unaffected. In the employed sub-lethal concentration the ZnO-NTP nanoparticles rather enhanced the bacterial growth, as observed for other ZnO NP types before. Yet illumination did not lead to any bactericidal effect, unlike for other ZnO nanoparticle types. This must be related with specific nanoparticle preparation using non-thermal plasma. This means that photoexcitation and PL spectra can be used as a robust sensing pathway in biological environment.

ACKNOWLEDGEMENTS

The presented work was supported financially by the project GACR 19-02858J.

REFERENCES

[1] KORTSHAGEN, U. R., SANKARAN, R. M., PEREIRA, R. N., GIRSHICK, S. L., WU, J. J. and AYDIL, E. S. Nonthermal Plasma Synthesis of Nanocrystals: Fundamental Principles, Materials, and Applications. *Chem. Rev.* 2016, vol. 116, pp. 11061-127.

[2] CHIANG, W-H, MARIOTTI, D., SANKARAN, R. M., EDEN, J. G. and OSTRIKOV, K. (Ken) Microplasmas for Advanced Materials and Devices. *Advanced Materials.* 2020, vol. 32, 1905508.

[3] Anon 2019 Material Safety Data Sheet (MSDS) - ZnO

[4] FELBIER, P., YANG, J., THEIS, J., LIPTAK, R. W., WAGNER, A., LORKE, A., BACHER, G. and KORTSHAGEN, U. Highly Luminescent ZnO Quantum Dots Made in a Nonthermal Plasma. *Advanced Functional Materials.* 2014, vol. 24, pp. 1988-93

[5] TSUMAKI, M., SHIMIZU, Y. and ITO, T. Size-controlled sub-micrometer spheroidized ZnO particles synthesis via plasma-induced processing in microdroplets. *Materials Letters.* 2016, vol. 166, pp. 81-4.

[6] JAIN, G., MACIAS-MONTERO, M., VELUSAMY, T., MAGUIRE, P. and MARIOTTI, D. Porous zinc oxide nanocrystalline film deposition by atmospheric pressure plasma: Fabrication and energy band estimation. *Plasma Process Polym.* 2017, vol. 14, 1700052.

[7] JAIN, G., ROCKS, C., MAGUIRE, P. and MARIOTTI, D. One-step synthesis of strongly confined, defect-free and hydroxy-terminated ZnO quantum dots. *Nanotechnology.* 2020, vol. 31, 215707.

[8] REZEK, B., HEMATIAN, H., JÍRA, J., RUTHERFORD, D., KULIČEK, J., UKRAINTSEV, E. and REMEŠ, Z. Microscopic Study of Bovine Serum Albumin Adsorption on Zinc Oxide (0001). Surface *physica status solidi (a).* 2021, vol. 218, 2000558.

[9] ISLAM, S. and AKYILDIZ, H. I. Immobilization of ZnO thin films onto fibrous glass substrates via atomic layer deposition and investigation of photocatalytic activity *J Mater Sci: Mater Electron.* 2021, vol. 32, pp. 27027-43.

[10] DJURIŠIĆ, A. B., LEUNG, Y. H., TAM, K. H., HSU, Y. F., DING, L., GE, W. K., ZHONG, Y. C., WONG, K. S., CHAN, W. K., TAM, H. L., CHEAH, K. W., KWOK, W. M. and PHILLIPS, D. L. Defect emissions in ZnO nanostructures. *Nanotechnology.* 2007, vol. 18, 095702.

[11] BHUNIA, A. K., KAMILYA, T. and SAHA, S. Temperature Dependent and Kinetic Study of the Adsorption of Bovine Serum Albumin to ZnO Nanoparticle Surfaces. *ChemistrySelect.* 2016, vol. 1, pp. 2872-82.

[12] HANSDA, C., MAITI, P., SINGHA, T., PAL, M., HUSSAIN, S. A., PAUL, S. and PAUL, P. K. Photophysical study of the interaction between ZnO nanoparticles and globular protein bovine serum albumin in solution and in a layer-by-layer self-assembled film. *Journal of Physics and Chemistry of Solids.* 2018, vol. 121, pp. 110-20.

[13] BHUNIA, A. K., KAMILYA, T. and SAHA, S. Synthesis, Characterization of ZnO Nanorods and its Interaction with Albumin Protein *Materials Today: Proceedings.* 2016, vol. **3**, pp. 592-7.

[14] PENG, Y.-H., TSO, C., TSAI, Y., ZHUANG, C. and SHIH, Y. The effect of electrolytes on the aggregation kinetics of three different ZnO nanoparticles in water. *Science of The Total Environment.* 2015, vol. 530-531, pp. 183-90.

[15] KAUSHIK, M., NIRANJAN, R., THANGAM, R., MADHAN, B., PANDIYARASAN, V., RAMACHANDRAN, C., OH, D-H. and VENKATASUBBU, G. D. Investigations on the antimicrobial activity and wound healing potential of ZnO nanoparticles. *Applied Surface Science.* 2019, vol. 479, pp. 1169-77.

[16] RUTHERFORD, D., JÍRA, J., KOLÁŘOVÁ, K., MATOLÍNOVÁ, I., MIČOVÁ, J., REMEŠ, Z. and REZEK, B. Growth Inhibition of Gram-Positive and Gram-Negative Bacteria by Zinc Oxide Hedgehog Particles. *IJN.* 2021, vol. 16, pp. 3541-54.

[17] SAPKOTA, A., ANCENO, A. J., BARUAH, S., SHIPIN, O. V. and DUTTA, J. Zinc oxide nanorod mediated visible light photoinactivation of model microbes in water. *Nanotechnology*. 2011, vol. 22, 215703.

FULLY-AUTOMATIC BENCHTOP MACHINE FOR THE PRODUCTION OF NANOPARTICLE SUSPENSIONS

[1]Tobias BESSEL, [1]Stephan BARCIKOWSKI, [1]Friedrich WAAG

[1]Technical Chemistry I, Center for Nanointegration Duisburg-Essen (CENIDE), University of Duisburg, Essen, Germany, EU, friedrich.waag@uni-due.de

https://doi.org/10.37904/nanocon.2021.4317

Abstract

Developing and exploiting applications of functional nanomaterials often require colloidal nanoparticles with defined properties. Their production suffers from batch-to-batch variability as well as poor shelf life, and distributors demand high prices. The conditions for innovative application research must be improved by easier and more robust access to colloidal nanoparticles. A compact and fully automatic benchtop machine has been developed for quick, easy and reliable production of colloidal nanoparticles. The machine uses the method of laser synthesis of colloids, converting almost any solid of choice into colloidal nanoparticles. In addition, different dispersants and stabilizers can be dosed, if desired. An almost infinite range of colloids becomes available with one machine. We provide an insight into the operation of the machine and report in particular on the high performance efficiency of the laser-based synthesis. Furthermore, we highlight the benefits of using the device in R&D of colloidal nanoparticles made possible by the technology.

Keywords: Ablation, productivity, PLAL, LASiS

1. INTRODUCTION

Inorganic nanoparticles represent an universal and essential resource with wide-ranging application relevance [1-3]. Many applications require non-agglomerated primary nanoparticles with a large and freely accessible surface area. These are generally prepared as suspensions by wet chemical methods using sometimes hazardous chemicals. Usually, surface-active agents are required to protect the particles from agglomeration, which in turn render access to the nanoparticle´s surface (which determine their functionality) more difficult in applications. The syntheses are also labor-intensive and error-prone due to many closely timed steps in the procedure. All this makes application-oriented research with nanoparticles difficult and lengthy if one does the synthesis work oneself, or expensive and limited if one purchases the particles from distributors.

Automation enables more reliable and simpler manufacturing processes in many fields of technology. In the synthesis of nanoparticle suspensions, however, no complete commercial solution exists to date. We are therefore working on a fully automated laboratory machine, which will provide researchers and developers with the best access to colloidal nanoparticles.

In the machine, we use the well-established synthesis technique of laser ablation of solids in liquids [4]. The method offers three immense advantages over wet chemical methods: 1) It allows the preparation of highly pure nanoparticle colloids, 2) gives access to diverse combinations of nanoparticle material and dispersant, and 3) benefits from the high automation compatibility of laser technology. Until recently, the disadvantages of the method were the limited control of the particle size distribution, in particular also batch-to-batch-variations and the limited laser-power-specific ablation efficiency.

In this work, we report on the use of microchip lasers in laser-based colloid synthesis. These enable a compact size of the laboratory machine as well as sufficiently high productivity for the use of the device in research and development. This makes possible a significantly higher power efficiency in synthesis compared to other laser systems usually used for the synthesis of colloidal nanoparticles. Dittrich et al. demonstrated 5-times higher power-specific mass yield for an infrared, low-power, single-nanosecond microchip laser compared to a standard manufacturing laser that is typically used in laser-based nanoparticles synthesis [5].

We applied another microchip laser system in the synthesis of colloidal nanoparticles and compared the power efficiencies with previously reported ones. The laser system tested differs in particular in its laser wavelength (532 nm) from previously studied systems (1.064 nm). It is known from the literature that shorter wavelengths can generate smaller nanoparticles with narrower particle size distribution [6-9]. However, due to stronger attenuation effects by the nanoparticles on the incident laser beam, shorter laser wavelength usually decreases productivity and thus power efficiency [10,11].

2. EXPERIMENTAL

A compact laser system (ONDA 532 nm, BrightSolutions, Prado, Italy) with a wavelength of 532 nm, a laser pulse length of 2 ns, a repetition rate of 15 kHz, and a power of 3.6 W was used to synthesize colloidal nanoparticles. The power was measured behind the galvanometer scanner (PT-A20, PhenixTechnology, Guangzhou, China) used. The scanner was controlled by custom software. After the scanner, the laser beam was focused by a lens system with an absolute focal length of 50 mm. The focused laser beam was directed onto the surface of rectangular metal plates. The platelets were placed in a chamber behind an anti-reflective coated glass plate followed by a flowing liquid layer.

Metal plates of Ag, Au and Pt (99.9 %, R. Götze, Berlin, Germany) were used. In the case of Ag, the liquid used was an aqueous, 0.5 mM trisodium citrate solution (analytical grade), and in the case of Au and Pt, an aqueous, 0.5 mM sodium chloride solution (analytical grade) was used. The fluids flowed through the chamber against gravity at a volume flow rate of 5 ml/min. The generated nanoparticles were thus continuously carried out of the chamber and subsequently collected. For the performance efficiency study, the distance between the lens system and the target surface was adjusted to achieve maximum nanoparticle concentration. To determine productivity, we weighted the metal platelets after the experiment. The mass difference was then divided by the experimental duration of 20 min. To determine the power efficiency, the productivity was divided by the measured laser power of 3.6 W.

The optimal distance between the focusing lens system and the metal platelet for high productivity was determined by varying the distance in 0.25 mm steps. The nanoparticle concentration was determined qualitatively using a UV-Vis extinction spectrometer (Evolution 201, Thermo Fisher Scientific, Waltham, US). For this purpose, the colloids were diluted 1:2 with the appropriate dispersant and placed in quartz cuvettes. The absorbance was measured in the range of 230 and 900 nm. All measurements were performed as differential measurements against cuvettes filled with the corresponding dispersant.

3. RESULTS AND DISCUSSION

We have recently reported performance efficiencies of the current prototype of our laboratory machine (**Figure 1**) in the synthesis of colloidal noble metal nanoparticles when using an infrared laser (1,064 nm) [12]. Below, we show corresponding data recorded using a green-wavelength laser beam source.

Since we worked with a focused laser beam, we first had to determine the distance between the lens system and the target surface at which the highest productivity was present. For this purpose, this distance was successively varied in 0.25 mm steps during the synthesis of the nanoparticles. At five different distances per metal, colloid samples were taken and analyzed with respect to their concentration by UV-Vis spectroscopy. **Figure 2** (left) shows the corresponding spectra. The spectra show different extinctions as a function of distance from the productivity optimum. In the case of Au and Ag, the characteristic surface plasmon resonance is also noticeable. From the spectra, the absorbance at a characteristic high-energy wavelength was extracted in the next step. **Figure 2** (right) shows the data as a function of the distance between lens and target, normalized to the distance of maximized productivity. In the case of Au and Pt, the best productivity was found at the same absolute position. For Ag, the position was 0.25 mm closer to the lens.

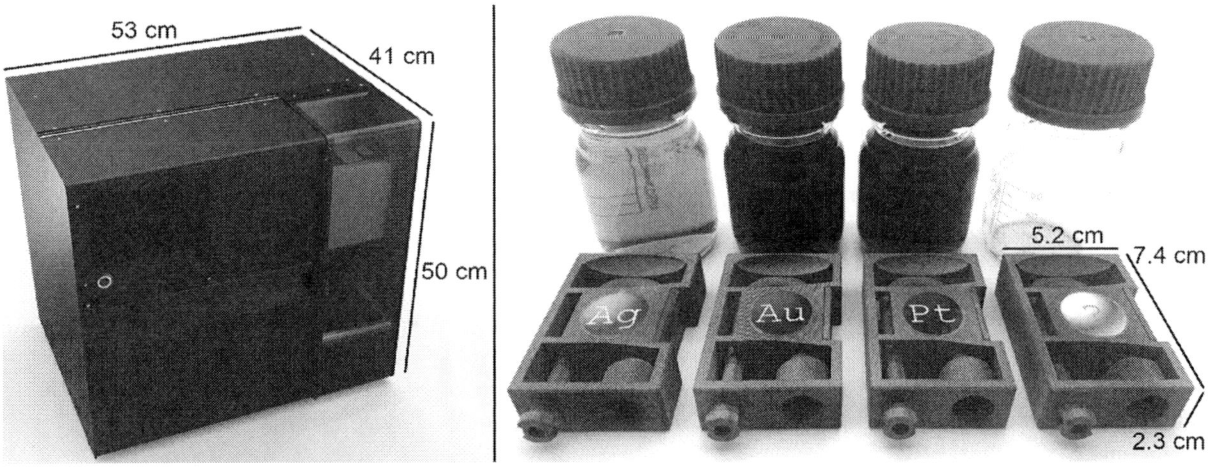

Figure 1 Laboratory machine prototype including its equipment: 3D model of the compact laboratory machine for the synthesis of colloidal nanoparticles (left) and colloidal nanoparticles of Ag, Au, and Pt as well as corresponding capsules for colloid production (right).

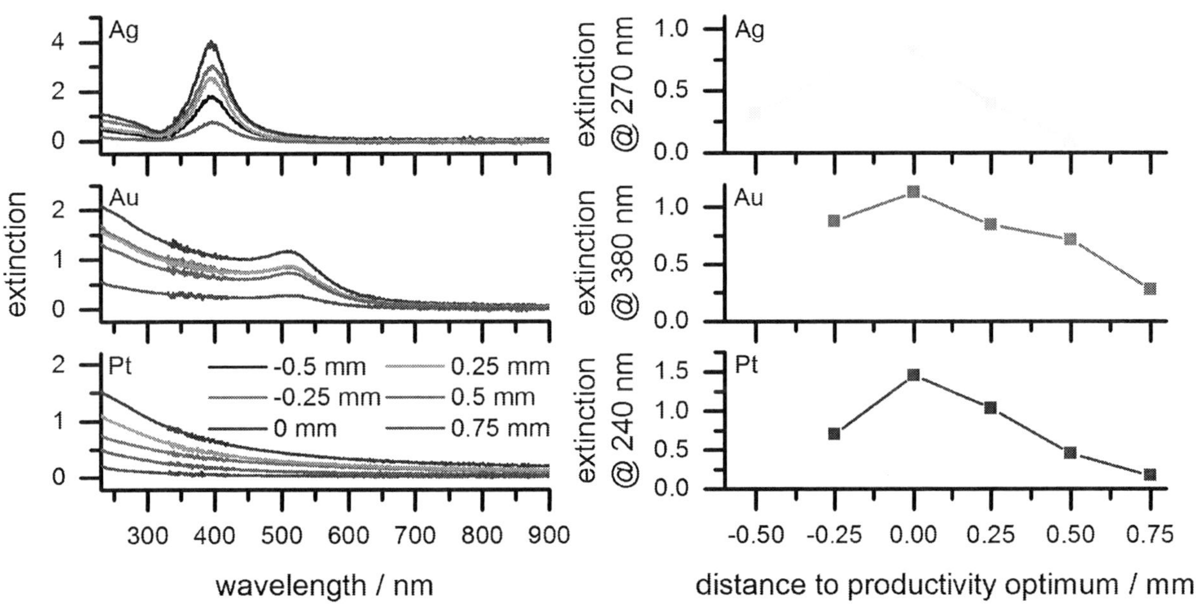

Figure 2 Dilution-corrected UV-Vis extinction spectra of Ag, Au, and Pt nanoparticle colloids produced at different distances to the optimum position for maximum productivity (left) and extracted extinctions of specific wavelengths versus the distance to the productivity optimum (right). More positive distance values represent shorter distances between lens and target.

We conducted ablation experiments with Ag, Au, and Pt at the respective optimal distance to determine the productivity and efficiency of nanoparticle generation with the compact laser system. **Figure 3** (left) shows the achieved productivities of 33 mg/h (Ag), 50 mg/h (Au) and 54 mg/h (Pt) compared to literature values. The productivities exceed those we were able to achieve in a previous study with an identical experimental setup but lower power and infrared laser system by factors of 3 to 4 [12]. However, the power of the green laser was about 7-times higher than the power of the infrared laser. Accordingly, we were able to achieve higher power-specific efficiencies with the infrared laser system, as also shown in **Figure 3** (right). In the case of Ag, the efficiency is comparable when scattering is taken into account. For Au and Pt, however, the efficiency of the green laser system was significantly lower than that of the infrared one. This was probably because of the optical properties of the colloids. While the absorption and scattering properties of all colloids should have been negligible at 1.064 nm, there were significant differences in the optical properties at 532 nm. In particular, Au nanoparticles absorb green light due to their surface plasmon resonance. As **Figure 2** (left) shows, the absorbance of the Pt colloid was 2-times higher than that of the Ag colloid. Despite the loss of efficiency due to the green laser wavelength, the green microchip laser achieved a 2-times higher efficiency in the synthesis of Au nanoparticles compared to a typical metal processing laser system, putting it in the same efficiency range as a high-end laser system designed specifically for the synthesis of colloidal nanoparticles.

The stronger interaction of laser light and nanoparticles in the case of the green laser could open up new possibilities in the manipulation of the particle size distribution [13-17]. Consequently, in the next step, it needs to be investigated to which extend in-process laser fragmentation as well as melting (or laser.induced aggregation) under the operating conditions of the automated synthesis system.

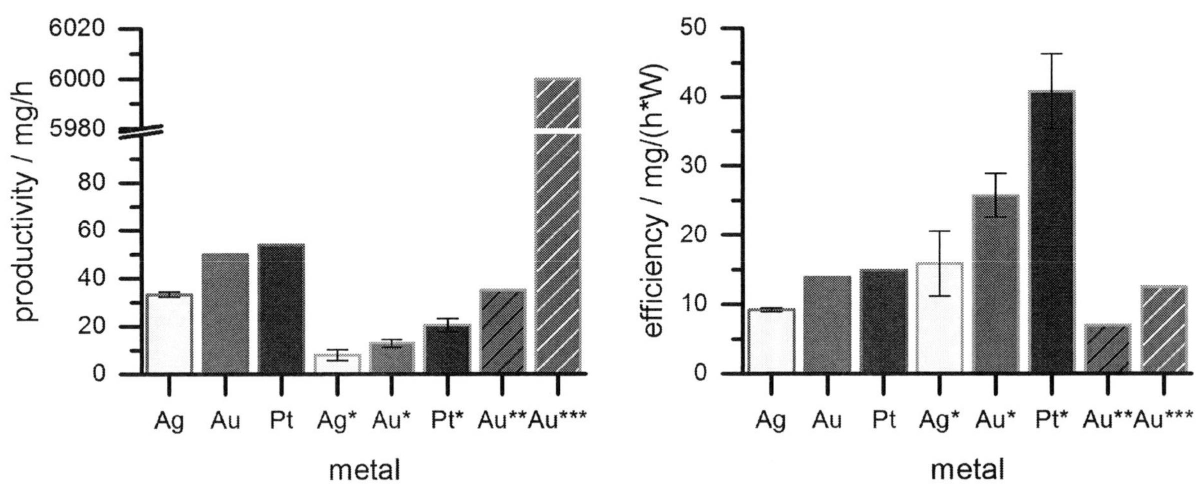

Figure 3 Productivity (left) and efficiency (right) of the compact laser-based synthesis of colloidal nanoparticles of Ag, Au, and Pt. The error bar of the Ag sample represents the absolute deviation of two experiments to their average. *: Data of [12] recorded using a microchip laser system (1,064 nm, 1 ns, 0.5 W). **: Data of [5] recorded using a standard laser system (1,064 nm, 5 ns, 5 W). ***: Data of [5] recorded using a high-end laser system (1,030 nm, 3 ps, 480 W).

4. CONCLUSION

By an automated laser synthesis system, nanoparticles for research and development can be produced easily, quickly, and reliably at the push of a button using a compact (i.e., benchtop) laboratory machine. The device can thus significantly accelerate application-oriented research based on nanoparticles in particular. The developed laser-based synthesis technology makes the automated machine a green and sustainable production device thanks to the greatly reduced use of chemicals compared to conventional synthesis methods. Note that atom yield (the consumed material that is converted into nanoparticles) of laser synthesis is 100%, unreached by wet chemical synthesis. Further advantages for research and development result from the wide-ranging possibilities in material selection. For example, the machine can not only reliably produce nanoparticle colloids of standard materials, such as Au nanoparticles in aqueous suspension, but can also make innovative materials accessible, such as alloy nanoparticles for catalysis or biomedicine. The previous shortcoming of low efficiency of the laser-based method could be eliminated by using power-efficient microchip lasers. In the future, the possibilities of in-process size manipulation using green laser radiation will be investigated in detail.

ACKNOWLEDGEMENTS

We thank the European Union and the Ministry of Economic Affairs, Innovation, Digitalization and Energy of the German State of North Rhine-Westphalia for funding the development of the automated colloid maker to market maturity through the European Regional Development Fund in the START-UP transfer.NRW program (AutoProNANO, EFRE-0400339). The START-UP transfer.NRW program supports researchers at public institutions in developing technologies to market maturity and thus in preparing a spin-off company. Please visit www.autopronano.eu for more information on our EU-funded project. We would also like to thank the German Federal Ministry for Economic Affairs and Energy for funding the prototype development in the Central Innovation Program for SMEs (Nanovollautomat, ZF4044504).

REFERENCES

[1] EALIA, S. A. M., SARAVANAKUMAR, M. P. A review on the classification, characterisation, synthesis of nanoparticles and their application. *IOP Conference Series: Materials Science and Engineering*. 2017, vol. 263, 032019.

[2] IJAZ, I., GILANI, E., NAZIR, A., BUKHARI, A. Detail review on chemical, physical and green synthesis, classification, characterizations and applications of nanoparticles. *Green Chemistry Letters and Reviews*. 2020, vol. 13, pp. 223-245.

[3] SOLANO, R., PATINO-RUIZ, D., TEJEDA-BENITEZ, L., HERRERA, A. Metal- and metal/oxide-based engineered nanoparticles and nanostructures: a review on the applications, nanotoxicological effects, and risk control strategies. *Environmental Science and Pollution Research*. 2021, vol. 28, pp. 16962-16981.

[4] KIM, M., OSONE, S., KIM, T., HIGASHI, H., SETO, T. Synthesis of Nanoparticles by Laser Ablation: A Review. *KONA Powder and Particle Journal*. 2017, vol. 34, pp. 80-90.

[5] DITTRICH, S., STREUBEL, R., MCDONNELL, C., HUBER, H. P., BARCIKOWSKI, S., GÖKCE, B. Comparison of the productivity and ablation efficiency of different laser classes for laser ablation of gold in water and air. *Applied Physics A*. 2019, vol. 125, p. 432.

[6] TSUJI, T., IRYO, K., WATANABE, N., TSUJI, M. Preparation of silver nanoparticles by laser ablation in solution: influence of laser wavelength on particle size. Applied Surface Science. 2002, vol. 202, pp. 80-85.

[7] KIM, J., REDDY, D. A., MA, R., KIM, T. K. The influence of laser wavelength and fluence on palladium nanoparticles produced by pulsed laser ablation in deionized water. *Solid State Sciences*. 2014, vol. 37, pp. 96-102.

[8] CHEWCHINDA, P., TSUGE, T., FUNAKUBO, H., ODAWARA, O., WADA, H. Laser Wavelength Effect on Size and Morphology of Silicon Nanoparticles Prepared by Laser Ablation in Liquid. *Japanese Journal of Applied Physics*. 2014, vol. 53, 039201.

[9] GIORGETTI, E., MUNIZ-MIRANDA, M., MARSILI, P., SCARPELLINI, D., GIAMMANCO, F. Stable gold nanoparticles obtained in pure acetone by laser ablation with different wavelengths. *Journal of Nanoparticle Research*. 2012, vol. 14, p. 648.

[10] NICHOLS, W. T., SASAKI, T., KOSHIZAKI, N. Laser ablation of a platinum target in water. II. Ablation rate and nanoparticle size distributions. *Journal of Applied Physics*. 2006, vol. 100, 114912.

[11] SCHWENKE, A., WAGENER, P., NOLTE, S., BARCIKOWSKI, S. Influence of processing time on nanoparticle generation during picosecond-pulsed fundamental and second harmonic laser ablation of metals in tetrahydrofuran. *Applied Physics A*. 2011, vol. 104, pp. 77-82.

[12] BESSEL, T., DITTRICH, S., GÖKCE, B., BARCIKOWSKI, S., WAAG, F. Automated synthesis of colloidal nanoparticles powered by microchip lasers. In: Lasers in Manufacturing (LiM 2021). Munich: Wissenschaftliche Gesellschaft Lasertechnik, 2021.

[13] WERNER, D., HASHIMOTO, S. Controlling the Pulsed-Laser-Induced Size Reduction of Au and Ag Nanoparticles via Changes in the External Pressure, Laser Intensity, and Excitation Wavelength. *Langmuir.* 2013, vol. 29, pp. 1295-1302.

[14] WERNER, D., HASHIMOTO, S. Improved Working Model for Interpreting the Excitation Wavelength- and Fluence-Dependent Response in Pulsed Laser-Induced Size Reduction of Aqueous Gold Nanoparticles. *The Journal of Physical Chemistry C.* 2011, vol. 115, pp. 5063-5072.

[15] CHUBILLEAU, C., LENOIR, B., MIGOT, S., DAUSCHER, A. Laser fragmentation in liquid medium: A new way for the synthesis of PbTe nanoparticles. *Journal of Colloid and Interface Science*. 2011, vol. 357, pp. 13-17.

[16] ISHIKAWA, Y., KOSHIZAKI, N., PYATENKO, A. Submicrometer-Sized Spherical Iron Oxide Particles Fabricated by Pulsed Laser Melting in Liquid. *Electronic and Communications in Japan*. 2016, vol. 99, pp. 37-42.

[17] PYATENKO, A., WANG, H., KOSHIZAKI, N., TSUJI, T. Mechanism of pulse laser interaction with colloidal nanoparticles. *Laser & Photonics Reviews*. 2013, vol. 7, pp. 596-604.

STUDY OF DYNAMIC GAS RESPONSE OF MICROWAVE PLASMA ENHANCED CHEMICAL VAPOR DEPOSITION SYSTEMS BY OPTICAL EMISSION SPECTROSCOPY

[1,2]Nicolas LAMBERT, [1,3]Vincent MORTET

[1]FZU - Institute of Physics of the Czech Academy of Sciences, Prague, Czech Republic, EU, mortetv@fzu.cz

[2]Faculty of Electrical Engineering, Czech Technical University in Prague, Prague, Czech Republic, EU, lambert@fzu.cz

[3]Faculty of Biomedical Engineering, Czech Technical University in Prague, Kladno, Czech Republic, EU, mortetv@fzu.cz

https://doi.org/10.37904/nanocon.2021.4379

Abstract

Microwave Plasma Enhanced Chemical Vapor Deposition (MWPECVD) is widely used for the growth and the study of synthetic doped diamond and its electronic and electrochemical applications. Recent results show the possible enhancement of dopant incorporation in diamond using pulsed gas injection in the reactor. Therefore, understanding the gas dynamic of the MWPECVD reactors is crucial to the optimization of the gas flows and diamond doping. In this work, we present a method to determine the gas dynamic response of a MWPECVD reactor. We determined the dynamic response of injected gas in a NIRIM type and a linear antenna type MWPECVD reactor by recording light emission from a hydrogen plasma during the instantaneous injection of a small fraction of nitrogen using either Optical Emission Spectroscopy (OES) or the combination of a photodiode and an optical band filter. Using linear system theory, we developed a program to determine the dynamic response of the reactor to an arbitrary shaped gas flow input from its pulse response.

Keywords: Pulse response, MWPECVD reactor, OES, gas dynamics, linear system theory

INTRODUCTION

Microwave Plasma Enhanced Chemical Vapor Deposition (MWPECVD) is widely used for the growth synthetic doped diamond used in electronic and electrochemical applications. Recent results show the possible enhancement of dopant incorporation using pulsed gas injection [1]. Therefore, understanding the gas dynamic of the MWPECVD reactors is crucial to the optimization of the gas flows and diamond doping.

In this work, we present and compare two methods for measuring the time response of two different MWPECVD reactors, with different volumes and different operating pressures, to a short N_2 gas pulse: a NIRIM type system [2] and a linear antenna MWPECVD system [3]. The time response is obtained by monitoring the intensity of the emission line at 337 nm attributed N_2 [4] in a hydrogen plasma by recording the voltage across a photodiode combined to a filter with a bandwidth centered around 340 nm and by Optical Emission Spectroscopy (OES) using a spectrometer with a wavelength range from 200 to 1100 nm. Both methods are compared and the advantages and flaws of each are discussed. Data were fitted using the impulse response equation proposed by H. Pendar et al. [5]. Finally, using linear system theory, it was possible to predict the dynamic response of the reactor to an arbitrary shaped gas flow input from its pulse response [6,7].

1. EXPERIMENTAL SETUP

The conditions of gas flow, pressure, and the volume of both the NIRIM and linear antenna MWPECVD system are summarized in **Table 1**.

Table 1 Plasma conditions

Reactor	H_2 gas flow (sccm)	Pressure (mbar)	Chamber volume (L)
NIRIM	200	70	1.4
Linear antenna MWPECVD	200	0.42	160

The measurement setup is described in **Figure 1**. H_2 flow was controlled by a MKS mass flow controller. In order to generate the N_2 gas pulse in the plasma a small volume of N_2 (ca. 1 ml) was trapped between two ball valves at a pressure of 2.6 bar and then released in the chamber.

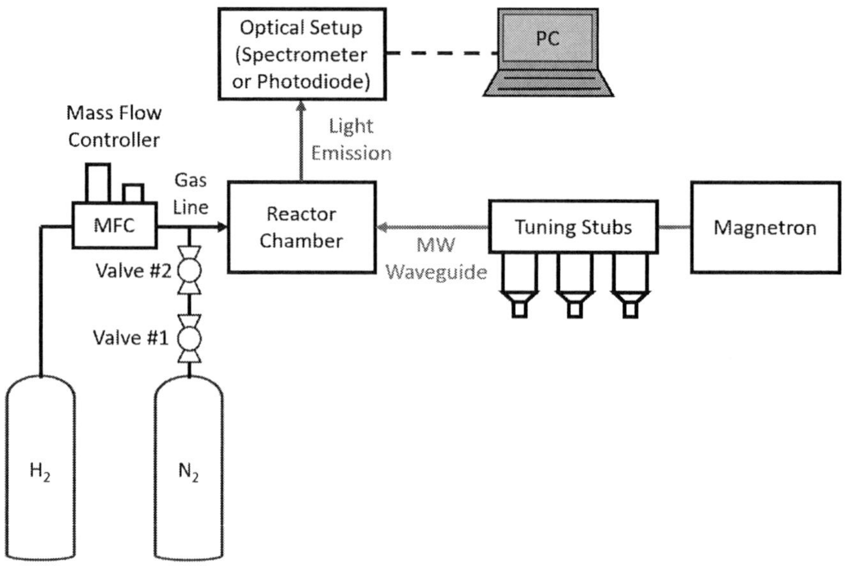

Figure 1 Measurement setup

To measure the time response of the reactors to the N_2 gas pulse, the light emission from the plasma was monitored using the following setups. In the first setup, the emission line at 337 nm attributed to molecular N_2 was monitored using a Thorlabs SM05PD3A photodiode with a wavelength range from 320 to 1100 nm in combination with a Thorlabs FB340-10 band pass filter centered at 340 nm with a full width half maximum of 10 nm. The photodiode voltage was recorded as a function of time using a Keysight 34461A multimeter and saved on a computer using Keysight BenchVue software. A homemade brass holder was designed to combine the photodiode and filter in a closed package and isolate it from any parasitic light sources. The main advantages of this setup are its simplicity and its operation speed with a sampling rate of about 20 Hz. However, its major flaw is its poor selectivity as it is impossible to measure and remove possible variation of the continuous background light coming from the hydrogen plasma during the gas pulse, which can overshadow the variation of the measured emission line. In the second setup, we used a Stellarnet BLACK-Comet-SR spectrometer, which can record spectra from 200 to 1100 nm. An integration time of 25 ms was used to retain a high sampling rate comparable to the first setup. The acquired spectra were analyzed using an algorithm in Origin® software to remove the variation of the continuous background from the N_2 emission peak at 337 nm.

2. EXPERIMENTAL RESULTS

The experimental dynamic responses of both reactors to a N_2 pulse measured with the photodiode and the spectrometer are presented in **Figure 2**. The time responses recorded on the MWPECVD linear antenna system have a very similar shape contrary to the ones recorded on the NIRIM reactor. The signal measured in the NIRIM reactor using the photodiode decreases below its reference value after the signal reaches its maximum. However, since the data extracted from the second measurement setup do not exhibit this unrealistic variation, we conclude that this anomaly is due to a change in the background signal during the pulse.

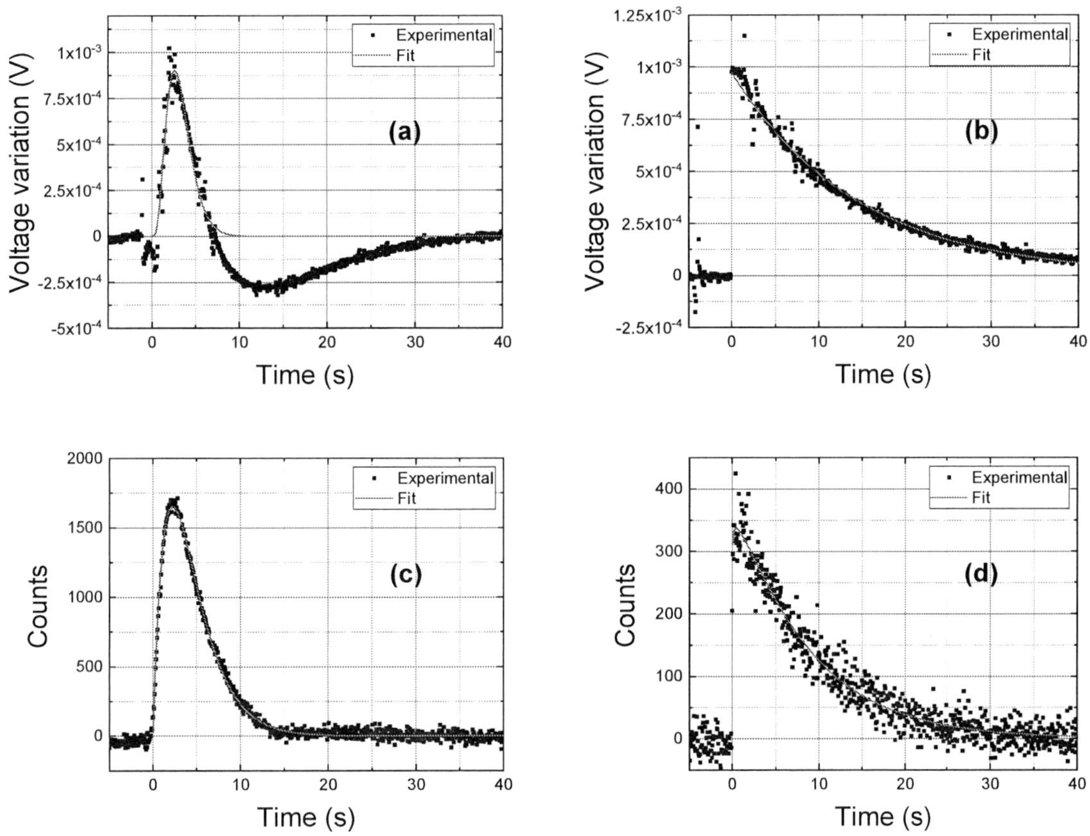

Figure 2 Experimental dynamic pulse responses and fitting curves measured on (a) the NIRIM reactor and (b) the MWPECVD linear antenna reactor using the photodiode and experimental dynamic pulse responses and fitting curves measured on (c) the NIRIM reactor and (d) the MWPECVD linear antenna reactor using the spectrometer

The time responses were fitted using the equation proposed by H. Pendar et al. [5]:

$$f(t) = \alpha t^m e^{-\beta t} \tag{1}$$

where t is the time and α, m, and β are fitting parameters. β is characteristic of the exponential decay of the signal which dominates for large value of time while m characterizes the rising portion of the signal for small value of time. The curve reaches its maximum for $t = \frac{m}{\beta}$ with a value of:

$$f\left(\frac{m}{\beta}\right) = \alpha \left(\frac{m}{\beta}\right)^m e^{-m} \tag{2}$$

Equation (1) fits well the experimental data as shown in **Figure 2** except in **Figure 2a**, which corresponds to the measurement carried out with the photodiode on the NIRIM reactor due to the variation of the background intensity, which cannot be taken into account in the model. The fitting parameters for each condition are presented in **Table 2**. Despite the differences in shape of the recorded signals on the NIRIM system, the time responses reach their maximum at a similar time of $t = \frac{m}{\beta} = 2.7$ s (photodiode) and $t = \frac{m}{\beta} = 2.2$ s (spectrometer). In the case of the MWPECVD linear antenna system, a larger value of β was obtained using the spectrometer compared to the photodiode, which reflects the faster decay of the signal when the time response is measured using the oscilloscope. This difference can be attributed to a change in the background, which affects the signal recorded with the photodiode similarly to what was already observed on the NIRIM system. The position of the maximum calculated from the fitting parameters is $t = \frac{m}{\beta} = 2.3 \times 10^{-7}$ s (photodiode) and $t = \frac{m}{\beta} = 4.6 \times 10^{-1}$ s (spectrometer). Even though the results largely differ, this difference is attributed to the limited time resolution of the measurement methods.

The measurements done with the photodiode are less noisy than the ones obtained using the spectrometer. It is particularly visible in **Figure 2b** and **Figure 2d**. However, the photodiode is unfortunately sensitive to background changes during the measurement as observed in **Figure 2a**. For this reason, the setup using the photodiode is less reliable for measuring the time response of the reactors in comparison to the one using the spectrometer, which should be preferred.

The fitting parameters for the two reactors determined from the spectrometer setup data analysis, show a much faster rising part (i.e. a smaller $\frac{m}{\beta}$ value) and a slower decay (i.e. a larger β value) for the linear antenna MWPECVD reactor compared to the NIRIM reactor. Those results are attributed to the different volumes of the two reactors [5].

Table 2 Fitting parameters

Measurement Method	Reactor	β (s^{-1})	m
Photodiode	NIRIM	1.5	4.0
	Linear antenna	6.9×10^{-2}	1.6×10^{-8}
Spectrometer	NIRIM	4.5×10^{-1}	1.0
	Linear antenna	1.2×10^{-1}	5.5×10^{-2}

Using linear system theory, it is possible to determine the time response of the reactors to an arbitrary shaped gas flow input from its pulse response [6,7]. The dynamic response is given by equation (3):

$$c(t) = \int_0^t u(\tau)h(t-\tau)d\tau \quad (3)$$

Where $u(t)$ is an arbitrary input signal and $h(t)$ is the impulse time response of the reactor.

In order to test this model, we have compared the experimental time response of the NIRIM reactor to a one-minute-long, square shaped flow of N_2 with an amplitude of 2 sccm injected into the hydrogen plasma using a MKS mass flow controller in the same conditions as described in **Table 1**, and its modeled response. The experimental response was recorded using the spectrometer and the calculated response was obtained using equations (1) and (3) and the fitting parameters in **Table 2**. The curves presented in **Figure 3** compare the experimental results to the calculated responses using both the photodiode or the spectrometer fitting parameters. It is clearly visible that the modeled response using the spectrometer fitting parameters more closely matches the experimental data. The discrepancy observed in the rising part of the curve is attributed to the slow response time of the mass flow controller (500 ms), which is smaller than the rising time of the NIRIM reactor dynamic response of 2.7 s, but not negligible.

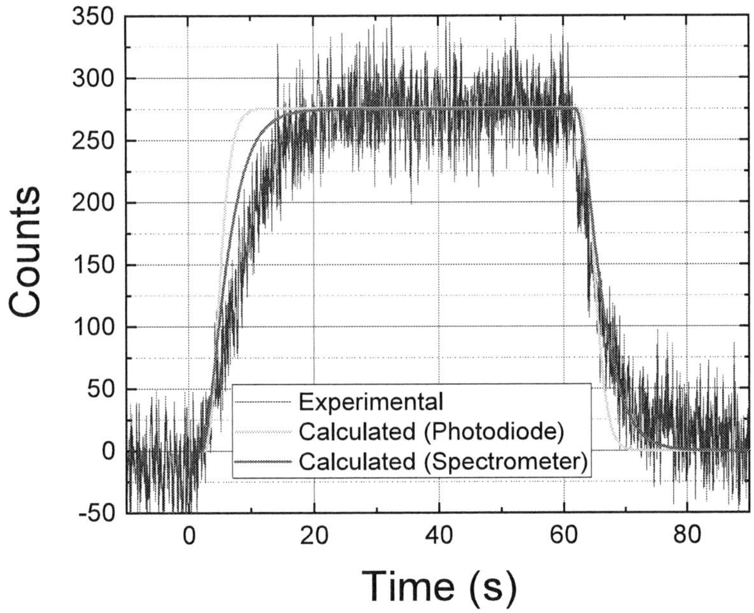

Figure 3 Recorded and simulated response of the NIRIM reactor to a one-minute-long, square shaped flow of N_2 with an amplitude of 2 sccm using the plasma conditions described in Table 1

3. CONCLUSION

Two different measurement setups were used to measure the time response of a NIRIM reactor and a linear antenna MWPECVD reactor to a N_2 gas pulse: one simple setup based on the voltage measurement across a photodiode combined with an optical filter, and another using a compact spectrometer. Despite a better signal to noise ratio compared to the spectrometer, the photodiode has the disadvantage of being sensitive to background changes during the experiment, which ultimately makes it unsuitable for measurement of the dynamic response of MWPECVD reactors. The time response depends on the physical characteristic of the reactor and operating pressure: a faster rising time and a slower decay were observed for the linear antenna MWPECVD reactor compared to the NIRIM system. Those differences can be explained by the large difference in volume between the two reactor chambers. Using linear system theory and the fitting curve of the time response obtained with the spectrometer, it was also possible to successfully predict the response of the NIRIM system to a one-minute-long square N_2 pulse.

ACKNOWLEDGEMENTS

The financial support from the following sources is gratefully acknowledged: The Czech Science Foundation grant ID: GACR 21-03538S, the SGS Student Grant ID: SGS21/057/OHK3/1T/13, and the Operational Program Research, Development, and Education financed by European Structural and Investment Funds and the Czech Ministry of Education, Youth and Sports (Project No. SOLID21 - CZ.02.1.01/0.0/0.0/16_019/0000760).

REFERENCES

[1] LAMBERT, N., WEISS, Z., KLIMŠA, L., FEKETE, L. and MORTET, V. Highly phosphorus-doped nanocrystalline diamond growth and properties. In: *31st International Conference on Diamond and Carbon Materials*. Sep. 2021.

[2] KAMO, M., SATO, Y., MATSUMOTO, S. and SETAKA, N. Diamond synthesis from gas phase in microwave plasma. *Journal of Crystal Growth.* [online]. 1983, vol. 62, no. 3, pp. 642-644. Available from: https://doi.org/10.1016/0022-0248(83)90411-6.

[3] FENDRYCH F. *et al.* Growth and characterization of nanodiamond layers prepared using the plasma-enhanced linear antennas microwave CVD system. *J. Phys. D: Appl. Phys.* [online]. 2010, vol. 43, no. 37, p. 374018. Available from: https://doi.org/10.1088/0022-3727/43/37/374018.

[4] QAYYUM, A., ZEB, S., NAVEED, M. A., REHMAN, N. U., GHAURI, S. A. and ZAKAULLAH, M. Optical emission spectroscopy of Ar-N2 mixture plasma. *Journal of Quantitative Spectroscopy and Radiative Transfer.* [online]. 2007, vol. 107, no. 3, pp. 361-371. Available from: https://doi.org/10.1016/j.jqsrt.2007.02.008.

[5] PENDAR H. and SOCHA, J. J. Estimation of Instantaneous Gas Exchange in Flow-Through Respirometry Systems: A Modern Revision of Bartholomew's Z-Transform Method. *PLoS ONE*. 2015, vol. 10, no. 10, p. e0139508. Available from: https://doi.org/10.1371/journal.pone.0139508.

[6] PORAT, B. *A course in digital signal processing.* New York: John Wiley, 1997.

[7] HAMILTON, J. D. *Time series analysis.* Princeton, N.J: Princeton University Press, 1994.

LIGHT SCATTERING TECHNIQUES AS A KEY IN PROVIDING INSIGHTS INTO THE INTERACTIONS AND STABILITY OF ALKALI-ACTIVATED MATERIALS BASED ON BLAST FURNACE SLAG

Michal KALINA, Petr HRUBÝ, Jiří SMILEK, Lukáš KALINA, Vlastimil BÍLEK Jr.

Brno University of Technology, Faculty of Chemistry, Brno, Czech Republic, EU, kalina-m@fch.vut.cz

https://doi.org/10.37904/nanocon.2021.4376

Abstract

Light scattering techniques represent a group of methods providing the absolute information on the molecular weight, particle size, conformation, polydispersity and the stability of analyzed dispersion systems. These techniques are primarily used for the characterization of biopolymers, synthetic polymers and nanoparticles. The utilization in the area of alkali-activated materials based on blast furnace slag is more challenging, mainly due to their highly alkaline nature and also due to the generally used high content of solid fractions dispersed in liquid media. This contribution was focused either on the optimization of sample preparation as well as on the definition of the optimal setting of measuring parameters to be able to properly use the electrophoretic light scattering (Zetasizer Nano ZS; Malvern Panalytical Ltd.) for the description of the stability of these complex samples. For these purposes, we studied dispersion systems of blast furnace slag in activating solutions (sodium hydroxide, sodium carbonate and sodium silicate). The obtained zeta potential provided crucial insight into the individual sample stability. The results are indicating that the proper technique of sample dispersion and the maintenance of the precise content and ratios of individual components represent important parameters managing the stability of alkali-activated materials based on blast furnace slag.

Keywords: Alkali-activated materials, blast furnace slag, conductivity, stability, zeta potential

1. INTRODUCTION

The development of modern building materials is nowadays closely connected to government restriction as well as by the environmental issues such as tendencies to reduce CO_2 emissions and the efforts to utilize the waste materials and by-products from other industries to protect nature and natural raw materials [1]. Alkali-activated materials (AAM) represent the non-traditional cementitious materials, which fulfil these ideas and currently are studied as materials that could serve as the potential substitute to the ordinary Portland cement in some applications [2]. The typical example of AAM is alkali-activated blast furnace slag (AAS) [3]. Generally, these materials show interesting chemical [4,5] and temperature resistance [6] and high early strength [7] of the material but on the other hand the high autogenous and drying shrinkage must be mentioned as the main disadvantage of the broader practical application of AAM [8]. The published research also indicates the necessity to design a specific admixture exclusively for the individually used AAM to overcome this issue [9], otherwise, AAMs will never fulfil their potential and will never be competitive to Portland cement [10].

At this point, the need for a technique allowing the simple description of interactions among blast furnace slag (BFS) and other components in the systems (activation solutions, admixtures) seems to be crucial. One of the methods how to determine these characteristics is the method of electrophoretic light scattering allowing to determine the zeta potential of the electrokinetic potential in the interfacial double layer of BFS grains [11]. As can be found in the literature [12] this important parameter characterizing the surface chemistry of BFS is dependent on pH, the nature of the used alkaline activator and also on the addition of admixture used for further processing of BFS dispersions. Moreover, the values of zeta potential determined for AAS are also influenced by the concentration of analysed particles. Kashani et al. [12] highlighted that the values of zeta potential of AAS have a direct connection to the rheological properties of slag-based suspensions containing sodium or potassium hydroxides and sodium silicate. The lowest yield stress was observed for slag suspension containing sodium silicate, followed by slag in water without activator, while it markedly increased for sodium hydroxide. Higher yield stress values are in accordance with the structure-making/structure-breaking theory of structural forces as well as with the determined values of zeta potential that were usually negative for slag in water, which was induced by the partial deprotonation of silanol groups on the slag surfaces. Contrary to alkali hydroxide activators, increasing dose of sodium silicate induced a shift to the more negative values of zeta potential and thus contribution of the repulsive electric double layer forces increased.

The zeta potential in cementitious suspensions can be determined using the electroacoustic method [13], enabling the measurement of concentrated suspensions [14], but an issue of too high alkalinity can still arise. In addition, often only the electrophoretic light scattering is available. However, it inherently requires a diluted sample, which may affect the results obtained. Therefore, our experimental work focused on optimizing the electrophoretic light scattering method for the investigation of the interactions and stabilization of BFS in various used AAS. For these purposes, we have used the dilution approach, where we studied the direct relation of BFS to Na^+ concentration and the total content of these components in the system to be able to predict the behaviour of original non-diluted samples.

2. EXPERIMENTAL

2.1. Materials

The individual samples of alkali-activated pastes were prepared using a blast furnace slag (BFS) with Blaine fineness of 400 $m^2 \cdot kg^{-1}$ and density of 2.88 $g \cdot cm^{-3}$. The chemical composition of BFS as was obtained by X-ray fluorescence analysis (XRF) is given in **Table 1**. The individual studied samples of alkali-activated slag (AAS) were prepared using the activation solutions of sodium hydroxide (Lach-Ner s.r.o), sodium carbonate (Lach-Ner, s.r.o) and sodium silicate having the molar ratio of SiO_2/Na_2O (Ms = 1.50) (Vodní sklo, a.s.).

Table 1 Chemical composition of blast furnace slag obtained by XRF analyses

material	Chemical composition (wt.%)									
	SiO_2	Al_2O_3	CaO	MgO	SO_3	Na_2O	K_2O	TiO_2	MnO	Fe_2O_3
blast furnace slag	34.7	9.1	41.1	10.5	1.4	0.4	0.9	1.0	0.6	0.3

2.2. Sample preparation

Concentration series of all activating solutions were prepared in advance prior to their mixing with slag. The specific range of concentrations was given by their achievability with respect to the original composition of the activators: 0-5 mol $Na^+ \cdot dm^{-3}$ for sodium carbonate solution, 0-9.93 mol $Na^+ \cdot dm^{-3}$ for sodium silicate solution and 0-15 mol $Na^+ \cdot dm^{-3}$ for sodium hydroxide solution. Then each activating solution was mixed with slag (1:1 by volume) to obtain 10 cm^3 of the whole suspension (paste) and homogenized with a spatula for 3 minutes. After another two minutes of rest, the paste was diluted by demineralized water in the range 1:10, 1:50, 1:100, 1:250, 1:500 and 1:1000 and immediately characterized by electrophoretic light scattering (section 2.4).

2.3. Physico-chemical characterization of activation solutions

The individual alkali-activated solutions were characterized through potentiometric pH determination using the Titrando 888 module equipped with a combined Unitrode electrode for a highly alkaline pH region with an integrated Pt1000 temperature sensor (Metrohm AG, Switzerland). The data shown further in the text represent the mean values of five repeated measurements (results in the form of mean values ± SD, n = 5). The data processing was performed using Tiamo software (version 2.5; Metrohm AG, Switzerland).

Dynamic viscosity of all the used alkali-activated solutions was measured employing Höppler falling ball viscometer AMVn (Anton Paar GmbH, Austria). The data shown further in the text represent mean values of three repeated measurements (results in the form of mean values ± SD, n = 3).

2.4. Zeta potential measurement

The effect of the type of alkaline activation (sodium hydroxide, sodium carbonate, sodium silicate with Ms = 1.5), their concentration in the sample (range from 0 to 15 mol·dm^{-3} concentration of Na^+) and the dilution ratio (dilution by volume with distilled water in the range of 1:10, 1:50, 1:100, 1:250, 1:500 and 1:1000) of the individual alkali-activated BFS dispersions was investigated by the method of electrophoretic light scattering using Zetasizer Nano ZS (Malvern Panalytical Ltd., UK). Simultaneously the average scattered light intensity, conductivity and the zeta potential of the suspensions were measured in three repeated measurements of three individually taken samples (results are presented in the form of mean values ± SD, n = 9). The obtained data from ELS analysis were processed by the Zetasizer software (version 7.11; Malvern Panalytical Ltd., UK).

3. RESULTS AND DISCUSSION

The principal task of this work was the investigation of the effect of the type of used alkali-activated solution and their concentration in the sample on the interactions with BFS. For these purposes, the measurement of zeta potential using the method of electrophoretic light scattering was selected. This method is used for investigation of the surface chemistry of colloidal dispersions, description of their surface charge with the corresponding stability of particles against mutual aggregation and for investigation of the interactions among particles in the system [5]. Due to the experimental limitations (mainly the limits for the pH, conductivity and concentration), it was not possible to study the sample directly, but we had to perform a dilution with distilled water to be able to subsequently extrapolate the zeta potential to the original samples. For these purposes, the dilution ratios between 1:50 and 1:1000 (by volume) were used.

3.1. Physico-chemical characterization

The physico-chemical characterization of all the used activation solutions by measurement of their pH, conductivity and dynamic viscosity was realized as the initial experiments necessary for subsequent discussion of zeta potential results obtained on AAS in a broader context in Section 3.2. The results of the measured pH and dynamic viscosity for all the used activation solutions are shown in **Figure 1**.

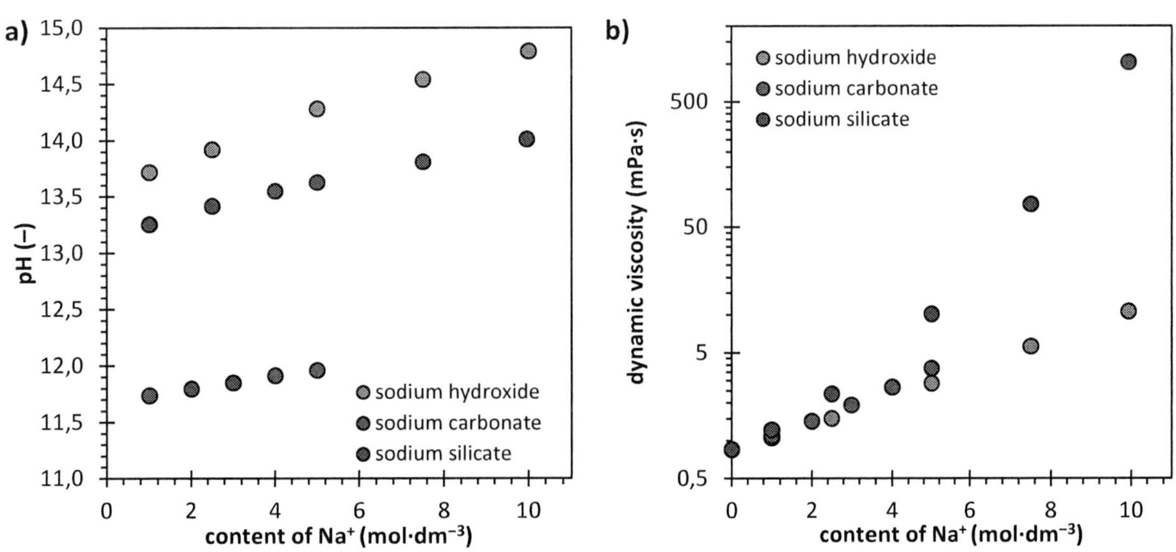

Figure 1 a) pH and b) dynamic viscosity of activation solutions recalculated on the real content of Na^+ ions

The determined pH values are indicating expected dependences on the real content of Na^+ ions. Especially for sodium hydroxide solutions, the data need further corrections on activity coefficients. The highest pH values were measured for sodium hydroxide solutions and on the other hand in agreement with our expectation, the lowest pH values were measured for sodium carbonate solutions. The data from the measurement of dynamic viscosity of the individual activating solutions (**Figure 1b**) are indicating similar trends of sodium hydroxide and sodium carbonate (logarithm of dynamic viscosity has almost linear dependence on the concentration of Na^+ in the solution). The different behaviour was observed in the sodium silicate activation solution, where the dependence of logarithm of dynamic viscosity on the concentration of Na^+ was much higher and indicated an almost polynomial trend in the data. These results are in good correlation with published literature [15].

3.2. Zeta potential measurement

The zeta potential measurement by using the method of ELS was used as the principal experimental approach for the investigation of the interactions and stability of dispersions of BFS in alkali-activated solutions (sodium hydroxide, sodium silicate and sodium carbonate). Due to the experimental limitations of ELS (pH and conductivity and concentration limits) the measurement was performed under the broad range of used dilution ratios (diluted by volume with distilled water 1:50, 1:100, 1:250, 1:500 and 1:1000). For this reason, the lowest used dilution (1:10) was not reliably measured and thus the results are not shown.

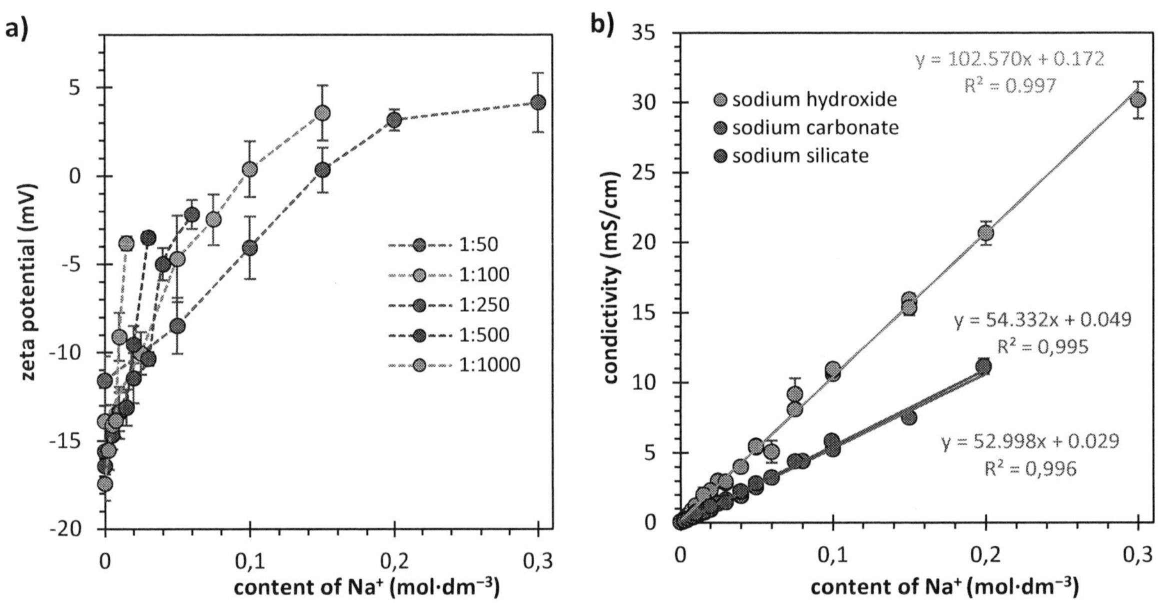

Figure 2 a) Development of zeta potential of various used dilution ratios of BFS dispersions in sodium hydroxide recalculated on the real content of Na+ ion in the samples, b) Measured conductivities of BFS dispersions in all used alkaline activators recalculated on the real content of Na+ ion in the samples.

The main task in this part of the experiments was the utilization of the dilution approach as the possibility to get an overview of the trend in zeta potential for individual samples in respect to the used dilution ratio, helping to predict and extrapolate the behaviour of original undiluted samples. The example of obtained zeta potential dependencies on the concentration of Na^+ ions in the samples as were obtained for sodium hydroxide alkaline activation for the individual used ranges of dilution are shown in **Figure 2a**. The results are indicating that the dilution has a minimal effect on the zeta potentials of BFS in distilled water without the addition of activators, where the average zeta potential (-15.7±2.2) mV was determined. With the increasing concentration of Na^+ ions in the samples the difference from the original dispersions of BFS in the water started to be more pronounced. The trends in our data are also indicating that from the concentration of Na^+ ions 0.05 mol·dm^{-3}, sample high conductivity and pH are starting to be the key features tailoring and affecting the final value of measured zeta potential of BFS dispersion. **Figure 2b** experimentally confirmed the direct relation of the concentration of Na^{+} ions in BFS dispersions and its measured conductivity. Our dilution approach also indicated that at the concentration of Na^{+} ions 0.1 $mol{\cdot}dm^{-3}$, BFS dispersions reached zero zeta potential, which is indicating their low electrostatic stabilization against aggregation. Further addition of Na^{+} ions gradually shifted zeta potentials into the positive region of the scale. These results are in good agreement with published literature [12].

The comparison of two selected dilution ranges (1:50 and 1:1000 by volume) for the other two studied alkaline activators is shown in **Figure 3**. The presented data are indicating an opposite effect observed for the activation with sodium carbonate and sodium silicate in comparison to sodium hydroxide alkaline activation of BFS. In all studied dilution ranges, we have observed a significant shift to the more negative values of BFS dispersions zeta potential in comparison with original BFS dispersion in water. Again, this finding is in good correlation with published data [12]. This is indicating a more pronounced negative charge of individual particles in these samples with further addition of both these activating solutions and also an increase in their colloidal stability. These results are really interesting, especially considering the observed dependences of conductivities of BFS dispersions on Na^{+} ion concentration showed in **Figure 2b**.

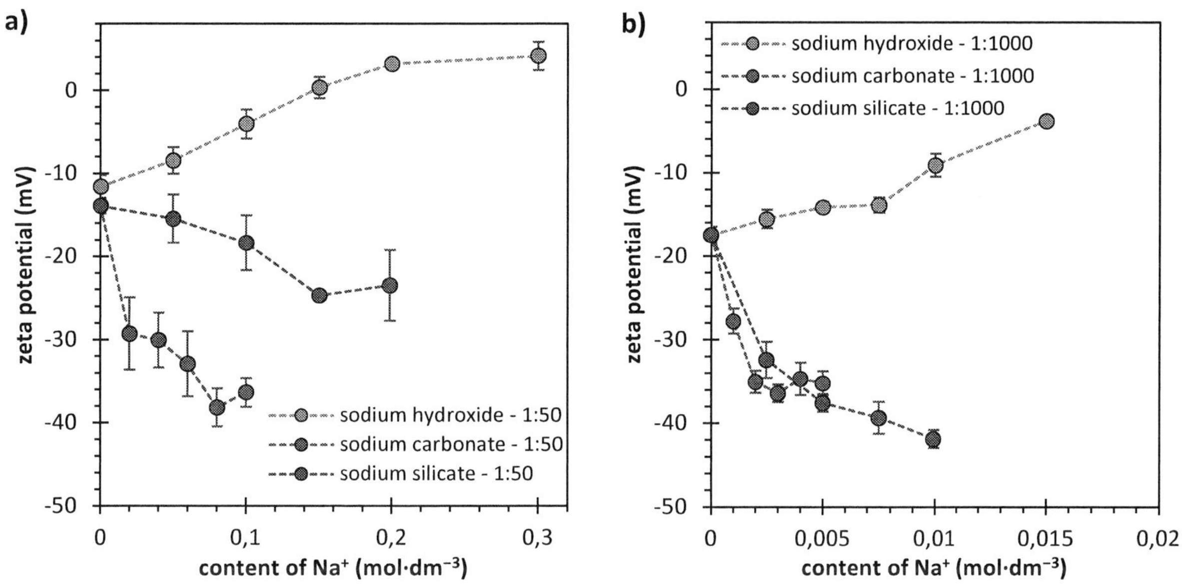

Figure 3 Development of zeta potential of BFS in used alkaline activators recalculated on the real content of Na+ in the samples - a) 1:50 dilution (by volume) with water; b) 1:1000 dilution (by volume) with water

The conductivities of sodium silicate and sodium carbonate dispersions are showing similar trends as the data for sodium hydroxide. These experimental data are indicating a completely different mode of operation of sodium carbonate and sodium silicate compared to the classical BFS activation using sodium hydroxide. The literature [9,12] explain this behaviour by the adsorption of carbonate or silicate anions onto the BFS grains.

The comparison of the dilution of sodium carbonate and sodium silicate is indicating an almost negligible effect in the case of sodium carbonate. This AAS showed the lowest pH in comparison with sodium silicate and sodium hydroxide solutions and due to this fact, the further dilution of these BFS dispersions has an almost negligible effect on the measured zeta potential and corresponding stability of these samples. On the other hand, sodium silicate solutions showed more alkaline pH and also significantly higher dynamic viscosity in comparison to sodium carbonate, which are probably the main reasons for the observed more significant role of dilution on the measured zeta potentials of BFS dispersions.

Although the trend seems similar for both dilutions presented in **Figure 3**, lower dilution gives values closer to zero (for NaOH even to positive values) and it is thus questionable, what situation is in the highly concentrated suspensions prior to their dilution with higher ionic strength etc. Therefore, the results illustrate that the dilution extent together with the nature of the alkaline activator are crucial parameters for the obtained absolute values of zeta potential and one should be cautious when trying to predict the behaviour of the real concentrated suspension. The comparison of the values with those of concentrated suspensions could be very interesting but unfortunately has not been done due to the unavailability of the electroacoustic method able to measure concentrated suspensions [11,13,14] at our department.

4. CONCLUSION

The main task of our work was the investigation of the effect of the type and the concentration of used alkaline activators on the stability and interactions between colloidal particles of BFS. For these purposes, we have used zeta potential as a crucial feature of analyzed samples together with the sample-dilution approach allowing us to predict further behaviour of these systems in a more concentrated state. These experiments were also supplemented by the physicochemical characterization of activation solutions as well as prepared BFS dispersions. The outcomes of our experiments are straightforward. The dilution approach was successful for modelling and extrapolation of possible behaviour and stability of BFS dispersions in all used alkali-activated solutions (sodium hydroxide, sodium silicate and sodium carbonate). We have also observed a different mode of action in the case of sodium hydroxide, where the addition of activation solution increased the zeta potentials from negative values (for zeta potential of slag in water observed average value (-15.7 ± 2.2) mV) to 0 mV and further to positive values. Contrary to this, in the case of sodium silicate and sodium carbonated the addition of AAS into the BFS slag dispersions caused a gradual shift toward more negative values of measured zeta potentials, which is indicating more significant colloidal stabilization of BFS particles in these systems. The results of our work are indicating that the proper technique of sample dispersion and the maintenance of the precise content and ratios of individual components represent crucial parameters managing the stability of alkali-activated materials based on blast furnace slag.

ACKNOWLEDGEMENTS

The work was supported by project GA20-26896S from the Science Foundation of Czech Republic.

REFERENCES

[1] PROVIS, J. L. Geopolymers and other alkali-activated materials: why, how, and what? *Materials and Structures*. 2014, vol. 47.

[2] PROVIS, J. L. Alkali-activated materials. Cement and Concrete Research. 2018, vol. 114.

[3] KALINA, L., BILEK, BRADOVA, L., TOPOLAR, L. Blast furnace Hybrid Cement with Waste Water Glass Activator: Alkali-Silica Reaction Study. *Materials*. 2020, vol.13.

[4] BAKHAREV, T., J. G. SANJAYAN AND Y. B. CHENG Resistance of alkali-activated slag concrete to acid attack. Cement and Concrete Research. 2003, vol. 33.

[5] BERNAL, S. A., E. D. RODRÍGUEZ, R. MEJÍA DE GUTIÉRREZ AND J. L. PROVIS Performance of alkali-activated slag mortars exposed to acids. *Journal of Sustainable Cement-Based Materials*. 2012, vol. 1.

[6] RASHAD, A. M., Y. BAI, P. A. M. BASHEER, N. C. COLLIER, et al. Chemical and mechanical stability of sodium sulfate activated slag after exposure to elevated temperature. *Cement and Concrete Research*. 2012, vol. 42.

[7] PROVIS, J. L. and J. S. J. VAN DEVENTER Alkali Activated Materials State-of-the-Art Report, RILEM TC 224-AAM. Edtion ed.: Springer, 2014. ISBN 978-94-007-7671-5.

[8] LI, Z., T. LU, X. LIANG, H. DONG, et al. Mechanisms of autogenous shrinkage of alkali-activated slag and fly ash pastes. *Cement and Concrete Research*. 2020, vol. 135.

[9] KALINA, L., BILEK, V., BARTONICKOVA, E., KALINA, M., HAJZLER, J., NOVOTNY, R. Doubts over capillary pressure theory in context with drying and autogenous shrinkage of alkali-activated materials. *Construction and Building Materials*. 2020, vol. 248.

[10] HEWELETT, P. C., LISKA, M. *Lea's Chemistry of Cement and Concrete*. 5th ed. Oxford, United Kingdom: Elsevier, 2019.

[11] HUNTER, R. J. *Zeta potential in colloid science: principles and applications*. London: Academic Press, 1988.

[12] KASHANI, A, PROVIS, J. L., QIAO, G. G., VAN DEVENTER, S. J. The interrelationship between surface chemistry and rheology in alkali-activated slag paste. *Construction and Building Materials*. 2014, vol. 65.

[13] FERRARI, L., J. KAUFMANN, F. WINNEFELD AND J. PLANK Interaction of cement model systems with superplasticizers investigated by atomic force microscopy, zeta potential, and adsorption measurements. *J Colloid Interface Sci*. 2010, vol. 347.

[14] DUKHIN, A. S. AND P. J. GOETZ Acoustic and electroacoustic spectroscopy for characterizing concentrated dispersions and emulsions. *Advances in Colloid and Interface Science*. 2001, vol. 92.

[15] VAIL, J.G. *Soluble Silicates - Their Properties and Uses*. Reinhold Book Corporation. 1952.

EMERGENCE OF DARK ZnO NANORODS BY HYDROGEN PLASMA TREATMENT

[1]Zdenek REMES, [1]Maksym BURYI, [1]Dhananjay K. SHARMA, [1]Anna ARTEMENKO, [2]Julia MICOVA, [3]Bohuslav REZEK, [4]Ales PORUBA, [5]Hua Shu HSU, [1]Stepan POTOCKY, [1]Vladimir BABIN

[1]*Institute of Physics, Czech Academy of Sciences, Prague, Czech Republic, EU,* remes@fzu.cz, buryi@fzu.cz, dhananjay@fzu.cz, artemenko@fzu.cz, potocky@fzu.cz, babin@fzu.cz

[2]*Institute of Chemistry, Slovak Academy of Sciences, Bratislava, Slovakia, EU,* chemjumi@savba.sk

[3]*Faculty of Electrical Engineering, Czech Technical University, Prague, Czech Republic, EU,* rezekboh@fel.cvut.cz

[4]*SVCS Process Innovation s.r.o., Valasské Mezirici, Czech Republic, EU,* aporuba@svcs.cz

[5]*Department of Applied Physics, National Pingtung University, Taiwan,* hshsu@mail.nptu.edu.tw

https://doi.org/10.37904/nanocon.2021.4374

Abstract

Plasma hydrogenation of hydrothermally grown ZnO nanorods was done at 13.56 MHz in inductively coupled plasma (ICP) with H_2/Ar gas mixture. Plasma was monitored by optical emission spectroscopy (OES) and by measuring self-bias potential of the stainless-steel sample holder isolated from the ground by a capacitor. The ZnO surface chemical composition was studied by XPS. The exciton-related UV photoluminescence of ZnO nanorods and optical absorption increased significantly after plasma treatment and diminished after plasma oxidation. Increasing Ar in hydrogen plasma caused more negative self-bias, less efficient hydrogenation and lower exciton photoluminescence. We attribute changes of optical properties to changes of surface and related electronic states giving also rise to its black color visually.

Keywords: Optical materials, ZnO, ICP, plasma hydrogenation, OES, XPS, photoluminescence

1. INTRODUCTION

The hydrothermally grown zinc oxide (ZnO) crystalline powder is a low cost and environmentally friendly material with unique optical properties and variety of nano and micro-structures. ZnO nanorods (ZnO NRs) with their high surface-to-volume ratio and related size effects impose challenges for energy conversion, scintillators, photocatalytic wastewater treatment, electrochemical energy storage or sensing applications [1]. For instance, black ZnO nanoparticles prepared by heating in hydrogen atmosphere at 400 °C for 2h displayed long-wavelength absorption and improved photocatalytic performance [2].

We have built inductively coupled plasma (ICP) reactor with up to 300 W discharge power operating at the radio frequency 13.56 MHz for plasma hydrogenation of powder samples. Plasma contains energetic electrons, ions, excited radicals and molecules providing further adjustment of surface properties while keeping a sample at a relatively low temperature [3]. Plasma hydrogenation of ZnO NRs leads to a significant increase of exciton-related UV luminescence [4]. Their surface properties were examined by X-ray photoelectron spectroscopy. To further study such effects, we elucidate in this paper the key role of self-bias on the optical properties of ZnO NRs exposed to plasma in an ICP reactor with an ungrounded stainless-steel sample holder isolated from the ground by a blocking capacitor.

2. EXPERIMENTAL

2.1. Hydrothermal growth of ZnO nanorods

The synthesis of ZnO NRs was based on the reaction of an equimolar mixture of 25 mM analytical grade zinc nitrate hexahydrate $Zn(NO_3)_2{\cdot}6H_2O$ (250 mL) and hexamethylenetetramine $C_6H_{12}N_4$ (HMTA, 250 mL) without further purification. Both solutions were filtered through a Whatman 2 filter and mixed with vigorous stirring (700 rpm, 15 min). The hydrothermal growth was performed at 90 °C for 3 h. After cooling to room temperature, ZnO NRs were freed of residual salts by washing 5 times with distilled water, separated by centrifugation and lyophilized for 24 h. Finally, ZnO NRs were annealed in air at 350 °C for 15 min to prepare anhydride powder with improved crystal quality.

2.2. Inductively coupled plasma (ICP)

An inductively coupled plasma (ICP) reactor with quartz walls, developed in collaboration with the Czech company SVCS Process Innovation s. r. o. operates at the radio frequency (RF) 13.56 MHz with up to 300 W discharge power. It employs H_2, O_2, N_2, NH_3 and Ar process gasses with a flow rate up to 50 sccm. To repeal plasma from the reactor walls, the capacitive coupled plasma (CCP) mode was reduced by an electrically grounded electrostatic shield installed outside the reactor between the quartz walls and the RF coil. Plasma treatment of ZnO NR was performed using a stainless steel holder similar to an rotating cradle isolated from the ground by a blocking capacitor with an RF impedance of 50 Ω. Prior to plasma treatment, the chamber was evacuated for 30 min to a base pressure of less than 0.1 Pa and purged for 10 minutes with 50 sccm of process gas flow. The plasma hydrogenation was done for 30 min with RF power 100 W, 10 sccm H_2 and up to 2 sccm Ar gas flow. The plasma oxidation was done for 5 min with RF power 100 W and 10 sccm O_2. The self-bias potential of the holder was measured using an electrometer Keithley 6517A in a guarded DC mode.

2.3. Characterization

Optical emission spectra (OES) were measured in 400-900 nm spectral range with 1 nm spectral resolution by spectrally calibrated fiber coupled CCD spectrometer (B&W Tek BTC112E) optically focused in the middle of the quartz chamber just above the sample holder. The exposition time varied from 0.3 to 3 s depending on plasma luminosity. Each spectrum was averaged 10 times and the spectra were divided by the exposition time. Spectra obtained in dark were subtracted.

The X-ray photoelectron spectra were acquired at AXIS Supra photoelectron spectrometer (Kratos Analytical Ltd., UK) equipped with a monochromated Al Kα X-ray source (1486.6 eV) and hemispherical energy analyzer. ZnO NRs were pressed in pellets and fixed by Cu tape on the sample holder. The XPS survey spectra were collected at a pass energy of 80 eV whereas the high resolution (HR) spectra were acquired at the pass energy of 20 eV. XPS spectra were obtained at a constant take-off angle of 90° from the analysis area of 0.3×0.7 mm^2. The XPS spectra were measured with a charge neutralization system. Peak fitting of the measured high resolution spectra was performed by the CasaXPS software using Shirley background and Gaussian/Lorentzian line shapes (GL(30), 70% Gaussian and 30% Lorentzian) without fixing of peak width (full width at half maximum, FWHM). Binding energies of photoelectron lines were determined with an accuracy of ± 0.2 eV and referenced to C-C bonds in C 1s component at 285.0 eV [5].

Photoluminescence emission spectra (PL) were excited by pulsed, narrow band-pass UV LED (340 nm, 100 mW/cm^2, 333 Hz) and measured at room temperature in the 350-780 nm spectral range using a perpendicular geometry, long pass optical filters (LP350 and LP400 nm), a spectrally calibrated double gratings monochromator SPEX1672, a cooled multi-alkali photomultiplier tube, current preamplifier with a gain 10 μA/V and a dual-phase lock-in amplifier referenced to the UV LED frequency.

3. RESULTS AND DISCUSSION

Figure 1 compares OES of hydrogen plasma (10 sccm H_2) upon addition of Ar (0, 0.5, 1 and 2 sccm Ar). The OES spectra clearly show the presence of atomic hydrogen (H peaks at 656 and 486 nm) and several Ar peaks in near IR (750-850 nm) with increasing intensity correlating with Ar flow [6]. OES shows that Ar reduces luminosity of H peaks due to a plasma shield around the sample holder, see also **Table 1**. This is because the asymmetry of charging and discharging processes in RF plasma induces on ungrounded holder a negative steady-state self-biased floating potential that repels electrons from the holder and equalizes the electron and ion flows [7].

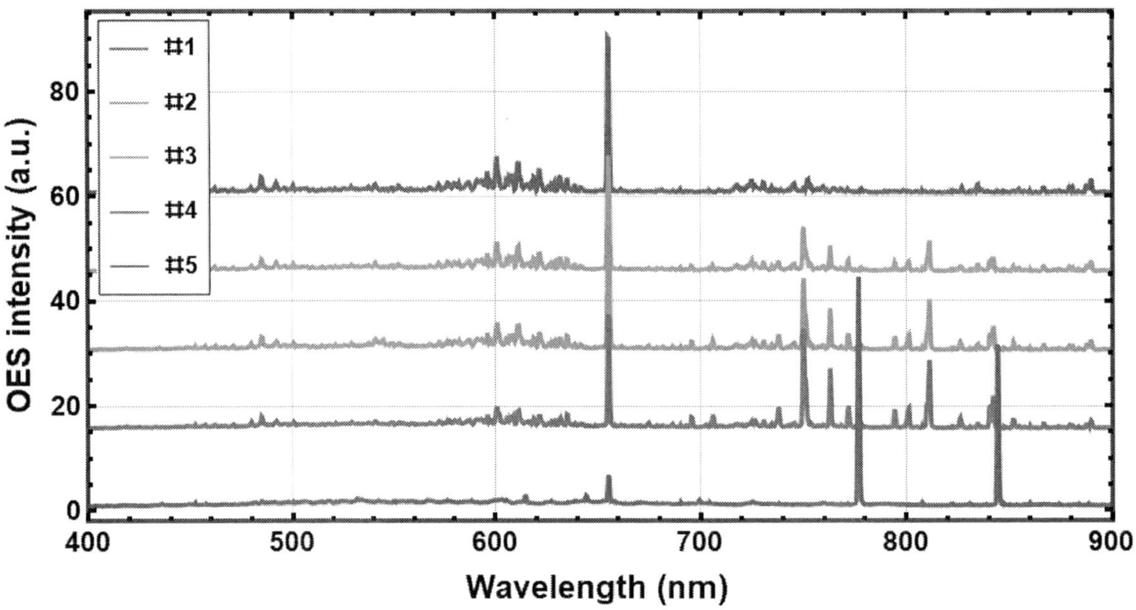

Figure 1 OES of hydrogen plasma upon addition of Ar (#1, #2, #3 and #4) compared to oxygen plasma OES (#5) measured with inserted hydrogenated ZnO sample.

Table 1 Gas flow, pressure, self-bias potential, integral intensity of H related OES at 653-659 nm, integral intensity of exciton-related UV PL (360-420 nm) and defect-related visible PL (480-780 nm). *Sample #5 was hydrogenated as #1 prior oxygen plasma treatment.

sample	H_2 flow (sccm)	Ar flow (sccm)	O_2 flow (sccm)	Gas pressure (Pa)	Self-bias (V)	OES @ 656 nm (a.u.)	UV PL (a.u.)	vis PL (a.u.)
#0	-	-	-	-	-	-	2 363	11 464
#1	10	0	0	6	-23.5	935	44 052	7 749
#2	10	0.5	0	11	-25.5	701	31 565	7 653
#3	10	1	0	9	-26.4	762	21 623	6 743
#4	10	2	0	11	-29.8	688	23 232	8 752
#5	0(10)*	0	10	9	-9.8	325	356	65 564

Two OES peaks at 635 & 637 nm are related to Si etched by hydrogen from quartz walls of the chamber. Oxygen plasma is dominated by strong IR peaks at 777 and 845 nm, but OES also shows that oxygen plasma releases some hydrogen from the previously hydrogenated ZnO NRs as demonstrated by the presence of the

656 nm peak (see #5 in **Figure 1**). No O peaks were detected in pure hydrogen plasma but small amount of O appears in H_2 & Ar plasma probably due to scattering of ZnO by heavy Ar ions.

Figure 2 SEM image of hydrothermally grown ZnO NRs prior plasma treatment (MAIA3, TESCAN, 10 kV).

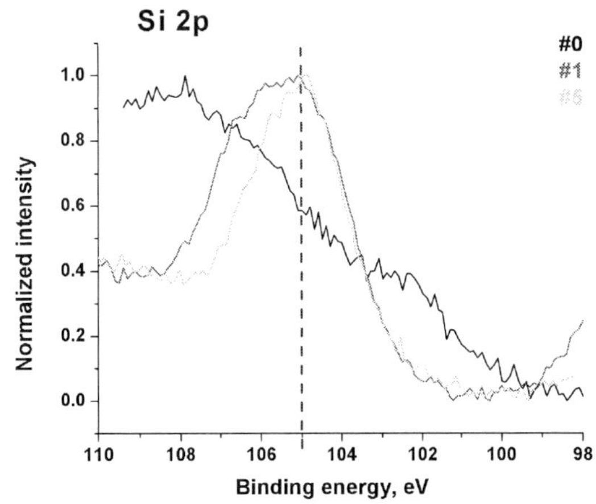

Figure 3 The comparison of normalized Si 2p XPS peaks of reference (#0), hydrogenated (#1) and oxidized (#5) ZnO.

Scanning electron microscopy (SEM) image of the synthetized ZnO NRs revealed crystallites with the hexagonal shape, the length about 2 µm and the diameter about 0.2 µm, see **Figure 2**.

Table 2 The atomic concentration of chemical elements calculated from HR XPS spectra of the plasma untreated sample (#0) and plasma hydrogenated sample (#1) subsequently plasma oxidized (#5).

sample	Atomic concentration of chemical elements, at.%						
	O	C	Zn	Si	F	Cl	Ca
#0	45.2	15.2	38.2	-	-	1.4	-
#1	52.0	11.6	19.3	11.0	2.9	2.0	1.2
#5	50.6	22.6	17.1	8.7	-	0.5	0.5

For the purpose of XPS studies, the reference sample (#0) was compared with the hydrogenated sample (#1) that was subsequently exposed for 5 min to oxygen plasma (#5). The Si 2p peak at about 104.9 eV were detected for plasma treated samples, see **Figure 3**. According to the Ref. [8], the Si signal can be attributed to SiF_4/Si. However, no F signal was observed in O_2 plasma treated ZnO NRs. **Table 2** shows that plasma hydrogenation caused increase of O content up to 52.0 at.% on the ZnO surface in contrast to reference sample (45.2 at.%) and the decrease of Zn to ~19 or even 17.1 at.% (after plasma oxidation) in comparison with initial values (38.2 at.%). On the other hand, hydrogen plasma treatment led to contamination of the ZnO surface with Si and Ca that comes probably from the quartz walls of the reactor chamber. The F signal (2.9 at.%) was detected after hydrogen plasma treatment (probably from the vacuum seals). Contamination by Si, Ca and F was reduced by plasma oxidation, thus it must be indeed surface related. Moreover, XPS analysis revealed that all ZnO samples were contaminated with C and Cl (probably a contamination from air or residual reactants used in hydrothermal growth of ZnO NRs).

The originally white reference sample became darker after plasma hydrogenation and white again after plasma oxidation as visualized by macro photographs, see **Figure 4**. Despite dark color, the PL spectrum of plasma hydrogenated ZnO NRs shows a strong exciton-related UV PL with the maximum at 380 nm and a low intensity of deep defect-related broad band in the red spectral region [9], see **Figure 5**. Plasma oxidation of the previously hydrogenated sample suppresses the exciton-related PL and strongly enhances the defect related

PL. This is a strong evidence that the observed optical properties after plasma treatment are caused by a complex changes of surface electronic states [10].

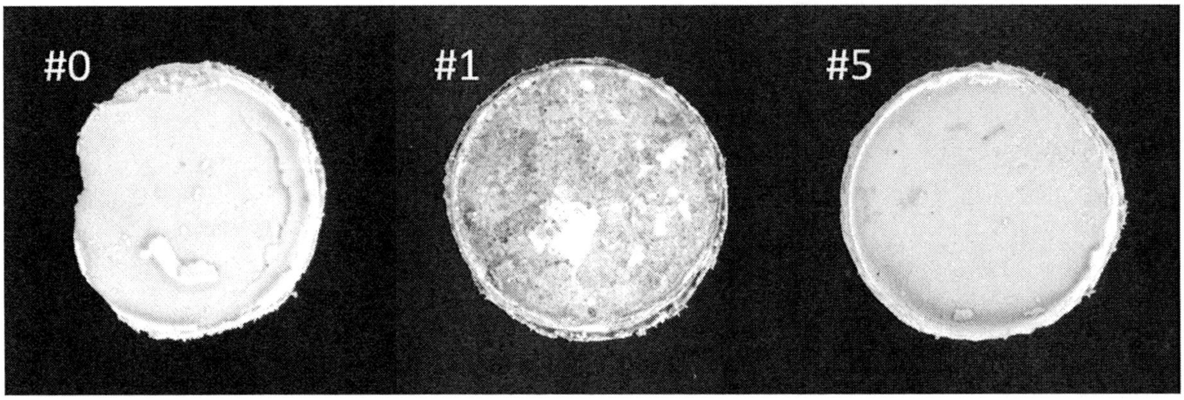

Figure 4 Macro photographs of plasma untreated (#0), plasma hydrogenated (#1) and plasma oxidized (#5) ZnO pellets (Ø 3 mm, 4 mg) taken with a macro objective (Lumix G Macro 30 mm, F2.8 ASPH)

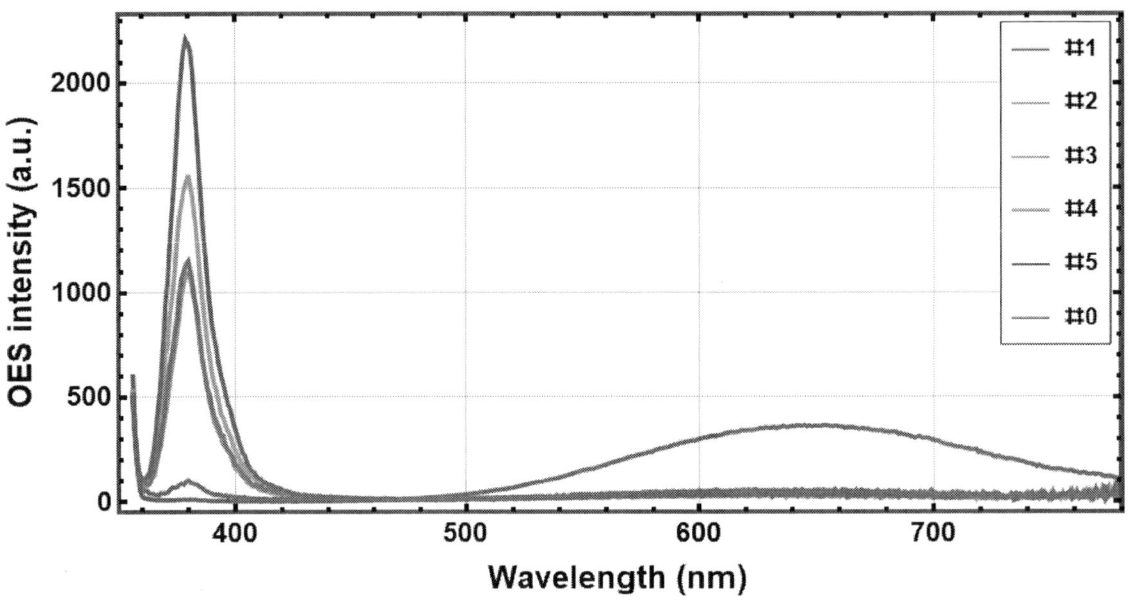

Figure 5 PL spectra of plasma hydrogenated ZnO upon addition of Ar (#1, #2, #3 and #4) compared to plasma oxidized (#5) and plasma untreated (#0) sample

4. CONCLUSION

Hydrothermally grown ZnO NRs were exposed to a hydrogen-argon plasma using a stainless-steel sample holder electrically isolated from the ground by a blocking capacitor. The PL spectra of plasma hydrogenated ZnO NRs show a strong exciton-related UV PL with a maximum at 380 nm, reduced red PL of defect states and increased absorption giving rise to its dark color visually. Changes in optical properties were attributed to complex changes in surface electronic states of ZnO NRs. Addition of Ar into hydrogen plasma caused more negative self-bias around the sample holder and less efficient plasma hydrogenation of ZnO NRs. Plasma oxygen treatment applied on hydrogenated ZnO NRs resulted in significantly reduced exciton-related UV PL and enhanced red PL associated with defects as well as reduced surface contamination by Si, Ca and F from quartz walls and vacuum seals of the ICP reactor. Further optimization of gas pressure, holder capacitance, and in-situ oxygen plasma treatment is required to obtain an optimally hydrogenated ZnO NRs with minimal contamination.

ACKNOWLEDGEMENTS

This work was supported by the Czech Science Foundation (GACR) project No. 19-02858J, CAS Mobility Plus project SAV-AVCR-21-09, Operational Program for Research, Development and Education project CZ.02.1.01/0.0/0.0/16_019/0000760 (Solid21), the Slovak Scientific Grant Agency (VEGA) grant 2/0157/20, the Ministry of Science and Technology of Taiwan (MOST) under the grant No. 108-2923-M-153-002-MY3, by the CzechNanoLab Research Infrastructure (MEYS CR, project no. LM2018110) and by the project Strategy AV21 (program Diagnostic Methods and Techniques). We thank R. Jackivová, J. Macíček and Yu Ying Chang for technical help.

REFERENCES

[1] KLINGSHIRN, C. F., ed. *Zinc Oxide: from fundamental properties towards novel applications*. Springer series in materials science. Heidelberg ; London: Springer, 2010, vol. 120. ISBN 978-3-642-10576-0.

[2] XIA, T., WALLENMEYER, P., ANDERSON, A., MUROWCHICK, J., LIU, L., CHEN, X. Hydrogenated black ZnO nanoparticles with enhanced photocatalytic performance. *RSC Adv.* 2014, vol. 4, pp. 41654-41658

[3] GRILL, A. *Cold plasma in material fabrication: from fundamentals to applications*. New York: The Institute of Electrical and Electronics Engineers, Inc., 1994.

[4] REMEŠ, Z., AUBRECHTOVÁ DRAGOUNOVÁ, K., MIČOVÁ, J. Plasma hydrogenation of hydrothermally grown ZnO micropods. In: *NANOCON 2020, 12th International Conference on Nanomaterials - Research & Application: NANOCON 2020- Conference Proceedings*. 2020, pp. 512-517

[5] MIČOVÁ, J., BURYI, M., ŠIMEK, D., DRAHOKOUPIL, J., NEYKOVA, N., CHANG, Y. Y., REMEŠ, Z., POP-GEORGIEVSKI, O., Jan SVOBODA, J., IM, Ch. Synthesis of zinc oxide nanostructures and comparison of their crystal quality. *Applied Surface Science*. 2018, vol. 461, pp. 190-195

[6] KRAMIDA, Alexander a Yuri RALCHENKO. *NIST Atomic Spectra Database, NIST Standard Reference Database 78*. National Institute of Standards and Technology. [online]. 1999. Available from: https://doi.org/10.18434/T4W30F

[7] BUTLER, H. S., KINO, G. S. Plasma Sheath Formation by Radio-Frequency Fields. *Physics of Fluids*. 1963, vol. 6, p. 1346.

[8] POWELL, C. *X-ray Photoelectron Spectroscopy Database XPS, Version 4.1, NIST Standard Reference Database 20*. National Institute of Standards and Technology. [online]. 1989. Available from: https://doi.org/10.18434/T4T88K

[9] BURYI, M., REMEŠ, Z., BABIN, V., NOVOTNÝ, M., VANĚČEK, V., AUBRECHTOVÁ DRAGOUNOVÁ, K., MIČOVÁ, J., LANDOVÁ, L., KUČERKOVÁ, R., MORE-CHEVALIER, J., CHERTOPALOV, S., FITL, P., KMJEČ, T. Influence of Mo doping on the luminescence properties and defect states in ZnO nanorods. Comparison with ZnO:Mo thin films. *Appl. Surf. Sci.* 2021, p. 149679.

[10] CHEN, X., LIU, L., LIU, Z., MARCUS, M. A., WANG, W.-Ch., OYLER, N. A., GRASS, M. E., MAO, B., GLANS, P. A., YU, P. Y., GUO, J., MAO, S. S. Properties of Disorder-Engineered Black Titanium Dioxide Nanoparticles through Hydrogenation. *Scientific Reports*. 2013, vol. 3, pp.1-7

SEQUENTIAL CHEMICAL FRACTIONATION AS A TOOL FOR ACCESSING THE ORGANIC MATTER IN BIOCHAR

[1]David ŠIRŮČEK, [1]Michal KALINA, [1]Martina KLUČÁKOVÁ

[1]*Brno University of Technology, Faculty of Chemistry, Brno, Czech Republic, EU*

https://doi.org/10.37904/nanocon.2021.4372

Abstract

This contribution is focused on utilization of a sequential chemical fractionation method for determination of organic matter content, its distribution in biochar and the possible use of biochar as a soil conditioner in agriculture. For these purposes the optimized chemical fractionation procedure was used to reveal the content of organic matter in different biochar samples with different properties (samples with European biochar certification for usage in agriculture) which depend significantly on the temperature used during a pyrolysis of biomass residues in the process of biochar production. In parallel, these samples were also fractionated by classic alkaline extraction to obtain the so-called extractable fraction of organic matter (NOM).

Biochar is one of the important soil conditioners, known for having positive effect on crop yield, soil quality, nutrient cycle and carbon sequestration due to the transfer of organic carbon from it to the soil. However, the effect depends on the properties of the biochar, its doses to the soil, but also on the properties of the soil itself. Surprisingly, some authors point to the fact that biochar does not always have a positive effect on soil, plants or microfauna. Therefore, it is necessary to perform its depth characterization to be able to predicts its role in soil and its optimal application dose.

Individual fractions obtained from sequential chemical fractionation as well as NOM samples were characterized by methods of elemental analysis (determination of organic elements content), thermogravimetry (contents of ash, organic matter and moisture). The results showed that the method of sequential chemical fractionation gives higher yields of organic matter compared to classical alkaline extraction. Moreover, the obtained fractions divided according to solubility and strength of binding to residual inorganic matter of biochar can be better characterized by physical-chemical methods, which provides more detailed information about organic matter in biochar and its possible effect on soil properties.

Keywords: Biochar, organic matter, chemical fractionation, elemental analysis, thermogravimetric analysis

1. INTRODUCTION

Soil organic matter (SOM) is one of the soil components consisting of three basic types - living organisms (worms, insects, microorganisms), dead microorganisms and finally matter formed from organic residues of plants, animals and humus. The decomposition of the first two types creates nutrients for plants (nitrogen, potassium, phosphorus) [1-3]. The main reactive part of SOM is humic substances - natural substances arising from the chemical and biological degradation of plants and animal remains. They are crucial in ecosystems because they regulate global carbon and nitrogen cycles, the growth of plants and microorganisms, the transport of anthropogenic compounds and heavy metals, and the stabilization of soil structure. To date, however, no technologies have been developed to control the activity of humic substances due to their high molecular complexity. There is a broad variety of analytical techniques for their characterization, however, none of these techniques alone is sufficient to elucidate the structural complexity of particles of humic molecules [4,5].

In general, humic substances are supramolecular associations of heterogeneous, relatively small molecules in nanoscale range that are held together by weak non-covalent interactions, such as hydrogen bonds and hydrophobic forces. The complexity of the humic molecule can be reduced by the gradual destruction of intramolecular interactions that stabilize these complex suprastructures (composed of nanoparticles of organic matter) and thus individual fractions of colloidal size of humic substances (organic matter) can be isolated. These individual fractions of organic matter obtained from sequential fractionation are called humeomics. It is a gradual separation of molecules from humic suprastructures by controlled cleavage of intermolecular bonds. In terms of particle size, these particles obtained by sequential fractionation are colloidal in size which significantly improve the resolution of various analytical methods often used for characterization [6,7].

In recent years, increasing demands have been placed on crop yields which also brings with it increased demands on soil fertility. For these reasons, various substances are often used to increase fertility, which are added artificially. They are most often fertilizers. Recently, however, research has also focused on the application of alternative additives to soils in order not only to increase fertility but also to improve the physicochemical properties of the soil and to increase the content and stabilization of soil organic matter. These additives are often included in a group called - soil supportive substances (SSS). Substances that are widely used for application include, for example, lignite fly ash, biochar and a mixture of fractions of humic substances. It is also very important to apply a suitable form with a given particle size, internal structure and morphology (porous structure) [8,9].

As mentioned before, one of the soil supportive substances is a biochar which is kind of a charcoal mostly produced by pyrolysis of biomass (any substance that has an organic character and comes from plants and animals) in absence of oxygen. Biomass can be considered an excellent source material, as it is renewable and includes many things that are often considered waste. It is a stable, solid and carbon rich substance that can endure in soil for thousands of years. It is characterized by its physical and chemical properties, such as adsorption capacity or composition which vary depending on the starting material and production conditions (temperature). Its main benefit can be found in the agricultural sector where it is used as a soil supplement which significantly improves soil quality. Besides its effect on physico-chemical properties of soil (soil structure, compactness, water and ion sorption capacity, pH, conductivity) tt contributes to the sustainability of agriculture and makes a positive influence to the transfer of atmospheric carbon to the soil which has a positive impact on flora, water and air [10,11].

2. EXPERIMENTAL

2.1. Materials and Methods

For this purpose, a sample of biochar from the Czech company BIOUHEL.CZ s.r.o. was used. But before it could be used, large particles and other impurities (incompletely pyrolyzed biomass or tiny stones) were removed from the material using a sieve. For study purposes, the method of sequential chemical fractionation was used in the work. Organic matter contained in biochar affects the key properties of the soil such as fertility, ability to retain water, transport nutrients etc. In order to be able to describe the effect of biochar, it is necessary to precisely characterize the organic matter which must first be separated from the inorganic part. In order to do that, the method previously mentioned - sequential chemical fractionation - was performed and for comparison purposes, the classical procedure - alkaline extraction - was performed too. Using the sequential chemical fractionation allowed us to separate all the different-strongly bound fractions of similar types of molecules. The alkaline extraction only allowed us to see a mixture of organic molecules contained but overall, this method wasn`t really good to use in order to separate the organic matter from biochar. Three replicates of sequential chemical fractionation were performed for each of these three samples.

2.2. Sequential Chemical Fractionation

The procedure of sequential chemical fractionation was adopted from the publications of Nebbioso and Drossos [6,7]. The details of the optimization of this chemical extraction procedure of organic matter for the samples of soil and lignite are summarized in chapters 2.2.1 - 2.3 and can be found in our previous publication [12].

2.2.1. Water soluble fraction 1 (AQU1)

First step was to dry used organic material in an oven at 45 °C to evaporate all the water. Then 50 g of dried material was transferred to Erlenmeyer flask and 150 ml of 0,1 M HCl was added. Flasks were covered with a double layer of parafilm and left on a shaker overnight. The next day samples were filtered under reduced pressure. Filtrate was frozen and all the water was evaporated in a freeze drier.

2.2.2. Organic soluble fraction 1 (ORG1)

Dried solid material after AQU1 was mixed with 100 ml DCM and 50 ml MeOH in Erlenmeyer flask and covered with a double layer of aluminium foil and left on a shaker overnight. The next day samples were filtered under reduced pressure. All the solvent was evaporated on a rotary evaporator at 50 °C. The residue after evaporation was transferred to vials left to evaporate the solvent.

2.2.3. Water and organic soluble fractions 2 (AQU2 and ORG2)

Dried solid material after ORG1 separation was transferred to high-pressure sealable teflon flasks and 50 ml of a 12% solution of boron trifluoride in methanol was added. Vials were tightly sealed, clamped in a metal rack and placed in an oven at 85 °C for 8 hours. Then vials were removed from the oven, let to cool down and additional 50 ml of the solution was added. Vials were sealed and clamped again and placed in the oven for another 8 hours under same conditions. Then after cooling down the solution was filtered under normal pressure and the solid material was dried. The filtrates were L/L extracted (liquid to liquid) with chloroform. The obtained AQU2 and ORG2 fractions were purified and dried in the same way as was described for AQU1 and ORG1.

2.2.4. Water and organic soluble fractions 3 (AQU3 and ORG3)

Dried solid material after the previous step was covered with 50 ml of 1 M solution of KOH in MeOH (for the lignite sample a 100 ml of the solution was used). A magnetic stirrer was added, the sample flask was attached to a direct condenser and heater was set to 75 °C. The mixture was refluxed under nitrogen atmosphere for 2 hours. Then all the content in the flask was filtrated under reduced pressure and the solid material was dried. The filtrates were extracted with dichloromethane using L/L extraction. The obtained AQU3 and ORG3 fractions were purified and dried in the same way as was described for AQU1 and ORG1.

2.2.5. Water soluble fraction 4 (AQU4)

Dried solid material after the previous step was covered with 50 ml of a 47% solution of HI in water (for the lignite sample a 100 ml of the solution was used). A magnetic stirrer was added, the sample flask was attached to a direct condenser and heater was set to 75 °C. The mixture was refluxed under nitrogen atmosphere for 2 days. Then all the content in the flask was filtrated under reduced pressure and the solid material was dried. The filtrates were neutralized with $NaHCO_3$ to pH 7 and the iodine contained was reduced with $Na_2S_2O_3$. Then the filtrates were dialyzed against distilled water and freeze-dried.

2.2.6. Residual extractable organic matter fraction (ResOM)

To isolate the residual organic matter, 5 g of dried solid material after AQU4 fractionation and 45 ml of extractant (0.5 M NaOH and 0.1 M $Na_4P_2O_7$) were mixed. Samples were left on a shaker overnight. The samples were filtered and the filtrates were dialyzed against distilled water and freeze-dried.

2.3. Natural organic matter fraction isolation (NOM)

In parallel, a conventional alkaline extraction was performed. Thus 5 g of the prime material was covered with 45 ml of extractant (0.5 M NaOH and 0.1 M $Na_4P_2O_7$). After overnight shaking, the samples were filtered under reducer pressure and the filtrates were dialyzed against distilled water and freeze-dried.

3. RESULTS AND DISCUSSION

3.1. Yield comparison depending on the method used

As mentioned before, yields of extractable organic matter from biochar were obtained using two methods - sequential chemical fractionation (humeomics) and alkaline extraction. Based on the obtained yields (**Figure 1**) it can be said that the largest fraction is AQU4, however, a majority of it are $NaHCO_3$ and $Na_2S_2O_3$ used during the fractionation. On the other hand, the AQU1 fraction, which consists of unbound or only weakly bound molecules, is significantly higher than other samples obtained. A similar trend can be observed with the ORG (organo-soluble) samples. Comparing the humeomics (ΣHUMEO) and the alkaline extraction (AE) we notice that AE yield is several times smaller but still we have to take in account the elemental analysis and thermogravimetry data to calculate the organic matter content which is much more significant and eliminates the content of salts introduced into the individual fractions during the fractionation procedures.

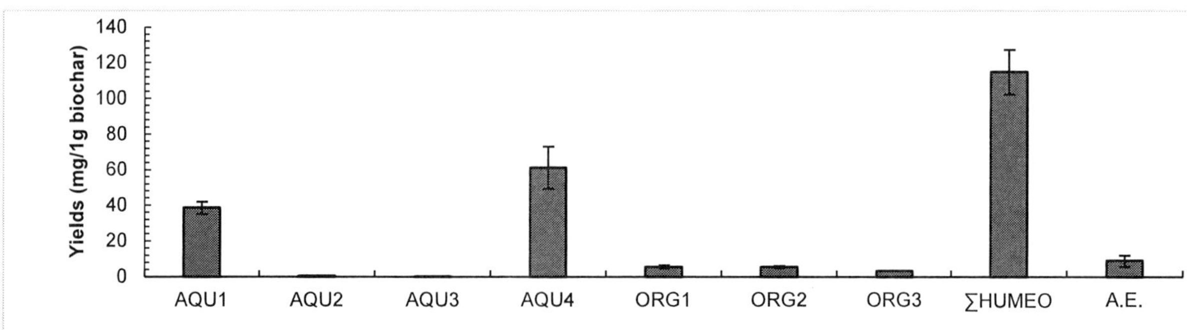

Figure 1 Average yields obtained by both sequential chemical fractionation and alkaline extraction related to 1 g of biochar used during both processes.

3.2. Elemental analysis and thermogravimetry application

Using elemental analyses, the content of carbon, hydrogen, nitrogen and sulfur was determined and the content of oxygen could be later calculated after the determination of moisture and ash content using thermogravimetry. Here we only focus on the carbon and nitrogen content as these are the most important in connection with the application of biochar to soil to improve its fertility and improve conditions for growth. The source biochar sample was also subjected to elemental analysis in order to determine the content or carbon and nitrogen. The total amount of organic carbon contained in biochar was (476 ± 50) mg and nitrogen was (26 ± 4) mg per 1 g of biochar. Using the thermogravimetry data shows that the sample of biochar used contains approximately (699 ± 45) mg per 1 g of organic matter which means that 70 % of the sample is only organic matter.

In general, organo-soluble fractions (ORG) contain more carbon than water-soluble fractions (AQU). As we can see on **Figure 2**, the yield of organic carbon and nitrogen is higher towards more weakly bound fractions and that AQU2 and AQU3 fractions almost do not contain neither one of them. Overall, we were only able to extract around 2.4 % of total organic carbon and slightly more, around 4.6 %, of total nitrogen contained. Both yields are relatively small so it can be deduced that the investigated biochar has a solid structure that does not allow us to extract much of its organic content. The alkaline extraction has even lower yields - about 0.9 % of carbon and 1.2 % of nitrogen.

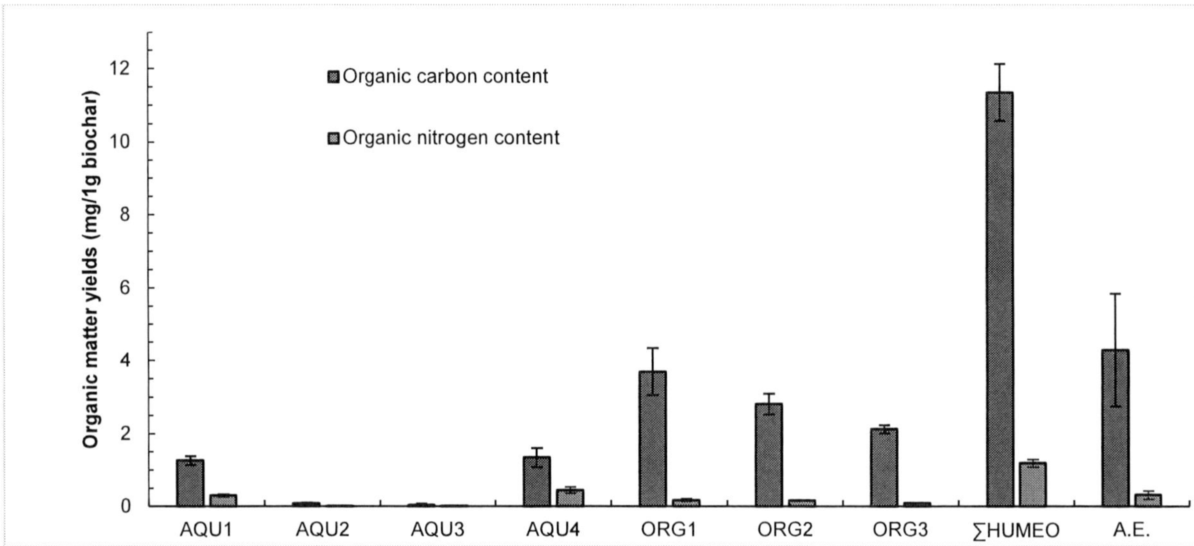

Figure 2 Organic matter content. The blue color represents the organic carbon content and the green color represents the organic nitrogen content.

So, as we can see above on **Figure 2**, the total amount of isolated organic carbon was about 12 mg per 1 g of source biochar which is really a low amount considering that it contains (476 ± 50) mg per 1 g. Nevertheless, it was much more than we were able to isolate by the alkaline extraction which makes the sequential chemical fractionation more appropriate method to use. It is more laborious but also more than twice as effective as alkaline extraction. In case of nitrogen it is more than 4 times as effective. Similar facts can be found in the work of Drosos et. al. [7]. Moreover, separating the extracted organic matter into individual fractions allows us to more easily characterize those fractions using advanced spectroscopic and chromatographic techniques which provides us with more comprehensive information about the biochar sample and will be important for us to predict its effect on soil after its application into the soil.

4. CONCLUSION

This contribution was focused on describing the organic matter of biochar in order to better understand the effect of its use as a soil supplement. The biochar was first fractionated using the sequential chemical fractionation. Fractions containing organic matter with different solubilities (water and organo-soluble) were prepared and the individual fractions were separated based on their binding force to the residual biochar matrix. The other method used was alkaline extraction. All the samples were subsequently characterized by elemental analysis which determined the organic elements (carbon, hydrogen, nitrogen and sulfur) content and, in order to determine the total organic and inorganic matter content, the thermogravimetric analysis was used.

The results showed how much and which organic elements are contained in the biochar sample related to 1 g of it. The assumptions about the solid structure were confirmed according to the fact that, despite the use of strong and relatively aggressive solvents, it has been impossible for us to extract a significant amount of total organic matter contained. Nevertheless, the biochar has a great potential for use in agricultural sector as a soil supplement as biochar is likely to be retained in the soil matrix for a relatively long time and thus contribute to the stabilization of soil organic matter.

Overall, the analyzes show that the investigated biochar is a really rigid and durable material that, even after using of aggressive reagents and high temperatures, still contains a high amount of organic matter. The structure couldn`t be cleaved at a level where strongly bound organic matter could be extracted, which is indicating the high resistance of organic matter in biochar structure.

To be able to make more detailed description we will perform chromatographic analysis (GC-MS and LC-MS) which will provide us information on the molecular level of fractions or structural analysis methods (e.g. FTIR, NMR). Another follow-up research will include using these separation methods to characterize a biochar that has been cultivated and treated with soil microorganisms. That will make it possible to better assess their potential effect on biochar in the soil. This experimental set-up will bring crucial insights into the study on the bioavailability of organic matter of biochar for the surrounding soil environment.

ACKNOWLEDGEMENTS

This work has been supported by project GJ20-28208Y from Czech Science Foundation of the Czech Republic.

REFERENCES

[1] *Agronomy Fact Sheet no. 41: Soil organic matter.* Cornell University Publications, 2008. [cit. 2014-01-21].

[2] PAUL, E. A., CLARK, F. E. *Soil Microbiology and Biochemistry.* San Diego: Academic Press, 1996, pp. 93, 320.

[3] TAN, K.H. *Humic Matter in Soil and the Environment. Principles and Controversies.* New York: Marcel Dekker, Inc., 2003, p. 386.

[4] GUSTAFSON, A. F. *Soils and soil management.* London: McGraw-Hill Book Company, 1941. McGraw-Hill publications in the agricultural sciences.

[5] BOHLEN, P. J.; EDWARDS, C. A. Earthworm effects on N dynamics and soil respiration in microcosms receiving organic and inorganic nutrients. *Soil Biology and 50 Biochemistry.* 1995, vol. 27, pp. 341-348. [cit. 2017-02-26].

[6] NEBBIOSO, A., PICCOLO, A. Basis of a Humeomics Science: Chemical Fractionation and Molecular Characterization of Humic Biosuprastructures. *Biomacromolecules.* 2011, vol. 12, no. 4, pp. 1187-1199. [cit. 2019-03-31].

[7] DROSOS, M., NEBBIOSO, A., MAZZEI, P., VINCI, G., SPACCINI, R., PICCOLO, A. A molecular zoom into soil Humeome by a direct sequential chemical fractionation of soil. *Science of The Total Environment.* 2017, vol. 586, pp. 807-816.

[8] MASTO, R. ebhin, Md. a. ANSARI, J. GEORGE, V.a. SELVI a L.c. RAM. Co-application of biochar and lignite fly ash on soil nutrients and biological parameters at 77 different crop growth stages of Zea mays. *Ecological Engineering.* Elsevier B.V. 2013, vol. 58, pp. 314-322. [cit. 2019-04-05].

[9] RAM, L., SRIVASTAVA, N., JHA, S., SINHA, A., MASTO, R., SELVI, V. Management of Lignite Fly Ash for Improving Soil Fertility and Crop Productivity. *Environmental Management.* New York: Springer-Verlag, 2007, vol. 40, no. 3, pp. 438-452. [cit. 2019-04-07].

[10] LEHMANN, Johannes, RILLIG, Matthias C., THIES, Janice, MASIELLO, Caroline A., HOCKADAY, William C. and CROWLEY, David. Biochar effects on soil biota - A review. *Soil Biology and Biochemistry.* [online]. 2011, vol. 43, no. 9, pp. 1812-1836. Available from: https://doi.org/10.1016/j.soilbio.2011.04.022. ISSN 00380717.

[11] TAN, Zhongxin, LIN, Carol S.K., JIn Xiaoyan, RAINEY, Thomas J. Returning biochar to fields: A review. *Applied Soil Ecology.* [online]. 2017, vol. 116, pp. 1-11. Available from: https://doi.org/10.1016/j.apsoil.2017.03.017. ISSN 09291393.

[12] ŠIRŮČEK, David, KALINA Michal, KLUČÁKOVÁ Martina. Influence of the Application of lignite on the distribution of organic carbon in soil. In: *NANOCON 2020 12th International Conference on Nanomaterials - Research and Application.* [online]. Ostrava: TANGER, 2020, p. 489-494. ISBN 978-80-87294-98-7. ISSN 2694-930X. Available from: https://doi.org/10.37904/nanocon.2020.3765

A SAMPLER OF DIFFRACTION AND REFRACTION OPTICALLY VARIABLE IMAGE ELEMENTS

[1*]Miroslav HORÁČEK, [1]Stanislav KRÁTKÝ, [1]Milan MATĚJKA, [1]Jana CHLUMSKÁ, [1]Petr MELUZÍN, [2]Jiří PIRUNČÍK, [2]Ivo AUBRECHT, [2]Marek KOTRLÝ, [1]Vladimír KOLAŘÍK

[1]*Institute of Scientific Instruments of the CAS, v. v. i., Brno, Czech Republic, EU,* [*]*mih@isibrno.cz*

[2]*Institute of Criminalistics of the Police of the Czech Republic, v. v. i., Prague, Czech Republic, EU*

https://doi.org/10.37904/nanocon.2021.4371

Abstract

Diffraction and refraction optically variable image elements are basic building blocks of planar structures for advanced security of documents and valuables. A sampler formed by an array of 36 diffraction structures - binary, tertiary, quaternary and blazed gratings (period range 400 nm - 20,000 nm) - represents a cross-section throughout technological steps - mastering, galvanic replication and embossing. Electron-beam writing technology with Gaussian beam and electron energy of 100 keV, with very small forward scattering of high energy electrons and with the possibilities to create a linear grating with the minimal period of 100 nm, was used to create the master. An important advantage of high-resolution electron-beam lithography is its substantial flexibility in combining possible planar structures with significantly different parameters, such as very dense and relatively shallow structures together with deep structures (approx. 10 microns) with precise shapes (micro-lenses or Fresnel structures). For protection of documents and valuables, interesting results are induced with planar optical structures consisting of non-periodic arrangements, which are characterized by high robustness to counterfeiting and imitation.

While the origination process is available for grating period down to 100 nm, the mass replication technology appears to be a bottleneck of the entire technological process. Measurement of topology and profiles of the structures by atomic forces microscope and documenting the quality of technological process of the three steps of replication of planar optically variable elements was performed for all 36 structure types of sampler.

Keywords: Diffraction, refraction, optically variable image element, e-beam writer, mastering, galvanic replication, embossing, security, valuables

1. INTRODUCTION

Various sophisticated techniques are involved in protection of documents and valuables against counterfeiting or tampering and the most wide-spread non-printing types are the ones that rely on spectral and directional redistribution of incident white light by surface-relief structures embossed in a foil. These structures are commonly denoted as variable image elements and may be characterized with varying appearance that depends on geometry of illumination and observation or even on polarization state.

While the electron-beam writing origination process is available for grating period down to 100 nm, the mass replication technology appears to be a bottleneck of the entire technological process - mastering, galvanic replication and embossing. Problems of individual technological steps of the production process are demonstrated on a sampler of 36 mesh diffraction structures - binary (2L), tertiary (3L), quaternary (4L) (with range of periods from 400 nm to 4,320 nm) and blazed gratings (16L) (with range of periods from 2,000 nm to 20,000 nm), see **Figure 1**.

2. MASTERING

Electron-beam lithography (EBL) is recognized to be a versatile origination tool as regards the possibility of creating very fine relief features and a general design. Electron-beam writer with Gaussian beam and electron energy of 100 keV was used to create the master. With very small forward scattering of high energy electrons, it is possible to create a linear grating with a period of down to 100 nm. Advantage of this versatility consist in combining possible planar structures with significantly different parameters: Very dense though relatively shallow structures may be combined together with deep structures (approx. 10 microns) with precise shapes. Micro lenses, Fresnel structures and non-periodic arrangements may serve as examples of the latter. Hence, EBL induced interesting results for protection of documents and valuables, which are characterized by high robustness to counterfeiting and imitation.

With the EBL the sampler was designed as an array of 6 x 6 structures and realized in a resist layer on a silicon substrate. The total sampler size is 35 mm x 35 mm and each diffraction structure occupies an area of 4 mm x 4 mm separated from each other with gaps.

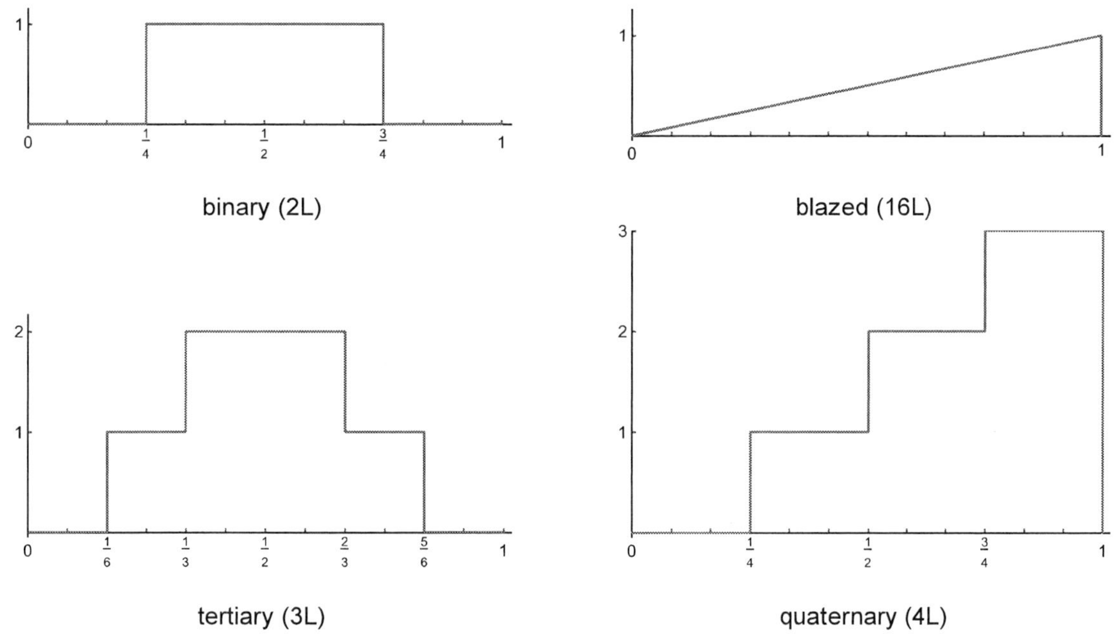

Figure 1 Binary (2L), blazed (16L), tertiary (3L) and quaternary (4L) profiles used in sampler

3. ELECTROFORMING AND EMBOSSING

As it is common practice in manufacturing micro- and nanostructures, a master is first copied into a sturdier material before it is mass replicated. These materials are often required to endure higher pressures, higher temperatures and to be good heat conductors which is fulfilled with variety of metals. Such an ideal material is nickel that is widely used for shims for embossing processes as it is also the case of our contribution. Main steps used in our production processes are shown in schematic **Figure 2**. First step after successful completion of the master is a vacuum metallization. From this essential step, the non-conducting photoresist of master is covered with thin conducting layer of silver which is necessary for following electroforming. The electroforming generally produces metal layers from assembling molecules from an electrolyte on conducting surfaces. Since a sturdy sheet is required for further applications, thickness of created nickel layers ranges from tenths of microns to hundreds of microns. The galvanic process itself (electroforming) actually comprises successive production of three nickel sheets. The first one is a galvanic replica of the master (a negative), the second replicates the first (a positive), and the third nickel sheet, being the replica of the second, is the final galvanic replica of the master (the negative again), which is a shim used for embossing (to create the positive imprint again). The structures were embossed in polymer foils (metallized or coated with dielectrics of high refractive index and with special layers facilitating transfer of relief patterns from the shim). The embossing process is performed on a laboratory-made roll-to-roll system that is designed for a narrow (around 5 cm) foil transport. A shim is clamped to a heated central roll, to which a foil is pneumatically pressed with two pressure rolls. With the system, only an embossing period of 250 mm is allowed which is given by used shims length 125 mm and transport of the foil that is set to continually emboss from the one pressure roll and in between from the second pressure roll.

The production steps were accompanied with AFM analyses of 3D reliefs and profiles of all the structures of the sampler, namely of the master, the third galvanic replica and the final pressing in a foil. The measurements of the master have to be done before the galvanic replication because the master is destroyed during it. Moreover in principle, it is not possible to make scanning electron microscope (SEM) images to analyze cross-sectional cuts of the structures of the master because it is subsequently used in the galvanic replication.

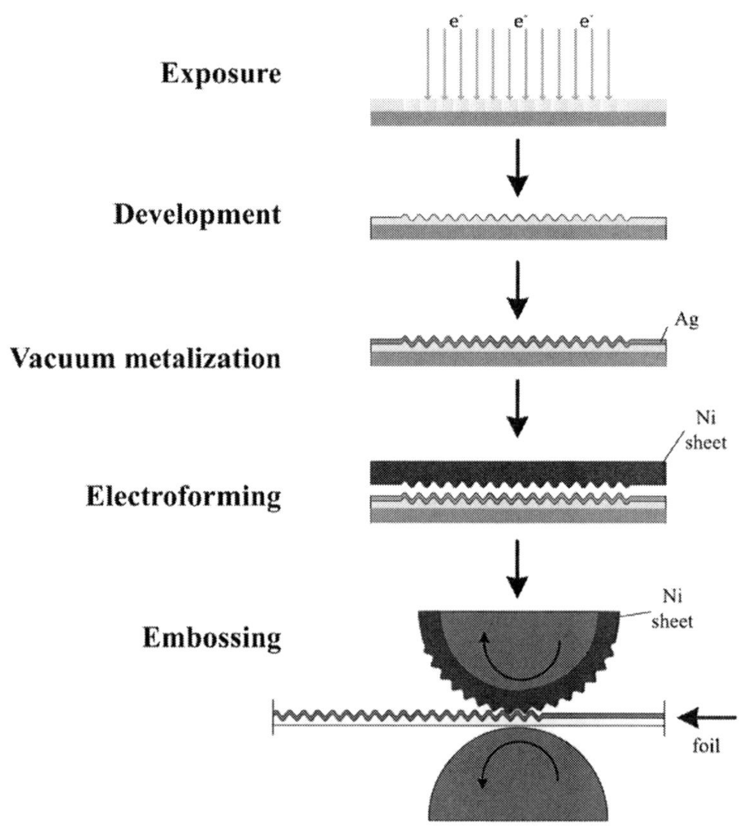

Figure 2 Realization steps of the sampler of diffraction structures

4. RESULTS AND DISCUSSIONS

Deviations of the profile of a final embossed structure from that of the master are almost ubiquitous in practice and can generally consists in a loss of modulation depth, a change of the period and a deformation of its shape. Throughout the structures on the sampler, these deviations were carefully studied and analyzed according to previous AFM measurements. Results of the analyses of the studied structures in a form of their profiles for a few periods are presented in **Figure 3**. Although one example of the profile is shown for each structure type from each part of the sampler, the following evaluations of the fidelity of profiles is based on the results of all the structures from the sampler.

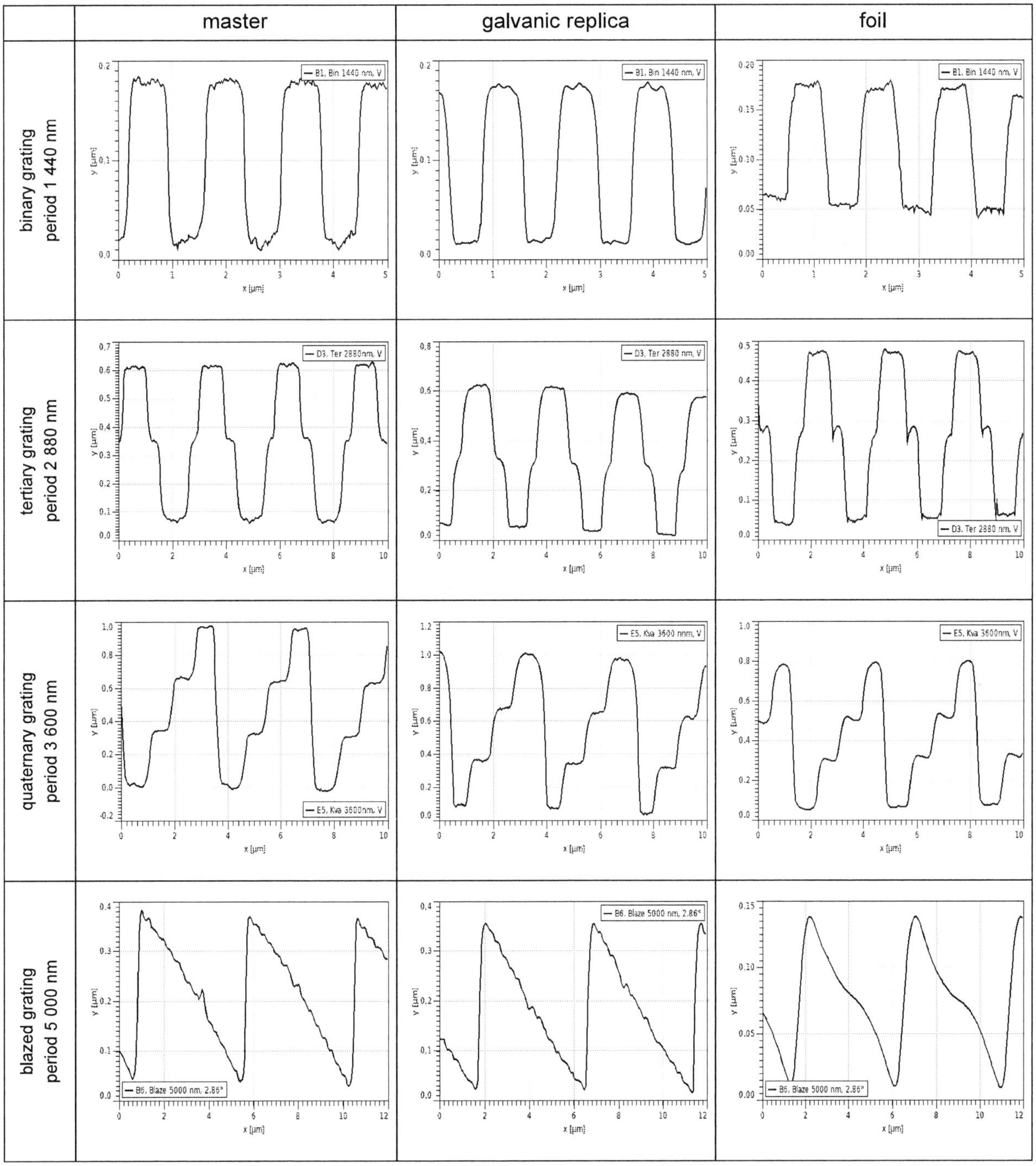

Figure 3 Profiles of the selected binary (2L), tertiary (3L), quaternary (4L) and blazed (16L) gratings from master, galvanic replica and foil

The electroforming process seems to cause some deformations of the profile shape including a certain decrease of the modulation depth and a very small change of a grating period may occur. As can be seen from the AFM images of grating profiles that were measured (see **Figure 3**) on the sampler's master and its Ni-replica grating structures, electroforming causes certain deformations of corresponding profile shapes. Especially, the tertiary and quaternary ones are affected. Generally, the more significant the deformations are, the shorter the grating period and this phenomenon does not seem to be essential for periods about approximately 2 μm or larger. There is a drop in modulation depth of the tertiary gratings and quaternary gratings for shortest periods (720 nm) of approximately 50 %. The absolute depths of other grating structures of 150 nm - 1,000 nm roughly matches the assumptions.

On the other hand, deformations of the embossed profiles is more significant compared to the corresponding master ones for large periods (above 2 μm), which could be seen best in the case of the binary shape. An explanation of apexes and valleys occurring at the bottom and top parts of the profile might be sought in different technological conditions at the edges and central part of both cavities and protrusions of the Ni-replica [4-7] but more detailed inquiries into this subject would be beyond the scope of this contribution. In the case of tertiary and quaternary profiles, deformations is rather difficult to describe but again it appears more evident with the increase of the grating period. There is a drop in modulation depth between Ni-replica and foil of approximately 20 % - 25 %. The more pronounced drop in modulation depth occurred for the shortest periods (720 nm) of the tertiary gratings and quaternary gratings and for largest periods (4,320 nm) for all the binary, tertiary and quaternary gratings.

Deformations of the blazed profiles, which were generated as 16-level stairs-like structures for the periods in the range 2,000 nm - 20,000 nm, deserves a special attention. There is a 30 % - 40 % drop of the modulation depth (compared to the master) in the whole period range. For periods larger than a certain limit, the embossed profiles are much less inclined in its central part than its master or shim counterpart while the profile inclination increases towards both the profile minimum and maximum. The limit is affected by modulation depths and approximately, it is lower for higher modulations. While deformations becomes significant for the periods above 5,000 nm for the design depth of about 250 nm - 300 nm, with a line of division around the period of 5,000 nm (see the profile in **Figure 3** at the bottom right), a deeper profile (with the design depth around 1,000 nm) is much more deformed for the same grating period and it is relatively undeformed though shallower, for the grating period of 2,000 nm.

5. CONCLUSIONS

The sampler of 36 mesh diffraction structures - binary, tertiary, quaternary and blazed gratings - was realized, measured and analyzed. The electron-beam writing technology was used to create the master, followed by the three steps of galvanic replication and embossing in polymer foils on a roll-to-roll system.

For the sake of this short contribution, we used the most straightforward quantification of grating profiles fidelity of the copies in comparison to their master. This means only those deformations that are clearly recognizable by an eye from measured profiles (**Figure 3**) were considered significant and discussed. A more rigorous approach is certainly possible (e.g. employment of convolution operators) but its scope would be a matter of standalone analyses.

The first reproduction method, electroforming of nickel shims, does not cause significant deviations of shapes or modulation depths in the whole studied range of periods of the binary profile (720 nm to 4,320 nm), of the blazed profile (2,000 nm to 20,000 nm). The same applies to deviations of the tertiary profile in the range from about 1,400 nm to 4,320 nm, of the quaternary profile in the range from about 2,160 nm to 4,320 nm.

The second reproduction method, embossing, causes significant deformations not only for the smallest period but in a much more significant manner for the longest ones as well. The latter can be seen best for the binary and blazed profile but deformations occur for the tertiary and quaternary profiles too.

Overall, faithful reproduction of the master engages two important aspects. One is limited "resolution" of both the electroforming and the embossing that plays a role mainly with profiles of the smallest periods. The other is related to rheological properties of liquefied foil materials during embossing and thus it is appearing more significantly for the larger periods. However, deviation of the modulation depth of the embossed profiles alone may not completely follow faithful reproduction with any mentioned aspect.

ACKNOWLEDGEMENTS

The research was partially supported by the Ministry of the Interior of the Czech Republic (project No. VI20192022147), and by CAS (RVO:68081731).

REFERENCES

[1] RENESSE, R. L. van. *Optical Document Security*. 3rd edition. Boston/London: Artech House, 2005, pp. 386. ISBN 1-58053-258-6.

[2] MELUZÍN, P. *et al.* Some Other Gratings: Benchmarks for Large-Area E-Beam Nanopatterning. In: *NANOCON 2014: 6th Int'l Conference Proceedings*. Ostrava: TANGER, 2014, pp. 246-251.

[3] KOLAŘÍK, V. *et al.*, Structural Color of Metallic Surfaces. In: *METAL 2014: 23rd International Conference on Metallurgy and Materials*. Ostrava: TANGER, 2014, pp. 962-967.

[4] SCHIFT, H. *et al.*, Pattern formation in hot embossing of thin polymer films. *Nanotechnology*. 2001, vol. 12, pp. 173-177.

[5] HEYDEMAN, L. J. *et al.* Flow behavior of thin polymer films used for hot embossing lithography. *Microelectronic Engineering* 2000, vol. 54, no. 3-4.

[6] TORRES, C. M. S. *Alternative Lithography*. US: Springer, 2003.

[7] PENG, L. *et al.* Micro hot embossing of thermoplastic polymers: a review. *J. Micromech. Microeng.* [online]. 2014, vol. 24, pp. 013001-013023. Available from: https://doi.org/10.1088/0960-1317/24/1/013001.

STUDY OF W-X BONDS IN AMORPHOUS W-B-C

[1]Jaroslav ŽENÍŠEK, [1]Pavel SOUČEK, [1,2]Pavel ONDRAČKA, [3]Petr VAŠINA

[1]*Masaryk University, Faculty of Science, Brno, Czech Republic, EU,*
jzenisek@mail.muni.cz, soucek@mail.muni.cz, pavel.ondracka@gmail.com

[2]*Materials Chemistry, RWTH Aachen University, Aachen, Germany, EU,* ondracka@mch.rwth-aachen.de

[3]*Masaryk University, Faculty of Science, Brno, Czech Republic, EU,* vasina@physics.muni.cz

https://doi.org/10.37904/nanocon.2021.4380

Abstract

We examine the validity of assumptions commonly used for the analysis of bonds (bond fractions) in amorphous W-B-C systems by a deconvolution procedure applied on measured XPS profiles. We focus on the relationship between the chemical shift of 4f electronic states of W atoms and the relative amounts of W-X bonds (X=W, B, C). The bond fraction analysis of amorphous W-B-C assumes that the formation of binary W-B and W-C bonds leads to a shift of 4f electron binding energies to lower and higher values, respectively, due to the differences in the Pauling electronegativity of W, B and C. Different W-B-C compositions also lead to different relative amounts of W-W bonds but their effect on the chemical shift is usually neglected. To test these assumptions we generated amorphous W-B-C atomic model of W40B30C30 composition (atomic percent), calculated the W 4f core electron binding energies from *ab initio* methods and examined relationships between the W-X bonds and the chemical shift of 4f electron binding energies. Our analysis confirms the expected chemical shifts in W-B-C caused by W-C bonds. Additionally, we found a correlation between W-W bonds and W 4f electronic states and discuss its influence on the bond fraction analysis.

Keywords: XPS, bonds, amorphous materials, atomistic model

1. INTRODUCTION

Amorphous WBC coatings were shown to be both hard and fracture resistant [1] which renders them as possible candidates for protection of cutting and forming tools. Previous studies, e.g. in [2,3], also correlated the mechanical properties with calculated bond fractions as determined from the analysis of X-ray Phototelectron Spectroscopy (XPS) measurements, specifically from the deconvolution of the W 4f spectra. This work discusses and reviews the assumptions so far commonly used in the XPS bonding analysis of amorphous W-B-C materials with different compositions. In the W-B-C systems, the formation of binary W-B and W-C bonds leads to a shift in core-level energy to lower and higher binding energies, respectively, of W 4f core electrons due to the differences in the Pauling electronegativity of W, B and C.

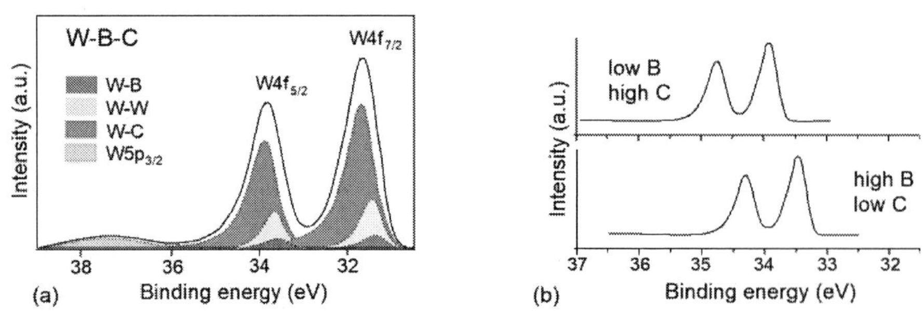

Figure 1 Schematically deconvolution of W 4f peaks to contributions due to W-X bonds (a) and the effect of relative amounts of B and C atoms on chemical shift in W-B-C materials (b)

The bond analysis of the WBC systems based on the deconvolution of XPS W4f peaks (**Figure 1**(a)) is usually based on the assumption that W-C and W-B bonds present in the WBC systems shift the W 4f peak position compared to the pure W (W-W bonds only). The B and C atoms should have an opposite (in direction, not in magnitude) effect on the peak position, as schematically depicted in **Figure 1**(b). It is also assumed that the variation in the number of W-W bonds does not change the peak position. To verify these presumptions, we generate an amorphous W-B-C atomistic model, calculate the W 4f core electron binding energies from *ab initio* methods and examine the relationship between the W 4f electron binding energies and the amount of W-X bonds in the local atomic configurations surrounding W atoms.

2. ATOMISTIC MODELLING OF AMORPHOUS WBC

2.1. Ab initio methodology

Density functional theory (DFT) [4,5] was used to create W40B30C30 atomistic amorphous-like model and calculate the W 4f core electron binding energies. First, 40 W, 30 B, and 30 C atoms were randomly (with the constraint of keeping physically appropriate interatomic distances) placed in a starting cubic unit cell, with size corresponding to the mass density of 13 g/cm3. That cell was used as a starting point for a "melt and quench" *ab initio* molecular dynamics (AIMD), where it was at first kept at 4000 K for 1000 2 fs steps, and then cooled to 300 K over 5000 2 fs MD steps. The temperature was controlled with the Nosé-Hoover thermostat [6]. The resulting structure was fully relaxed at 0 K with respect to the atomic positions, cell size and shape, so that the remaining forces on the atoms were smaller than 0.01 eV/Å. The binding energies of electrons in states $4f_{5/2}$ were calculated for each W atom in the W-B-C model by the Slater´s Transition State method [7] in which half of an electron is moved from the specific core level to the valence band and the binding energy is calculated as the difference between the energy eigenvalue of the state with the half-occupancy and a Fermi level. We used three different DFT codes in order to maintain optimal trade-off between speed and precision. AIMD was done with the OpenMX package using norm-conserving pseudopotentials [8], pseudo-atomic localized basis functions [9,10] (C6.0-s2p2d1, B7.0-s2p2d1, W7.0-s3p2d2), 200 eV cutoff energy and 3x3x3 k-point grid. Structral relaxation at 0 K was done with the VASP code [11], PAW pseudopotentials [12,13], 500 eV planewave energy cutoff, 0.2 eV Gaussian smearing and a 4x4x4 Monkhorst-Pack [14] *k*-point grid. For the calculation of W 4f core electron binding energies we used all-electron full-potential Wien2k code [15]. The calculations were carried out with the basis set size parameter RKmax=7 and k-point grid of 4x4x4. The 4f states were modelled as core-states (for all atoms at least 98 % of the corresponding 4f electronic charge around W atoms was inside the muffin-tin region). Core density superposition was done each self-consistent cycle. Muffin tin radii were set to almost touching spheres, corresponding to approximately 2.1, 1.4, 1.4 Bohr radii for W,B and C respectively. PBE exchange-correlation functional [16] was used for all calculations.

2.2. Analysis of local environments of W atoms

The W atoms are sorted by their $4f_{5/2}$ electron binding energies (from the lowest to the highest) and divided in groups corresponding to 0.2 eV binding energy intervals. The number of atoms X within an interval of distances from W is determined for each W from a binding energy interval. An example of this is schematically depicted in **Figure 2** for B atoms within the distance interval (2.0-2.5) Å from W atoms with $4f_{5/2}$ electron binding energies from (34.0-34.2)eV interval. **Figure 2**(a) shows a structural model (unit cell) of an amorphous W-B-C system in which counting of the atomic neighbours to W atoms is performed. In this particular case, we have five W atoms whose $4f_{5/2}$ electron binding energies are within the (34.0-34.2) eV interval. **Figure 2**(b) shows only those W atoms together with their neighbouring B atoms in the distance interval (2.0-2.5) Å. We see that we have (1+2+0+1+2)=6B atoms, the volume of the shell specified by the radii of 2.0 Å and 2.5 Å is 32 $Å^3$ so the average number density is 6/(5*32)=0.04 B atoms per $Å^3$. This value is shown with a gray column in **Figure 2**(c). The same procedure is then applied for W atoms from (34.2-34.4)eV and further intervals as indicated by the dashed columns in **Figure 2**(c). The average number densities of W and C in the selected distances from W atoms grouped according to their $4f_{5/2}$ electron binding energies are determined analogically. The identification of atoms surrounding W atoms is performed with the use of periodic boundary conditions. An example of this is marked with arrows in **Figure 2**(a) and (b). The B atom close to W with 34.02eV in **Figure 2**(b) finds itself actually slightly below the bottom side of the unit cell. It is the periodic image of the B atom inside unit cell marked with arrow in **Figure 2**(a).

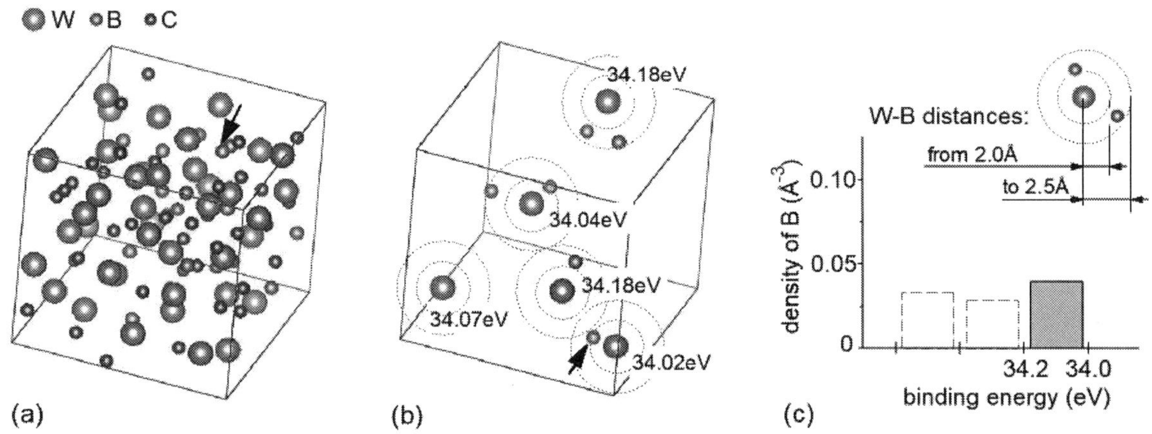

Figure 2 Analysis of the average number density of B atoms within distances (2.0-2.5) Å from W atoms whose $4f_{5/2}$ electron binding energies are within (34.0-34.2) eV interval

The results of analysis described above are shown below as columnar plots. The analysis itself is similar to the determination of radial distribution function whose schematic example is depicted in the upper part of **Figure 3**. If the thickness of a shell specifying the examined zone around W atoms goes to zero and the number density of atoms X is evaluated for all shells centred at W atoms together regardless of W $4f_{5/2}$ electron binding energies, then a point on radial distribution function is in principle obtained (other steps like normalizing are omitted now for simplicity). Some features of the columnar plots can be therefore predicted from the shape of a radial distribution function and this is shown in the bottom part of **Figure 3**. Shells too close to the central W atoms will be empty and no columns will be present in the plots.

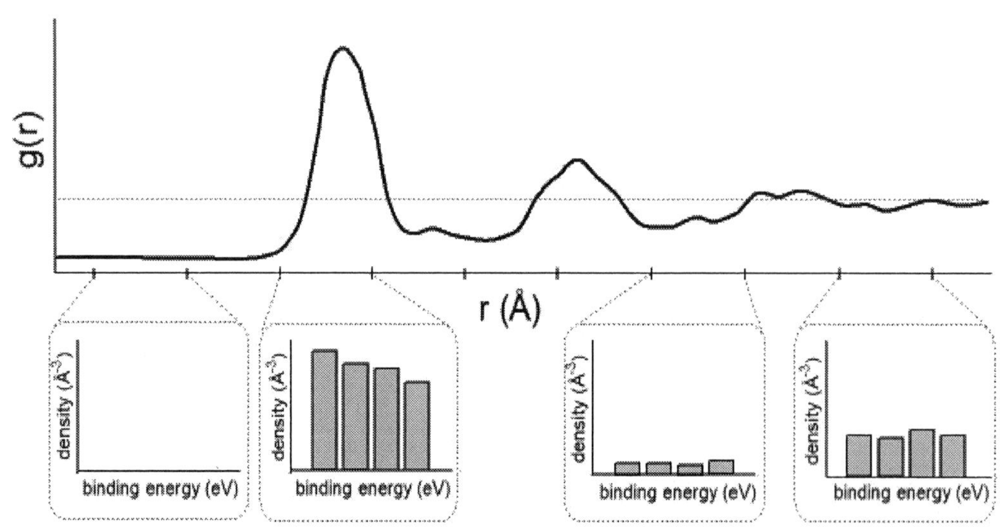

Figure 3 Expected relative column heights for different distance intervals of atoms X from central W atoms from comparison with a radial distribution function (schematic example)

The closest atomic neighbours X to the W atoms form the first peak in the radial distribution function. Corresponding columnar plot should show relatively high columns. If there is any dependency between the local density of X in the bonding distances to W (relative amount of W-X bonds) and the W $4f_{5/2}$ electron binding energies, then the column height in the plot should vary. When moving further away from the W-X bonding distances, no correlation between the density of X in the corresponding shell and the W $4f_{5/2}$ electron binding energies is expected anymore. Therefore, the column height within one plot should become equal (within some fluctuations) but some variations in the column height are possible for various W-X distances.

3. RESULTS

Our analysis tests the dependency between local chemical composition near W atoms and the W $4f_{5/2}$ electron binding energies. The vicinities of W atoms are evaluated inside selected concentric shells 0.5 Å thick. Local composition in a shell is described by number densities of W, B, and C atoms. The number densities are given as functions of W $4f_{5/2}$ electron binding energies in **Figure 4**. Densities of B and C atoms are examined in the interval of distances (2.0-2.5) Å from W atoms, densities of neighbouring W atoms are examined in distances (2.5-3.0) Å from central W atoms. These intervals of distances correspond approximately to the distances where W-X bonds can form and consequently influence the W $4f_{5/2}$ electron binding energies. The upper limits of the distance intervals (2.5 Å for W-B and W-C and 3.0 Å for W-W) are reasonably close to the bonding distances estimated from the sum of the covalent radii of elements W and X [17].

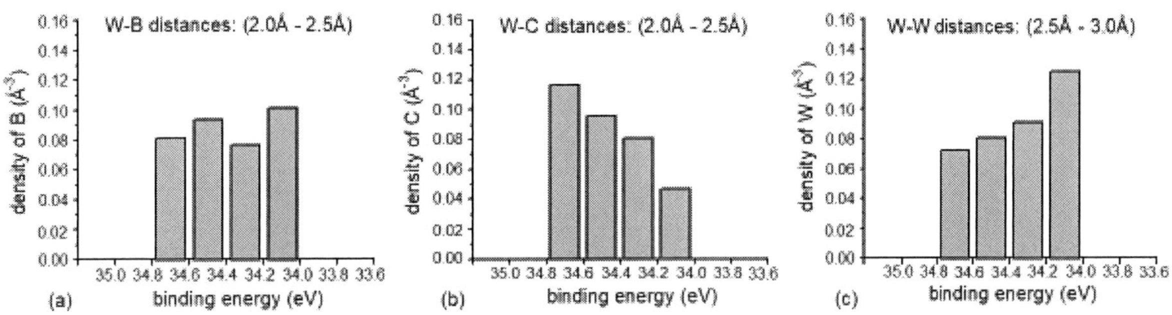

Figure 4 Relationships between the number densities of W, B, and C atoms near central W atoms and the $4f_{5/2}$ electron binding energies of the central W atoms

4. DISCUSSION

Plots from **Figure 4** are used in **Figure 5** where some remarks have been added to discuss the findings. **Figure 5**(a) shows that higher amount of C in the bonding distances to W shifts the $4f_{5/2}$ electron binding energy to higher values. Shift in the opposite direction holds for lower C amounts. This is consistent with the assumption for the bond fraction analysis in W-B-C materials schematically depicted in **Figure 1**(b) in Introduction. The effect of B shows an opposite trend than that of C, i.e., shift to higher binding energies with higher B density. However the effect is much weaker and exhibits some fluctuations, so while this is qualitatively consistent with the experimental bonding analysis assumptions, a larger dataset and/or analysis of more W-B-C compositions is needed for unequivocal validation.

The W-B-C bond fraction analysis (e.g. in [3]) neglects the effect of changing amount of W-W bonds on the chemical shift. If formation of W-W bonds does not affect the W $4f_{5/2}$ electron binding energy, then an equal column height (marked by the dashed line in **Figure 5**(b)) would be expected. However, there is a clearly visible trend marked by the full line. This indicates that at lower $4f_{5/2}$ electron binding energies the W atoms might be closely surrounded not only with the greater amount of B relative to C but also rather higher number of W atoms. This means that the shift to lower binding energies is not only due to forming a relatively higher amount of W-B, but also by forming more W-W bonds. This is an important finding since if one assumes that changing the W concentration (and consequently the amount of W-W bonds) in the W-B-C systems has no impact on the W 4f peak positions in the measured XPS profiles, then the deconvolution-based bond analysis might overestimate the W-B bond fractions at the expense of W-W. Conversely, W 4f peaks shifted to the higher binding energies correspond to W atoms surrounded by greater amount of C atoms and lower amount of B and also W atoms. If one again assumes that changing of W concentration in the W-B-C systems has no impact on the W 4f peak positions, then the deconvolution-based bond analysis might also underestimate the W-C bond fractions in favour of W-W.

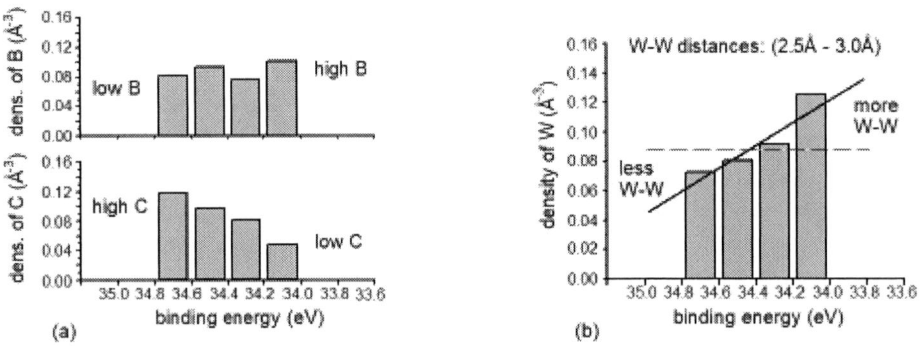

Figure 5 Verification of the expected effect of relative amounts of B and C atoms on chemical shift in W-B-C materials (a). Effect of varying density of W atoms in the bonding distances to the central W atoms on the $4f_{5/2}$ electron binding energies of the central W atoms (b).

5. CONCLUSION

An atomistic model of amorphous W-B-C models was generated and used to analyze relationships between the local atomic environment of W atoms and their $4f_{5/2}$ electron binding energies. It was discovered that the W 4f binding energies depend not only on the local density of C atoms, and therefore W-C bonds in the system (one of the assumptions commonly utilized in the bond fraction analysis) but also on the local density of W atoms (W-W bonds), which should be accounted for in future analyses.

ACKNOWLEDGEMENTS

This research has been supported by project LM2018097 funded by the Ministry of Education, Youth and Sports of the Czech Republic and project GA19-03899S financed by the Grant Agency of the Czech Republic. Computational resources were supplied by the project "e-Infrastruktura CZ" (e-INFRA LM2018140) provided within the program Projects of Large Research, Development and Innovations Infrastructures and by JARA-HPC from RWTH Aachen University under project JARA0131.

REFERENCES

[1] ALISHAHI, M., MIRZAEI, S., SOUČEK, P., ZÁBRANSKY, L., BURŠÍKOVÁ, V., STUPAVSKÁ, M., PEŘINA, V., BALÁSZI, K., CZIGÁNI, Z., VAŠINA, P. Evolution of structure and mechanical properties of hard yet fracture resistant W-B-C coatings with varying B/W ratio. *Surface and Coatings Technology*. 2018, vol. 340, pp. 103-111.

[2] MIRZAEI, S., ALISHAHI, M., SOUČEK, P., ZÁBRANSKÝ, L., BURŠÍKOVÁ, V., STUPAVSKÁ, M., PEŘINA, V., BALÁSZI, K., CZIGÁNI, Z., VAŠINA, P. Effect of bonding structure on hardness and fracture resistance of W-B-C coatings with varying B/W ratio. *Surface and Coatings Technology*. 2019, vol. 358, pp. 843-849.

[3] MIRZAEI, S., ALISHAHI, M., SOUČEK, P., ŽENÍŠEK, J., HOLEC, D., KOUTNÁ, N., BURŠÍKOVÁ, V., STUPAVSKÁ, M., ZÁBRANSKÝ, L., BURMEISTER, F., BLUG, B., CZIGÁNI, Z., BALÁSZI, K., MIKŠOVÁ, R., VAŠINA, P. The effect of chemical composition on the structure, chemistry and mechanical properties of magnetron sputtered W-B-C coatings: Modeling and experiments. *Surface and Coatings Technology.* 2020, vol. 383, pp. 125274-125284.

[4] HOHENBERG, P., KOHN, W. Inhomogeneous electron gas. *Physical Review.* 1964, vol. 136, pp. 864-871.

[5] KOHN, W., SHAM, L. J. Self-consistent equations including exchange and correlation effects. *Physical Review.* 1965, vol. 140, pp. 1133-1138.

[6] NOSÉ, S. A unified formulation of the constant temperature molecular dynamics methods. *The Journal of Chemical Physics.* 1984, vol. 81, pp. 511-519.

[7] SLATER, J.C. Statistical exchange-correlation in the self-consistent field. *Advances in Quantum Chemistry.* 1972, vol. 6, pp. 1-92.

[8] BLOECHL, P.E. Generalized separable potentials for electronic-structure calculations. *Physical Review B.* 1990, vol. 41, pp. 5414-5416.

[9] OZAKI, T. Variationally optimized atomic orbitals for large-scale electronic structures. *Physical Review B.* 2003, vol. 67, pp. 155108-155112.

[10] OZAKI, T., KINO, H. Numerical atomic basis orbitals from H to Kr. *Physical Review B.* 2004, vol. 69, pp. 195113-195131.

[11] KRESSE, G., FURTHMUELLER, J. Efficient iterative schemes for ab initio total-energy calculations using a plane-wave basis set. *Physical Review B.* 1996, vol. 54, pp. 11169-11186.

[12] BLOECHL, P.E. Projector augmented-wave method. *Physical Review B.* 1994, vol. 50, pp. 17953-17979.

[13] KRESSE, G., JOUBERT, J. From ultrasoft pseudopotentials to the projector augmented wave method. *Physical Review B.* 1999, vol. 59, pp. 1758-1775.

[14] MONKHORST, H.J., PACK, J.D. Special points for Brillouin-zone integrations. *Physical Review B.* 1976, vol. 13, pp. 5188-5192.

[15] BLAHA, P., SCHWARZ, K., TRAN, F., LASKOWSKI, R., MADSEN, G.K.H., MARKS, L.D. WIEN2k: an APW+lo program for calculating the properties of solids. *The Journal of Chemical Physics.* 2020, vol. 152, pp. 074101-1 - 074101-30.

[16] PERDEW, J.P., BURKE, K., ERNZERNHOF, M. Generalized gradient approximation made simple. *Physical Review Letters.* 1996, vol. 77, pp. 3865-3868.

[17] CORDERO, B., GÓMEZ, V., PLATERO-PRATS, A.E., REVÉS, M., ECHEVERRÍA, J., CREMADES, E., BARRAGÁN, F., ALVAREZ, S. Covalent radii revisited. *Dalton Transactions.* 2008, issue 21, pp. 2832-2838.

DIMINISHING OF STEEL CORROSION IN ACID ENVIRONMENT USING THIN BI-LAYER SURFACES OF MONO-CARBOXYL-SUBSTITUTED A_3B PORPHYRIN AND $MNTA_2O_6$

[1]Mihaela BIRDEANU, [2]Aurel-Valentin BIRDEANU, [3]Ion FRATILESCU, [3*]Eugenia FAGADAR-COSMA

[1]National Institute for Research and Development in Electrochemistry and Condensed Matter, Timisoara, Romania, mihaelabirdeanu@gamil.com

[2]National Research & Development Institute for Welding and Material Testing - ISIM, Timisoara, Romania, valentinbirdeanu@yahoo.com

[3]Institute of Chemistry "Coriolan Dragulescu", Timisoara, Romania, efagadar@yahoo.com, ion.fratilescu@gmail.com

https://doi.org/10.37904/nanocon.2021.4373

Abstract

It is already known, that porphyrin-bases are offering better protection against steel corrosion than metalloporphyrins due to their ability to form very stable metallic complexes by linking in their inner core the iron contained in different steels. In this respect, 5-(4-carboxy-phenyl)-10,15,20-tris-(4-phenoxy-phenyl)-porphyrin was synthesized by multicomponent Adler-Longo synthesis, using two different aldehydes and pyrrole, characterized to certify the structure and further used for formulating double thin layers in all possible combinations with $MnTa_2O_6$. The pseudo-binary oxide nanomaterials $MnTa_2O_6$ were synthesized by solid-state method. The corrosion protection exhibited by these nanostructures alone or in composites after deposition on carbon steel was comparatively assessed in acid environment. Different electrochemical techniques, such as: open circuit potential measurement and potentiodynamic polarization technique with Tafel representation were used. The performed tests reveal that corrosion protection is fairly good, with the best results indicating that the electrode covered with 5-(4-carboxy-phenyl)-10,15,20-tris-(4-phenoxy-phenyl)-porphyrin / $MnTa_2O_6$ has offered the best corrosion inhibition efficiency of 86.05

Keywords: Mono-carboxyl-substituted A_3B porphyrins, $MnTa_2O_6$, steel, corrosion inhibition, Tafel curves

1. INTRODUCTION

Corrosion protection plays a prominent role in several sectors, such as: marine infrastructures, oil and gas industries, aircrafts and transportations systems [1,2]. Corrosion direct impact on different industrial sectors is magnified by the higher additional costs [3-6]. The use of inhibitors is one of the most practical methods for protection against corrosion in aggressive media as acid medium [7]. Metal corrosion leading to the degradation of metal properties is due to the interaction of metals with certain elements in the environment, this being a normal and inevitable process [8]. During time, porphyrins and/or pseudo-binary oxides demonstrated their qualities as corrosion inhibitors [9] Recent attempts have been made to clarify the influence of the nature of pseudo-binary oxide and/or porphyrins, used for protective thin films, and the deposition order onto steel with the purpose to finally improve the corrosion inhibition of steel in different environments [9-11]. For a long time manganese has been extensively used in various steels, batteries, non-ferrous metal alloys and scientific research [12], due to its versatility. Tantalum has attracted attention for corrosion inhibition applications destined for chemical and biomedical industries due to its excellent results obtained in several aggressive environments [13,14].

The present study it's about investigations concerning the inhibition of corrosion in 1M HCl media for $MnTa_2O_6$ pseudo-binary oxide and 5-(4-carboxy-phenyl)-10,15,20-tris-(4-phenoxy-phenyl)-porphyrin thin layers deposited by drop-casting technique on modified steel electrodes, in single or sandwich type structures. The porphyrin used in this study was chosen due to its already proven self-organization properties that provide multi lamellar triangles, uniformly in shape, dimensions and orientation that uniformly cover the surface.

2. MATERIALS AND METHODS

The pseudo-binary oxide nanomaterials $MnTa_2O_6$ were synthesized by solid-state method. The starting materials used during the synthesis were: tantalum (V) oxide - Ta_2O_5 (99.99%, Merck), and manganese (VI) oxide - MnO_2 (99.99%, Sigma-Aldrich). The mixture of oxides was afterwards heated at 1200 °C for 3 h soaking time.

The A_3B mono substituted COOH-porphyrin, 5-(4-carboxyphenyl)-5,10,15-tris(4-phenoxyphenyl)-porphyrin, might be obtained in several ways: by hydrolysis of ester type Zn-metalloporphyrins [15], or by applying Adler -Longo multicomponent reaction based on the condensation of pyrrole and two different structured benzaldehydes 4-carboxylmethylbenzaldehyde and 4-phenoxybenzaldehyde in a particular ratio of 4:3:1, as previously reported [16] The methyl ester was hydrolyzed in basic condition followed by neutralization with diluted HCl [17].

Both tested materials, $MnTa_2O_6$ pseudo-binary oxide and 5-(4-carboxy-phenyl)-10,15,20-tris-(4-phenoxy-phenyl)-porphyrin were used to obtain single and sandwich type structures by drop casting technique on carbon steel (OL) disks electrodes (10 mm diameter and 2 mm thick). The deposition order and the combinations of the layers in the sandwich structures are presented in **Table 1**.

Table 1 Drop casting thin film depositions

Sample	Deposition order	Deposition Mode
a	$MnTa_2O_6$	Single structure
b	5-(4-carboxy-phenyl)-10,15,20-tris-(4-phenoxy-phenyl) porphyrin	Single structure
c	5-(4-carboxy-phenyl)-10,15,20-tris-(4-phenoxy-phenyl) porphyrin / $MnTa_2O_6$	Sandwich structure
d	$MnTa_2O_6$ / 5-(4-carboxy-phenyl)-10,15,20-tris-(4-phenoxy-phenyl) porphyrin	Sandwich structure

The morphology and the particle dimensions of the samples were investigated by SEM / EDAX (Model INSPECT S) and AFM (Model Nanosurf EasyScan2 Advanced Research) using the noncontact mode cantilever (scan size of 1.1 µm × 1.1 µm).

The electrochemical measurements were realized using a Voltalab potentiostat (Model PGZ 402) coupled with a three electrode electrochemical cell, comprised of: bare (OL) or drop casting modified alloy steel tubes disk as working electrode, a platinum wire as counter electrode and a satured calomel electrode as reference electrode (SCE). The working electrodes were mounted in a Teflon body to ensure a controlled active surface. The potentiodynamic polarization measurements were recorded at 23 °C, by sweeping the potential in the -1.3 V ÷ -0.6 V range at a scan rate of 1 mV /s. The open circuit potential (OCP) of the modified electrodes was monitored for 30 minutes before polarization. The electrolyte solution used for the corrosion tests was 1M HCl. The degree of inhibition efficiency (IE%) were calculated based on equation (1) [18].

$$IE\% = \left(1 - \frac{i_{corr}}{i^0_{corr}}\right) \times 100 \quad (1)$$

where i_{corr}^{0} and i_{corr} represent the values of the current density in the absence and in the presence of the pseudo-binary oxide/porphyrin films, respectively.

Disks of bare electrode carbon steel (OL) containing (wt%) 0.10-0.35 % Si, 0.07-0.13 % S, 0.70-1.10 % Cr, 0.12-0.18 % C, 0.03 % P and 98.21-98.98 % Fe were employed as working electrodes.

The carbon steel surfaces were polished to a mirror-like surface finishing using emery paper (granulation diameter 2000 μm), rinsed with distilled water and degreased in ethanol, before each thin film deposition by drop casting.

3. RESULTS AND DISCUSSION

The morphological aspects of the resulting thin films were comparatively examined by SEM, as shown in **Figure 1**, recorded for various single and sandwich structures based on combinations of pseudo-binary oxides and porphyrins before the corrosion tests. At the surface of the thin films can be noticed the formation of some irregular agglomerates heavily compacted structures of particles with dimensions in the range of 500 ÷1000 nm (**Figures 1 a** and **b**). Also, for the sandwich structures (**Figures 1 c** and **d**) is observed a larger density of the particles at the surface of the thin films than for the single layers.

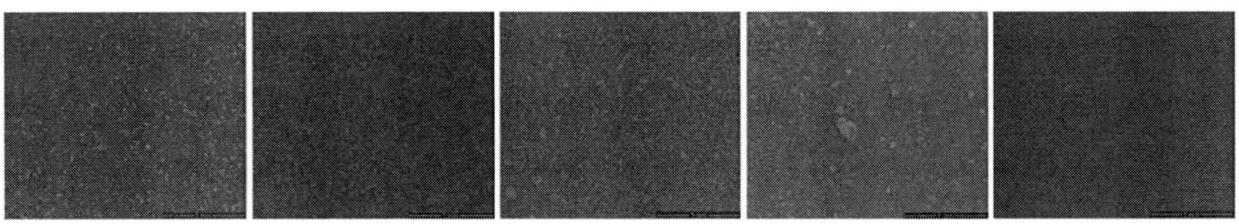

a) b) c) d) e)

Figure 1 The SEM images of: a) $MnTa_2O_6$; b) 5-(4-carboxy-phenyl)-10,15,20-tris-(4-phenoxy-phenyl) porphyrin; c) 5-(4-carboxy-phenyl)-10,15,20-tris-(4-phenoxy-phenyl) porphyrin / $MnTa_2O_6$; d) $MnTa_2O_6$ / 5-(4-carboxy-phenyl)-10,15,20-tris-(4-phenoxy-phenyl) porphyrin; e) OL(bare electrode)

Topographic analysis for the obtained thin films of $MnTa_2O_6$ pseudo-binary oxide with 5-(4-carboxy-phenyl)-10,15,20-tris-(4-phenoxy-phenyl)-porphyrin was perfomed using atomic force microscopy (AFM). The recorded 3D images are presented for each thin film obtained by the drop casting method according with **Figures 2a-d**. **Figure 2 e** shows the carbon steel electrode (OL) before deposition. Besides particle size analysis, AFM measurements were also used to determine the surface roughness. Using the NanoSurf EasyScan 2 computer program, the surface roughness for each sample was calculated according to [18], using the equations for the average roughness (S_a) and the mean square root roughness (S_q) -for each deposited layer onto steel electrodes.

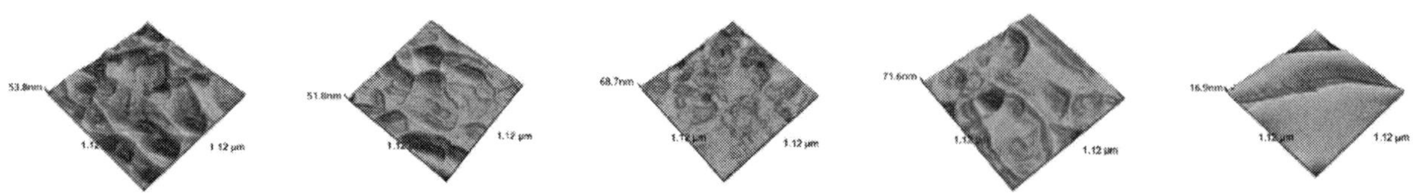

a) b) c) d) e)

Figure 2 3D AFM images for the depositions of: a) $MnTa_2O_6$; b) 5-(4-carboxy-phenyl)-10,15,20-tris-(4-phenoxy-phenyl) porphyrin; c) 5-(4-carboxy-phenyl)-10,15,20-tris-(4-phenoxy-phenyl) porphyrin / $MnTa_2O_6$; d) $MnTa_2O_6$ / 5-(4-carboxy-phenyl)-10,15,20-tris-(4-phenoxy-phenyl) porphyrin; e) OL(bare electrode)

The particle dimensions and the nanorugosities were calculated for all the depositions (**Table 2**) using the AFM software.

From **Table 2** it was observed that the smallest roughness value was obtained for $MnTa_2O_6$ single layer and the values were increasing in case of depositing sandwich layers (larger values being obtained for 5-(4-carboxy-phenyl)-10,15,20-tris-(4-phenoxy-phenyl) porphyrin / $MnTa_2O_6$ sandwich layers of 13.90 nm).

It was found that the size of the particles at surface (**Table 2**) significantly varies between 14.26 and 43.1 nm. The largest size for particles was obtained at the surface of the OL modified with $MnTa_2O_6$ layers.

Table 2 The particle dimension and calculated nanorugosities at the depositions' surfaces

Sample	**Area (pm2)**	**S_a (nm)**	**S_q (nm)**	**S_y (µm)**	**Particles dimension (nm)**
OL		1.43	2.37	-	-
a		3.64	4.65	27.29	43.1
b	1.326	7.54	9.52	66.72	26.9
c		13.90	17.96	113.4	14.26
d		10.10	12.57	81.56	20.2

The evolution of the OCP function of time was analyzed for the bare and modified steel disk electrodes, in 1M HCl electrolyte solution, for 30 min and are presented in **Figure 3a**. The drop- casting deposited layers determined the shifting the OCP of the electrodes toward more positive values, compared with the bare OL electrode. This analyze provides preliminary information on the nature of the processes occurring at the protected metal interface with electrolyte. The OCP measurements (**Figure 3a**) showed that the thin films obtained with porphyrins (5-(4-carboxy-phenyl)-10,15,20-tris-(4-phenoxy-phenyl) porphyrin, $MnTa_2O_6$ / 5-(4-carboxy-phenyl)-10,15,20-tris-(4-phenoxy-phenyl) porphyrin and 5-(4-carboxy-phenyl)-10,15,20-tris-(4-phenoxy-phenyl) porphyrin / $MnTa_2O_6$) were stabilized around 500 s, while the thin film obtained from pseudo-binary oxide ($MnTa_2O_6$) stabilizes around 800 s.

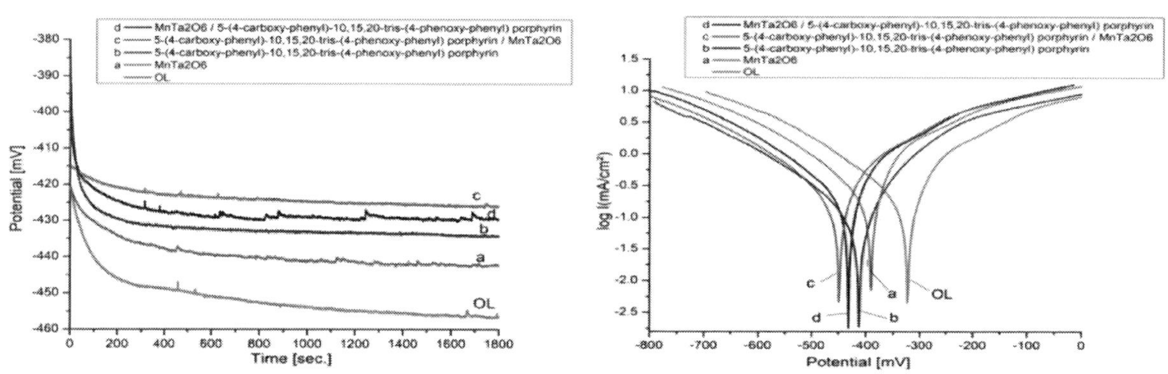

a b

Figure 3 a) Evolution of OCP and b) Tafel representation of polarization curves recorded in 1 M HCl for the investigated electrodes for: OL (bare electrode), a) $MnTa_2O_6$; b) 5-(4-carboxy-phenyl)-10,15,20-tris-(4-phenoxy-phenyl) porphyrin; c) 5-(4-carboxy-phenyl)-10,15,20-tris-(4-phenoxy-phenyl) porphyrin / $MnTa_2O_6$; d) $MnTa_2O_6$ / 5-(4-carboxy-phenyl)-10,15,20-tris-(4-phenoxy-phenyl) porphyrin;

The potentiodynamic cathodic and anodic polarization curves measured in 1M HCl solution, after 30 minutes stabilization of OCP, are presented in **Figure 3b** and the electrochemical parameters obtained from the Tafel extrapolation are summarized in **Table 3**. It is noticed that the corrosion rates decrease for all modified steel electrodes comparing with the corrosion rate for bare electrode. The degree of inhibition efficiency IE (%) has been calculated on the basis of equation reported in [19]. By corroborating all the obtained results, it was observed that the highest inhibition efficiency of 86.1%, that is a fair result, was obtained in the case of the thickest film deposition 5-(4-carboxy-phenyl)-10,15,20-tris-(4-phenoxy-phenyl) porphyrin / $MnTa_2O_6$ (see **Table 2**) which also presents the highest value of the roughness (see **Table 2**), meaning that defects in the protective layers do not create centers for corrosion. The values of the inhibition efficiencies are obviously influenced by the thickness of the protective layers, meaning that the mechanism of protection is mainly of physical nature.

Table 3 The Tafel parameters calculated for the investigated electrodes, after 30 min immersion in 1 M HCl solution

Sample	E_{corr} (mV)	Rp (Ω cm^2)	i_{corr} (mA/cm^2)	v_{cor} (mm/Y^1)	IE (%)
OL	-406.3	101.35	0.9548	3.256	-
a	-410.7	110.22	0.3781	2.635	60.4
b	-416.1	120.03	0.1623	1.669	83
c	-418.6	256.73	0.1331	1.345	86.1
d	-415.4	137.71	0.1557	1.629	83.7

It was found that the corrosion potential E_{corr} decrease from -406.3mV (the corresponding current density i_{corr} = 0.9548 mA / cm^2) for the bare OL to -418.6 mV (i_{corr} = 0.1331 mA / cm^2) for the depositions with 5-(4-carboxy-phenyl)-10,15,20-tris-(4-phenoxy-phenyl) porphyrin as the first layer and $MnTa_2O_6$ as the second layer. The highest polarization resistance (*Rp*) increase from 101.35 $\Omega\cdot$cm^2 for bare OL to 256.73 $\Omega\cdot$cm^2 which corresponds to the sandwich structure with the highest *IE*.

4. CONCLUSION

The purpose of this study was to realize new combinations of materials (porphyrins and pseudo-binary oxides) with efficiency in corrosion inhibition of steel in acid medium. An A_3B porphyrin, 5-(4-carboxy-phenyl)-10,15,20-tris-(4-phenoxy-phenyl)-porphyrin was synthesized by multicomponent Adler-Longo synthesis. $MnTa_2O_6$ pseudo-binary oxide nanomaterials were obtained by solid-state method. Both materials were deposited by drop casting method on carbon steel electrodes, in order to realize synergistic protective properties. Corrosion tests were performed by electrochemical measurements in 0.1 M HCl acid solution. For all the modified electrodes with drop-casted layers an improvement of the corrosion inhibition was observed. The OL modified with deposited layers of 5-(4-carboxy-phenyl)-10,15,20-tris-(4-phenoxy-phenyl) porphyrin / $MnTa_2O_6$ showed the best corrosion inhibition efficiency (over 86 %). So, the sandwich layer with the lower particle dimension offered the best result.

ACKNOWLEDGEMENTS

This work was supported by grant of the Romanian Ministry of Research, Innovation and Digitization/UEFISCDI, project number PN-III-P2-2.1-PED-2019-0487, 528PED/2020, CeraPor and partially by Romanian Academy through Programme 3/2021 from Institute of Chemistry "Coriolan Dragulescu".

REFERENCES

[1] AGHILI, M., YAZDI, M. K., RANJBAR, Z., JAFARI, S. H. Anticorrosion performance of electro-deposited epoxy/ amine functionalized graphene oxide nanocomposite coatings. *Corrosion Science*. 2021, vol. 179, p.109143.

[2] CHANG, C.-H., HUANG, T.-C., PENG, C.-W., YEH, T.-C., LU, H.-I, HUNG, W.-I, WENG, C.-J., YANG, T.-I, YEH, J.-M. Novel anticorrosion coatings prepared from polyaniline/graphene composites. *Carbon.* 2012, vol. 50, no. 14, pp. 5044-5051.

[3] International, N. Assessment of the global cost of corrosion. Available from: http://impact.nace.org/economic-impact.aspx.

[4] NIKPOUR, B., RAMEZANZADEH, B., BAHLAKEH, G., MAHDAVIAN, M. Synthesis of graphene oxide nanosheets functionalized by green corrosion inhibitive compounds to fabricate a protective system. *Corros. Sci.* 2017, vol. 127, pp. 240-259.

[5] MALIK, M. A., HASHIM, M. A., NABI, F., AL-THABAITI, S. A., KHAN, Z. Anti-corrosion Ability of Surfactants: A Review. *Int. J. Electrochem. Sci.* 2011, vol. 6, pp. 1927 - 1948.

[6] LU, X., LIU, L., XIE, X., CUI, Y., OGUZIE, E. E., WANG, F. Synergetic effect of graphene and Co (OH) 2 as cocatalysts of TiO2 nanotubes for enhanced photogenerated cathodic protection. *J. Mater. Sci. Technol.* 2020, vol 37, pp. 55-63.

[7] SARI, M.G., RAMEZANZADEH, B. Epoxy composite coating corrosion protection properties reinforcement through the addition of hydroxyl-terminated hyperbranched polyamide non-covalently assembled graphene oxide platforms. *Constr. Build. Mater.* 2020, vol. 234, p. 117421.

[8] TAN, B., HE, J., ZHANG, S., XU, C., CHEN, S., LIU, H., LI, W. Insight into anti-corrosion nature of Betel leaves water extracts as the novel and eco-friendly inhibitors. *Journal of Colloid and Interface Science.* 2021, vol. 585, pp. 287-301.

[9] BÎRDEANU, A. V., BIRDEANU, M., FAGADAR-COSMA, E. Corrosion protection characteristics of ceramics, porphyrins and hybrid ceramics/porphyrins, deposited as single and sandwich layers, by pulsed laser deposition (PLD) *J. Alloys Compds.* 2017, vol. 706, pp. 220-226.

[10] BIRDEANU, M., BÎRDEANU, A.V., VAIDA, M., MILOVANOVIC, D., LASCU, A., FAGADAR-COSMA, E. Corrosion behaviour of ZnTa2O6 pseudo-binary oxide, zinc meso-tetra (4-pyridyl) porphyrin (ZnTPyP) and hybrid ZnTa2O6/ZnTPyP layers deposited by PLD. *Phys. Scr.* 2019, vol. 94, p. 075702

[11] BIRDEANU, M., BIRDEANU, A. - V., VAIDA, M., ORHA, C., FAGADAR-COSMA, E. Anticorrosive properties of ZnTa2O6 and ZnV2O6 nanomaterials - deposited as sandwich structures by drop casting method - in NaCl. In: *Conference Proceedings 9th International Conference on Nanomaterials - Research & Application, NANOCON 2017*, Brno, Czeh Republic: Tanger, 2018, pp. 359-365,

[12] CAO, Z.-F., WANG, J., QIU, P., YANG, F., WANG, S., LIU, G., ZHONG, H. Hydrophobic coatings for improving corrosion resistance of manganese substrate. *Surface & Coatings Technology.* 2018, vol. 347, pp. 235-244.

[13] LU, F., CAO, Z.-F., YANG, F., WANG, S., ZHONGA, H. Fabrication of hydrophobic coating on electrolytic manganese surface for enhancing corrosion resistance. *Progress in Organic Coatings.* 2019, vol. 132, pp. 379-387.

[14] HU, W., XU, J., LU, X., HU, D., TAO, H., MUNROE, P., XIE, Z.-H. Corrosion and wear behaviours of a reactive-sputter-deposited Ta2O5 nanoceramic coating. *Applied Surface Science.* 2016, vol. 368, pp. 177-190.

[15] BELINE, T., DE ALMEIDA, A. B., NETO, N. F. A., MATOS, A. O., RICOMINI-FILHO, A. P., SUKOTJO, C., SMEETS, P. J. M., DA SILVA, J. H. D., NOCITI, JR F. H., BARAO, V. A. R. β-Ta2O5 thin film for implant surface modification triggers superior anticorrosion performance and cytocompatibility of titanium. *Applied Surface Science.* 2020, vol. 520, p. 146326.

[16] BAKAR, M.B., OELGEMOLLER, M., SENGE, M.O. Lead Structures for Applications in Photodynamic Therapy. Part 2 Synthetic Studies for Photo-Triggered Release Systems of Bioconjugate Porphyrin Photosensitizers. *Tetrahedron.* 2009, vol. 65, pp. 7064-7078.

[17] MAK, C. A., PERICAS, M. A., FAGADAR-COSMA, E. Functionalization of A3B-type porphyrin with Fe3O4 MNPs. Supramolecular assemblies, gas sensor and catalytic applications. *Catalysis Today.* 2018, vol. 306, pp. 268-275.

[18] AHMAD Z. *Principles of Corrosion Engineering and Corrosion Control.* Butterworth-Heinemann/IChemE Series. Elsevier, Amsterdam, 2006, vol. 377.

[19] KAPAKLIS, V., POULOPOULOS, P., KAROUTSOS, V., MANOURAS, TH., POLITIS, C. Growth of thin Ag films produced by radio frequency magnetron sputtering. *Thin Solid Films.* 2006, vol. 510, p. 138.

CORRELATION BETWEEN THE ONSET OF BEADLESS CHARACTER OF NANOFIBROUS WEBS AND RHEOLOGICAL CHARACTERISTICS OF ELECTROSPUN POLYMER SOLUTIONS

Jana ZELENKOVA, Petra PEER, Petr FILIP

Institute of Hydrodynamics, Czech Academy Sciences, Prague, Czech Republic, EU,
zelenkova@ih.cas.cz, peer@ih.cas.cz, filip@ih.cas.cz

https://doi.org/10.37904/nanocon.2021.4375

Abstract

An overwhelming majority of applications of nanofibrous webs requires sufficiently smooth character of nanofibres surface. This character is apart other parameters strongly influenced by a concentration of chosen polymeric material in the solvents. Qualitative attributes roughly depend on the prepared concentration, specifically whether it belongs to one of the four basic regions: dilute, semidilute unentangled, semidilute entangled and concentrated. A concentration separating the latter two regions is often taken as a zero-th approximation indicating an onset of beadless nanofibrous webs. The present contribution uses as a more precise indicator behaviour of so-called phase angle relating viscous and elastic moduli, in other words mutual participation of viscous and elastic components. To this aim three frequently used polymers were used: copolymer of poly(vinylidene fluoride) and hexafluoropropylene (PVDF-*co*-HFP), poly(ethylene oxide) (PEO) and poly(vinyl butyral) (PVB). For materials exhibiting first a constant behaviour or moderate decrease in a phase angle for lower concentrations, an approximation of the starting concentration is given by a concentration value where a phase angle curve starts to decrease, apparently reflecting a more progressive viscoelastic nature. As shown, such approximation provides relatively very good approximation enabling to eliminate a traditional trial-and-error method.

Keywords: Nanofibres, beadless morphology, PVDF-*co*-HFP, polymer concentration, rheology

1. INTRODUCTION

In past decades the application of nanofibrous materials has shifted from classical filters to other uses such as protective clothing, antibacterial wound dressing, antibacterial filtration membranes, and many others. The electrospinning process [1,2] represents one way to produce nanofibres in a relatively cheap manner. This process is based on applying a high voltage (in orders of tens kV) to polymer solutions or melts when ejected nanofibres are deposited on a grounded collector. Quality of the resulting nanofibres is also closely related with an appearance of so-called beads along the individual nanofibres which is mostly an unwanted phenomenon. Bead formation strongly depends on various factors such as voltage, tip-to-collector distance, additives as well as a choice of solvents.

One of the crucial parameters is represented by the polymer concentration. Its value contributes to distinguish the visual properties of the final product to three distinct cases: electrospraying (discrete drops), continuous nanofibres and the creation of spots produced by spinning due to incomplete solvent evaporation. De Gennes [3] set four consecutive regions of polymer solutions differing in concentration: dilute, semidilute unentangled, semidilute entangled and concentrated. The first two regions are separated by the so called overlap concentration c^*, the semidilute regions by the so called entanglement concentration c_e. For individual materials the multiplicative coefficients k of entanglement concentration c_e (where a semidilute entanglement region starts) are determined in such a way that for concentrations lower than $k.c_e$ the electrospun nanofibers exhibit singularities (appearance of beads along the nanofibres) and for concentration higher than $k.c_e$ the beads are suppressed and the nanofibres exhibit a relatively smooth surface. The question is whether this method of evaluating an onset of concentrations for which beadless nanofibres are produced cannot be simplified. To this aim we chose three frequently used materials introduced in the Abstract. These materials are biocompatible and their application as e.g. antibacterial filtration membranes seems to be promising [4].

Based on oscillatory rheological measurements the goal of this contribution is to show a close correlation between the onset of a starting concentration for which a presence of beadless nanofibres dominates and a location of phase angle decrease, i.e. a sudden decrease from the constant course or only a moderate decrease in the curve phase angle vs. the concentration of polymer solution.

2. MATERIALS AND METHODS

Kynar Flex® 2801 (copolymer poly(vinylidene fluoride)-*co*-hexafluoropropylene) was purchased from Arkema (France), poly(vinyl butyral) Mowital B 75H was purchased from Kuraray Specialities Europe and poly(ethylene oxide) was purchased from Sigma Aldrich (USA). The solvents N,N´- dimethylformamide (DMF) (p.a.) and ethanol (quality of p.a.) were purchased from P-LAB, a.s. (Czech Republic) and Penta (Czech Republic) respectively. All chemicals were used as obtained without further refinement.

Polymer materials were dissolved in corresponding solvents (**Table 1**) using a magnetic stirrer MR Hei-Tec (Heidolph, Germany) with the help of a teflon-coated magnetic cross under these conditions: mixing rate was set to 250 rpm, temperature 25 °C and time of mixing was 48 hours (24 hours in the case of Kynar Flex® 2801). The sufficiently wide and dense concentration range ensured that the full palette of nanofibres was at disposal and moreover, it ensured an acceptable approximation of a location where a phase angle started to decrease.

Table 1 Ranges of concentrations of the individual polymer solution

Polymer	Solvent	Molecular weight [g/mol]	Concentration [wt.%]
PVDF-*co*-HFP	DMF	455,000[a]	8, 10, 12, 13, 15, 17, 18, 20 - 29 in steps of 1 %
PVB	ethanol	75,000[a]	3 - 10 in steps of 1 %, 12, 14
PEO	distilled water	637,500[b]	0.3, 0.5, 0.6, 0.8, 1, 1.2, 1,5, 1.8, 2 - 6 in steps of 0.5 %, 7, 7.5, 8, 9

[a] molecular weight provided by the producer, [b] molecular weight determined by SEC (Size Exclusion Chromatography)

The nanofibres were spun (see **Table 2** presenting the conditions) using our laboratory needleless device equipped with a high-voltage power supply SL70PN150 (Spellman, USA), a carbon steel stick (10 mm in diameter) with a semispherical hole for depositing of 0.2 mL of polymer solution and a motionless flat metal collector.

Table 2 Summary of the entry parameters of the electrospinning process

Material	Voltage [kV]	Tip-to-collector distance [cm]	Temperature [°C]	Humidity [%]
PVDF-*co*-HFP / DMF	18	10	23 ± 1	34 ± 1
PVB / ethanol	20	10	21 ± 1	40 ± 1
PEO / distilled water	12 - 25	20	22 ± 1	39 ± 2

A rotational rheometer Physica MCR 501 (Anton Paar, Austria) equipped with the concentric cylinder geometry (the inner and outer diameters were 26.6 and 28.9 mm, respectively) was used both for oscillatory measurements (frequency sweep within 0.1 - 100 Hz at strain 1 %) providing elastic *G*' and viscous *G*" moduli, and for shear viscosity measurements (a range 0.01-300 s^{-1}). The value of shear rate $\dot{\gamma}$ = 0.12 s^{-1} belonging to a linear viscoelastic region was chosen for measurement of shear viscosity of polymer solutions with different polymer concentrations, and consequently applied to a determination of specific viscosity. The phase angle δ ($\tan \delta = G''/G'$) was determined at frequency of 10 Hz. Temperature was set to 25 °C. Each measurement was carried out at least three times and the individual runs were more or less identical.

A high resolution scanning electron microscope Vega 3 (Tescan, Czech Republic) was used for characterization of nanofibrous mats. Prior to imaging, the samples were sputtered by a conductive coating layer using a sputter Quorum Q150R (Quorum Technologies Ltd, UK).

3. RESULTS

Based on measurement of shear viscosities for individual concentrations of PVDF-*co*-HFP, the specific viscosities were calculated using the relation $\eta_{sp} = (\eta_0 - \eta_s)/\eta_s$, where η_0 is the zero shear rate viscosity (measured at $\dot{\gamma}$ = 0.12 s^{-1}) and η_s is the solvent (DMF) viscosity (= 0.807 mPa.s). Consequently, the first three concentration regions according to de Gennes [3] were determined separated by the overlap concentration c^* (≈ 12 wt.%) and entanglement concentration c_e (≈19 wt.%). These two concentrations correspond to the intersection points of linear segments optimized with respect to the experimental points, see **Figure 1**.

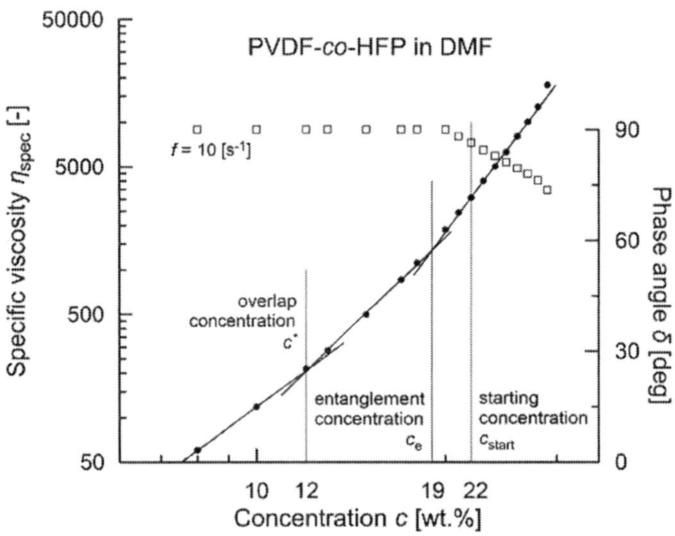

Figure 1 Correlation between an onset of viscoelasticity (a decrease of phase angle, open circles) and a starting concentration corresponding to an onset of beadless PVDF-*co*-HFP nanofibres

Based on elastic (storage) *G*' and viscous (loss) *G*" moduli, a course of the phase angle δ ($\tan\delta = G''/G'$) is depicted in **Figure 1**. A value of concentration for which the phase angle starts to decrease (slightly more than 20 wt.%) corresponds to the formation of network structures among the polymer chains and exceeds a location of the entanglement concentration (≈ 19 wt.%).

The morphology of PVDF-*co*-HFP nanofibres was correlated with the rheological characteristics of measured solutions. **Figure 2** documents that up to the overlap concentration (c^* ≈ 12 wt.%) no nanofibres are formed and as a result only a disordered set of blobs are received. Starting with the overlap concentration the primary forms of nanofibres are indicated mixed with the blobs, and finally a passage from bead to bead-free nanofibres is apparent round the entanglement concentration. It seems that an adequate quality of nanofibres is achieved for c_{start} = 22 wt.% and higher. This value closely corresponds to a concentration for which the phase angle starts to decrease as elasticity begins to manifest.

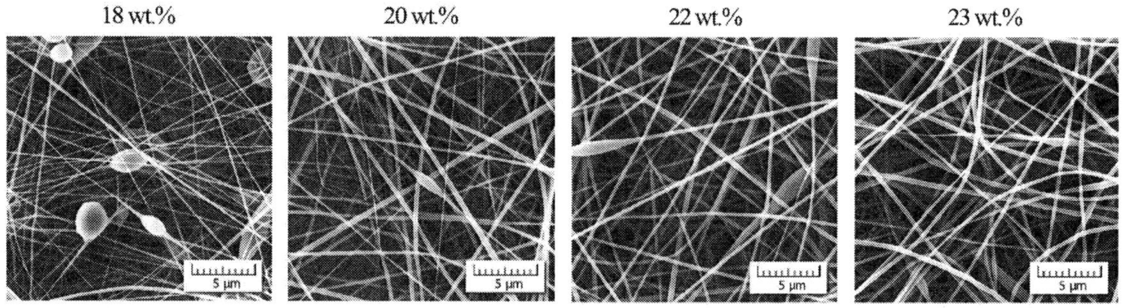

Figure 2 Morphology of individual nanofibres in dependence on PVDF-*co*-HFP concentration

The analogous experimental approach was carried out with PVB dissolved in ethanol (a measured value η_s = 1.087 mPa.s). The results are introduced in **Figures 3** and **4**. In this case the starting concentration is slightly lower than the entanglement concentration.

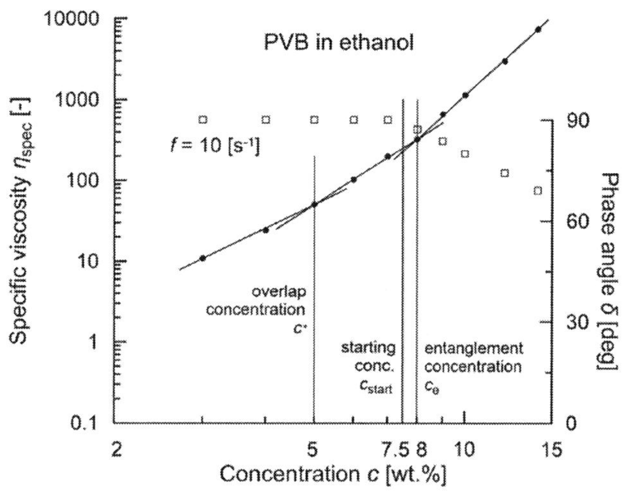

Figure 3 Correlation between an onset of viscoelasticity (a decrease of phase angle, open circles) and a starting concentration corresponding to an onset of beadless PVB nanofibres

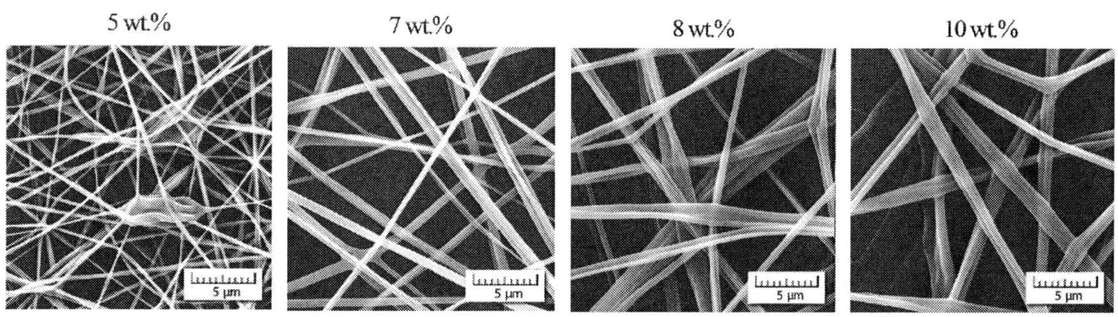

Figure 4 Morphology of individual nanofibres in dependence on PVB concentration

The analogous experimental approach was also carried out with PEO dissolved in distilled water (η_s = 0. 890 mPa.s). The results are summarized in **Figures 5** and **6**. In this case (M_w = 637,500 g/mol) the polymer solutions exhibit viscoelastic behaviour also for lower concentrations due to higher molecular weight. First, a decrease in the relation phase angle vs. concentration is rather moderate. Then, at the concentration of approximately 6 wt.% a sudden drop is apparent. Hence, this value (c_{start}) indicates the region from which the nanofibres have dominantly bead-free character.

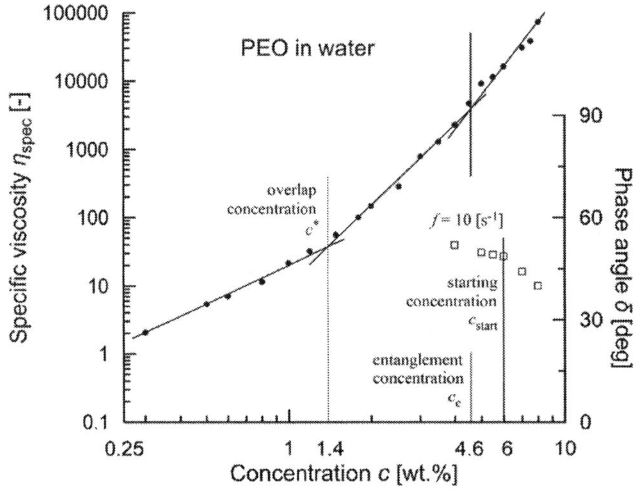

Figure 5 Correlation between a sudden decrease of phase angle (open circles) and a starting concentration corresponding to an onset of beadless PEO nanofibers

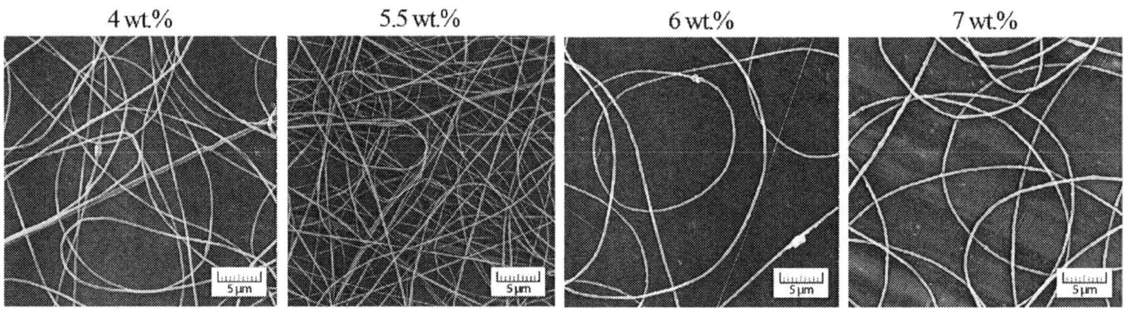

Figure 6 Morphology of individual nanofibres in dependence on PEO concentration

The following **Table 3** summarizes the results for all three materials. The differences between the individual polymers also subject to topology of macromolecular chains and their possible entanglement.

Table 3 Summary of the overlap, entanglement and starting concentrations for used polymeric materials

Material	Overlap concentration c^* [wt.%]	Entanglement concentration c_e [wt.%]	Starting concentration c_{start} [wt.%]
PVDF-*co*-HFP / DMF	12	19	22
PVB / ethanol	5	8	7.5
PEO / distilled water	1.4	4.6	6

4. DISCUSSION

The determination of the starting concentration for the PEO solution is not so strict as for the remaining two materials as PEO solutions exhibit viscoelastic behaviour also for lower concentrations due to relatively high molecular weight. A more pronounced decrease of the phase angle starts from already mildly decreasing curve phase angle vs. concentration which contrasts to the PVDF-*co*-HFP and PVB solutions for which the phase angle attains a constant value (90°) for lower concentrations. This also results in relatively gradual passage of PEO nanofibres from beaded to beadless character.

Possible electrospinnability of polymeric materials and quality of the resulting nanofibres is subject to a number of entry parameters which can be roughly distributed into four categories: polymer characteristics (molecular weight, viscosity, etc.), solvent characteristics (rheological parameters, etc.), polymer solution characteristics (concentration, Hansen solubility parameters, etc.), and process characteristics (e.g. voltage, tip-to collector distance, temperature, humidity).

The topic of this study is a participation of polymer concentration on the appearance of singularities (beads) along the electrospun nanofibres. It is well known that characterization of polymeric materials differs from batch to batch [5]. It results in a variable value of the concentration separating beaded and beadless nanofibres. The primary factor causing these discrepancies is represented by variable molecular weight modifying rheological characteristics. It means that the experimental findings valid for one batch cannot be automatically applied for the other and the whole evaluation process should be repeated. The rheological procedure presented in the preceding section substantially accelerates acquiring new data (starting concentrations) with new batches eliminating hitherto necessity to repeat the individual experiments using a spinning device.

5. CONCLUSION

Determination of a starting concentration of electrospinnable polymer solutions for which the obtained electrospun nanofibres exhibit a bead-free surface is usually carried out by a trial-and-error method. The proposed method using oscillatory rheological measurements provides a very good approximation of the starting concentrations. For materials exhibiting first a constant behaviour or moderate decrease of a phase angle for lower concentrations, an approximation of the starting concentration is given by a concentration value where a phase angle curve starts to decrease apparently reflecting more progressive viscoelastic nature. It was documented using solutions of three often used polymers: PVDF-*co*-HFP, DMF and PEO.

ACKNOWLEDGEMENTS

This research was supported by the Ministry of Education, Youth and Sports of the Czech Republic (Project LTC 19034). This work was carried out in the frame of the COST Actions CA17107.

REFERENCES

[1] RENEKER, D.H., YARIN, A.L. Electrospinning jets and polymer nanofibers. *Polymer.* 2008, vol. 49, pp. 2387-2425.

[2] XUE, J., WU, T., DAI, Y., XIA, Y. Electrospinning and electrospun nanofibers: Methods, materials, and applications. *Chem. Rev.* 2019, vol. 119, pp. 5298-5415.

[3] de GENNES, P.G. *Scaling Concepts in Polymer Physics.* Ithaca, NY: Cornell University Press, 1979.

[4] PEER P., JANALIKOVA, M., SEDLARIKOVA, J., PLEVA, P., FILIP P., ZELENKOVA J., OPALKOVA-SISKOVA, A. Antibacterial filtration membranes based on PVDF-co-HFP nanofibers with the addition of medium-chain 1-monoacylglycerols. *ACS Appl. Mater. Interfaces* (accepted). 2021, vol. 13, no. 34, pp. 41021-41033.

[5] ZATLOUKAL, M.; KOPYTKO, W.; VLCEK, J. The effect of different batches of the same polymer on the flow in flat coextrusion dies. In: *Proc. Annual Technical Conf. Soc. Plast. Engnrs.* Nashville, TN, USA, 4-8 May 2003; pp. 3509-3513.

THE EFFECT OF MULTI-WALLED CARBON NANOTUBES ADDICTION ON THE PROPERTIES OF SHEAR THICKENING FLUIDS

Paulina NAKONIECZNA-DABROWSKA, Marcin LEONOWICZ

Warsaw University of Technology, Faculty of Material Science, Poland, EU,
paulina.nakonieczna.dokt@pw.edu.pl

https://doi.org/10.37904/nanocon.2021.4378

Abstract

Shear thickening fluid (STF) - often referred to as dilatant fluid - is a non-Newtonian suspension of a solid component in a liquid carrier, with an unique ability to dissipate the impact energy. The viscosity of STFs increases abruptly with a growing shear rate. Beyond the critical value of the shear rate, the fluid transforms from a liquid-like to a solid-like state. High-grade dilatant fluids, besides the appropriate rheological properties and protecting capability, have to present high structural stability. The unique properties of the STFs make them suitable for human body protection. The STFs provide lots of opportunities for creating composites with great energy absorption ability, for example, in smart armors or sports protectors.

The properties of shear thickening fluid, modified by the addition of multiwalled carbon nanotubes (MWCNT), were characterized. The addition of a small amount of carbon nanotubes to shear thickening fluid leads to an increase of maximal viscosity, from 2128 to 12213 Pa·s. To show the differences between various compositions, the microstructures of fluids were observed by scanning electron microscopy (SEM). The pronounced influence of the MWCNT on the ability of impact force absorption was noticed. The protective structure containing 55 and 0.25 vol.% of fumed silica and MWCNT, respectively, is able to absorb up to 74 % of impact force.

Keywords: Carbon nanotubes, shear thickening fluids, scanning electron microscopy, nanomaterials

1. INTRODUCTION

A fluid is defined as a substance that flows, that is, deforms continuously over time [1]. The fluids include gases and liquids, as well as those solids that flow at certain, easily achievable conditions. Newtonian fluids comply with Newton's law - their viscosity is constant, independent of the shear rate, and the flow curve passes through the origin of the coordinate system. All fluids that do not comply with this law are called non-Newtonian - their viscosity depends on the speed and time of shearing [2]. Non-Newtonian fluids are divided into two groups: rheologically stable ones, which do not change rheological properties during shearing, and rheologically unstable ones whose rheological properties are a function of shear time [3], [4]. There are many rheological models, which is a more or less precise way, allow to describe the behavior of a given rheological system in conditions of low or high shear rates.

Shear thickening fluids (STF) display a dilation phenomenon which means that at a sufficiently high shear rate, their properties change from typical for liquids to characteristic for solids [5]. The viscosity of such liquids changes as a function of the shear rate. If the force acts at low speed, the liquid has a relatively low viscosity and does not put up much resistance. If the force acts rapidly, the viscosity of the liquid will also rise sharply, and the resistance will significantly increase [6-7]. The properties of such liquids can be controlled through their composition, which results in the change in the shear rate at which the viscosity (critical shear rate) and the maximum and initial viscosity increase [8].

STFs are promising for various applications, such as liquid body armor, sports protectors, such as smart knee pads, and others [9-12]. A variety of modifications have been introduced for the fabrication of composite STFs with different rheological properties. Ge et al. modified STFs with SiC nanowires [13]. Other studies have reported a minor addition of graphene and carbon nanotubes [14-16], in which the level of maximal viscosities obtained was below 1000 Pa·s. Carbon additives seem to be the best fillers that have improved the maximal viscosity of STFs and also their stability [17]. Therefore, this topic should be investigated thoroughly by synthesizing new fluids whose properties may be unique and significantly better than the results already presented in this field of research.

2. EXPERIMENTAL

2.1. Determination of static rheological properties of the developed dilatant fluids

The ARES rotational rheometer was used to define the rheological properties of the fluids developed at room temperature in the parallel plate system using a 0.3 mm gap and 25 mm diameter plates. In the rotational rheometer, sample shearing occurs between two surfaces, where one is rotating, and the other is stationary. In static mode, the dependence of viscosity on the shear rate for all fluids developed was determined. The critical shear rate and the initial and maximum viscosity values were also established.

For the synthesis of shear thickening fluids, an amorphous silica KE-P50, with diameters ranging from 500 to 600 nm, from Nippon Shokubai (Osaka, Japan), were mixed, in appropriate proportions, with polypropylene glycol PPG2000 from Acros Organics (Geel, Belgium), having a molar mass 2000 g/mol. The content of silica in the basic fluid was 50, 53 and 55 vol.%. The composition of STFs was modified by the addition of the multi-walled carbon nanotubes (MWCNT, Nanocyl NC7000, Sambreville, Belgium), with the amounts of 0.05, 0.15 and 0.25 vol%.

2.2. Determination of the microstructure of shear thickening fluids through their observation using scanning electron microscope

The test consisted of proper preparation of the fluid so that observation of the structure of shear thickening fluids using the SEM microscope is possible. The first step comprised a proper degassing, removing liquid phase to show the structure of the solid components. Degassing was performed by heating of the samples in vacuum. The resultant structure kept the solids intact so that visual observation of the interaction between the silica and the carbon fillers could be realized.

2.3. Defining the ability to absorb impact force

The ability to absorb the impact force was tested in accordance with the developed test procedure, which was prepared on the basis of the BS 7971-4:2002: Protective clothing and equipment for use in violent situations and in training. Part 4 - Limb protectors. Requirements and test methods. The fluids were closed in silicone molds. In the test, the hammer was dropped onto the sample with energy of 5J. This energy is obtained by means of the appropriately selected weight of the hammer and the height from which it is dropped. The sample is attached to the table with four two-kilo lead weights. Under the work table, there is a force sensor that records the change in the impact force overtime. Before the test, a reference measurement is carried out by dropping the hammer directly on the table and recording the maximum impact force. Subsequently, samples are placed on the table, and the hammer with the energy of 5J is dropped.

3. RESULTS AND DISCUSSION

3.1. Static rheological properties

In **Figure 1**, the rheological properties of STFs with various additives are presented.

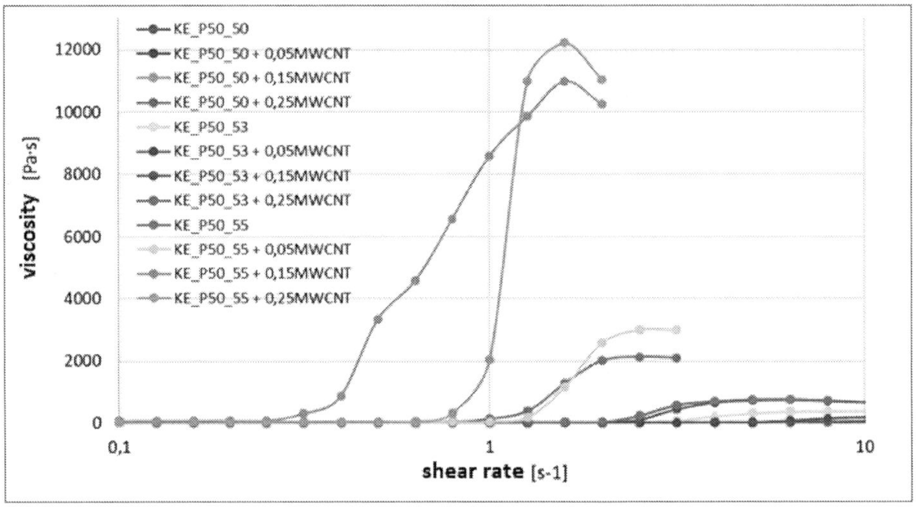

Figure 1 Viscosity vs. shear rate for the STFs with different contents of MWCNT

The highest viscosity and the critical shear rate, at which a drastic increase of the viscosity begins, depend on concentrations of both the silica and MWCNT. Increasing the volume fraction of MWCNT results in an increase in the value of the highest viscosity. The highest viscosity, 2213 Pa·s was recorded for the fluid containing 55 vol% of silica and 0.25 vol.% of MWCNT, with the critical shear rate below 2 s^{-1}.

3.2. Microstructure of shear thickening fluids

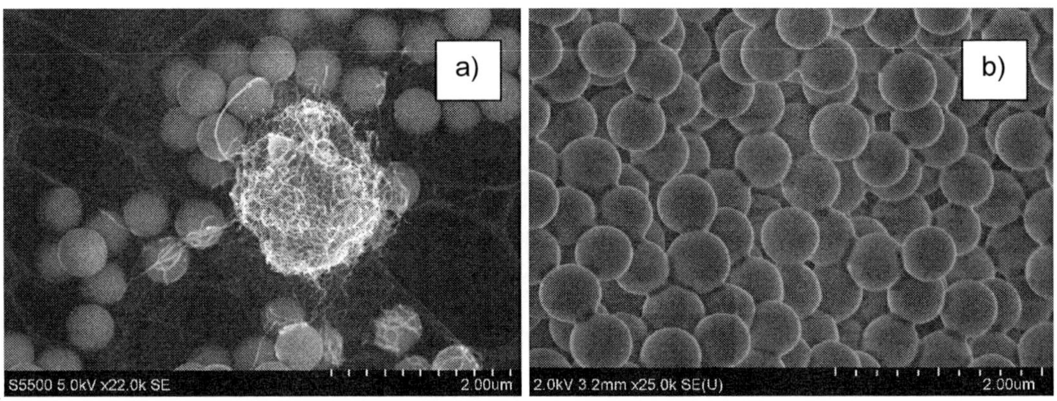

Figure 2 SEM images of a) the starting mixture of silica and agglomerates of MWCNT, b) the arrangement of solid components in the STF

The solid substrates of the STFs were subjected to SEM observations in **Figures 2a)** and **2b)**. Both the silica and the carbon nanotubes had a tendency to agglomeration. In **Figure 2b**, the distribution of silica and carbon nanofillers in the STFs is shown. The agglomerates disappeared, and only silica particles were visible. Mixing silica with MWCNT in a carrier liquid leads to the breaking of the agglomerates of the nanotubes.

3.3. Ability to absorb the impact force

The shear thickening fluids were sealed in silicone moulds. During the tests, none of the samples were destroyed. The samples were deformed and then they returned to their initial shape. **Figure 3** presents the values of the absorbed force for STFs sealed in silicone form after one, two, and three strikes at the same point with the addition of MWCNT.

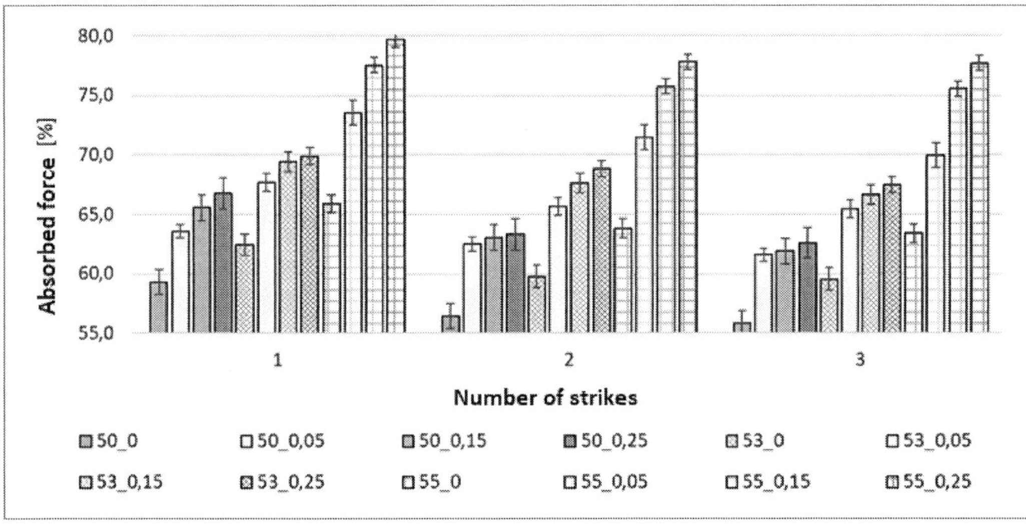

Figure 3 The percentage of absorbed force for STFs sealed in silicone forms after one, two, and three strikes at the same point, with various additions of MWCNT

The higher addition of MWCNT provides an increase of the absorbed force. The samples containing 53 and 55 vol% silica and 0.25 vol% MWCNT show the highest absorbing properties (above 70% for the first strike). One can clearly see that the STFs modified by MWCNT have a greater ability to absorb impact force than the ones based on fumed silica only.

4. CONCLUSION

Shear thickening fluids based on amorphous silica and multi-walled carbon nanotubes were produced. The results show that the addition of the MWCNT has a significant influence on the maximum viscosity, critical share rate and impact force absorbing efficiency. The STFs containing MWCNT have great potential for application in smart protective structures, such as sports protectors, helmets, liquid body armor, and many other applications.

ACKNOWLEDGEMENTS

The research was financed from the Subsidy Research Funds of the Warsaw University of Technology.

The publication was financed from the "Excellence initiative - research university" (IDUB), program "Mobilność Doktorantów Studiów Doktoranckich"

REFERENCES

[1] YOUNG, D.F.; MUNSON, B.R.; OKIISHI, T.H.; HUEBSCH, W.W. *A Brief Introduction to Fluid Mechanics*, 4th ed.; NJ, USA, Wiley: Hoboken, 2007.

[2] MEWIS J., WAGNER N. J. *Colloidal suspension rheology.* Cambridge University Press, 2012.

[3] JANMEY P. A., WILLIAMS N., Quick guide: Rheology. *Current Biology.* 2010, pp. 639-641.

[4] BINGHAM E. C. Some Fundamental Definitions of Rheology. *Journal of Rheology.* 2005, vol. 1, pp. 5.

[5] FREUNDLICH H., RÖDER H. L. Dilatancy and its relation to thixotropy. *Trans. Faraday Soc.* 1938, vol. 34, pp. 308-316.

[6] BOSCO A., CALADO V., MAIA J. Rheological parameters of shear-thickening fluids using an experimental design. *Mater. Res.* 2019, vol. 22, 5, pp. 1-9.

[7] BARNES, H.A. Shear-Thickening ("Dilatancy") in Suspensions of Nonaggregating Solid Particles Dispersed in Newtonian Liquids. *J. Rheol.* 1989, vol. 33, pp. 329-366.

[8] WAGNER N.J., BRADY J. F., Shear thickening in colloidal dispersions. *Phys. Today.* 2009, vol. 62, p. 27.

[9] LEE, Y.S., WETZEL, E.D., WAGNER, N.J. The ballistic impact characteristics of Kevlar® woven fabrics impregnated with a colloidal shear thickening fluid. *J. Mater. Sci.* 2003, vol. 38, pp. 2825-2833.

[10] QIN, J.; ZHANG, G.; ZHOU, L.; LI, J.; SHI, X. Dynamic/quasi-static stab-resistance and mechanical properties of soft body armour composites constructed from Kevlar fabrics and shear thickening fluids. *RSC Adv.* 2017, vol. 7, pp. 39803-39813.

[11] TIAN, T.; NAKANO, M.; LI, W. Applications of shear thickening fluids: A review. *Int. J. Hydromechatron.* 2018, vol. 1, pp. 238-257.

[12] KUCSHAN, M.C.; GÜRGEN, S.; ÜNALIR, T.; ÇEVIK, S. A novel approach for armor applications of shear thickening fluids in aviation and defense industry. *Int. Conf. Sci.* 2014, vol. 22, pp. 4.

[13] GE, J.; TAN, Z.; LI, W.; ZHANG, H. The rheological properties of shear thickening fluid reinforced with SiC nanowires. *Results Phys.* 2017, vol. 7, pp. 3369-3372.

[14] EGRES R. G., WAGNER N. J., The rheology and microstructure of acicular precipitated calcium carbonate colloidal suspension through the shear thickening transition. *Journal of Rheology.* 2005, vol. 49, pp. 719.

[15] YU K., CAO H., QIAN K., SHA X., CHEN Y. Shear-thickening behavior of modified silica nanoparticles in polyethylene glycol. *J. Nanopart. Res.* 2021, vol. 14, p. 747.

[16] MAJUMDAR S., KRISHNASWAMY R., SOOD A. K., Discontinuous shear thickening in confined dilute carbon nanotube suspensions. *Proc. Natl. Acad. Sci.* 2011, vol. 108, pp. 8996-9001.

[17] WEI M., Y. SUN Lv, L., SUN H. Rheological properties of multi-walled carbon nanotubes / silica shear thickening fluid suspensions Rheological properties of multi-walled carbon nanotubes / silica shear thickening fluid suspensions. *Colloid Polym. Sci.* 2020, vol. 298, pp. 243-250.

CALIBRATING THE STIFFNESS OF AFM CANTILEVERS VIA INSTRUMENTED INDENTATION DEVICES

[1]Anna CHARVÁTOVÁ CAMPBELL, [1,2]Marek HAVLÍČEK, [1,2]Petr KLAPETEK

[1]*Czech Metrology Institute, Brno, Czech Republic, EU, acharvatovacampbell@cmi.cz*
[2]*Central European Institute of Technology, Brno, Czech Republic, EU*

https://doi.org/10.37904/nanocon.2021.4377

Abstract

In recent years the study of mechanical properties using atomic force microscopy (AFM) has become very popular. As it uses much lower forces than nanoindentation methods it allows to focus on smaller volumes which makes it an excellent tool for the study of thin films and the creation of high-resolution maps. Although it gives a quantitative as well as a qualitative analysis some open problems remain in the quantitative aspects. One of the problems encountered is related to the stiffness of the AFM cantilever. Most microscopes offer built-in methods for the calibration of cantilever stiffness. Unfortunately, these are often insufficient from the point of view of metrology. Uncertainties and traceability are rarely discussed.

Alternatively, the stiffness of a cantilever can be determined using a nanoindentation device. This well-known method offers simpler uncertainty analysis and traceability. In this contribution we explore the possibilities of this method from the metrological point of view. We shall present an uncertainty budget with focus on repeatability and we shall also discuss traceability issues.

Keywords: AFM, mechanical properties, calibration

1. INTRODUCTION

The study of mechanical properties using atomic force microscopy (AFM) has become very popular in recent years. While it has become a standard tool for the study of biological samples (e.g. [1-5]), stiffer samples are studied less frequently.

For a quantitative analysis the value of the cantilever spring constant must be known. However, this value is not trivial to achieve. The estimates given by probe manufacturers may vary up to hundreds of percent making them almost useless for quantitative analysis. Several methods for the calibration of the spring constant have been developed, for a review see e.g. [6,7]. AFM instruments usually have some method built-in, often Sader's method or the thermal vibration method. However, these methods work well only for soft cantilevers, as used in biological applications, and are not suitable for stiff cantilevers. Reference cantilevers can be used but are often not stiff enough for applications in material sciences. The traceability of these methods is rarely studied but is crucial for the quantitative analyses: without traceability the comparison of results between different instruments or laboratories is highly questionable.

One of the methods which seems most promising in terms of traceability is the method proposed by Clifford and Seah [8] which uses an indentation device to bend an AFM cantilever. Indentation devices are constructed for well-controlled application of load while recording precisely both the applied load and displacement. The traceability of the cantilever then follows from the traceability of the load and displacement sensors.

In this contribution we present a detailed case study of the determination of the spring constant of a stiff AFM cantilever. The uncertainty budget of all steps in the analyses is presented, as well as repeatability and traceability issues. The applicability of this method for a given cantilever depends also on the ranges of the indentation device. Indentations instead of bending might occur or the limits of the mathematical model this method is based on may be reached. Some suggestions how to estimate the impact of these effects are made.

2. METHODS

In this work we follow the method described in [8]. We shall briefly present its main idea. It is based on the well-known model of the small elastic deformations on an ideal rectangular homogenous beam with a fixed end (e.g. [9,10]). Also, the beam is considered massless [9]. In this case the deformation Δz is proportional to the applied force F as

$$\Delta z = \frac{L^3}{3\,E\,J_z} F, \tag{1}$$

where L is the length of the beam, E its Young's modulus and J_z the moment of inertia about the neutral axis (along the length). For a rectangular cross section this is $J_z = wt^3/12$.

Equation (1) holds for a general position of the load allowing to obtain the stiffness from multiple points by fitting the relation

$$X = p\left(\frac{dF}{dz}\right)^{-\frac{1}{3}} + X_0\ . \tag{2}$$

Here X_0 is the possible difference between the apparent fixed end and the true position. The stiffness can be obtained from this as

$$k = \frac{p^3}{(L-X_0)^3}. \tag{3}$$

Solutions using for large deformations exist in literature as well, see e.g. [11,12]. Unfortunately, it is not possible to find a simple relation between deformation and load. Instead, numerical solutions involving inversions of elliptic integrals are used [13].

The procedure in [8] does not take into account the displacement into the cantilever due to indentation. The total compliance measured by the indentation device, i.e., the inverse of the derivative of the loading curve, is the sum of the compliance of the cantilever and the compliance of contact. Assuming the surface of the cantilever is homogenous along its length we can measure the compliance of contact on the fixed part of the cantilever.

Experiments were performed using the Ultra Hardness Nano Tester from Anton Paar. The instrument has been calibrated traceably as described in [14]. The resulting linear calibration constants for the force and depth sensor have a relative uncertainty of 12 % and 1 %. The optical camera and the mechanical moving stage allow placements with an uncertainty approx. 0.5 um. Both were calibrated using a calibration grating.

One measurement of the cantilever's spring constant requires the determination of the coordinates of the points defined in **Figure 1**.

A reference indent I on the fixed part of the cantilever was made in order to align precisely the position of the indenter with respect to the optical camera. A set of loading curves was measured at different positions $X_1, \ldots, X_N$. Further indentations $I_1, \ldots, I_M$ were performed on the fixed part of the cantilever to assess the contact stiffness. In all cases a linear loading, linear unloading scheme was used. For bending we used very low maximum loads, usually 50 or 70 μN which are close to the lower limit of our instrument. For indentations we used the same values in order to be able to use directly the contact stiffness. For the reference indent we used 2 mN, which was one of the lowest values for which the resulting indent could be reliably identified in the optical microscope.

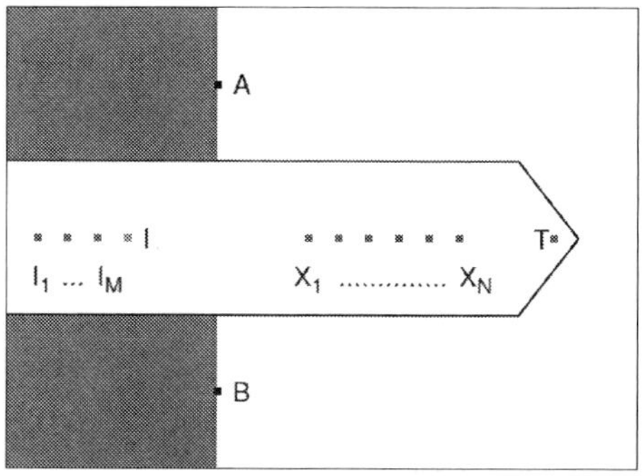

Figure 1 Plan of the measuring points on the cantilever. T - position of tip, A, B - positions on the edge, X_1, ..., X_N - position of bending measurements, I_1, ..., I_M - positions of indentation measurements

Data processing was based on the opensource Niget [15] (force-distance curve processing) and the opensource algorithm OEFPIL, obtainable at https://cran.r-project.org/package=OEFPIL. The OEFPIL - "Optimal Estimate for Function Parameters by Iterated Linearization" algorithm allows to take into account a general covariance matrix, in contrast to ordinary least squares minimization.

Different commercial cantilevers produced mainly by Bruker have been used for testing. The main case study in this work was performed on a stiff cantilever RTESPA525 from Bruker.

3. RESULTS AND DISCUSSION

The contact stiffness was found to be (87.9 ± 2.5) nm/mN at 50 μN. For the following quantitative analysis we chose to use the same loading force as used for the bending. This allows us to avoid the conversion to the Young's modulus which involves the tip shape and would lead to an unnecessary increase of the uncertainties. An example of the set of acquired force distance curves is shown in **Figure 2**. The corresponding slopes both corrected and uncorrected are plotted together with the distance from the edge in the same figure.

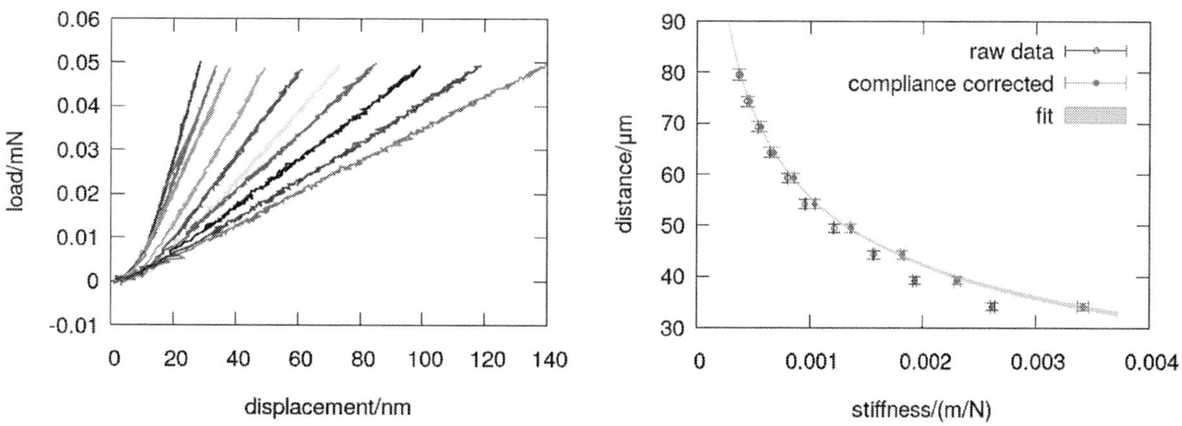

Figure 2 Example of unloading curves measured at different locations (left), corresponding plot of the distance versus the measured stiffness (right)

The resulting spring constant of the cantilever for this particular dataset is (140 ± 18) N/m.

The apparent compliance measured at the point closest to the fixed edge was in this particular case 380 nm/mN, thus the contact compliance is more than 20% of the measured compliance. We can see that although the force-distance curves seem linear at first sight the correction for the contact stiffness has a noticeable impact. It shifts the cantilever spring constant from (146 ± 18) N/m without the correction to (140 ± 18) N/m, thus ignoring the indentation introduces an overestimate of approx. 4 %.

The uncertainty of the slopes of the unloading curves was around 1.5 %, usually well below 3 %.

The cantilever length and the distance from the edge were calculated from the coordinates measured according to **Figure 1**. The corresponding uncertainties were calculated assuming an uncertainty in all coordinates equal to 0.5 µm, with the exception of the coordinates of the tip which are assumed to be 1.0 µm. This is necessary as the shape of the tip is not visible in the optical microscope. The average uncertainty of the length was 1.7 µm. For the total case study, the length was (115.5 ± 0.4) µm.

The spring constant of the cantilever is being calibrated with a factor $c = c_F/c_h$, given by the calibration constants of the sensors. For our instrument it is 1.01 ± 0.12. This is due to a large uncertainty in the calibration of the force sensor which will be the focus of our future calibration studies.

For our case study the fitting parameters (p, X_0) were found to have uncertainties approx. 2-3 % for p and around 15-20 % for X_0. They are strongly correlated, with a correlation around -0.98.

The uncertainty of the spring constant is given by the combination of the uncertainties due to the uncertainties of the calibration factor, length and fitting parameters

$$\mathrm{u(k)} = \sqrt{\left(\frac{\partial k}{\partial c}\right)^2 u(c)^2 + \left(\frac{\partial k}{\partial L}\right)^2 u(L)^2 + \left(\frac{\partial k}{\partial p}\right)^2 u(p)^2 + \left(\frac{\partial k}{\partial X_0}\right)^2 u(X_0)^2 + \left(\frac{\partial k}{\partial p}\right)\left(\frac{\partial k}{\partial X_0}\right) corr(p, X_0)}. \quad (4)$$

Here corr(p, X_0) is the correlation between the two fitting parameters. In all cases we encountered it was negative thus lowering the total uncertainty.

The uncertainty budget for our example is shown in **Table 1**.

Table 1 Example uncertainty budget

Variable	Value of variable x_i	Value of standard uncertainty $u(x_i)$	Sensitivity coefficient $\frac{\partial k}{\partial x_i}$	Contribution to uncertainty (N/m)
Calibration factor c	1.01	0.12	$\frac{p^3}{(L-X_0)^3}$	16
Length L	115.3 µm	1.7 µm	$-3c\frac{p^3}{(L-X_0)^4}$	6
Fitting parameter p	640 $N^{1/3} \cdot m^{2/3}$	13 $N^{1/3} \cdot m^{2/3}$	$3c\frac{p^2}{(L-X_0)^3}$	4
Fitting parameter X_0	-8.6 µm	1.3 µm	$3c\frac{p^3}{(L-X_0)^4}$	

We can see from **Table 1** that the largest contribution to the uncertainty is due to the calibration factor. Its uncertainty is fairly high and future work will attempt to decrease it. The uncertainty due to the fitting parameters includes mainly the uncertainties due to sensor noise and positioning, the effect of the uncertainty of the contact stiffness is negligible. There is little room for improvement in positioning and sensor noise. The contributions due to length and fitting make up for a relative uncertainty of roughly 5 %. The total uncertainty is around 15 % currently so in order to decrease our total spring constant uncertainty below 10 % we need to achieve a calibration uncertainty approx. 8 %.

The case study included a number of experiments. We evaluated the repeatability on a set of measurements between which the instrument had been restarted or the cantilever moved. We found a standard deviation of 6 N/m which corresponds to 4 %. This is significantly smaller than the now dominant uncertainty due to the instrument calibration but comparable with the uncertainty due to length. Based on our long-term experience with this instrument we assume that this reflects the small differences in the attachment of the cantilever as well as time instabilities in the calibration constants.

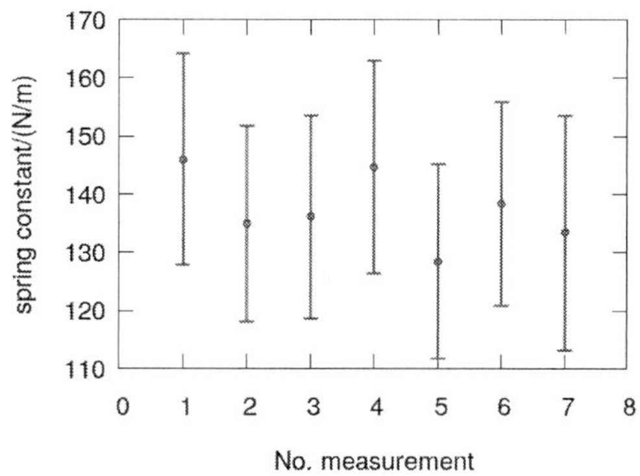

Figure 3 Repeatability of the measurement of the spring constant

The number and location of the points where the bending was performed obviously influences the fit, and leads to differences in the fitting parameters and their covariance matrix. Obviously, it is important to choose the range of displacements as large as possible, as omitting the end points has a larger impact than omitting points in the middle. Numerical tests were performed by omitting combinations of points from the measured data set. However, even when using only 4 points for the fit the differences compared to the value obtained from the full data didn't exceed 7 %. The relative uncertainties ranged from 13 %, same as for the full data, to 18 % in the worst case. Larger differences and uncertainties occurred when several neighboring points were omitted. This shows that an insufficient number of measuring points may introduce errors on the order of a few percent.

The limits of this model should be kept in mind. Notably, the assumption of small deformations should be checked. The large deformation model [11,12] uses a dimensionless parameter α

$$\alpha = \sqrt{\frac{FL^2}{EJ_z}} \qquad (5)$$

Comparing the model with the large deformation model [11,12] it can be seen that in order to keep the deviation of the deformation below 1 %, α must be smaller than approx. 0.55. Since the exact parameters, such as thickness and Young's modulus, are not known well, only very rough estimates can be made. We use nominal values for length, width and thickness and literature values for Young's modulus for the estimates. In order to keep α low, the lowest load at which the indentation instrument can operate should be used; for our instrument this is 50 μN. While for a typical stiff cantilever with nominal stiffness 200 N/m we find for α the value 0.28, the value is 2.04 for a typical soft cantilever. These estimates refer to the application of load at the end of the cantilever. If the load is applied closer to the fixed side the α value is reduced. Obviously for soft cantilevers care must be taken to choose appropriate positions to apply load. If the deformations become too large, the large deformation model would have to be used which requires significantly more complex data processing.

4. CONCLUSION

Although calibration of AFM cantilevers using indentation devices is a long-established method, details on the uncertainties and other aspects are rarely discussed in literature. General recommendations are difficult to give as they depend on the technical parameters of the instrument and the stiffness of the cantilever. In this contribution we illustrated some metrological aspects such as uncertainty analysis, corrections due to indentation and repeatability on a case study using a commercial cantilever. For our instrument the uncertainties in the calibration of the force and displacement sensors were the dominant contribution to the total uncertainty. It was also necessary to correct for the contact stiffness which could be measured on the fixed part of the cantilever. Repeatability has been checked as the attachment of the cantilever may vary and sensors can show time dependencies. However, differences were within margins of uncertainty. An uncertainty around 15-20 % can be achieved by this method but this includes full traceability. The disadvantages are its demands for time and cantilever manipulation.

ACKNOWLEDGEMENTS

This work is supported by the Czech Science Foundation project GA19-15240S. "Multifunctional Nanocomposite Polymer Thin Films with Controlled Surface and Mechanical Properties Deposited in RF Dusty Plasma".

REFERENCES

[1] SANTOS, N.C., CASTANHO, M.A.R.B. An overview of the biophysical applications of atomic force microscopy. *Biophysical Chemistry*. 2004, vol. 107, pp. 133-149.

[2] PANG, D., THIERRY, A. R., DRITSCHILLO, A. DNA studies using atomic force microscopy: capabilities for measurement of short DNA fragments. *Frontiers in Molecular Biosciences*. 2015, vol. 2.

[3] HILAL, N., BOWEN, W.R., ALKHATIB, L., OGUNBIYI, O. A Review of Atomic Force Microscopy Applied to Cell Interactions with Membranes. *Chemical Engineering Research and Design*. 2006, vol. 84, pp. 282-292.

[4] MAIN, K.H.S., PROVAN, J.I., HAYNES, P.J. et al. Atomic Force Microscopy - A tool for structural and translational DNA research. *APL Bioengineering*. 2021, vol. 5.

[5] HAASE, K., PELLING, A.E., Investigating Cell Mechanics With Atomic Force Microscopy. *J. R. Soc.Interface*. 2014, vol. 12.

[6] BURNHAM, N. A., CHEN, X., H ODGES, C. S., MATEI, G. A. THORESON, E. J., ROBERTS, C. J., DAVIES. M. C., TENDLER, S. J. B. Comparison of calibration methods for atomic-force microscopy cantilevers. *Nanotechnology*. 2003, vol. 14, pp. 1-6.

[7] GIBSON, C.T., WATSON, G.S., MYHRA, S. Scanning force microscopy - calibrative procedures for best practice. *Scanning*. 2006, vol 19, pp. 564 - 581.

[8] CLIFFORD, C.A., SEAH, M.P. Improved methods and uncertainty analysis in the calibration of the spring constant of an atomic force microscope cantilever using static experimental methods. *Measurement Science Technology*. 2009, vol. 20, p. 125501.

[9] FEYNMAN, R., LEIGHTON, R., SANDS, M., *The Feynman lectures on Physics: mainly Electromagnetism and Matter*. Reading: Addison-Wesley, 1989.

[10] TIMOSHENKO, S.P. *History of the Strength of Materials*. New York: Dover, 1983.

[11] BISSHOPP, K. E., DRUCKER, D. C. Large deflections of cantilever beams. *Q. Appl. Math*. 1945, vol. 3, pp. 272-275.

[12] BELENDEZ, T., NEIPP, C., BELENDEZ, A. Large and small deflections of a cantilever beam. *European Journal of Physics*. 2002, vol. 23, pp. 371-379.

[13] CHEN, L. An Integral Approach for Large Deflection Cantilever Beams. *International Journal of Non-Linear Mechanics*. 2010, vol.45, pp. 301 - 305.

[14] CAMPBELLOVÁ, A., VALTR, M., ZŮDA, J. and KLAPETEK, P. Traceable measurements of small forces and local mechanical properties. *Meas. Sci. Technol*. 2011, vol. 22, pp. 094007-094013.

[15] CHARVÁTOVÁ CAMPBELL, A., GROLICH P., ŠLESINGER, R. Niget: Nanoindentation general evaluation tool. *Software X*. 2019, vol. 9, pp. 248-254.